▶ Course Calendar

Study Plan ─────────────────────────────────── Legend 🖨 ⑦

Click a chapter below to start practicing, or follow these steps to create a personalized study plan.

① To determine what you need to study, do work on the following material: Quizzes, Tests, Sample Tests

② Practice the questions in the topics you need to study (✏).

③ When you have answered all questions correctly (✓), prove mastery (🎓) by again working on the following material: ▶ Learn More
Quizzes, Tests, Sample Tests

[Show All] ✏ Show What I Need to Study ➡ Jump to where I worked last

Book Contents for All Topics		Correct	Worked	Questions	Time Spent
⊕ Ch. 0: Orientation Questions for Students				8	
⊕ Ch. R: Algebra Reference				114	
⊕ Ch. 1: Linear Functions				45	
⊖ Ch. 2: Systems of Linear Equations and Matrices	✏			90	
⊕ 2.1 Solution of Linear Systems by the Echelon Method	✏			15	
⊕ 2.2 Solution of Linear Systems by the Gauss-Jordan Method	✏			21	
⊕ 2.3 Addition and Subtraction of Matrices	✏			13	
⊕ 2.4 Multiplication of Matrices	✏			14	
⊕ 2.5 Matrix Inverses	✏			18	
⊕ 2.6 Input-Output Models	✏			9	

Test Prep When You Need It
Take a practice test to gauge your readiness. MyMathLab then generates a personalized Study Plan with links to interactive tutorial exercises for those skills you need to review.

Pearson
Tutor Center

Tutors When You Need Them
MyMathLab users have access to tutoring from the Pearson Math Tutor Center.* The Tutor Center is staffed by qualified math instructors who provide one-on-one tutoring. For more details, visit
www.pearsontutorservices.com

Proven Success
Since 2001, more than 10.3 million students at more than 2,000 colleges have used MyMathLab, and a related product, MathXL.® MyMathLab and MathXL provide you with a personalized, interactive learning environment where you can learn at your own pace, measure your progress, and improve your success in this course.

Student Purchasing Options
Talk to your instructor about using
MyMathLab or MathXL for this course.

Finite Mathematics

TENTH EDITION

Margaret L. Lial
American River College

Raymond N. Greenwell
Hofstra University

Nathan P. Ritchey
Youngstown State University

PEARSON

Boston Columbus Indianapolis New York San Francisco Upper Saddle River
Amsterdam Cape Town Dubai London Madrid Milan Munich Paris Montréal Toronto
Delhi Mexico City São Paulo Sydney Hong Kong Seoul Singapore Taipei Tokyo

Editor in Chief: Deirdre Lynch
Executive Editor: Jennifer Crum
Executive Content Editor: Christine O'Brien
Senior Project Editor: Rachel S. Reeve
Editorial Assistant: Joanne Wendelken
Senior Managing Editor: Karen Wernholm
Senior Production Project Manager: Patty Bergin
Associate Director of Design, USHE North and West: Andrea Nix
Senior Designer: Heather Scott
Digital Assets Manager: Marianne Groth
Media Producer: Jean Choe
Software Development: Mary Durnwald and Bob Carroll
Executive Marketing Manager: Jeff Weidenaar
Marketing Coordinator: Caitlin Crain
Senior Author Support/Technology Specialist: Joe Vetere
Rights and Permissions Advisor: Michael Joyce
Image Manager: Rachel Youdelman
Senior Manufacturing Buyer: Carol Melville
Senior Media Buyer: Ginny Michaud
Production Coordination and Composition: Nesbitt Graphics, Inc.
Illustrations: Nesbitt Graphics, Inc. and IllustraTech
Cover Design: Heather Scott
Cover Image: iStock Photo/mmbirdy

Credits appear on page C-1, which constitutes a continuation of the copyright page.

Many of the designations used by manufacturers and sellers to distinguish their products are claimed as trademarks. Where those designations appear in this book, and Pearson was aware of a trademark claim, the designations have been printed in initial caps or all caps.

Library of Congress Cataloging-in-Publication Data
Lial, Margaret L.
    Finite mathematics / Margaret L. Lial, Raymond N. Greenwell, Nathan P.
    Ritchey. — 10th ed.
        p.cm.
Includes bibliographical references and index.
    ISBN-13: 978-0-321-74899-7 (student ed.)
    ISBN-10: 0-321-74899-9 (student ed.)
1. Mathematics—Textbooks.   I. Greenwell, Raymond N.   II. Ritchey, Nathan P.   III. Title.
QA37.3.L532012
510—dc22

                                                            2010029154

1 2 3 4 5 6 7 8 9 10—QG—15 14 13 12 11

www.pearsonhighered.com

ISBN-10: 0-321-74899-9
ISBN-13: 978-0-321-74899-7

Contents

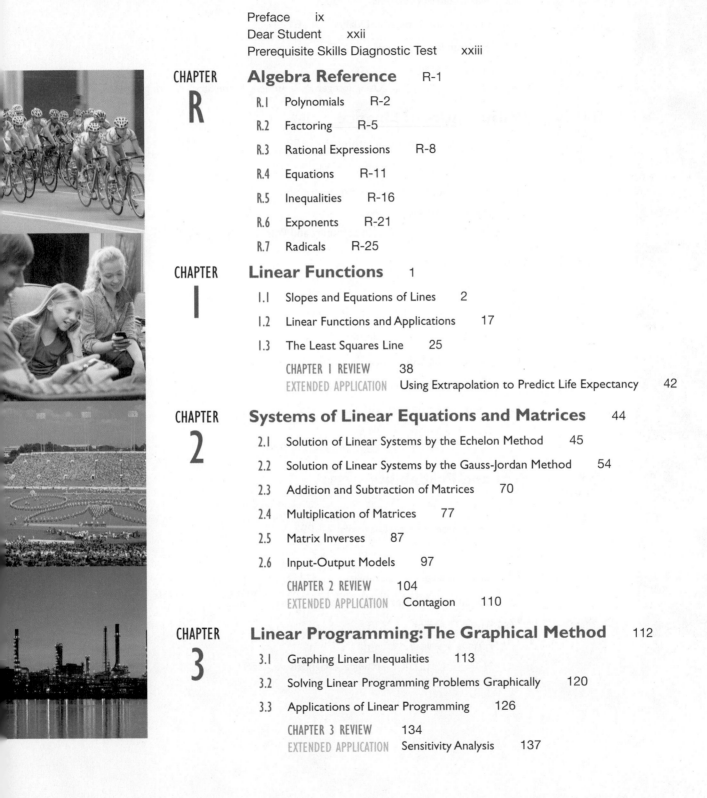

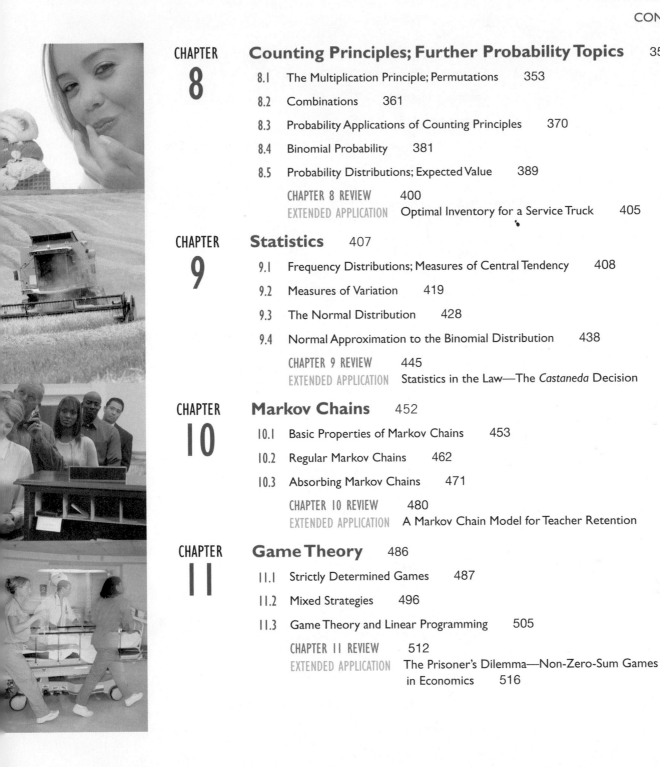

Special Topics to Accompany Finite Mathematics

The following material is provided free to adopters at www.pearsonhighered.com/mathstatsresources:

Digraphs and Networks
Graphs and Digraphs
Dominance Graphs
Communication Graphs
Networks
Review Exercises

Preface

Finite Mathematics is a thorough, application-oriented text for students majoring in business, management, economics, or the life or social sciences. In addition to its clear exposition, this text consistently connects the mathematics to career and everyday-life situations. A prerequisite of two to three semesters of high school algebra is assumed. A renewed focus on quick and effective assessments, new applications and exercises, as well as other new learning tools make this 10th edition an even richer learning resource for students.

Our Approach

Our main goal is to present finite mathematics in a concise and meaningful way so that students can understand the full picture of the concepts they are learning and apply it to real-life situations. This is done through a variety of ways.

Focus on Applications Making this course meaningful to students is critical to their success. Applications of the mathematics are integrated throughout the text in the exposition, the examples, the exercise sets, and the supplementary resources. *Finite Mathematics* presents students with a myriad of opportunities to relate what they're learning to career situations through the *Apply It* questions, the applied examples, and the *Extended Applications*. To get a sense of the breadth of applications presented, look at the Index of Applications in the back of the book or the extended list of sources of real-world data on www.pearsonhighered.com/math-statsresources.

Pedagogy to Support Students Students need careful explanations of the mathematics along with examples presented in a clear and consistent manner. Additionally students and instructors should have a means to assess the basic prerequisite skills. This can now be done with the *Prerequisite Skills Diagnostic Test* located just before Chapter R. In addition, the students need a mechanism to check their understanding as they go and resources to help them remediate if necessary. *Finite Mathematics* has this support built into the pedagogy of the text through fully developed and annotated examples, *Your Turn* exercises, *For Review* references, and supplementary material.

Beyond the Textbook Students today take advantage of a variety of resources and delivery methods for instruction. As such, we have developed a robust MyMathLab course for *Finite Mathematics*. MyMathLab has a well-established and well-documented track record of helping students succeed in mathematics. The MyMathLab online course for *Finite Mathematics* contains over 1300 exercises to challenge students and provides help when they need it. Students who learn best by seeing and hearing can view section- and example-level videos within MyMathLab or on the book-specific DVD-Rom. These and other resources are available to students as a unified and reliable tool for their success.

New to the Tenth Edition

Based on the authors' experience in the classroom along with feedback from many instructors across the country, the focus of this revision is to improve the clarity of the presentation and provide students with more opportunities to learn, practice, and apply what they've learned on their own. This is done in both the presentation of the content and in new features added to the text.

New and Revised Content

- **Chapter R**   The flow of the material was improved by reordering some exercises and examples. Exercises were added to Section R.1 (on performing algebraic operations) and Section R.5 (on solving inequalities).

- **Chapter 1**   Changes in the presentation were made throughout to increase clarity, including adding some examples and rewriting others. Terminology in Section 1.2 was adjusted to be more consistent with usage in economics.

- **Chapter 2**   Section 2.1 was changed so that only systems of two equations are solved by the echelon method, while systems with three or more equations are solved using the Gauss-Jordan method in Section 2.2. The discussion of subtraction of matrices in Section 2.3 was simplified.

- **Chapter 3**   The concept of bounded and unbounded regions was moved from Section 3.2 to Section 3.1, where such regions are first encountered. An *Extended Application* on sensitivity analysis was added to this chapter.

- **Chapter 4**   Exercises 25 through 30 in Section 4.1 were modified to clarify the role of slack variables. Exercise 30 in Section 4.2 was modified to amplify how multiple solutions may occur. The method for handling ties in nonstandard problems in Section 4.4 was improved.

- **Chapter 5**   In Section 5.1, examples and accompanying exercises were added covering how to solve for the interest rate and how to find the compounding time, both with a graphing calculator and with logarithms. An appendix on logarithms was added to the end of the book to help students with this approach. The explanation of the rule of 70 and the rule of 72 was improved. Material on continuous compounding was also added to Section 5.1. In Section 5.3, an example and accompanying exercises were added on how a loan can be paid off early.

- **Chapter 6**   Many exercises in this chapter were revised so that the information would be more relevant to students. For example, tax references include scholarships, tuition, paychecks, reporting tips, filing taxes, inheritances, and tuition deductions. Law references include car accidents, contracts, lawsuits, driver's licenses, and marriage, and warranty references cover iPhones and eBay. In Section 6.5, applications were revised to give more diversity in topics.

- **Chapter 7**   Empirical probability was moved from Section 7.4 to 7.3 so that methods for determining probability are contained in the same section. In Section 7.4, probability distributions are emphasized more and a probability distribution example was added. The introduction to Bayes' Theorem was rewritten for brevity and clarity in Section 7.6.

- **Chapter 8**   The notation for combinations was changed from $\binom{n}{r}$ to $C(n, r)$ to be more current and consistent with our notation throughout the book. Section 8.3 now includes an example illustrating probabilities using permutations and the multiplication principle.

- **Chapter 9**   In Section 9.1, a new example was added illustrating a case in which the median is a truer representation of data than the mean.

- **Chapter 10**   The procedures for finding powers of transition matrices, for finding absorbing states, and determining if a Markov chain is absorbing were clarified.

- **Chapter 11**   Changes in presentation were made throughout to increase clarity.

Prerequisite Skills Diagnostic Test

The Prerequisite Skills Diagnostic Test gives students and instructors a means to assess the basic prerequisite skills needed to be successful in this course. In addition, the answers to the test include references to specific content in Chapter R as applicable so students can zero in on where they need improvement. Solutions to the questions in this test are in Appendix A.

More Applications and Exercises

This text is used in large part because of the enormous amounts of real data used in examples and exercises throughout the text. This 10th edition will not disappoint in this area. We have added or updated 25% of the applications and 43% of the examples throughout the text and added or updated nearly 500 exercises.

Reference Tables for Exercises

The answers to odd-numbered exercises in the back of the textbook now contain a table referring students to a specific example in the section for help with most exercises. For the review exercises, the table refers to the section in the chapter where the topic of that exercise is first discussed.

Annotated Instructor's Edition

The annotated instructor's edition is filled with valuable teaching tips in the margins for those instructors who are new to teaching this course. In addition, answers to most exercises are provided directly on the exercise set page to make assigning and checking homework easier.

New to MyMathLab

Available now with *Finite Mathematics* are the following resources within MyMathLab that will benefit students in this course.

- "Getting Ready for Finite Mathematics" chapter covers basic prerequisite skills
- Personalized Homework allows you to create homework assignments based on the results of student assessments
- Videos for every section of the textbook
- Application labels within exercise sets (e.g., "Business/Econ") make it easy for you to find types of applications appropriate to your students
- Additional graphing calculator and Excel spreadsheet help

A detailed description of the overall capabilities of MyMathLab is provided on page xvii.

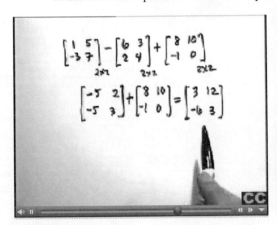

Source Lines

Sources for the exercises are now written in an abbreviated format within the actual exercise so that students immediately see that the problem comes from, or pulls data from, actual research or industry. The complete references are available at www.pearsonhighered.com/mathstatresources as well as on page S-1.

Other New Features

We have worked hard to meet the needs of today's students through this revision. In addition to the new content and resources listed above, there are many new features to this 10th edition including **new and enhanced examples, Your Turn** exercises, the inclusion of and instruction for **new technology**, and **new and updated Extended Applications**. You can view these new features in context in the following *Quick Walk-Through of Finite Mathematics, 10e.*

A Quick Walk-Through of *Finite Mathematics, 10e*

5 Mathematics of Finance

5.1 Simple and Compound Interest

5.2 Future Value of an Annuity

5.3 Present Value of an Annuity; Amortization

Chapter 5 Review

Extended Application: Time, Money, and Polynomials

Buying a car usually requires both some savings for a down payment and a loan for the balance. An exercise in Section 2 calculates the regular deposits that would be needed to save up the full purchase price, and other exercises and examples in this chapter compute the payments required to amortize a loan.

◀ Chapter Opener

Each chapter opens with a quick introduction that relates to an application presented in the chapter. In addition, a section-level table of contents is included.

Apply It ▶

An **Apply It** question, typically at the start of a section, asks students to consider how to solve a real-life situation related to the math they are about to learn. The **Apply It** question is answered in an application within the section or the exercise set. ("Apply It" was labeled "Think About It" in the previous edition.)

NEW!
Teaching Tips ▶

Teaching Tips are provided in the margins of the Annotated Instructor's Edition for those who are new to teaching this course. In addition, answers to most exercises are provided directly on the exercise set page making it easier to assign and check homework.

For Review ▶

For Review boxes are provided in the margin as appropriate, giving students just-in-time help with skills they should already know but may have forgotten. **For Review** comments sometimes include an explanation while others refer students back to earlier parts of the book for a more thorough review.

5.3 Present Value of an Annuity; Amortization

APPLY IT What monthly payment will pay off a $17,000 car loan in 36 monthly payments at 6% annual interest?

The answer to this question is given in Example 2 in this section. We shall see that it involves finding the present value of an annuity.

Suppose that at the end of each year, for the next 10 years, $500 is deposited in a savings account paying 7% interest compounded annually. This is an example of an ordinary annuity. The **present value of an annuity** is the amount that would have to be deposited in one lump sum today (at the same compound interest rate) in order to produce exactly the same balance at the end of 10 years. We can find a formula for the present value of an annuity as follows.

Suppose deposits of R dollars are made at the end of each period for n periods at interest rate i per period. Then the amount in the account after n periods is the future value of this annuity:

$$S = R \cdot s_{\overline{n}|i} = R\left[\frac{(1 + i)^n - 1}{i}\right].$$

Teaching Tip: This section extends the ideas of the previous section to show how to calculate payments on a loan and create an amortization table.

On the other hand, if P dollars are deposited today at the same compound interest rate i, then at the end of n periods, the amount in the account is $P(1 + i)^n$. If P is the present value of the annuity, this amount must be the same as the amount S in the formula above; that is,

$$P(1 + i)^n = R\left[\frac{(1 + i)^n - 1}{i}\right].$$

To solve this equation for P, multiply both sides by $(1 + i)^{-n}$.

$$P = R(1 + i)^{-n}\left[\frac{(1 + i)^n - 1}{i}\right]$$

Use the distributive property; also recall that $(1 + i)^{-n}(1 + i)^n = 1$.

$$P = R\left[\frac{(1 + i)^{-n}(1 + i)^n - (1 + i)^{-n}}{i}\right] = R\left[\frac{1 - (1 + i)^{-n}}{i}\right]$$

The amount P is the *present value of the annuity*. The quantity in brackets is abbreviated as $a_{\overline{n}|i}$, so

$$a_{\overline{n}|i} = \frac{1 - (1 + i)^{-n}}{i}.$$

(The symbol $a_{\overline{n}|i}$ is read "a-angle-n at i." Compare this quantity with $s_{\overline{n}|i}$ in the previous section.) The formula for the present value of an annuity is summarized on the next page.

FOR REVIEW

Recall from Section R.6 that for any nonzero number a, $a^0 = 1$. Also, by the product rule for exponents, $a^x \cdot a^y = a^{x+y}$. In particular, if a is any nonzero number $a^n \cdot a^{-n} = a^{n+(-n)} = a^0 = 1$.

Caution ▶

Caution boxes provide students with a quick "heads-up" to common student difficulties and errors.

NEW!
"Your Turn" Exercises ▶

The **Your Turn** exercises, following selected examples, provide students with an easy way to quickly stop and check their understanding of the skill or concept being presented. Answers are provided at the end of the section's exercises.

Apply It ▶

The solution to the **Apply It** question often falls in the body of the text where it can be seen in context with the mathematics.

CAUTION Don't confuse the formula for the present value of an annuity with the one for the future value of an annuity. Notice the difference: the numerator of the fraction in the present value formula is $1 - (1 + i)^{-n}$, but in the future value formula, it is $(1 + i)^n - 1$.

TECHNOLOGY NOTE The financial feature of the TI-84 Plus calculator can be used to find the present value of an annuity by choosing that option from the menu and entering the required information. If your calculator does not have this built-in feature, it will be useful to store a program to calculate present value of an annuity in your calculator. A program is given in the *Graphing Calculator and Excel Spreadsheet Manual* available with this book.

EXAMPLE 1 Present Value of an Annuity

John Cross and Wendy Mears are both graduates of the Brisbane Institute of Technology (BIT). They both agree to contribute to the endowment fund of BIT. John says that he will give $500 at the end of each year for 9 years. Wendy prefers to give a lump sum today. What lump sum can she give that will equal the present value of John's annual gifts, if the endowment fund earns 7.5% compounded annually?

SOLUTION Here, $R = 500$, $n = 9$, and $i = 0.075$, and we have

$$P = R \cdot a_{\overline{9}|0.075} = 500\left[\frac{1 - (1.075)^{-9}}{0.075}\right] \approx 3189.44.$$

Therefore, Wendy must donate a lump sum of $3189.44 today. **TRY YOUR TURN 1**

YOUR TURN 1 Find the present value of an annuity of $120 at the end of each month put into an account yielding 4.8% compounded monthly for 5 years.

One of the most important uses of annuities is in determining the equal monthly payments needed to pay off a loan, as illustrated in the next example.

EXAMPLE 2 Car Payments

A car costs $19,000. After a down payment of $2000, the balance will be paid off in 36 equal monthly payments with interest of 6% per year on the unpaid balance. Find the amount of each payment.

SOLUTION A single lump sum payment of $17,000 today would pay off the loan. So, $17,000 is the present value of an annuity of 36 monthly payments with interest of $6\%/12 = 0.5\%$ per month. Thus, $P = 17,000$, $n = 36$, $i = 0.005$, and we must find the monthly payment R in the formula

$$P = R\left[\frac{1 - (1 + i)^{-n}}{i}\right]$$

$$17,000 = R\left[\frac{1 - (1.005)^{-36}}{0.005}\right]$$

$$R \approx 517.17.$$

A monthly payment of $517.17 will be needed. **TRY YOUR TURN 2**

YOUR TURN 2 Find the car payment in Example 2 if there are 48 equal monthly payments and the interest rate is 5.4%.

TECHNOLOGY NOTE A graphing calculator program to produce an amortization schedule is available in the *Graphing Calculator and Excel Spreadsheet Manual* available with this book. The TI-84 Plus includes a built-in program to find the amortization payment. Spreadsheets are another useful tool for creating amortization tables. Microsoft Excel has a built-in feature for calculating monthly payments. Figure 14 shows an Excel amortization table for Example 5. For more details, see the *Graphing Calculator and Excel Spreadsheet Manual*, also available with this book.

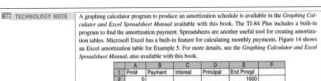

FIGURE 14

EXAMPLE 6 Paying Off a Loan Early

Suppose that in Example 2, the car owner decides that she can afford to make payments of $700 rather than $517.17. How much earlier would she pay off the loan? How much interest would she save?

SOLUTION Putting $R = 700$, $P = 17,000$, and $i = 0.005$ into the formula for the present value of an annuity gives

$$17,000 = 700\left[\frac{1 - 1.005^{-n}}{0.005}\right].$$

Multiply both sides by 0.005 and divide by 700 to get

$$\frac{85}{700} = 1 - 1.005^{-n},$$

or

$$1.005^{-n} = 1 - \frac{85}{700}.$$

Solve this using either logarithms or a graphing calculator, as in Example 11 in Section 5.1, to get $n = 25.956$. This means that 25 payments of $700, plus a final, smaller payment, would be sufficient to pay off the loan. Create an amortization table to verify that the final payment would be $669.47 (the sum of the principal after the penultimate payment plus the interest on that principal for the final month). The loan would then be paid off after 26 months, or 10 months early.

The original loan required 36 payments of $517.17, or $36(517.17) = \$18,618.12$, although the amount is actually $18,618.24 because the final payment was $517.29, as an amortization table would show. With larger payments, the car owner paid $25(700) + 669.47 = \$18,169.47$. Therefore, the car owner saved $18,618.24 - \$18,169.47 = \$448.77$ in interest by making larger payments each month.

NEW!
◀ Recognizing New Technology

Material on graphing calculators or Microsoft Excel™ is now set off to make it easier for instructors to use this material or not. All of the figures depicting graphing calculator screens have been redrawn to create a more accurate depiction of the math. In addition, this edition references and provides students with a transition to the new Math-Print™ operating system of the TI-84 Plus through the technology notes, a new appendix, and the *Graphing Calculator and Excel Spreadsheet Manual*.

NEW!
◀ Enhanced Examples

Most learning from a textbook takes place within the examples of the text. The authors have taken advantage of this by adding more detailed annotations to the already well-developed examples to guide students through new concepts and skills.

Note ▶

Note boxes highlight and emphasize important treatments and asides.

NOTE Compare this formula for compound interest with the formula for simple interest.

Compound interest $A = P(1 + r)^t$
Simple interest $A = P(1 + rt)$

Exercises ▶

Skill-based problems are followed by **application exercises**, which are grouped by subject with subheads indicating the specific topic.

Connection exercises integrate topics presented in different sections or chapters and are indicated with ⬤⬤.

5.3 EXERCISES

1. Explain the difference between the present value of an annuity and the future value of an annuity. For a given annuity, which is larger? Why?

2. What does it mean to amortize a loan?

Find the present value of each ordinary annuity.

3. Payments of $890 each year for 16 years at 6% compounded annually

4. Payments of $1400 each year for 8 years at 6% compounded annually

5. Payments of $10,000 semiannually for 15 years at 5% compounded semiannually

6. Payments of $50,000 quarterly for 10 years at 4% compounded quarterly

7. Payments of $15,806 quarterly for 3 years at 6.8% compounded quarterly

8. Payments of $18,579 every 6 months for 8 years at 5.4% compounded semiannually

Find the lump sum deposited today that will yield the same total amount as payments of $10,000 at the end of each year for 15 years at each of the given interest rates.

9. 4% compounded annually

10. 6% compounded annually

Find (a) the payment necessary to amortize each loan; (b) the total payments and the total amount of interest paid based on the calculated monthly payments, and (c) the total payments and total amount of interest paid based upon an amortization table.

11. $2500; 6% compounded quarterly; 6 quarterly payments

12. $41,000; 8% compounded semiannually; 10 semiannual payments

13. $90,000; 6% compounded annually; 12 annual payments

14. $140,000; 8% compounded quarterly; 15 quarterly payments

15. $7400; 6.2% compounded semiannually; 18 semiannual payments

16. $5500; 10% compounded monthly; 24 monthly payments

Suppose that in the loans described in Exercises 13–16, the borrower paid off the loan after the time indicated below. Calculate the amount needed to pay off the loan, using either of the two methods described in Example 4.

23. How much interest is paid in the first 4 months of the loan?

24. How much interest is paid in the last 4 months of the loan?

⬤⬤ 25. What sum deposited today at 5% compounded annually for 8 years will provide the same amount as $1000 deposited at the end of each year for 8 years at 6% compounded annually?

⬤⬤ 26. What lump sum deposited today at 8% compounded quarterly for 10 years will yield the same final amount as deposits of $4000 at the end of each 6-month period for 10 years at 6% compounded semiannually?

Find the monthly house payments necessary to amortize each loan. Then calculate the total payments and the total amount of interest paid.

27. $199,000 at 7.01% for 25 years

28. $175,000 at 6.24% for 30 years

29. $253,000 at 6.45% for 30 years

30. $310,000 at 5.96% for 25 years

Suppose that in the loans described in Exercises 13–16, the borrower made a larger payment, as indicated below. Calculate (a) the time needed to pay off the loan, (b) the total amount of the payments, and (c) the amount of interest saved, compared with part c of Exercises 13–16.

31. $16,000 in Exercise 13

32. $18,000 in Exercise 14

33. $850 in Exercise 15

34. $400 in Exercise 16

APPLICATIONS

Business and Economics

35. **House Payments** Calculate the monthly payment and total amount of interest paid in Example 3 with a 15-year loan, and then compare with the results of Example 3.

36. **Installment Buying** Stereo Shack sells a stereo system for $600 down and monthly payments of $30 for the next 3 years. If the interest rate is 1.25% per month on the unpaid balance, find

   a. the cost of the stereo system.

   b. the total amount of interest paid.

37. **Car Payments** Hong Le buys a car costing $14,000. He agrees to make payments at the end of each monthly period for 4 years. He pays 7% interest, compounded monthly.

   a. What is the amount of each payment?

   b. Find the total amount of interest Le will pay.

38. **Credit Card Debt** Tom Shaffer charged $8430 on his credit card to relocate for his first job. When he realized that the interest rate for the unpaid balance was 27% compounded monthly, he decided not to charge any more on that account. He wants to have this account paid off by the end of 3 years,

---

48. **Loan Payments** When Nancy Hart opened her law office, she bought $14,000 worth of law books and $7200 worth of office furniture. She paid $1200 down and agreed to amortize the balance with semiannual payments for 5 years, at 8% compounded semiannually.

   a. Find the amount of each payment.

   b. Refer to the text and Figure 13. When her loan had been reduced below $5000, Nancy received a large tax refund and decided to pay off the loan. How many payments were left at this time?

49. **House Payments** Ian Desrosiers buys a house for $285,000. He pays $60,000 down and takes out a mortgage at 6.5% on the balance. Find his monthly payment and the total amount of interest he will pay if the length of the mortgage is

   a. 15 years;

   b. 20 years;

   c. 25 years.

   d. Refer to the text and Figure 13. When will half the 20-year loan in part c be paid off?

50. **House Payments** The Chavara family buys a house for $225,000. They pay $50,000 down and take out a 30-year mortgage on the balance. Find their monthly payment and the total amount of interest they will pay if the interest rate is

   a. 6%;

   b. 6.5%;

   c. 7%.

   d. Refer to the text and Figure 13. When will half the 7% loan in part c be paid off?

51. **Refinancing a Mortgage** Fifteen years ago, the Budai family bought a home and financed $150,000 with a 30-year mortgage at 8.2%.

   a. Find their monthly payment, the total amount of their payments, and the total amount of interest they will pay over the life of this loan.

   b. The Budais made payments for 15 years. Estimate the unpaid balance using the formula

$$y = R\left[\frac{1 - (1+i)^{-(n-x)}}{i}\right],$$

   and then calculate the total of their remaining payments.

   c. Suppose interest rates have dropped since the Budai family took out their original loan. One local bank now offers a 30-year mortgage at 6.5%. The bank fees for refinancing are $3400. If the Budais pay this fee up front and refinance the balance of their loan, find their monthly payment. Including the refinancing fee, what is the total amount of their payments? Discuss whether or not the family should refinance with this option.

   d. A different bank offers the same 6.5% rate but on a 15-year mortgage. Their fee for financing is $4500. If the Budais pay this fee up front and refinance the balance of their loan, find their monthly payment. Including the refinancing fee, what is the total amount of their payments? Discuss whether or not the family should refinance with this option.

52. **Inheritance** Deborah Harden has inherited $25,000 from her grandfather's estate. She deposits the money in an account offering 6% interest compounded annually. She wants to make equal annual withdrawals from the account so that the money (principal and interest) lasts exactly 8 years.

   a. Find the amount of each withdrawal.

   b. Find the amount of each withdrawal if the money must last 12 years.

53. **Charitable Trust** The trustees of a college have accepted a gift of $150,000. The donor has directed the trustees to deposit the money in an account paying 6% per year, compounded semiannually. The trustees may make equal withdrawals at the end of each 6-month period; the money must last 5 years.

   a. Find the amount of each withdrawal.

   b. Find the amount of each withdrawal if the money must last 6 years.

Amortization Prepare an amortization schedule for each loan.

54. A loan of $37,948 with interest at 6.5% compounded annually, to be paid with equal annual payments over 10 years.

55. A loan of $4836 at 7.25% interest compounded semi-annually, to be repaid in 5 years in equal semiannual payments.

56. **Perpetuity** A *perpetuity* is an annuity in which the payments go on forever. We can derive a formula for the present value of a perpetuity by taking the formula for the present value of an annuity and looking at what happens when *n* gets larger and larger. Explain why the present value of a perpetuity is given by

$$P = \frac{R}{i}.$$

57. **Perpetuity** Using the result of Exercise 56, find the present value of perpetuities for each of the following.

   a. Payments of $1000 a year with 4% interest compounded annually

   b. Payments of $600 every 3 months with 6% interest compounded quarterly

YOUR TURN ANSWERS

1. $6389.86 2. $394.59

3. $1977.42, $135,935.60 4. $679.84

(partial column)
... Exercise 13
... Exercise 14
... ercise 15
... Exercise 16

... table in Example 5 to answer the ques-
...24.

... fourth payment is interest?

... eleventh payment is used to reduce the debt?

Technology exercises are labeled with 📈 for graphing calculator and ▦ for spreadsheet.

Writing exercises, labeled with ✎, provide students with an opportunity to explain important mathematical ideas.

Exercises that are particularly **challenging** are denoted with + in the Annotated Instructor's Edition only.

End-of-Chapter Summary ▶

End-of-Chapter Summary provides students with a quick summary of the key ideas of the chapter followed by a list of key definitions, terms, and examples.

5 CHAPTER REVIEW

SUMMARY

In this chapter we introduced the mathematics of finance. We first extended simple interest calculations to compound interest, which is interest earned on interest previously earned. We then developed the mathematics associated with the following financial concepts.

- In an annuity, money continues to be deposited at regular intervals, and compound interest is earned on that money as well.
- In an ordinary annuity, payments are made at the end of each time period and the compounding period is the same as the time between payments, which simplifies the calculations.

- An annuity due is slightly different, in that the payments are made at the beginning of each time period.
- A sinking fund is like an ordinary annuity; a fund is set up to receive periodic payments. The payments plus the compound interest will produce a desired sum by a certain date.
- The present value of an annuity is the amount that would have to be deposited today to produce the same amount as the annuity at the end of a specified time.
- An amortization table shows how a loan is paid back after a specified time. It shows the payments broken down into interest and principal.

We have presented a lot of new formulas in this chapter. By answering the following questions, you can decide which formula to use for a particular problem.

1. Is simple or compound interest involved?

Simple interest is normally used for investments or loans of a year or less; compound interest is normally used in all other cases.

2. If simple interest is being used, what is being sought: interest amount, future value, present value, or interest rate?

REVIEW EXERCISES

CONCEPT CHECK

Determine whether each of the following statements is true or false, and explain why.

1. For a particular interest rate, compound interest is always better than simple interest.

2. The sequence 1, 2, 4, 6, 8, . . . is a geometric sequence.

3. If a geometric sequence has first term 3 and common ratio 2, then the sum of the first 5 terms is $S_5 = 93$.

4. The value of a sinking fund should decrease over time.

5. For payments made on a mortgage, the (noninterest) portion of the payment applied on the principal increases over time.

6. On a 30-year conventional home mortgage, at recent interest rates, it is common to pay more money on the interest on the loan than the actual loan itself.

7. One can use the amortization payments formula to calculate the monthly payment of a car loan.

8. The effective rate formula can be used to calculate the present value of a loan.

9. The following calculation gives the monthly payment on a $25,000 loan, compounded monthly at a rate of 5% for a period of six years:

$$25,000 \left[\frac{(1 + 0.05/12)^{72} - 1}{0.05/12} \right]$$

10. The following calculation gives the present value of an annuity of $5,000 payments at the end of each year for 10 years. The fund earns 4.5% compounded annually.

$$5000 \left[\frac{1 - (1.045)^{-10}}{0.045} \right]$$

PRACTICE AND EXPLORATION

Find the simple interest for each loan.

11. $15,903 at 6% for 8 months

12. $4902 at 5.4% for 11 months

13. $42,368 at 5.22% for 7 months

14. $3478 at 6.8% for 88 days (assume a 360-day year)

◀ Chapter Review Exercises

Chapter Review Exercises have been slightly reorganized so that the Concept Check exercises fall within the Chapter Review Exercises. This provides students with a more complete review of both the skills and the concepts they should have mastered in this chapter. These exercises in their entirety provide a comprehensive review for a chapter-level exam.

...matics of Finance

15. For a given amount of money at a given interest rate for a given time period, does simple interest or compound interest produce more interest?

Find the compound amount in each loan.

16. $2800 at 7% compounded annually for 10 years

17. $19,456.11 at 8% compounded semiannually for 7 years

18. $312.45 at 5.6% compounded semiannually for 16 years

19. $57,809.34 at 6% compounded quarterly for 5 years

Find the amount of interest earned by each deposit.

20. $3954 at 8% compounded annually for 10 years

21. $12,699.36 at 5% compounded semiannually for 7 years

22. $12,903.45 at 6.4% compounded quarterly for 29 quarters

23. $34,677.23 at 4.8% compounded monthly for 32 months

24. What is meant by the present value of an amount A?

Find the present value of each amount.

25. $42,000 in 7 years, 6% compounded monthly

26. $17,650 in 4 years, 4% compounded quarterly

27. $1347.89 in 3.5 years, 6.77% compounded semiannually

28. $2388.90 in 44 months, 5.93% compounded monthly

29. Write the first five terms of the geometric sequence with a = 2...

43. $11,900 deposited at the beginning of each month for 13 months; money earns 6% compounded monthly.

44. What is the purpose of a sinking fund?

Find the amount of each payment that must be made into a sinking fund to accumulate each amount.

45. $6500; money earns 5% compounded annually for 6 years.

46. $57,000; money earns 4% compounded semiannually for $8\frac{1}{2}$ years.

47. $233,188; money earns 5.2% compounded quarterly for $7\frac{3}{4}$ years.

48. $1,056,788; money earns 7.2% compounded monthly for $4\frac{1}{2}$ years.

Find the present value of each ordinary annuity.

49. Deposits of $850 annually for 4 years at 6% compounded annually

50. Deposits of $1500 quarterly for 7 years at 5% compounded quarterly

51. Payments of $4210 semiannually for 8 years at 4.2% compounded semiannually

52. Payments of $877.34 monthly for 17 months at 6.4% compounded monthly

53. Give two examples of the types of loans that are commonly amortized.

Find the amount of the payment necessary to amortize each...

EXTENDED APPLICATION

TIME, MONEY, AND POLYNOMIALS*

A time line is often helpful for evaluating complex investments. For example, suppose you buy a $1000 CD at time t_0. After one year $2500 is added to the CD at t_1. By time t_2, after another year, your money has grown to $3851 with interest. What rate of interest, called *yield to maturity* (YTM), did your money earn? A time line for this situation is shown in Figure 15.

FIGURE 15

Assuming interest is compounded annually at a rate i, and using the...

To determine the yield to maturity, we must solve this equation for i. Since the quantity $1 + i$ is repeated, let $x = 1 + i$ and first solve the second-degree (quadratic) polynomial equation for x.

$$1000x^2 + 2500x - 3851 = 0$$

We can use the quadratic formula with $a = 1000$, $b = 2500$, and $c = -3851$.

$$x = \frac{-2500 \pm \sqrt{2500^2 - 4(1000)(-3851)}}{2(1000)}$$

We get $x = 1.0767$ and $x = -3.5767$. Since $x = 1 + i$, the two values for i are $0.0767 = 7.67\%$ and $-4.5767 = -457.67\%$. We reject the negative value because the final accumulation is greater than the sum of the deposits. In some applications, however, negative rates may be meaningful. By checking in the first equation, we see that the yield to maturity for the CD is 7.67%.

Now let us consider a more complex but realistic problem. Suppose Austin Caperton has contributed for 4 years to a retirement fund. He contributed $6000 at the beginning of the first year. At the beginning of the next 3 years, he contributed $5840, $4000, and $5200, respectively. At the end of the fourth year, he had $29,912.38 in his fund. The interest rate earned by the fund varied between 21% and −3%, so Caperton would like to know the YTM = i for his hard-earned retirement dollars. From a time line (see Figure 16), we set up...

◀ Extended Applications

Extended Applications are provided now at the end of **every** chapter as in-depth applied exercises to help stimulate student interest. These activities can be completed individually or as a group project.

xv

Flexible Syllabus

The flexibility of the text is indicated in the following chart of chapter prerequisites. As shown, the course could begin with either Chapter 1 or Chapter 7. Chapter 5 on the mathematics of finance and Chapter 6 on logic could be covered at any time, although Chapter 6 makes a nice introduction to ideas covered in Chapter 7.

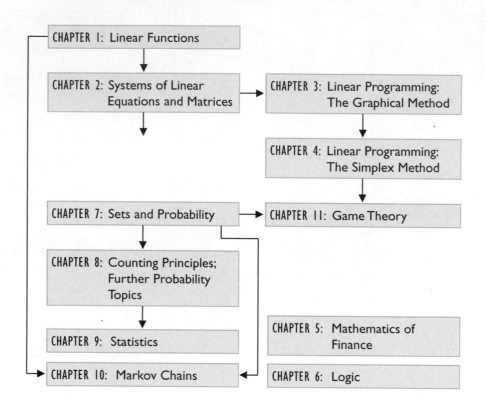

Supplements

STUDENT RESOURCES	INSTRUCTOR RESOURCES

Student Edition

- ISBN 0-321-74899-9 / 978-0-321-74899-7

Student's Solutions Manual

- Provides detailed solutions to all odd-numbered text exercises and sample chapter tests with answers.
- Authored by Elka Block and Frank Purcell, Twin Prime Editorial
- ISBN 0-321-74867-0 / 978-0-321-74867-6

Graphing Calculator and Excel Spreadsheet Manual

- Provides instructions and keystroke operations for the TI-83/84 Plus, the TI-84 Plus with the new operating system featuring MathPrint™, and the TI-89 as well as for the Excel spreadsheet program.
- Authored by Steve Ouellette
- ISBN 0-321-70966-7 / 978-0-321-70966-0

Annotated Instructor's Edition

- Numerous teaching tips
- Includes all the answers, usually on the same page as the exercises, for quick reference
- ISBN 0-321-70964-0 / 978-0-321-70964-6

Instructor's Resource Guide and Solutions Manual (download only)

- Provides complete solutions to all exercises, two versions of a pre-test and final exam, and teaching tips.
- Authored by Elka Block and Frank Purcell, Twin Prime Editorial
- Available to qualified instructors within MyMathLab or through the Pearson Instructor Resource Center, www.pearsonhighered.com/irc

STUDENT RESOURCES	INSTRUCTOR RESOURCES

Video Lectures on DVD-ROM with Optional Captioning

- Complete set of digitized videos, covering all sections of the text, for student use at home or on campus
- Ideal for distance learning or supplemental instruction
- ISBN 0-321-74612-0 / 978-0-321-74612-2

Supplementary Content

- Digraphs and Networks
- Additional Extended Applications
- Comprehensive source list
- Available at the Downloadable Student Resources site, www.pearsonhighered.com/mathstatsresources, and to qualified instructors within MyMathLab or through the Pearson Instructor Resource Center, www.pearsonhighered.com/irc

PowerPoint Lecture Presentation

- Newly revised and greatly improved
- Classroom presentation slides are geared specifically to the sequence and philosophy of this textbook.
- Includes lecture content and key graphics from the book
- Available to qualified instructors within MyMathLab or through the Pearson Instructor Resource Center, www.pearsonhighered.com/irc

Pearson Math Adjunct Support Center (www.pearsontutorservices.com/math-adjunct.html)

The Pearson Math Adjunct Support Center is staffed by qualified instructors with more than 100 years of combined experience at both the community college and university levels. Assistance is provided for faculty in the following areas:

- Suggested syllabus consultation
- Tips on using materials that accompany your book
- Book-specific content assistance
- Teaching suggestions, including advice on classroom strategies

Media Resources www.mymathlab.com

MyMathLab® Online Course (access code required)

MyMathLab® is a text-specific, easily customizable online course that integrates interactive multimedia instruction with textbook content. MyMathLab gives you the tools you need to deliver all or a portion of your course online, whether your students are in a lab setting or working from home.

- **Interactive homework exercises,** correlated to your textbook at the objective level, are algorithmically generated for unlimited practice and mastery. Most exercises are free-response and provide guided solutions, sample problems, and tutorial learning aids for extra help.

- **Personalized homework** assignments that you can design to meet the needs of your class. MyMathLab tailors the assignment for each student based on test or quiz scores. Each student receives a homework assignment that contains only the problems the student still needs to master.

- **Personalized Study Plan,** generated when students complete a test or quiz or homework, indicates which topics have been mastered and links to tutorial exercises for topics students have not mastered. You can customize the Study Plan so that the topics available match your course content.

- **Multimedia learning aids,** such as video lectures and podcasts, animations, and a complete multimedia textbook, help students independently improve their understanding and performance. You can assign these multimedia learning aids as homework to help your students grasp the concepts.

- **Homework and Test Manager** lets you assign homework, quizzes, and tests that are automatically graded. Select just the right mix of questions from the MyMathLab exercise bank, instructor-created custom exercises, and/or TestGen® test items.

- **Gradebook,** designed specifically for mathematics and statistics, automatically tracks students' results, lets you stay on top of student performance, and gives you control over how to calculate final grades. You can also add offline (paper-and-pencil) grades to the gradebook.

- **MathXL Exercise Builder** allows you to create static and algorithmic exercises for your online assignments. You can use the library of sample exercises as an easy starting point, or you can edit any course-related exercise.

- **Pearson Tutor Center** (www.pearsontutorservices.com) access is automatically included with MyMathLab. The Tutor Center is staffed by qualified math instructors who provide textbook-specific tutoring for students via toll-free phone, fax, email, and interactive Web sessions.

Students do their assignments in the Flash®-based MathXL Player, which is compatible with almost any browser (Firefox®, Safari™, or Internet Explorer®) on almost any platform (Macintosh® or Windows®). MyMathLab is powered by CourseCompass™, Pearson Education's online teaching and learning environment, and by MathXL®, our online homework, tutorial, and assessment system. MyMathLab is available to qualified adopters. For more information, visit www.mymathlab.com or contact your Pearson representative.

MathXL® Online Course (access code required)
www.mathxl.com

MathXL® is an online homework, tutorial, and assessment system that accompanies Pearson's textbooks in mathematics or statistics.

- **Interactive homework exercises**, correlated to your textbook at the objective level, are algorithmically generated for unlimited practice and mastery. Most exercises are free-response and provide guided solutions, sample problems, and learning aids for extra help.

- **Personalized homework** assignments are designed by the instructor to meet the needs of the class and then personalized for each student based on test or quiz results. As a result, each student receives a homework assignment in which the problems cover only the objectives the student has not mastered.

- **Personalized Study Plan**, generated when students complete a test or quiz or homework, indicates which topics have been mastered and links to tutorial exercises for topics students have not mastered. Instructors can customize the available topics in the study plan to match their course concepts.

- **Multimedia learning aids**, such as video lectures and animations, help students independently improve their understanding and performance. These may be assigned as homework to further encourage their use.

- **Gradebook**, designed specifically for mathematics and statistics, automatically tracks students' results, lets you stay on top of student performance, and gives you control over how to calculate final grades.

- **MathXL Exercise Builder** allows you to create static and algorithmic exercises for your online assignments. You can use the library of sample exercises as an easy starting point or use the Exercise Builder to edit any of the course-related exercises.

- **Homework and Test Manager** lets you create online homework, quizzes, and tests that are automatically graded. Select just the right mix of questions from the MathXL exercise bank, instructor-created custom exercises, and/or TestGen test items.

The new, Flash®-based MathXL Player is compatible with almost any browser (Firefox®, Safari™, or Internet Explorer®) on almost any platform (Macintosh® or Windows®). MathXL is available to qualified adopters. For more information, visit our website at www.mathxl.com, or contact your Pearson representative.

InterAct Math Tutorial Website
www.interactmath.com

Get practice and tutorial help online! This interactive tutorial website provides algorithmically generated practice exercises that correlate directly to the exercises in the textbook. Students can retry an exercise as many times as they would like with new values each time for unlimited practice and mastery. Every exercise is accompanied by an interactive guided solution that provides helpful feedback for incorrect answers, and students can view a worked-out sample problem that guides them through an exercise similar to the one in which they're working.

TestGen®
www.pearsoned.com/testgen

TestGen® enables instructors to build, edit, print, and administer tests using a computerized bank of questions developed to cover all the objectives of the text. TestGen is algorithmically based, allowing instructors to create multiple but equivalent versions of the same question or test with the click of a button. Instructors can also modify test bank questions or add new questions. The software and testbank are available for download from Pearson Education's online catalog.

Acknowledgments

We wish to thank the following professors for their contributions in reviewing portions of this text:

Lowell Abrams, *The George Washington University*
Nathan Borchelt, *Clayton State University*
Joanne Brunner, *Joliet Jr. College*
James K. Bryan, *Jr., Merced College*
James Carolan, *Wharton County Junior College*
Michelle DeDeo, *University of North Florida*
Lauren Fern, *University of Montana*
Sharda K. Gudehithlu, *Wilbur Wright College*
Yvette Hester, *Texas A & M University*
Lynette J. King, *Gadsden State Community College*
Donna S. Krichiver, *Johnson County Community College*
Lia Liu, *University of Illinois at Chicago*
Donna M. Lynch, *Clinton Community College*
Phillip Miller, *Indiana University Southeast*
Marna Mozeff, *Drexel University*
Bishnu Naraine, *St. Cloud State University*
Charles Odion, *Houston Community College*
Charles B. Pierre, *Clark Atlanta University*
Stela Pudar-Hozo, *Indiana University Northwest*
Nancy Ressler, *Oakton Community College*
Mary Alice Smeal, *Alabama State University*
Alexis Sternhell, *Delaware County Community College*

We also thank Elka Block and Frank Purcell of Twin Prime Editorial for doing an excellent job updating the *Student's Solutions Manual* and *Instructor's Resource Guide and Solutions Manual*, an enormous and time-consuming task. Further thanks go to our accuracy checkers Renato Mirollo, Jon Weerts, and Tom Wegleitner. We are grateful to Karla Harby and Mary Ann Ritchey for their editorial assistance. We especially appreciate the staff at Pearson, whose contributions have been very important in bringing this project to a successful conclusion.

Margaret L. Lial
Raymond N. Greenwell
Nathan P. Ritchey

Finite Mathematics

TENTH EDITION

Dear Student,

Hello! The fact that you're reading this preface is good news. One of the keys to success in a math class is to read the book. Another is to answer all the questions correctly on your professor's tests. You've already started doing the first; doing the second may be more of a challenge, but by reading this book and working out the exercises, you'll be in a much stronger position to ace the tests. One last essential key to success is to go to class and actively participate.

You'll be happy to discover that we've provided the answers to the odd-numbered exercises in the back of the book. As you begin the exercises, you may be tempted to immediately look up the answer in the back of the book, and then figure out how to get that answer. It is an easy solution that has a consequence—you won't learn to do the exercises without that extra hint. Then, when you take a test, you will be forced to answer the questions without knowing what the answer is. Believe us, this is a lot harder! The learning comes from figuring out the exercises. Once you have an answer, look in the back and see if your answer agrees with ours. If it does, you're on the right path. If it doesn't, try to figure out what you did wrong. Once you've discovered your error, continue to work out more exercises to master the concept and skill.

Equations are a mathematician's way of expressing ideas in concise shorthand. The problem in reading mathematics is unpacking the shorthand. One useful technique is to read with paper and pencil in hand so you can work out calculations as you go along. When you are baffled, and you wonder, "How did they get that result?" try doing the calculation yourself and see what you get. You'll be amazed (or at least mildly satisfied) at how often that answers your question. Remember, math is not a spectator sport. You don't learn math by passively reading it or watching your professor. You learn mathematics by doing mathematics.

Finally, if there is anything you would like to see changed in the book, feel free to write to us at matrng@hofstra.edu or npritchey@ysu.edu. We're constantly trying to make this book even better. If you'd like to know more about us, we have Web sites that we invite you to visit: http://people.hofstra.edu/rgreenwell and http://people.ysu.edu/~npritchey.

Marge Lial
Ray Greenwell
Nate Ritchey

Prerequisite Skills Diagnostic Test

Below is a very brief test to help you recognize which, if any, prerequisite skills you may need to remediate in order to be successful in this course. After completing the test, check your answers in the back of the book. In addition to the answers, we have also provided the solutions to these problems in Appendix A. These solutions should help remind you how to solve the problems. For problems 5-10, the answers are followed by references to sections within Chapter R where you can find guidance on how to solve the problem and/or additional instruction. Addressing any weak prerequisite skills now will make a positive impact on your success as you progress through this course.

1. What percent of 50 is 10?

2. Simplify $\dfrac{13}{7} - \dfrac{2}{5}$.

3. Let x be the number of apples and y be the number of oranges. Write the following statement as an algebraic equation: "The total number of apples and oranges is 75."

4. Let s be the number of students and p be the number of professors. Write the following statement as an algebraic equation: "There are at least four times as many students as professors."

5. Solve for k: $7k + 8 = -4(3 - k)$.

6. Solve for x: $\dfrac{5}{8}x + \dfrac{1}{16}x = \dfrac{11}{16} + x$.

7. Write in interval notation: $-2 < x \le 5$.

8. Using the variable x, write the following interval as an inequality: $[-\infty, -3]$.

9. Solve for y: $5(y - 2) + 1 \le 7y + 8$.

10. Solve for p: $\dfrac{2}{3}(5p - 3) > \dfrac{3}{4}(2p + 1)$.

R

Algebra Reference

In this chapter, we will review the most important topics in algebra. Knowing algebra is a fundamental prerequisite to success in higher mathematics. This algebra reference is designed for self-study; study it all at once or refer to it when needed throughout the course. Since this is a review, answers to all exercises are given in the answer section at the back of the book.

R.1 Polynomials

An expression such as $9p^4$ is a **term**; the number 9 is the **coefficient**, p is the **variable**, and 4 is the **exponent**. The expression p^4 means $p \cdot p \cdot p \cdot p$, while p^2 means $p \cdot p$, and so on. Terms having the same variable and the same exponent, such as $9x^4$ and $-3x^4$, are **like terms**. Terms that do not have both the same variable and the same exponent, such as m^2 and m^4, are **unlike terms**.

A **polynomial** is a term or a finite sum of terms in which all variables have whole number exponents, and no variables appear in denominators. Examples of polynomials include

$$5x^4 + 2x^3 + 6x, \qquad 8m^3 + 9m^2n - 6mn^2 + 3n^3, \qquad 10p, \qquad \text{and} \qquad -9.$$

Order of Operations

Algebra is a language, and you must be familiar with its rules to correctly interpret algebraic statements. The following order of operations have been agreed upon through centuries of usage.

- Expressions in **parentheses** are calculated first, working from the inside out. The numerator and denominator of a fraction are treated as expressions in parentheses.

- **Powers** are performed next, going from left to right.

- **Multiplication** and **division** are performed next, going from left to right.

- **Addition** and **subtraction** are performed last, going from left to right.

For example, in the expression $(6(x + 1)^2 + 3x - 22)^2$, suppose x has the value of 2. We would evaluate this as follows:

$$
\begin{aligned}
(6(2 + 1)^2 + 3(2) - 22)^2 &= (6(3)^2 + 3(2) - 22)^2 && \text{Evaluate the expression in the innermost parentheses.} \\
&= (6(9) + 3(2) - 22)^2 && \text{Evaluate 3 raised to a power.} \\
&= (54 + 6 - 22)^2 && \text{Perform the multiplication.} \\
&= (38)^2 && \text{Perform the addition and subtraction from left to right.} \\
&= 1444 && \text{Evaluate the power.}
\end{aligned}
$$

In the expression $\dfrac{x^2 + 3x + 6}{x + 6}$, suppose x has the value of 2. We would evaluate this as follows:

$$
\begin{aligned}
\frac{2^2 + 3(2) + 6}{2 + 6} &= \frac{16}{8} && \text{Evaluate the numerator and the denominator.} \\
&= 2 && \text{Simplify the fraction.}
\end{aligned}
$$

Adding and Subtracting Polynomials

The following properties of real numbers are useful for performing operations on polynomials.

Properties of Real Numbers

For all real numbers a, b, and c:

1. $a + b = b + a$;   **Commutative properties**
   $ab = ba$;

2. $(a + b) + c = a + (b + c)$;   **Associative properties**
   $(ab)c = a(bc)$;

3. $a(b + c) = ab + ac$.   **Distributive property**

EXAMPLE 1 **Properties of Real Numbers**

(a) $2 + x = x + 2$ Commutative property of addition

(b) $x \cdot 3 = 3x$ Commutative property of multiplication

(c) $(7x)x = 7(x \cdot x) = 7x^2$ Associative property of multiplication

(d) $3(x + 4) = 3x + 12$ Distributive property

One use of the distributive property is to add or subtract polynomials. Only like terms may be added or subtracted. For example,

$$12y^4 + 6y^4 = (12 + 6)y^4 = 18y^4,$$

and

$$-2m^2 + 8m^2 = (-2 + 8)m^2 = 6m^2,$$

but the polynomial $8y^4 + 2y^5$ cannot be further simplified. To subtract polynomials, we use the facts that $-(a + b) = -a - b$ and $-(a - b) = -a + b$. In the next example, we show how to add and subtract polynomials.

EXAMPLE 2 **Adding and Subtracting Polynomials**

Add or subtract as indicated.

(a) $(8x^3 - 4x^2 + 6x) + (3x^3 + 5x^2 - 9x + 8)$

     SOLUTION Combine like terms.

$$(8x^3 - 4x^2 + 6x) + (3x^3 + 5x^2 - 9x + 8)$$
$$= (8x^3 + 3x^3) + (-4x^2 + 5x^2) + (6x - 9x) + 8$$
$$= 11x^3 + x^2 - 3x + 8$$

(b) $2(-4x^4 + 6x^3 - 9x^2 - 12) + 3(-3x^3 + 8x^2 - 11x + 7)$

     SOLUTION Multiply each polynomial by the coefficient in front of the polynomial, and then combine terms as before.

$$2(-4x^4 + 6x^3 - 9x^2 - 12) + 3(-3x^3 + 8x^2 - 11x + 7)$$
$$= -8x^4 + 12x^3 - 18x^2 - 24 - 9x^3 + 24x^2 - 33x + 21$$
$$= -8x^4 + 3x^3 + 6x^2 - 33x - 3$$

(c) $(2x^2 - 11x + 8) - (7x^2 - 6x + 2)$

YOUR TURN 1 Perform the operation $3(x^2 - 4x - 5) - 4(3x^2 - 5x - 7)$.

     SOLUTION Distributing the minus sign and combining like terms yields

$$(2x^2 - 11x + 8) + (-7x^2 + 6x - 2)$$
$$= -5x^2 - 5x + 6.$$ **TRY YOUR TURN 1**

Multiplying Polynomials
The distributive property is also used to multiply polynomials, along with the fact that $a^m \cdot a^n = a^{m+n}$. For example,

$$x \cdot x = x^1 \cdot x^1 = x^{1+1} = x^2 \qquad \text{and} \qquad x^2 \cdot x^5 = x^{2+5} = x^7.$$

EXAMPLE 3 **Multiplying Polynomials**

Multiply.

(a) $8x(6x - 4)$

     SOLUTION Using the distributive property yields

$$8x(6x - 4) = 8x(6x) - 8x(4)$$
$$= 48x^2 - 32x.$$

(b) $(3p - 2)(p^2 + 5p - 1)$

SOLUTION Using the distributive property yields

$$(3p - 2)(p^2 + 5p - 1)$$
$$= 3p(p^2 + 5p - 1) - 2(p^2 + 5p - 1)$$
$$= 3p(p^2) + 3p(5p) + 3p(-1) - 2(p^2) - 2(5p) - 2(-1)$$
$$= 3p^3 + 15p^2 - 3p - 2p^2 - 10p + 2$$
$$= 3p^3 + 13p^2 - 13p + 2.$$

(c) $(x + 2)(x + 3)(x - 4)$

SOLUTION Multiplying the first two polynomials and then multiplying their product by the third polynomial yields

$$(x + 2)(x + 3)(x - 4)$$
$$= [(x + 2)(x + 3)](x - 4)$$
$$= (x^2 + 2x + 3x + 6)(x - 4)$$
$$= (x^2 + 5x + 6)(x - 4)$$
$$= x^3 + 5x^2 + 6x - 4x^2 - 20x - 24$$
$$= x^3 + x^2 - 14x - 24. \qquad \textbf{TRY YOUR TURN 2}$$

YOUR TURN 2 Perform the operation $(3y + 2)(4y^2 - 2y - 5)$.

A **binomial** is a polynomial with exactly two terms, such as $2x + 1$ or $m + n$. When two binomials are multiplied, the FOIL method (First, Outer, Inner, Last) is used as a memory aid.

EXAMPLE 4 Multiplying Polynomials

Find $(2m - 5)(m + 4)$ using the FOIL method.

SOLUTION

$$(2m - 5)(m + 4) = \overset{\text{F}}{(2m)(m)} + \overset{\text{O}}{(2m)(4)} + \overset{\text{I}}{(-5)(m)} + \overset{\text{L}}{(-5)(4)}$$
$$= 2m^2 + 8m - 5m - 20$$
$$= 2m^2 + 3m - 20$$

EXAMPLE 5 Multiplying Polynomials

Find $(2k - 5m)^3$.

SOLUTION Write $(2k - 5m)^3$ as $(2k - 5m)(2k - 5m)(2k - 5m)$. Then multiply the first two factors using FOIL.

$$(2k - 5m)(2k - 5m) = 4k^2 - 10km - 10km + 25m^2$$
$$= 4k^2 - 20km + 25m^2$$

Now multiply this last result by $(2k - 5m)$ using the distributive property, as in Example 3(b).

$$(4k^2 - 20km + 25m^2)(2k - 5m)$$
$$= 4k^2(2k - 5m) - 20km(2k - 5m) + 25m^2(2k - 5m)$$
$$= 8k^3 - 20k^2m - 40k^2m + 100km^2 + 50km^2 - 125m^3$$
$$= 8k^3 - 60k^2m + 150km^2 - 125m^3 \qquad \text{Combine like terms.}$$

Notice in the first part of Example 5, when we multiplied $(2k - 5m)$ by itself, that the product of the square of a binomial is the square of the first term, $(2k)^2$, plus twice the product of the two terms, $(2)(2k)(-5m)$, plus the square of the last term, $(-5k)^2$.

CAUTION Avoid the common error of writing $(x + y)^2 = x^2 + y^2$. As the first step of Example 5 shows, the square of a binomial has three terms, so

$$(x + y)^2 = x^2 + 2xy + y^2.$$

Furthermore, higher powers of a binomial also result in more than two terms. For example, verify by multiplication that

$$(x + y)^3 = x^3 + 3x^2y + 3xy^2 + y^3.$$

Remember, for any value of $n \neq 1$,

$$(x + y)^n \neq x^n + y^n.$$

R.1 EXERCISES

Perform the indicated operations.

1. $(2x^2 - 6x + 11) + (-3x^2 + 7x - 2)$

2. $(-4y^2 - 3y + 8) - (2y^2 - 6y - 2)$

3. $-6(2q^2 + 4q - 3) + 4(-q^2 + 7q - 3)$

4. $2(3r^2 + 4r + 2) - 3(-r^2 + 4r - 5)$

5. $(0.613x^2 - 4.215x + 0.892) - 0.47(2x^2 - 3x + 5)$

6. $0.5(5r^2 + 3.2r - 6) - (1.7r^2 - 2r - 1.5)$

7. $-9m(2m^2 + 3m - 1)$

8. $6x(-2x^3 + 5x + 6)$

9. $(3t - 2y)(3t + 5y)$

10. $(9k + q)(2k - q)$

11. $(2 - 3x)(2 + 3x)$

12. $(6m + 5)(6m - 5)$

13. $\left(\frac{2}{5}y + \frac{1}{8}z\right)\left(\frac{3}{5}y + \frac{1}{2}z\right)$

14. $\left(\frac{3}{4}r - \frac{2}{3}s\right)\left(\frac{5}{4}r + \frac{1}{3}s\right)$

15. $(3p - 1)(9p^2 + 3p + 1)$

16. $(3p + 2)(5p^2 + p - 4)$

17. $(2m + 1)(4m^2 - 2m + 1)$

18. $(k + 2)(12k^3 - 3k^2 + k + 1)$

19. $(x + y + z)(3x - 2y - z)$

20. $(r + 2s - 3t)(2r - 2s + t)$

21. $(x + 1)(x + 2)(x + 3)$

22. $(x - 1)(x + 2)(x - 3)$

23. $(x + 2)^2$

24. $(2a - 4b)^2$

25. $(x - 2y)^3$

26. $(3x + y)^3$

YOUR TURN ANSWERS

1. $-9x^2 + 8x + 13$
2. $12y^3 + 2y^2 - 19y - 10$

R.2 Factoring

Multiplication of polynomials relies on the distributive property. The reverse process, where a polynomial is written as a product of other polynomials, is called **factoring**. For example, one way to factor the number 18 is to write it as the product $9 \cdot 2$; both 9 and 2 are **factors** of 18. Usually, only integers are used as factors of integers. The number 18 can also be written with three integer factors as $2 \cdot 3 \cdot 3$.

The Greatest Common Factor
To factor the algebraic expression $15m + 45$, first note that both $15m$ and 45 are divisible by 15; $15m = 15 \cdot m$ and $45 = 15 \cdot 3$. By the distributive property,

$$15m + 45 = 15 \cdot m + 15 \cdot 3 = 15(m + 3).$$

Both 15 and $m + 3$ are factors of $15m + 45$. Since 15 divides into both terms of $15m + 45$ (and is the largest number that will do so), 15 is the **greatest common factor** for the polynomial $15m + 45$. The process of writing $15m + 45$ as $15(m + 3)$ is often called **factoring out** the greatest common factor.

EXAMPLE 1 Factoring

Factor out the greatest common factor.

(a) $12p - 18q$

SOLUTION Both $12p$ and $18q$ are divisible by 6. Therefore,

$$12p - 18q = 6 \cdot 2p - 6 \cdot 3q = 6(2p - 3q).$$

(b) $8x^3 - 9x^2 + 15x$

SOLUTION Each of these terms is divisible by x.

$$8x^3 - 9x^2 + 15x = (8x^2) \cdot x - (9x) \cdot x + 15 \cdot x$$
$$= x(8x^2 - 9x + 15) \quad \text{or} \quad (8x^2 - 9x + 15)x$$

TRY YOUR TURN 1

YOUR TURN 1 Factor $4z^4 + 4z^3 + 18z^2$.

One can always check factorization by finding the product of the factors and comparing it to the original expression.

CAUTION When factoring out the greatest common factor in an expression like $2x^2 + x$, be careful to remember the 1 in the second term.

$$2x^2 + x = 2x^2 + 1x = x(2x + 1), \quad \text{not } x(2x).$$

Factoring Trinomials

A polynomial that has no greatest common factor (other than 1) may still be factorable. For example, the polynomial $x^2 + 5x + 6$ can be factored as $(x + 2)(x + 3)$. To see that this is correct, find the product $(x + 2)(x + 3)$; you should get $x^2 + 5x + 6$. A polynomial such as this with three terms is called a **trinomial**. To factor the trinomial $x^2 + 5x + 6$, where the coefficient of x^2 is 1, we use FOIL backwards.

EXAMPLE 2 Factoring a Trinomial

Factor $y^2 + 8y + 15$.

SOLUTION Since the coefficient of y^2 is 1, factor by finding two numbers whose *product* is 15 and whose *sum* is 8. Since the constant and the middle term are positive, the numbers must both be positive. Begin by listing all pairs of positive integers having a product of 15. As you do this, also form the sum of each pair of numbers.

Products	Sums
$15 \cdot 1 = 15$	$15 + 1 = 16$
$5 \cdot 3 = 15$	$5 + 3 = 8$

The numbers 5 and 3 have a product of 15 and a sum of 8. Thus, $y^2 + 8y + 15$ factors as

$$y^2 + 8y + 15 = (y + 5)(y + 3).$$

The answer also can be written as $(y + 3)(y + 5)$.

If the coefficient of the squared term is *not* 1, work as shown below.

EXAMPLE 3 Factoring a Trinomial

Factor $4x^2 + 8xy - 5y^2$.

SOLUTION The possible factors of $4x^2$ are $4x$ and x or $2x$ and $2x$; the possible factors of $-5y^2$ are $-5y$ and y or $5y$ and $-y$. Try various combinations of these factors until one works (if, indeed, any work). For example, try the product $(x + 5y)(4x - y)$.

$$(x + 5y)(4x - y) = 4x^2 - xy + 20xy - 5y^2$$
$$= 4x^2 + 19xy - 5y^2$$

This product is not correct, so try another combination.

$$(2x - y)(2x + 5y) = 4x^2 + 10xy - 2xy - 5y^2$$
$$= 4x^2 + 8xy - 5y^2$$

YOUR TURN 2   Factor
$6a^2 + 5ab - 4b^2$.

Since this combination gives the correct polynomial,

$$4x^2 + 8xy - 5y^2 = (2x - y)(2x + 5y).$$ **TRY YOUR TURN 2**

Special Factorizations   Four special factorizations occur so often that they are listed here for future reference.

Special Factorizations

$x^2 - y^2 = (x + y)(x - y)$	**Difference of two squares**
$x^2 + 2xy + y^2 = (x + y)^2$	**Perfect square**
$x^3 - y^3 = (x - y)(x^2 + xy + y^2)$	**Difference of two cubes**
$x^3 + y^3 = (x + y)(x^2 - xy + y^2)$	**Sum of two cubes**

A polynomial that cannot be factored is called a **prime polynomial**.

EXAMPLE 4   Factoring Polynomials

Factor each polynomial, if possible.

(a) $64p^2 - 49q^2 = (8p)^2 - (7q)^2 = (8p + 7q)(8p - 7q)$ Difference of two squares

(b) $x^2 + 36$ is a prime polynomial.

(c) $x^2 + 12x + 36 = (x + 6)^2$ Perfect square

(d) $9y^2 - 24yz + 16z^2 = (3y - 4z)^2$ Perfect square

(e) $y^3 - 8 = y^3 - 2^3 = (y - 2)(y^2 + 2y + 4)$ Difference of two cubes

(f) $m^3 + 125 = m^3 + 5^3 = (m + 5)(m^2 - 5m + 25)$ Sum of two cubes

(g) $8k^3 - 27z^3 = (2k)^3 - (3z)^3 = (2k - 3z)(4k^2 + 6kz + 9z^2)$ Difference of two cubes

(h) $p^4 - 1 = (p^2 + 1)(p^2 - 1) = (p^2 + 1)(p + 1)(p - 1)$ Difference of two squares

CAUTION   In factoring, always look for a common factor first. Since $36x^2 - 4y^2$ has a common factor of 4,

$$36x^2 - 4y^2 = 4(9x^2 - y^2) = 4(3x + y)(3x - y).$$

It would be incomplete to factor it as

$$36x^2 - 4y^2 = (6x + 2y)(6x - 2y),$$

since each factor can be factored still further. To *factor* means to factor completely, so that each polynomial factor is prime.

R.2   EXERCISES

Factor each polynomial. If a polynomial cannot be factored, write *prime*. Factor out the greatest common factor as necessary.

1. $7a^3 + 14a^2$

2. $3y^3 + 24y^2 + 9y$

3. $13p^4q^2 - 39p^3q + 26p^2q^2$

4. $60m^4 - 120m^3n + 50m^2n^2$

5. $m^2 - 5m - 14$

6. $x^2 + 4x - 5$

7. $z^2 + 9z + 20$

8. $b^2 - 8b + 7$

9. $a^2 - 6ab + 5b^2$

10. $s^2 + 2st - 35t^2$

11. $y^2 - 4yz - 21z^2$

12. $3x^2 + 4x - 7$

13. $3a^2 + 10a + 7$

14. $15y^2 + y - 2$

15. $21m^2 + 13mn + 2n^2$

16. $6a^2 - 48a - 120$

17. $3m^3 + 12m^2 + 9m$

18. $4a^2 + 10a + 6$

19. $24a^4 + 10a^3b - 4a^2b^2$

20. $24x^4 + 36x^3y - 60x^2y^2$

21. $x^2 - 64$ **22.** $9m^2 - 25$

23. $10x^2 - 160$ **24.** $9x^2 + 64$

25. $z^2 + 14zy + 49y^2$ **26.** $s^2 - 10st + 25t^2$

27. $9p^2 - 24p + 16$ **28.** $a^3 - 216$

29. $27r^3 - 64s^3$ **30.** $3m^3 + 375$

31. $x^4 - y^4$ **32.** $16a^4 - 81b^4$

◼ YOUR TURN ANSWERS

1. $2z^2(2z^2 + 2z + 9)$ **2.** $(2a - b)(3a + 4b)$

R.3 Rational Expressions

Many algebraic fractions are **rational expressions**, which are quotients of polynomials with nonzero denominators. Examples include

$$\frac{8}{x - 1}, \qquad \frac{3x^2 + 4x}{5x - 6}, \qquad \text{and} \qquad \frac{2y + 1}{y^2}.$$

Next, we summarize properties for working with rational expressions.

Properties of Rational Expressions

For all mathematical expressions P, Q, R, and S, with $Q \neq 0$ and $S \neq 0$:

$$\frac{P}{Q} = \frac{PS}{QS} \qquad \textbf{Fundamental property}$$

$$\frac{P}{Q} + \frac{R}{Q} = \frac{P + R}{Q} \qquad \textbf{Addition}$$

$$\frac{P}{Q} - \frac{R}{Q} = \frac{P - R}{Q} \qquad \textbf{Subtraction}$$

$$\frac{P}{Q} \cdot \frac{R}{S} = \frac{PR}{QS} \qquad \textbf{Multiplication}$$

$$\frac{P}{Q} \div \frac{R}{S} = \frac{P}{Q} \cdot \frac{S}{R} \ (R \neq 0) \qquad \textbf{Division}$$

When writing a rational expression in lowest terms, we may need to use the fact that $\dfrac{a^m}{a^n} = a^{m-n}$. For example,

$$\frac{x^4}{3x} = \frac{1x^4}{3x} = \frac{1}{3} \cdot \frac{x^4}{x} = \frac{1}{3} \cdot x^{4-1} = \frac{1}{3}x^3.$$

EXAMPLE 1   **Reducing Rational Expressions**

Write each rational expression in lowest terms, that is, reduce the expression as much as possible.

(a) $\dfrac{8x + 16}{4} = \dfrac{8(x + 2)}{4} = \dfrac{4 \cdot 2(x + 2)}{4} = 2(x + 2)$

Factor both the numerator and denominator in order to identify any common factors, which have a quotient of 1. The answer could also be written as $2x + 4$.

YOUR TURN 1 Write
in lowest terms

$$\frac{z^2 + 5z + 6}{2z^2 + 7z + 3}.$$

(b) $\dfrac{k^2 + 7k + 12}{k^2 + 2k - 3} = \dfrac{(k + 4)(k + 3)}{(k - 1)(k + 3)} = \dfrac{k + 4}{k - 1}$

The answer cannot be further reduced.   **TRY YOUR TURN 1** ▬

> **CAUTION** One of the most common errors in algebra involves incorrect use of the fundamental property of rational expressions. Only common *factors* may be divided or "canceled." It is essential to factor rational expressions before writing them in lowest terms. In Example 1(b), for instance, it is not correct to "cancel" k^2 (or cancel k, or divide 12 by -3) because the additions and subtraction must be performed first. Here they cannot be performed, so it is not possible to divide. After factoring, however, the fundamental property can be used to write the expression in lowest terms.

EXAMPLE 2   Combining Rational Expressions

Perform each operation.

(a) $\dfrac{3y + 9}{6} \cdot \dfrac{18}{5y + 15}$

SOLUTION   Factor where possible, then multiply numerators and denominators and reduce to lowest terms.

$$\frac{3y + 9}{6} \cdot \frac{18}{5y + 15} = \frac{3(y + 3)}{6} \cdot \frac{18}{5(y + 3)}$$

$$= \frac{3 \cdot 18(y + 3)}{6 \cdot 5(y + 3)}$$

$$= \frac{3 \cdot \cancel{6} \cdot 3\cancel{(y + 3)}}{\cancel{6} \cdot 5\cancel{(y + 3)}} = \frac{3 \cdot 3}{5} = \frac{9}{5}$$

(b) $\dfrac{m^2 + 5m + 6}{m + 3} \cdot \dfrac{m}{m^2 + 3m + 2}$

SOLUTION   Factor where possible.

$$\frac{(m + 2)(m + 3)}{m + 3} \cdot \frac{m}{(m + 2)(m + 1)}$$

$$= \frac{m\cancel{(m + 2)}\cancel{(m + 3)}}{\cancel{(m + 3)}\cancel{(m + 2)}(m + 1)} = \frac{m}{m + 1}$$

(c) $\dfrac{9p - 36}{12} \div \dfrac{5(p - 4)}{18}$

SOLUTION   Use the division property of rational expressions.

$$\frac{9p - 36}{12} \cdot \frac{18}{5(p - 4)}$$ Invert and multiply.

$$= \frac{9\cancel{(p - 4)}}{\cancel{6} \cdot 2} \cdot \frac{\cancel{6} \cdot 3}{5\cancel{(p - 4)}} = \frac{27}{10}$$ Factor and reduce to lowest terms.

(d) $\dfrac{4}{5k} - \dfrac{11}{5k}$

SOLUTION   As shown in the list of properties, to subtract two rational expressions that have the same denominators, subtract the numerators while keeping the same denominator.

$$\frac{4}{5k} - \frac{11}{5k} = \frac{4 - 11}{5k} = -\frac{7}{5k}$$

(e) $\dfrac{7}{p} + \dfrac{9}{2p} + \dfrac{1}{3p}$

SOLUTION These three fractions cannot be added until their denominators are the same. A **common denominator** into which p, $2p$, and $3p$ all divide is $6p$. Note that $12p$ is also a common denominator, but $6p$ is the **least common denominator**. Use the fundamental property to rewrite each rational expression with a denominator of $6p$.

$$\frac{7}{p} + \frac{9}{2p} + \frac{1}{3p} = \frac{6 \cdot 7}{6 \cdot p} + \frac{3 \cdot 9}{3 \cdot 2p} + \frac{2 \cdot 1}{2 \cdot 3p}$$

$$= \frac{42}{6p} + \frac{27}{6p} + \frac{2}{6p}$$

$$= \frac{42 + 27 + 2}{6p}$$

$$= \frac{71}{6p}$$

(f) $\dfrac{x + 1}{x^2 + 5x + 6} - \dfrac{5x - 1}{x^2 - x - 12}$

SOLUTION To find the least common denominator, we first factor each denominator. Then we change each fraction so they all have the same denominator, being careful to multiply only by quotients that equal 1.

$$\frac{x + 1}{x^2 + 5x + 6} - \frac{5x - 1}{x^2 - x - 12}$$

$$= \frac{x + 1}{(x + 2)(x + 3)} - \frac{5x - 1}{(x + 3)(x - 4)}$$

$$= \frac{x + 1}{(x + 2)(x + 3)} \cdot \frac{(x - 4)}{(x - 4)} - \frac{5x - 1}{(x + 3)(x - 4)} \cdot \frac{(x + 2)}{(x + 2)}$$

$$= \frac{(x^2 - 3x - 4) - (5x^2 + 9x - 2)}{(x + 2)(x + 3)(x - 4)}$$

$$= \frac{-4x^2 - 12x - 2}{(x + 2)(x + 3)(x - 4)}$$

$$= \frac{-2(2x^2 + 6x + 1)}{(x + 2)(x + 3)(x - 4)}$$

Because the numerator cannot be factored further, we leave our answer in this form. We could also multiply out the denominator, but factored form is usually more useful.

TRY YOUR TURN 2

YOUR TURN 2 Perform each of the following operations.

(a) $\dfrac{z^2 + 5z + 6}{2z^2 - 5z - 3} \cdot \dfrac{2z^2 - z - 1}{z^2 + 2z - 3}$.

(b) $\dfrac{a - 3}{a^2 + 3a + 2} + \dfrac{5a}{a^2 - 4}$.

R.3 EXERCISES

Write each rational expression in lowest terms.

1. $\dfrac{5v^2}{35v}$

2. $\dfrac{25p^3}{10p^2}$

3. $\dfrac{8k + 16}{9k + 18}$

4. $\dfrac{2(t - 15)}{(t - 15)(t + 2)}$

5. $\dfrac{4x^3 - 8x^2}{4x^2}$

6. $\dfrac{36y^2 + 72y}{9y}$

7. $\dfrac{m^2 - 4m + 4}{m^2 + m - 6}$

8. $\dfrac{r^2 - r - 6}{r^2 + r - 12}$

9. $\dfrac{3x^2 + 3x - 6}{x^2 - 4}$

10. $\dfrac{z^2 - 5z + 6}{z^2 - 4}$

11. $\dfrac{m^4 - 16}{4m^2 - 16}$

12. $\dfrac{6y^2 + 11y + 4}{3y^2 + 7y + 4}$

Perform the indicated operations.

13. $\dfrac{9k^2}{25} \cdot \dfrac{5}{3k}$

14. $\dfrac{15p^3}{9p^2} \div \dfrac{6p}{10p^2}$

15. $\dfrac{3a + 3b}{4c} \cdot \dfrac{12}{5(a + b)}$

16. $\dfrac{a - 3}{16} \div \dfrac{a - 3}{32}$

17. $\dfrac{2k-16}{6} \div \dfrac{4k-32}{3}$

18. $\dfrac{9y-18}{6y+12} \cdot \dfrac{3y+6}{15y-30}$

19. $\dfrac{4a+12}{2a-10} \div \dfrac{a^2-9}{a^2-a-20}$

20. $\dfrac{6r-18}{9r^2+6r-24} \cdot \dfrac{12r-16}{4r-12}$

21. $\dfrac{k^2+4k-12}{k^2+10k+24} \cdot \dfrac{k^2+k-12}{k^2-9}$

22. $\dfrac{m^2+3m+2}{m^2+5m+4} \div \dfrac{m^2+5m+6}{m^2+10m+24}$

23. $\dfrac{2m^2-5m-12}{m^2-10m+24} \div \dfrac{4m^2-9}{m^2-9m+18}$

24. $\dfrac{4n^2+4n-3}{6n^2-n-15} \cdot \dfrac{8n^2+32n+30}{4n^2+16n+15}$

25. $\dfrac{a+1}{2} - \dfrac{a-1}{2}$

26. $\dfrac{3}{p} + \dfrac{1}{2}$

27. $\dfrac{6}{5y} - \dfrac{3}{2}$

28. $\dfrac{1}{6m} + \dfrac{2}{5m} + \dfrac{4}{m}$

29. $\dfrac{1}{m-1} + \dfrac{2}{m}$

30. $\dfrac{5}{2r+3} - \dfrac{2}{r}$

31. $\dfrac{8}{3(a-1)} + \dfrac{2}{a-1}$

32. $\dfrac{2}{5(k-2)} + \dfrac{3}{4(k-2)}$

33. $\dfrac{4}{x^2+4x+3} + \dfrac{3}{x^2-x-2}$

34. $\dfrac{y}{y^2+2y-3} - \dfrac{1}{y^2+4y+3}$

35. $\dfrac{3k}{2k^2+3k-2} - \dfrac{2k}{2k^2-7k+3}$

36. $\dfrac{4m}{3m^2+7m-6} - \dfrac{m}{3m^2-14m+8}$

37. $\dfrac{2}{a+2} + \dfrac{1}{a} + \dfrac{a-1}{a^2+2a}$

38. $\dfrac{5x+2}{x^2-1} + \dfrac{3}{x^2+x} - \dfrac{1}{x^2-x}$

YOUR TURN ANSWERS

1. $(z+2)/(2z+1)$

2a. $(z+2)/(z-3)$

2b. $6(a^2+1)/[(a-2)(a+2)(a+1)]$

R.4 Equations

Linear Equations

Equations that can be written in the form $ax + b = 0$, where a and b are real numbers, with $a \neq 0$, are **linear equations**. Examples of linear equations include $5y + 9 = 16$, $8x = 4$, and $-3p + 5 = -8$. Equations that are *not* linear include absolute value equations such as $|x| = 4$. The following properties are used to solve linear equations.

Properties of Equality

For all real numbers a, b, and c:

1. **If $a = b$, then $a + c = b + c$.**   **Addition property of equality**
   (The same number may be added to both sides of an equation.)

2. **If $a = b$, then $ac = bc$.**   **Multiplication property of equality**
   (Both sides of an equation may be multiplied by the same number.)

EXAMPLE 1   Solving Linear Equations

Solve the following equations.

(a) $x - 2 = 3$

SOLUTION   The goal is to isolate the variable. Using the addition property of equality yields

$$x - 2 + 2 = 3 + 2, \quad \text{or} \quad x = 5.$$

(b) $\dfrac{x}{2} = 3$

SOLUTION Using the multiplication property of equality yields

$$2 \cdot \frac{x}{2} = 2 \cdot 3, \quad \text{or} \quad x = 6.$$

The following example shows how these properties are used to solve linear equations. The goal is to isolate the variable. The solutions should always be checked by substitution in the original equation.

EXAMPLE 2   Solving a Linear Equation

Solve $2x - 5 + 8 = 3x + 2(2 - 3x)$.

SOLUTION

$2x - 5 + 8 = 3x + 4 - 6x$	Distributive property
$2x + 3 = -3x + 4$	Combine like terms.
$5x + 3 = 4$	Add $3x$ to both sides.
$5x = 1$	Add -3 to both sides.
$x = \dfrac{1}{5}$	Multiply both sides by $\frac{1}{5}$.

YOUR TURN 1 Solve $3x - 7 = 4(5x + 2) - 7x$.

Check by substituting in the original equation. The left side becomes $2(1/5) - 5 + 8$ and the right side becomes $3(1/5) + 2[2 - 3(1/5)]$. Verify that both of these expressions simplify to $17/5$. **TRY YOUR TURN 1**

Quadratic Equations
An equation with 2 as the highest exponent of the variable is a *quadratic equation*. A **quadratic equation** has the form $ax^2 + bx + c = 0$, where a, b, and c are real numbers and $a \neq 0$. A quadratic equation written in the form $ax^2 + bx + c = 0$ is said to be in **standard form**.

The simplest way to solve a quadratic equation, but one that is not always applicable, is by factoring. This method depends on the **zero-factor property**.

Zero-Factor Property
If a and b are real numbers, with $ab = 0$, then either

$$a = 0 \text{ or } b = 0 \quad \text{(or both).}$$

EXAMPLE 3   Solving a Quadratic Equation

Solve $6r^2 + 7r = 3$.

SOLUTION First write the equation in standard form.

$$6r^2 + 7r - 3 = 0$$

Now factor $6r^2 + 7r - 3$ to get

$$(3r - 1)(2r + 3) = 0.$$

By the zero-factor property, the product $(3r - 1)(2r + 3)$ can equal 0 if and only if

$$3r - 1 = 0 \quad \text{or} \quad 2r + 3 = 0.$$

YOUR TURN 2 Solve $2m^2 + 7m = 15$.

Solve each of these equations separately to find that the solutions are $1/3$ and $-3/2$. Check these solutions by substituting them in the original equation. **TRY YOUR TURN 2**

CAUTION Remember, the zero-factor property requires that the product of two (or more) factors be equal to *zero*, not some other quantity. It would be incorrect to use the zero-factor property with an equation in the form $(x + 3)(x - 1) = 4$, for example.

If a quadratic equation cannot be solved easily by factoring, use the *quadratic formula*. (The derivation of the quadratic formula is given in most algebra books.)

Quadratic Formula

The solutions of the quadratic equation $ax^2 + bx + c = 0$, where $a \neq 0$, are given by

$$x = \frac{-b \pm \sqrt{b^2 - 4ac}}{2a}.$$

EXAMPLE 4 **Quadratic Formula**

Solve $x^2 - 4x - 5 = 0$ by the quadratic formula.

SOLUTION The equation is already in standard form (it has 0 alone on one side of the equal sign), so the values of a, b, and c from the quadratic formula are easily identified. The coefficient of the squared term gives the value of a; here, $a = 1$. Also, $b = -4$ and $c = -5$. (Be careful to use the correct signs.) Substitute these values into the quadratic formula.

$$x = \frac{-(-4) \pm \sqrt{(-4)^2 - 4(1)(-5)}}{2(1)} \qquad a = 1, b = -4, c = -5$$

$$x = \frac{4 \pm \sqrt{16 + 20}}{2} \qquad (-4)^2 = (-4)(-4) = 16$$

$$x = \frac{4 \pm 6}{2} \qquad \sqrt{16 + 20} = \sqrt{36} = 6$$

The $\pm$ sign represents the two solutions of the equation. To find both of the solutions, first use $+$ and then use $-$.

$$x = \frac{4 + 6}{2} = \frac{10}{2} = 5 \qquad \text{or} \qquad x = \frac{4 - 6}{2} = \frac{-2}{2} = -1$$

The two solutions are 5 and -1.

CAUTION Notice in the quadratic formula that the square root is added to or subtracted from the value of $-b$ *before* dividing by $2a$.

EXAMPLE 5 **Quadratic Formula**

Solve $x^2 + 1 = 4x$.

SOLUTION First, add $-4x$ on both sides of the equal sign in order to get the equation in standard form.

$$x^2 - 4x + 1 = 0$$

Now identify the letters a, b, and c. Here $a = 1$, $b = -4$, and $c = 1$. Substitute these numbers into the quadratic formula.

$$x = \frac{-(-4) \pm \sqrt{(-4)^2 - 4(1)(1)}}{2(1)}$$

$$= \frac{4 \pm \sqrt{16 - 4}}{2}$$

$$= \frac{4 \pm \sqrt{12}}{2}$$

Simplify the solutions by writing $\sqrt{12}$ as $\sqrt{4 \cdot 3} = \sqrt{4} \cdot \sqrt{3} = 2\sqrt{3}$. Substituting $2\sqrt{3}$ for $\sqrt{12}$ gives

$$x = \frac{4 \pm 2\sqrt{3}}{2}$$

$$= \frac{2(2 \pm \sqrt{3})}{2} \qquad \text{Factor } 4 \pm 2\sqrt{3}.$$

$$= 2 \pm \sqrt{3}. \qquad \text{Reduce to lowest terms.}$$

The two solutions are $2 + \sqrt{3}$ and $2 - \sqrt{3}$.

The exact values of the solutions are $2 + \sqrt{3}$ and $2 - \sqrt{3}$. The $\sqrt{}$ key on a calculator gives decimal approximations of these solutions (to the nearest thousandth):

$$2 + \sqrt{3} \approx 2 + 1.732 = 3.732*$$
$$2 - \sqrt{3} \approx 2 - 1.732 = 0.268 \qquad \text{TRY YOUR TURN 3}$$

YOUR TURN 3 Solve $z^2 + 6 = 8z$.

NOTE Sometimes the quadratic formula will give a result with a negative number under the radical sign, such as $3 \pm \sqrt{-5}$. A solution of this type is a complex number. Since this text deals only with real numbers, such solutions cannot be used.

Equations with Fractions
When an equation includes fractions, first eliminate all denominators by multiplying both sides of the equation by a common denominator, a number that can be divided (with no remainder) by each denominator in the equation. When an equation involves fractions with variable denominators, it is *necessary* to check all solutions in the original equation to be sure that no solution will lead to a zero denominator.

EXAMPLE 6   Solving Rational Equations

Solve each equation.

(a) $\dfrac{r}{10} - \dfrac{2}{15} = \dfrac{3r}{20} - \dfrac{1}{5}$

SOLUTION   The denominators are 10, 15, 20, and 5. Each of these numbers can be divided into 60, so 60 is a common denominator. Multiply both sides of the equation by 60 and use the distributive property. (If a common denominator cannot be found easily, all the denominators in the problem can be multiplied together to produce one.)

$$\frac{r}{10} - \frac{2}{15} = \frac{3r}{20} - \frac{1}{5}$$

$$60\left(\frac{r}{10} - \frac{2}{15}\right) = 60\left(\frac{3r}{20} - \frac{1}{5}\right) \qquad \text{Multiply by the common denominator.}$$

$$60\left(\frac{r}{10}\right) - 60\left(\frac{2}{15}\right) = 60\left(\frac{3r}{20}\right) - 60\left(\frac{1}{5}\right) \qquad \text{Distributive property}$$

$$6r - 8 = 9r - 12$$

*The symbol $\approx$ means "is approximately equal to."

Add $-9r$ and 8 to both sides.

$$6r - 8 + (-9r) + 8 = 9r - 12 + (-9r) + 8$$
$$-3r = -4$$
$$r = \frac{4}{3} \qquad \text{Multiply each side by } -\tfrac{1}{3}.$$

Check by substituting into the original equation.

(b) $\dfrac{3}{x^2} - 12 = 0$

SOLUTION   Begin by multiplying both sides of the equation by x^2 to get $3 - 12x^2 = 0$. This equation could be solved by using the quadratic formula with $a = -12$, $b = 0$, and $c = 3$. Another method that works well for the type of quadratic equation in which $b = 0$ is shown below.

$$3 - 12x^2 = 0$$
$$3 = 12x^2 \qquad \text{Add } 12x^2.$$
$$\frac{1}{4} = x^2 \qquad \text{Multiply by } \tfrac{1}{12}.$$
$$\pm\frac{1}{2} = x \qquad \text{Take square roots.}$$

Verify that there are two solutions, $-1/2$ and $1/2$.

(c) $\dfrac{2}{k} - \dfrac{3k}{k+2} = \dfrac{k}{k^2 + 2k}$

SOLUTION   Factor $k^2 + 2k$ as $k(k + 2)$. The least common denominator for all the fractions is $k(k + 2)$. Multiplying both sides by $k(k + 2)$ gives the following:

$$k(k + 2) \cdot \left(\frac{2}{k} - \frac{3k}{k+2} \right) = k(k + 2) \cdot \frac{k}{k^2 + 2k}$$
$$2(k + 2) - 3k(k) = k$$
$$2k + 4 - 3k^2 = k \qquad \text{Distributive property}$$
$$-3k^2 + k + 4 = 0 \qquad \text{Add } -k; \text{ rearrange terms.}$$
$$3k^2 - k - 4 = 0 \qquad \text{Multiply by } -1.$$
$$(3k - 4)(k + 1) = 0 \qquad \text{Factor.}$$
$$3k - 4 = 0 \qquad \text{or} \qquad k + 1 = 0$$
$$k = \frac{4}{3} \qquad k = -1$$

YOUR TURN 4   Solve
$\dfrac{1}{x^2 - 4} + \dfrac{2}{x - 2} = \dfrac{1}{x}$.

Verify that the solutions are $4/3$ and -1.   **TRY YOUR TURN 4**

CAUTION   It is possible to get, as a solution of a rational equation, a number that makes one or more of the denominators in the original equation equal to zero. That number is not a solution, so it is *necessary* to check all potential solutions of rational equations. These introduced solutions are called **extraneous solutions**.

EXAMPLE 7   Solving a Rational Equation

Solve $\dfrac{2}{x - 3} + \dfrac{1}{x} = \dfrac{6}{x(x - 3)}$.

SOLUTION The common denominator is $x(x-3)$. Multiply both sides by $x(x-3)$ and solve the resulting equation.

$$x(x-3)\cdot\left(\frac{2}{x-3}+\frac{1}{x}\right)=x(x-3)\cdot\left[\frac{6}{x(x-3)}\right]$$

$$2x+x-3=6$$

$$3x=9$$

$$x=3$$

Checking this potential solution by substitution in the original equation shows that 3 makes two denominators 0. Thus, 3 cannot be a solution, so there is no solution for this equation. ▬

R.4 EXERCISES

Solve each equation.

1. $2x+8=x-4$

2. $5x+2=8-3x$

3. $0.2m-0.5=0.1m+0.7$

4. $\frac{2}{3}k-k+\frac{3}{8}=\frac{1}{2}$

5. $3r+2-5(r+1)=6r+4$

6. $5(a+3)+4a-5=-(2a-4)$

7. $2[3m-2(3-m)-4]=6m-4$

8. $4[2p-(3-p)+5]=-7p-2$

Solve each equation by factoring or by using the quadratic formula. If the solutions involve square roots, give both the exact solutions and the approximate solutions to three decimal places.

9. $x^2+5x+6=0$

10. $x^2=3+2x$

11. $m^2=14m-49$

12. $2k^2-k=10$

13. $12x^2-5x=2$

14. $m(m-7)=-10$

15. $4x^2-36=0$

16. $z(2z+7)=4$

17. $12y^2-48y=0$

18. $3x^2-5x+1=0$

19. $2m^2-4m=3$

20. $p^2+p-1=0$

21. $k^2-10k=-20$

22. $5x^2-8x+2=0$

23. $2r^2-7r+5=0$

24. $2x^2-7x+30=0$

25. $3k^2+k=6$

26. $5m^2+5m=0$

Solve each equation.

27. $\dfrac{3x-2}{7}=\dfrac{x+2}{5}$

28. $\dfrac{x}{3}-7=6-\dfrac{3x}{4}$

29. $\dfrac{4}{x-3}-\dfrac{8}{2x+5}+\dfrac{3}{x-3}=0$

30. $\dfrac{5}{p-2}-\dfrac{7}{p+2}=\dfrac{12}{p^2-4}$

31. $\dfrac{2m}{m-2}-\dfrac{6}{m}=\dfrac{12}{m^2-2m}$

32. $\dfrac{2y}{y-1}=\dfrac{5}{y}+\dfrac{10-8y}{y^2-y}$

33. $\dfrac{1}{x-2}-\dfrac{3x}{x-1}=\dfrac{2x+1}{x^2-3x+2}$

34. $\dfrac{5}{a}+\dfrac{-7}{a+1}=\dfrac{a^2-2a+4}{a^2+a}$

35. $\dfrac{5}{b+5}-\dfrac{4}{b^2+2b}=\dfrac{6}{b^2+7b+10}$

36. $\dfrac{2}{x^2-2x-3}+\dfrac{5}{x^2-x-6}=\dfrac{1}{x^2+3x+2}$

37. $\dfrac{4}{2x^2+3x-9}+\dfrac{2}{2x^2-x-3}=\dfrac{3}{x^2+4x+3}$

▬ **YOUR TURN ANSWERS**

1. $-3/2$

2. $3/2, -5$

3. $4\pm\sqrt{10}$

4. $-1, -4$

R.5 Inequalities

To write that one number is greater than or less than another number, we use the following symbols.

Inequality Symbols

$<$ means *is less than*

$>$ means *is greater than*

$\le$ means *is less than or equal to*

$\ge$ means *is greater than or equal to*

Linear Inequalities
An equation states that two expressions are equal; an **inequality** states that they are unequal. A **linear inequality** is an inequality that can be simplified to the form $ax < b$. (Properties introduced in this section are given only for $<$, but they are equally valid for $>$, $\leq$, or $\geq$.) Linear inequalities are solved with the following properties.

Properties of Inequality
For all real numbers a, b, and c:

1. If $a < b$, then $a + c < b + c$.
2. If $a < b$ and if $c > 0$, then $ac < bc$.
3. If $a < b$ and if $c < 0$, then $ac > bc$.

Pay careful attention to property 3; it says that if both sides of an inequality are multiplied by a negative number, the direction of the inequality symbol must be reversed.

EXAMPLE 1 Solving a Linear Inequality

Solve $4 - 3y \leq 7 + 2y$.

SOLUTION Use the properties of inequality.

$$4 - 3y + (-4) \leq 7 + 2y + (-4) \qquad \text{Add } -4 \text{ to both sides.}$$
$$-3y \leq 3 + 2y$$

Remember that *adding* the same number to both sides never changes the direction of the inequality symbol.

$$-3y + (-2y) \leq 3 + 2y + (-2y) \qquad \text{Add } -2y \text{ to both sides.}$$
$$-5y \leq 3$$

Multiply both sides by $-1/5$. Since $-1/5$ is negative, change the direction of the inequality symbol.

$$-\frac{1}{5}(-5y) \geq -\frac{1}{5}(3)$$

$$y \geq -\frac{3}{5}$$

YOUR TURN 1 Solve $3z - 2 > 5z + 7$.

TRY YOUR TURN 1

| CAUTION | It is a common error to forget to reverse the direction of the inequality sign when multiplying or dividing by a negative number. For example, to solve $-4x \leq 12$, we must multiply by $-1/4$ on both sides *and* reverse the inequality symbol to get $x \geq -3$. |

The solution $y \geq -3/5$ in Example 1 represents an interval on the number line. **Interval notation** often is used for writing intervals. With interval notation, $y \geq -3/5$ is written as $[-3/5, \infty)$. This is an example of a **half-open interval**, since one endpoint, $-3/5$, is included. The **open interval** $(2, 5)$ corresponds to $2 < x < 5$, with neither endpoint included. The **closed interval** $[2, 5]$ includes both endpoints and corresponds to $2 \leq x \leq 5$.

The **graph** of an interval shows all points on a number line that correspond to the numbers in the interval. To graph the interval $[-3/5, \infty)$, for example, use a solid circle at $-3/5$, since $-3/5$ is part of the solution. To show that the solution includes all real numbers greater than or equal to $-3/5$, draw a heavy arrow pointing to the right (the positive direction). See Figure 1.

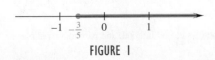

FIGURE 1

EXAMPLE 2 Graphing a Linear Inequality

Solve $-2 < 5 + 3m < 20$. Graph the solution.

SOLUTION The inequality $-2 < 5 + 3m < 20$ says that $5 + 3m$ is *between* -2 and 20. Solve this inequality with an extension of the properties given above. Work as follows, first adding -5 to each part.

$$-2 + (-5) < 5 + 3m + (-5) < 20 + (-5)$$
$$-7 < 3m < 15$$

Now multiply each part by $1/3$.

$$-\frac{7}{3} < m < 5$$

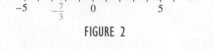

FIGURE 2

A graph of the solution is given in Figure 2; here open circles are used to show that $-7/3$ and 5 are *not* part of the graph.*

Quadratic Inequalities

A **quadratic inequality** has the form $ax^2 + bx + c > 0$ (or $<$, or $\le$, or $\ge$). The highest exponent is 2. The next few examples show how to solve quadratic inequalities.

EXAMPLE 3 Solving a Quadratic Inequality

Solve the quadratic inequality $x^2 - x < 12$.

SOLUTION Write the inequality with 0 on one side, as $x^2 - x - 12 < 0$. This inequality is solved with values of x that make $x^2 - x - 12$ negative (<0). The quantity $x^2 - x - 12$ changes from positive to negative or from negative to positive at the points where it equals 0. For this reason, first solve the *equation* $x^2 - x - 12 = 0$.

$$x^2 - x - 12 = 0$$
$$(x - 4)(x + 3) = 0$$
$$x = 4 \quad \text{or} \quad x = -3$$

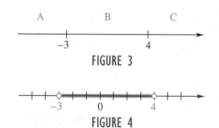

FIGURE 3

FIGURE 4

YOUR TURN 2 Solve $3y^2 \le 16y + 12$.

Locating -3 and 4 on a number line, as shown in Figure 3, determines three intervals A, B, and C. Decide which intervals include numbers that make $x^2 - x - 12$ negative by substituting any number from each interval in the polynomial. For example,

choose -4 from interval A: $(-4)^2 - (-4) - 12 = 8 > 0$;

choose 0 from interval B: $0^2 - 0 - 12 = -12 < 0$;

choose 5 from interval C: $5^2 - 5 - 12 = 8 > 0$.

Only numbers in interval B satisfy the given inequality, so the solution is $(-3, 4)$. A graph of this solution is shown in Figure 4. **TRY YOUR TURN 2**

EXAMPLE 4 Solving a Polynomial Inequality

Solve the inequality $x^3 + 2x^2 - 3x \ge 0$.

SOLUTION This is not a quadratic inequality because of the x^3 term, but we solve it in a similar way by first factoring the polynomial.

$$x^3 + 2x^2 - 3x = x(x^2 + 2x - 3) \quad \text{Factor out the common factor.}$$
$$= x(x - 1)(x + 3) \quad \text{Factor the quadratic.}$$

*Some textbooks use brackets in place of solid circles for the graph of a closed interval, and parentheses in place of open circles for the graph of an open interval.

Now solve the corresponding equation.

$$x(x - 1)(x + 3) = 0$$

$$x = 0 \quad \text{or} \quad x - 1 = 0 \quad \text{or} \quad x + 3 = 0$$

$$x = 1 \qquad\qquad x = -3$$

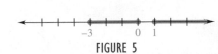

FIGURE 5

These three solutions determine four intervals on the number line: $(-\infty, -3)$, $(-3, 0)$, $(0, 1)$, and $(1, \infty)$. Substitute a number from each interval into the original inequality to determine that the solution consists of the numbers between -3 and 0 (including the endpoints) and all numbers that are greater than or equal to 1. See Figure 5. In interval notation, the solution is

$$[-3, 0] \cup [1, \infty).*$$

Inequalities with Fractions

Inequalities with fractions are solved in a similar manner as quadratic inequalities.

EXAMPLE 5   Solving a Rational Inequality

Solve $\dfrac{2x - 3}{x} \geq 1.$

SOLUTION   First solve the corresponding equation.

$$\frac{2x - 3}{x} = 1$$

$$2x - 3 = x$$

$$x = 3$$

The solution, $x = 3$, determines the intervals on the number line where the fraction may change from greater than 1 to less than 1. This change also may occur on either side of a number that makes the denominator equal 0. Here, the x-value that makes the denominator 0 is $x = 0$. Test each of the three intervals determined by the numbers 0 and 3.

$$\text{For } (-\infty, 0), \text{ choose } -1: \frac{2(-1) - 3}{-1} = 5 \geq 1.$$

$$\text{For } (0, 3), \quad \text{choose } \quad 1: \frac{2(1) - 3}{1} = -1 \ngeq 1.$$

$$\text{For } (3, \infty), \quad \text{choose } \quad 4: \frac{2(4) - 3}{4} = \frac{5}{4} \geq 1.$$

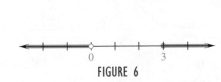

FIGURE 6

The symbol $\ngeq$ means "is *not* greater than or equal to." Testing the endpoints 0 and 3 shows that the solution is $(-\infty, 0) \cup [3, \infty)$, as shown in Figure 6.

> **CAUTION**   A common error is to try to solve the inequality in Example 5 by multiplying both sides by x. The reason this is wrong is that we don't know in the beginning whether x is positive or negative. If x is negative, the $\geq$ would change to $\leq$ according to the third property of inequality listed at the beginning of this section.

*The symbol $\cup$ indicates the *union* of two sets, which includes all elements in either set.

EXAMPLE 6 **Solving a Rational Inequality**

Solve $\dfrac{(x-1)(x+1)}{x} \leq 0$.

SOLUTION   We first solve the corresponding equation.

$$\frac{(x-1)(x+1)}{x} = 0$$

$$(x-1)(x+1) = 0 \qquad \text{Multiply both sides by } x.$$

$$x = 1 \qquad \text{or} \qquad x = -1 \qquad \text{Use the zero-factor property.}$$

Setting the denominator equal to 0 gives $x = 0$, so the intervals of interest are $(-\infty, -1)$, $(-1, 0)$, $(0, 1)$, and $(1, \infty)$. Testing a number from each region in the original inequality and checking the endpoints, we find the solution is

$$(-\infty, -1] \cup (0, 1],$$

FIGURE 7

as shown in Figure 7.

CAUTION Remember to solve the equation formed by setting the *denominator* equal to zero. Any number that makes the denominator zero always creates two intervals on the number line. For instance, in Example 6, substituting $x = 0$ makes the denominator of the rational inequality equal to 0, so we know that there may be a sign change from one side of 0 to the other (as was indeed the case).

EXAMPLE 7 **Solving a Rational Inequality**

Solve $\dfrac{x^2 - 3x}{x^2 - 9} < 4$.

SOLUTION   Solve the corresponding equation.

$$\frac{x^2 - 3x}{x^2 - 9} = 4$$

$$x^2 - 3x = 4x^2 - 36 \qquad \text{Multiply by } x^2 - 9.$$

$$0 = 3x^2 + 3x - 36 \qquad \text{Get 0 on one side.}$$

$$0 = x^2 + x - 12 \qquad \text{Multiply by } \tfrac{1}{3}.$$

$$0 = (x + 4)(x - 3) \qquad \text{Factor.}$$

$$x = -4 \qquad \text{or} \qquad x = 3$$

Now set the denominator equal to 0 and solve that equation.

$$x^2 - 9 = 0$$

$$(x - 3)(x + 3) = 0$$

$$x = 3 \qquad \text{or} \qquad x = -3$$

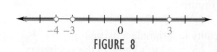

FIGURE 8

The intervals determined by the three (different) solutions are $(-\infty, -4)$, $(-4, -3)$, $(-3, 3)$, and $(3, \infty)$. Testing a number from each interval in the given inequality shows that the solution is

YOUR TURN 3 Solve

$$(-\infty, -4) \cup (-3, 3) \cup (3, \infty),$$

$\dfrac{k^2 - 35}{k} \geq 2.$

as shown in Figure 8. For this example, none of the endpoints are part of the solution because $x = 3$ and $x = -3$ make the denominator zero and $x = -4$ produces an equality.

TRY YOUR TURN 3

R.5 EXERCISES

Write each expression in interval notation. Graph each interval.

1. $x < 4$

2. $x \geq -3$

3. $1 \leq x < 2$

4. $-2 \leq x \leq 3$

5. $-9 > x$

6. $6 \leq x$

Using the variable x, write each interval as an inequality.

7. $[-7, -3]$

8. $[4, 10)$

9. $(-\infty, -1]$

10. $(3, \infty)$

11.
$$-2 \quad 0 \qquad 6$$

12.
$$0 \qquad 8$$

13.
$$-4 \quad 0 \quad 4$$

14.
$$0 \qquad 3$$

Solve each inequality and graph the solution.

15. $6p + 7 \leq 19$

16. $6k - 4 < 3k - 1$

17. $m - (3m - 2) + 6 < 7m - 19$

18. $-2(3y - 8) \geq 5(4y - 2)$

19. $3p - 1 < 6p + 2(p - 1)$

20. $x + 5(x + 1) > 4(2 - x) + x$

21. $-11 < y - 7 < -1$

22. $8 \leq 3r + 1 \leq 13$

23. $-2 < \dfrac{1 - 3k}{4} \leq 4$

24. $-1 \leq \dfrac{5y + 2}{3} \leq 4$

25. $\dfrac{3}{5}(2p + 3) \geq \dfrac{1}{10}(5p + 1)$

26. $\dfrac{8}{3}(z - 4) \leq \dfrac{2}{9}(3z + 2)$

Solve each quadratic inequality. Graph each solution.

27. $(m - 3)(m + 5) < 0$

28. $(t + 6)(t - 1) \geq 0$

29. $y^2 - 3y + 2 < 0$

30. $2k^2 + 7k - 4 > 0$

31. $x^2 - 16 > 0$

32. $2k^2 - 7k - 15 \leq 0$

33. $x^2 - 4x \geq 5$

34. $10r^2 + r \leq 2$

35. $3x^2 + 2x > 1$

36. $3a^2 + a > 10$

37. $9 - x^2 \leq 0$

38. $p^2 - 16p > 0$

39. $x^3 - 4x \geq 0$

40. $x^3 + 7x^2 + 12x \leq 0$

41. $2x^3 - 14x^2 + 12x < 0$

42. $3x^3 - 9x^2 - 12x > 0$

Solve each inequality.

43. $\dfrac{m - 3}{m + 5} \leq 0$

44. $\dfrac{r + 1}{r - 1} > 0$

45. $\dfrac{k - 1}{k + 2} > 1$

46. $\dfrac{a - 5}{a + 2} < -1$

47. $\dfrac{2y + 3}{y - 5} \leq 1$

48. $\dfrac{a + 2}{3 + 2a} \leq 5$

49. $\dfrac{2k}{k - 3} \leq \dfrac{4}{k - 3}$

50. $\dfrac{5}{p + 1} > \dfrac{12}{p + 1}$

51. $\dfrac{2x}{x^2 - x - 6} \geq 0$

52. $\dfrac{8}{p^2 + 2p} > 1$

53. $\dfrac{z^2 + z}{z^2 - 1} \geq 3$

54. $\dfrac{a^2 + 2a}{a^2 - 4} \leq 2$

YOUR TURN ANSWERS

1. $z < -9/2$ **2.** $[-2/3, 6]$ **3.** $[-5, 0) \cup [7, \infty)$

R.6 Exponents

Integer Exponents

Recall that $a^2 = a \cdot a$, while $a^3 = a \cdot a \cdot a$, and so on. In this section, a more general meaning is given to the symbol a^n.

Definition of Exponent

If n is a natural number, then

$$a^n = a \cdot a \cdot a \cdot \cdots \cdot a,$$

where a appears as a factor n times.

In the expression a^n, the power n is the **exponent** and a is the **base**. This definition can be extended by defining a^n for zero and negative integer values of n.

Zero and Negative Exponents

If a is any nonzero real number, and if n is a positive integer, then

$$a^0 = 1 \quad \text{and} \quad a^{-n} = \frac{1}{a^n}.$$

(The symbol 0^0 is meaningless.)

EXAMPLE 1 Exponents

(a) $6^0 = 1$

(b) $(-9)^0 = 1$

(c) $3^{-2} = \dfrac{1}{3^2} = \dfrac{1}{9}$

(d) $9^{-1} = \dfrac{1}{9^1} = \dfrac{1}{9}$

(e) $\left(\dfrac{3}{4}\right)^{-1} = \dfrac{1}{(3/4)^1} = \dfrac{1}{3/4} = \dfrac{4}{3}$

The following properties follow from the definitions of exponents given above.

Properties of Exponents

For any integers m and n, and any real numbers a and b for which the following exist:

1. $a^m \cdot a^n = a^{m+n}$

2. $\dfrac{a^m}{a^n} = a^{m-n}$

3. $(a^m)^n = a^{mn}$

4. $(ab)^m = a^m \cdot b^m$

5. $\left(\dfrac{a}{b}\right)^m = \dfrac{a^m}{b^m}$

Note that $(-a)^n = a^n$ if n is an even integer, but $(-a)^n = -a^n$ if n is an odd integer.

EXAMPLE 2 Simplifying Exponential Expressions

Use the properties of exponents to simplify each expression. Leave answers with positive exponents. Assume that all variables represent positive real numbers.

(a) $7^4 \cdot 7^6 = 7^{4+6} = 7^{10}$ (or 282,475,249)   Property 1

(b) $\dfrac{9^{14}}{9^6} = 9^{14-6} = 9^8$ (or 43,046,721)   Property 2

(c) $\dfrac{r^9}{r^{17}} = r^{9-17} = r^{-8} = \dfrac{1}{r^8}$ Property 2

(d) $(2m^3)^4 = 2^4 \cdot (m^3)^4 = 16m^{12}$ Properties 3 and 4

(e) $(3x)^4 = 3^4 \cdot x^4 = 81x^4$ Property 4

(f) $\left(\dfrac{x^2}{y^3}\right)^6 = \dfrac{(x^2)^6}{(y^3)^6} = \dfrac{x^{2\cdot6}}{y^{3\cdot6}} = \dfrac{x^{12}}{y^{18}}$ Properties 3 and 5

(g) $\dfrac{a^{-3}b^5}{a^4b^{-7}} = \dfrac{b^{5-(-7)}}{a^{4-(-3)}} = \dfrac{b^{5+7}}{a^{4+3}} = \dfrac{b^{12}}{a^7}$ Property 2

(h) $p^{-1} + q^{-1} = \dfrac{1}{p} + \dfrac{1}{q} = \dfrac{1}{p} \cdot \dfrac{q}{q} + \dfrac{1}{q} \cdot \dfrac{p}{p} = \dfrac{q}{pq} + \dfrac{p}{pq} = \dfrac{p+q}{pq}$

(i) $\dfrac{x^{-2} - y^{-2}}{x^{-1} - y^{-1}} = \dfrac{\dfrac{1}{x^2} - \dfrac{1}{y^2}}{\dfrac{1}{x} - \dfrac{1}{y}}$ Definition of a^{-n}

$= \dfrac{\dfrac{y^2 - x^2}{x^2 y^2}}{\dfrac{y - x}{xy}}$ Get common denominators and combine terms.

$= \dfrac{y^2 - x^2}{x^2 y^2} \cdot \dfrac{xy}{y - x}$ Invert and multiply.

$= \dfrac{(y - x)(y + x)}{x^2 y^2} \cdot \dfrac{xy}{y - x}$ Factor.

$= \dfrac{x + y}{xy}$ Simplify.

YOUR TURN 1
Simplify
$\left(\dfrac{y^2 z^{-4}}{y^{-3} z^4} \right)^{-2}.$

TRY YOUR TURN 1

CAUTION If Example 2(e) were written $3x^4$, the properties of exponents would not apply. When no parentheses are used, the exponent refers only to the factor closest to it. Also notice in Examples 2(c), 2(g), 2(h), and 2(i) that a negative exponent does *not* indicate a negative number.

Roots
For *even* values of n and nonnegative values of a, the expression $a^{1/n}$ is defined to be the **positive nth root** of a or the **principal nth root** of a. For example, $a^{1/2}$ denotes the positive second root, or **square root**, of a, while $a^{1/4}$ is the positive fourth root of a. When n is *odd*, there is only one nth root, which has the same sign as a. For example, $a^{1/3}$, the **cube root** of a, has the same sign as a. By definition, if $b = a^{1/n}$, then $b^n = a$. On a calculator, a number is raised to a power using a key labeled x^y, y^x, or $\wedge$. For example, to take the fourth root of 6 on a TI-84 Plus calculator, enter $6 \wedge (1/4)$ to get the result 1.56508458.

EXAMPLE 3 **Calculations with Exponents**

(a) $121^{1/2} = 11$, since 11 is positive and $11^2 = 121$.

(b) $625^{1/4} = 5$, since $5^4 = 625$.

(c) $256^{1/4} = 4$

(d) $64^{1/6} = 2$

(e) $27^{1/3} = 3$

(f) $(-32)^{1/5} = -2$

(g) $128^{1/7} = 2$

(h) $(-49)^{1/2}$ is not a real number.

Rational Exponents
In the following definition, the domain of an exponent is extended to include all rational numbers.

Definition of $a^{m/n}$

For all real numbers a for which the indicated roots exist, and for any rational number m/n,

$$a^{m/n} = (a^{1/n})^m.$$

EXAMPLE 4 Calculations with Exponents

(a) $27^{2/3} = (27^{1/3})^2 = 3^2 = 9$

(b) $32^{2/5} = (32^{1/5})^2 = 2^2 = 4$

(c) $64^{4/3} = (64^{1/3})^4 = 4^4 = 256$

(d) $25^{3/2} = (25^{1/2})^3 = 5^3 = 125$

NOTE $27^{2/3}$ could also be evaluated as $(27^2)^{1/3}$, but this is more difficult to perform without a calculator because it involves squaring 27 and then taking the cube root of this large number. On the other hand, when we evaluate it as $(27^{1/3})^2$, we know that the cube root of 27 is 3 without using a calculator, and squaring 3 is easy.

All the properties for integer exponents given in this section also apply to any rational exponent on a nonnegative real-number base.

EXAMPLE 5 Simplifying Exponential Expressions

(a) $\dfrac{y^{1/3}y^{5/3}}{y^3} = \dfrac{y^{1/3+5/3}}{y^3} = \dfrac{y^2}{y^3} = y^{2-3} = y^{-1} = \dfrac{1}{y}$

(b) $m^{2/3}(m^{7/3} + 2m^{1/3}) = m^{2/3+7/3} + 2m^{2/3+1/3} = m^3 + 2m$

(c) $\left(\dfrac{m^7 n^{-2}}{m^{-5} n^2}\right)^{1/4} = \left(\dfrac{m^{7-(-5)}}{n^{2-(-2)}}\right)^{1/4} = \left(\dfrac{m^{12}}{n^4}\right)^{1/4} = \dfrac{(m^{12})^{1/4}}{(n^4)^{1/4}} = \dfrac{m^{12/4}}{n^{4/4}} = \dfrac{m^3}{n}$

In calculus, it is often necessary to factor expressions involving fractional exponents.

EXAMPLE 6 Simplifying Exponential Expressions

Factor out the smallest power of the variable, assuming all variables represent positive real numbers.

(a) $4m^{1/2} + 3m^{3/2}$

SOLUTION   The smallest exponent is $1/2$. Factoring out $m^{1/2}$ yields

$$4m^{1/2} + 3m^{3/2} = m^{1/2}(4m^{1/2-1/2} + 3m^{3/2-1/2})$$
$$= m^{1/2}(4 + 3m).$$

Check this result by multiplying $m^{1/2}$ by $4 + 3m$.

(b) $9x^{-2} - 6x^{-3}$

SOLUTION   The smallest exponent here is -3. Since 3 is a common numerical factor, factor out $3x^{-3}$.

$$9x^{-2} - 6x^{-3} = 3x^{-3}(3x^{-2-(-3)} - 2x^{-3-(-3)}) = 3x^{-3}(3x - 2)$$

Check by multiplying. The factored form can be written without negative exponents as

$$\dfrac{3(3x - 2)}{x^3}.$$

(c) $(x^2 + 5)(3x - 1)^{-1/2}(2) + (3x - 1)^{1/2}(2x)$

SOLUTION   There is a common factor of 2. Also, $(3x - 1)^{-1/2}$ and $(3x - 1)^{1/2}$ have a common factor. Always factor out the quantity to the *smallest* exponent. Here $-1/2 < 1/2$, so the common factor is $2(3x - 1)^{-1/2}$ and the factored form is

$$2(3x - 1)^{-1/2}[(x^2 + 5) + (3x - 1)x] = 2(3x - 1)^{-1/2}(4x^2 - x + 5).$$

YOUR TURN 2 Factor $5z^{1/3} + 4z^{-2/3}$.

TRY YOUR TURN 2

R.6 EXERCISES

Evaluate each expression. Write all answers without exponents.

1. 8^{-2}

2. 3^{-4}

3. 5^0

4. $\left(-\dfrac{3}{4}\right)^0$

5. $-(-3)^{-2}$

6. $-(-3^{-2})$

7. $\left(\dfrac{1}{6}\right)^{-2}$

8. $\left(\dfrac{4}{3}\right)^{-3}$

Simplify each expression. Assume that all variables represent positive real numbers. Write answers with only positive exponents.

9. $\dfrac{4^{-2}}{4}$

10. $\dfrac{8^9 \cdot 8^{-7}}{8^{-3}}$

11. $\dfrac{10^8 \cdot 10^{-10}}{10^4 \cdot 10^2}$

12. $\left(\dfrac{7^{-12} \cdot 7^3}{7^{-8}}\right)^{-1}$

13. $\dfrac{x^4 \cdot x^3}{x^5}$

14. $\dfrac{y^{10} \cdot y^{-4}}{y^6}$

15. $\dfrac{(4k^{-1})^2}{2k^{-5}}$

16. $\dfrac{(3z^2)^{-1}}{z^5}$

17. $\dfrac{3^{-1} \cdot x \cdot y^2}{x^{-4} \cdot y^5}$

18. $\dfrac{5^{-2}m^2y^{-2}}{5^2m^{-1}y^{-2}}$

19. $\left(\dfrac{a^{-1}}{b^2}\right)^{-3}$

20. $\left(\dfrac{c^3}{7d^{-2}}\right)^{-2}$

Simplify each expression, writing the answer as a single term without negative exponents.

21. $a^{-1} + b^{-1}$

22. $b^{-2} - a$

23. $\dfrac{2n^{-1} - 2m^{-1}}{m + n^2}$

24. $\left(\dfrac{m}{3}\right)^{-1} + \left(\dfrac{n}{2}\right)^{-2}$

25. $(x^{-1} - y^{-1})^{-1}$

26. $(x \cdot y^{-1} - y^{-2})^{-2}$

Write each number without exponents.

27. $121^{1/2}$

28. $27^{1/3}$

29. $32^{2/5}$

30. $-125^{2/3}$

31. $\left(\dfrac{36}{144}\right)^{1/2}$

32. $\left(\dfrac{64}{27}\right)^{1/3}$

33. $8^{-4/3}$

34. $625^{-1/4}$

35. $\left(\dfrac{27}{64}\right)^{-1/3}$

36. $\left(\dfrac{121}{100}\right)^{-3/2}$

Simplify each expression. Write all answers with only positive exponents. Assume that all variables represent positive real numbers.

37. $3^{2/3} \cdot 3^{4/3}$

38. $27^{2/3} \cdot 27^{-1/3}$

39. $\dfrac{4^{9/4} \cdot 4^{-7/4}}{4^{-10/4}}$

40. $\dfrac{3^{-5/2} \cdot 3^{3/2}}{3^{7/2} \cdot 3^{-9/2}}$

41. $\left(\dfrac{x^6 y^{-3}}{x^{-2} y^5}\right)^{1/2}$

42. $\left(\dfrac{a^{-7}b^{-1}}{b^{-4}a^2}\right)^{1/3}$

43. $\dfrac{7^{-1/3} \cdot 7r^{-3}}{7^{2/3} \cdot (r^{-2})^2}$

44. $\dfrac{12^{3/4} \cdot 12^{5/4} \cdot y^{-2}}{12^{-1} \cdot (y^{-3})^{-2}}$

45. $\dfrac{3k^2 \cdot (4k^{-3})^{-1}}{4^{1/2} \cdot k^{7/2}}$

46. $\dfrac{8p^{-3} \cdot (4p^2)^{-2}}{p^{-5}}$

47. $\dfrac{a^{4/3} \cdot b^{1/2}}{a^{2/3} \cdot b^{-3/2}}$

48. $\dfrac{x^{3/2} \cdot y^{4/5} \cdot z^{-3/4}}{x^{5/3} \cdot y^{-6/5} \cdot z^{1/2}}$

49. $\dfrac{k^{-3/5} \cdot h^{-1/3} \cdot t^{2/5}}{k^{-1/5} \cdot h^{-2/3} \cdot t^{1/5}}$

50. $\dfrac{m^{7/3} \cdot n^{-2/5} \cdot p^{3/8}}{m^{-2/3} \cdot n^{3/5} \cdot p^{-5/8}}$

Factor each expression.

51. $3x^3(x^2 + 3x)^2 - 15x(x^2 + 3x)^2$

52. $6x(x^3 + 7)^2 - 6x^2(3x^2 + 5)(x^3 + 7)$

53. $10x^3(x^2 - 1)^{-1/2} - 5x(x^2 - 1)^{1/2}$

54. $9(6x + 2)^{1/2} + 3(9x - 1)(6x + 2)^{-1/2}$

55. $x(2x + 5)^2(x^2 - 4)^{-1/2} + 2(x^2 - 4)^{1/2}(2x + 5)$

56. $(4x^2 + 1)^2(2x - 1)^{-1/2} + 16x(4x^2 + 1)(2x - 1)^{1/2}$

YOUR TURN ANSWERS

1. z^{16}/y^{10}

2. $z^{-2/3}(5z + 4)$

R.7 Radicals

We have defined $a^{1/n}$ as the positive or principal nth root of a for appropriate values of a and n. An alternative notation for $a^{1/n}$ uses radicals.

Radicals

If n is an even natural number and $a > 0$, or n is an odd natural number, then

$$a^{1/n} = \sqrt[n]{a}.$$

The symbol $\sqrt[n]{}$ is a **radical sign**, the number a is the **radicand**, and n is the **index** of the radical. The familiar symbol $\sqrt{a}$ is used instead of $\sqrt[2]{a}$.

EXAMPLE 1   Radical Calculations

(a) $\sqrt[4]{16} = 16^{1/4} = 2$

(b) $\sqrt[5]{-32} = -2$

(c) $\sqrt[3]{1000} = 10$

(d) $\sqrt[6]{\dfrac{64}{729}} = \dfrac{2}{3}$

With $a^{1/n}$ written as $\sqrt[n]{a}$, the expression $a^{m/n}$ also can be written using radicals.

$$a^{m/n} = (\sqrt[n]{a})^m \quad \text{or} \quad a^{m/n} = \sqrt[n]{a^m}$$

The following properties of radicals depend on the definitions and properties of exponents.

Properties of Radicals

For all real numbers a and b and natural numbers m and n such that $\sqrt[n]{a}$ and $\sqrt[n]{b}$ are real numbers:

1. $(\sqrt[n]{a})^n = a$

2. $\sqrt[n]{a^n} = \begin{cases} |a| & \text{if } n \text{ is even} \\ a & \text{if } n \text{ is odd} \end{cases}$

3. $\sqrt[n]{a} \cdot \sqrt[n]{b} = \sqrt[n]{ab}$

4. $\dfrac{\sqrt[n]{a}}{\sqrt[n]{b}} = \sqrt[n]{\dfrac{a}{b}} \quad (b \neq 0)$

5. $\sqrt[m]{\sqrt[n]{a}} = \sqrt[mn]{a}$

Property 3 can be used to simplify certain radicals. For example, since $48 = 16 \cdot 3$,

$$\sqrt{48} = \sqrt{16 \cdot 3} = \sqrt{16} \cdot \sqrt{3} = 4\sqrt{3}.$$

To some extent, simplification is in the eye of the beholder, and $\sqrt{48}$ might be considered as simple as $4\sqrt{3}$. In this textbook, we will consider an expression to be simpler when we have removed as many factors as possible from under the radical.

EXAMPLE 2   Radical Calculations

(a) $\sqrt{1000} = \sqrt{100 \cdot 10} = \sqrt{100} \cdot \sqrt{10} = 10\sqrt{10}$

(b) $\sqrt{128} = \sqrt{64 \cdot 2} = 8\sqrt{2}$

(c) $\sqrt{2} \cdot \sqrt{18} = \sqrt{2 \cdot 18} = \sqrt{36} = 6$

(d) $\sqrt[3]{54} = \sqrt[3]{27 \cdot 2} = \sqrt[3]{27} \cdot \sqrt[3]{2} = 3\sqrt[3]{2}$

(e) $\sqrt{288m^5} = \sqrt{144 \cdot m^4 \cdot 2m} = 12m^2\sqrt{2m}$

(f) $2\sqrt{18} - 5\sqrt{32} = 2\sqrt{9 \cdot 2} - 5\sqrt{16 \cdot 2}$

$$= 2\sqrt{9} \cdot \sqrt{2} - 5\sqrt{16} \cdot \sqrt{2}$$

$$= 2(3)\sqrt{2} - 5(4)\sqrt{2} = -14\sqrt{2}$$

YOUR TURN 1
Simplify $\sqrt{28x^9y^5}$.

(g) $\sqrt{x^5} \cdot \sqrt[3]{x^5} = x^{5/2} \cdot x^{5/3} = x^{5/2+5/3} = x^{25/6} = \sqrt[6]{x^{25}} = x^4\sqrt[6]{x}$ **TRY YOUR TURN 1**

When simplifying a square root, keep in mind that $\sqrt{x}$ is nonnegative by definition. Also, $\sqrt{x^2}$ is not x, but $|x|$, the **absolute value of x**, defined as

$$|x| = \begin{cases} x & \text{if } x \geq 0 \\ -x & \text{if } x < 0. \end{cases}$$

For example, $\sqrt{(-5)^2} = |-5| = 5$. It is correct, however, to simplify $\sqrt{x^4} = x^2$. We need not write $|x^2|$ because x^2 is always nonnegative.

EXAMPLE 3 **Simplifying by Factoring**

Simplify $\sqrt{m^2 - 4m + 4}$.

SOLUTION Factor the polynomial as $m^2 - 4m + 4 = (m - 2)^2$. Then by property 2 of radicals and the definition of absolute value,

$$\sqrt{(m-2)^2} = |m - 2| = \begin{cases} m - 2 & \text{if } m - 2 \geq 0 \\ -(m - 2) = 2 - m & \text{if } m - 2 < 0. \end{cases}$$

CAUTION Avoid the common error of writing $\sqrt{a^2 + b^2}$ as $\sqrt{a^2} + \sqrt{b^2}$. We must add a^2 and b^2 *before* taking the square root. For example, $\sqrt{16 + 9} = \sqrt{25} = 5$, *not* $\sqrt{16} + \sqrt{9} = 4 + 3 = 7$. This idea applies as well to higher roots. For example, in general,

$$\sqrt[3]{a^3 + b^3} \neq \sqrt[3]{a^3} + \sqrt[3]{b^3},$$

$$\sqrt[4]{a^4 + b^4} \neq \sqrt[4]{a^4} + \sqrt[4]{b^4}.$$

Also, $\sqrt{a + b} \neq \sqrt{a} + \sqrt{b}.$

Rationalizing Denominators
The next example shows how to *rationalize* (remove all radicals from) the denominator in an expression containing radicals.

EXAMPLE 4 **Rationalizing Denominators**

Simplify each expression by rationalizing the denominator.

(a) $\dfrac{4}{\sqrt{3}}$

SOLUTION To rationalize the denominator, multiply by $\sqrt{3}/\sqrt{3}$ (or 1) so that the denominator of the product is a rational number.

$$\frac{4}{\sqrt{3}} \cdot \frac{\sqrt{3}}{\sqrt{3}} = \frac{4\sqrt{3}}{3} \qquad \sqrt{3} \cdot \sqrt{3} = \sqrt{9} = 3$$

(b) $\dfrac{2}{\sqrt[3]{x}}$

SOLUTION Here, we need a perfect cube under the radical sign to rationalize the denominator. Multiplying by $\sqrt[3]{x^2}/\sqrt[3]{x^2}$ gives

$$\frac{2}{\sqrt[3]{x}} \cdot \frac{\sqrt[3]{x^2}}{\sqrt[3]{x^2}} = \frac{2\sqrt[3]{x^2}}{\sqrt[3]{x^3}} = \frac{2\sqrt[3]{x^2}}{x}.$$

(c) $\dfrac{1}{1 - \sqrt{2}}$

SOLUTION The best approach here is to multiply both numerator and denominator by the number $1 + \sqrt{2}$. The expressions $1 + \sqrt{2}$ and $1 - \sqrt{2}$ are conjugates,* and their product is $1^2 - (\sqrt{2})^2 = 1 - 2 = -1$. Thus,

$$\dfrac{1}{1 - \sqrt{2}} = \dfrac{1(1 + \sqrt{2})}{(1 - \sqrt{2})(1 + \sqrt{2})} = \dfrac{1 + \sqrt{2}}{1 - 2} = -1 - \sqrt{2}.$$

TRY YOUR TURN 2

YOUR TURN 2 Rationalize the denominator in $\dfrac{5}{\sqrt{x} - \sqrt{y}}$.

Sometimes it is advantageous to rationalize the *numerator* of a rational expression. The following example arises in calculus when evaluating a *limit*.

EXAMPLE 5 **Rationalizing Numerators**

Rationalize each numerator.

(a) $\dfrac{\sqrt{x} - 3}{x - 9}$.

SOLUTION Multiply the numerator and denominator by the conjugate of the numerator, $\sqrt{x} + 3$.

$$\dfrac{\sqrt{x} - 3}{x - 9} \cdot \dfrac{\sqrt{x} + 3}{\sqrt{x} + 3} = \dfrac{(\sqrt{x})^2 - 3^2}{(x - 9)(\sqrt{x} + 3)} \qquad (a - b)(a + b) = a^2 - b^2$$

$$= \dfrac{x - 9}{(x - 9)(\sqrt{x} + 3)}$$

$$= \dfrac{1}{\sqrt{x} + 3}$$

(b) $\dfrac{\sqrt{3} + \sqrt{x + 3}}{\sqrt{3} - \sqrt{x + 3}}$

SOLUTION Multiply the numerator and denominator by the conjugate of the numerator, $\sqrt{3} - \sqrt{x + 3}$.

$$\dfrac{\sqrt{3} + \sqrt{x + 3}}{\sqrt{3} - \sqrt{x + 3}} \cdot \dfrac{\sqrt{3} - \sqrt{x + 3}}{\sqrt{3} - \sqrt{x + 3}} = \dfrac{3 - (x + 3)}{3 - 2\sqrt{3}\sqrt{x + 3} + (x + 3)}$$

$$= \dfrac{-x}{6 + x - 2\sqrt{3(x + 3)}}$$

R.7 EXERCISES

Simplify each expression by removing as many factors as possible from under the radical. Assume that all variables represent positive real numbers.

1. $\sqrt[3]{125}$

2. $\sqrt[4]{1296}$

3. $\sqrt[5]{-3125}$

4. $\sqrt{50}$

5. $\sqrt{2000}$

6. $\sqrt{32y^5}$

7. $\sqrt{27} \cdot \sqrt{3}$

8. $\sqrt{2} \cdot \sqrt{32}$

9. $7\sqrt{2} - 8\sqrt{18} + 4\sqrt{72}$

10. $4\sqrt{3} - 5\sqrt{12} + 3\sqrt{75}$

11. $4\sqrt{7} - \sqrt{28} + \sqrt{343}$

*If a and b are real numbers, the *conjugate* of $a + b$ is $a - b$.

12. $3\sqrt{28} - 4\sqrt{63} + \sqrt{112}$

13. $\sqrt[3]{2} - \sqrt[3]{16} + 2\sqrt[3]{54}$

14. $2\sqrt[3]{5} - 4\sqrt[3]{40} + 3\sqrt[3]{135}$

15. $\sqrt{2x^3y^2z^4}$ **16.** $\sqrt{160r^7s^9t^{12}}$

17. $\sqrt[3]{128x^3y^8z^9}$ **18.** $\sqrt[4]{x^8y^7z^{11}}$

19. $\sqrt{a^3b^5} - 2\sqrt{a^7b^3} + \sqrt{a^3b^9}$

20. $\sqrt{p^7q^3} - \sqrt{p^5q^9} + \sqrt{p^9q}$

21. $\sqrt{a} \cdot \sqrt[3]{a}$

22. $\sqrt{b^3} \cdot \sqrt[4]{b^3}$

Simplify each root, if possible.

23. $\sqrt{16 - 8x + x^2}$

24. $\sqrt{9y^2 + 30y + 25}$

25. $\sqrt{4 - 25z^2}$

26. $\sqrt{9k^2 + h^2}$

Rationalize each denominator. Assume that all radicands represent positive real numbers.

27. $\dfrac{5}{\sqrt{7}}$ **28.** $\dfrac{5}{\sqrt{10}}$

29. $\dfrac{-3}{\sqrt{12}}$ **30.** $\dfrac{4}{\sqrt{8}}$

31. $\dfrac{3}{1 - \sqrt{2}}$ **32.** $\dfrac{5}{2 - \sqrt{6}}$

33. $\dfrac{6}{2 + \sqrt{2}}$ **34.** $\dfrac{\sqrt{5}}{\sqrt{5} + \sqrt{2}}$

35. $\dfrac{1}{\sqrt{r} - \sqrt{3}}$ **36.** $\dfrac{5}{\sqrt{m} - \sqrt{5}}$

37. $\dfrac{y - 5}{\sqrt{y} - \sqrt{5}}$ **38.** $\dfrac{\sqrt{z} - 1}{\sqrt{z} - \sqrt{5}}$

39. $\dfrac{\sqrt{x} + \sqrt{x + 1}}{\sqrt{x} - \sqrt{x + 1}}$ **40.** $\dfrac{\sqrt{p} + \sqrt{p^2 - 1}}{\sqrt{p} - \sqrt{p^2 - 1}}$

Rationalize each numerator. Assume that all radicands represent positive real numbers.

41. $\dfrac{1 + \sqrt{2}}{2}$ **42.** $\dfrac{3 - \sqrt{3}}{6}$

43. $\dfrac{\sqrt{x} + \sqrt{x + 1}}{\sqrt{x} - \sqrt{x + 1}}$ **44.** $\dfrac{\sqrt{p} - \sqrt{p - 2}}{\sqrt{p}}$

■■■YOUR TURN ANSWERS

1. $2x^4y^2\sqrt{7xy}$ **2.** $5(\sqrt{x} + \sqrt{y})/(x - y)$

Linear Functions

Over short time intervals, many changes in the economy are well modeled by linear functions. In an exercise in the first section of this chapter, we will examine a linear model that predicts the number of cellular telephone users in the United States. Such predictions are important tools for cellular telephone company executives and planners.

Before using mathematics to solve a real-world problem, we must usually set up a **mathematical model,** a mathematical description of the situation. In this chapter we look at some mathematics of *linear* models, which are used for data whose graphs can be approximated by straight lines. Linear models have an immense number of applications, because even when the underlying phenomenon is not linear, a linear model often provides an approximation that is sufficiently accurate and much simpler to use.

1.1   Slopes and Equations of Lines

APPLY IT

How fast has tuition at public colleges been increasing in recent years, and how well can we predict tuition in the future?
In Example 14 of this section, we will answer these questions using the equation of a line.

There are many everyday situations in which two quantities are related. For example, if a bank account pays 6% simple interest per year, then the interest I that a deposit of P dollars would earn in one year is given by

$$I = 0.06 \cdot P, \qquad \text{or} \qquad I = 0.06P.$$

The formula $I = 0.06P$ describes the relationship between interest and the amount of money deposited.

Using this formula, we see, for example, that if $P = \$100$, then $I = \$6$, and if $P = \$200$, then $I = \$12$. These corresponding pairs of numbers can be written as **ordered pairs**, $(100, 6)$ and $(200, 12)$, whose order is important. The first number denotes the value of P and the second number the value of I.

Ordered pairs are graphed with the perpendicular number lines of a **Cartesian coordinate system**, shown in Figure 1.* The horizontal number line, or **x-axis**, represents the first components of the ordered pairs, while the vertical or **y-axis** represents the second components. The point where the number lines cross is the zero point on both lines; this point is called the **origin**.

Each point on the *xy*-plane corresponds to an ordered pair of numbers, where the *x*-value is written first. From now on, we will refer to the point corresponding to the ordered pair (x, y) as "the point (x, y)."

Locate the point $(-2, 4)$ on the coordinate system by starting at the origin and counting 2 units to the left on the horizontal axis and 4 units upward, parallel to the vertical axis. This point is shown in Figure 1, along with several other sample points. The number -2 is the **x-coordinate** and the number 4 is the **y-coordinate** of the point $(-2, 4)$.

The *x*-axis and *y*-axis divide the plane into four parts, or **quadrants**. For example, quadrant I includes all those points whose *x*- and *y*-coordinates are both positive. The quadrants are numbered as shown in Figure 1. The points on the axes themselves belong to no quadrant. The set of points corresponding to the ordered pairs of an equation is the **graph** of the equation.

The *x*- and *y*-values of the points where the graph of an equation crosses the axes are called the **x-intercept** and **y-intercept**, respectively.** See Figure 2.

---

*The name "Cartesian" honors René Descartes (1596–1650), one of the greatest mathematicians of the seventeenth century. According to legend, Descartes was lying in bed when he noticed an insect crawling on the ceiling and realized that if he could determine the distance from the bug to each of two perpendicular walls, he could describe its position at any given moment. The same idea can be used to locate a point in a plane.
**Some people prefer to define the intercepts as ordered pairs, rather than as numbers.

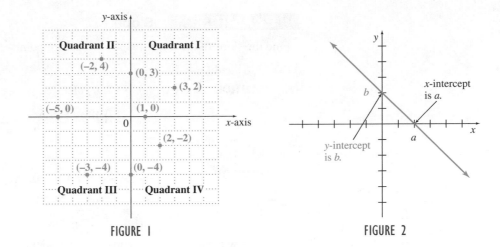

FIGURE 1   FIGURE 2

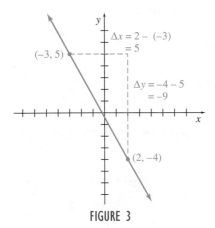

FIGURE 3

Slope of a Line

An important characteristic of a straight line is its *slope*, a number that represents the "steepness" of the line. To see how slope is defined, look at the line in Figure 3. The line goes through the points $(x_1, y_1) = (-3, 5)$ and $(x_2, y_2) = (2, -4)$. The difference in the two *x*-values,

$$x_2 - x_1 = 2 - (-3) = 5$$

in this example, is called the **change in x**. The symbol Δx (read "delta *x*") is used to represent the change in *x*. In the same way, Δy represents the **change in y**. In our example,

$$\Delta y = y_2 - y_1$$
$$= -4 - 5$$
$$= -9.$$

These symbols, Δx and Δy, are used in the following definition of slope.

Slope of a Line
The **slope** of a line is defined as the vertical change (the "rise") over the horizontal change (the "run") as one travels along the line. In symbols, taking two different points (x_1, y_1) and (x_2, y_2) on the line, the slope is

$$m = \frac{\text{Change in } y}{\text{Change in } x} = \frac{\Delta y}{\Delta x} = \frac{y_2 - y_1}{x_2 - x_1},$$

where $x_1 \neq x_2$.

By this definition, the slope of the line in Figure 3 is

$$m = \frac{\Delta y}{\Delta x} = \frac{-4 - 5}{2 - (-3)} = -\frac{9}{5}.$$

The slope of a line tells how fast *y* changes for each unit of change in *x*.

NOTE Using similar triangles, it can be shown that the slope of a line is independent of the choice of points on the line. That is, the same slope will be obtained for *any* choice of two different points on the line.

EXAMPLE 1 Slope

Find the slope of the line through each pair of points.

(a) $(7, 6)$ and $(-4, 5)$

SOLUTION Let $(x_1, y_1) = (7, 6)$ and $(x_2, y_2) = (-4, 5)$. Use the definition of slope.

$$m = \frac{\Delta y}{\Delta x} = \frac{5 - 6}{-4 - 7} = \frac{-1}{-11} = \frac{1}{11}$$

(b) $(5, -3)$ and $(-2, -3)$

SOLUTION Let $(x_1, y_1) = (5, -3)$ and $(x_2, y_2) = (-2, -3)$. Then

$$m = \frac{-3 - (-3)}{-2 - 5} = \frac{0}{-7} = 0.$$

Lines with zero slope are horizontal (parallel to the x-axis).

(c) $(2, -4)$ and $(2, 3)$

SOLUTION Let $(x_1, y_1) = (2, -4)$ and $(x_2, y_2) = (2, 3)$. Then

$$m = \frac{3 - (-4)}{2 - 2} = \frac{7}{0},$$

YOUR TURN 1 Find the slope of the line through $(1, 5)$ and $(4, 6)$.

which is undefined. This happens when the line is vertical (parallel to the y-axis).

TRY YOUR TURN 1

CAUTION The phrase "no slope" should be avoided; specify instead whether the slope is zero or undefined.

In finding the slope of the line in Example 1(a), we could have let $(x_1, y_1) = (-4, 5)$ and $(x_2, y_2) = (7, 6)$. In that case,

$$m = \frac{6 - 5}{7 - (-4)} = \frac{1}{11},$$

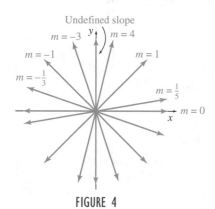

FIGURE 4

the same answer as before. The order in which coordinates are subtracted does not matter, as long as it is done consistently.

Figure 4 shows examples of lines with different slopes. Lines with positive slopes go up from left to right, while lines with negative slopes go down from left to right.

It might help you to compare slope with the percent grade of a hill. If a sign says a hill has a 10% grade uphill, this means the slope is 0.10, or 1/10, so the hill rises 1 foot for every 10 feet horizontally. A 15% grade downhill means the slope is -0.15.

Equations of a Line
An equation in two first-degree variables, such as $4x + 7y = 20$, has a line as its graph, so it is called a **linear equation**. In the rest of this section, we consider various forms of the equation of a line.

Suppose a line has a slope m and y-intercept b. This means that it goes through the point $(0, b)$. If (x, y) is any other point on the line, then the definition of slope tells us that

$$m = \frac{y - b}{x - 0}.$$

FOR REVIEW

For review on solving a linear equation, see Section R.4.

We can simplify this equation by multiplying both sides by x and adding b to both sides. The result is

$$y = mx + b,$$

which we call the *slope-intercept* form of a line. This is the most common form for writing the equation of a line.

Slope-Intercept Form

If a line has slope m and y-intercept b, then the equation of the line in **slope-intercept form** is

$$y = mx + b.$$

When $b = 0$, we say that y is **proportional** to x.

EXAMPLE 2   **Equation of a Line**

Find an equation in slope-intercept form for each line.

(a) Through $(0, -3)$ with slope $3/4$

SOLUTION   We recognize $(0, -3)$ as the y-intercept because it's the point with 0 as its x-coordinate, so $b = -3$. The slope is $3/4$, so $m = 3/4$. Substituting these values into $y = mx + b$ gives

$$y = \frac{3}{4}x - 3.$$

(b) With x-intercept 7 and y-intercept 2

SOLUTION   Notice that $b = 2$. To find m, use the definition of slope after writing the x-intercept as $(7, 0)$ (because the y-coordinate is 0 where the line crosses the x-axis) and the y-intercept as $(0, 2)$.

$$m = \frac{0 - 2}{7 - 0} = -\frac{2}{7}$$

YOUR TURN 2 Find the equation of the line with x-intercept -4 and y-intercept 6.

Substituting these values into $y = mx + b$, we have

$$y = -\frac{2}{7}x + 2.$$

TRY YOUR TURN 2

EXAMPLE 3   **Finding the Slope**

Find the slope of the line whose equation is $3x - 4y = 12$.

SOLUTION   To find the slope, solve the equation for y.

$$3x - 4y = 12$$
$$-4y = -3x + 12 \qquad \text{Subtract } 3x \text{ from both sides.}$$
$$y = \frac{3}{4}x - 3 \qquad \text{Divide both sides by } -4.$$

YOUR TURN 3 Find the slope of the line whose equation is $8x + 3y = 5$.

The coefficient of x is $3/4$, which is the slope of the line. Notice that this is the same line as in Example 2 (a).   TRY YOUR TURN 3

The slope-intercept form of the equation of a line involves the slope and the y-intercept. Sometimes, however, the slope of a line is known, together with one point (perhaps *not* the y-intercept) that the line goes through. The *point-slope form* of the equation of a line is used to find the equation in this case. Let (x_1, y_1) be any fixed point on the line, and let (x, y) represent any other point on the line. If m is the slope of the line, then by the definition of slope,

$$\frac{y - y_1}{x - x_1} = m,$$

or

$$y - y_1 = m(x - x_1). \qquad \text{Multiply both sides by } x - x_1.$$

> **Point-Slope Form**
>
> If a line has slope m and passes through the point (x_1, y_1), then an equation of the line is given by
>
> $$y - y_1 = m(x - x_1),$$
>
> the **point-slope form** of the equation of a line.

EXAMPLE 4   Point-Slope Form

Find an equation of the line that passes through the point $(3, -7)$ and has slope $m = 5/4$.

SOLUTION   Use the point-slope form.

$$y - y_1 = m(x - x_1)$$

$$y - (-7) = \frac{5}{4}(x - 3) \qquad y_1 = -7, m = \tfrac{5}{4}, x_1 = 3$$

$$y + 7 = \frac{5}{4}(x - 3)$$

$$4y + 28 = 5(x - 3) \qquad \text{Multiply both sides by 4.}$$

$$4y + 28 = 5x - 15 \qquad \text{Distribute.}$$

$$4y = 5x - 43 \qquad \text{Combine constants.}$$

$$y = \frac{5}{4}x - \frac{43}{4} \qquad \text{Divide both sides by 4.}$$

--- FOR REVIEW ---

See Section R.4 for details on eliminating denominators in an equation.

The equation of the same line can be given in many forms. To avoid confusion, the linear equations used in the rest of this section will be written in slope-intercept form, $y = mx + b$, which is often the most useful form.

The point-slope form also can be useful to find an equation of a line if we know two different points that the line goes through, as in the next example.

EXAMPLE 5   Using Point-Slope Form to Find an Equation

Find an equation of the line through $(5, 4)$ and $(-10, -2)$.

SOLUTION   Begin by using the definition of slope to find the slope of the line that passes through the given points.

$$\text{Slope} = m = \frac{-2 - 4}{-10 - 5} = \frac{-6}{-15} = \frac{2}{5}$$

Either $(5, 4)$ or $(-10, -2)$ can be used in the point-slope form with $m = 2/5$. If $(x_1, y_1) = (5, 4)$, then

$$y - y_1 = m(x - x_1)$$

$$y - 4 = \frac{2}{5}(x - 5) \qquad y_1 = 4, m = \tfrac{2}{5}, x_1 = 5$$

$$5y - 20 = 2(x - 5) \qquad \text{Multiply both sides by 5.}$$

$$5y - 20 = 2x - 10 \qquad \text{Distributive property}$$

$$5y = 2x + 10 \qquad \text{Add 20 to both sides.}$$

$$y = \frac{2}{5}x + 2 \qquad \text{Divide by 5 to put in slope-intercept form.}$$

YOUR TURN 4 Find the equation of the line through $(2, 9)$ and $(5, 3)$. Put your answer in slope-intercept form.

Check that the same result is found if $(x_1, y_1) = (-10, -2)$.   **TRY YOUR TURN 4**

EXAMPLE 6   Horizontal Line

Find an equation of the line through $(8, -4)$ and $(-2, -4)$.

SOLUTION   Find the slope.

$$m = \frac{-4 - (-4)}{-2 - 8} = \frac{0}{-10} = 0$$

Choose, say, $(8, -4)$ as (x_1, y_1).

$$y - y_1 = m(x - x_1)$$
$$y - (-4) = 0(x - 8) \qquad y_1 = -4, m = 0, x_1 = 8$$
$$y + 4 = 0 \qquad\qquad 0(x - 8) = 0$$
$$y = -4$$

Plotting the given ordered pairs and drawing a line through the points show that the equation $y = -4$ represents a horizontal line. See Figure 5(a). Every horizontal line has a slope of zero and an equation of the form $y = k$, where k is the y-value of all ordered pairs on the line.

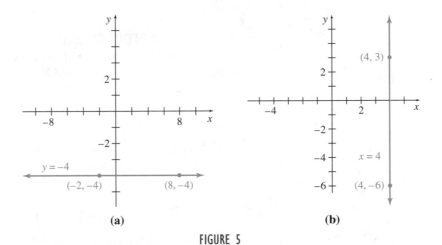

(a)   (b)

FIGURE 5

EXAMPLE 7   Vertical Line

Find an equation of the line through $(4, 3)$ and $(4, -6)$.

SOLUTION   The slope of the line is

$$m = \frac{-6 - 3}{4 - 4} = \frac{-9}{0},$$

which is undefined. Since both ordered pairs have x-coordinate 4, the equation is $x = 4$. Because the slope is undefined, the equation of this line cannot be written in the slope-intercept form.

Again, plotting the given ordered pairs and drawing a line through them show that the graph of $x = 4$ is a vertical line. See Figure 5(b).

> ## Slope of Horizontal and Vertical Lines
> The slope of a horizontal line is 0.
>
> The slope of a vertical line is undefined.

The different forms of linear equations discussed in this section are summarized below. The slope-intercept and point-slope forms are equivalent ways to express the equation of a nonvertical line. The slope-intercept form is simpler for a final answer, but you may find the point-slope form easier to use when you know the slope of a line and a point through which the line passes. The slope-intercept form is often considered the standard form. Any line that is not vertical has a unique slope-intercept form but can have many point-slope forms for its equation.

Equations of Lines

Equation	Description
$y = mx + b$	**Slope-intercept form:** slope m, y-intercept b
$y - y_1 = m(x - x_1)$	**Point-slope form:** slope m, line passes through (x_1, y_1)
$x = k$	**Vertical line:** x-intercept k, no y-intercept (except when $k = 0$), undefined slope
$y = k$	**Horizontal line:** y-intercept k, no x-intercept (except when $k = 0$), slope 0

Parallel and Perpendicular Lines
One application of slope involves deciding whether two lines are parallel, which means that they never intersect. Since two parallel lines are equally "steep," they should have the same slope. Also, two lines with the same "steepness" are parallel.

Parallel Lines
Two lines are **parallel** if and only if they have the same slope, or if they are both vertical.

EXAMPLE 8 Parallel Line

Find the equation of the line that passes through the point $(3, 5)$ and is parallel to the line $2x + 5y = 4$.

SOLUTION The slope of $2x + 5y = 4$ can be found by writing the equation in slope-intercept form.

$$2x + 5y = 4$$

$$y = -\frac{2}{5}x + \frac{4}{5} \qquad \text{Subtract } 2x \text{ from both sides and divide both sides by 5.}$$

This result shows that the slope is $-2/5$. Since the lines are parallel, $-2/5$ is also the slope of the line whose equation we want. This line passes through $(3, 5)$. Substituting $m = -2/5$, $x_1 = 3$, and $y_1 = 5$ into the point-slope form gives

$$y - y_1 = m(x - x_1)$$

$$y - 5 = -\frac{2}{5}(x - 3) = -\frac{2}{5}x + \frac{6}{5}$$

$$y = -\frac{2}{5}x + \frac{6}{5} + 5$$

$$y = -\frac{2}{5}x + \frac{31}{5}. \qquad \text{Multiply 5 by 5/5 to get a common denominator.}$$

YOUR TURN 5 Find the equation of the line that passes through the point $(4, 5)$ and is parallel to the line $3x - 6y = 7$. Put your answer in slope-intercept form.

TRY YOUR TURN 5

As already mentioned, two nonvertical lines are parallel if and only if they have the same slope. Two lines having slopes with a product of -1 are perpendicular. A proof of this fact, which depends on similar triangles from geometry, is given as Exercise 43 in this section.

> **Perpendicular Lines**
>
> Two lines are **perpendicular** if and only if the product of their slopes is -1, or if one is vertical and the other horizontal.

EXAMPLE 9   Perpendicular Line

Find the equation of the line L passing through the point $(3, 7)$ and perpendicular to the line having the equation $5x - y = 4$.

SOLUTION   To find the slope, write $5x - y = 4$ in slope-intercept form:

$$y = 5x - 4.$$

The slope is 5. Since the lines are perpendicular, if line L has slope m, then

$$5m = -1$$

$$m = -\frac{1}{5}.$$

Now substitute $m = -1/5$, $x_1 = 3$, and $y_1 = 7$ into the point-slope form.

$$y - 7 = -\frac{1}{5}(x - 3)$$

$$y - 7 = -\frac{1}{5}x + \frac{3}{5}$$

$$y = -\frac{1}{5}x + \frac{3}{5} + 7 \cdot \frac{5}{5} \qquad \text{Add 7 to both sides and get a common denominator.}$$

$$y = -\frac{1}{5}x + \frac{38}{5} \qquad\qquad\qquad \textbf{TRY YOUR TURN 6}$$

YOUR TURN 6 Find the equation of the line passing through the point $(3, 2)$ and perpendicular to the line having the equation $2x + 3y = 4$.

The next example uses the equation of a line to analyze real-world data. In this example, we are looking at how one variable changes over time. To simplify the arithmetic, we will *rescale* the variable representing time, although computers and calculators have made rescaling less important than in the past. Here it allows us to work with smaller numbers, and, as you will see, find the y-intercept of the line more easily. We will use rescaling on many examples throughout this book. When we do, it is important to be consistent.

EXAMPLE 10   Prevalence of Cigarette Smoking

In recent years, the percentage of the U.S. population age 18 and older who smoke has decreased at a roughly constant rate, from 24.1% in 1998 to 20.6% in 2008. *Source: Centers for Disease Control and Prevention.*

(a) Find the equation describing this linear relationship.

SOLUTION   Let t represent time in years, with $t = 0$ representing 1990. With this rescaling, the year 1998 corresponds to $t = 8$ and the year 2008 corresponds to $t = 2008 - 1990 = 18$. Let y represent the percentage of the population who smoke. The two ordered pairs representing the given information are then $(8, 24.1)$ and $(18, 20.6)$. The slope of the line through these points is

$$m = \frac{20.6 - 24.1}{18 - 8} = \frac{-3.5}{10} = -0.35.$$

This means that, on average, the percentage of the adult population who smoke is decreasing by about 0.35% per year.

Using $m = -0.35$ in the point-slope form, and choosing $(t_1, y_1) = (8, 24.1)$, gives the required equation.

$$y - 24.1 = -0.35\,(t - 8)$$
$$y - 24.1 = -0.35\,t + 2.8$$
$$y = -0.35\,t + 26.9$$

We could have used the other point $(18, 20.6)$ and found the same answer. Instead, we'll use this to check our answer by observing that $-0.35(18) + 26.9 = 20.6$, which agrees with the y-value at $t = 18$.

(b) One objective of Healthy People 2010 (a campaign of the U.S. Department of Health and Human Services) was to reduce the percentage of U.S. adults who smoke to 12% or less by the year 2010. If this decline in smoking continued at the same rate, did they meet this objective?

SOLUTION Using the same rescaling, $t = 20$ corresponds to the year 2010. Substituting this value into the above equation gives

$$y = -0.35(20) + 26.9 = 19.9.$$

Continuing at this rate, an estimated 19.9% of the adult population still smoked in 2010, and the objective of Healthy People 2010 was not met. ■

Notice that if the formula from part (b) of Example 10 is valid for all nonnegative t, then eventually y becomes 0:

$$-0.35\,t + 26.9 = 0$$
$$-0.35\,t = -26.9 \qquad \text{Subtract 26.9 from both sides.}$$
$$t = \frac{-26.9}{-0.35} = 76.857 \approx 77^*, \qquad \text{Divide both sides by } -0.35.$$

which indicates that 77 years from 1990 (in the year 2067), 0% of the U.S. adult population will smoke. Of course, it is still possible that in 2067 there will be adults who smoke; the trend of recent years may not continue. Most equations are valid for some specific set of numbers. It is highly speculative to extrapolate beyond those values.

On the other hand, people in business and government often need to make some prediction about what will happen in the future, so a tentative conclusion based on past trends may be better than no conclusion at all. There are also circumstances, particularly in the physical sciences, in which theoretical reasons imply that the trend will continue.

Graph of a Line

We can graph the linear equation defined by $y = x + 1$ by finding several ordered pairs that satisfy the equation. For example, if $x = 2$, then $y = 2 + 1 = 3$, giving the ordered pair $(2, 3)$. Also, $(0, 1)$, $(4, 5)$, $(-2, -1)$, $(-5, -4)$, $(-3, -2)$, among many others, satisfy the equation.

To graph $y = x + 1$, we begin by locating the ordered pairs obtained above, as shown in Figure 6(a). All the points of this graph appear to lie on a straight line, as in Figure 6(b). This straight line is the graph of $y = x + 1$.

It can be shown that every equation of the form $ax + by = c$ has a straight line as its graph, assuming a and b are not both 0. Although just two points are needed to determine a line, it is a good idea to plot a third point as a check. It is often convenient to use the x- and y-intercepts as the two points, as in the following example.

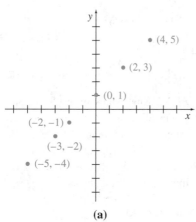

(a)

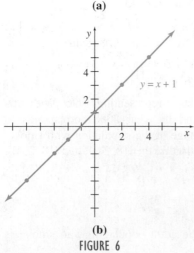

(b)

FIGURE 6

*The symbol $\approx$ means "is approximately equal to."

EXAMPLE 11 **Graph of a Line**

Graph $3x + 2y = 12$.

SOLUTION To find the y-intercept, let $x = 0$.

$$3(0) + 2y = 12$$
$$2y = 12 \qquad \text{Divide both sides by 2.}$$
$$y = 6$$

Similarly, find the x-intercept by letting $y = 0$, which gives $x = 4$. Verify that when $x = 2$, the result is $y = 3$. These three points are plotted in Figure 7(a). A line is drawn through them in Figure 7(b).

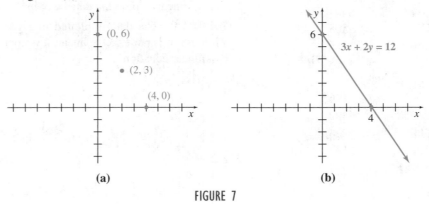

(a)   **(b)**

FIGURE 7

Not every line has two distinct intercepts; the graph in the next example does not cross the x-axis, and so it has no x-intercept.

EXAMPLE 12 **Graph of a Horizontal Line**

Graph $y = -3$.

SOLUTION The equation $y = -3$, or equivalently, $y = 0x - 3$, always gives the same y-value, -3, for any value of x. Therefore, no value of x will make $y = 0$, so the graph has no x-intercept. As we saw in Example 6, the graph of such an equation is a horizontal line parallel to the x-axis. In this case the y-intercept is -3, as shown in Figure 8.

The graph in Example 12 has only one intercept. Another type of linear equation with coinciding intercepts is graphed in Example 13.

EXAMPLE 13 **Graph of a Line Through the Origin**

Graph $y = -3x$.

SOLUTION Begin by looking for the x-intercept. If $y = 0$, then

$$y = -3x$$
$$0 = -3x \qquad \text{Let } y = 0.$$
$$0 = x. \qquad \text{Divide both sides by } -3.$$

We have the ordered pair $(0, 0)$. Starting with $x = 0$ gives exactly the same ordered pair, $(0, 0)$. Two points are needed to determine a straight line, and the intercepts have led to only one point. To get a second point, we choose some other value of x (or y). For example, if $x = 2$, then

$$y = -3x = -3(2) = -6, \qquad \text{Let } x = 2.$$

giving the ordered pair $(2, -6)$. These two ordered pairs, $(0, 0)$ and $(2, -6)$, were used to get the graph shown in Figure 9.

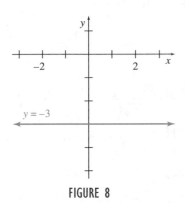

FIGURE 8

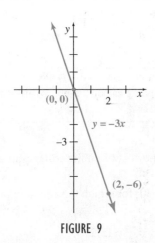

FIGURE 9

Linear equations allow us to set up simple mathematical models for real-life situations. In almost every case, linear (or any other reasonably simple) equations provide only approximations to real-world situations. Nevertheless, these are often remarkably useful approximations.

EXAMPLE 14 Tuition

APPLY IT

The table on the left lists the average annual cost (in dollars) of tuition and fees at public four-year colleges for selected years. *Source: The College Board.*

(a) Plot the cost of public colleges by letting $t = 0$ correspond to 2000. Are the data *exactly* linear? Could the data be *approximated* by a linear equation?

SOLUTION   The data is plotted in Figure 10(a) in a figure known as a **scatterplot**. Although it is not exactly linear, it is approximately linear and could be approximated by a linear equation.

Cost of Public College	
Year	Tuition and Fees
2000	3508
2001	3766
2002	4098
2003	4645
2004	5126
2005	5492
2006	5804
2007	6191
2008	6591
2009	7020

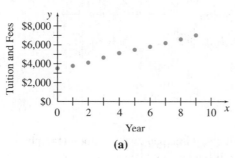

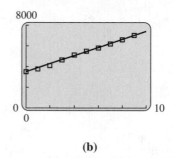

(a)

(b)

FIGURE 10

(b) Use the points (0, 3508) and (9, 7020) to determine an equation that models the data.

SOLUTION   We first find the slope of the line as follows:

$$m = \frac{7020 - 3508}{9 - 0} = \frac{3512}{9} \approx 390.2.$$

We have rounded to four digits, noting that we cannot expect more accuracy in our answer than in our data, which is accurate to four digits. Using the slope-intercept form of the line, $y = mt + b$, with $m = 390.2$ and $b = 3508$, gives

$$y = 390.2t + 3508.$$

TECHNOLOGY NOTE   A graphing calculator plot of this line and the data points are shown in Figure 10(b). Notice that the points closely fit the line. More details on how to construct this graphing calculator plot are given at the end of this example.

(c) Discuss the accuracy of using this equation to estimate the cost of public colleges in the year 2030.

SOLUTION   The year 2030 corresponds to the year $t = 30$, for which the equation predicts a cost of

$$y = 390.2(30) + 3508 = 15,214, \quad \text{or} \quad \$15,214.$$

The year 2030 is many years in the future, however. Many factors could affect the tuition, and the actual figure for 2030 could turn out to be very different from our prediction. ▬

TECHNOLOGY NOTE   You can plot data with a TI-84 Plus graphing calculator using the following steps.

1. Store the data in lists.

2. Define the stat plot.

3. Turn off $Y =$ functions (unless you also want to graph a function).

4. Turn on the plot you want to display.

5. Define the viewing window.

6. Display the graph.

Consult the calculator's instruction booklet or the *Graphing Calculator and Excel Spreadsheet Manual*, available with this book, for specific instructions. See the calculator-generated graph in Figure 10(b), which includes the points and line from Example 14. Notice how the line closely approximates the data.

1.1   EXERCISES

Find the slope of each line.

1. Through $(4, 5)$ and $(-1, 2)$

2. Through $(5, -4)$ and $(1, 3)$

3. Through $(8, 4)$ and $(8, -7)$

4. Through $(1, 5)$ and $(-2, 5)$

5. $y = x$ 6. $y = 3x - 2$

7. $5x - 9y = 11$ 8. $4x + 7y = 1$

9. $x = 5$ 10. The x-axis

11. $y = 8$ 12. $y = -6$

13. A line parallel to $6x - 3y = 12$

14. A line perpendicular to $8x = 2y - 5$

In Exercises 15–24, find an equation in slope-intercept form for each line.

15. Through $(1, 3)$, $m = -2$

16. Through $(2, 4)$, $m = -1$

17. Through $(-5, -7)$, $m = 0$

18. Through $(-8, 1)$, with undefined slope

19. Through $(4, 2)$ and $(1, 3)$

20. Through $(8, -1)$ and $(4, 3)$

21. Through $(2/3, 1/2)$ and $(1/4, -2)$

22. Through $(-2, 3/4)$ and $(2/3, 5/2)$

23. Through $(-8, 4)$ and $(-8, 6)$

24. Through $(-1, 3)$ and $(0, 3)$

In Exercises 25–34, find an equation for each line in the form $ax + by = c$, where a, b, and c are integers with no factor common to all three and $a \geq 0$.

25. x-intercept -6, y-intercept -3

26. x-intercept -2, y-intercept 4

27. Vertical, through $(-6, 5)$

28. Horizontal, through $(8, 7)$

29. Through $(-4, 6)$, parallel to $3x + 2y = 13$

30. Through $(2, -5)$, parallel to $2x - y = -4$

31. Through $(3, -4)$, perpendicular to $x + y = 4$

32. Through $(-2, 6)$, perpendicular to $2x - 3y = 5$

33. The line with y-intercept 4 and perpendicular to $x + 5y = 7$

34. The line with x-intercept $-2/3$ and perpendicular to $2x - y = 4$

35. Do the points $(4, 3)$, $(2, 0)$, and $(-18, -12)$ lie on the same line? Explain why or why not. (*Hint:* Find the slopes between the points.)

36. Find k so that the line through $(4, -1)$ and $(k, 2)$ is
    a. parallel to $2x + 3y = 6$,
    b. perpendicular to $5x - 2y = -1$.

37. Use slopes to show that the quadrilateral with vertices at $(1, 3)$, $(-5/2, 2)$, $(-7/2, 4)$, and $(2, 1)$ is a parallelogram.

38. Use slopes to show that the square with vertices at $(-2, 5)$, $(4, 5)$, $(4, -1)$, and $(-2, -1)$ has diagonals that are perpendicular.

For the lines in Exercises 39 and 40, which of the following is closest to the slope of the line? (a) 1   (b) 2   (c) 3   (d) 21   (e) 22   (f) −3

39.

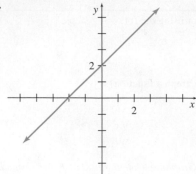

40.

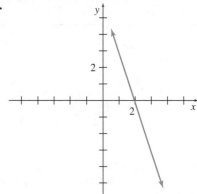

In Exercises 41 and 42, estimate the slope of the lines.

41.

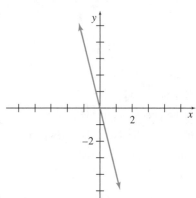

42.

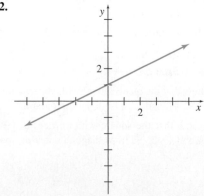

43. To show that two perpendicular lines, neither of which is vertical, have slopes with a product of −1, go through the following steps. Let line L_1 have equation $y = m_1x + b_1$, and let L_2 have equation $y = m_2x + b_2$, with $m_1 > 0$ and $m_2 < 0$. Assume that L_1 and L_2 are perpendicular, and use right triangle *MPN* shown in the figure. Prove each of the following statements.

a. *MQ* has length m_1.

b. *QN* has length $-m_2$.

c. Triangles *MPQ* and *PNQ* are similar.

d. $m_1/1 = 1/(-m_2)$ and $m_1m_2 = -1$

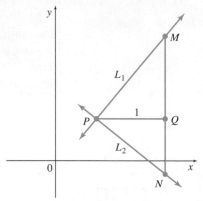

44. Consider the equation $\dfrac{x}{a} + \dfrac{y}{b} = 1$.

a. Show that this equation represents a line by writing it in the form $y = mx + b$.

b. Find the *x*- and *y*-intercepts of this line.

c. Explain in your own words why the equation in this exercise is known as the intercept form of a line.

Graph each equation.

45. $y = x - 1$ **46.** $y = 4x + 5$

47. $y = -4x + 9$ **48.** $y = -6x + 12$

49. $2x - 3y = 12$ **50.** $3x - y = -9$

51. $3y - 7x = -21$ **52.** $5y + 6x = 11$

53. $y = -2$ **54.** $x = 4$

55. $x + 5 = 0$ **56.** $y + 8 = 0$

57. $y = 2x$ **58.** $y = -5x$

59. $x + 4y = 0$ **60.** $3x - 5y = 0$

APPLICATIONS

Business and Economics

61. Sales   The sales of a small company were $27,000 in its second year of operation and $63,000 in its fifth year. Let *y* represent sales in the *x*th year of operation. Assume that the data can be approximated by a straight line.

a. Find the slope of the sales line, and give an equation for the line in the form $y = mx + b$.

b. Use your answer from part a to find out how many years must pass before the sales surpass $100,000.

62. Use of Cellular Telephones The following table shows the subscribership of cellular telephones in the United States (in millions) for even-numbered years between 2000 and 2008. *Source: Time Almanac 2010.*

Year	2000	2002	2004	2006	2008
Subscribers (in millions)	109.48	140.77	182.14	233.04	270.33

  a. Plot the data by letting $t = 0$ correspond to 2000. Discuss how well the data fit a straight line.

  b. Determine a linear equation that approximates the number of subscribers using the points $(0, 109.48)$ and $(8, 270.33)$.

  c. Repeat part b using the points $(2, 140.77)$ and $(8, 270.33)$.

  d. Discuss why your answers to parts b and c are similar but not identical.

  e. Using your equations from parts b and c, approximate the number of cellular phone subscribers in the year 2007. Compare your result with the actual value of 255.40 million.

63. Consumer Price Index The Consumer Price Index (CPI) is a measure of the change in the cost of goods over time. The index was 100 for the three-year period centered on 1983. For simplicity, we will assume that the CPI was exactly 100 in 1983. Then the CPI of 215.3 in 2008 indicates that an item that cost $1.00 in 1983 would cost $2.15 in 2008. The CPI has been increasing approximately linearly over the last few decades. *Source: Time Almanac 2010.*

  a. Use this information to determine an equation for the CPI in terms of t, which represents the years since 1980.

  b. Based on the answer to part a, what was the predicted value of the CPI in 2000? Compare this estimate with the actual CPI of 172.2.

  c. Describe the rate at which the annual CPI is changing.

Life Sciences

64. HIV Infection The time interval between a person's initial infection with HIV and that person's eventual development of AIDS symptoms is an important issue. The method of infection with HIV affects the time interval before AIDS develops. One study of HIV patients who were infected by intravenous drug use found that 17% of the patients had AIDS after 4 years, and 33% had developed the disease after 7 years. The relationship between the time interval and the percentage of patients with AIDS can be modeled accurately with a linear equation. *Source: Epidemiologic Review.*

  a. Write a linear equation $y = mt + b$ that models this data, using the ordered pairs $(4, 0.17)$ and $(7, 0.33)$.

  b. Use your equation from part a to predict the number of years before half of these patients will have AIDS.

65. Exercise Heart Rate To achieve the maximum benefit for the heart when exercising, your heart rate (in beats per minute) should be in the target heart rate zone. The lower limit of this zone is found by taking 70% of the difference between 220 and your age. The upper limit is found by using 85%. *Source: Physical Fitness.*

  a. Find formulas for the upper and lower limits (u and l) as linear equations involving the age x.

  b. What is the target heart rate zone for a 20-year-old?

  c. What is the target heart rate zone for a 40-year-old?

  d. Two women in an aerobics class stop to take their pulse and are surprised to find that they have the same pulse. One woman is 36 years older than the other and is working at the upper limit of her target heart rate zone. The younger woman is working at the lower limit of her target heart rate zone. What are the ages of the two women, and what is their pulse?

  e. Run for 10 minutes, take your pulse, and see if it is in your target heart rate zone. (After all, this is listed as an exercise!)

66. Ponies Trotting A 1991 study found that the peak vertical force on a trotting Shetland pony increased linearly with the pony's speed, and that when the force reached a critical level, the pony switched from a trot to a gallop. For one pony, the critical force was 1.16 times its body weight. It experienced a force of 0.75 times its body weight at a speed of 2 meters per second and a force of 0.93 times its body weight at 3 meters per second. At what speed did the pony switch from a trot to a gallop? *Source: Science.*

67. Life Expectancy Some scientists believe there is a limit to how long humans can live. One supporting argument is that during the last century, life expectancy from age 65 has increased more slowly than life expectancy from birth, so eventually these two will be equal, at which point, according to these scientists, life expectancy should increase no further. In 1900, life expectancy at birth was 46 yr, and life expectancy at age 65 was 76 yr. In 2004, these figures had risen to 77.8 and 83.7, respectively. In both cases, the increase in life expectancy has been linear. Using these assumptions and the data given, find the maximum life expectancy for humans. *Source: Science.*

Social Sciences

68. Child Mortality Rate The mortality rate for children under 5 years of age around the world has been declining in a roughly linear fashion in recent years. The rate per 1000 live births was 90 in 1990 and 65 in 2008. *Source: World Health Organization.*

  a. Determine a linear equation that approximates the mortality rate in terms of time t, where t represents the number of years since 1900.

  b. If this trend continues, in what year will the mortality rate first drop to 50 or below per 1000 live births?

69. Health Insurance The percentage of adults in the United States without health insurance increased at a roughly linear rate from 1999, when it was 17.2%, to 2008, when it was 20.3%. *Source: The New York Times.*

  a. Determine a linear equation that approximates the percentage of adults in the United States without health insurance in terms of time t, where t represents the number of years since 1990.

  b. If this trend were to continue, in what year would the percentage of adults without health insurance be at least 25%?

70. Marriage The following table lists the U.S. median age at first marriage for men and women. The age at which both groups marry for the first time seems to be increasing at a roughly linear rate in recent decades. Let t correspond to the number of years since 1980. *Source: U.S. Census Bureau.*

Age at First Marriage						
Year	1980	1985	1990	1995	2000	2005
Men	24.7	25.5	26.1	26.9	26.8	27.1
Women	22.0	23.3	23.9	24.5	25.1	25.3

a. Find a linear equation that approximates the data for men, using the data for the years 1980 and 2005.

b. Repeat part a using the data for women.

c. Which group seems to have the faster increase in median age at first marriage?

d. In what year will the men's median age at first marriage reach 30?

e. When the men's median age at first marriage is 30, what will the median age be for women?

71. Immigration In 1950, there were 249,187 immigrants admitted to the United States. In 2008, the number was 1,107,126. *Source: 2008 Yearbook of Immigration Statistics.*

a. Assuming that the change in immigration is linear, write an equation expressing the number of immigrants, y, in terms of t, the number of years after 1900.

b. Use your result in part a to predict the number of immigrants admitted to the United States in 2015.

c. Considering the value of the y-intercept in your answer to part a, discuss the validity of using this equation to model the number of immigrants throughout the entire 20th century.

Physical Sciences

72. Global Warming In 1990, the Intergovernmental Panel on Climate Change predicted that the average temperature on Earth would rise 0.3°C per decade in the absence of international controls on greenhouse emissions. Let t measure the time in years since 1970, when the average global temperature was 15°C. *Source: Science News.*

a. Find a linear equation giving the average global temperature in degrees Celsius in terms of t, the number of years since 1970.

b. Scientists have estimated that the sea level will rise by 65 cm if the average global temperature rises to 19°C. According to your answer to part a, when would this occur?

73. Galactic Distance The table lists the distances (in megaparsecs where 1 megaparsec $\approx 3.1 \times 10^{19}$ km) and velocities (in kilometers per second) of four galaxies moving rapidly away from Earth. *Source: Astronomical Methods and Calculations, and Fundamental Astronomy.*

Galaxy	Distance	Velocity
Virga	15	1600
Ursa Minor	200	15,000
Corona Borealis	290	24,000
Bootes	520	40,000

a. Plot the data points letting x represent distance and y represent velocity. Do the points lie in an approximately linear pattern?

b. Write a linear equation $y = mx$ to model this data, using the ordered pair $(520, 40{,}000)$.

c. The galaxy Hydra has a velocity of 60,000 km per sec. Use your equation to approximate how far away it is from Earth.

d. The value of m in the equation is called the *Hubble constant*. The Hubble constant can be used to estimate the age of the universe A (in years) using the formula

$$A = \frac{9.5 \times 10^{11}}{m}.$$

Approximate A using your value of m.

General Interest

74. News/Talk Radio From 2001 to 2007, the number of stations carrying news/talk radio increased at a roughly linear rate, from 1139 in 2001 to 1370 in 2007. *Source: State of the Media.*

a. Find a linear equation expressing the number of stations carrying news/talk radio, y, in terms of t, the years since 2000.

b. Use your answer from part a to predict the number of stations carrying news/talk radio in 2008. Compare with the actual number of 2046. Discuss how the linear trend from 2001 to 2007 might have changed in 2008.

75. Tuition The table lists the annual cost (in dollars) of tuition and fees at private four-year colleges for selected years. (See Example 14.) *Source: The College Board.*

Year	Tuition and Fees
2000	16,072
2002	18,060
2004	20,045
2006	22,308
2008	25,177
2009	26,273

a. Sketch a graph of the data. Do the data appear to lie roughly along a straight line?

b. Let $t = 0$ correspond to the year 2000. Use the points $(0, 16{,}072)$ and $(9, 26{,}273)$ to determine a linear equation that models the data. What does the slope of the graph of the equation indicate?

c. Discuss the accuracy of using this equation to estimate the cost of private college in 2025.

YOUR TURN ANSWERS

1. $1/3$

2. $y = (3/2)x + 6$

3. $-8/3$

4. $y = -2x + 13$

5. $y = (1/2)x + 3$

6. $y = (3/2)x - 5/2$

1.2 Linear Functions and Applications

APPLY IT   How many units must be sold for a firm to break even?
In Example 6 in this section, this question will be answered using a linear function.

As we saw in the previous section, many situations involve two variables related by a linear equation. For such a relationship, when we express the variable y in terms of x, we say that y is a **linear function** of x. This means that for any allowed value of x (the **independent variable**), we can use the equation to find the corresponding value of y (the **dependent variable**). Examples of equations defining linear functions include $y = 2x + 3$, $y = -5$, and $2x - 3y = 7$, which can be written as $y = (2/3)x - (7/3)$. Equations in the form $x = k$, where k is a constant, do not define linear functions. All other linear equations define linear functions.

$f(x)$ Notation
Letters such as f, g, or h are often used to name functions. For example, f might be used to name the function defined by

$$y = 5 - 3x.$$

To show that this function is named f, it is common to replace y with $f(x)$ (read "f of x") to get

$$f(x) = 5 - 3x.$$

By choosing 2 as a value of x, $f(x)$ becomes $5 - 3 \cdot 2 = 5 - 6 = -1$, written

$$f(2) = -1.$$

The corresponding ordered pair is $(2, -1)$. In a similar manner,

$$f(-4) = 5 - 3(-4) = 17, \qquad f(0) = 5, \qquad f(-6) = 23,$$

and so on.

EXAMPLE 1   Function Notation

Let $g(x) = -4x + 5$. Find $g(3)$, $g(0)$, $g(-2)$, and $g(b)$.

SOLUTION   To find $g(3)$, substitute 3 for x.

$$g(3) = -4(3) + 5 = -12 + 5 = -7$$

Similarly,

$$g(0) = -4(0) + 5 = 0 + 5 = 5,$$
$$g(-2) = -4(-2) + 5 = 8 + 5 = 13,$$

and

$$g(b) = -4b + 5.$$ **TRY YOUR TURN 1**

YOUR TURN 1
Calculate $g(-5)$.

We summarize the discussion below.

Linear Function
A relationship f defined by

$$y = f(x) = mx + b,$$

for real numbers m and b, is a **linear function**.

Supply and Demand
Linear functions are often good choices for supply and demand curves. Typically, as the price of an item increases, consumers are less likely to buy an increasingly expensive item, and so the demand for the item decreases. On the other

hand, as the price of an item increases, producers are more likely to see a profit in selling the item, and so the supply of the item increases. The increase in the quantity supplied and decrease in the quantity demanded can eventually result in a surplus, which causes the price to fall. These countervailing trends tend to move the price, as well as the quantity supplied and demanded toward an equilibrium value.

For example, during the late 1980s and early 1990s, the consumer demand for cranberries (and all of their healthy benefits) soared. The quantity demanded surpassed the quantity supplied, causing a shortage, and cranberry prices rose dramatically. As prices increased, growers wanted to increase their profits, so they planted more acres of cranberries. Unfortunately, cranberries take 3 to 5 years from planting until they can first be harvested. As growers waited and prices increased, consumer demand decreased. When the cranberries were finally harvested, the supply overwhelmed the demand and a huge surplus occurred, causing the price of cranberries to drop in the late 1990s. *Source: Agricultural Marketing Resource Center.* Other factors were involved in this situation, but the relationship between price, supply, and demand was nonetheless typical.

Although economists consider price to be the independent variable, they have the unfortunate habit of plotting price, usually denoted by p, on the vertical axis, while everyone else graphs the independent variable on the horizontal axis. This custom was started by the English economist Alfred Marshall (1842–1924). In order to abide by this custom, we will write p, the price, as a function of q, the quantity produced, and plot p on the vertical axis. But remember, it is really *price* that determines how much consumers demand and producers supply, not the other way around.

Supply and demand functions are not necessarily linear, the simplest kind of function. Yet most functions are approximately linear if a small enough piece of the graph is taken, allowing applied mathematicians to often use linear functions for simplicity. That approach will be taken in this chapter.

EXAMPLE 2   Supply and Demand

Suppose that Greg Tobin, manager of a giant supermarket chain, has studied the supply and demand for watermelons. He has noticed that the demand increases as the price decreases. He has determined that the quantity (in thousands) demanded weekly, q, and the price (in dollars) per watermelon, p, are related by the linear function

$$p = D(q) = 9 - 0.75q. \qquad \text{Demand function}$$

(a) Find the quantity demanded at a price of $5.25 per watermelon and at a price of $3.75 per watermelon.

SOLUTION   To find the quantity demanded at a price of $5.25 per watermelon, replace p in the demand function with 5.25 and solve for q.

$$5.25 = 9 - 0.75q$$
$$-3.75 = -0.75q \qquad \text{Subtract 9 from both sides.}$$
$$5 = q \qquad \text{Divide both sides by } -0.75.$$

Thus, at a price of $5.25, the quantity demanded is 5000 watermelons.

Similarly, replace p with 3.75 to find the demand when the price is $3.75. Verify that this leads to $q = 7$. When the price is lowered from $5.25 to $3.75 per watermelon, the quantity demanded increases from 5000 to 7000 watermelons.

(b) Greg also noticed that the quantity of watermelons supplied decreased as the price decreased. Price p and supply q are related by the linear function

$$p = S(q) = 0.75q. \qquad \text{Supply function}$$

Find the quantity supplied at a price of $5.25 per watermelon and at a price of $3.00 per watermelon.

SOLUTION Substitute 5.25 for p in the supply function, $p = 0.75q$, to find that $q = 7$, so the quantity supplied is 7000 watermelons. Similarly, replacing p with 3 in the supply equation gives a quantity supplied of 4000 watermelons. If the price decreases from $5.25 to $3.00 per watermelon, the quantity supplied also decreases, from 7000 to 4000 watermelons.

(c) Graph both functions on the same axes.

YOUR TURN 2 Find the quantity of watermelon demanded and supplied at a price of $3.30 per watermelon.

SOLUTION The results of part (a) are written as the ordered pairs (5, 5.25) and (7, 3.75). The line through those points is the graph of the demand function, $p = 9 - 0.75q$, shown in red in Figure 11(a). We used the ordered pairs (7, 5.25) and (4, 3) from the work in part (b) to graph the supply function, $p = 0.75q$, shown in blue in Figure 11(a).

TRY YOUR TURN 2

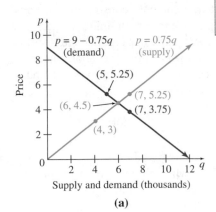

Supply and demand (thousands)

(a)

(b)

FIGURE 11

▓ **TECHNOLOGY NOTE** A calculator-generated graph of the lines representing the supply and demand functions in Example 2 is shown in Figure 11(b). To get this graph, the equation of each line, using x and y instead of q and p, was entered, along with an appropriate window. A special menu choice gives the coordinates of the intersection point, as shown at the bottom of the graph.

NOTE Not all supply and demand problems will have the same scale on both axes. It helps to consider the intercepts of both the supply graph and the demand graph to decide what scale to use. For example, in Figure 11, the y-intercept of the demand function is 9, so the scale should allow values from 0 to at least 9 on the vertical axis. The x-intercept of the demand function is 12, so the values on the x-axis must go from 0 to 12.

As shown in the graphs of Figure 11, both the supply graph and the demand graph pass through the point (6, 4.5). If the price of a watermelon is more than $4.50, the quantity supplied will exceed the quantity demanded and there will be a **surplus** of watermelons. At a price less than $4.50, the quantity demanded will exceed the quantity supplied and there will be a **shortage** of watermelons. Only at a price of $4.50 will quantity demanded and supplied be equal. For this reason, $4.50 is called the *equilibrium price*. When the price is $4.50, quantity demanded and supplied both equal 6000 watermelons, the *equilibrium quantity*. In general, the **equilibrium price** of the commodity is the price found at the point where the supply and demand graphs for that commodity intersect. The **equilibrium quantity** is the quantity demanded and supplied at that same point. Figure 12 illustrates a general supply and demand situation.

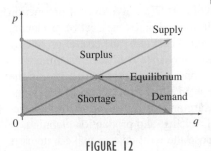

FIGURE 12

YOUR TURN 3 Repeat Example 3 using the demand equation $D(q) = 10 - 0.85q$ and the supply equation $S(q) = 0.4q$.

EXAMPLE 3 Equilibrium Quantity

Use algebra to find the equilibrium quantity and price for the watermelons in Example 2.

SOLUTION The equilibrium quantity is found when the prices from both supply and demand are equal. Set the two expressions for p equal to each other and solve.

$$9 - 0.75q = 0.75q$$
$$9 = 1.5q \qquad \text{Add 0.75q to both sides.}$$
$$6 = q$$

The equilibrium quantity is 6000 watermelons, the same answer found earlier.

The equilibrium price can be found by plugging the value of $q = 6$ into either the demand or the supply function. Using the demand function,

$$p = D(6) = 9 - 0.75(6) = 4.5.$$

The equilibrium price is $4.50, as we found earlier. Check your work by also plugging $q = 6$ into the supply function. **TRY YOUR TURN 3**

⊞ TECHNOLOGY NOTE You may prefer to find the equilibrium quantity by solving the equation with your calculator. Or, if your calculator has a TABLE feature, you can use it to find the value of q that makes the two expressions equal.

Another important issue is how, in practice, the equations of the supply and demand functions can be found. Data need to be collected, and if they lie perfectly along a line, then the equation can easily be found with any two points. What usually happens, however, is that the data are scattered, and there is no line that goes through all the points. In this case we must find a line that approximates the linear trend of the data as closely as possible (assuming the points lie approximately along a line) as in Example 14 in the previous section. This is usually done by the *method of least squares*, also referred to as *linear regression*. We will discuss this method in Section 1.3.

Cost Analysis

The cost of manufacturing an item commonly consists of two parts. The first is a **fixed cost** for designing the product, setting up a factory, training workers, and so on. Within broad limits, the fixed cost is constant for a particular product and does not change as more items are made. The second part is a *cost per item* for labor, materials, packing, shipping, and so on. The total value of this second cost *does* depend on the number of items made.

EXAMPLE 4 Cost Analysis

A small company decides to produce video games. The owners find that the fixed cost for creating the game is $5000, after which they must spend $12 to produce each individual copy of the game. Find a formula $C(x)$ for the cost as a linear function of x, the number of games produced.

SOLUTION Notice that $C(0) = 5000$, since $5000 must be spent even if no games are produced. Also, $C(1) = 5000 + 12 = 5012$, and $C(2) = 5000 + 2 \cdot 12 = 5024$. In general,

$$C(x) = 5000 + 12x,$$

because every time x increases by 1, the cost should increase by $12. The number 12 is also the slope of the graph of the cost function; the slope gives us the cost to produce one additional item.

In economics, **marginal cost** is the rate of change of cost $C(x)$ at a level of production x and is equal to the slope of the cost function at x. It approximates the cost of producing one additional item. In fact, some books define the marginal cost to be the cost of producing one additional item. With *linear functions*, these two definitions are equivalent, and the marginal cost, which is equal to the slope of the cost function, is *constant*. For instance, in the video game example, the marginal cost of each game is $12. For other types of functions, these two definitions are only approximately equal. Marginal cost is important to management in making decisions in areas such as cost control, pricing, and production planning.

The work in Example 4 can be generalized. Suppose the total cost to make x items is given by the linear cost function $C(x) = mx + b$. The fixed cost is found by letting $x = 0$:

$$C(0) = m \cdot 0 + b = b;$$

thus, the fixed cost is b dollars. The additional cost of each additional item, the marginal cost, is m, the slope of the line $C(x) = mx + b$.

Linear Cost Function
In a cost function of the form $C(x) = mx + b$, the m represents the marginal cost and b the fixed cost. Conversely, if the fixed cost of producing an item is b and the marginal cost is m, then the **linear cost function** $C(x)$ for producing x items is $C(x) = mx + b$.

EXAMPLE 5   Cost Function

The marginal cost to make x batches of a prescription medication is \$10 per batch, while the cost to produce 100 batches is \$1500. Find the cost function $C(x)$, given that it is linear.

SOLUTION   Since the cost function is linear, it can be expressed in the form $C(x) = mx + b$. The marginal cost is \$10 per batch, which gives the value for m. Using $x = 100$ and $C(x) = 1500$ in the point-slope form of the line gives

$$C(x) - 1500 = 10(x - 100)$$
$$C(x) - 1500 = 10x - 1000$$
$$C(x) = 10x + 500. \qquad \text{Add 1500 to both sides.}$$

The cost function is given by $C(x) = 10x + 500$, where the fixed cost is \$500.

TRY YOUR TURN 4

YOUR TURN 4 Repeat Example 5, using a marginal cost of \$15 per batch and a cost of \$1930 to produce 80 batches.

Break-Even Analysis
The **revenue** $R(x)$ from selling x units of an item is the product of the price per unit p and the number of units sold (demand) x, so that

$$R(x) = px.$$

The corresponding **profit** $P(x)$ is the difference between revenue $R(x)$ and cost $C(x)$. That is,

$$P(x) = R(x) - C(x).$$

A company can make a profit only if the revenue received from its customers exceeds the cost of producing and selling its goods and services. The number of units at which revenue just equals cost is the **break-even quantity**; the corresponding ordered pair gives the **break-even point**.

EXAMPLE 6   Break-Even Analysis

APPLY IT   A firm producing poultry feed finds that the total cost $C(x)$ in dollars of producing and selling x units is given by

$$C(x) = 20x + 100.$$

Management plans to charge \$24 per unit for the feed.

(a) How many units must be sold for the firm to break even?

SOLUTION   The firm will break even (no profit and no loss) as long as revenue just equals cost, or $R(x) = C(x)$. From the given information, since $R(x) = px$ and $p = \$24$,

$$R(x) = 24x.$$

Substituting for $R(x)$ and $C(x)$ in the equation $R(x) = C(x)$ gives

$$24x = 20x + 100,$$

from which $x = 25$. The firm breaks even by selling 25 units, which is the break-even quantity. The graphs of $C(x) = 20x + 100$ and $R(x) = 24x$ are shown in Figure 13.

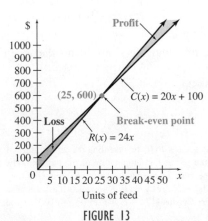

FIGURE 13

The break-even point (where $x = 25$) is shown on the graph. If the company sells more than 25 units (if $x > 25$), it makes a profit. If it sells fewer than 25 units, it loses money.

(b) What is the profit if 100 units of feed are sold?

SOLUTION Use the formula for profit $P(x)$.

$$P(x) = R(x) - C(x)$$
$$= 24x - (20x + 100)$$
$$= 4x - 100$$

Then $P(100) = 4(100) - 100 = 300$. The firm will make a profit of $300 from the sale of 100 units of feed.

(c) How many units must be sold to produce a profit of $900?

SOLUTION Let $P(x) = 900$ in the equation $P(x) = 4x - 100$ and solve for x.

YOUR TURN 5 Repeat Example 6(c), using a cost function $C(x) = 35x + 250$, a charge of $58 per unit, and a profit of $8030.

$$900 = 4x - 100$$
$$1000 = 4x$$
$$x = 250$$

Sales of 250 units will produce $900 profit.   **TRY YOUR TURN 5**

Temperature   One of the most common linear relationships found in everyday situations deals with temperature. Recall that water freezes at $32°$ Fahrenheit and $0°$ Celsius, while it boils at $212°$ Fahrenheit and $100°$ Celsius.* The ordered pairs $(0, 32)$ and $(100, 212)$ are graphed in Figure 14 on axes showing Fahrenheit (F) as a function of Celsius (C). The line joining them is the graph of the function.

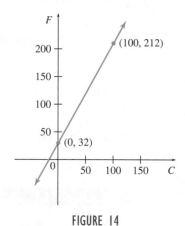

FIGURE 14

EXAMPLE 7   Temperature

Derive an equation relating F and C.

SOLUTION   To derive the required linear equation, first find the slope using the given ordered pairs, $(0, 32)$ and $(100, 212)$.

$$m = \frac{212 - 32}{100 - 0} = \frac{9}{5}$$

*Gabriel Fahrenheit (1686–1736), a German physicist, invented his scale with $0°$ representing the temperature of an equal mixture of ice and ammonium chloride (a type of salt), and $96°$ as the temperature of the human body. (It is often said, erroneously, that Fahrenheit set $100°$ as the temperature of the human body. Fahrenheit's own words are quoted in *A History of the Thermometer and Its Use in Meteorology* by W. E. Knowles, Middleton: The Johns Hopkins Press, 1966, p. 75.) The Swedish astronomer Anders Celsius (1701–1744) set $0°$ and $100°$ as the freezing and boiling points of water.

The *F*-intercept of the graph is 32, so by the slope-intercept form, the equation of the line is

$$F = \frac{9}{5}C + 32.$$

With simple algebra this equation can be rewritten to give *C* in terms of *F*:

$$C = \frac{5}{9}(F - 32).$$

1.2 EXERCISES

For Exercises 1–10, let $f(x) = 7 - 5x$ and $g(x) = 2x - 3$. Find the following.

1. $f(2)$

2. $f(4)$

3. $f(-3)$

4. $f(-1)$

5. $g(1.5)$

6. $g(2.5)$

7. $g(-1/2)$

8. $g(-3/4)$

9. $f(t)$

10. $g(k^2)$

In Exercises 11–14, decide whether the statement is true or false.

11. To find the *x*-intercept of the graph of a linear function, we solve $y = f(x) = 0$, and to find the *y*-intercept, we evaluate $f(0)$.

12. The graph of $f(x) = -5$ is a vertical line.

13. The slope of the graph of a linear function cannot be undefined.

14. The graph of $f(x) = ax$ is a straight line that passes through the origin.

15. Describe what fixed costs and marginal costs mean to a company.

16. In a few sentences, explain why the price of a commodity not already at its equilibrium price should move in that direction.

17. Explain why a linear function may not be adequate for describing the supply and demand functions.

18. In your own words, describe the break-even quantity, how to find it, and what it indicates.

Write a linear cost function for each situation. Identify all variables used.

19. A Lake Tahoe resort charges a snowboard rental fee of $10 plus $2.25 per hour.

20. An Internet site for downloading music charges a $10 registration fee plus 99 cents per downloaded song.

21. A parking garage charges 2 dollars plus 75 cents per half-hour.

22. For a one-day rental, a car rental firm charges $44 plus 28 cents per mile.

Assume that each situation can be expressed as a linear cost function. Find the cost function in each case.

23. Fixed cost: $100; 50 items cost $1600 to produce.

24. Fixed cost: $35; 8 items cost $395 to produce.

25. Marginal cost: $75; 50 items cost $4300 to produce.

26. Marginal cost: $120; 700 items cost $96,500 to produce.

APPLICATIONS

Business and Economics

27. Supply and Demand   Suppose that the demand and price for a certain model of a youth wristwatch are related by

$$p = D(q) = 16 - 1.25q,$$

where *p* is the price (in dollars) and *q* is the quantity demanded (in hundreds). Find the price at each level of demand.

a. 0 watches   **b.** 400 watches   **c.** 800 watches

Find the quantity demanded for the watch at each price.

d. $8   **e.** $10   **f.** $12

g. Graph $p = 16 - 1.25q$.

Suppose the price and supply of the watch are related by

$$p = S(q) = 0.75q,$$

where *p* is the price (in dollars) and *q* is the quantity supplied (in hundreds) of watches. Find the quantity supplied at each price.

h. $0   **i.** $10   **j.** $20

k. Graph $p = 0.75q$ on the same axis used for part g.

l. Find the equilibrium quantity and the equilibrium price.

28. Supply and Demand   Suppose that the demand and price for strawberries are related by

$$p = D(q) = 5 - 0.25q,$$

where *p* is the price (in dollars) and *q* is the quantity demanded (in hundreds of quarts). Find the price at each level of demand.

a. 0 quarts   **b.** 400 quarts   **c.** 840 quarts

Find the quantity demanded for the strawberries at each price.

d. $4.50   **e.** $3.25   **f.** $2.40

g. Graph $p = 5 - 0.25q$.

Suppose the price and supply of strawberries are related by

$$p = S(q) = 0.25q,$$

where p is the price (in dollars) and q is the quantity supplied (in hundreds of quarts) of strawberries. Find the quantity supplied at each price.

h. $0   **i.** $2   **j.** $4.50

k. Graph $p = 0.75q$ on the same axis used for part g.

l. Find the equilibrium quantity and the equilibrium price.

29. Supply and Demand Let the supply and demand functions for butter pecan ice cream be given by

$$p = S(q) = \frac{2}{5}q \quad \text{and} \quad p = D(q) = 100 - \frac{2}{5}q,$$

where p is the price in dollars and q is the number of 10-gallon tubs.

a. Graph these on the same axes.

b. Find the equilibrium quantity and the equilibrium price. (*Hint:* The way to divide by a fraction is to multiply by its reciprocal.)

30. Supply and Demand Let the supply and demand functions for sugar be given by

$$p = S(q) = 1.4q - 0.6 \quad \text{and}$$
$$p = D(q) = -2q + 3.2,$$

where p is the price per pound and q is the quantity in thousands of pounds.

a. Graph these on the same axes.

b. Find the equilibrium quantity and the equilibrium price.

31. Supply and Demand Suppose that the supply function for honey is $p = S(q) = 0.3q + 2.7$, where p is the price in dollars for an 8-oz container and q is the quantity in barrels. Suppose also that the equilibrium price is $4.50 and the demand is 2 barrels when the price is $6.10. Find an equation for the demand function, assuming it is linear.

32. Supply and Demand Suppose that the supply function for walnuts is $p = S(q) = 0.25q + 3.6$, where p is the price in dollars per pound and q is the quantity in bushels. Suppose also that the equilibrium price is $5.85, and the demand is 4 bushels when the price is $7.60. Find an equation for the demand function, assuming it is linear.

33. T-Shirt Cost Joanne Wendelken sells silk-screened T-shirts at community festivals and crafts fairs. Her marginal cost to produce one T-shirt is $3.50. Her total cost to produce 60 T-shirts is $300, and she sells them for $9 each.

a. Find the linear cost function for Joanne's T-shirt production.

b. How many T-shirts must she produce and sell in order to break even?

c. How many T-shirts must she produce and sell to make a profit of $500?

34. Publishing Costs Alfred Juarez owns a small publishing house specializing in Latin American poetry. His fixed cost to produce a typical poetry volume is $525, and his total cost to produce 1000 copies of the book is $2675. His books sell for $4.95 each.

a. Find the linear cost function for Alfred's book production.

b. How many poetry books must he produce and sell in order to break even?

c. How many books must he produce and sell to make a profit of $1000?

35. Marginal Cost of Coffee The manager of a restaurant found that the cost to produce 100 cups of coffee is $11.02, while the cost to produce 400 cups is $40.12. Assume the cost $C(x)$ is a linear function of x, the number of cups produced.

a. Find a formula for $C(x)$.

b. What is the fixed cost?

c. Find the total cost of producing 1000 cups.

d. Find the total cost of producing 1001 cups.

e. Find the marginal cost of the 1001st cup.

f. What is the marginal cost of *any* cup and what does this mean to the manager?

36. Marginal Cost of a New Plant In deciding whether to set up a new manufacturing plant, company analysts have decided that a linear function is a reasonable estimation for the total cost $C(x)$ in dollars to produce x items. They estimate the cost to produce 10,000 items as $547,500, and the cost to produce 50,000 items as $737,500.

a. Find a formula for $C(x)$.

b. Find the fixed cost.

c. Find the total cost to produce 100,000 items.

d. Find the marginal cost of the items to be produced in this plant and what does this mean to the manager?

37. Break-Even Analysis Producing x units of tacos costs $C(x) = 5x + 20$; revenue is $R(x) = 15x$, where $C(x)$ and $R(x)$ are in dollars.

a. What is the break-even quantity?

b. What is the profit from 100 units?

c. How many units will produce a profit of $500?

38. Break-Even Analysis To produce x units of a religious medal costs $C(x) = 12x + 39$. The revenue is $R(x) = 25x$. Both $C(x)$ and $R(x)$ are in dollars.

a. Find the break-even quantity.

b. Find the profit from 250 units.

c. Find the number of units that must be produced for a profit of $130.

Break-Even Analysis You are the manager of a firm. You are considering the manufacture of a new product, so you ask the accounting department for cost estimates and the sales department for sales estimates. After you receive the data, you must decide whether to go ahead with production of the new product. Analyze the data in Exercises 39–42 (find a break-even quantity) and then decide what you would do in each case. Also write the profit function.

39. $C(x) = 85x + 900$; $R(x) = 105x$; no more than 38 units can be sold.

40. $C(x) = 105x + 6000$; $R(x) = 250x$; no more than 400 units can be sold.

41. $C(x) = 70x + 500$; $R(x) = 60x$ (*Hint*: What does a negative break-even quantity mean?)

42. $C(x) = 1000x + 5000$; $R(x) = 900x$

43. Break-Even Analysis Suppose that the fixed cost for a product is $400 and the break-even quantity is 80. Find the marginal profit (the slope of the linear profit function).

44. Break-Even Analysis Suppose that the fixed cost for a product is $650 and the break-even quantity is 25. Find the marginal profit (the slope of the linear profit function).

Physical Sciences

45. Temperature Use the formula for conversion between Fahrenheit and Celsius derived in Example 7 to convert each temperature.

a. 58°F to Celsius

b. −20°F to Celsius

c. 50°C to Fahrenheit

46. Body Temperature You may have heard that the average temperature of the human body is 98.6°. Recent experiments show that the actual figure is closer to 98.2°. The figure of 98.6 comes from experiments done by Carl Wunderlich in 1868. But Wunderlich measured the temperatures in degrees Celsius and rounded the average to the nearest degree, giving 37°C as the average temperature. *Source: Science News.*

a. What is the Fahrenheit equivalent of 37°C?

b. Given that Wunderlich rounded to the nearest degree Celsius, his experiments tell us that the actual average human body temperature is somewhere between 36.5°C and 37.5°C. Find what this range corresponds to in degrees Fahrenheit.

47. Temperature Find the temperature at which the Celsius and Fahrenheit temperatures are numerically equal.

General Interest

48. Education Cost The 2009–2010 budget for the California State University system projected a fixed cost of $486,000 at each of five off-campus centers, plus a marginal cost of $1140 per student. *Source: California State University.*

a. Find a formula for the cost at each center, $C(x)$, as a linear function of x, the number of students.

b. The budget projected 500 students at each center. Calculate the total cost at each center.

c. Suppose, due to budget cuts, that each center is limited to $1 million. What is the maximum number of students that each center can then support?

YOUR TURN ANSWERS

1. 25   **2.** 7600 and 4400
3. 8000 watermelons and $3.20 per watermelon
4. $C(x) = 15x + 730$ **5.** 360

1.3 The Least Squares Line

APPLY IT How has the accidental death rate in the United States changed over time? *In Example 1 in this section, we show how to answer such questions using the method of least squares.*

We use past data to find trends and to make tentative predictions about the future. The only assumption we make is that the data are related linearly—that is, if we plot pairs of data, the resulting points will lie close to some line. This method cannot give exact answers. The best we can expect is that, if we are careful, we will get a reasonable approximation.

The table lists the number of accidental deaths per 100,000 people in the United States through the past century. *Source: National Center for Health Statistics.* If you were a manager at an insurance company, these data could be very important. You might need to make some predictions about how much you will pay out next year in accidental death benefits, and even a very tentative prediction based on past trends is better than no prediction at all.

The first step is to draw a scatterplot, as we have done in Figure 15 on the next page. Notice that the points lie approximately along a line, which means that a linear function may give a good approximation of the data. If we select two points and find the line that passes through them, as we did in Section 1.1, we will get a different line for each pair of points, and in some cases the lines will be very different. We want to draw one line that is simultaneously close to all the points on the graph, but many such lines are possible, depending upon how we define the phrase "simultaneously close to all the points." How do we decide on the best possible line? Before going on, you might want to try drawing the line you think is best on Figure 15.

Accidental Death Rate	
Year	**Death Rate**
1910	84.4
1920	71.2
1930	80.5
1940	73.4
1950	60.3
1960	52.1
1970	56.2
1980	46.5
1990	36.9
2000	34.0

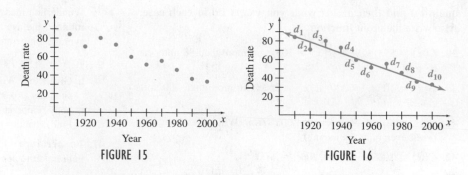

FIGURE 15   FIGURE 16

The line used most often in applications is that in which the sum of the squares of the vertical distances from the data points to the line is as small as possible. Such a line is called the **least squares line**. The least squares line for the data in Figure 15 is drawn in Figure 16. How does the line compare with the one you drew on Figure 15? It may not be exactly the same, but should appear similar.

In Figure 16, the vertical distances from the points to the line are indicated by d_1, d_2, and so on, up through d_{10} (read "d-sub-one, d-sub-two, d-sub-three," and so on). For n points, corresponding to the n pairs of data, the least squares line is found by minimizing the sum $(d_1)^2 + (d_2)^2 + (d_3)^2 + \cdots + (d_n)^2$.

We often use **summation notation** to write the sum of a list of numbers. The Greek letter sigma, Σ, is used to indicate "the sum of." For example, we write the sum $x_1 + x_2 + \cdots + x_n$, where n is the number of data points, as

$$x_1 + x_2 + \cdots + x_n = \Sigma x.$$

Similarly, Σxy means $x_1 y_1 + x_2 y_2 + \cdots + x_n y_n$, and so on.

> **CAUTION**   Note that Σx^2 means $x_1^2 + x_2^2 + \cdots + x_n^2$, which is *not* the same as squaring Σx. When we square Σx, we write it as $(\Sigma x)^2$.

For the least squares line, the sum of the distances we are to minimize, $d_1^2 + d_2^2 + \cdots + d_n^2$, is written as

$$d_1^2 + d_2^2 + \cdots + d_n^2 = \Sigma d^2.$$

To calculate the distances, we let $(x_1, y_1), (x_2, y_2), \cdots, (x_n, y_n)$ be the actual data points and we let the least squares line be $Y = mx + b$. We use Y in the equation instead of y to distinguish the predicted values (Y) from the y-value of the given data points. The predicted value of Y at x_1 is $Y_1 = mx_1 + b$, and the distance, d_1, between the actual y-value y_1 and the predicted value Y_1 is

$$d_1 = |Y_1 - y_1| = |mx_1 + b - y_1|.$$

Likewise,

$$d_2 = |Y_2 - y_2| = |mx_2 + b - y_2|,$$

and

$$d_n = |Y_n - y_n| = |mx_n + b - y_n|.$$

The sum to be minimized becomes

$$\Sigma d^2 = (mx_1 + b - y_1)^2 + (mx_2 + b - y_2)^2 + \cdots + (mx_n + b - y_n)^2$$
$$= \Sigma(mx + b - y)^2,$$

where $(x_1, y_1), (x_2, y_2), \ldots, (x_n, y_n)$ are known and m and b are to be found.

The method of minimizing this sum requires advanced techniques and is not given here. To obtain the equation for the least squares line, a system of equations must be solved, producing the following formulas for determining the slope m and y-intercept b.*

Least Squares Line

The **least squares line** $Y = mx + b$ that gives the best fit to the data points $(x_1, y_1), (x_2, y_2), \dots, (x_n, y_n)$ has slope m and y-intercept b given by

$$m = \frac{n(\Sigma xy) - (\Sigma x)(\Sigma y)}{n(\Sigma x^2) - (\Sigma x)^2} \quad \text{and} \quad b = \frac{\Sigma y - m(\Sigma x)}{n}.$$

EXAMPLE 1   Least Squares Line

APPLY IT   Calculate the least squares line for the accidental death rate data.

SOLUTION

Method 1
Calculating by Hand

To find the least squares line for the given data, we first find the required sums. To reduce the size of the numbers, we rescale the year data. Let x represent the years since 1900, so that, for example, $x = 10$ corresponds to the year 1910. Let y represent the death rate. We then calculate the values in the xy, x^2, and y^2 columns and find their totals. (The column headed y^2 will be used later.) Note that the number of data points is $n = 10$.

Least Squares Calculations				
x	y	xy	x^2	y^2
10	84.4	844	100	7123.36
20	71.2	1424	400	5069.44
30	80.5	2415	900	6480.25
40	73.4	2936	1600	5387.56
50	60.3	3015	2500	3636.09
60	52.1	3126	3600	2714.41
70	56.2	3934	4900	3158.44
80	46.5	3720	6400	2162.25
90	36.9	3321	8100	1361.61
100	34.0	3400	10,000	1156.00
$\Sigma x = 550$	$\Sigma y = 595.5$	$\Sigma xy = 28{,}135$	$\Sigma x^2 = 38{,}500$	$\Sigma y^2 = 38{,}249.41$

Putting the column totals into the formula for the slope m, we get

$$m = \frac{n(\Sigma xy) - (\Sigma x)(\Sigma y)}{n(\Sigma x^2) - (\Sigma x)^2} \qquad \text{Formula for } m$$

$$= \frac{10(28{,}135) - (550)(595.5)}{10(38{,}500) - (550)^2} \qquad \text{Substitute from the table.}$$

$$= \frac{281{,}350 - 327{,}525}{385{,}000 - 302{,}500} \qquad \text{Multiply.}$$

$$= \frac{-46{,}175}{82{,}500} \qquad \text{Subtract.}$$

$$= -0.5596970 \approx -0.560.$$

*See Exercise 9 at the end of this section.

The significance of m is that the death rate per 100,000 people is tending to drop (because of the negative) at a rate of 0.560 per year.

Now substitute the value of m and the column totals in the formula for b:

$$b = \frac{\Sigma y - m(\Sigma x)}{n} \qquad \text{Formula for } b$$

$$= \frac{595.5 - (-0.559697)(550)}{10} \qquad \text{Substitute.}$$

$$= \frac{595.5 - (-307.83335)}{10} \qquad \text{Multiply.}$$

$$= \frac{903.33335}{10} = 90.333335 \approx 90.3$$

Substitute m and b into the least squares line, $Y = mx + b$; the least squares line that best fits the 10 data points has equation

$$Y = -0.560x + 90.3.$$

This gives a mathematical description of the relationship between the year and the number of accidental deaths per 100,000 people. The equation can be used to predict y from a given value of x, as we will show in Example 2. As we mentioned before, however, caution must be exercised when using the least squares equation to predict data points that are far from the range of points on which the equation was modeled.

CAUTION In computing m and b, we rounded the final answer to three digits because the original data were known only to three digits. It is important, however, *not* to round any of the intermediate results (such as Σx^2) because round-off error may have a detrimental effect on the accuracy of the answer. Similarly, it is important not to use a rounded-off value of m when computing b.

Method 2
Graphing Calculator

The calculations for finding the least squares line are often tedious, even with the aid of a calculator. Fortunately, many calculators can calculate the least squares line with just a few keystrokes. For purposes of illustration, we will show how the least squares line in the previous example is found with a TI-84 Plus graphing calculator.

We begin by entering the data into the calculator. We will be using the first two lists, called L_1 and L_2. Choosing the STAT menu, then choosing the fourth entry ClrList, we enter L_1, L_2, to indicate the lists to be cleared. Now we press STAT again and choose the first entry EDIT, which brings up the blank lists. As before, we will only use the last two digits of the year, putting the numbers in L_1. We put the death rate in L_2, giving the two screens shown in Figure 17.

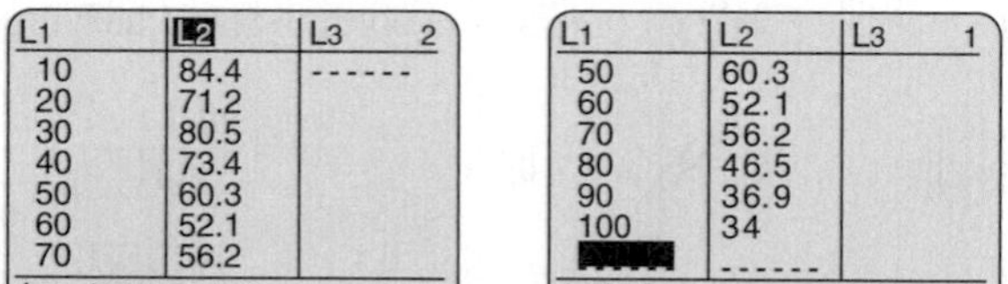

L1	**L2**	L3	2
10	84.4	------	
20	71.2		
30	80.5		
40	73.4		
50	60.3		
60	52.1		
70	56.2		
L2 =(84.4, 71.2, 8...			

L1	L2	L3	1
50	60.3		
60	52.1		
70	56.2		
80	46.5		
90	36.9		
100	34		
▮▮▮▮▮	------		
L1(11)=			

FIGURE 17

Quit the editor, press STAT again, and choose CALC instead of EDIT. Then choose item 4 LinReg($ax + b$) to get the values of a (the slope) and b (the y-intercept) for the least squares line, as shown in Figure 18. With a and b rounded to three decimal places, the least squares line is $Y = -0.560x + 90.3$. A graph of the data points and the line is shown in Figure 19.

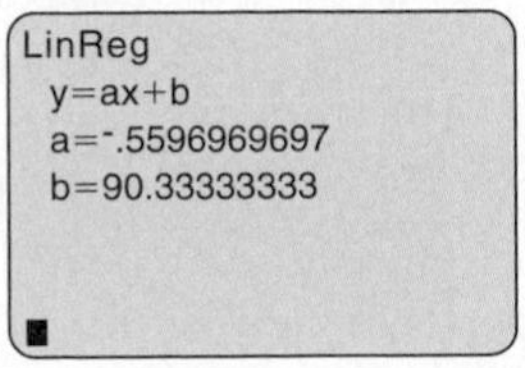

LinReg
  y=ax+b
  a=⁻.5596969697
  b=90.33333333

FIGURE 18 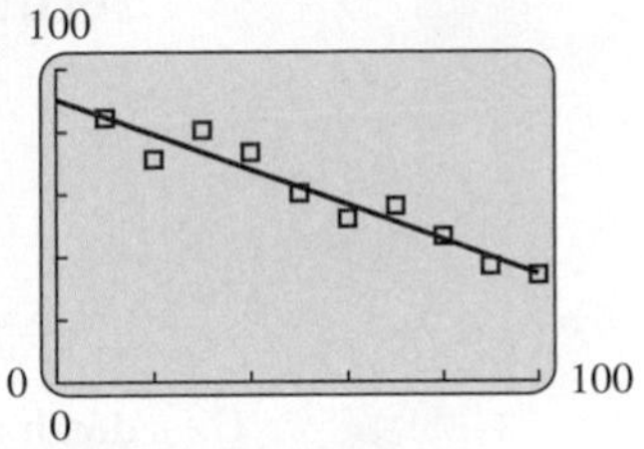   FIGURE 19

For more details on finding the least squares line with a graphing calculator, see the *Graphing Calculator and Excel Spreadsheet Manual* available with this book.

Method 3
Spreadsheet

YOUR TURN 1 Repeat Example 1, deleting the last pair of data (100, 34.0) and changing the second to last pair to (90, 40.2).

Many computer spreadsheet programs can also find the least squares line. Figure 20 shows the scatterplot and least squares line for the accidental death rate data using an Excel spreadsheet. The scatterplot was found using the Marked Scatter chart from the Gallery and the line was found using the Add Trendline command under the Chart menu. For details, see the *Graphing Calculator and Excel Spreadsheet Manual* available with this book.

TRY YOUR TURN 1

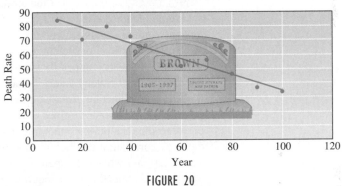

Accidental Deaths

FIGURE 20

EXAMPLE 2   Least Squares Line

What do we predict the accidental death rate to be in 2012?

SOLUTION Use the least squares line equation given above with $x = 112$.

$$Y = -0.560x + 90.3$$
$$= -0.56(112) + 90.3$$
$$= 27.6$$

The accidental death rate in 2012 is predicted to be about 27.6 per 100,000 population. In this case, we will have to wait until the 2012 data become available to see how accurate our prediction is. We have observed, however, that the accidental death rate began to go up after 2000 and was 40.6 per 100,000 population in 2006. This illustrates the danger of extrapolating beyond the data.

EXAMPLE 3   Least Squares Line

In what year is the death rate predicted to drop below 26 per 100,000 population?

SOLUTION Let $Y = 26$ in the equation above and solve for x.

$$26 = -0.560x + 90.3$$

$$-64.3 = -0.560x \qquad \text{Subtract 90.3 from both sides.}$$

$$x = 115 \qquad \text{Divide both sides by } -0.560.$$

This corresponds to the year 2015 (115 years after 1900), when our equation predicts the death rate to be $-0.560(115) + 90.3 = 25.9$ per 100,000 population.

Correlation

Although the least squares line can always be found, it may not be a good model. For example, if the data points are widely scattered, no straight line will model the data accurately. One measure of how well the original data fits a straight line is the **correlation coefficient,** denoted by r, which can be calculated by the following formula.

Correlation Coefficient

$$r = \frac{n(\sum xy) - (\sum x)(\sum y)}{\sqrt{n(\sum x^2) - (\sum x)^2} \cdot \sqrt{n(\sum y^2) - (\sum y)^2}}$$

Although the expression for r looks daunting, remember that each of the summations, $\sum x$, $\sum y$, $\sum xy$, and so on, are just the totals from a table like the one we prepared for the data on accidental deaths. Also, with a calculator, the arithmetic is no problem! Furthermore, statistics software and many calculators can calculate the correlation coefficient for you.

The correlation coefficient measures the strength of the linear relationship between two variables. It was developed by statistics pioneer Karl Pearson (1857–1936). The correlation coefficient r is between 1 and -1 or is equal to 1 or -1. Values of exactly 1 or -1 indicate that the data points lie *exactly* on the least squares line. If $r = 1$, the least squares line has a positive slope; $r = -1$ gives a negative slope. If $r = 0$, there is no linear correlation between the data points (but some *nonlinear* function might provide an excellent fit for the data). A correlation coefficient of zero may also indicate that the data fit a horizontal line. To investigate what is happening, it is always helpful to sketch a scatterplot of the data. Some scatterplots that correspond to these values of r are shown in Figure 21.

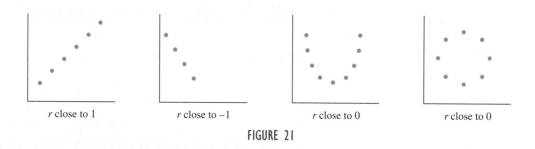

r close to 1 r close to -1 r close to 0 r close to 0

FIGURE 21

A value of r close to 1 or -1 indicates the presence of a linear relationship. The exact value of r necessary to conclude that there is a linear relationship depends upon n, the number of data points, as well as how confident we want to be of our conclusion. For details, consult a text on statistics.*

*For example, see *Introductory Statistics*, 8th edition, by Neil A. Weiss, Boston, Mass.: Pearson, 2008.

| EXAMPLE 4 | **Correlation Coefficient** |

Find r for the data on accidental death rates in Example 1.

SOLUTION

Method 1
Calculating by Hand

From the table in Example 1,

$\Sigma x = 550$, $\Sigma y = 595.5$, $\Sigma xy = 28,135$, $\Sigma x^2 = 38,500$,

$\Sigma y^2 = 38,249.41$,   and $n = 10$.

Substituting these values into the formula for r gives

$$r = \frac{n(\Sigma xy)-(\Sigma x)(\Sigma y)}{\sqrt{n(\Sigma x^2)-(\Sigma x)^2}\cdot\sqrt{n(\Sigma y^2)-(\Sigma y)^2}} \qquad \text{Formula for } r$$

$$= \frac{10(28,135)-(550)(595.5)}{\sqrt{10(38,500)-(550)^2}\cdot\sqrt{10(38,249.41)-(595.5)^2}} \qquad \text{Substitute.}$$

$$= \frac{281,350-327,525}{\sqrt{385,000-302,500}\cdot\sqrt{382,494.1-354,620.25}} \qquad \text{Multiply.}$$

$$= \frac{-46,175}{\sqrt{82,500}\cdot\sqrt{27,873.85}} \qquad \text{Subtract.}$$

$$= \frac{-46,175}{47,954.06787} \qquad \text{Take square roots and multiply.}$$

$$= -0.9629005849 \approx -0.963.$$

This is a high correlation, which agrees with our observation that the data fit a line quite well.

Method 2
Graphing Calculator

Most calculators that give the least squares line will also give the correlation coefficient. To do this on the TI-84 Plus, press the second function CATALOG and go down the list to the entry DiagnosticOn. Press ENTER at that point, then press STAT, CALC, and choose item 4 to get the display in Figure 22. The result is the same as we got by hand. The command DiagnosticOn need only be entered once, and the correlation coefficient will always appear in the future.

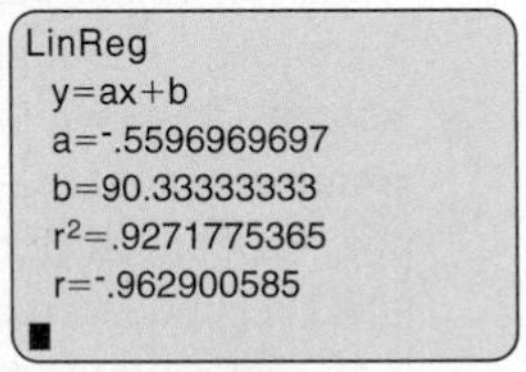

```
LinReg
 y=ax+b
 a=-.5596969697
 b=90.33333333
 r²=.9271775365
 r=-.962900585
```

FIGURE 22

Method 3
Spreadsheet

Many computer spreadsheet programs have a built-in command to find the correlation coefficient. For example, in Excel, use the command "= CORREL(A1:A10,B1:B10)" to find the correlation of the 10 data points stored in columns A and B. For more details, see the *Graphing Calculator and Excel Spreadsheet Manual* available with this text.

YOUR TURN 2 Repeat Example 4, deleting the last pair of data (100, 34.0) and changing the second to last pair to (90, 40.2).

TRY YOUR TURN 2

The square of the correlation coefficient gives the fraction of the variation in y that is explained by the linear relationship between x and y. Consider Example 4, where $r^2 = (-0.963)^2 = 0.927$. This means that 92.7% of the variation in y is explained by the linear relationship found earlier in Example 1. The remaining 7.3% comes from the scattering of the points about the line.

EXAMPLE 5 Average Expenditure per Pupil Versus Test Scores

Many states and school districts debate whether or not increasing the amount of money spent per student will guarantee academic success. The following scatterplot shows the average eighth grade reading score on the National Assessment of Education Progress (NAEP) for the 50 states and the District of Columbia in 2007 plotted against the average expenditure per pupil in 2007. Explore how the correlation coefficient is affected by the inclusion of the District of Columbia, which spent $14,324 per pupil and had a NAEP score of 241. *Source: U.S. Census Bureau and National Center for Education Statistics.*

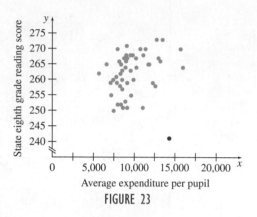

FIGURE 23

SOLUTION A spreadsheet was used to create a plot of the points shown in Figure 23. Washington D.C. corresponds to the red point in the lower right, which is noticeably separate from all the other points. Using the original data, the correlation coefficient when Washington D.C. is included is 0.1981, indicating that there is not a strong linear correlation. Excluding Washington D.C. raises the correlation coefficient to 0.3745, which is a somewhat stronger indication of a linear correlation. This illustrates that one extreme data point that is separate from the others, known as an **outlier**, can have a strong effect on the correlation coefficient.

Even if the correlation between average expenditure per pupil and reading score in Example 5 was high, this would not prove that spending more per pupil causes high reading scores. To prove this would require further research. It is a common statistical fallacy to assume that correlation implies causation. Perhaps the correlation is due to a third underlying variable. In Example 5, perhaps states with wealthier families spend more per pupil, and the students read better because wealthier families have greater access to reading material. Determining the truth requires careful research methods that are beyond the scope of this textbook.

1.3 EXERCISES

1. Suppose a positive linear correlation is found between two quantities. Does this mean that one of the quantities increasing causes the other to increase? If not, what does it mean?

2. Given a set of points, the least squares line formed by letting x be the independent variable will not necessarily be the same as the least squares line formed by letting y be the independent variable. Give an example to show why this is true.

3. For the following table of data,

x	1	2	3	4	5	6	7	8	9	10
y	0	0.5	1	2	2.5	3	3	4	4.5	5

  a. draw a scatterplot.

  b. calculate the correlation coefficient.

c. calculate the least squares line and graph it on the scatterplot.

d. predict the y-value when x is 11.

The following problem is reprinted from the November 1989 *Actuarial Examination on Applied Statistical Methods*. Source: Society of Actuaries.

4. You are given

X	6.8	7.0	7.1	7.2	7.4
Y	0.8	1.2	0.9	0.9	1.5

Determine r^2, the coefficient of determination for the regression of Y on X. Choose one of the following. (*Note:* The coefficient of determination is defined as the square of the correlation coefficient.)

a. 0.3   **b.** 0.4   **c.** 0.5   **d.** 0.6   **e.** 0.7

5. Consider the following table of data.

x	1	1	2	2	9
y	1	2	1	2	9

a. Calculate the least squares line and the correlation coefficient.

b. Repeat part a, but this time delete the last point.

c. Draw a graph of the data, and use it to explain the dramatic difference between the answers to parts a and b.

6. Consider the following table of data.

x	1	2	3	4	9
y	1	2	3	4	−20

a. Calculate the least squares line and the correlation coefficient.

b. Repeat part a, but this time delete the last point.

c. Draw a graph of the data, and use it to explain the dramatic difference between the answers to parts a and b.

7. Consider the following table of data.

x	1	2	3	4
y	1	1	1	1.1

a. Calculate the correlation coefficient.

b. Sketch a graph of the data.

c. Based on how closely the data fits a straight line, is your answer to part a surprising? Discuss the extent to which the correlation coefficient describes how well the data fit a horizontal line.

8. Consider the following table of data.

x	0	1	2	3	4
y	4	1	0	1	4

a. Calculate the least squares line and the correlation coefficient.

b. Sketch a graph of the data.

c. Comparing your answers to parts a and b, does a correlation coefficient of 0 mean that there is no relationship between the x and y values? Would some curve other than a line fit the data better? Explain.

9. The formulas for the least squares line were found by solving the system of equations
$$nb + (\Sigma x)m = \Sigma y$$
$$(\Sigma x)b + (\Sigma x^2)m = \Sigma xy.$$
Solve the above system for b and m to show that
$$m = \frac{n(\Sigma xy) - (\Sigma x)(\Sigma y)}{n(\Sigma x^2) - (\Sigma x)^2} \quad \text{and}$$
$$b = \frac{\Sigma y - m(\Sigma x)}{n}.$$

APPLICATIONS

Business and Economics

10. **Consumer Durable Goods** The total value of consumer durable goods has grown at an approximately linear rate in recent years. The annual data for the years 2002 through 2008 can be summarized as follows, where x represents the years since 2000 and y the total value of consumer durable goods in trillions of dollars. *Source: Bureau of Economic Analysis.*

$n = 7$ $\Sigma x = 35$ $\Sigma x^2 = 203$
$\Sigma y = 28.4269$ $\Sigma y^2 = 116.3396$ $\Sigma xy = 147.1399$

a. Find an equation for the least squares line.

b. Use your result from part a to predict the total value of consumer durable goods in the year 2015.

c. If this growth continues linearly, in what year will the total value of consumer durable goods first reach at least 6 trillion dollars?

d. Find and interpret the correlation coefficient.

11. **Decrease in Banks** The number of banks in the United States has been dropping steadily since 1984, and the trend in recent years has been roughly linear. The annual data for the years 1999 through 2008 can be summarized as follows, where x represents the years since 1990 and y the number of banks, in thousands, in the United States. *Source: FDIC.*

$n = 10$ $\Sigma x = 235$ $\Sigma x^2 = 5605$
$\Sigma y = 77.564$ $\Sigma y^2 = 603.60324$ $\Sigma xy = 1810.095$

a. Find an equation for the least squares line.

b. Use your result from part a to predict the number of U.S. banks in the year 2020.

c. If this trend continues linearly, in what year will the number of U.S. banks drop below 6000?

d. Find and interpret the correlation coefficient.

12. **Digital Cable Subscribers** The number of subscribers to digital cable television has been growing steadily, as shown by the

following table. *Source: National Cable and Telecommunications Association.*

Year	2000	2002	2004	2006	2008
Customers (in millions)	8.5	19.3	25.4	32.6	40.4

a. Find an equation for the least squares line, letting x equal the number of years since 2000.

b. Based on your answer to part a, at approximately what rate is the number of subscribers to digital cable television growing per year?

c. Use your result from part a to predict the number of digital cable subscribers in the year 2012.

d. If this trend continues linearly, in what year will the number of digital cable subscribers first exceed 70 million?

e. Find and interpret the correlation coefficient.

13. Consumer Credit The total amount of consumer credit has been increasing steadily in recent years. The following table gives the total U.S. outstanding consumer credit. *Source: Federal Reserve.*

Year	2004	2005	2006	2007	2008
Consumer credit (in billions of dollars)	2219.5	2319.8	2415.0	2551.9	2592.1

a. Find an equation for the least squares line, letting x equal the number of years since 2000.

b. Based on your answer to part a, at approximately what rate is the consumer credit growing per year?

c. Use your result from part a to predict the amount of consumer credit in the year 2015.

d. If this trend continues linearly, in what year will the total debt first exceed $4000 billion?

e. Find and interpret the correlation coefficient.

14. New Car Sales New car sales have increased at a roughly linear rate. Sales, in millions of vehicles, from 2000 to 2007, are given in the table below. *Source: National Automobile Dealers Association.* Let x represent the number of years since 2000.

Year	Sales
2000	17.3
2001	17.1
2002	16.8
2003	16.6
2004	16.9
2005	16.9
2006	16.5
2007	16.1

a. Find the equation of the least squares line and the correlation coefficient.

b. Find the equation of the least squares line using only the data for every other year starting with 2000, 2002, and so on. Find the correlation coefficient.

c. Compare your results for parts a and b. What do you find? Why do you think this happens?

15. Air Fares In 2006, for passengers who made early reservations, American Airlines offered lower prices on one-way fares from New York to various cities. Fourteen of the cities are listed in the following table, with the distances from New York to the cities included. *Source: American Airlines.*

a. Plot the data. Do the data points lie in a linear pattern?

b. Find the correlation coefficient. Combining this with your answer to part a, does the cost of a ticket tend to go up with the distance flown?

c. Find the equation of the least squares line, and use it to find the approximate marginal cost per mile to fly.

d. For similar data in a January 2000 *New York Times* ad, the equation of the least squares line was $Y = 113 + 0.0243x$. *Source: The New York Times.* Use this information and your answer to part b to compare the cost of flying American Airlines for these two time periods.

e. Identify the outlier in the scatterplot. Discuss the reason why there would be a difference in price to this city.

City	Distance (x) (miles)	Price (y) (dollars)
Boston	206	95
Chicago	802	138
Denver	1771	228
Kansas City	1198	209
Little Rock	1238	269
Los Angeles	2786	309
Minneapolis	1207	202
Nashville	892	217
Phoenix	2411	109
Portland	2885	434
Reno	2705	399
St. Louis	948	206
San Diego	2762	239
Seattle	2815	329

Life Sciences

16. Bird Eggs The average length and width of various bird eggs are given in the following table. *Source: National Council of Teachers of Mathematics.*

Bird Name	Width (cm)	Length (cm)
Canada goose	5.8	8.6
Robin	1.5	1.9
Turtledove	2.3	3.1
Hummingbird	1.0	1.0
Raven	3.3	5.0

a. Plot the points, putting the length on the *y*-axis and the width on the *x*-axis. Do the data appear to be linear?

b. Find the least squares line, and plot it on the same graph as the data.

c. Suppose there are birds with eggs even smaller than those of hummingbirds. Would the equation found in part b continue to make sense for all positive widths, no matter how small? Explain.

d. Find the correlation coefficient.

 17. Size of Hunting Parties   In the 1960s, the famous researcher Jane Goodall observed that chimpanzees hunt and eat meat as part of their regular diet. Sometimes chimpanzees hunt alone, while other times they form hunting parties. The following table summarizes research on chimpanzee hunting parties, giving the size of the hunting party and the percentage of successful hunts. *Source: American Scientist and Mathematics Teacher.*

Number of Chimps in Hunting Party	Percentage of Successful Hunts
1	20
2	30
3	28
4	42
5	40
6	58
7	45
8	62
9	65
10	63
12	75
13	75
14	78
15	75
16	82

a. Plot the data. Do the data points lie in a linear pattern?

b. Find the correlation coefficient. Combining this with your answer to part a, does the percentage of successful hunts tend to increase with the size of the hunting party?

c. Find the equation of the least squares line, and graph it on your scatterplot.

18. Crickets Chirping   Biologists have observed a linear relationship between the temperature and the frequency with which a cricket chirps. The following data were measured for the striped ground cricket. *Source: The Song of Insects.*

Temperature °F (*x*)	Chirps Per Second (*y*)
88.6	20.0
71.6	16.0
93.3	19.8
84.3	18.4
80.6	17.1
75.2	15.5
69.7	14.7
82.0	17.1
69.4	15.4
83.3	16.2
79.6	15.0
82.6	17.2
80.6	16.0
83.5	17.0
76.3	14.4

a. Find the equation for the least squares line for the data.

b. Use the results of part a to determine how many chirps per second you would expect to hear from the striped ground cricket if the temperature were 73°F.

c. Use the results of part a to determine what the temperature is when the striped ground crickets are chirping at a rate of 18 times per sec.

d. Find the correlation coefficient.

Social Sciences

19. Pupil-Teacher Ratios   The following table gives the national average pupil-teacher ratio in public schools over selected years. *Source: National Center for Education Statistics.*

Year	Ratio
1990	17.4
1994	17.7
1998	16.9
2002	16.2
2006	15.8

a. Find the equation for the least squares line. Let *x* correspond to the number of years since 1990 and let *y* correspond to the average number of pupils per 1 teacher.

b. Use your answer from part a to predict the pupil-teacher ratio in 2020. Does this seem realistic?

c. Calculate and interpret the correlation coefficient.

20. Poverty Levels The following table lists how poverty level income cutoffs (in dollars) for a family of four have changed over time. *Source: U.S. Census Bureau.*

Year	Income
1980	8414
1985	10,989
1990	13,359
1995	15,569
2000	17,604
2005	19,961
2008	22,207

Let x represent the year, with $x = 0$ corresponding to 1980 and y represent the income in thousands of dollars.

a. Plot the data. Do the data appear to lie along a straight line?

b. Calculate the correlation coefficient. Does your result agree with your answer to part a?

c. Find the equation of the least squares line.

d. Use your answer from part c to predict the poverty level in the year 2018.

21. Ideal Partner Height In an introductory statistics course at Cornell University, 147 undergraduates were asked their own height and the ideal height for their ideal spouse or partner. For this exercise, we are including the data for only a representative sample of 10 of the students, as given in the following table. All heights are in inches. *Source: Chance.*

Height	Ideal Partner's Height
59	66
62	71
66	72
68	73
71	75
67	63
70	63
71	67
73	66
75	66

a. Find the regression line and correlation coefficient for this data. What strange phenomenon do you observe?

b. The first five data pairs are for female students and the second five for male students. Find the regression line and correlation coefficient for each set of data.

c. Plot all the data on one graph, using different types of points to distinguish the data for the males and for the females. Using this plot and the results from part b, explain the strange phenomenon that you observed in part a.

22. SAT Scores At Hofstra University, all students take the math SAT before entrance, and most students take a mathematics placement test before registration. Recently, one professor collected the following data for 19 students in his Finite Mathematics class:

Math SAT	Placement Test	Math SAT	Placement Test	Math SAT	Placement Test
540	20	580	8	440	10
510	16	680	15	520	11
490	10	560	8	620	11
560	8	560	13	680	8
470	12	500	14	550	8
600	11	470	10	620	7
540	10				

a. Find an equation for the least squares line. Let x be the math SAT and y be the placement test score.

b. Use your answer from part a to predict the mathematics placement test score for a student with a math SAT score of 420.

c. Use your answer from part a to predict the mathematics placement test score for a student with a math SAT score of 620.

d. Calculate the correlation coefficient.

e. Based on your answer to part d, what can you conclude about the relationship between a student's math SAT and mathematics placement test score?

Physical Sciences

23. Length of a Pendulum Grandfather clocks use pendulums to keep accurate time. The relationship between the length of a pendulum L and the time T for one complete oscillation can be determined from the data in the table.* *Source: Gary Rockswold.*

L (ft)	T (sec)
1.0	1.11
1.5	1.36
2.0	1.57
2.5	1.76
3.0	1.92
3.5	2.08
4.0	2.22

a. Plot the data from the table with L as the horizontal axis and T as the vertical axis.

b. Find the least squares line equation and graph it simultaneously, if possible, with the data points. Does it seem to fit the data?

c. Find the correlation coefficient and interpret it. Does it confirm your answer to part b?

24. Air Conditioning While shopping for an air conditioner, Adam Bryer consulted the following table, which gives a machine's BTUs and the square footage (ft²) that it would cool.

*The actual relationship is $L = 0.81T^2$, which is not a linear relationship. This illustration that even if the relationship is not linear, a line can give a good approximation.

ft²(x)	BTUs (y)
150	5000
175	5500
215	6000
250	6500
280	7000
310	7500
350	8000
370	8500
420	9000
450	9500

a. Find the equation for the least squares line for the data.

b. To check the fit of the data to the line, use the results from part a to find the BTUs required to cool a room of 150 ft², 280 ft², and 420 ft². How well does the actual data agree with the predicted values?

c. Suppose Adam's room measures 230 ft². Use the results from part a to decide how many BTUs it requires. If air conditioners are available only with the BTU choices in the table, which would Adam choose?

d. Why do you think the table gives ft² instead of ft³, which would give the volume of the room?

General Interest

25. Football   The following data give the expected points for a football team with first down and 10 yards to go from various points on the field. *Source: Operations Research.* (*Note:* $\sum x = 500$, $\sum x^2 = 33,250$, $\sum y = 20.668$, $\sum y^2 = 91.927042$, $\sum xy = 399.16$.)

Yards from Goal (x)	Expected Points (y)
5	6.041
15	4.572
25	3.681
35	3.167
45	2.392
55	1.538
65	0.923
75	0.236
85	−0.637
95	−1.245

a. Calculate the correlation coefficient. Does there appear to be a linear correlation?

b. Find the equation of the least squares line.

c. Use your answer from part a to predict the expected points when a team is at the 50-yd line.

26. Athletic Records   The table shows the men's and women's outdoor world records (in seconds) in the 800-m run. *Source: Nature, Track and Field Athletics, Statistics in Sports, and The World Almanac and Book of Facts.*

Year	Men's Record	Women's Record
1905	113.4	—
1915	111.9	—
1925	111.9	144
1935	109.7	135.6
1945	106.6	132
1955	105.7	125
1965	104.3	118
1975	103.7	117.48
1985	101.73	113.28
1995	101.11	113.28
2005	101.11	113.28

Let x be the year, with x = 0 corresponding to 1900.

a. Find the equation for the least squares line for the men's record (y) in terms of the year (x).

b. Find the equation for the least squares line for the women's record.

c. Suppose the men's and women's records continue to improve as predicted by the equations found in parts a and b. In what year will the women's record catch up with the men's record? Do you believe that will happen? Why or why not?

d. Calculate the correlation coefficient for both the men's and the women's record. What do these numbers tell you?

e. Draw a plot of the data, and discuss to what extent a linear function describes the trend in the data.

27. Running If you think a marathon is a long race, consider the Hardrock 100, a 100.5 mile running race held in southwestern Colorado. The chart at right lists the times that the 2008 winner, Kyle Skaggs, arrived at various mileage points along the way. *Source: www.run100s.com.*

a. What was Skagg's average speed?

b. Graph the data, plotting time on the x-axis and distance on the y-axis. You will need to convert the time from hours and minutes into hours. Do the data appear to lie approximately on a straight line?

c. Find the equation for the least squares line, fitting distance as a linear function of time.

d. Calculate the correlation coefficient. Does it indicate a good fit of the least squares line to the data?

e. Based on your answer to part d, what is a good value for Skagg's average speed? Compare this with your answer to part a. Which answer do you think is better? Explain your reasoning.

Time (hr:min)	Miles
0	0
2:19	11.5
3:43	18.9
5:36	27.8
7:05	32.8
7:30	36.0
8:30	43.9
10:36	51.5
11:56	58.4
15:14	71.8
17:49	80.9
18:58	85.2
20:50	91.3
23:23	100.5

YOUR TURN ANSWERS

1. $Y = -0.535x + 89.5$ **2.** -0.949

CHAPTER REVIEW

SUMMARY

In this chapter we studied linear functions, whose graphs are straight lines. We developed the slope-intercept and point-slope formulas, which can be used to find the equation of a line, given a point and the slope or given two points. We saw that lines have many applications in virtually every discipline. Lines are used through the rest of this book, so fluency in their use is important. We concluded the chapter by introducing the method of least squares, which is used to find an equation of the line that best fits a given set of data.

Slope of a Line The slope of a line is defined as the vertical change (the "rise") over the horizontal change (the "run") as one travels along the line. In symbols, taking two different points (x_1, y_1) and (x_2, y_2) on the line, the slope is

$$m = \frac{y_2 - y_1}{x_2 - x_1},$$

where $x_1 \neq x_2$.

Equations of Lines

Equation	Description
$y = mx + b$	Slope intercept form: slope m and y-intercept b.
$y - y_1 = m(x - x_1)$	Point-slope form: slope m and line passes through (x_1, y_1).
$x = k$	Vertical line: x-intercept k, no y-intercept (except when $k = 0$), undefined slope.
$y = k$	Horizontal line: y-intercept k, no x-intercept (except when $k = 0$), slope 0.

Parallel Lines Two lines are parallel if and only if they have the same slope, or if they are both vertical.

Perpendicular Lines Two lines are perpendicular if and only if the product of their slopes is -1, or if one is vertical and the other horizontal.

Linear Function A relationship f defined by

$$y = f(x) = mx + b,$$

for real numbers m and b, is a linear function.

Linear Cost Function In a cost function of the form $C(x) = mx + b$, the m represents the marginal cost and b represents the fixed cost.

Least Squares Line The least squares line $Y = mx + b$ that gives the best fit to the data points $(x_1, y_1), (x_2, y_2), \ldots,$ (x_n, y_n) has slope m and y-intercept b given by the equations

$$m = \frac{n(\Sigma xy) - (\Sigma x)(\Sigma y)}{n(\Sigma x^2) - (\Sigma x)^2}$$

$$b = \frac{\Sigma y - m(\Sigma x)}{n}$$

Correlation Coefficient $$r = \frac{n(\Sigma xy) - (\Sigma x)(\Sigma y)}{\sqrt{n(\Sigma x^2) - (\Sigma x)^2}\ \sqrt{n(\Sigma y^2) - (\Sigma y)^2}}$$

KEY TERMS

To understand the concepts presented in this chapter, you should know the meaning and use of the following terms.
For easy reference, the section in the chapter where a word (or expression) was first used is provided.

mathematical model	slope	independent variable	profit
1.1	linear equation	dependent variable	break-even quantity
ordered pair	slope-intercept form	surplus	break-even point
Cartesian coordinate system	proportional	shortage	**1.3**
axes	point-slope form	equilibrium price	least squares line
origin	parallel	equilibrium quantity	summation notation
coordinates	perpendicular	fixed cost	correlation coefficient
quadrants	scatterplot	marginal cost	outlier
graph	**1.2**	linear cost function	
intercepts	linear function	revenue	

REVIEW EXERCISES

CONCEPT CHECK

Determine whether each statement is true or false, and explain why.

1. A given line can have more than one slope.

2. The equation $y = 3x + 4$ represents the equation of a line with slope 4.

3. The line $y = -2x + 5$ intersects the point $(3, -1)$.

4. The line that intersects the points $(2, 3)$ and $(2, 5)$ is a horizontal line.

5. The line that intersects the points $(4, 6)$ and $(5, 6)$ is a horizontal line.

6. The x-intercept of the line $y = 8x + 9$ is 9.

7. The function $f(x) = \pi x + 4$ represents a linear function.

8. The function $f(x) = 2x^2 + 3$ represents a linear function.

9. The lines $y = 3x + 17$ and $y = -3x + 8$ are perpendicular.

10. The lines $4x + 3y = 8$ and $4x + y = 5$ are parallel.

11. A correlation coefficient of zero indicates a perfect fit with the data.

12. It is not possible to get a correlation coefficient of -1.5 for a set of data.

PRACTICE AND EXPLORATIONS

13. What is marginal cost? Fixed cost?

14. What six quantities are needed to compute a correlation coefficient?

Find the slope for each line that has a slope.

15. Through $(-3, 7)$ and $(2, 12)$

16. Through $(4, -1)$ and $(3, -3)$

17. Through the origin and $(11, -2)$

18. Through the origin and $(0, 7)$

19. $4x + 3y = 6$ **20.** $4x - y = 7$

21. $y + 4 = 9$ **22.** $3y - 1 = 14$

23. $y = 5x + 4$ **24.** $x = 5y$

Find an equation in the form $y = mx + b$ for each line.

25. Through $(5, -1)$; slope $= 2/3$

26. Through $(8, 0)$; slope $= -1/4$

27. Through $(-6, 3)$ and $(2, -5)$

28. Through $(2, -3)$ and $(-3, 4)$

29. Through $(2, -10)$, perpendicular to a line with undefined slope

30. Through $(-2, 5)$; slope $= 0$

Find an equation for each line in the form $ax + by = c$, where a, b, and c are integers with no factor common to all three and $a \geq 0$.

31. Through $(3, -4)$, parallel to $4x - 2y = 9$

32. Through $(0, 5)$, perpendicular to $8x + 5y = 3$

33. Through $(-1, 4)$; undefined slope

34. Through $(7, -6)$, parallel to a line with undefined slope

35. Through $(3, -5)$, parallel to $y = 4$

36. Through $(-3, 5)$, perpendicular to $y = -2$

Graph each linear equation defined as follows.

37. $y = 4x + 3$ **38.** $y = 6 - 2x$

39. $3x - 5y = 15$ **40.** $4x + 6y = 12$

41. $x - 3 = 0$ **42.** $y = 1$

43. $y = 2x$ **44.** $x + 3y = 0$

APPLICATIONS

Business and Economics

45. Profit To manufacture x thousand computer chips requires fixed expenditures of $352 plus $42 per thousand chips. Receipts from the sale of x thousand chips amount to $130 per thousand.

    a. Write an expression for expenditures.

    b. Write an expression for receipts.

    c. For profit to be made, receipts must be greater than expenditures. How many chips must be sold to produce a profit?

46. Supply and Demand The supply and demand for crabmeat in a local fish store are related by the equations

$$\text{Supply: } p = S(q) = 6q + 3$$

and

$$\text{Demand: } p = D(q) = 19 - 2q,$$

where p represents the price in dollars per pound and q represents the quantity of crabmeat in pounds per day. Find the quantity supplied and demanded at each of the following prices.

    a. $10   **b.** $15   **c.** $18

    d. Graph both the supply and the demand functions on the same axes.

    e. Find the equilibrium price.

    f. Find the equilibrium quantity.

47. Supply For a new diet pill, 60 pills will be supplied at a price of $40, while 100 pills will be supplied at a price of $60. Write a linear supply function for this product.

48. Demand The demand for the diet pills in Exercise 47 is 50 pills at a price of $47.50 and 80 pills at a price of $32.50. Determine a linear demand function for these pills.

49. Supply and Demand Find the equilibrium price and quantity for the diet pills in Exercises 47 and 48.

Cost **In Exercises 50–53, find a linear cost function.**

50. Eight units cost $300; fixed cost is $60.

51. Fixed cost is $2000; 36 units cost $8480.

52. Twelve units cost $445; 50 units cost $1585.

53. Thirty units cost $1500; 120 units cost $5640.

54. Break-Even Analysis The cost of producing x cartons of CDs is $C(x)$ dollars, where $C(x) = 200x + 1000$. The CDs sell for $400 per carton.

    a. Find the break-even quantity.

    b. What revenue will the company receive if it sells just that number of cartons?

55. Break-Even Analysis The cost function for flavored coffee at an upscale coffeehouse is given in dollars by $C(x) = 3x + 160$, where x is in pounds. The coffee sells for $7 per pound.

    a. Find the break-even quantity.

    b. What will the revenue be at that point?

56. U.S. Imports from China The United States is China's largest export market. Imports from China have grown from about 102 billion dollars in 2001 to 338 billion dollars in 2008. This growth has been approximately linear. Use the given data pairs to write a linear equation that describes this growth in imports over the years. Let $t = 1$ represent 2001 and $t = 8$ represent 2008. *Source: TradeStats Express™.*

57. U.S. Exports to China U.S. exports to China have grown (although at a slower rate than imports) since 2001. In 2001, about 19.1 billion dollars of goods were exported to China. By 2008, this amount had grown to 69.7 billion dollars. Write a linear equation describing the number of exports each year, with $t = 1$ representing 2001 and $t = 8$ representing 2008. *Source: TradeStats Express™.*

58. Median Income The U.S. Census Bureau reported that the median income for all U.S. households in 2008 was $50,303. In 1988, the median income (in 2008 dollars) was $47,614. The median income is approximately linear and is a function of time. Find a formula for the median income, I, as a function of the year t, where t is the number of years since 1900. *Source: U.S Census Bureau.*

59. New Car Cost The average new car cost (in dollars) for selected years from 1980 to 2005 is given in the table. *Source: Chicago Tribune and National Automobile Dealers Association.*

Year	1980	1985	1990	1995	2000	2005
Cost	7500	12,000	16,000	20,450	24,900	28,400

a. Find a linear equation for the average new car cost in terms of x, the number of years since 1980, using the data for 1980 and 2005.

b. Repeat part a, using the data for 1995 and 2005.

c. Find the equation of the least squares line using all the data.

d. Use a graphing calculator to plot the data and the three lines from parts a-c.

e. Discuss which of the three lines found in parts a–c best describes the data, as well as to what extent a linear model accurately describes the data.

f. Calculate the correlation coefficient.

Life Sciences

60. World Health In general, people tend to live longer in countries that have a greater supply of food. Listed below is the 2003–2005 daily calorie supply and 2005 life expectancy at birth for 10 randomly selected countries. *Source: Food and Agriculture Organization.*

Country	Calories (x)	Life Expectancy (y)
Belize	2818	75.4
Cambodia	2155	59.4
France	3602	80.4
India	2358	62.7
Mexico	3265	75.5
New Zealand	3235	79.8
Peru	2450	72.5
Sweden	3120	80.5
Tanzania	2010	53.7
United States	3826	78.7

a. Find the correlation coefficient. Do the data seem to fit a straight line?

b. Draw a scatterplot of the data. Combining this with your results from part a, do the data seem to fit a straight line?

c. Find the equation of the least squares line.

d. Use your answer from part c to predict the life expectancy in the United Kingdom, which has a daily calorie supply of 3426. Compare your answer with the actual value of 79.0 years.

e. Briefly explain why countries with a higher daily calorie supply might tend to have a longer life expectancy. Is this trend likely to continue to higher calorie levels? Do you think that an American who eats 5000 calories a day is likely to live longer than one who eats 3600 calories a day? Why or why not?

f. (For the ambitious!) Find the correlation coefficient and least squares line using the data for a larger sample of countries, as

found in an almanac or other reference. Is the result in general agreement with the previous results?

61. Blood Sugar and Cholesterol Levels The following data show the connection between blood sugar levels and cholesterol levels for eight different patients.

Patient	Blood Sugar Level (x)	Cholesterol Level (y)
1	130	170
2	138	160
3	142	173
4	159	181
5	165	201
6	200	192
7	210	240
8	250	290

For the data given in the preceding table, $\Sigma x = 1394$, $\Sigma y = 1607$, $\Sigma xy = 291,990$, $\Sigma x^2 = 255,214$, and $\Sigma y^2 = 336,155$.

a. Find the equation of the least squares line.

b. Predict the cholesterol level for a person whose blood sugar level is 190.

c. Find the correlation coefficient.

Social Sciences

62. Beef Consumption The per capita consumption of beef in the United States decreased from 115.7 lb in 1974 to 92.9 lb in 2007. Assume a linear function describes the decrease. Write a linear equation defining the function. Let t represent the number of years since 1950 and y represent the number of pounds of red meat consumed. *Source: U.S. Department of Agriculture.*

63. Marital Status More people are staying single longer in the United States. In 1995, the number of never-married adults, age 15 and over, was 55.0 million. By 2009, it was 72.1 million. Assume the data increase linearly, and write an equation that defines a linear function for this data. Let t represent the number of years since 1990. *Source: U.S. Census Bureau.*

64. Poverty The following table gives the number of families under the poverty level in the U.S. in recent years. *Source: U.S. Census Bureau.*

Year	Families Below Poverty Level (in thousands)
2000	6400
2001	6813
2002	7229
2003	7607
2004	7623
2005	7657
2006	7668
2007	7835
2008	8147

a. Find a linear equation for the number of families below poverty level (in thousands) in terms of x, the number of years since 2000, using the data for 2000 and 2008.

b. Repeat part a, using the data for 2004 and 2008.

c. Find the equation of the least squares line using all the data. Then plot the data and the three lines from parts a–c on a graphing calculator.

d. Discuss which of the three lines found in parts a–c best describes the data, as well as to what extent a linear model accurately describes the data.

e. Calculate the correlation coefficient.

65. Governors' Salaries In general, the larger a state's population, the more the governor earns. Listed in the table below are the estimated 2008 populations (in millions) and the salary of the governor (in thousands of dollars) for eight randomly selected states. *Source: U.S. Census Bureau and Alaska Department of Administration.*

State	AZ	DE	MD	MA	NY	PA	TN	WY
Population (x)	6.50	0.88	5.54	6.45	19.30	12.39	5.92	0.53
Governor's Salary (y)	95	133	150	141	179	170	160	105

a. Find the correlation coefficient. Do the data seem to fit a straight line?

b. Draw a scatterplot of the data. Compare this with your answer from part a.

c. Find the equation for the least squares line.

d. Based on your answer to part c, how much does a governor's salary increase, on average, for each additional million in population?

e. Use your answer from part c to predict the governor's salary in your state. Based on your answers from parts a and b, would this prediction be very accurate? Compare with the actual salary, as listed in an almanac or other reference.

f. (For the ambitious!) Find the correlation coefficient and least squares line using the data for all 50 states, as found in an almanac or other reference. Is the result in general agreement with the previous results?

66. Movies A mathematician exploring the relationship between ratings of movies, their year of release, and their length discovered a paradox. Rather than list the data set of 100 movies in the original research, we have created a sample of size 10 that captures the properties of the original dataset. In the following table, the rating is a score from 1 to 10, and the length is in minutes. *Source: Journal of Statistics Education.*

Year	Rating	Length
2001	10	120
2003	5	85
2004	3	100
2004	6	105
2005	4	110
2005	8	115
2006	6	135
2007	2	105
2007	5	125
2008	6	130

a. Find the correlation coefficient between the years since 2000 and the length.

b. Find the correlation coefficient between the length and the rating.

c. Given that you found a positive correlation between the year and the length in part a, and a positive correlation between the length and the rating in part b, what would you expect about the correlation between the year and the rating? Calculate this correlation. Are you surprised?

d. Discuss the paradoxical result in part c. Write out in words what each correlation tells you. Try to explain what is happening. You may want to look at a scatterplot between the year and the rating, and consider which points on the scatterplot represent movies of length no more than 110 minutes, and which represent movies of length 115 minutes or more.

EXTENDED APPLICATION

USING EXTRAPOLATION TO PREDICT LIFE EXPECTANCY

One reason for developing a mathematical model is to make predictions. If your model is a least squares line, you can predict the y-value corresponding to some new x by substituting this x into an equation of the form $Y = mx + b$. (We use a capital Y to remind us that we're getting a predicted value rather than an actual data value.) Data analysts distinguish between two very different kinds of prediction, *interpolation*, and *extrapolation*. An interpolation uses a new x inside the x range of your original data. For example, if you have inflation data at 5-year intervals from 1950 to 2000, estimating the rate of inflation in 1957 is an interpolation problem. But if you use the same data to estimate what the inflation rate was in 1920, or what it will be in 2020, you are extrapolating.

In general, interpolation is much safer than extrapolation, because data that are approximately linear over a short interval may be nonlinear over a larger interval. One way to detect nonlinearity is to look at *residuals*, which are the differences between the actual data values and the values predicted by the line of best fit. Here is a simple example:

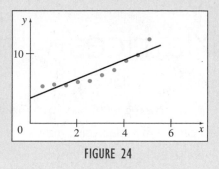

FIGURE 24

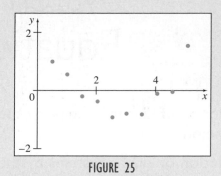

FIGURE 25

The regression equation for the linear fit in Figure 24 is $Y = 3.431 + 1.334x$. Since the r-value for this regression line is 0.93, our linear model fits the data very well. But we might notice that the predictions are a bit low at the ends and high in the middle. We can get a better look at this pattern by plotting the residuals. To find them, we put each value of the independent variable into the regression equation, calculate the predicted value Y, and subtract it from the actual y-value. The residual plot is shown in Figure 25, with the vertical axis rescaled to exaggerate the pattern. The residuals indicate that our data have a nonlinear, U-shaped component that is not captured by the linear fit. Extrapolating from this data set is probably not a good idea; our linear prediction for the value of y when x is 10 may be much too low.

EXERCISES

The following table gives the life expectancy at birth of females born in the United States in various years from 1970 to 2005. **Source: National Center for Health Statistics.**

Year of Birth	Life Expectancy (years)
1970	74.7
1975	76.6
1980	77.4
1985	78.2
1990	78.8
1995	78.9
2000	79.3
2005	79.9

1. Find an equation for the least squares line for these data, using year of birth as the independent variable.

2. Use your regression equation to guess a value for the life expectancy of females born in 1900.

3. Compare your answer with the actual life expectancy for females born in 1900, which was 48.3 years. Are you surprised?

4. Find the life expectancy predicted by your regression equation for each year in the table, and subtract it from the actual value in the second column. This gives you a table of residuals. Plot your residuals as points on a graph.

5. Now look at the residuals as a fresh data set, and see if you can sketch the graph of a smooth function that fits the residuals well. How easy do you think it will be to predict the life expectancy at birth of females born in 2015?

6. What will happen if you try linear regression on the *residuals*? If you're not sure, use your calculator or software to find the regression equation for the residuals. Why does this result make sense?

7. Since most of the females born in 1995 are still alive, how did the Public Health Service come up with a life expectancy of 78.9 years for these women?

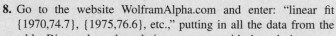 8. Go to the website WolframAlpha.com and enter: "linear fit {1970,74.7}, {1975,76.6}, etc.," putting in all the data from the table. Discuss how the solution compares with the solutions provided by a graphing calculator and by Microsoft Excel.

DIRECTIONS FOR GROUP PROJECT

Assume that you and your group (3–5 students) are preparing a report for a local health agency that is interested in using linear regression to predict life expectancy. Using the questions above as a guide, write a report that addresses the spirit of each question and any issues related to that question. The report should be mathematically sound, grammatically correct, and professionally crafted. Provide recommendations as to whether the health agency should proceed with the linear equation or whether it should seek other means of making such predictions.

2

Systems of Linear Equations and Matrices

The synchronized movements of band members marching on a field can be modeled using matrix arithmetic. An exercise in Section 5 in this chapter shows how multiplication by a matrix inverse transforms the original positions of the marchers into their new coordinates as they change direction.

M any mathematical models require finding the solutions of two or more equations. The solutions must satisfy *all* of the equations in the model. A set of equations related in this way is called a **system of equations**. In this chapter we will discuss systems of equations, introduce the idea of a *matrix*, and then show how matrices are used to solve systems of equations.

2.1 Solution of Linear Systems by the Echelon Method

APPLY IT How much of each ingredient should be used in an animal feed to meet dietary requirements?

APPLY IT   Suppose that an animal feed is made from two ingredients: corn and soybeans. One serving of each ingredient provides the number of grams of protein and fiber shown in the table. For example, the entries in the first column, 3 and 11, indicate that one serving of corn provides 3 g of protein and 11 g of fiber.

Nutritional Content of Ingredients		
	Corn	Soybeans
Protein	3	10
Fiber	11	4

Now suppose we want to know how many servings of corn and soybeans should be used to make a feed that contains 115 g of protein and 95 g of fiber. Let x represent the number of servings of corn used and y the number of servings of soybeans. Each serving of corn provides 3 g of protein, so the amount of protein provided by x servings of corn is $3x$. Similarly, the amount of protein provided by y servings of soybeans is $10y$. Since the total amount of protein is to be 115 g,

$$3x + 10y = 115.$$

The feed must supply 95 g of fiber, so

$$11x + 4y = 95.$$

Solving this problem means finding values of x and y that satisfy this system of equations. Verify that $x = 5$ and $y = 10$ is a solution of the system, since these numbers satisfy both equations. In fact, this is the only solution of this system. Many practical problems lead to such a system of *first-degree equations*.

A **first-degree equation in n unknowns** is any equation of the form

$$a_1x_1 + a_2x_2 + \cdots + a_nx_n = k,$$

where $a_1, a_2, \ldots, a_n$ and k are real numbers and $x_1, x_2, \ldots, x_n$ represent variables.* Each of the two equations from the animal feed problem is a first-degree equation. For example, the first equation

$$3x + 10y = 115$$

*a_1 is read "a-sub-one." The notation $a_1, a_2, \ldots, a_n$ represents n real-number coefficients (some of which may be equal), and the notation $x_1, x_2, \ldots, x_n$ represents n different variables, or unknowns.

is a first-degree equation with $n = 2$ where

$$a_1 = 3, \qquad a_2 = 10, \qquad k = 115,$$

and the variables are x and y. We use x_1, x_2, etc., rather than x, y, etc., in the general case because we might have any number of variables. When n is no more than 4, we usually use x, y, z, and w to represent x_1, x_2, x_3, and x_4.

A *solution* of the first-degree equation

$$a_1x_1 + a_2x_2 + \cdots + a_nx_n = k$$

is a sequence of numbers $s_1, s_2, \ldots, s_n$ such that

$$a_1s_1 + a_2s_2 + \cdots + a_ns_n = k.$$

A solution of an equation is usually written in parentheses as $(s_1, s_2, \ldots, s_n)$. For example, $(1, 6, 2)$ is a solution of the equation $3x_1 + 2x_2 - 4x_3 = 7$, since $3(1) + 2(6) - 4(2) = 7$. This is an extension of the idea of an ordered pair, which was introduced in Chapter 1. A solution of a first-degree equation in two unknowns is an ordered pair, and the graph of the equation is a straight line. For this reason, all first-degree equations are also called linear equations.

Because the graph of a linear equation in two unknowns is a straight line, there are three possibilities for the solutions of a system of two linear equations in two unknowns.

Types of Solutions for Two Equations in Two Unknowns

1. The two graphs are lines intersecting at a single point. The system has a **unique solution**, and it is given by the coordinates of this point. See Figure 1(a).

2. The graphs are distinct parallel lines. When this is the case, the system is **inconsistent**; that is, there is no solution common to both equations. See Figure 1(b).

3. The graphs are the same line. In this case, the equations are said to be **dependent**, since any solution of one equation is also a solution of the other. There are infinitely many solutions. See Figure 1(c).

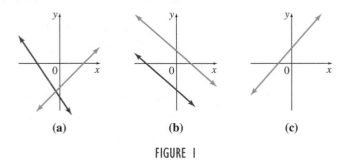

(a)   (b)   (c)

FIGURE 1

In larger systems, with more variables or more equations (or both, as is usually the case), there also may be exactly one solution, no solution, or infinitely many solutions. With more than two variables, the geometrical interpretation becomes more complicated, so we will only consider the geometry of two variables in two unknowns. We will, however, look at the algebra of two equations in three unknowns. In the next section, we will show how to handle any number of equations in any number of unknowns.

In Chapter 1, we used the graphing method and the substitution method to find the solution to a pair of equations describing the supply and demand for a commodity. The graphing method does not necessarily give an exact answer, and the substitution method becomes awkward with larger systems. In this section we will demonstrate a method to determine solutions using the addition property of equality to eliminate variables. This method forms the basis of the more general method presented in the next section.

Transformations

To solve a linear system of equations, we use properties of algebra to change, or transform, the system into a simpler *equivalent* system. An **equivalent system** is one that has the same solution(s) as the given system. Algebraic properties are the basis of the following transformations.

Transformations of a System

The following transformations can be applied to a system of equations to get an equivalent system:

1. exchanging any two equations;
2. multiplying both sides of an equation by any nonzero real number;
3. replacing any equation by a nonzero multiple of that equation plus a nonzero multiple of any other equation.

Use of these transformations leads to an equivalent system because each transformation can be reversed or "undone," allowing a return to the original system.

The Echelon Method

A systematic approach for solving systems of equations using the three transformations is called the **echelon method**. The goal of the echelon method is to use the transformations to rewrite the equations of the system until the system has a triangular form.

For a system of two equations in two variables, for example, the system should be transformed into the form

$$x + ay = b$$
$$y = c,$$

where a, b, and c are constants. Then the value for y from the second equation can be substituted into the first equation to find x. This is called **back substitution**. In a similar manner, a system of three equations in three variables should be transformed into the form

$$x + ay + bz = c$$
$$y + dz = e,$$
$$z = f.$$

EXAMPLE 1 **Solving a System of Equations with a Unique Solution**

Solve the system of equations from the animal feed example that began this section:

$$3x + 10y = 115 \tag{1}$$
$$11x + 4y = 95. \tag{2}$$

SOLUTION We first use transformation 3 to eliminate the x-term from equation (2). We multiply equation (1) by 11 and add the results to -3 times equation (2).

FOR REVIEW

For review of how to combine like terms using properties of real numbers, see Section R.1.

$$11(3x + 10y) = 11 \cdot 115$$
$$-3(11x + 4y) = -3 \cdot 95 \qquad \rightarrow$$

$$33x + 110y = 1265$$
$$\underline{-33x - 12y = -285}$$
$$98y = 980$$

We will indicate this process by the notation $11R_1 + (-3)R_2 \rightarrow R_2$. (R stands for the row.) The new system is

$$3x + 10y = 115 \tag{1}$$
$$11R_1 + (-3)R_2 \rightarrow R_2 \qquad 98y = 980 \tag{3}$$

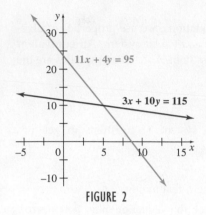

FIGURE 2

YOUR TURN 1

Solve the system

$2x + 3y = 12$

$3x - 4y = 1.$

┌─ FOR REVIEW ────────────

For review on graphing lines, see Section 1.1.

└─────────────────────────

Now we use transformation 2 to make the coefficient of the first term in each row equal to 1. Here, we must multiply equation (1) by 1/3 and equation (3) by 1/98 to accomplish this.

We get the system

$$\tfrac{1}{3}R_1 \rightarrow R_1 \qquad x + \frac{10}{3}y = \frac{115}{3}$$

$$\tfrac{1}{98}R_2 \rightarrow R_2 \qquad\qquad y = 10.$$

Back-substitution gives

$$x + \frac{10}{3}(10) = \frac{115}{3} \qquad \text{Substitute } y = 10.$$

$$x + \frac{100}{3} = \frac{115}{3}$$

$$x = \frac{115}{3} - \frac{100}{3}$$

$$= \frac{15}{3} = 5.$$

The solution of the system is (5, 10). The graphs of the two equations in Figure 2 suggest that (5, 10) satisfies both equations in the system. **TRY YOUR TURN 1**

The echelon method, as illustrated in Example 1, is a specific case of a procedure known as the *elimination method*. The elimination method allows for any variable to be eliminated. For example, the variable y in Equations (1) and (2) could be eliminated by calculating $2R_1 - 5R_2$. In the echelon method, the goal is to get the system into the specific form described just before Example 1.

▬▬ **EXAMPLE 2** **Solving a System of Equations with No Solution**

Solve the system

$$2x - 3y = 6 \tag{1}$$

$$-4x + 6y = 8. \tag{2}$$

SOLUTION Eliminate x in equation (2) to get the system

$$2x - 3y = 6 \tag{1}$$

$$2R_1 + R_2 \rightarrow R_2 \qquad 0 = 20. \tag{3}$$

In equation (3), both variables have been eliminated, leaving a *false statement*. This means it is impossible to have values of x and y that satisfy both equations, because this leads to a contradiction. To see why this occurs, see Figure 3, where we have shown that the graph of the system consists of two parallel lines. A solution would be a point lying on both lines, which is impossible. We conclude that the system is inconsistent and has no solution. ▬▬

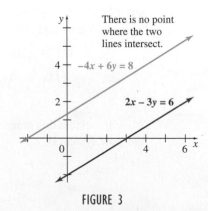

FIGURE 3

EXAMPLE 3 Solving a System of Equations with an Infinite Number of Solutions

Solve the system

$$3x - y = 4 \tag{1}$$
$$-6x + 2y = -8. \tag{2}$$

SOLUTION We use transformation 3 to eliminate x in equation (2), getting the system

$$3x - y = 4 \tag{1}$$
$$2R_1 + R_2 \rightarrow R_2 \qquad 0 = 0. \tag{3}$$

The system becomes

$$\tfrac{1}{3}R_1 \rightarrow R_1 \qquad x - \frac{1}{3}y = \frac{4}{3} \tag{4}$$
$$0 = 0. \tag{3}$$

In equation (3), both variables have been eliminated, leaving a *true statement*. If we graph the original equations of the system on the same axes, as shown in Figure 4, we see that the graphs are the same line, and any point on the line will satisfy the system. This indicates that one equation was simply a multiple of the other. This system is dependent and has an infinite number of solutions.

We will express the solutions in terms of y, where y can be any real number. The variable y in this case is called a **parameter**. (We could also let x be the parameter. In this text, we will follow the common practice of letting the rightmost variable be the parameter.) Solving equation (4) for x gives $x = (1/3)y + 4/3 = (y + 4)/3$, and all ordered pairs of the form

$$\left(\frac{y + 4}{3}, y \right)$$

are solutions. For example, if we let $y = 5$, then $x = (5 + 4)/3 = 3$ and one solution is $(3, 5)$. Similarly, letting $y = -10$ and $y = 3$ gives the solutions $(-2, -10)$ and $(7/3, 3)$.

Note that the original two equations are solved not only by the particular solutions like $(3, 5)$, $(-2, -10)$, and $(7/3, 3)$ but also by the general solution $(x, y) = ((y + 4)/3, y)$. For example, substituting this general solution into the first equation gives

$$3\left(\frac{y + 4}{3} \right) - y = y + 4 - y = 4,$$

which verifies that this general solution is indeed a solution. ▬

In some applications, x and y must be nonnegative integers. For instance, in Example 3, if x and y represent the number of male and female workers in a factory, it makes no sense to have $x = 7/3$ or $x = -2$. To make both x and y nonnegative, we solve the inequalities

$$\frac{y + 4}{3} \geq 0 \qquad \text{and} \qquad y \geq 0,$$

yielding

$$y \geq -4 \qquad \text{and} \qquad y \geq 0.$$

To make these last two inequalities true, we require $y \geq 0$, from which $y \geq -4$ automatically follows. Furthermore, to ensure $(y + 4)/3$ is an integer, it is necessary that y be 2 more than a whole-number multiple of 3. Therefore, the possible values of y are 2, 5, 8, 11, and so on, and the corresponding values of x are 2, 3, 4, 5, and so on.

The echelon method can be generalized to systems with more equations and unknowns. Because systems with three or more equations are more complicated, however,

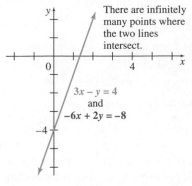

There are infinitely many points where the two lines intersect.

$3x - y = 4$
and
$-6x + 2y = -8$

FIGURE 4

we will postpone those to the next section, where we will show a procedure for solving them based on the echelon system. Meanwhile, we summarize the echelon method here. To solve a linear system in n variables, perform the following steps using the three transformations given earlier.

Echelon Method of Solving a Linear System

1. If possible, arrange the equations so that there is an x_1-term in the first equation, an x_2-term in the second equation, and so on.
2. Eliminate the x_1-term in all equations after the first equation.
3. Eliminate the x_2-term in all equations after the second equation.
4. Eliminate the x_3-term in all equations after the third equation.
5. Continue in this way until the last equation has the form $ax_n = k$, for constants a and k, if possible.
6. Multiply each equation by the reciprocal of the coefficient of its first term.
7. Use back-substitution to find the value of each variable.

Applications
The mathematical techniques in this text will be useful to you only if you are able to apply them to practical problems. To do this, always begin by reading the problem carefully. Next, identify what must be found. Let each unknown quantity be represented by a variable. (It is a good idea to *write down* exactly what each variable represents.) Now reread the problem, looking for all necessary data. Write those down, too. Finally, look for one or more sentences that lead to equations or inequalities. The next example illustrates these steps.

EXAMPLE 4   Flight Time

A flight leaves New York at 8 P.M. and arrives in Paris at 9 A.M. (Paris time). This 13-hour difference includes the flight time plus the change in time zones. The return flight leaves Paris at 1 P.M. and arrives in New York at 3 P.M. (New York time). This 2-hour difference includes the flight time *minus* time zones, plus an extra hour due to the fact that flying westward is against the wind. Find the actual flight time eastward and the difference in time zones.

SOLUTION   Let x be the flight time eastward and y be the difference in time zones. For the trip east, the flight time plus the change in time zones is 13 hours, so

$$x + y = 13.$$

For the trip west, the flight time (which is $x + 1$ hours due to the wind) minus the time zone is 2 hours, so

$$(x + 1) - y = 2.$$

Subtract 1 from both sides of this equation, and then solve the system

$$x + y = 13$$
$$x - y =\ \ 1$$

using the echelon method.

YOUR TURN 2 Solve Example 4 for the case in which there is a 16-hour time difference on the trip from New York, a 2-hour difference on the return trip, and no wind.

$$\begin{array}{r} x + y = 13 \\ R_1 + (-1)R_2 \to R_2 \quad\quad 2y = 12 \end{array}$$

Dividing the last equation by 2 gives $y = 6$. Substituting this into the first equation gives $x + 6 = 13$, so $x = 7$. Therefore, the flight time eastward is 7 hours, and the difference in time zones is 6 hours.   **TRY YOUR TURN 2**

EXAMPLE 5   Integral Solutions

A restaurant owner orders a replacement set of knives, forks, and spoons. The box arrives containing 40 utensils and weighing 141.3 oz (ignoring the weight of the box). A knife, fork, and spoon weigh 3.9 oz, 3.6 oz, and 3.0 oz, respectively.

(a) How many solutions are there for the number of knives, forks, and spoons in the box?

SOLUTION   Let

$$x = \text{the number of knives};$$
$$y = \text{the number of forks};$$
$$z = \text{the number of spoons}.$$

A chart is useful for organizing the information in a problem of this type.

Number and Weight of Utensils			
	Knives	Forks	Spoons	Total
Number	x	y	z	40
Weight	3.9	3.6	3.0	141.3

Because the box contains 40 utensils,

$$x + y + z = 40.$$

The x knives weigh $3.9x$ ounces, the y forks weigh $3.6y$ ounces, and the z spoons weigh $3.0z$ ounces. Since the total weight is 141.3 oz, we have the system

$$x + \quad y + \quad z = \quad 40$$
$$3.9x + 3.6y + 3.0z = 141.3.$$

Solve using the echelon method.

$$x + \quad y + \quad z = 40$$
$$3.9R_1 + (-1)R_2 \rightarrow R_2 \qquad 0.3y + 0.9z = 14.7$$

We do not have a third equation to solve for z, as we did in Example 4. We could thus let z be any real number, and then use this value of z in the second equation to find a corresponding value of y. These values of y and z could be put into the first equation to find a corresponding value of x. This system, then, has an infinite number of solutions. Letting z be the parameter, solve the second equation for y to get

$$y = \frac{14.7 - 0.9z}{0.3} = 49 - 3z.$$

Substituting this into the first equation, we get

$$x + (49 - 3z) + z = 40.$$

Solving this for x gives

$$x = 2z - 9.$$

Thus, the solutions are $(2z - 9, 49 - 3z, z)$, where z is any real number.

Now that we have solved for x and y in terms of z, let us investigate what values z can take on. This application demands that the solutions be nonnegative integers. The number of forks cannot be negative, so set

$$49 - 3z \geq 0.$$

Solving for z gives

$$z \leq \frac{49}{3} \approx 16.33.$$

Also, the number of knives cannot be negative, so set

$$2z - 9 \geq 0.$$

Solving for z gives

$$z \geq \frac{9}{2} = 4.5.$$

Therefore, the permissible values of z are 5, 6, 7, . . . , 16, for a total of 12 solutions.

YOUR TURN 3 Find the solution with the largest number of spoons.

(b) Find the solution with the smallest number of spoons.

SOLUTION The smallest value of z is $z = 5$, from which we find $x = 2(5) - 9 = 1$ and $y = 49 - 3(5) = 34$. This solution has 1 knife, 34 forks, and 5 spoons.

TRY YOUR TURN 3

2.1 EXERCISES

Use the echelon method to solve each system of two equations in two unknowns. Check your answers.

1. $x + y = 5$
$2x - 2y = 2$

2. $4x + y = 9$
$3x - y = 5$

3. $3x - 2y = -3$
$5x - y = 2$

4. $2x + 7y = -8$
$-2x + 3y = -12$

5. $3x + 2y = -6$
$5x - 2y = -10$

6. $-3x + y = 4$
$2x - 2y = -4$

7. $6x - 2y = -4$
$3x + 4y = 8$

8. $4m + 3n = -1$
$2m + 5n = 3$

9. $5p + 11q = -7$
$3p - 8q = 25$

10. $12s - 5t = 9$
$3s - 8t = -18$

11. $6x + 7y = -2$
$7x - 6y = 26$

12. $3a - 8b = 14$
$a - 2b = 2$

13. $3x + 2y = 5$
$6x + 4y = 8$

14. $9x - 5y = 1$
$-18x + 10y = 1$

15. $3x - 2y = -4$
$-6x + 4y = 8$

16. $3x + 5y + 2 = 0$
$9x + 15y + 6 = 0$

17. $x - \frac{3y}{2} = \frac{5}{2}$
$\frac{4x}{3} + \frac{2y}{3} = 6$

18. $\frac{x}{5} + 3y = 31$
$2x - \frac{y}{5} = 8$

19. $\frac{x}{2} + y = \frac{3}{2}$
$\frac{x}{3} + y = \frac{1}{3}$

20. $\frac{x}{9} + \frac{y}{6} = \frac{1}{3}$
$2x + \frac{8y}{5} = \frac{2}{5}$

21. An inconsistent system has _____ solutions.

22. The solution of a system with two dependent equations in two variables is _____.

For each of the following systems of equations in echelon form, tell how many solutions there are in nonnegative integers.

23. $x + 2y + 3z = 90$
$3y + 4z = 36$

24. $x - 7y + 4z = 75$
$2y + 7z = 60$

25. $3x + 2y + 4z = 80$
$y - 3z = 10$

26. $4x + 2y + 3z = 72$
$2y - 3z = 12$

27. Describe what a parameter is and why it is used in systems of equations with an infinite number of solutions.

28. In your own words, describe the echelon method as used to solve a system of two equations in three variables.

Solve each system of equations. Let z be the parameter.

29. $2x + 3y - z = 1$
$3x + 5y + z = 3$

30. $3x + y - z = 0$
$2x - y + 3z = -7$

31. $x + 2y + 3z = 11$
$2x - y + z = 2$

32. $-x + y - z = -7$
$2x + 3y + z = 7$

33. In Exercise 9 in Section 1.3, you were asked to solve the system of least squares line equations

$$nb + (\Sigma x)m = \Sigma y$$
$$(\Sigma x)b + (\Sigma x^2)m = \Sigma xy$$

by the method of substitution. Now solve the system by the echelon method to get

$$m = \frac{n(\Sigma xy) - (\Sigma x)(\Sigma y)}{n(\Sigma x^2) - (\Sigma x)^2}$$

$$b = \frac{\Sigma y - m(\Sigma x)}{n}.$$

34. The examples in this section did not use the first transformation. How might this transformation be used in the echelon method?

APPLICATIONS

Business and Economics

35. Groceries If 20 lb of rice and 10 lb of potatoes cost $16.20, and 30 lb of rice and 12 lb of potatoes cost $23.04, how much will 10 lb of rice and 50 lb of potatoes cost?

36. Sales An apparel shop sells skirts for $45 and blouses for $35. Its entire stock is worth $51,750. But sales are slow and only half the skirts and two-thirds of the blouses are sold, for a total of $30,600. How many skirts and blouses are left in the store?

37. Sales A theater charges $8 for main floor seats and $5 for balcony seats. If all seats are sold, the ticket income is $4200. At one show, 25% of the main floor seats and 40% of the balcony seats were sold and ticket income was $1200. How many seats are on the main floor and how many are in the balcony?

38. Stock Lorri Morgan has $16,000 invested in Disney and Exxon stock. The Disney stock currently sells for $30 a share and the Exxon stock for $70 a share. Her stockbroker points out that if Disney stock goes up 50% and Exxon stock goes up by $35 a share, her stock will be worth $25,500. Is this possible? If so, tell how many shares of each stock she owns. If not, explain why not.

39. Production A company produces two models of bicycles, model 201 and model 301. Model 201 requires 2 hours of assembly time and model 301 requires 3 hours of assembly time. The parts for model 201 cost $18 per bike and the parts for model 301 cost $27 per bike. If the company has a total of 34 hours of assembly time and $335 available per day for these two models, how many of each should be made in a day to use up all available time and money? If it is not possible, explain why not.

40. Banking A bank teller has a total of 70 bills in five-, ten-, and twenty-dollar denominations. The total value of the money is $960.

   a. Find the total number of solutions.

   b. Find the solution with the smallest number of five-dollar bills.

   c. Find the solution with the largest number of five-dollar bills.

41. Production Felsted Furniture makes dining room furniture. A buffet requires 30 hours for construction and 10 hours for finishing. A chair requires 10 hours for construction and 10 hours for finishing. A table requires 10 hours for construction and 30 hours for finishing. The construction department has 350 hours of labor and the finishing department has 150 hours of labor available each week. How many pieces of each type of furniture should be produced each week if the factory is to run at full capacity?

42. Rug Cleaning Machines Kelly Karpet Kleaners sells rug cleaning machines. The EZ model weighs 10 lb and comes in a 10-cubic-ft box. The compact model weighs 20 lb and comes in an 8-cubic-ft box. The commercial model weighs 60 lb and comes in a 28-cubic-ft box. Each of their delivery vans has 248 cubic ft of space and can hold a maximum of 440 lb. In order for a van to be fully loaded, how many of each model should it carry?

43. Production Turley Tailor, Inc. makes long-sleeve, short-sleeve, and sleeveless blouses. A long-sleeve blouse requires 1.5 hours of cutting and 1.2 hours of sewing. A short-sleeve blouse requires 1 hour of cutting and 0.9 hour of sewing. A sleeveless blouse requires 0.5 hour of cutting and 0.6 hour of sewing. There are 380 hours of labor available in the cutting department each day and 330 hours in the sewing department. If the plant is to run at full capacity, how many of each type of blouse should be made each day?

44. Broadway Economics When Neil Simon opens a new play, he has to decide whether to open the show on Broadway or Off Broadway. For example, in his play *London Suite*, he decided to open it Off Broadway. From information provided by Emanuel Azenberg, his producer, the following equations were developed:

$$43,500x - y = 1,295,000$$
$$27,000x - y = 440,000,$$

where x represents the number of weeks that the show has run and y represents the profit or loss from the show (first equation is for Broadway and second equation is for Off Broadway). *Source: The Mathematics Teacher.*

   a. Solve this system of equations to determine when the profit/loss from the show will be equal for each venue. What is the profit at that point?

   b. Discuss which venue is favorable for the show.

Life Sciences

45. Birds The date of the first sighting of robins has been occurring earlier each spring over the past 25 years at the Rocky Mountain Biological Laboratory. Scientists from this laboratory have developed two linear equations that estimate the date of the first sighting of robins:

$$y = 759 - 0.338x$$
$$y = 1637 - 0.779x,$$

where x is the year and y is the estimated number of days into the year when a robin can be expected. *Source: Proceedings of the National Academy of Science.*

   a. Compare the date of first sighting in 2000 for each of these equations. (*Hint:* 2000 was a leap year.)

   b. Solve this system of equations to find the year in which the two estimates agree.

Physical Sciences

46. Stopping Distance The stopping distance of a car traveling 25 mph is 61.7 ft, and for a car traveling 35 mph it is 106 ft. The stopping distance in feet can be described by the equation $y = ax^2 + bx$, where x is the speed in mph. *Source: National Traffic Safety Institute.*

   a. Find the values of a and b.

   b. Use your answers from part a to find the stopping distance for a car traveling 55 mph.

General Interest

47. Basketball Wilt Chamberlain holds the record for the highest number of points scored in a single NBA basketball game.

Chamberlain scored 100 points for Philadelphia against the New York Knicks on March 2, 1962. This is an amazing feat, considering he scored all of his points without the help of three-point shots. Chamberlain made a total of 64 baskets, consisting of field goals (worth two points) and foul shots (worth one point). Find the number of field goals and the number of foul shots that Chamberlain made. *Source: ESPN.*

48. **The 24® Game** The object of the 24® Game, created by Robert Sun, is to combine four numbers, using addition, subtraction, multiplication, and/or division, to get the number 24. For example, the numbers 2, 5, 5, 4 can be combined as $2(5 + 5) + 4 = 24$. For the algebra edition of the game and the game card shown to the right, the object is to find single-digit positive integer values x and y so the four numbers $x + y$, $3x + 2y$, 8, and 9 can be combined to make 24. *Source: Suntex.*

    a. Using the game card, write a system of equations that, when solved, can be used to make 24 from the game card. What is the solution to this system, and how can it be used to make 24 on the game card?

b. Repeat part a and develop a second system of equations.

YOUR TURN ANSWERS
1. $(3, 2)$
2. Flight time is 9 hours, difference in time zones is 7 hours.
3. 23 knives, 1 fork, and 16 spoons

2.2 Solution of Linear Systems by the Gauss-Jordan Method

APPLY IT   How can an auto manufacturer with more than one factory and several dealers decide how many cars to send to each dealer from each factory? *Questions like this are called transportation problems; they frequently lead to a system of equations that must be satisfied. In Exercise 52 in this section we use a further refinement of the echelon method to answer this question.*

When we use the echelon method, since the variables are in the same order in each equation, we really need to keep track of just the coefficients and the constants. For example, consider the following system of three equations in three unknowns.

$$2x + y - z = 2$$
$$x + 3y + 2z = 1$$
$$x + y + z = 2$$

This system can be written in an abbreviated form as

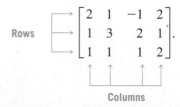

$$\text{Rows} \begin{bmatrix} 2 & 1 & -1 & 2 \\ 1 & 3 & 2 & 1 \\ 1 & 1 & 1 & 2 \end{bmatrix}.$$

Columns

Such a rectangular array of numbers enclosed by brackets is called a **matrix** (plural: **matrices**).* Each number in the array is an **element** or **entry**. To separate the constants in

---

*The word matrix, Latin for "womb," was coined by James Joseph Sylvester (1814–1897) and made popular by his friend Arthur Cayley (1821–1895). Both mathematicians were English, although Sylvester spent much of his life in the United States.

the last column of the matrix from the coefficients of the variables, we use a vertical line, producing the following **augmented matrix**.

$$\begin{bmatrix} 2 & 1 & -1 & | & 2 \\ 1 & 3 & 2 & | & 1 \\ 1 & 1 & 1 & | & 2 \end{bmatrix}$$

The rows of the augmented matrix can be transformed in the same way as the equations of the system, since the matrix is just a shortened form of the system. The following **row operations** on the augmented matrix correspond to the transformations of systems of equations given earlier.

Row Operations

For any augmented matrix of a system of equations, the following operations produce the augmented matrix of an equivalent system:

1. interchanging any two rows;

2. multiplying the elements of a row by any nonzero real number;

3. adding a nonzero multiple of the elements of one row to the corresponding elements of a nonzero multiple of some other row.

In steps 2 and 3, we are replacing a row with a new, modified row, which the old row helped to form, just as we replaced an equation with a new, modified equation in the previous section.

Row operations, like the transformations of systems of equations, are reversible. If they are used to change matrix A to matrix B, then it is possible to use row operations to transform B back into A. In addition to their use in solving equations, row operations are very important in the simplex method to be described in Chapter 4.

In the examples in this section, we will use the same notation as in Section 1 to show the row operation used. For example, the notation R_1 indicates row 1 of the previous matrix, and $-3R_1 + R_2$ means that row 1 is multiplied by -3 and added to row 2.

By the first row operation, interchanging two rows, the matrix

$$\begin{bmatrix} 0 & 1 & 2 & | & 3 \\ -2 & -6 & -10 & | & -12 \\ 2 & 1 & -2 & | & -5 \end{bmatrix} \quad \text{becomes} \quad \begin{bmatrix} -2 & -6 & -10 & | & -12 \\ 0 & 1 & 2 & | & 3 \\ 2 & 1 & -2 & | & -5 \end{bmatrix} \quad \begin{array}{l} \text{Interchange } \mathbf{R_1} \\ \text{and } \mathbf{R_2} \end{array}$$

by interchanging the first two rows. Row 3 is left unchanged.

The second row operation, multiplying a row by a number, allows us to change

$$\begin{bmatrix} -2 & -6 & -10 & | & -12 \\ 0 & 1 & 2 & | & 3 \\ 2 & 1 & -2 & | & -5 \end{bmatrix} \quad \text{to} \quad \begin{bmatrix} 1 & 3 & 5 & | & 6 \\ 0 & 1 & 2 & | & 3 \\ 2 & 1 & -2 & | & -5 \end{bmatrix} \quad (-1/2)\mathbf{R_1} \rightarrow \mathbf{R_1}$$

by multiplying the elements of row 1 of the original matrix by $-1/2$. Note that rows 2 and 3 are left unchanged.

Using the third row operation, adding a multiple of one row to another, we change

$$\begin{bmatrix} 1 & 3 & 5 & | & 6 \\ 0 & 1 & 2 & | & 3 \\ 2 & 1 & -2 & | & -5 \end{bmatrix} \quad \text{to} \quad \begin{bmatrix} 1 & 3 & 5 & | & 6 \\ 0 & 1 & 2 & | & 3 \\ 0 & -5 & -12 & | & -17 \end{bmatrix} \quad -2\mathbf{R_1} + \mathbf{R_3} \rightarrow \mathbf{R_3}$$

by first multiplying each element in row 1 of the original matrix by -2 and then adding the results to the corresponding elements in the third row of that matrix. Work as follows.

$$
\begin{bmatrix}
1 & 3 & 5 & 6 \\
0 & 1 & 2 & 3 \\
(-2)1+2 & (-2)3+1 & (-2)5-2 & (-2)6-5
\end{bmatrix}
=
\begin{bmatrix}
1 & 3 & 5 & 6 \\
0 & 1 & 2 & 3 \\
0 & -5 & -12 & -17
\end{bmatrix}
$$

Notice that rows 1 and 2 are left unchanged, *even though the elements of row 1 were used to transform row 3.*

The Gauss-Jordan Method

The **Gauss-Jordan method** is an extension of the echelon method of solving systems.* Before the Gauss-Jordan method can be used, the system must be in proper form: the terms with variables should be on the left and the constants on the right in each equation, with the variables in the same order in each equation.

The system is then written as an augmented matrix. Using row operations, the goal is to transform the matrix so that it has zeros above and below a diagonal of 1's on the left of the vertical bar. Once this is accomplished, the final solution can be read directly from the last matrix. The following example illustrates the use of the Gauss-Jordan method to solve a system of equations.

EXAMPLE 1 Gauss-Jordan Method

Solve the system

$$3x - 4y = 1 \tag{1}$$
$$5x + 2y = 19. \tag{2}$$

SOLUTION

Method 1
1's on Diagonal

The system is already in the proper form to use the Gauss-Jordan method. Our goal is to transform this matrix, if possible, into the form

$$
\begin{bmatrix}
1 & 0 & m \\
0 & 1 & n
\end{bmatrix},
$$

where m and n are real numbers. To begin, we change the 3 in the first row to 1 using the second row operation. (Notice that the same notation is used to indicate each transformation, as in the previous section.)

$$
\begin{bmatrix}
3 & -4 & 1 \\
5 & 2 & 19
\end{bmatrix}
\quad \text{Augmented matrix}
$$

$$
\tfrac{1}{3}R_1 \rightarrow R_1 \quad
\begin{bmatrix}
1 & -\tfrac{4}{3} & \tfrac{1}{3} \\
5 & 2 & 19
\end{bmatrix}
$$

Using the third row operation, we change the 5 in row 2 to 0.

$$
-5R_1 + R_2 \rightarrow R_2 \quad
\begin{bmatrix}
1 & -\tfrac{4}{3} & \tfrac{1}{3} \\
0 & \tfrac{26}{3} & \tfrac{52}{3}
\end{bmatrix}
$$

We now change 26/3 in row 2 to 1 to complete the diagonal of 1's.

$$
\tfrac{3}{26}R_2 \rightarrow R_2 \quad
\begin{bmatrix}
1 & -\tfrac{4}{3} & \tfrac{1}{3} \\
0 & 1 & 2
\end{bmatrix}
$$

*The great German mathematician Carl Friedrich Gauss (1777–1855), sometimes referred to as the "Prince of Mathematicians," originally developed his elimination method for use in finding least squares coefficients. (See Section 1.3.) The German geodesist Wilhelm Jordan (1842–1899) improved his method and used it in surveying problems. Gauss's method had been known to the Chinese at least 1800 years earlier and was described in the *Jiuahang Suanshu (Nine Chapters on the Mathematical Art).*

The final transformation is to change the $-4/3$ in row 1 to 0.

$$\tfrac{4}{3}R_2 + R_1 \rightarrow R_1 \quad \begin{bmatrix} 1 & 0 & | & 3 \\ 0 & 1 & | & 2 \end{bmatrix}$$

The last matrix corresponds to the system

$$x = 3$$
$$y = 2,$$

so we can read the solution directly from the last column of the final matrix. Check that $(3, 2)$ is the solution by substitution in the equations of the original matrix.

Method 2
Fraction-Free

An alternate form of Gauss-Jordan is to first transform the matrix so that it contains zeros above and below the main diagonal. Then, use the second transformation to get the required 1's. When doing calculations by hand, this second method simplifies the calculations by avoiding fractions and decimals. We will use this method when doing calculations by hand throughout the remainder of this chapter.

To begin, we change the 5 in row 2 to 0.

$$\begin{bmatrix} 3 & -4 & | & 1 \\ 5 & 2 & | & 19 \end{bmatrix} \quad \text{Augmented matrix}$$

$$5R_1 + (-3)R_2 \rightarrow R_2 \quad \begin{bmatrix} 3 & -4 & | & 1 \\ 0 & -26 & | & -52 \end{bmatrix}$$

We change the -4 in row 1 to 0.

$$-4R_2 + 26R_1 \rightarrow R_1 \quad \begin{bmatrix} 78 & 0 & | & 234 \\ 0 & -26 & | & -52 \end{bmatrix}$$

Then we change the first nonzero number in each row to 1.

$$\begin{array}{c} \tfrac{1}{78}R_1 \rightarrow R_1 \\ -\tfrac{1}{26}R_2 \rightarrow R_2 \end{array} \quad \begin{bmatrix} 1 & 0 & | & 3 \\ 0 & 1 & | & 2 \end{bmatrix}$$

YOUR TURN 1 Use the Gauss-Jordan method to solve the system
$$4x + 5y = 10$$
$$7x + 8y = 19.$$

The solution is read directly from this last matrix: $x = 3$ and $y = 2$, or $(3, 2)$.
TRY YOUR TURN 1

NOTE If your solution does not check, the most efficient way to find the error is to substitute back through the equations that correspond to each matrix, starting with the last matrix. When you find a system that is not satisfied by your (incorrect) answers, you have probably reached the matrix just before the error occurred. Look for the error in the transformation to the next matrix. For example, if you erroneously wrote the 2 as -2 in the final matrix of the fraction-free method of Example 2, you would find that $(3, -2)$ was not a solution of the system represented by the previous matrix because $-26(-2) \neq -52$, telling you that your error occurred between this matrix and the final one.

When the Gauss-Jordan method is used to solve a system, the final matrix always will have zeros above and below the diagonal of 1's on the left of the vertical bar. To transform the matrix, it is best to work column by column from left to right. Such an orderly method avoids confusion and going around in circles. For each column, first perform the steps that give the zeros. When all columns have zeros in place, multiply each row by the reciprocal of the coefficient of the remaining nonzero number in that row to get the required 1's. With dependent equations or inconsistent systems, it will not be possible to get the complete diagonal of 1's.

We will demonstrate the method in an example with three variables and three equations, after which we will summarize the steps.

EXAMPLE 2 Gauss-Jordan Method

Use the Gauss-Jordan method to solve the system

$$\begin{aligned} x + 5z &= -6 + y \\ 3x + 3y &= 10 + z \\ x + 3y + 2z &= 5. \end{aligned}$$

Method 1
Calculating by Hand

SOLUTION

First, rewrite the system in proper form, as follows.

$$\begin{aligned} x - y + 5z &= -6 \\ 3x + 3y - z &= 10 \\ x + 3y + 2z &= 5 \end{aligned}$$

Begin to find the solution by writing the augmented matrix of the linear system.

$$\begin{bmatrix} 1 & -1 & 5 & | & -6 \\ 3 & 3 & -1 & | & 10 \\ 1 & 3 & 2 & | & 5 \end{bmatrix}$$

Row transformations will be used to rewrite this matrix in the form

$$\begin{bmatrix} 1 & 0 & 0 & | & m \\ 0 & 1 & 0 & | & n \\ 0 & 0 & 1 & | & p \end{bmatrix},$$

where m, n, and p are real numbers (if this form is possible). From this final form of the matrix, the solution can be read: $x = m$, $y = n$, $z = p$, or (m, n, p).

In the first column, we need zeros in the second and third rows. Multiply the first row by -3 and add to the second row to get a zero there. Then multiply the first row by -1 and add to the third row to get that zero.

$$\begin{matrix} \\ -3R_1 + R_2 \rightarrow R_2 \\ -1R_1 + R_3 \rightarrow R_3 \end{matrix} \begin{bmatrix} 1 & -1 & 5 & | & -6 \\ 0 & 6 & -16 & | & 28 \\ 0 & 4 & -3 & | & 11 \end{bmatrix}$$

Now get zeros in the second column in a similar way. We want zeros in the first and third rows. Row 2 will not change.

$$\begin{matrix} R_2 + 6R_1 \rightarrow R_1 \\ \\ 2R_2 + (-3)R_3 \rightarrow R_3 \end{matrix} \begin{bmatrix} 6 & 0 & 14 & | & -8 \\ 0 & 6 & -16 & | & 28 \\ 0 & 0 & -23 & | & 23 \end{bmatrix}$$

In transforming the third row, you may have used the operation $4R_2 + (-6)R_3 \rightarrow R_3$ instead of $2R_2 + (-3)R_3 \rightarrow R_3$. This is perfectly fine; the last row would then have -46 and 46 in place of -23 and 23. To avoid errors, it helps to keep the numbers as small as possible. We observe at this point that all of the numbers can be reduced in size by multiplying each row by an appropriate constant. This next step is not essential, but it simplifies the arithmetic.

$$\begin{matrix} \frac{1}{2}R_1 \rightarrow R_1 \\ \frac{1}{2}R_2 \rightarrow R_2 \\ -\frac{1}{23}R_3 \rightarrow R_3 \end{matrix} \begin{bmatrix} 3 & 0 & 7 & | & -4 \\ 0 & 3 & -8 & | & 14 \\ 0 & 0 & 1 & | & -1 \end{bmatrix}$$

Next, we want zeros in the first and second rows of the third column. Row 3 will not change.

$$\begin{matrix} -7R_3 + R_1 \rightarrow R_1 \\ 8R_3 + R_2 \rightarrow R_2 \end{matrix} \begin{bmatrix} 3 & 0 & 0 & | & 3 \\ 0 & 3 & 0 & | & 6 \\ 0 & 0 & 1 & | & -1 \end{bmatrix}$$

Finally, get 1's in each row by multiplying the row by the reciprocal of (or dividing the row by) the number in the diagonal position.

$$\begin{array}{c} \frac{1}{3}R_1 \to R_1 \\ \frac{1}{3}R_2 \to R_2 \end{array} \left[\begin{array}{ccc|c} 1 & 0 & 0 & 1 \\ 0 & 1 & 0 & 2 \\ 0 & 0 & 1 & -1 \end{array}\right]$$

The linear system associated with the final augmented matrix is

$$x = 1$$
$$y = 2$$
$$z = -1,$$

and the solution is $(1, 2, -1)$. Verify that this is the solution to the original system of equations.

> **CAUTION** Notice that we have performed two or three operations on the same matrix in one step. This is permissible as long as we do not use a row that we are changing as part of another row operation. For example, when we changed row 2 in the first step, we could not use row 2 to transform row 3 in the same step. To avoid difficulty, use *only* row 1 to get zeros in column 1, row 2 to get zeros in column 2, and so on.

**Method 2
Graphing Calculator**

The row operations of the Gauss-Jordan method can also be done on a graphing calculator. For example, Figure 5 shows the result when the augmented matrix is entered into a TI-84 Plus. Figures 6 and 7 show how row operations can be used to get zeros in rows 2 and 3 of the first column.

Calculators typically do not allow any multiple of a row to be added to any multiple of another row, such as in the operation $2R_2 + 6R_1 \to R_1$. They normally allow a multiple of a row to be added only to another unmodified row. To get around this restriction, we can convert the diagonal element to a 1 before changing the other elements in the column to 0, as we did in the first method of Example 1. In this example, we change the 6 in row 2, column 2, to a 1 by dividing by 6. The result is shown in Figure 8. (The right side of the matrix is not visible but can be seen by pressing the right arrow key.) Notice that this operation introduces decimals. Converting to fractions is preferable on calculators that have that option; 1/3 is certainly more concise than 0.3333333333. Figure 9 shows such a conversion on the TI-84 Plus.

```
[A]
[1  -1   5   -6]
[3   3  -1   10]
[1   3   2    5]
```

FIGURE 5

```
*row+(-3, [A], 1, 2) → [A]
[1  -1    5   -6]
[0   6  -16   28]
[1   3    2    5]
```

FIGURE 6

```
*row+(-1, [A], 1, 3)→[A]
[1  -1    5   -6]
[0   6  -16   28]
[0   4   -3   11]
```

FIGURE 7

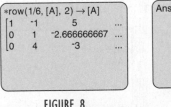

```
*row(1/6, [A], 2) → [A]
[1  -1        5        ...]
[0   1  -2.666666667  ...]
[0   4       -3        ...]
```

FIGURE 8

```
Ans▶Frac
[1  -1   5    -6 ]
[0   1  -8/3  14/3]
[0   4  -3    11 ]
```

FIGURE 9

When performing row operations without a graphing calculator, it is best to avoid fractions and decimals, because these make the operations more difficult and more prone to error. A calculator, on the other hand, encounters no such difficulties.

Continuing in the same manner, the solution $(1, 2, -1)$ is found as shown in Figure 10.

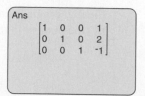

FIGURE 10

Some calculators can do the entire Gauss-Jordan process with a single command; on the TI-84 Plus, for example, this is done with the `rref` command. This is very useful in practice, although it does not show any of the intermediate steps. For more details, see the *Graphing Calculator and Excel Spreadsheet Manual* available with this book.

**Method 3
Spreadsheet**

The Gauss-Jordan method can be done using a spreadsheet either by using a macro or by developing the pivot steps using formulas with the copy and paste commands. However, spreadsheets also have built-in methods to solve systems of equations. Although these solvers do not usually employ the Gauss-Jordan method for solving systems of equations, they are, nonetheless, efficient and practical to use.

The Solver included with Excel can solve systems of equations that are both linear and nonlinear. The Solver is located in the Tools menu and requires that cells be identified ahead of time for each variable in the problem. It also requires that the left-hand side of each equation be placed in the spreadsheet as a formula. For example, to solve the above problem, we could identify cells A1, B1, and C1 for the variables x, y, and z, respectively. The Solver requires that we place a guess for the answer in these cells. It is convenient to place a zero in each of these cells. The left-hand side of each equation must be placed in a cell. We could choose A3, A4, and A5 to hold each of these formulas. Thus, in cell A3, we would type "=A1 − B1 + 5*C1" and put the other two equations in cells A4 and A5.

We now click on the Tools menu and choose Solver. (In some versions, it may be necessary to install the Solver. For more details, see the *Graphing Calculator and Excel Spreadsheet Manual* available with this book.) Since this solver attempts to find a solution that is best in some way, we are required to identify a cell with a formula in it that we want to optimize. In this case, it is convenient to use the cell with the left-hand side of the first constraint in it, A3. Figure 11 illustrates the Solver box and the items placed in it.

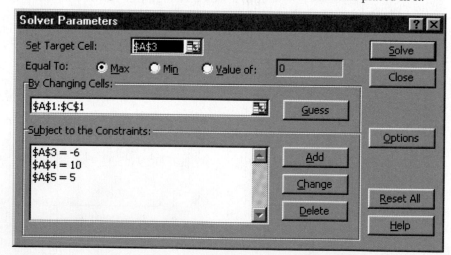

FIGURE 11

YOUR TURN 2 Use the Gauss-Jordan method to solve the system

$$x + 2y + 3z = 2$$
$$2x + 2y - 3z = 27$$
$$3x + 2y + 5z = 10.$$

To obtain a solution, click on Solve. The approximate solution is located in cells A1, B1, and C1, and these correspond to x, y, and z, respectively.

In summary, the Gauss-Jordan method of solving a linear system requires the following steps.

Gauss-Jordan Method of Solving a Linear System

1. Write each equation so that variable terms are in the same order on the left side of the equal sign and constants are on the right.

2. Write the augmented matrix that corresponds to the system.

3. Use row operations to transform the first column so that all elements except the element in the first row are zero.

4. Use row operations to transform the second column so that all elements except the element in the second row are zero.

5. Use row operations to transform the third column so that all elements except the element in the third row are zero.

6. Continue in this way, when possible, until the last row is written in the form

$$[0 \quad 0 \quad 0 \quad \cdots \quad 0 \quad j \mid k],$$

where j and k are constants with $j \neq 0$. When this is not possible, continue until every row has more zeros on the left than the previous row (except possibly for any rows of all zero at the bottom of the matrix), and the first nonzero entry in each row is the only nonzero entry in its column.

7. Multiply each row by the reciprocal of the nonzero element in that row.

Systems without a Unique Solution

In the previous examples, we were able to get the last row in the form $[0 \quad 0 \quad 0 \quad \cdots \quad 0 \quad j \mid k]$, where j and k are constants with $j \neq 0$. We will now look at examples where this is not the case.

EXAMPLE 3   Solving a System of Equations with No Solution

Use the Gauss-Jordan method to solve the system

$$x - 2y = 2$$
$$3x - 6y = 5.$$

SOLUTION Begin by writing the augmented matrix.

$$\begin{bmatrix} 1 & -2 & 2 \\ 3 & -6 & 5 \end{bmatrix}$$

To get a zero for the second element in column 1, multiply the numbers in row 1 by -3 and add the results to the corresponding elements in row 2.

$$-3R_1 + R_2 \rightarrow R_2 \quad \begin{bmatrix} 1 & -2 & 2 \\ 0 & 0 & -1 \end{bmatrix}$$

This matrix corresponds to the system

$$x - 2y = 2$$
$$0x + 0y = -1.$$

Since the second equation is $0 = -1$, the system is inconsistent and, therefore, has no solution. The row $[0 \quad 0 \mid k]$ for any nonzero k is a signal that the given system is inconsistent.

EXAMPLE 4 Solving a System of Equations with an Infinite Number of Solutions

Use the Gauss-Jordan method to solve the system

$$
\begin{aligned}
x + 2y - z &= 0 \\
3x - y + z &= 6 \\
-2x - 4y + 2z &= 0.
\end{aligned}
$$

SOLUTION The augmented matrix is

$$
\left[\begin{array}{ccc|c}
1 & 2 & -1 & 0 \\
3 & -1 & 1 & 6 \\
-2 & -4 & 2 & 0
\end{array}\right].
$$

We first get zeros in the second and third rows of column 1.

$$
\begin{array}{c}
-3R_1 + R_2 \rightarrow R_2 \\
2R_1 + R_3 \rightarrow R_3
\end{array}
\left[\begin{array}{ccc|c}
1 & 2 & -1 & 0 \\
0 & -7 & 4 & 6 \\
0 & 0 & 0 & 0
\end{array}\right]
$$

To continue, we get a zero in the first row of column 2 using the second row, as usual.

$$
2R_2 + 7R_1 \rightarrow R_1
\left[\begin{array}{ccc|c}
7 & 0 & 1 & 12 \\
0 & -7 & 4 & 6 \\
0 & 0 & 0 & 0
\end{array}\right]
$$

We cannot get a zero for the first-row, third-column element without changing the form of the first two columns. We must multiply each of the first two rows by the reciprocal of the first nonzero number.

$$
\begin{array}{c}
\frac{1}{7}R_1 \rightarrow R_1 \\
-\frac{1}{7}R_2 \rightarrow R_2
\end{array}
\left[\begin{array}{ccc|c}
1 & 0 & \frac{1}{7} & \frac{12}{7} \\
0 & 1 & -\frac{4}{7} & -\frac{6}{7} \\
0 & 0 & 0 & 0
\end{array}\right]
$$

To complete the solution, write the equations that correspond to the first two rows of the matrix.

$$
x + \frac{1}{7}z = \frac{12}{7}
$$

$$
y - \frac{4}{7}z = -\frac{6}{7}
$$

Because both equations involve z, let z be the parameter. There are an infinite number of solutions, corresponding to the infinite number of values of z. Solve the first equation for x and the second for y to get

$$
x = \frac{12 - z}{7} \quad \text{and} \quad y = \frac{4z - 6}{7}.
$$

As shown in the previous section, the general solution is written

$$
\left(\frac{12 - z}{7}, \frac{4z - 6}{7}, z \right),
$$

where z is any real number. For example, $z = 2$ and $z = 12$ lead to the solutions $(10/7, 2/7, 2)$ and $(0, 6, 12)$.

EXAMPLE 5 Solving a System of Equations with an Infinite Number of Solutions

Consider the following system of equations.

$$x + 2y + 3z - w = 4$$
$$2x + 3y \qquad + w = -3$$
$$3x + 5y + 3z \qquad = 1$$

(a) Set this up as an augmented matrix, and verify that the result after the Gauss-Jordan method is

$$\begin{bmatrix} 1 & 0 & -9 & 5 & | & -18 \\ 0 & 1 & 6 & -3 & | & 11 \\ 0 & 0 & 0 & 0 & | & 0 \end{bmatrix}$$

(b) Find the solution to this system of equations.

SOLUTION   To complete the solution, write the equations that correspond to the first two rows of the matrix.

$$x \quad - 9z + 5w = -18$$
$$y + 6z - 3w = \quad 11$$

Because both equations involve both z and w, let z and w be parameters. There are an infinite number of solutions, corresponding to the infinite number of values of z and w. Solve the first equation for x and the second for y to get

$$x = -18 + 9z - 5w \qquad \text{and} \qquad y = 11 - 6z + 3w.$$

In an analogous manner to problems with a single parameter, the general solution is written

$$(-18 + 9z - 5w, 11 - 6z + 3w, z, w),$$

where z and w are any real numbers. For example, $z = 1$ and $w = -2$ leads to the solution $(1, -1, 1, -2)$.   **TRY YOUR TURN 3**

YOUR TURN 3   Use the Gauss-Jordan method to solve the system

$$2x - 2y + 3z - 4w = 6$$
$$3x + 2y + 5z - 3w = 7$$
$$4x + y + 2z - 2w = 8.$$

Although the examples have used only systems with two equations in two unknowns, three equations in three unknowns, or three equations in four unknowns, the Gauss-Jordan method can be used for any system with n equations and m unknowns. The method becomes tedious with more than three equations in three unknowns; on the other hand, it is very suitable for use with graphing calculators and computers, which can solve fairly large systems quickly. Sophisticated computer programs modify the method to reduce round-off error. Other methods used for special types of large matrices are studied in a course on numerical analysis.

EXAMPLE 6 Soda Sales

A convenience store sells 23 sodas one summer afternoon in 12-, 16-, and 20-oz cups (small, medium, and large). The total volume of soda sold was 376 oz.

(a) Suppose that the prices for a small, medium, and large soda are $1, $1.25, and $1.40, respectively, and that the total sales were $28.45. How many of each size did the store sell?

SOLUTION   As in Example 6 of the previous section, we will organize the information in a table.

	Soda Sales			
	Small	Medium	Large	Total
Number	x	y	z	23
Volume	12	16	20	376
Price	1.00	1.25	1.40	28.45

The three rows of the table lead to three equations: one for the total number of sodas, one for the volume, and one for the price.

$$
\begin{aligned}
x + \quad y + \quad z &= \quad 23 \\
12x + \quad 16y + \quad 20z &= \quad 376 \\
1.00x + 1.25y + 1.40z &= 28.45
\end{aligned}
$$

Set this up as an augmented matrix, and verify that the result after the Gauss-Jordan method is

$$
\begin{bmatrix}
1 & 0 & 0 & 6 \\
0 & 1 & 0 & 9 \\
0 & 0 & 1 & 8
\end{bmatrix}.
$$

The store sold 6 small, 9 medium, and 8 large sodas.

(b) Suppose the prices for small, medium, and large sodas are changed to $1, $2, and $3, respectively, but all other information is kept the same. How many of each size did the store sell?

SOLUTION   Change the third equation to

$$
x + 2y + 3z = 28.45
$$

and go through the Gauss-Jordan method again. The result is

$$
\begin{bmatrix}
1 & 0 & -1 & -2 \\
0 & 1 & 2 & 25 \\
0 & 0 & 0 & -19.55
\end{bmatrix}.
$$

(If you do the row operations in a different order in this example, you will have different numbers in the last column.) The last row of this matrix says that $0 = -19.55$, so the system is inconsistent and has no solution. (In retrospect, this is clear, because each soda sells for a whole number of dollars, and the total amount of money is not a whole number of dollars. In general, however, it is not easy to tell whether a system of equations has a solution or not by just looking at it.)

(c) Suppose the prices are the same as in part (b), but the total revenue is $48. Now how many of each size did the store sell?

SOLUTION   The third equation becomes

$$
x + 2y + 3z = 48,
$$

and the Gauss-Jordan method leads to

$$
\begin{bmatrix}
1 & 0 & -1 & -2 \\
0 & 1 & 2 & 25 \\
0 & 0 & 0 & 0
\end{bmatrix}.
$$

The system is dependent, similar to Example 4. Let z be the parameter, and solve the first two equations for x and y, yielding

$$
x = z - 2 \qquad \text{and} \qquad y = 25 - 2z.
$$

Remember that in this problem, x, y, and z must be nonnegative integers. From the equation for x, we must have

$$
z \geq 2,
$$

and from the equation for y, we must have

$$
25 - 2z \geq 0,
$$

from which we find

$$z \leq 12.5.$$

Therefore, we have 11 solutions corresponding to $z = 2, 3, \ldots, 12$.

(d) Give the solutions from part (c) that have the smallest and largest numbers of large sodas.

SOLUTION   For the smallest number of large sodas, let $z = 2$, giving $x = 2 - 2 = 0$ and $y = 25 - 2(2) = 21$. There are 0 small, 21 medium, and 2 large sodas.

For the largest number of large sodas, let $z = 12$, giving $x = 12 - 2 = 10$ and $y = 25 - 2(12) = 1$. There are 10 small, 1 medium, and 12 large sodas.   ▬

2.2   EXERCISES

Write the augmented matrix for each system. Do not solve.

1. $3x + y = 6$
   $2x + 5y = 15$

2. $4x - 2y = 8$
   $-7y = -12$

3. $2x + y + z = 3$
   $3x - 4y + 2z = -7$
   $x + y + z = 2$

4. $2x - 5y + 3z = 4$
   $-4x + 2y - 7z = -5$
   $3x - y = 8$

Write the system of equations associated with each augmented matrix.

5. $\begin{bmatrix} 1 & 0 & | & 2 \\ 0 & 1 & | & 3 \end{bmatrix}$

6. $\begin{bmatrix} 1 & 0 & | & 5 \\ 0 & 1 & | & -3 \end{bmatrix}$

7. $\begin{bmatrix} 1 & 0 & 0 & | & 4 \\ 0 & 1 & 0 & | & -5 \\ 0 & 0 & 1 & | & 1 \end{bmatrix}$

8. $\begin{bmatrix} 1 & 0 & 0 & | & 4 \\ 0 & 1 & 0 & | & 2 \\ 0 & 0 & 1 & | & 3 \end{bmatrix}$

9. _____ on a matrix correspond to transformations of a system of equations.

10. Describe in your own words what $2R_1 + R_3 \to R_3$ means.

Use the indicated row operations to change each matrix.

11. Replace R_2 by $R_1 + (-3)R_2$.

$\begin{bmatrix} 3 & 7 & 4 & | & 10 \\ 1 & 2 & 3 & | & 6 \\ 0 & 4 & 5 & | & 11 \end{bmatrix}$

12. Replace R_3 by $(-1)R_1 + 3R_3$.

$\begin{bmatrix} 3 & 2 & 6 & | & 18 \\ 2 & -2 & 5 & | & 7 \\ 1 & 0 & 5 & | & 20 \end{bmatrix}$

13. Replace R_1 by $(-2)R_2 + R_1$.

$\begin{bmatrix} 1 & 6 & 4 & | & 7 \\ 0 & 3 & 2 & | & 5 \\ 0 & 5 & 3 & | & 7 \end{bmatrix}$

14. Replace R_1 by $R_3 + (-3)R_1$.

$\begin{bmatrix} 1 & 0 & 4 & | & 21 \\ 0 & 6 & 5 & | & 30 \\ 0 & 0 & 12 & | & 15 \end{bmatrix}$

15. Replace R_1 by $\frac{1}{3}R_1$.

$\begin{bmatrix} 3 & 0 & 0 & | & 18 \\ 0 & 5 & 0 & | & 9 \\ 0 & 0 & 4 & | & 8 \end{bmatrix}$

16. Replace R_3 by $\frac{1}{6}R_3$.

$\begin{bmatrix} 1 & 0 & 0 & | & 30 \\ 0 & 1 & 0 & | & 17 \\ 0 & 0 & 6 & | & 162 \end{bmatrix}$

Use the Gauss-Jordan method to solve each system of equations.

17. $x + y = 5$
   $3x + 2y = 12$

18. $x + 2y = 5$
   $2x + y = -2$

19. $x + y = 7$
   $4x + 3y = 22$

20. $4x - 2y = 3$
   $-2x + 3y = 1$

21. $2x - 3y = 2$
   $4x - 6y = 1$

22. $2x + 3y = 9$
   $4x + 6y = 7$

23. $6x - 3y = 1$
   $-12x + 6y = -2$

24. $x - y = 1$
   $-x + y = -1$

25. $y = x - 3$
   $y = 1 + z$
   $z = 4 - x$

26. $x = 1 - y$
   $2x = z$
   $2z = -2 - y$

27. $2x - 2y = -5$
   $2y + z = 0$
   $2x + z = -7$

28. $x - z = -3$
   $y + z = 9$
   $-2x + 3y + 5z = 33$

29. $4x + 4y - 4z = 24$
   $2x - y + z = -9$
   $x - 2y + 3z = 1$

30. $x + 2y - 7z = -2$
   $-2x - 5y + 2z = 1$
   $3x + 5y + 4z = -9$

31. $3x + 5y - z = 0$
   $4x - y + 2z = 1$
   $7x + 4y + z = 1$

32. $3x - 6y + 3z = 11$
   $2x + y - z = 2$
   $5x - 5y + 2z = 6$

33.
$$5x - 4y + 2z = 6$$
$$5x + 3y - z = 11$$
$$15x - 5y + 3z = 23$$

34.
$$3x + 2y - z = -16$$
$$6x - 4y + 3z = 12$$
$$5x - 2y + 2z = 4$$

35.
$$2x + 3y + z = 9$$
$$4x + 6y + 2z = 18$$
$$-\frac{1}{2}x - \frac{3}{4}y - \frac{1}{4}z = -\frac{9}{4}$$

36.
$$3x - 5y - 2z = -9$$
$$-4x + 3y + z = 11$$
$$8x - 5y + 4z = 6$$

37.
$$x + 2y - w = 3$$
$$2x + 4z + 2w = -6$$
$$x + 2y - z = 6$$
$$2x - y + z + w = -3$$

38.
$$x + 3y - 2z - w = 9$$
$$2x + 4y + 2w = 10$$
$$-3x - 5y + 2z - w = -15$$
$$x - y - 3z + 2w = 6$$

39.
$$x + y - z + 2w = -20$$
$$2x - y + z + w = 11$$
$$3x - 2y + z - 2w = 27$$

40.
$$4x - 3y + z + w = 21$$
$$-2x - y + 2z + 7w = 2$$
$$10x - 5z - 20w = 15$$

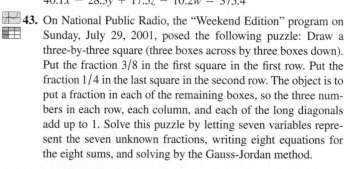

41.
$$10.47x + 3.52y + 2.58z - 6.42w = 218.65$$
$$8.62x - 4.93y - 1.75z + 2.83w = 157.03$$
$$4.92x + 6.83y - 2.97z + 2.65w = 462.3$$
$$2.86x + 19.10y - 6.24z - 8.73w = 398.4$$

42.
$$28.6x + 94.5y + 16.0z - 2.94w = 198.3$$
$$16.7x + 44.3y - 27.3z + 8.9w = 254.7$$
$$12.5x - 38.7y + 92.5z + 22.4w = 562.7$$
$$40.1x - 28.3y + 17.5z - 10.2w = 375.4$$

43. On National Public Radio, the "Weekend Edition" program on Sunday, July 29, 2001, posed the following puzzle: Draw a three-by-three square (three boxes across by three boxes down). Put the fraction 3/8 in the first square in the first row. Put the fraction 1/4 in the last square in the second row. The object is to put a fraction in each of the remaining boxes, so the three numbers in each row, each column, and each of the long diagonals add up to 1. Solve this puzzle by letting seven variables represent the seven unknown fractions, writing eight equations for the eight sums, and solving by the Gauss-Jordan method.

APPLICATIONS

Business and Economics

44. Surveys The president of Sam's Supermarkets plans to hire two public relations firms to survey 500 customers by phone, 750 by mail, and 250 by in-person interviews. The Garcia firm has personnel to do 10 phone surveys, 30 mail surveys, and 5 interviews per hour. The Wong firm can handle 20 phone surveys, 10 mail surveys, and 10 interviews per hour. For how many hours should each firm be hired to produce the exact number of surveys needed?

45. Investments Katherine Chong invests $10,000 received from her grandmother in three ways. With one part, she buys U.S. savings bonds at an interest rate of 2.5% per year. She uses the second part, which amounts to twice the first, to buy mutual funds that offer a return of 6% per year. She puts the rest of the money into a money market account paying 4.5% annual interest. The first year her investments bring a return of $470. How much did she invest in each way?

46. Office Technology Pyro-Tech, Inc. is upgrading office technology by purchasing inkjet printers, LCD monitors, and additional memory chips. The total number of pieces of hardware purchased is 46. The cost of each inkjet printer is $109, the cost of each LD monitor is $129, and the cost of each memory chip is $89. The total amount of money spent on new hardware came to $4774. They purchased two times as many memory chips as they did LCD monitors. Determine the number of each that was purchased. *Source: Nathan Borchelt.*

47. Manufacturing Fred's Furniture Factory has 1950 machine hours available each week in the cutting department, 1490 hours in the assembly department, and 2160 in the finishing department. Manufacturing a chair requires 0.2 hours of cutting, 0.3 hours of assembly, and 0.1 hours of finishing. A cabinet requires 0.5 hours of cutting, 0.4 hours of assembly, and 0.6 hours of finishing. A buffet requires 0.3 hours of cutting, 0.1 hours of assembly, and 0.4 hours of finishing. How many chairs, cabinets, and buffets should be produced in order to use all the available production capacity?

48. Manufacturing Nadir, Inc. produces three models of television sets: deluxe, super-deluxe, and ultra. Each deluxe set requires 2 hours of electronics work, 3 hours of assembly time, and 5 hours of finishing time. Each super-deluxe requires 1, 3, and 2 hours of electronics, assembly, and finishing time, respectively. Each ultra requires 2, 2, and 6 hours of the same work, respectively.

a. There are 54 hours available for electronics, 72 hours available for assembly, and 148 hours available for finishing per week. How many of each model should be produced each week if all available time is to be used?

b. Suppose everything is the same as in part a, but a super-deluxe set requires 1, rather than 2, hours of finishing time. How many solutions are there now?

c. Suppose everything is the same as in part b, but the total hours available for finishing changes from 148 hours to 144 hours. Now how many solutions are there?

49. Transportation An electronics company produces three models of stereo speakers, models A, B, and C, and can deliver them by truck, van, or SUV. A truck holds 2 boxes of model A, 2 of model B, and 3 of model C. A van holds 3 boxes of model A, 4 boxes of model B, and 2 boxes of model C. An SUV holds 3 boxes of model A, 5 boxes of model B, and 1 box of model C.

a. If 25 boxes of model A, 33 boxes of model B, and 22 boxes of model C are to be delivered, how many vehicles of each type should be used so that all operate at full capacity?

b. Model C has been discontinued. If 25 boxes of model A and 33 boxes of model B are to be delivered, how many vehicles of each type should be used so that all operate at full capacity?

50. **Truck Rental** The U-Drive Rent-A-Truck company plans to spend $7 million on 200 new vehicles. Each commercial van will cost $35,000, each small truck $30,000, and each large truck $50,000. Past experience shows that they need twice as many vans as small trucks. How many of each type of vehicle can they buy?

51. **Loans** To get the necessary funds for a planned expansion, a small company took out three loans totaling $25,000. Company owners were able to get interest rates of 8%, 9%, and 10%. They borrowed $1000 more at 9% than they borrowed at 10%. The total annual interest on the loans was $2190.

a. How much did they borrow at each rate?

b. Suppose we drop the condition that they borrowed $1000 more at 9% than at 10%. What can you say about the amount borrowed at 10%? What is the solution if the amount borrowed at 10% is $5000?

c. Suppose the bank sets a maximum of $10,000 at the lowest interest rate of 8%. Is a solution possible that still meets all of the original conditions?

d. Explain why $10,000 at 8%, $8000 at 9%, and $7000 at 10% is not a feasible solution for part c.

52. **APPLY IT** **Transportation** An auto manufacturer sends cars from two plants, I and II, to dealerships A and B located in a midwestern city. Plant I has a total of 28 cars to send, and plant II has 8. Dealer A needs 20 cars, and dealer B needs 16. Transportation costs per car, based on the distance of each dealership from each plant, are $220 from I to A, $300 from I to B, $400 from II to A, and $180 from II to B. The manufacturer wants to limit transportation costs to $10,640. How many cars should be sent from each plant to each of the two dealerships?

53. **Transportation** A manufacturer purchases a part for use at both of its plants — one at Roseville, California, the other at Akron, Ohio. The part is available in limited quantities from two suppliers. Each supplier has 75 units available. The Roseville plant needs 40 units, and the Akron plant requires 75 units. The first supplier charges $70 per unit delivered to Roseville and $90 per unit delivered to Akron. Corresponding costs from the second supplier are $80 and $120. The manufacturer wants to order a total of 75 units from the first, less expensive supplier, with the remaining 40 units to come from the second supplier. If the company spends $10,750 to purchase the required number of units for the two plants, find the number of units that should be sent from each supplier to each plant.

54. **Packaging** A company produces three combinations of mixed vegetables that sell in 1-kg packages. Italian style combines 0.3 kg of zucchini, 0.3 of broccoli, and 0.4 of carrots. French style combines 0.6 kg of broccoli and 0.4 of carrots. Oriental style combines 0.2 kg of zucchini, 0.5 of broccoli, and 0.3 of carrots. The company has a stock of 16,200 kg of zucchini, 41,400 kg of broccoli, and 29,400 kg of carrots. How many packages of each style should it prepare to use up existing supplies?

55. **Tents** L.L. Bean makes three sizes of Ultra Dome tents: two-person, four-person, and six-person models, which cost $129, $179, and $229, respectively. A two-person tent provides 40 ft^2 of floor space, while a four-person and a six-person model provide 64 ft^2 and 88 ft^2 of floor space, respectively. *Source: L. L. Bean.* A recent order by an organization that takes children camping ordered enough tents to hold 200 people and provide 3200 ft^2 of floor space. The total cost was $8950, and we wish to know how many tents of each size were ordered.

a. How many solutions are there to this problem?

b. What is the solution with the most four-person tents?

c. What is the solution with the most two-person tents?

d. Discuss the company's pricing strategy that led to a system of equations that is dependent. Do you think that this is a coincidence or an example of logical thinking?

Life Sciences

56. **Animal Breeding** An animal breeder can buy four types of food for Vietnamese pot-bellied pigs. Each case of Brand A contains 25 units of fiber, 30 units of protein, and 30 units of fat. Each case of Brand B contains 50 units of fiber, 30 units of protein, and 20 units of fat. Each case of Brand C contains 75 units of fiber, 30 units of protein, and 20 units of fat. Each case of Brand D contains 100 units of fiber, 60 units of protein, and 30 units of fat. How many cases of each should the breeder mix together to obtain a food that provides 1200 units of fiber, 600 units of protein, and 400 units of fat?

57. **Dietetics** A hospital dietician is planning a special diet for a certain patient. The total amount per meal of food groups A, B, and C must equal 400 grams. The diet should include one-third as much of group A as of group B, and the sum of the amounts of group A and group C should equal twice the amount of group B.

a. How many grams of each food group should be included?

b. Suppose we drop the requirement that the diet include one-third as much of group A as of group B. Describe the set of all possible solutions.

c. Suppose that, in addition to the conditions given in the original problem, foods A and B cost 2 cents per gram and food C costs 3 cents per gram, and that a meal must cost $8. Is a solution possible?

58. **Bacterial Food Requirements** Three species of bacteria are fed three foods, I, II, and III. A bacterium of the first species consumes 1.3 units each of foods I and II and 2.3 units of food III each day. A bacterium of the second species consumes 1.1 units of food I, 2.4 units of food II, and 3.7 units of food III each day. A bacterium of the third species consumes 8.1 units of I, 2.9 units of II, and 5.1 units of III each day. If 16,000 units of I, 28,000 units of II, and 44,000 units of III are supplied each day, how many of each species can be maintained in this environment?

59. **Fish Food Requirements** A lake is stocked each spring with three species of fish, A, B, and C. Three foods, I, II, and III, are available in the lake. Each fish of species A requires an average of 1.32 units of food I, 2.9 units of food II, and 1.75 units of food III each day. Species B fish each require 2.1 units of food I, 0.95 unit of food II, and 0.6 unit of food III daily. Species C fish require 0.86, 1.52, and 2.01 units of I, II, and III per day, respectively. If 490 units of food I, 897 units of food II, and

653 units of food III are available daily, how many of each species should be stocked?

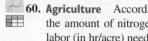

60. Agriculture According to data from a Texas agricultural report, the amount of nitrogen (in lb/acre), phosphate (in lb/acre), and labor (in hr/acre) needed to grow honeydews, yellow onions, and lettuce is given by the following table. *Source: The AMATYC Review.*

	Honeydews	Yellow Onions	Lettuce
Nitrogen	120	150	180
Phosphate	180	80	80
Labor	4.97	4.45	4.65

a. If the farmer has 220 acres, 29,100 lb of nitrogen, 32,600 lb of phosphate, and 480 hours of labor, is it possible to use all resources completely? If so, how many acres should he allot for each crop?

b. Suppose everything is the same as in part a, except that 1061 hours of labor are available. Is it possible to use all resources completely? If so, how many acres should he allot for each crop?

61. Archimedes' Problem Bovinum Archimedes is credited with the authorship of a famous problem involving the number of cattle of the sun god. A simplified version of the problem is stated as follows:

> The sun god had a herd of cattle consisting of bulls and cows, one part of which was white, a second black, a third spotted, and a fourth brown.
>
> Among the bulls, the number of white ones was one-half plus one-third the number of the black greater than the brown; the number of the black, one-quarter plus one-fifth the number of the spotted greater than the brown; the number of the spotted, one-sixth and one-seventh the number of the white greater than the brown.
>
> Among the cows, the number of white ones was one-third plus one-quarter of the total black cattle; the number of the black, one-quarter plus one-fifth the total of the spotted cattle; the number of the spotted, one-fifth plus one-sixth the total of the brown cattle; the number of the brown, one-sixth plus one-seventh the total of the white cattle.
>
> What was the composition of the herd?

Source: 100 Great Problems of Elementary Mathematics.
The problem can be solved by converting the statements into two systems of equations, using X, Y, Z, and T for the number of white, black, spotted, and brown bulls, respectively, and x, y, z, and t for the number of white, black, spotted, and brown cows, respectively. For example, the first statement can be written as $X = (1/2 + 1/3)Y + T$ and then reduced. The result is the following two systems of equations:

$$6X - 5Y = 6T \qquad\qquad 12x - 7y = 7Y$$
$$20Y - 9Z = 20T \quad \text{and} \quad 20y - 9z = 9Z$$
$$42Z - 13X = 42T \qquad\qquad 30z - 11t = 11T$$
$$\qquad\qquad\qquad\qquad -13x + 42t = 13X$$

a. Show that these two systems of equations represent Archimedes' Problem Bovinum.

b. If it is known that the number of brown bulls, T, is 4,149,387, use the Gauss-Jordan method to first find a solution to the 3×3 system and then use these values and the Gauss-Jordan method to find a solution to the 4×4 system of equations.

62. Health The U.S. National Center for Health Statistics tracks the major causes of death in the United States. After a steady increase, the death rate by cancer has decreased since the early 1990s. The table lists the age-adjusted death rate per 100,000 people for 4 years. *Source: The New York Times 2010 Almanac.*

Year	Rate
1980	183.9
1990	203.3
2000	196.5
2006	187.0

a. If the relationship between the death rate R and the year t is expressed as $R = at^2 + bt + c$, where $t = 0$ corresponds to 1980, use data from 1980, 1990, and 2000 and a linear system of equations to determine the constants a, b, and c.

b. Use the equation from part a to predict the rate in 2006, and compare the result with the actual data.

c. If the relationship between the death rate R and the year t is expressed as $R = at^3 + bt^2 + ct + d$, where $t = 0$ corresponds to 1980, use all four data points and a linear system of equations to determine the constants a, b, c, and d.

d. Discuss the appropriateness of the functions used in parts a and c to model this data.

Social Sciences

63. Modeling War One of the factors that contribute to the success or failure of a particular army during war is its ability to get new troops ready for service. It is possible to analyze the rate of change in the number of troops of two hypothetical armies with the following simplified model,

Rate of increase (RED ARMY) $= 200{,}000 - 0.5r - 0.3b$

Rate of increase (BLUE ARMY) $= 350{,}000 - 0.5r - 0.7b$,

where r is the number of soldiers in the Red Army at a given time and b is the number of soldiers in the Blue Army at a given time. The factors 0.5 and 0.7 represent each army's efficiency of bringing new soldiers to the fight. *Source: Journal of Peace Research.*

a. Solve this system of equations to determine the number of soldiers in each army when the rate of increase for each is zero.

b. Describe what might be going on in a war when the rate of increase is zero.

64. Traffic Control At rush hours, substantial traffic congestion is encountered at the traffic intersections shown in the figure on the next page. (The streets are one-way, as shown by the arrows.)

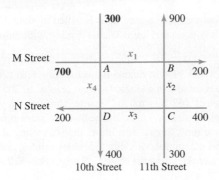

The city wishes to improve the signals at these corners so as to speed the flow of traffic. The traffic engineers first gather data. As the figure shows, 700 cars per hour come down M Street to intersection A, and 300 cars per hour come down 10th Street to intersection A. A total of x_1 of these cars leave A on M Street, and x_4 cars leave A on 10th Street. The number of cars entering A must equal the number leaving, so that

$$x_1 + x_4 = 700 + 300$$

or

$$x_1 + x_4 = 1000.$$

For intersection B, x_1 cars enter on M Street and x_2 on 11th Street. The figure shows that $\overline{900}$ cars leave B on 11th and 200 on M. Thus,

$$x_1 + x_2 = 900 + 200$$
$$x_1 + x_2 = 1100.$$

a. Write two equations representing the traffic entering and leaving intersections C and D.

b. Use the four equations to set up an augmented matrix, and solve the system by the Gauss-Jordan method, using x_4 as the parameter.

c. Based on your solution to part b, what are the largest and smallest possible values for the number of cars leaving intersection A on 10th Street?

d. Answer the question in part c for the other three variables.

e. Verify that you could have discarded any one of the four original equations without changing the solution. What does this tell you about the original problem?

General Interest

65. Snack Food According to the nutrition labels on the package, a single serving of Oreos® has 10 g of fat, 36 g of carbohydrates, 2 g of protein, and 240 calories. The figures for a single serving of Twix® are 14 g, 37 g, 3 g, and 280 calories. The figures for a single serving of trail mix are 20 g, 23 g, 11 g, and 295 calories. How many calories are in each gram of fat, carbohydrate, and protein? *Source: The Mathematics Teacher.**

*For a discussion of some complications that can come up in solving linear systems, read the original article: Szydlik, Stephen D., "The Problem with the Snack Food Problem," *The Mathematics Teacher,* Vol. 103, No. 1, Aug. 2009, p. 18–28.

66. Basketball Kobe Bryant has the second highest single game point total in the NBA. Bryant scored 81 points for the Los Angeles Lakers on January 22, 2006, against the Toronto Raptors. Bryant made a total of 46 baskets, including foul shots (worth one point), field goals (worth two points), and three-point shots (worth three points). The number of field goal shots he made is equal to three times the number of three pointers he made. Find the number of foul shots, field goals, and three pointers Bryant made. *Source: ESPN.*

67. Toys One hundred toys are to be given out to a group of children. A ball costs $2, a doll costs $3, and a car costs $4. A total of $295 was spent on the toys.

a. A ball weighs 12 oz, a doll 16 oz, and a car 18 oz. The total weight of all the toys is 1542 oz. Find how many of each toy there are.

b. Now suppose the weight of a ball, doll, and car are 11, 15, and 19 oz, respectively. If the total weight is still 1542 oz, how many solutions are there now?

c. Keep the weights as in part b, but change the total weight to 1480 oz. How many solutions are there?

d. Give the solution to part c that has the smallest number of cars.

e. Give the solution to part c that has the largest number of cars.

68. Ice Cream Researchers have determined that the amount of sugar contained in ice cream helps to determine the overall "degree of like" that a consumer has toward that particular flavor. They have also determined that too much or too little sugar will have the same negative affect on the "degree of like" and that this relationship follows a quadratic function. In an experiment conducted at Pennsylvania State University, the following condensed table was obtained. *Source: Journal of Food Science.*

Percentage of Sugar	Degree of Like
8	5.4
13	6.3
18	5.6

a. Use this information and the Gauss-Jordan method to determine the coefficients $a, b, c,$ of the quadratic equation

$$y = ax^2 + bx + c,$$

where y is the "degree of like" and x is the percentage of sugar in the ice cream mix.

b. Repeat part a by using the quadratic regression feature on a graphing calculator. Compare your answers.

69. Lights Out The Tiger Electronics' game, Lights Out, consists of five rows of five lighted buttons. When a button is pushed, it changes the on/off status of it and the status of all of its vertical and horizontal neighbors. For any given situation where some of the lights are on and some are off, the goal of the game is to push buttons until all of the lights are turned off. It turns out that for any given array of lights, solving a system of equations can be used to develop a strategy for turning the lights out. The following system of equations can be used to solve the problem for a simplified version of the game with 2 rows of 2 buttons where all of the lights are initially turned on:

$$x_{11} + x_{12} + x_{21} = 1$$
$$x_{11} + x_{12} + x_{22} = 1$$
$$x_{11} + x_{21} + x_{22} = 1$$
$$x_{12} + x_{21} + x_{22} = 1,$$

where $x_{ij} = 1$ if the light in row i, column j, is on and $x_{ij} = 0$ when it is off. The order in which the buttons are pushed does not matter, so we are only seeking which buttons should be pushed. *Source: Mathematics Magazine.*

a. Solve this system of equations and determine a strategy to turn the lights out. (*Hint:* While doing row operations, if an odd number is found, immediately replace this value with a 1; if an even number is found, then immediately replace that number with a zero. This is called modulo 2 arithmetic, and it is necessary in problems dealing with on/off switches.)

b. Resolve the equation with the right side changed to (0, 1, 1, 0).

70. Baseball   Ichiro Suzuki holds the American League record for the most hits in a single baseball season. In 2004, Suzuki had a total of 262 hits for the Seattle Mariners. He hit three fewer triples than home runs, and he hit three times as many doubles as home runs. Suzuki also hit 45 times as many singles as triples. Find the number of singles, doubles, triples, and home runs hit by Suzuki during the season. *Source: Baseball Almanac.*

YOUR TURN ANSWERS

1. $(5, -2)$ 2. $(7, 2, -3)$
3. $(17/9 + w/3, -4/9 - 2w/3, 4/9 + 2w/3, w)$

2.3 Addition and Subtraction of Matrices

APPLY IT   A company sends monthly shipments to its warehouses in several cities. How might the company keep track of the shipments to each warehouse most efficiently?

In the previous section, matrices were used to store information about systems of linear equations. In this section, we begin to study calculations with matrices, which we will use in Examples 1, 5, and 7 to answer the question posed above.

The use of matrices has gained increasing importance in the fields of management, natural science, and social science because matrices provide a convenient way to organize data, as Example 1 demonstrates.

EXAMPLE 1   **Furniture Shipments**

The EZ Life Company manufactures sofas and armchairs in three models, A, B, and C. The company has regional warehouses in New York, Chicago, and San Francisco. In its August shipment, the company sends 10 model-A sofas, 12 model-B sofas, 5 model-C sofas, 15 model-A chairs, 20 model-B chairs, and 8 model-C chairs to each warehouse. Use a matrix to organize this information.

APPLY IT   **SOLUTION**   To organize this data, we might tabulate the data in a chart.

Number of Furniture Pieces				
		Model		
		A	B	C
Furniture Type	Sofas	10	12	5
	Chairs	15	20	8

With the understanding that the numbers in each row refer to the furniture type (sofa, chair) and the numbers in each column refer to the model (A, B, C), the same information can be given by a matrix, as follows.

$$M = \begin{bmatrix} 10 & 12 & 5 \\ 15 & 20 & 8 \end{bmatrix}$$

Matrices often are named with capital letters, as in Example 1. Matrices are classified by **size**; that is, by the number of rows and columns they contain. For example, matrix M above has two rows and three columns. This matrix is a 2×3 (read "2 by 3") matrix. By definition, a matrix with m rows and n columns is an $m \times n$ matrix. The number of rows is always given first.

EXAMPLE 2   Matrix Size

(a) The matrix $\begin{bmatrix} -3 & 5 \\ 2 & 0 \\ 5 & -1 \end{bmatrix}$ is a 3×2 matrix.

(b) $\begin{bmatrix} 0.5 & 8 & 0.9 \\ 0 & 5.1 & -3 \\ -4 & 0 & 5 \end{bmatrix}$ is a 3×3 matrix.

(c) $\begin{bmatrix} 1 & 6 & 5 & -2 & 5 \end{bmatrix}$ is a 1×5 matrix.

(d) $\begin{bmatrix} 3 \\ -5 \\ 0 \\ 2 \end{bmatrix}$ is a 4×1 matrix.

A matrix with the same number of rows as columns is called a **square matrix**. The matrix in Example 2(b) is a square matrix.

A matrix containing only one row is called a **row matrix** or a **row vector**. The matrix in Example 2(c) is a row matrix, as are

$$\begin{bmatrix} 5 & 8 \end{bmatrix}, \quad \begin{bmatrix} 6 & -9 & 2 \end{bmatrix}, \quad \text{and} \quad \begin{bmatrix} -4 & 0 & 0 & 0 \end{bmatrix}.$$

A matrix of only one column, as in Example 2(d), is a **column matrix** or a **column vector**.

Equality for matrices is defined as follows.

Matrix Equality
Two matrices are equal if they are the same size and if each pair of corresponding elements is equal.

By this definition,

$$\begin{bmatrix} 2 & 1 \\ 3 & -5 \end{bmatrix} \quad \text{and} \quad \begin{bmatrix} 1 & 2 \\ -5 & 3 \end{bmatrix}$$

are not equal (even though they contain the same elements and are the same size) since the corresponding elements differ.

EXAMPLE 3   Matrix Equality

(a) From the definition of matrix equality given above, the only way that the statement

$$\begin{bmatrix} 2 & 1 \\ p & q \end{bmatrix} = \begin{bmatrix} x & y \\ -1 & 0 \end{bmatrix}$$

can be true is if $2 = x$, $1 = y$, $p = -1$, and $q = 0$.

(b) The statement

$$\begin{bmatrix} x \\ y \end{bmatrix} = \begin{bmatrix} 1 \\ -3 \\ 0 \end{bmatrix}$$

can never be true, since the two matrices are different sizes. (One is 2×1 and the other is 3×1.)

Addition

The matrix given in Example 1,

$$M = \begin{bmatrix} 10 & 12 & 5 \\ 15 & 20 & 8 \end{bmatrix},$$

shows the August shipment from the EZ Life plant to each of its warehouses. If matrix N below gives the September shipment to the New York warehouse, what is the total shipment of each item of furniture to the New York warehouse for these two months?

$$N = \begin{bmatrix} 45 & 35 & 20 \\ 65 & 40 & 35 \end{bmatrix}$$

If 10 model-A sofas were shipped in August and 45 in September, then altogether $10 + 45 = 55$ model-A sofas were shipped in the two months. The other corresponding entries can be added in a similar way to get a new matrix Q, which represents the total shipment for the two months.

$$Q = \begin{bmatrix} 55 & 47 & 25 \\ 80 & 60 & 43 \end{bmatrix}$$

It is convenient to refer to Q as the sum of M and N.

The way these two matrices were added illustrates the following definition of addition of matrices.

Adding Matrices

The sum of two $m \times n$ matrices X and Y is the $m \times n$ matrix $X + Y$ in which each element is the sum of the corresponding elements of X and Y.

CAUTION It is important to remember that only matrices that are the same size can be added.

EXAMPLE 4 Adding Matrices

Find each sum, if possible.

SOLUTION

YOUR TURN 1 Find each sum, if possible.

(a) $\begin{bmatrix} 3 & 4 & 5 & 6 \\ 1 & 2 & 3 & 4 \end{bmatrix} + \begin{bmatrix} 1 & -2 & 4 \\ -2 & -4 & 8 \end{bmatrix}$

(b) $\begin{bmatrix} 3 & 4 & 5 \\ 1 & 2 & 3 \end{bmatrix} + \begin{bmatrix} 1 & -2 & 4 \\ -2 & -4 & 8 \end{bmatrix}$

(a) $\begin{bmatrix} 5 & -6 \\ 8 & 9 \end{bmatrix} + \begin{bmatrix} -4 & 6 \\ 8 & -3 \end{bmatrix} = \begin{bmatrix} 5 + (-4) & -6 + 6 \\ 8 + 8 & 9 + (-3) \end{bmatrix} = \begin{bmatrix} 1 & 0 \\ 16 & 6 \end{bmatrix}$

(b) The matrices

$$A = \begin{bmatrix} 5 & -8 \\ 6 & 2 \end{bmatrix} \quad \text{and} \quad B = \begin{bmatrix} 3 & -9 & 1 \\ 4 & 2 & -5 \end{bmatrix}$$

are different sizes. Therefore, the sum $A + B$ does not exist.   **TRY YOUR TURN 1**

EXAMPLE 5 Furniture Shipments

The September shipments from the EZ Life Company to the New York, San Francisco, and Chicago warehouses are given in matrices N, S, and C below.

$$N = \begin{bmatrix} 45 & 35 & 20 \\ 65 & 40 & 35 \end{bmatrix} \quad S = \begin{bmatrix} 30 & 32 & 28 \\ 43 & 47 & 30 \end{bmatrix} \quad C = \begin{bmatrix} 22 & 25 & 38 \\ 31 & 34 & 35 \end{bmatrix}$$

What was the total amount shipped to the three warehouses in September?

APPLY IT

SOLUTION   The total of the September shipments is represented by the sum of the three matrices N, S, and C.

$$N + S + C = \begin{bmatrix} 45 & 35 & 20 \\ 65 & 40 & 35 \end{bmatrix} + \begin{bmatrix} 30 & 32 & 28 \\ 43 & 47 & 30 \end{bmatrix} + \begin{bmatrix} 22 & 25 & 38 \\ 31 & 34 & 35 \end{bmatrix}$$

$$= \begin{bmatrix} 97 & 92 & 86 \\ 139 & 121 & 100 \end{bmatrix}$$

For example, this sum shows that the total number of model-C sofas shipped to the three warehouses in September was 86.

Subtraction   Subtraction of matrices is defined similarly to addition.

Subtracting Matrices
The difference of two $m \times n$ matrices X and Y is the $m \times n$ matrix $X - Y$ in which each element is found by subtracting the corresponding elements of X and Y.

EXAMPLE 6   Subtracting Matrices

Subtract each pair of matrices, if possible.

SOLUTION

(a) $\begin{bmatrix} 8 & 6 & -4 \\ -2 & 7 & 5 \end{bmatrix} - \begin{bmatrix} 3 & 5 & -8 \\ -4 & 2 & 9 \end{bmatrix} = \begin{bmatrix} 8-3 & 6-5 & -4-(-8) \\ -2-(-4) & 7-2 & 5-9 \end{bmatrix} = \begin{bmatrix} 5 & 1 & 4 \\ 2 & 5 & -4 \end{bmatrix}.$

(b) The matrices

$$\begin{bmatrix} -2 & 5 \\ 0 & 1 \end{bmatrix} \quad \text{and} \quad \begin{bmatrix} 3 \\ 5 \end{bmatrix}$$

YOUR TURN 2   Calculate

$\begin{bmatrix} 3 & 4 & 5 \\ 1 & 2 & 3 \end{bmatrix} - \begin{bmatrix} 1 & -2 & 4 \\ -2 & -4 & 8 \end{bmatrix}.$

are different sizes and cannot be subtracted.   **TRY YOUR TURN 2**

EXAMPLE 7   Furniture Shipments

APPLY IT   During September the Chicago warehouse of the EZ Life Company shipped out the following numbers of each model.

$$K = \begin{bmatrix} 5 & 10 & 8 \\ 11 & 14 & 15 \end{bmatrix}$$

What was the Chicago warehouse inventory on October 1, taking into account only the number of items received and sent out during the month?

Method 1
Calculating by Hand

SOLUTION   The number of each kind of item received during September is given by matrix C from Example 5; the number of each model sent out during September is given by matrix K. The October 1 inventory will be represented by the matrix $C - K$:

[A]−[B]
$\begin{bmatrix} 17 & 15 & 30 \\ 20 & 20 & 20 \end{bmatrix}$

$$\begin{bmatrix} 22 & 25 & 38 \\ 31 & 34 & 35 \end{bmatrix} - \begin{bmatrix} 5 & 10 & 8 \\ 11 & 14 & 15 \end{bmatrix} = \begin{bmatrix} 17 & 15 & 30 \\ 20 & 20 & 20 \end{bmatrix}.$$

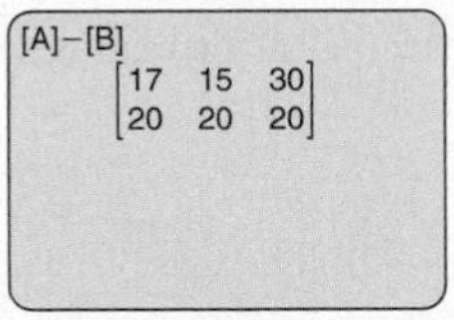

FIGURE 12

Method 2
Graphing Calculator

Matrix operations are easily performed on a graphing calculator. Figure 12 shows the previous operation; the matrices A and B were already entered into the calculator.

TECHNOLOGY NOTE   Spreadsheet programs are designed to effectively organize data that can be represented in rows and columns. Accordingly, matrix operations are also easily performed on spreadsheets. See the *Graphing Calculator and Excel Spreadsheet Manual* available with this book for details.

2.3 EXERCISES

Decide whether each statement is true or false. If false, tell why.

1. $\begin{bmatrix} 1 & 3 \\ 5 & 7 \end{bmatrix} = \begin{bmatrix} 1 & 5 \\ 3 & 7 \end{bmatrix}$

2. $\begin{bmatrix} 1 \\ 2 \\ 3 \end{bmatrix} = \begin{bmatrix} 1 & 2 & 3 \end{bmatrix}$

3. $\begin{bmatrix} x \\ y \end{bmatrix} = \begin{bmatrix} -2 \\ 8 \end{bmatrix}$ if $x = -2$ and $y = 8$.

4. $\begin{bmatrix} 3 & 5 & 2 & 8 \\ 1 & -1 & 4 & 0 \end{bmatrix}$ is a 4×2 matrix.

5. $\begin{bmatrix} 1 & 9 & -4 \\ 3 & 7 & 2 \\ -1 & 1 & 0 \end{bmatrix}$ is a square matrix.

6. $\begin{bmatrix} 2 & 4 & -1 \\ 3 & 7 & 5 \\ 0 & 0 & 0 \end{bmatrix} = \begin{bmatrix} 2 & 4 & -1 \\ 3 & 7 & 5 \end{bmatrix}$

Find the size of each matrix. Identify any square, column, or row matrices.

7. $\begin{bmatrix} -4 & 8 \\ 2 & 3 \end{bmatrix}$

8. $\begin{bmatrix} 2 & -3 & 7 \\ 1 & 0 & 4 \end{bmatrix}$

9. $\begin{bmatrix} -6 & 8 & 0 & 0 \\ 4 & 1 & 9 & 2 \\ 3 & -5 & 7 & 1 \end{bmatrix}$

10. $\begin{bmatrix} 8 & -2 & 4 & 6 & 3 \end{bmatrix}$

11. $\begin{bmatrix} -7 \\ 5 \end{bmatrix}$

12. $\begin{bmatrix} -9 \end{bmatrix}$

13. The sum of an $n \times m$ matrix and an $m \times n$ matrix, where $m \neq n$, is _____.

14. If A is a 5×2 matrix and $A + K = A$, what do you know about K?

Find the values of the variables in each equation.

15. $\begin{bmatrix} 3 & 4 \\ -8 & 1 \end{bmatrix} = \begin{bmatrix} 3 & x \\ y & z \end{bmatrix}$

16. $\begin{bmatrix} -5 \\ y \end{bmatrix} = \begin{bmatrix} -5 \\ 8 \end{bmatrix}$

17. $\begin{bmatrix} s - 4 & t + 2 \\ -5 & 7 \end{bmatrix} = \begin{bmatrix} 6 & 2 \\ -5 & r \end{bmatrix}$

18. $\begin{bmatrix} 9 & 7 \\ r & 0 \end{bmatrix} = \begin{bmatrix} m - 3 & n + 5 \\ 8 & 0 \end{bmatrix}$

19. $\begin{bmatrix} a + 2 & 3b & 4c \\ d & 7f & 8 \end{bmatrix} + \begin{bmatrix} -7 & 2b & 6 \\ -3d & -6 & -2 \end{bmatrix} = \begin{bmatrix} 15 & 25 & 6 \\ -8 & 1 & 6 \end{bmatrix}$

20. $\begin{bmatrix} a + 2 & 3z + 1 & 5m \\ 4k & 0 & 3 \end{bmatrix} + \begin{bmatrix} 3a & 2z & 5m \\ 2k & 5 & 6 \end{bmatrix} = \begin{bmatrix} 10 & -14 & 80 \\ 10 & 5 & 9 \end{bmatrix}$

Perform the indicated operations, where possible.

21. $\begin{bmatrix} 2 & 4 & 5 & -7 \\ 6 & -3 & 12 & 0 \end{bmatrix} + \begin{bmatrix} 8 & 0 & -10 & 1 \\ -2 & 8 & -9 & 11 \end{bmatrix}$

22. $\begin{bmatrix} 1 & 5 \\ 2 & -3 \\ 3 & 7 \end{bmatrix} + \begin{bmatrix} 2 & 3 \\ 8 & 5 \\ -1 & 9 \end{bmatrix}$

23. $\begin{bmatrix} 1 & 3 & -2 \\ 4 & 7 & 1 \end{bmatrix} + \begin{bmatrix} 3 & 0 \\ 6 & 4 \\ -5 & 2 \end{bmatrix}$

24. $\begin{bmatrix} 8 & 0 & -3 \\ 1 & 19 & -5 \end{bmatrix} - \begin{bmatrix} 1 & -5 & 2 \\ 3 & 9 & -8 \end{bmatrix}$

25. $\begin{bmatrix} 2 & 8 & 12 & 0 \\ 7 & 4 & -1 & 5 \\ 1 & 2 & 0 & 10 \end{bmatrix} - \begin{bmatrix} 1 & 3 & 6 & 9 \\ 2 & -3 & -3 & 4 \\ 8 & 0 & -2 & 17 \end{bmatrix}$

26. $\begin{bmatrix} 2 & 1 \\ 5 & -3 \\ -7 & 2 \\ 9 & 0 \end{bmatrix} + \begin{bmatrix} 1 & -8 & 0 \\ 5 & 3 & 2 \\ -6 & 7 & -5 \\ 2 & -1 & 0 \end{bmatrix}$

27. $\begin{bmatrix} 2 & 3 \\ -2 & 4 \end{bmatrix} + \begin{bmatrix} 4 & 3 \\ 7 & 8 \end{bmatrix} - \begin{bmatrix} 3 & 2 \\ 1 & 4 \end{bmatrix}$

28. $\begin{bmatrix} 4 & 3 \\ 1 & 2 \end{bmatrix} - \begin{bmatrix} 1 & 1 \\ 1 & 0 \end{bmatrix} + \begin{bmatrix} 1 & 1 \\ 1 & 4 \end{bmatrix}$

29. $\begin{bmatrix} 2 & -1 \\ 0 & 13 \end{bmatrix} - \begin{bmatrix} 4 & 8 \\ -5 & 7 \end{bmatrix} + \begin{bmatrix} 12 & 7 \\ 5 & 3 \end{bmatrix}$

30. $\begin{bmatrix} 5 & 8 \\ -3 & 1 \end{bmatrix} + \begin{bmatrix} 0 & 1 \\ -2 & -2 \end{bmatrix} + \begin{bmatrix} -5 & -8 \\ 6 & 1 \end{bmatrix}$

31. $\begin{bmatrix} -4x + 2y & -3x + y \\ 6x - 3y & 2x - 5y \end{bmatrix} + \begin{bmatrix} -8x + 6y & 2x \\ 3y - 5x & 6x + 4y \end{bmatrix}$

32. $\begin{bmatrix} 4k - 8y \\ 6z - 3x \\ 2k + 5a \\ -4m + 2n \end{bmatrix} - \begin{bmatrix} 5k + 6y \\ 2z + 5x \\ 4k + 6a \\ 4m - 2n \end{bmatrix}$

33. For matrices $X = \begin{bmatrix} x & y \\ z & w \end{bmatrix}$ and $0 = \begin{bmatrix} 0 & 0 \\ 0 & 0 \end{bmatrix}$, find the matrix $0 - X$.

Using matrices $O = \begin{bmatrix} 0 & 0 \\ 0 & 0 \end{bmatrix}$, $P = \begin{bmatrix} m & n \\ p & q \end{bmatrix}$, $T = \begin{bmatrix} r & s \\ t & u \end{bmatrix}$, **and** $X = \begin{bmatrix} x & y \\ z & w \end{bmatrix}$, **verify the statements in Exercises 34–37.**

34. $X + T = T + X$ (commutative property of addition of matrices)

35. $X + (T + P) = (X + T) + P$ (associative property of addition of matrices)

36. $X - X = O$ (inverse property of addition of matrices)

37. $P + O = P$ (identity property of addition of matrices)

38. Which of the above properties are valid for matrices that are not square?

APPLICATIONS

Business and Economics

39. Management A toy company has plants in Boston, Chicago, and Seattle that manufacture toy phones and calculators. The following matrix gives the production costs (in dollars) for each item at the Boston plant:

	Phones	Calculators
Material	4.27	6.94
Labor	3.45	3.65

a. In Chicago, a phone costs $4.05 for material and $3.27 for labor; a calculator costs $7.01 for material and $3.51 for labor. In Seattle, material costs are $4.40 for a phone and $6.90 for a calculator; labor costs are $3.54 for a phone and $3.76 for a calculator. Write the production cost matrices for Chicago and Seattle.

b. Suppose labor costs increase by $0.11 per item in Chicago and material costs there increase by $0.37 for a phone and $0.42 for a calculator. What is the new production cost matrix for Chicago?

40. Management There are three convenience stores in Folsom. This week, store I sold 88 loaves of bread, 48 qt of milk, 16 jars of peanut butter, and 112 lb of cold cuts. Store II sold 105 loaves of bread, 72 qt of milk, 21 jars of peanut butter, and 147 lb of cold cuts. Store III sold 60 loaves of bread, 40 qt of milk, no peanut butter, and 50 lb of cold cuts.

a. Use a 4×3 matrix to express the sales information for the three stores.

b. During the following week, sales on these products at store I increased by 25%; sales at store II increased by 1/3; and sales at store III increased by 10%. Write the sales matrix for that week.

c. Write a matrix that represents total sales over the two-week period.

Life Sciences

41. Dietetics A dietician prepares a diet specifying the amounts a patient should eat of four basic food groups: group I, meats; group II, fruits and vegetables; group III, breads and starches; group IV, milk products. Amounts are given in "exchanges" that represent 1 oz (meat), 1/2 cup (fruits and vegetables), 1 slice (bread), 8 oz (milk), or other suitable measurements.

a. The number of "exchanges" for breakfast for each of the four food groups, respectively, are 2, 1, 2, and 1; for lunch, 3, 2, 2, and 1; and for dinner, 4, 3, 2, and 1. Write a 3×4 matrix using this information.

b. The amounts of fat, carbohydrates, and protein (in appropriate units) in each food group, respectively, are as follows.

Fat: 5, 0, 0, 10

Carbohydrates: 0, 10, 15, 12

Protein: 7, 1, 2, 8

Use this information to write a 4×3 matrix.

c. There are 8 calories per exchange of fat, 4 calories per exchange of carbohydrates, and 5 calories per exchange of protein. Summarize this data in a 3×1 matrix.

42. Animal Growth At the beginning of a laboratory experiment, five baby rats measured 5.6, 6.4, 6.9, 7.6, and 6.1 cm in length, and weighed 144, 138, 149, 152, and 146 g, respectively.

a. Write a 2×5 matrix using this information.

b. At the end of two weeks, their lengths (in centimeters) were 10.2, 11.4, 11.4, 12.7, and 10.8 and their weights (in grams) were 196, 196, 225, 250, and 230. Write a 2×5 matrix with this information.

c. Use matrix subtraction and the matrices found in parts a and b to write a matrix that gives the amount of change in length and weight for each rat.

d. During the third week, the rats grew by the amounts shown in the matrix below.

$$\begin{array}{c} \text{Length} \\ \text{Weight} \end{array} \begin{bmatrix} 1.8 & 1.5 & 2.3 & 1.8 & 2.0 \\ 25 & 22 & 29 & 33 & 20 \end{bmatrix}$$

What were their lengths and weights at the end of this week?

43. Testing Medication A drug company is testing 200 patients to see if Painfree (a new headache medicine) is effective. Half the patients receive Painfree and half receive a placebo. The data on the first 50 patients is summarized in this matrix:

Pain Relief Obtained

$$\begin{array}{c} \\ \text{Painfree} \\ \text{Placebo} \end{array} \begin{array}{cc} \text{Yes} & \text{No} \end{array} \\ \begin{bmatrix} 22 & 3 \\ 8 & 17 \end{bmatrix}.$$

a. Of those who took the placebo, how many got relief?

b. Of those who took the new medication, how many got no relief?

c. The test was repeated on three more groups of 50 patients each, with the results summarized by these matrices.

$$\begin{bmatrix} 21 & 4 \\ 6 & 19 \end{bmatrix} \quad \begin{bmatrix} 19 & 6 \\ 10 & 15 \end{bmatrix} \quad \begin{bmatrix} 23 & 2 \\ 3 & 22 \end{bmatrix}$$

Find the total results for all 200 patients.

d. On the basis of these results, does it appear that Painfree is effective?

44. Motorcycle Helmets The following table shows the percentage of motorcyclists in various regions of the country who used helmets compliant with federal safety regulations and the percentage who used helmets that were noncompliant in two recent years. *Source: NHTSA.*

2008	Compliant	Noncompliant
Northeast	45	8
Midwest	67	16
South	61	14
West	71	5

2009	Compliant	Noncompliant
Northeast	61	15
Midwest	67	8
South	65	6
West	83	4

a. Write two matrices for the 2008 and 2009 helmet usage.

b. Use the two matrices from part a to write a matrix showing the change in helmet usage from 2008 to 2009.

c. Analyze the results from part b and discuss the extent to which changes from 2008 to 2009 differ from one region to another.

45. Life Expectancy The following table gives the life expectancy of African American males and females and white American males and females at the beginning of each decade since 1970. *Source: The New York Times 2010 Almanac.*

Year	African American		White American	
	Male	Female	Male	Female
1970	60.0	68.3	68.0	75.6
1980	63.8	72.5	70.7	78.1
1990	64.5	73.6	72.7	79.4
2000	68.3	75.2	74.9	80.1

a. Write a matrix for the life expectancy of African Americans.

b. Write a matrix for the life expectancy of white Americans.

c. Use the matrices from parts a and b to write a matrix showing the difference between the two groups.

d. Analyze the results from part c and discuss any noticeable trends.

46. Educational Attainment The table below gives the educational attainment of the U.S. population 25 years and older since 1970. *Source: U. S. Census Bureau.*

Year	Male		Female	
	Percentage with 4 Years of High School or More	Percentage with 4 Years of College or More	Percentage with 4 Years of High School or More	Percentage with 4 Years of College or More
1970	55.0	14.1	55.4	8.2
1980	69.2	20.9	68.1	13.6
1990	77.7	24.4	77.5	18.4
2000	84.2	27.8	84.0	23.6
2008	85.9	30.1	87.2	28.8

a. Write a matrix for the educational attainment of males.

b. Write a matrix for the educational attainment of females.

c. Use the matrices from parts a and b to write a matrix showing the difference in educational attainment between males and females since 1970.

47. Educational Attainment The following table gives the educational attainment of African Americans and Hispanic Americans 25 years and older since 1980. *Source: U. S. Census Bureau.*

a. Write a matrix for the educational attainment of African Americans.

b. Write a matrix for the educational attainment of Hispanic Americans.

Year	African American		Hispanic American	
	Percentage with 4 Years of High School or More	Percentage with 4 Years of College or More	Percentage with 4 Years of High School or More	Percentage with 4 Years of College or More
1980	51.2	7.9	45.3	7.9
1985	59.8	11.1	47.9	8.5
1990	66.2	11.3	50.8	9.2
1995	73.8	13.2	53.4	9.3
2000	78.5	16.5	57.0	10.6
2008	83.0	19.6	62.3	13.3

c. Use the matrices from parts a and b to write a matrix showing the difference in educational attainment between African and Hispanic Americans.

General Interest

48. Animal Interactions When two kittens named Cauchy and Cliché were introduced into a household with Jamie (an older cat) and Musk (a dog), the interactions among animals were complicated. The two kittens liked each other and Jamie, but didn't like Musk. Musk liked everybody, but Jamie didn't like any of the other animals.

a. Write a 4 × 4 matrix in which rows (and columns) 1, 2, 3, and 4 refer to Musk, Jamie, Cauchy, and Cliché. Make an element a 1 if the animal for that row likes the animal for that column, and otherwise make the element a 0. Assume every animal likes herself.

b. Within a few days, Cauchy and Cliché decided that they liked Musk after all. Write a 4 × 4 matrix, as you did in part a, representing the new situation.

YOUR TURN ANSWERS

1. (a) Does not exist **(b)** $\begin{bmatrix} 4 & 2 & 9 \\ -1 & -2 & 11 \end{bmatrix}$

2. $\begin{bmatrix} 2 & 6 & 1 \\ 3 & 6 & -5 \end{bmatrix}$

2.4 Multiplication of Matrices

What is a contractor's total cost for materials required for various types of model homes?
Matrix multiplication will be used to answer this question in Example 6.

We begin by defining the product of a real number and a matrix. In work with matrices, a real number is called a **scalar**.

> **Product of a Matrix and a Scalar**
> The product of a scalar k and a matrix X is the matrix kX, each of whose elements is k times the corresponding element of X.

EXAMPLE 1   Scalar Product of a Matrix

Calculate $-5A$, where $A = \begin{bmatrix} 3 & 4 \\ 0 & -1 \end{bmatrix}$.

SOLUTION

$$(-5)\begin{bmatrix} 3 & 4 \\ 0 & -1 \end{bmatrix} = \begin{bmatrix} -15 & -20 \\ 0 & 5 \end{bmatrix}.$$

Finding the product of two matrices is more involved, but such multiplication is important in solving practical problems. To understand the reasoning behind matrix multiplication, it may be helpful to consider another example concerning EZ Life Company discussed in the previous section. Suppose sofas and chairs of the same model are often sold as sets. Matrix W shows the number of sets of each model in each warehouse.

$$\begin{array}{c} \\ \text{New York} \\ \text{Chicago} \\ \text{San Francisco} \end{array} \begin{array}{ccc} A & B & C \\ \begin{bmatrix} 10 & 7 & 3 \\ 5 & 9 & 6 \\ 4 & 8 & 2 \end{bmatrix} \end{array} = W$$

If the selling price of a model-A set is $1000, of a model-B set $1200, and of a model-C set $1400, the total value of the sets in the New York warehouse is found as follows.

Value of Furniture					
Type	Number of Sets		Price of Set		Total
A	10	×	$1000	=	$10,000
B	7	×	$1200	=	$8400
C	3	×	$1400	=	$4200
	(Total for New York)				$22,600

The total value of the three kinds of sets in New York is $22,600.

The work done in the table above is summarized as follows:

$$10(\$1000) + 7(\$1200) + 3(\$1400) = \$22,\!600.$$

In the same way, we find that the Chicago sets have a total value of

$$5(\$1000) + 9(\$1200) + 6(\$1400) = \$24,\!200,$$

and in San Francisco, the total value of the sets is

$$4(\$1000) + 8(\$1200) + 2(\$1400) = \$16,\!400.$$

The selling prices can be written as a column matrix P, and the total value in each location as another column matrix, V.

$$\begin{bmatrix} 1000 \\ 1200 \\ 1400 \end{bmatrix} = P \qquad \begin{bmatrix} 22{,}600 \\ 24{,}200 \\ 16{,}400 \end{bmatrix} = V$$

Look at the elements of W and P below; multiplying the first, second, and third elements of the first row of W by the first, second, and third elements, respectively, of the column matrix P and then adding these products gives the first element in V. Doing the same thing with the second row of W gives the second element of V; the third row of W leads to the third element of V, suggesting that it is reasonable to write the product of matrices

$$W = \begin{bmatrix} 10 & 7 & 3 \\ 5 & 9 & 6 \\ 4 & 8 & 2 \end{bmatrix} \qquad \text{and} \qquad P = \begin{bmatrix} 1000 \\ 1200 \\ 1400 \end{bmatrix}$$

as

$$WP = \begin{bmatrix} 10 & 7 & 3 \\ 5 & 9 & 6 \\ 4 & 8 & 2 \end{bmatrix} \begin{bmatrix} 1000 \\ 1200 \\ 1400 \end{bmatrix} = \begin{bmatrix} 22{,}600 \\ 24{,}200 \\ 16{,}400 \end{bmatrix} = V.$$

The product was found by multiplying the elements of *rows* of the matrix on the left and the corresponding elements of the *column* of the matrix on the right, and then finding the sum of these separate products. Notice that the product of a 3×3 matrix and a 3×1 matrix is a 3×1 matrix. Notice also that each element of the product of two matrices is a sum of products. This is exactly what you do when you go to the store and buy 8 candy bars at \$0.80 each, 4 bottles of water at \$1.25 each, and so forth, and calculate the total as $8 \times 0.80 + 4 \times 1.25 + \cdots$.

The product AB of an $m \times n$ matrix A and an $n \times k$ matrix B is found as follows. Multiply each element of the first row of A by the corresponding element of the *first column* of B. The sum of these n products is the *first-row, first-column* element of AB. Similarly, the sum of the products found by multiplying the elements of the *first row* of A by the corresponding elements of the *second column* of B gives the *first-row, second-column* element of AB, and so on.

Product of Two Matrices

Let A be an $m \times n$ matrix and let B be an $n \times k$ matrix. To find the element in the ith row and jth column of the **product matrix** AB, multiply each element in the ith row of A by the corresponding element in the jth column of B, and then add these products. The product matrix AB is an $m \times k$ matrix.

EXAMPLE 2   Matrix Product

Find the product AB of matrices

$$A = \begin{bmatrix} 2 & 3 & -1 \\ 4 & 2 & 2 \end{bmatrix} \qquad \text{and} \qquad B = \begin{bmatrix} 1 \\ 8 \\ 6 \end{bmatrix}.$$

SOLUTION   Since A is 2×3 and B is 3×1, we can find the product matrix AB.

Step 1 Multiply the elements of the first row of A and the corresponding elements of the column of B.

$$\begin{bmatrix} 2 & 3 & -1 \\ 4 & 2 & 2 \end{bmatrix} \begin{bmatrix} 1 \\ 8 \\ 6 \end{bmatrix} \qquad 2 \cdot 1 + 3 \cdot 8 + (-1) \cdot 6 = 20$$

Thus, 20 is the first-row entry of the product matrix AB.

Step 2 Multiply the elements of the second row of A and the corresponding elements of B.

$$\begin{bmatrix} 2 & 3 & -1 \\ 4 & 2 & 2 \end{bmatrix} \begin{bmatrix} 1 \\ 8 \\ 6 \end{bmatrix} \quad 4 \cdot 1 + 2 \cdot 8 + 2 \cdot 6 = 32$$

The second-row entry of the product matrix AB is 32.

Step 3 Write the product as a column matrix using the two entries found above.

$$AB = \begin{bmatrix} 2 & 3 & -1 \\ 4 & 2 & 2 \end{bmatrix} \begin{bmatrix} 1 \\ 8 \\ 6 \end{bmatrix} = \begin{bmatrix} 20 \\ 32 \end{bmatrix}$$

Note that the product of a 2×3 matrix and a 3×1 matrix is a 2×1 matrix.

EXAMPLE 3   Matrix Product

Find the product CD of matrices

$$C = \begin{bmatrix} -3 & 4 & 2 \\ 5 & 0 & 4 \end{bmatrix} \quad \text{and} \quad D = \begin{bmatrix} -6 & 4 \\ 2 & 3 \\ 3 & -2 \end{bmatrix}.$$

SOLUTION   Since C is 2×3 and D is 3×2, we can find the product matrix CD.

Step 1
$$\begin{bmatrix} -3 & 4 & 2 \\ 5 & 0 & 4 \end{bmatrix} \begin{bmatrix} -6 & 4 \\ 2 & 3 \\ 3 & -2 \end{bmatrix} \quad (-3) \cdot (-6) + 4 \cdot 2 + 2 \cdot 3 = 32$$

Step 2
$$\begin{bmatrix} -3 & 4 & 2 \\ 5 & 0 & 4 \end{bmatrix} \begin{bmatrix} -6 & 4 \\ 2 & 3 \\ 3 & -2 \end{bmatrix} \quad (-3) \cdot 4 + 4 \cdot 3 + 2 \cdot (-2) = -4$$

Step 3
$$\begin{bmatrix} -3 & 4 & 2 \\ 5 & 0 & 4 \end{bmatrix} \begin{bmatrix} -6 & 4 \\ 2 & 3 \\ 3 & -2 \end{bmatrix} \quad 5 \cdot (-6) + 0 \cdot 2 + 4 \cdot 3 = -18$$

Step 4
$$\begin{bmatrix} -3 & 4 & 2 \\ 5 & 0 & 4 \end{bmatrix} \begin{bmatrix} -6 & 4 \\ 2 & 3 \\ 3 & -2 \end{bmatrix} \quad 5 \cdot 4 + 0 \cdot 3 + 4 \cdot (-2) = 12$$

Step 5 The product is

YOUR TURN 1 Calculate the product AB where $A = \begin{bmatrix} 3 & 4 \\ 1 & 2 \end{bmatrix}$ and $B = \begin{bmatrix} 1 & -2 \\ -2 & -4 \end{bmatrix}$.

$$CD = \begin{bmatrix} -3 & 4 & 2 \\ 5 & 0 & 4 \end{bmatrix} \begin{bmatrix} -6 & 4 \\ 2 & 3 \\ 3 & -2 \end{bmatrix} = \begin{bmatrix} 32 & -4 \\ -18 & 12 \end{bmatrix}.$$

Here the product of a 2×3 matrix and a 3×2 matrix is a 2×2 matrix.

TRY YOUR TURN 1

NOTE   One way to avoid errors in matrix multiplication is to lower the first matrix so it is below and to the left of the second matrix, and then write the product in the space between the two matrices. For example, to multiply the matrices in Example 2, we could rewrite the product as shown on the following page.

$$\downarrow$$

$$\begin{bmatrix} -6 & 4 \\ 2 & 3 \\ 3 & -2 \end{bmatrix}$$

$$\rightarrow \begin{bmatrix} -3 & 4 & 2 \\ 5 & 0 & 4 \end{bmatrix} \begin{bmatrix} \\ * \end{bmatrix}$$

To find the entry where the * is, for example, multiply the row and the column indicated by the arrows: $5 \cdot (-6) + 0 \cdot 2 + 4 \cdot 3 = -18$.

As the definition of matrix multiplication shows,

> the product AB of two matrices A and B can be found only if the number of columns of A is the same as the number of rows of B.

The final product will have as many rows as A and as many columns as B.

EXAMPLE 4 Matrix Product

Suppose matrix A is 2×2 and matrix B is 2×4. Can the products AB and BA be calculated? If so, what is the size of each product?

SOLUTION   The following diagram helps decide the answers to these questions.

Matrix A size 2×2 Matrix B size 2×4
must match
Size of AB
is 2×4

The product of A and B can be found because A has two columns and B has two rows. The size of the product is 2×4.

Matrix B size 2×4 Matrix A size 2×2
do not match

The product BA cannot be found because B has 4 columns and A has 2 rows.

EXAMPLE 5 Comparing Matrix Products *AB* and *BA*

Find AB and BA, given

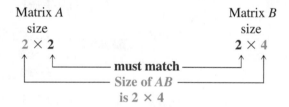

$$A = \begin{bmatrix} 1 & -3 \\ 7 & 2 \\ -2 & 5 \end{bmatrix} \quad \text{and} \quad B = \begin{bmatrix} 1 & 0 & -1 \\ 3 & 1 & 4 \end{bmatrix}.$$

SOLUTION

Method 1
Calculating by Hand

$$AB = \begin{bmatrix} 1 & -3 \\ 7 & 2 \\ -2 & 5 \end{bmatrix} \begin{bmatrix} 1 & 0 & -1 \\ 3 & 1 & 4 \end{bmatrix}$$

$$= \begin{bmatrix} -8 & -3 & -13 \\ 13 & 2 & 1 \\ 13 & 5 & 22 \end{bmatrix}$$

YOUR TURN 2 Calculate each product AB and BA, if possible, where $A = \begin{bmatrix} 3 & 5 & -1 \\ 2 & 4 & -2 \end{bmatrix}$ and $B = \begin{bmatrix} 3 & -4 \\ -5 & -3 \end{bmatrix}$.

$$BA = \begin{bmatrix} 1 & 0 & -1 \\ 3 & 1 & 4 \end{bmatrix} \begin{bmatrix} 1 & -3 \\ 7 & 2 \\ -2 & 5 \end{bmatrix}$$

$$= \begin{bmatrix} 3 & -8 \\ 2 & 13 \end{bmatrix}$$

Method 2
Graphing Calculator

Matrix multiplication is easily performed on a graphing calculator. Figure 13 in the margin below shows the results. The matrices A and B were already entered into the calculator.

TRY YOUR TURN 2

TECHNOLOGY NOTE Matrix multiplication can also be easily done with a spreadsheet. See the *Graphing Calculator and Excel Spreadsheet Manual* available with this textbook for details.

```
[A]*[B]
   [-8  -3  -13]
   [13   2    1]
   [13   5   22]
[B]*[A]
        [3  -8]
        [2  13]
```

FIGURE 13

Notice in Example 5 that $AB \neq BA$; matrices AB and BA aren't even the same size. In Example 4, we showed that they may not both exist. This means that matrix multiplication is *not* commutative. Even if both A and B are square matrices, in general, matrices AB and BA are not equal. (See Exercise 31.) Of course, there may be special cases in which they are equal, but this is not true in general.

CAUTION Since matrix multiplication is not commutative, always be careful to multiply matrices in the correct order.

Matrix multiplication *is* associative, however. For example, if

$$C = \begin{bmatrix} 3 & 2 \\ 0 & -4 \\ -1 & 1 \end{bmatrix},$$

then $(AB)C = A(BC)$, where A and B are the matrices given in Example 5. (Verify this.) Also, there is a distributive property of matrices such that, for appropriate matrices A, B, and C,

$$A(B + C) = AB + AC.$$

(See Exercises 32 and 33.) Other properties of matrix multiplication involving scalars are included in the exercises. Multiplicative inverses and multiplicative identities are defined in the next section.

EXAMPLE 6 **Home Construction**

A contractor builds three kinds of houses, models A, B, and C, with a choice of two styles, Spanish and contemporary. Matrix P shows the number of each kind of house planned for a new 100-home subdivision. The amounts for each of the exterior materials depend primarily on the style of the house. These amounts are shown in matrix Q. (Concrete is in cubic yards, lumber in units of 1000 board feet, brick in 1000s, and shingles in units of 100 ft².) Matrix R gives the cost in dollars for each kind of material.

$$
\begin{array}{c}
\hspace{2.5em} \text{Spanish} \quad \text{Contemporary} \\
\begin{array}{c}
\text{Model A} \\
\text{Model B} \\
\text{Model C}
\end{array}
\left[
\begin{array}{cc}
0 & 30 \\
10 & 20 \\
20 & 20
\end{array}
\right] = P
\end{array}
$$

$$
\begin{array}{c}
\hspace{4em} \text{Concrete} \quad \text{Lumber} \quad \text{Brick} \quad \text{Shingles} \\
\begin{array}{c}
\text{Spanish} \\
\text{Contemporary}
\end{array}
\left[
\begin{array}{cccc}
10 & 2 & 0 & 2 \\
50 & 1 & 20 & 2
\end{array}
\right] = Q
\end{array}
$$

$$
\begin{array}{c}
\hspace{2em} \text{Cost per Unit} \\
\begin{array}{c}
\text{Concrete} \\
\text{Lumber} \\
\text{Brick} \\
\text{Shingles}
\end{array}
\left[
\begin{array}{c}
20 \\
180 \\
60 \\
25
\end{array}
\right] = R
\end{array}
$$

(a) What is the total cost of these materials for each model?

APPLY IT

SOLUTION To find the cost for each model, first find PQ, which shows the amount of each material needed for each model.

$$
PQ = \left[
\begin{array}{cc}
0 & 30 \\
10 & 20 \\
20 & 20
\end{array}
\right]
\left[
\begin{array}{cccc}
10 & 2 & 0 & 2 \\
50 & 1 & 20 & 2
\end{array}
\right]
$$

$$
\begin{array}{c}
\hspace{2em} \text{Concrete} \quad \text{Lumber} \quad \text{Brick} \quad \text{Shingles} \\
= \left[
\begin{array}{cccc}
1500 & 30 & 600 & 60 \\
1100 & 40 & 400 & 60 \\
1200 & 60 & 400 & 80
\end{array}
\right]
\begin{array}{c}
\text{Model A} \\
\text{Model B} \\
\text{Model C}
\end{array}
\end{array}
$$

Now multiply PQ and R, the cost matrix, to get the total cost of the exterior materials for each model.

$$
\left[
\begin{array}{cccc}
1500 & 30 & 600 & 60 \\
1100 & 40 & 400 & 60 \\
1200 & 60 & 400 & 80
\end{array}
\right]
\left[
\begin{array}{c}
20 \\
180 \\
60 \\
25
\end{array}
\right]
=
\begin{array}{c}
\text{Cost} \\
\left[
\begin{array}{c}
72{,}900 \\
54{,}700 \\
60{,}800
\end{array}
\right]
\end{array}
\begin{array}{c}
\text{Model A} \\
\text{Model B} \\
\text{Model C}
\end{array}
$$

The total cost of materials is \$72,900 for model A, \$54,700 for model B, and \$60,800 for model C.

(b) How much of each of the four kinds of material must be ordered?

SOLUTION The totals of the columns of matrix PQ will give a matrix whose elements represent the total amounts of each material needed for the subdivision. Call this matrix T, and write it as a row matrix.

$$
T = \left[
\begin{array}{cccc}
3800 & 130 & 1400 & 200
\end{array}
\right]
$$

Thus, 3800 yd^3 of concrete, 130,000 board feet of lumber, 1,400,000 bricks, and 20,000 ft^2 of shingles are needed.

(c) What is the total cost for exterior materials?

SOLUTION For the total cost of all the exterior materials, find the product of matrix T, the matrix showing the total amount of each material, and matrix R, the cost matrix. (To multiply these and get a 1×1 matrix representing total cost, we need a 1×4 matrix multiplied by a 4×1 matrix. This is why T was written as a row matrix in (b) above.)

$$TR = \begin{bmatrix} 3800 & 130 & 1400 & 200 \end{bmatrix} \begin{bmatrix} 20 \\ 180 \\ 60 \\ 25 \end{bmatrix} = [188,400]$$

The total cost for exterior materials is $188,400.

(d) Suppose the contractor builds the same number of homes in five subdivisions. Calculate the total amount of each exterior material for each model for all five subdivisions.

SOLUTION   Multiply PQ by the scalar 5, as follows.

$$5(PQ) = 5\begin{bmatrix} 1500 & 30 & 600 & 60 \\ 1100 & 40 & 400 & 60 \\ 1200 & 60 & 400 & 80 \end{bmatrix} = \begin{bmatrix} 7500 & 150 & 3000 & 300 \\ 5500 & 200 & 2000 & 300 \\ 6000 & 300 & 2000 & 400 \end{bmatrix}$$

The total amount of concrete needed for model A homes, for example, is 7500 yd³.

Meaning of a Matrix Product

It is helpful to use a notation that keeps track of the quantities a matrix represents. We will use the notation

meaning of the rows/meaning of the columns,

that is, writing the meaning of the rows first, followed by the meaning of the columns. In Example 6, we would use the notation models/styles for matrix P, styles/materials for matrix Q, and materials/cost for matrix R. In multiplying PQ, we are multiplying models/styles by styles/materials. The result is models/materials. Notice that styles, the common quantity in both P and Q, was eliminated in the product PQ. By this method, the product $(PQ)R$ represents models/cost.

In practical problems this notation helps us decide in which order to multiply matrices so that the results are meaningful. In Example 6(c) either RT or TR can be calculated. Since T represents subdivisions/materials and R represents materials/cost, the product TR gives subdivisions/cost, while the product RT is meaningless.

2.4   EXERCISES

Let $A = \begin{bmatrix} -2 & 4 \\ 0 & 3 \end{bmatrix}$ and $B = \begin{bmatrix} -6 & 2 \\ 4 & 0 \end{bmatrix}$. Find each value.

1. $2A$
2. $-3B$
3. $-6A$
4. $5B$
5. $-4A + 5B$
6. $7B - 3A$

In Exercises 7–12, the sizes of two matrices A and B are given. Find the sizes of the product AB and the product BA, whenever these products exist.

7. A is 2×2, and B is 2×2.
8. A is 3×3, and B is 3×3.
9. A is 3×4, and B is 4×4.
10. A is 4×3, and B is 3×6.
11. A is 4×2, and B is 3×4.
12. A is 3×2, and B is 1×3.

13. To find the product matrix AB, the number of _____ of A must be the same as the number of _____ of B.

14. The product matrix AB has the same number of _____ as A and the same number of _____ as B.

Find each matrix product, if possible.

15. $\begin{bmatrix} 2 & -1 \\ 5 & 8 \end{bmatrix}\begin{bmatrix} 3 \\ -2 \end{bmatrix}$

16. $\begin{bmatrix} -1 & 5 \\ 7 & 0 \end{bmatrix}\begin{bmatrix} 6 \\ 2 \end{bmatrix}$

17. $\begin{bmatrix} 2 & -1 & 7 \\ -3 & 0 & -4 \end{bmatrix}\begin{bmatrix} 5 \\ 10 \\ 2 \end{bmatrix}$

18. $\begin{bmatrix} 5 & 2 \\ 7 & 6 \\ 1 & 0 \end{bmatrix}\begin{bmatrix} 1 & 4 & 0 \\ 2 & -1 & 2 \end{bmatrix}$

19. $\begin{bmatrix} 2 & -1 \\ 3 & 6 \end{bmatrix}\begin{bmatrix} -1 & 0 & 4 \\ 5 & -2 & 0 \end{bmatrix}$

20. $\begin{bmatrix} 6 & 0 & -4 \\ 1 & 2 & 5 \\ 10 & -1 & 3 \end{bmatrix}\begin{bmatrix} 1 \\ 2 \\ 0 \end{bmatrix}$

21. $\begin{bmatrix} 2 & 2 & -1 \\ 3 & 0 & 1 \end{bmatrix}\begin{bmatrix} 0 & 2 \\ -1 & 4 \\ 0 & 2 \end{bmatrix}$

22. $\begin{bmatrix} -3 & 1 & 0 \\ 6 & 0 & 8 \end{bmatrix}\begin{bmatrix} 3 \\ -1 \\ -2 \end{bmatrix}$

23. $\begin{bmatrix} 1 & 2 \\ 3 & 4 \end{bmatrix}\begin{bmatrix} -1 & 5 \\ 7 & 0 \end{bmatrix}$

24. $\begin{bmatrix} 2 & 8 \\ -7 & 5 \end{bmatrix}\begin{bmatrix} 1 & 0 \\ 0 & 1 \end{bmatrix}$

25. $\begin{bmatrix} -2 & -3 & 7 \\ 1 & 5 & 6 \end{bmatrix}\begin{bmatrix} 1 \\ 2 \\ 3 \end{bmatrix}$

26. $\begin{bmatrix} 2 \\ -9 \\ 12 \end{bmatrix}\begin{bmatrix} 1 & 0 & -1 \end{bmatrix}$

27. $\left(\begin{bmatrix} 2 & 1 \\ -3 & -6 \\ 4 & 0 \end{bmatrix}\begin{bmatrix} 1 & -2 \\ 2 & -1 \end{bmatrix}\right)\begin{bmatrix} 3 \\ 1 \end{bmatrix}$

28. $\begin{bmatrix} 2 & 1 \\ -3 & -6 \\ 4 & 0 \end{bmatrix}\left(\begin{bmatrix} 1 & -2 \\ 2 & -1 \end{bmatrix}\begin{bmatrix} 3 \\ 1 \end{bmatrix}\right)$

29. $\begin{bmatrix} 2 & -2 \\ 1 & -1 \end{bmatrix}\left(\begin{bmatrix} 4 & 3 \\ 1 & 2 \end{bmatrix}+\begin{bmatrix} 7 & 0 \\ -1 & 5 \end{bmatrix}\right)$

30. $\begin{bmatrix} 2 & -2 \\ 1 & -1 \end{bmatrix}\begin{bmatrix} 4 & 3 \\ 1 & 2 \end{bmatrix}+\begin{bmatrix} 2 & -2 \\ 1 & -1 \end{bmatrix}\begin{bmatrix} 7 & 0 \\ -1 & 5 \end{bmatrix}$

31. Let $A = \begin{bmatrix} -2 & 4 \\ 1 & 3 \end{bmatrix}$ and $B = \begin{bmatrix} -2 & 1 \\ 3 & 6 \end{bmatrix}$.

  a. Find AB.

  b. Find BA.

  c. Did you get the same answer in parts a and b?

  d. In general, for matrices A and B such that AB and BA both exist, does AB always equal BA?

Given matrices $P = \begin{bmatrix} m & n \\ p & q \end{bmatrix}$, $X = \begin{bmatrix} x & y \\ z & w \end{bmatrix}$, and $T = \begin{bmatrix} r & s \\ t & u \end{bmatrix}$, verify that the statements in Exercises 32–35 are true. The statements are valid for any matrices whenever matrix multiplication and addition can be carried out. This, of course, depends on the size of the matrices.

32. $(PX)T = P(XT)$ (associative property: see Exercises 27 and 28)

33. $P(X + T) = PX + PT$ (distributive property: see Exercises 29 and 30)

34. $k(X + T) = kX + kT$ for any real number k.

35. $(k + h)P = kP + hP$ for any real numbers k and h.

36. Let I be the matrix $I = \begin{bmatrix} 1 & 0 \\ 0 & 1 \end{bmatrix}$, and let matrices P, X, and T be defined as for Exercises 32–35.

  a. Find IP, PI, and IX.

  b. Without calculating, guess what the matrix IT might be.

  c. Suggest a reason for naming a matrix such as I an *identity matrix*.

37. Show that the system of linear equations

$$2x_1 + 3x_2 + x_3 = 5$$
$$x_1 - 4x_2 + 5x_3 = 8$$

can be written as the matrix equation

$$\begin{bmatrix} 2 & 3 & 1 \\ 1 & -4 & 5 \end{bmatrix}\begin{bmatrix} x_1 \\ x_2 \\ x_3 \end{bmatrix} = \begin{bmatrix} 5 \\ 8 \end{bmatrix}.$$

38. Let $A = \begin{bmatrix} 1 & 2 \\ -3 & 5 \end{bmatrix}$, $X = \begin{bmatrix} x_1 \\ x_2 \end{bmatrix}$, and $B = \begin{bmatrix} -4 \\ 12 \end{bmatrix}$. Show that the equation $AX = B$ represents a linear system of two equations in two unknowns. Solve the system and substitute into the matrix equation to check your results.

Use a computer or graphing calculator and the following matrices to find the matrix products and sums in Exercises 39–41.

$$A = \begin{bmatrix} 2 & 3 & -1 & 5 & 10 \\ 2 & 8 & 7 & 4 & 3 \\ -1 & -4 & -12 & 6 & 8 \\ 2 & 5 & 7 & 1 & 4 \end{bmatrix}$$

$$B = \begin{bmatrix} 9 & 3 & 7 & -6 \\ -1 & 0 & 4 & 2 \\ -10 & -7 & 6 & 9 \\ 8 & 4 & 2 & -1 \\ 2 & -5 & 3 & 7 \end{bmatrix}$$

$$C = \begin{bmatrix} -6 & 8 & 2 & 4 & -3 \\ 1 & 9 & 7 & -12 & 5 \\ 15 & 2 & -8 & 10 & 11 \\ 4 & 7 & 9 & 6 & -2 \\ 1 & 3 & 8 & 23 & 4 \end{bmatrix}$$

$$D = \begin{bmatrix} 5 & -3 & 7 & 9 & 2 \\ 6 & 8 & -5 & 2 & 1 \\ 3 & 7 & -4 & 2 & 11 \\ 5 & -3 & 9 & 4 & -1 \\ 0 & 3 & 2 & 5 & 1 \end{bmatrix}$$

39. a. Find AC.   **b.** Find CA.   **c.** Does $AC = CA$?

40. a. Find CD.   **b.** Find DC.   **c.** Does $CD = DC$?

41. a. Find $C + D$.   **b.** Find $(C + D)B$.   **c.** Find CB.

  d. Find DB.   **e.** Find $CB + DB$.

  f. Does $(C + D)B = CB + DB$?

42. Which property of matrices does Exercise 41 illustrate?

APPLICATIONS

Business and Economics

43. Cost Analysis The four departments of Spangler Enterprises need to order the following amounts of the same products.

	Paper	Tape	Binders	Memo Pads	Pens
Department 1	10	4	3	5	6
Department 2	7	2	2	3	8
Department 3	4	5	1	0	10
Department 4	0	3	4	5	5

The unit price (in dollars) of each product is given below for two suppliers.

	Supplier A	Supplier B
Paper	2	3
Tape	1	1
Binders	4	3
Memo Pads	3	3
Pens	1	2

a. Use matrix multiplication to get a matrix showing the comparative costs for each department for the products from the two suppliers.

b. Find the total cost over all departments to buy products from each supplier. From which supplier should the company make the purchase?

44. Cost Analysis The Mundo Candy Company makes three types of chocolate candy: Cheery Cherry, Mucho Mocha, and Almond Delight. The company produces its products in San Diego, Mexico City, and Managua using two main ingredients: chocolate and sugar.

a. Each kilogram of Cheery Cherry requires 0.5 kg of sugar and 0.2 kg of chocolate; each kilogram of Mucho Mocha requires 0.4 kg of sugar and 0.3 kg of chocolate; and each kilogram of Almond Delight requires 0.3 kg of sugar and 0.3 kg of chocolate. Put this information into a 2×3 matrix called A, labeling the rows and columns.

b. The cost of 1 kg of sugar is $4 in San Diego, $2 in Mexico City, and $1 in Managua. The cost of 1 kg of chocolate is $3 in San Diego, $5 in Mexico City, and $7 in Managua. Put this information into a matrix called C in such a way that when you multiply it with your matrix from part a, you get a matrix representing the ingredient cost of producing each type of candy in each city.

c. Only one of the two products AC and CA is meaningful. Determine which one it is, calculate the product, and describe what the entries represent.

d. From your answer to part c, what is the combined sugar-and-chocolate cost to produce 1 kg of Mucho Mocha in Managua?

e. Mundo Candy needs to quickly produce a special shipment of 100 kg of Cheery Cherry, 200 kg of Mucho Mocha, and 500 kg of Almond Delight, and it decides to select one factory to fill the entire order. Use matrix multiplication to determine in which city the total sugar-and-chocolate cost to produce the order is the smallest.

45. Management In Exercise 39 from Section 2.3, consider the matrices $\begin{bmatrix} 4.27 & 6.94 \\ 3.45 & 3.65 \end{bmatrix}$, $\begin{bmatrix} 4.05 & 7.01 \\ 3.27 & 3.51 \end{bmatrix}$, and $\begin{bmatrix} 4.40 & 6.90 \\ 3.54 & 3.76 \end{bmatrix}$ for the production costs at the Boston, Chicago, and Seattle plants, respectively.

a. Assume each plant makes the same number of each item. Write a matrix that expresses the average production costs for all three plants.

b. In part b of Exercise 39 in Section 2.3, cost increases for the Chicago plant resulted in a new production cost matrix $\begin{bmatrix} 4.42 & 7.43 \\ 3.38 & 3.62 \end{bmatrix}$. Following those cost increases the Boston plant was closed and production divided evenly between the Chicago and Seattle plants. What is the matrix that now expresses the average production cost for the entire country?

46. House Construction Consider the matrices P, Q, and R given in Example 6.

a. Find and interpret the matrix product QR.

b. Verify that $P(QR)$ is equal to $(PQ)R$ calculated in Example 5.

47. Shoe Sales Sal's Shoes and Fred's Footwear both have outlets in California and Arizona. Sal's sells shoes for $80, sandals for $40, and boots for $120. Fred's prices are $60, $30, and $150 for shoes, sandals, and boots, respectively. Half of all sales in California stores are shoes, 1/4 are sandals, and 1/4 are boots. In Arizona the fractions are 1/5 shoes, 1/5 sandals, and 3/5 boots.

a. Write a 2×3 matrix called P representing prices for the two stores and three types of footwear.

b. Write a 3×2 matrix called F representing the fraction of each type of footwear sold in each state.

c. Only one of the two products PF and FP is meaningful. Determine which one it is, calculate the product, and describe what the entries represent.

d. From your answer to part c, what is the average price for a pair of footwear at an outlet of Fred's in Arizona?

48. Management In Exercise 40 from Section 2.3, consider the matrix

$$\begin{bmatrix} 88 & 105 & 60 \\ 48 & 72 & 40 \\ 16 & 21 & 0 \\ 112 & 147 & 50 \end{bmatrix}$$

expressing the sales information for the three stores.

a. Write a 3×1 matrix expressing the factors by which sales in each store should be multiplied to reflect the fact that sales increased during the following week by 25%, 1/3, and 10% in stores I, II, and III, respectively, as described in part b of Exercise 40 from Section 2.3.

b. Multiply the matrix expressing sales information by the matrix found in part a of this exercise to find the sales for all three stores in the second week.

Life Sciences

49. Dietetics In Exercise 41 from Section 2.3, label the matrices

$$\begin{bmatrix} 2 & 1 & 2 & 1 \\ 3 & 2 & 2 & 1 \\ 4 & 3 & 2 & 1 \end{bmatrix}, \quad \begin{bmatrix} 5 & 0 & 7 \\ 0 & 10 & 1 \\ 0 & 15 & 2 \\ 10 & 12 & 8 \end{bmatrix}, \quad \text{and} \quad \begin{bmatrix} 8 \\ 4 \\ 5 \end{bmatrix}$$

found in parts a, b, and c, respectively, X, Y, and Z.

a. Find the product matrix XY. What do the entries of this matrix represent?

b. Find the product matrix YZ. What do the entries represent?

c. Find the products $(XY)Z$ and $X(YZ)$ and verify that they are equal. What do the entries represent?

50. Motorcycle Helmets In Exercise 44 from Section 2.3, you constructed matrices that represented usage of motorcycle helmets for 2008 and 2009. Use matrix operations to combine these two matrices to form one matrix that represents the average of the two years.

51. Life Expectancy In Exercise 45 from Section 2.3, you constructed matrices that represent the life expectancy of African American and white American males and females. Use matrix operations to combine these two matrices to form one matrix that represents the combined life expectancy of both races at the beginning of each decade since 1970. Use the fact that of the combined African and white American population, African Americans are about one-sixth of the total and white Americans about five-sixths. (*Hint:* Multiply the matrix for African Americans by 1/6 and the matrix for the white Americans by 5/6, and then add the results.)

52. Northern Spotted Owl Population In an attempt to save the endangered northern spotted owl, the U.S. Fish and Wildlife Service imposed strict guidelines for the use of 12 million acres of Pacific Northwest forest. This decision led to a national debate between the logging industry and environmentalists. Mathematical ecologists have created a mathematical model to analyze population dynamics of the northern spotted owl by dividing the female owl population into three categories: juvenile (up to 1 year old), subadult (1 to 2 years), and adult (over 2 years old). By analyzing these three subgroups, it is possible to use the number of females in each subgroup at time n to estimate the number of females in each group at any time $n + 1$ with the following matrix equation:

$$\begin{bmatrix} j_{n+1} \\ s_{n+1} \\ a_{n+1} \end{bmatrix} = \begin{bmatrix} 0 & 0 & 0.33 \\ 0.18 & 0 & 0 \\ 0 & 0.71 & 0.94 \end{bmatrix} \begin{bmatrix} j_n \\ s_n \\ a_n \end{bmatrix},$$

where j_n is the number of juveniles, s_n is the number of subadults, and a_n is the number of adults at time n. **Source: Conservation Biology.**

a. If there are currently 4000 female northern spotted owls made up of 900 juveniles, 500 subadults, and 2600 adults, use a graphing calculator or spreadsheet and matrix operations to determine the total number of female owls for each of the next 5 years. (*Hint:* Round each answer to the nearest whole number after each matrix multiplication.)

b. With advanced techniques from linear algebra, it is possible to show that in the long run, the following holds.

$$\begin{bmatrix} j_{n+1} \\ s_{n+1} \\ a_{n+1} \end{bmatrix} \approx 0.98359 \begin{bmatrix} j_n \\ s_n \\ a_n \end{bmatrix}$$

What can we conclude about the long-term survival of the northern spotted owl?

c. Notice that only 18 percent of the juveniles become subadults. Assuming that, through better habitat management, this number could be increased to 40 percent, rework part a. Discuss possible reasons why only 18 percent of the juveniles become subadults. Under the new assumption, what can you conclude about the long-term survival of the northern spotted owl?

Social Sciences

53. World Population The 2010 birth and death rates per million for several regions and the world population (in millions) by region are given in the following tables. **Source: U. S. Census Bureau.**

	Births	**Deaths**
Africa	0.0346	0.0118
Asia	0.0174	0.0073
Latin America	0.0189	0.0059
North America	0.0135	0.0083
Europe	0.0099	0.0103

			Population (millions)		
Year	**Africa**	**Asia**	**Latin America**	**North America**	**Europe**
1970	361	2038	286	227	460
1980	473	2494	362	252	484
1990	627	2978	443	278	499
2000	803	3435	524	314	511
2010	1013	3824	591	344	522

a. Write the information in each table as a matrix.

b. Use the matrices from part a to find the total number (in millions) of births and deaths in each year (assuming birth and death rates for all years are close to these in 2010).

c. Using the results of part b, compare the number of births in 1970 and in 2010. Also compare the birth rates from part a. Which gives better information?

d. Using the results of part b, compare the number of deaths in 1980 and in 2010. Discuss how this comparison differs from a comparison of death rates from part a.

YOUR TURN ANSWERS

1. $\begin{bmatrix} -5 & -22 \\ -3 & -10 \end{bmatrix}$ **2.** *AB* does not exist; $BA = \begin{bmatrix} 1 & -1 & 5 \\ -21 & -37 & 11 \end{bmatrix}$

2.5 Matrix Inverses

APPLY IT

One top leader needs to get an important message to one of her agents. How can she encrypt the message to ensure secrecy?
This question is answered in Example 7.

In this section, we introduce the idea of a matrix inverse, which is comparable to the reciprocal of a real number. This will allow us to solve a matrix equation.

The real number 1 is the *multiplicative* identity for real numbers: for any real number a, we have $a \cdot 1 = 1 \cdot a = a$. In this section, we define a *multiplicative identity matrix I* that has properties similar to those of the number 1. We then use the definition of matrix I to find the *multiplicative inverse* of any square matrix that has an inverse.

If I is to be the identity matrix, both of the products AI and IA must equal A. This means that an identity matrix exists only for square matrices. The 2×2 **identity matrix** that satisfies these conditions is

$$I = \begin{bmatrix} 1 & 0 \\ 0 & 1 \end{bmatrix}.$$

To check that I, as defined above, is really the 2×2 identity, let

$$A = \begin{bmatrix} a & b \\ c & d \end{bmatrix}.$$

Then AI and IA should both equal A.

$$AI = \begin{bmatrix} a & b \\ c & d \end{bmatrix} \begin{bmatrix} 1 & 0 \\ 0 & 1 \end{bmatrix} = \begin{bmatrix} a(1) + b(0) & a(0) + b(1) \\ c(1) + d(0) & c(0) + d(1) \end{bmatrix} = \begin{bmatrix} a & b \\ c & d \end{bmatrix} = A$$

$$IA = \begin{bmatrix} 1 & 0 \\ 0 & 1 \end{bmatrix} \begin{bmatrix} a & b \\ c & d \end{bmatrix} = \begin{bmatrix} 1(a) + 0(c) & 1(b) + 0(d) \\ 0(a) + 1(c) & 0(b) + 1(d) \end{bmatrix} = \begin{bmatrix} a & b \\ c & d \end{bmatrix} = A$$

This verifies that I has been defined correctly.

It is easy to verify that the identity matrix I is unique. Suppose there is another identity; call it J. Then IJ must equal I, because J is an identity, and IJ must also equal J, because I is an identity. Thus $I = J$.

The identity matrices for 3×3 matrices and 4×4 matrices, respectively, are

$$I = \begin{bmatrix} 1 & 0 & 0 \\ 0 & 1 & 0 \\ 0 & 0 & 1 \end{bmatrix} \quad \text{and} \quad I = \begin{bmatrix} 1 & 0 & 0 & 0 \\ 0 & 1 & 0 & 0 \\ 0 & 0 & 1 & 0 \\ 0 & 0 & 0 & 1 \end{bmatrix}.$$

By generalizing, we can find an $n \times n$ identity matrix for any value of n.

Recall that the multiplicative inverse of the nonzero real number a is $1/a$. The product of a and its multiplicative inverse $1/a$ is 1. Given a matrix A, can a **multiplicative inverse matrix A^{-1}** (read "A-inverse") that will satisfy both

$$AA^{-1} = I \quad \text{and} \quad A^{-1}A = I$$

be found? For a given matrix, we often can find an inverse matrix by using the row operations of Section 2.2.

NOTE A^{-1} does not mean $1/A$; here, A^{-1} is just the notation for the multiplicative inverse of matrix A. Also, only square matrices can have inverses because both $A^{-1}A$ and AA^{-1} must exist and be equal to an identity matrix of the same size.

EXAMPLE 1 Inverse Matrices

Verify that the matrices $A = \begin{bmatrix} 1 & 3 \\ 2 & 5 \end{bmatrix}$ and $B = \begin{bmatrix} -5 & 3 \\ 2 & -1 \end{bmatrix}$ are inverses of each other.

SOLUTION Multiply A times B as in the previous section:

$$AB = \begin{bmatrix} 1 & 3 \\ 2 & 5 \end{bmatrix}\begin{bmatrix} -5 & 3 \\ 2 & -1 \end{bmatrix} = \begin{bmatrix} 1 & 0 \\ 0 & 1 \end{bmatrix}.$$

Similarly,

$$BA = \begin{bmatrix} -5 & 3 \\ 2 & -1 \end{bmatrix}\begin{bmatrix} 1 & 3 \\ 2 & 5 \end{bmatrix} = \begin{bmatrix} 1 & 0 \\ 0 & 1 \end{bmatrix}.$$

Since $AB = BA = I$, A and B are inverses of each other. ▬

If an inverse exists, it is unique. That is, any given square matrix has no more than one inverse. The proof of this is left to Exercise 50 in this section.

As an example, let us find the inverse of

$$A = \begin{bmatrix} 1 & 3 \\ -1 & 2 \end{bmatrix}.$$

Let the unknown inverse matrix be

$$A^{-1} = \begin{bmatrix} x & y \\ z & w \end{bmatrix}.$$

By the definition of matrix inverse, $AA^{-1} = I$, or

$$AA^{-1} = \begin{bmatrix} 1 & 3 \\ -1 & 2 \end{bmatrix}\begin{bmatrix} x & y \\ z & w \end{bmatrix} = \begin{bmatrix} 1 & 0 \\ 0 & 1 \end{bmatrix}.$$

By matrix multiplication,

$$\begin{bmatrix} x + 3z & y + 3w \\ -x + 2z & -y + 2w \end{bmatrix} = \begin{bmatrix} 1 & 0 \\ 0 & 1 \end{bmatrix}.$$

Setting corresponding elements equal gives the system of equations

$$x \qquad + 3z \qquad\quad = 1 \tag{1}$$
$$\qquad y \qquad + 3w = 0 \tag{2}$$
$$-x \qquad + 2z \qquad\quad = 0 \tag{3}$$
$$\qquad -y \qquad + 2w = 1. \tag{4}$$

Since equations (1) and (3) involve only x and z, while equations (2) and (4) involve only y and w, these four equations lead to two systems of equations,

$$\begin{matrix} x + 3z = 1 \\ -x + 2z = 0 \end{matrix} \quad\text{and}\quad \begin{matrix} y + 3w = 0 \\ -y + 2w = 1. \end{matrix}$$

Writing the two systems as augmented matrices gives

$$\left[\begin{array}{cc|c} 1 & 3 & 1 \\ -1 & 2 & 0 \end{array}\right] \quad\text{and}\quad \left[\begin{array}{cc|c} 1 & 3 & 0 \\ -1 & 2 & 1 \end{array}\right].$$

Each of these systems can be solved by the Gauss-Jordan method. Notice, however, that the elements to the left of the vertical bar are identical. The two systems can be combined into the single matrix

$$\begin{bmatrix} 1 & 3 & | & 1 & 0 \\ -1 & 2 & | & 0 & 1 \end{bmatrix}.$$

This is of the form $[A|I]$. It is solved simultaneously as follows.

$$R_1 + R_2 \rightarrow R_2 \quad \begin{bmatrix} 1 & 3 & | & 1 & 0 \\ 0 & 5 & | & 1 & 1 \end{bmatrix} \quad \begin{array}{l} \text{Get 0 in the second-row,} \\ \text{first-column position.} \end{array}$$

$$-3R_2 + 5R_1 \rightarrow R_1 \quad \begin{bmatrix} 5 & 0 & | & 2 & -3 \\ 0 & 5 & | & 1 & 1 \end{bmatrix} \quad \begin{array}{l} \text{Get 0 in the first-row,} \\ \text{second-column position.} \end{array}$$

$$\begin{array}{l} \frac{1}{5}R_1 \rightarrow R_1 \\ \frac{1}{5}R_2 \rightarrow R_2 \end{array} \quad \begin{bmatrix} 1 & 0 & | & \frac{2}{5} & -\frac{3}{5} \\ 0 & 1 & | & \frac{1}{5} & \frac{1}{5} \end{bmatrix} \quad \begin{array}{l} \text{Get 1's down the} \\ \text{diagonal.} \end{array}$$

The numbers in the first column to the right of the vertical bar give the values of x and z. The second column gives the values of y and w. That is,

$$\begin{bmatrix} 1 & 0 & | & x & y \\ 0 & 1 & | & z & w \end{bmatrix} = \begin{bmatrix} 1 & 0 & | & \frac{2}{5} & -\frac{3}{5} \\ 0 & 1 & | & \frac{1}{5} & \frac{1}{5} \end{bmatrix},$$

so that

$$A^{-1} = \begin{bmatrix} x & y \\ z & w \end{bmatrix} = \begin{bmatrix} \frac{2}{5} & -\frac{3}{5} \\ \frac{1}{5} & \frac{1}{5} \end{bmatrix}.$$

To check, multiply A by A^{-1}. The result should be I.

$$AA^{-1} = \begin{bmatrix} 1 & 3 \\ -1 & 2 \end{bmatrix} \begin{bmatrix} \frac{2}{5} & -\frac{3}{5} \\ \frac{1}{5} & \frac{1}{5} \end{bmatrix} = \begin{bmatrix} \frac{2}{5} + \frac{3}{5} & -\frac{3}{5} + \frac{3}{5} \\ -\frac{2}{5} + \frac{2}{5} & \frac{3}{5} + \frac{2}{5} \end{bmatrix} = \begin{bmatrix} 1 & 0 \\ 0 & 1 \end{bmatrix} = I$$

Verify that $A^{-1}A = I$, also.

Finding a Multiplicative Inverse Matrix

To obtain A^{-1} for any $n \times n$ matrix A for which A^{-1} exists, follow these steps.

1. Form the augmented matrix $[A|I]$, where I is the $n \times n$ identity matrix.
2. Perform row operations on $[A|I]$ to get a matrix of the form $[I|B]$, if this is possible.
3. Matrix B is A^{-1}.

EXAMPLE 2   Inverse Matrix

Find A^{-1} if $A = \begin{bmatrix} 1 & 0 & 1 \\ 2 & -2 & -1 \\ 3 & 0 & 0 \end{bmatrix}$.

Method 1
Calculating by Hand

SOLUTION   Write the augmented matrix $[A \mid I]$.

$$[A|I] = \begin{bmatrix} 1 & 0 & 1 & | & 1 & 0 & 0 \\ 2 & -2 & -1 & | & 0 & 1 & 0 \\ 3 & 0 & 0 & | & 0 & 0 & 1 \end{bmatrix}$$

Begin by selecting the row operation that produces a zero for the first element in row 2.

$$\begin{array}{l} -2R_1 + R_2 \rightarrow R_2 \\ -3R_1 + R_3 \rightarrow R_3 \end{array} \quad \begin{bmatrix} 1 & 0 & 1 & | & 1 & 0 & 0 \\ 0 & -2 & -3 & | & -2 & 1 & 0 \\ 0 & 0 & -3 & | & -3 & 0 & 1 \end{bmatrix} \quad \text{Get 0's in the first column.}$$

Column 2 already has zeros in the required positions, so work on column 3.

$$\begin{matrix} R_3 + 3R_1 \rightarrow R_1 \\ R_3 + (-1)R_2 \rightarrow R_2 \\ \\ \end{matrix} \left[\begin{array}{ccc|ccc} 3 & 0 & 0 & 0 & 0 & 1 \\ 0 & 2 & 0 & -1 & -1 & 1 \\ 0 & 0 & -3 & -3 & 0 & 1 \end{array} \right] \quad \text{Get 0's in the third column.}$$

Now get 1's down the main diagonal.

$$\begin{matrix} \frac{1}{3}R_1 \rightarrow R_1 \\ \frac{1}{2}R_2 \rightarrow R_2 \\ -\frac{1}{3}R_3 \rightarrow R_3 \end{matrix} \left[\begin{array}{ccc|ccc} 1 & 0 & 0 & 0 & 0 & \frac{1}{3} \\ 0 & 1 & 0 & -\frac{1}{2} & -\frac{1}{2} & \frac{1}{2} \\ 0 & 0 & 1 & 1 & 0 & -\frac{1}{3} \end{array} \right] \quad \text{Get 1's down the diagonal.}$$

YOUR TURN 1 Find A^{-1} if
$A = \begin{bmatrix} 2 & 3 & 1 \\ 1 & -2 & -1 \\ 3 & 3 & 2 \end{bmatrix}$.

From the last transformation, the desired inverse is

$$A^{-1} = \begin{bmatrix} 0 & 0 & \frac{1}{3} \\ -\frac{1}{2} & -\frac{1}{2} & \frac{1}{2} \\ 1 & 0 & -\frac{1}{3} \end{bmatrix}.$$

Confirm this by forming the products $A^{-1}A$ and AA^{-1}, both of which should equal I.

Method 2
Graphing Calculator

The inverse of A can also be found with a graphing calculator, as shown in Figure 14 in the margin below. (The matrix A had previously been entered into the calculator.) The entire answer can be viewed by pressing the right and left arrow keys on the calculator.

TRY YOUR TURN 1

TECHNOLOGY NOTE

Spreadsheets also have the capability of calculating the inverse of a matrix with a simple command. See the *Graphing Calculator and Excel Spreadsheet Manual* available with this book for details.

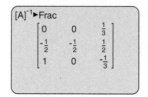

FIGURE 14

EXAMPLE 3 Inverse Matrix

Find A^{-1} if $A = \begin{bmatrix} 2 & -4 \\ 1 & -2 \end{bmatrix}$.

SOLUTION Using row operations to transform the first column of the augmented matrix

$$\left[\begin{array}{cc|cc} 2 & -4 & 1 & 0 \\ 1 & -2 & 0 & 1 \end{array} \right]$$

gives the following results.

$$R_1 + (-2)R_2 \rightarrow R_2 \quad \left[\begin{array}{cc|cc} 2 & -4 & 1 & 0 \\ 0 & 0 & 1 & -2 \end{array} \right]$$

Because the last row has all zeros to the left of the vertical bar, there is no way to complete the process of finding the inverse matrix. What is wrong? Just as the real number 0 has no multiplicative inverse, some matrices do not have inverses. Matrix A is an example of a matrix that has no inverse: there is no matrix A^{-1} such that $AA^{-1} = A^{-1}A = I$.

Solving Systems of Equations with Inverses
We used matrices to solve systems of linear equations by the Gauss-Jordan method in Section 2.2. Another way to use matrices to solve linear systems is to write the system as a matrix equation $AX = B$, where A is the matrix of the coefficients of the variables of the system, X is the matrix of the variables, and B is the matrix of the constants. Matrix A is called the **coefficient matrix**.

To solve the matrix equation $AX = B$, first see if A^{-1} exists. Assuming A^{-1} exists and using the facts that $A^{-1}A = I$ and $IX = X$ gives

$$AX = B$$
$$A^{-1}(AX) = A^{-1}B \quad \text{Multiply both sides by } A^{-1}.$$
$$(A^{-1}A)X = A^{-1}B \quad \text{Associative property}$$
$$IX = A^{-1}B \quad \text{Multiplicative inverse property}$$
$$X = A^{-1}B. \quad \text{Identity property}$$

CAUTION When multiplying by matrices on both sides of a matrix equation, be careful to multiply in the same order on both sides of the equation, since multiplication of matrices is not commutative (unlike multiplication of real numbers).

The work thus far leads to the following method of solving a system of equations written as a matrix equation.

Solving a System $AX = B$ Using Matrix Inverses

To solve a system of equations $AX = B$, where A is the square matrix of coefficients and A^{-1} exists, X is the matrix of variables, and B is the matrix of constants, first find A^{-1}. Then $X = A^{-1}B$.

This method is most practical in solving several systems that have the same coefficient matrix but different constants, as in Example 5 in this section. Then just one inverse matrix must be found.

EXAMPLE 4   Inverse Matrices and Systems of Equations

Use the inverse of the coefficient matrix to solve the linear system

$$2x - 3y = 4$$
$$x + 5y = 2.$$

SOLUTION   To represent the system as a matrix equation, use the coefficient matrix of the system together with the matrix of variables and the matrix of constants:

$$A = \begin{bmatrix} 2 & -3 \\ 1 & 5 \end{bmatrix}, \qquad X = \begin{bmatrix} x \\ y \end{bmatrix}, \qquad \text{and} \qquad B = \begin{bmatrix} 4 \\ 2 \end{bmatrix}.$$

The system can now be written in matrix form as the equation $AX = B$ since

$$AX = \begin{bmatrix} 2 & -3 \\ 1 & 5 \end{bmatrix} \begin{bmatrix} x \\ y \end{bmatrix} = \begin{bmatrix} 2x - 3y \\ x + 5y \end{bmatrix} = \begin{bmatrix} 4 \\ 2 \end{bmatrix} = B.$$

To solve the system, first find A^{-1}. Do this by using row operations on matrix $[A|I]$ to get

$$\begin{bmatrix} 1 & 0 & \frac{5}{13} & \frac{3}{13} \\ 0 & 1 & -\frac{1}{13} & \frac{2}{13} \end{bmatrix}.$$

From this result,

$$A^{-1} = \begin{bmatrix} \frac{5}{13} & \frac{3}{13} \\ -\frac{1}{13} & \frac{2}{13} \end{bmatrix}.$$

Next, find the product $A^{-1}B$.

$$A^{-1}B = \begin{bmatrix} \frac{5}{13} & \frac{3}{13} \\ -\frac{1}{13} & \frac{2}{13} \end{bmatrix} \begin{bmatrix} 4 \\ 2 \end{bmatrix} = \begin{bmatrix} 2 \\ 0 \end{bmatrix}.$$

YOUR TURN 2 Use the inverse of the coefficient matrix to solve the linear system
$$5x + 4y = 23$$
$$4x - 3y = 6.$$

Since $X = A^{-1}B$,

$$X = \begin{bmatrix} x \\ y \end{bmatrix} = \begin{bmatrix} 2 \\ 0 \end{bmatrix}.$$

The solution of the system is $(2, 0)$.

TRY YOUR TURN 2

EXAMPLE 5   Fertilizer

Three brands of fertilizer are available that provide nitrogen, phosphoric acid, and soluble potash to the soil. One bag of each brand provides the following units of each nutrient.

		Brand		
		Fertifun	**Big Grow**	**Soakem**
	Nitrogen	1	2	3
Nutrient	*Phosphoric Acid*	3	1	2
	Potash	2	0	1

For ideal growth, the soil on a Michigan farm needs 18 units of nitrogen, 23 units of phosphoric acid, and 13 units of potash per acre. The corresponding numbers for a California farm are 31, 24, and 11, and for a Kansas farm are 20, 19, and 15. How many bags of each brand of fertilizer should be used per acre for ideal growth on each farm?

SOLUTION Rather than solve three separate systems, we consider the single system

$$x + 2y + 3z = a$$
$$3x + y + 2z = b$$
$$2x \quad\;\; + z = c,$$

where a, b, and c represent the units of nitrogen, phosphoric acid, and potash needed for the different farms. The system of equations is then of the form $AX = B$, where

$$A = \begin{bmatrix} 1 & 2 & 3 \\ 3 & 1 & 2 \\ 2 & 0 & 1 \end{bmatrix} \quad \text{and} \quad X = \begin{bmatrix} x \\ y \\ z \end{bmatrix}.$$

B has different values for the different farms. We find A^{-1} first, then use it to solve all three systems.

To find A^{-1}, we start with the matrix

$$[A|I] = \begin{bmatrix} 1 & 2 & 3 & | & 1 & 0 & 0 \\ 3 & 1 & 2 & | & 0 & 1 & 0 \\ 2 & 0 & 1 & | & 0 & 0 & 1 \end{bmatrix}$$

and use row operations to get $[I|A^{-1}]$. The result is

$$A^{-1} = \begin{bmatrix} -\frac{1}{3} & \frac{2}{3} & -\frac{1}{3} \\ -\frac{1}{3} & \frac{5}{3} & -\frac{7}{3} \\ \frac{2}{3} & -\frac{4}{3} & \frac{5}{3} \end{bmatrix}.$$

Now we can solve each of the three systems by using $X = A^{-1}B$.

For the Michigan farm, $B = \begin{bmatrix} 18 \\ 23 \\ 13 \end{bmatrix}$, and

$$X = \begin{bmatrix} -\frac{1}{3} & \frac{2}{3} & -\frac{1}{3} \\ -\frac{1}{3} & \frac{5}{3} & -\frac{7}{3} \\ \frac{2}{3} & -\frac{4}{3} & \frac{5}{3} \end{bmatrix} \begin{bmatrix} 18 \\ 23 \\ 13 \end{bmatrix} = \begin{bmatrix} 5 \\ 2 \\ 3 \end{bmatrix}.$$

Therefore, $x = 5$, $y = 2$, and $z = 3$. Buy 5 bags of Fertifun, 2 bags of Big Grow, and 3 bags of Soakem.

For the California farm, $B = \begin{bmatrix} 31 \\ 24 \\ 11 \end{bmatrix}$, and

$$X = \begin{bmatrix} -\frac{1}{3} & \frac{2}{3} & -\frac{1}{3} \\ -\frac{1}{3} & \frac{5}{3} & -\frac{7}{3} \\ \frac{2}{3} & -\frac{4}{3} & \frac{5}{3} \end{bmatrix} \begin{bmatrix} 31 \\ 24 \\ 11 \end{bmatrix} = \begin{bmatrix} 2 \\ 4 \\ 7 \end{bmatrix}.$$

Buy 2 bags of Fertifun, 4 bags of Big Grow, and 7 bags of Soakem.

For the Kansas farm, $B = \begin{bmatrix} 20 \\ 19 \\ 15 \end{bmatrix}$. Verify that this leads to $x = 1$, $y = -10$, and

$z = 13$. We cannot have a negative number of bags, so this solution is impossible. In buying enough bags to meet all of the nutrient requirements, the farmer must purchase an excess of some nutrients. In the next two chapters, we will study a method of solving such problems at a minimum cost. ▬▬

In Example 5, using the matrix inverse method of solving the systems involved considerably less work than using row operations for each of the three systems.

EXAMPLE 6 **Solving an Inconsistent System of Equations**

Use the inverse of the coefficient matrix to solve the system

$$2x - 4y = 13$$
$$x - 2y = 1.$$

SOLUTION We saw in Example 3 that the coefficient matrix $\begin{bmatrix} 2 & -4 \\ 1 & -2 \end{bmatrix}$ does not have an inverse. This means that the given system either has no solution or has an infinite number of solutions. Verify that this system is inconsistent and has no solution. ▬▬

NOTE If a matrix has no inverse, then a corresponding system of equations might have either no solutions or an infinite number of solutions, and we instead use the echelon or Gauss-Jordan method. In Example 3, we saw that the matrix $\begin{bmatrix} 2 & -4 \\ 1 & -2 \end{bmatrix}$ had no inverse. The reason is that the first row of this matrix is double the second row. The system

$$2x - 4y = 2$$
$$x - 2y = 1$$

has an infinite number of solutions because the first equation is double the second equation. The system

$$2x - 4y = 13$$
$$x - 2y = 1$$

has no solutions because the left side of the first equation is double the left side of the second equation, but the right side is not.

EXAMPLE 7 **Cryptography**

Throughout the Cold War and as the Internet has grown and developed, the need for sophisticated methods of coding and decoding messages has increased. Although there are many methods of encrypting messages, one fairly sophisticated method uses matrix operations. This method first assigns a number to each letter of the alphabet. The simplest way to do this is to assign the number 1 to A, 2 to B, and so on, with the number 27 used to represent a space between words.

▬APPLY IT   For example, the message *math is cool* can be divided into groups of three letters and spaces each and then converted into numbers as follows

$$\begin{bmatrix} m \\ a \\ t \end{bmatrix} = \begin{bmatrix} 13 \\ 1 \\ 20 \end{bmatrix}.$$

The entire message would then consist of four 3×1 columns of numbers:

$$\begin{bmatrix} 13 \\ 1 \\ 20 \end{bmatrix}, \quad \begin{bmatrix} 8 \\ 27 \\ 9 \end{bmatrix}, \quad \begin{bmatrix} 19 \\ 27 \\ 3 \end{bmatrix}, \quad \begin{bmatrix} 15 \\ 15 \\ 12 \end{bmatrix}.$$

This code is easy to break, so we further complicate the code by choosing a matrix that has an inverse (in this case a 3×3 matrix) and calculate the products of the matrix and each of the column vectors above.

If we choose the coding matrix

$$A = \begin{bmatrix} 1 & 3 & 4 \\ 2 & 1 & 3 \\ 4 & 2 & 1 \end{bmatrix},$$

then the products of A with each of the column vectors above produce a new set of vectors

$$\begin{bmatrix} 96 \\ 87 \\ 74 \end{bmatrix}, \quad \begin{bmatrix} 125 \\ 70 \\ 95 \end{bmatrix}, \quad \begin{bmatrix} 112 \\ 74 \\ 133 \end{bmatrix}, \quad \begin{bmatrix} 108 \\ 81 \\ 102 \end{bmatrix}.$$

This set of vectors represents our coded message, and it will be transmitted as 96, 87, 74, 125 and so on.

When the intended person receives the message, it is divided into groups of three numbers, and each group is formed into a column matrix. The message is easily decoded if the receiver knows the inverse of the original matrix. The inverse of matrix A is

$$A^{-1} = \begin{bmatrix} -0.2 & 0.2 & 0.2 \\ 0.4 & -0.6 & 0.2 \\ 0 & 0.4 & -0.2 \end{bmatrix}.$$

Thus, the message is decoded by taking the product of the inverse matrix with each column vector of the received message. For example,

$$A^{-1} \begin{bmatrix} 96 \\ 87 \\ 74 \end{bmatrix} = \begin{bmatrix} 13 \\ 1 \\ 20 \end{bmatrix}.$$

YOUR TURN 3 Use the matrix A and its inverse A^{-1} in Example 7 to do the following.

a) Encode the message "Behold".
b) Decode the message 96, 87, 74, 141, 117, 114.

Unless the original matrix or its inverse is known, this type of code can be difficult to break. In fact, very large matrices can be used to encrypt data. It is interesting to note that many mathematicians are employed by the National Security Agency to develop encryption methods that are virtually unbreakable. **TRY YOUR TURN 3** ■

2.5 EXERCISES

Decide whether the given matrices are inverses of each other. (Check to see if their product is the identity matrix I.)

1. $\begin{bmatrix} 2 & 1 \\ 5 & 3 \end{bmatrix}$ and $\begin{bmatrix} 3 & -1 \\ -5 & 2 \end{bmatrix}$ **2.** $\begin{bmatrix} 1 & -4 \\ 2 & -7 \end{bmatrix}$ and $\begin{bmatrix} -7 & 4 \\ -2 & 1 \end{bmatrix}$

3. $\begin{bmatrix} 2 & 6 \\ 2 & 4 \end{bmatrix}$ and $\begin{bmatrix} -1 & 2 \\ 2 & -4 \end{bmatrix}$ **4.** $\begin{bmatrix} -1 & 2 \\ 3 & -5 \end{bmatrix}$ and $\begin{bmatrix} -5 & -2 \\ -3 & -1 \end{bmatrix}$

5. $\begin{bmatrix} 2 & 0 & 1 \\ 1 & 1 & 2 \\ 0 & 1 & 0 \end{bmatrix}$ and $\begin{bmatrix} 1 & 1 & -1 \\ 0 & 1 & 0 \\ -1 & -2 & 2 \end{bmatrix}$

6. $\begin{bmatrix} 0 & 1 & 0 \\ 0 & 0 & -2 \\ 1 & -1 & 0 \end{bmatrix}$ and $\begin{bmatrix} 1 & 0 & 1 \\ 1 & 0 & 0 \\ 0 & -1 & 0 \end{bmatrix}$

7. $\begin{bmatrix} 1 & 3 & 3 \\ 1 & 4 & 3 \\ 1 & 3 & 4 \end{bmatrix}$ and $\begin{bmatrix} 7 & -3 & -3 \\ -1 & 1 & 0 \\ -1 & 0 & 1 \end{bmatrix}$

8. $\begin{bmatrix} 1 & 0 & 0 \\ -1 & -2 & 3 \\ 0 & 1 & 0 \end{bmatrix}$ and $\begin{bmatrix} 1 & 0 & 0 \\ 0 & 0 & 1 \\ \frac{1}{3} & \frac{1}{3} & \frac{2}{3} \end{bmatrix}$

9. Does a matrix with a row of all zeros have an inverse? Why?

 10. Matrix A has A^{-1} as its inverse. What does $\left(A^{-1}\right)^{-1}$ equal? (*Hint:* Experiment with a few matrices to see what you get.)

Find the inverse, if it exists, for each matrix.

11. $\begin{bmatrix} 1 & -1 \\ 2 & 0 \end{bmatrix}$

12. $\begin{bmatrix} 1 & 1 \\ 2 & 3 \end{bmatrix}$

13. $\begin{bmatrix} 3 & -1 \\ -5 & 2 \end{bmatrix}$

14. $\begin{bmatrix} -3 & -8 \\ 1 & 3 \end{bmatrix}$

15. $\begin{bmatrix} 1 & -3 \\ -2 & 6 \end{bmatrix}$

16. $\begin{bmatrix} 5 & 10 \\ -3 & -6 \end{bmatrix}$

17. $\begin{bmatrix} 1 & 0 & 0 \\ 0 & -1 & 0 \\ 1 & 0 & 1 \end{bmatrix}$

18. $\begin{bmatrix} 1 & 3 & 0 \\ 0 & 2 & -1 \\ 1 & 0 & 2 \end{bmatrix}$

19. $\begin{bmatrix} -1 & -1 & -1 \\ 4 & 5 & 0 \\ 0 & 1 & -3 \end{bmatrix}$

20. $\begin{bmatrix} 2 & 1 & 0 \\ 0 & 3 & 1 \\ 4 & -1 & -3 \end{bmatrix}$

21. $\begin{bmatrix} 1 & 2 & 3 \\ -3 & -2 & -1 \\ -1 & 0 & 1 \end{bmatrix}$

22. $\begin{bmatrix} 2 & 0 & 4 \\ 1 & 0 & -1 \\ 3 & 0 & -2 \end{bmatrix}$

23. $\begin{bmatrix} 1 & 3 & -2 \\ 2 & 7 & -3 \\ 3 & 8 & -5 \end{bmatrix}$

24. $\begin{bmatrix} 4 & 1 & -4 \\ 2 & 1 & -1 \\ -2 & -4 & 5 \end{bmatrix}$

25. $\begin{bmatrix} 1 & -2 & 3 & 0 \\ 0 & 1 & -1 & 1 \\ -2 & 2 & -2 & 4 \\ 0 & 2 & -3 & 1 \end{bmatrix}$

26. $\begin{bmatrix} 1 & 1 & 0 & 2 \\ 2 & -1 & 1 & -1 \\ 3 & 3 & 2 & -2 \\ 1 & 2 & 1 & 0 \end{bmatrix}$

Solve each system of equations by using the inverse of the coefficient matrix if it exists and by the echelon method if the inverse doesn't exist.

27. $2x + 5y = 15$
$x + 4y = 9$

28. $-x + 2y = 15$
$-2x - y = 20$

29. $2x + y = 5$
$5x + 3y = 13$

30. $-x - 2y = 8$
$3x + 4y = 24$

31. $3x - 2y = 3$
$7x - 5y = 0$

32. $3x - 6y = 1$
$-5x + 9y = -1$

33. $-x - 8y = 12$
$3x + 24y = -36$

34. $2x + 7y = 14$
$4x + 14y = 28$

Solve each system of equations by using the inverse of the coefficient matrix if it exists and by the Gauss-Jordan method if the inverse doesn't exist. (The inverses for the first four problems were found in Exercises 19, 20, 23, and 24.)

35. $-x - y - z = 1$
$4x + 5y \phantom{{}- z} = -2$
$y - 3z = 3$

36. $2x + y \phantom{{}+ z} = 1$
$3y + z = 8$
$4x - y - 3z = 8$

37. $x + 3y - 2z = 4$
$2x + 7y - 3z = 8$
$3x + 8y - 5z = -4$

38. $4x + y - 4z = 17$
$2x + y - z = 12$
$-2x - 4y + 5z = 17$

39. $2x - 2y = 5$
$\phantom{2x - {}}4y + 8z = 7$
$x + 2z = 1$

40. $x + 2z = -1$
$y - z = 5$
$x + 2y = 7$

Solve each system of equations by using the inverse of the coefficient matrix. (The inverses were found in Exercises 25 and 26.)

41. $x - 2y + 3z \phantom{{}+ w} = 4$
$\phantom{-x + {}}y - z + w = -8$
$-2x + 2y - 2z + 4w = 12$
$\phantom{-2x + {}}2y - 3z + w = -4$

42. $x + y \phantom{{}+ 2z} + 2w = 3$
$2x - y + z - w = 3$
$3x + 3y + 2z - 2w = 5$
$x + 2y + z \phantom{{}- 2w} = 3$

Let $A = \begin{bmatrix} a & b \\ c & d \end{bmatrix}$ and $0 = \begin{bmatrix} 0 & 0 \\ 0 & 0 \end{bmatrix}$ in Exercises 43–48.

43. Show that $IA = A$.

44. Show that $AI = A$.

45. Show that $A \cdot O = O$.

46. Find A^{-1}.
(Assume $ad - bc \neq 0$.)

47. Show that $A^{-1}A = I$.

48. Show that $AA^{-1} = I$.

49. Using the definition and properties listed in this section, show that for square matrices A and B of the same size, if $AB = O$ and if A^{-1} exists, then $B = O$, where O is a matrix whose elements are all zeros.

50. Prove that, if it exists, the inverse of a matrix is unique. (*Hint:* Assume there are two inverses B and C for some matrix A, so that $AB = BA = I$ and $AC = CA = I$. Multiply the first equation by C and the second by B.)

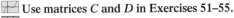 **Use matrices C and D in Exercises 51–55.**

$$C = \begin{bmatrix} -6 & 8 & 2 & 4 & -3 \\ 1 & 9 & 7 & -12 & 5 \\ 15 & 2 & -8 & 10 & 11 \\ 4 & 7 & 9 & 6 & -2 \\ 1 & 3 & 8 & 23 & 4 \end{bmatrix}, \quad D = \begin{bmatrix} 5 & -3 & 7 & 9 & 2 \\ 6 & 8 & -5 & 2 & 1 \\ 3 & 7 & -4 & 2 & 11 \\ 5 & -3 & 9 & 4 & -1 \\ 0 & 3 & 2 & 5 & 1 \end{bmatrix}$$

51. Find C^{-1}.

52. Find $(CD)^{-1}$.

53. Find D^{-1}.

54. Is $C^{-1}D^{-1} = (CD)^{-1}$?

55. Is $D^{-1}C^{-1} = (CD)^{-1}$?

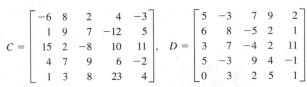 **Solve the matrix equation $AX = B$ for X by finding A^{-1}, given A and B as follows.**

56. $A = \begin{bmatrix} 2 & -5 & 7 \\ 4 & -3 & 2 \\ 15 & 2 & 6 \end{bmatrix}$, $B = \begin{bmatrix} -2 \\ 5 \\ 8 \end{bmatrix}$

57. $A = \begin{bmatrix} 2 & 5 & 7 & 9 \\ 1 & 3 & -4 & 6 \\ -1 & 0 & 5 & 8 \\ 2 & -2 & 4 & 10 \end{bmatrix}$, $B = \begin{bmatrix} 3 \\ 7 \\ -1 \\ 5 \end{bmatrix}$

58. $A = \begin{bmatrix} 3 & 2 & -1 & -2 & 6 \\ -5 & 17 & 4 & 3 & 15 \\ 7 & 9 & -3 & -7 & 12 \\ 9 & -2 & 1 & 4 & 8 \\ 1 & 21 & 9 & -7 & 25 \end{bmatrix}$, $B = \begin{bmatrix} -2 \\ 5 \\ 3 \\ -8 \\ 25 \end{bmatrix}$

APPLICATIONS

Business and Economics

Solve each exercise by using the inverse of the coefficient matrix to solve a system of equations.

59. Analysis of Orders The Bread Box Bakery sells three types of cakes, each requiring the amounts of the basic ingredients shown in the following matrix.

		Type of Cake		
		I	II	III
	Flour (in cups)	2	4	2
Ingredient	Sugar (in cups)	2	1	2
	Eggs	2	1	3

To fill its daily orders for these three kinds of cake, the bakery uses 72 cups of flour, 48 cups of sugar, and 60 eggs.

a. Write a 3×1 matrix for the amounts used daily.

b. Let the number of daily orders for cakes be a 3×1 matrix X with entries x_1, x_2, and x_3. Write a matrix equation that can be solved for X, using the given matrix and the matrix from part a.

c. Solve the equation from part b to find the number of daily orders for each type of cake.

60. Production Requirements An electronics company produces transistors, resistors, and computer chips. Each transistor requires 3 units of copper, 1 unit of zinc, and 2 units of glass. Each resistor requires 3, 2, and 1 units of the three materials, and each computer chip requires 2, 1, and 2 units of these materials, respectively. How many of each product can be made with the following amounts of materials?

a. 810 units of copper, 410 units of zinc, and 490 units of glass

b. 765 units of copper, 385 units of zinc, and 470 units of glass

c. 1010 units of copper, 500 units of zinc, and 610 units of glass

61. Investments An investment firm recommends that a client invest in AAA-, A-, and B-rated bonds. The average yield on AAA bonds is 6%, on A bonds 6.5%, and on B bonds 8%. The client wants to invest twice as much in AAA bonds as in B bonds. How much should be invested in each type of bond under the following conditions?

a. The total investment is $25,000, and the investor wants an annual return of $1650 on the three investments.

b. The values in part a are changed to $30,000 and $1985, respectively.

c. The values in part a are changed to $40,000 and $2660, respectively.

62. Production Pretzels cost $4 per lb, dried fruit $5 per lb, and nuts $9 per lb. The three ingredients are to be combined in a trail mix containing twice the weight of pretzels as dried fruit. How many pounds of each should be used to produce the following amounts at the given cost?

a. 140 lb at $6 per lb

b. 100 lb at $7.60 per lb

c. 125 lb at $6.20 per lb

Life Sciences

63. Vitamins Greg Tobin mixes together three types of vitamin tablets. Each Super Vim tablet contains, among other things, 15 mg of niacin and 12 I.U. of vitamin E. The figures for a Multitab tablet are 20 mg and 15 I.U., and for a Mighty Mix are 25 mg and 35 I.U. How many of each tablet are there if the total number of tablets, total amount of niacin, and total amount of vitamin E are as follows?

a. 225 tablets, 4750 mg of niacin, and 5225 I.U. of vitamin E

b. 185 tablets, 3625 mg of niacin, and 3750 I.U. of vitamin E

c. 230 tablets, 4450 mg of niacin, and 4210 I.U. of vitamin E

General Interest

64. Encryption Use the matrices presented in Example 7 of this section to do the following:

a. Encode the message, "All is fair in love and war."

b. Decode the message 138, 81, 102, 101, 67, 109, 162, 124, 173, 210, 150, 165.

65. Encryption Use the methods presented in Example 7 along with the given matrix B to do the following.

$$B = \begin{bmatrix} 2 & 4 & 6 \\ -1 & -4 & -3 \\ 0 & 1 & -1 \end{bmatrix}$$

a. Encode the message, "To be or not to be."

b. Find the inverse of B.

c. Use the inverse of B to decode the message 116, -60, -15, 294, -197, -2, 148, -92, -9, 96, -64, 4, 264, -182, -2.

66. Music During a marching band's half-time show, the band members generally line up in such a way that a common shape is recognized by the fans. For example, as illustrated in the figure, a band might form a letter T, where an x represents a member of the band. As the music is played, the band will either create a new shape or rotate the original shape. In doing this, each member of the band will need to move from one point on the field to another. For larger bands, keeping track of who goes where can be a daunting task. However, it is possible to use matrix inverses to make the process a bit easier. The entire process is calculated by knowing how three band members, all of whom cannot be in a straight line, will move from the current position to a new position. For example, in the figure, we can see that there are band members at $(50, 0)$, $(50, 15)$, and $(45, 20)$. We will assume that these three band members move to $(40, 10)$, $(55, 10)$, and $(60, 15)$, respectively. *Source: The College Mathematics Journal.*

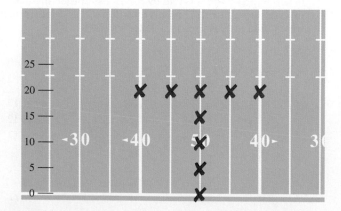

a. Find the inverse of $B = \begin{bmatrix} 50 & 50 & 45 \\ 0 & 15 & 20 \\ 1 & 1 & 1 \end{bmatrix}$.

b. Find $A = \begin{bmatrix} 40 & 55 & 60 \\ 10 & 10 & 15 \\ 1 & 1 & 1 \end{bmatrix} B^{-1}$.

c. Use the result of part b to find the new position of the other band members. What is the shape of the new position? (*Hint:* Multiply the matrix A by a 3×1 column vector with the first two components equal to the original position of each band member and the third component equal to 1. The new position of the band member is in the first two components of the product.)

YOUR TURN ANSWERS

1. $\begin{bmatrix} 1/8 & 3/8 & 1/8 \\ 5/8 & -1/8 & -3/8 \\ -9/8 & -3/8 & 7/8 \end{bmatrix}$ **2.** $(3, 2)$ **3. a.** $49, 33, 26, 67, 54, 88$ **b.** matrix

2.6 Input-Output Models

APPLY IT   **What production levels are needed to keep an economy going and to supply demands from outside the economy?**

A method for solving such questions is developed in this section and applied in Examples 3 and 4.

Wassily Leontief (1906–1999) developed an interesting and powerful application of matrix theory to economics and was recognized for this contribution with the Nobel prize in economics in 1973. His matrix models for studying the interdependencies in an economy are called *input-output* models. In practice these models are very complicated, with many variables. Only simple examples with a few variables are discussed here.

Input-output models are concerned with the production and flow of goods (and perhaps services). In an economy with n basic commodities, or sectors, the production of each commodity uses some (perhaps all) of the commodities in the economy as inputs. For example, oil is needed to run the machinery that plants and harvests the wheat, and wheat is used to feed the people who drill and refine the oil. The amounts of each commodity used in the production of one unit of each commodity can be written as an $n \times n$ matrix A, called the **technological matrix** or **input-output matrix** of the economy.

EXAMPLE 1   Input-Output Matrix

Suppose a simplified economy involves just three commodity categories: agriculture, manufacturing, and transportation, all in appropriate units. Production of 1 unit of agriculture requires 1/2 unit of manufacturing and 1/4 unit of transportation; production of 1 unit of manufacturing requires 1/4 unit of agriculture and 1/4 unit of transportation; and production of 1 unit of transportation requires 1/3 unit of agriculture and 1/4 unit of manufacturing. Give the input-output matrix for this economy.

SOLUTION

$$\begin{array}{c} \text{Agriculture} \\ \text{Manufacturing} \\ \text{Transportation} \end{array} \begin{array}{ccc} \text{Agriculture} & \text{Manufacturing} & \text{Transportation} \end{array} \\ \begin{bmatrix} 0 & \frac{1}{4} & \frac{1}{3} \\ \frac{1}{2} & 0 & \frac{1}{4} \\ \frac{1}{4} & \frac{1}{4} & 0 \end{bmatrix} = A$$

The first column of the input-output matrix represents the amount of each of the three commodities consumed in the production of 1 unit of agriculture. The second column gives the amounts required to produce 1 unit of manufacturing, and the last column gives the amounts required to produce 1 unit of transportation. (Although it is perhaps unrealistic that production of a unit of each commodity requires none of that commodity, the simpler matrix involved is useful for our purposes.)

NOTE Notice that for each commodity produced, the various units needed are put in a column. Each column corresponds to a commodity produced, and the rows correspond to what is needed to produce the commodity.

Another matrix used with the input-output matrix is the matrix giving the amount of each commodity produced, called the **production matrix**, or the matrix of gross output. In an economy producing n commodities, the production matrix can be represented by a column matrix X with entries $x_1, x_2, x_3, \dots, x_n$.

EXAMPLE 2   Production Matrix

In Example 1, suppose the production matrix is

$$X = \begin{bmatrix} 60 \\ 52 \\ 48 \end{bmatrix}.$$

Then 60 units of agriculture, 52 units of manufacturing, and 48 units of transportation are produced. Because 1/4 unit of agriculture is used for each unit of manufacturing produced, $1/4 \times 52 = 13$ units of agriculture must be used in the "production" of manufacturing. Similarly, $1/3 \times 48 = 16$ units of agriculture will be used in the "production" of transportation. Thus, $13 + 16 = 29$ units of agriculture are used for production in the economy. Look again at the matrices A and X. Since X gives the number of units of each commodity produced and A gives the amount (in units) of each commodity used to produce 1 unit of each of the various commodities, the matrix product AX gives the amount of each commodity used in the production process.

$$AX = \begin{bmatrix} 0 & \frac{1}{4} & \frac{1}{3} \\ \frac{1}{2} & 0 & \frac{1}{4} \\ \frac{1}{4} & \frac{1}{4} & 0 \end{bmatrix} \begin{bmatrix} 60 \\ 52 \\ 48 \end{bmatrix} = \begin{bmatrix} 29 \\ 42 \\ 28 \end{bmatrix}$$

From this result, 29 units of agriculture, 42 units of manufacturing, and 28 units of transportation are used to produce 60 units of agriculture, 52 units of manufacturing, and 48 units of transportation.

The matrix product AX represents the amount of each commodity used in the production process. The remainder (if any) must be enough to satisfy the demand for the various commodities from outside the production system. In an n-commodity economy, this demand can be represented by a **demand matrix** D with entries $d_1, d_2, \dots, d_n$. If no production is to remain unused, the difference between the production matrix X and the amount AX used in the production process must equal the demand D, or

$$D = X - AX.$$

In Example 2,

$$D = \begin{bmatrix} 60 \\ 52 \\ 48 \end{bmatrix} - \begin{bmatrix} 29 \\ 42 \\ 28 \end{bmatrix} = \begin{bmatrix} 31 \\ 10 \\ 20 \end{bmatrix},$$

so production of 60 units of agriculture, 52 units of manufacturing, and 48 units of transportation would satisfy a demand of 31, 10, and 20 units of each commodity, respectively.

In practice, A and D usually are known and X must be found. That is, we need to decide what amounts of production are needed to satisfy the required demands. Matrix algebra can be used to solve the equation $D = X - AX$ for X.

$$D = X - AX$$
$$D = IX - AX \qquad \text{Identity property}$$
$$D = (I - A)X \qquad \text{Distributive property}$$

If the matrix $I - A$ has an inverse, then

$$X = (I - A)^{-1}D.$$

---FOR REVIEW---

Recall that I is the identity matrix, a square matrix in which each element on the main diagonal is 1 and all other elements are 0.

TECHNOLOGY NOTE If the production matrix is large or complicated, we could use a graphing calculator. On the TI-84 Plus, for example, we would enter the command $(\text{identity}(3) - [A])^{-1}*[D]$ for a 3×3 matrix A.

TECHNOLOGY NOTE It is also practical to do these calculations on a spreadsheet. For more details, see the *Graphing Calculator and Excel Spreadsheet Manual* available with this book.

EXAMPLE 3 Demand Matrix

Suppose, in the three-commodity economy from Examples 1 and 2, there is a demand for 516 units of agriculture, 258 units of manufacturing, and 129 units of transportation. What should production of each commodity be?

APPLY IT   **SOLUTION**   The demand matrix is

$$D = \begin{bmatrix} 516 \\ 258 \\ 129 \end{bmatrix}.$$

To find the production matrix X, first calculate $I - A$.

$$I - A = \begin{bmatrix} 1 & 0 & 0 \\ 0 & 1 & 0 \\ 0 & 0 & 1 \end{bmatrix} - \begin{bmatrix} 0 & \frac{1}{4} & \frac{1}{3} \\ \frac{1}{2} & 0 & \frac{1}{4} \\ \frac{1}{4} & \frac{1}{4} & 0 \end{bmatrix} = \begin{bmatrix} 1 & -\frac{1}{4} & -\frac{1}{3} \\ -\frac{1}{2} & 1 & -\frac{1}{4} \\ -\frac{1}{4} & -\frac{1}{4} & 1 \end{bmatrix}$$

Use row operations to find the inverse of $I - A$ (the entries are rounded to three decimal places).

$$(I - A)^{-1} = \begin{bmatrix} 1.395 & 0.496 & 0.589 \\ 0.837 & 1.364 & 0.620 \\ 0.558 & 0.465 & 1.302 \end{bmatrix}$$

Since $X = (I - A)^{-1}D$,

$$X = \begin{bmatrix} 1.395 & 0.496 & 0.589 \\ 0.837 & 1.364 & 0.620 \\ 0.558 & 0.465 & 1.302 \end{bmatrix} \begin{bmatrix} 516 \\ 258 \\ 129 \end{bmatrix} = \begin{bmatrix} 924 \\ 864 \\ 576 \end{bmatrix}.$$

(Each entry in X has been rounded to the nearest whole number.)

The last result shows that production of 924 units of agriculture, 864 units of manufacturing, and 576 units of transportation are required to satisfy demands of 516, 258, and 129 units, respectively.

The entries in the matrix $(I - A)^{-1}$ are often called *multipliers*, and they have important economic interpretations. For example, every \$1 increase in total agricultural demand will result in an increase in agricultural production by about \$1.40, an increase in manufacturing production by about \$0.84, and an increase in transportation production by about \$0.56. Similarly, every \$3 increase in total manufacturing demand will result in an increase of about $3(0.50) = 1.50$, $3(1.36) = 4.08$, and $3(0.47) = 1.41$ dollars in agricultural production, manufacturing production, and transportation production, respectively.

TRY YOUR TURN 1

YOUR TURN 1 Find the production of each commodity for the economy in Example 3 if there is a demand for 322 units of agriculture, 447 units of manufacturing, and 133 units of transportation.

EXAMPLE 4 Wheat and Oil Production

An economy depends on two basic products, wheat and oil. To produce 1 metric ton of wheat requires 0.25 metric tons of wheat and 0.33 metric tons of oil. Production of 1 metric ton of oil consumes 0.08 metric tons of wheat and 0.11 metric tons of oil.

(a) Find the production that will satisfy a demand for 500 metric tons of wheat and 1000 metric tons of oil.

APPLY IT **SOLUTION** The input-output matrix is

$$A = \begin{bmatrix} 0.25 & 0.08 \\ 0.33 & 0.11 \end{bmatrix}.$$

Also,

$$I - A = \begin{bmatrix} 0.75 & -0.08 \\ -0.33 & 0.89 \end{bmatrix}.$$

Next, calculate $(I - A)^{-1}$.

$$(I - A)^{-1} = \begin{bmatrix} 1.3882 & 0.1248 \\ 0.5147 & 1.1699 \end{bmatrix} \quad \text{(rounded)}$$

To find the production matrix X, use the equation $X = (I - A)^{-1}D$, with

$$D = \begin{bmatrix} 500 \\ 1000 \end{bmatrix}.$$

The production matrix is

$$X = \begin{bmatrix} 1.3882 & 0.1248 \\ 0.5147 & 1.1699 \end{bmatrix} \begin{bmatrix} 500 \\ 1000 \end{bmatrix} \approx \begin{bmatrix} 819 \\ 1427 \end{bmatrix}.$$

Production of 819 metric tons of wheat and 1427 metric tons of oil is required to satisfy the indicated demand.

(b) Suppose the demand for wheat goes up from 500 to 600 metric tons. Find the increased production in wheat and oil that will be required to meet the new demand.

SOLUTION One way to solve this problem is using the multipliers for wheat, found in the first column of $(I - A)^{-1}$ from part (a). The element in the first row, 1.3882, is used to find the increased production in wheat, while the item in the second row, 0.5147, is used to find the increased production in oil. Since the increase in demand for wheat is 100 metric tons, the increased production in wheat must be $100(1.3882) \approx 139$ metric tons. Similarly, the increased production in oil is $100(0.5147) \approx 51$ metric tons.

Alternatively, we could have found the new production in wheat and oil with the equation $X = (I - A)^{-1}D$, giving

$$X = \begin{bmatrix} 1.3882 & 0.1248 \\ 0.5147 & 1.1699 \end{bmatrix} \begin{bmatrix} 600 \\ 1000 \end{bmatrix} \approx \begin{bmatrix} 958 \\ 1479 \end{bmatrix}.$$

We find the increased production by subtracting the answers found in part (a) from these answers. The increased production in wheat is $958 - 819 = 139$ metric tons, and the increased production in oil is $1479 - 1427 = 52$ metric tons. The slight difference here from the previous answer of 51 metric tons is due to rounding.

Closed Models

The input-output model discussed above is referred to as an **open model**, since it allows for a surplus from the production equal to D. In the **closed model**, all the production is consumed internally in the production process, so that $X = AX$. There is nothing left over to satisfy any outside demands from other parts of the economy or from other economies. In this case, the sum of each column in the input-output matrix equals 1.

To solve the closed model, set $D = O$ in the equation derived earlier, where O is a **zero matrix**, a matrix whose elements are all zeros.

$$(I - A)X = D = O$$

FOR REVIEW

Parameters were discussed in the first section of this chapter. As mentioned there, parameters are required when a system has infinitely many solutions.

The system of equations that corresponds to $(I - A)X = O$ does not have a single unique solution, but it can be solved in terms of a parameter. (It can be shown that if the columns of a matrix A sum to 1, then the equation $(I - A)X = O$ has an infinite number of solutions.)

EXAMPLE 5 Closed Input-Output Model

Use matrix A below to find the production of each commodity in a closed model.

$$A = \begin{bmatrix} \frac{1}{2} & \frac{1}{4} & \frac{1}{3} \\ 0 & \frac{1}{4} & \frac{1}{3} \\ \frac{1}{2} & \frac{1}{2} & \frac{1}{3} \end{bmatrix}$$

SOLUTION Find the value of $I - A$, then set $(I - A)X = O$ to find X.

$$I - A = \begin{bmatrix} \frac{1}{2} & -\frac{1}{4} & -\frac{1}{3} \\ 0 & \frac{3}{4} & -\frac{1}{3} \\ -\frac{1}{2} & -\frac{1}{2} & \frac{2}{3} \end{bmatrix}$$

$$(I - A)X = \begin{bmatrix} \frac{1}{2} & -\frac{1}{4} & -\frac{1}{3} \\ 0 & \frac{3}{4} & -\frac{1}{3} \\ -\frac{1}{2} & -\frac{1}{2} & \frac{2}{3} \end{bmatrix} \begin{bmatrix} x_1 \\ x_2 \\ x_3 \end{bmatrix} = \begin{bmatrix} 0 \\ 0 \\ 0 \end{bmatrix}$$

Multiply to get

$$\begin{bmatrix} \frac{1}{2}x_1 - \frac{1}{4}x_2 - \frac{1}{3}x_3 \\ 0x_1 + \frac{3}{4}x_2 - \frac{1}{3}x_3 \\ -\frac{1}{2}x_1 - \frac{1}{2}x_2 + \frac{2}{3}x_3 \end{bmatrix} = \begin{bmatrix} 0 \\ 0 \\ 0 \end{bmatrix}.$$

The last matrix equation corresponds to the following system.

$$\frac{1}{2}x_1 - \frac{1}{4}x_2 - \frac{1}{3}x_3 = 0$$
$$\frac{3}{4}x_2 - \frac{1}{3}x_3 = 0$$
$$-\frac{1}{2}x_1 - \frac{1}{2}x_2 + \frac{2}{3}x_3 = 0$$

YOUR TURN 2 Change the last column of matrix A in Example 5 to $\begin{bmatrix} \frac{1}{6} \\ \frac{1}{6} \\ \frac{2}{3} \end{bmatrix}$ and find the solution corresponding to $x_3 = 9$.

Solving the system with x_3 as the parameter gives the solution of the system

$$\left(\tfrac{8}{9}x_3, \tfrac{4}{9}x_3, x_3\right).$$

For example, if $x_3 = 9$ (a choice that eliminates fractions in the answer), then $x_1 = 8$ and $x_2 = 4$, so the production of the three commodities should be in the ratio 8:4:9.

TRY YOUR TURN 2

Finding a Production Matrix

To obtain the production matrix, X, for an open input-output model, follow these steps:

1. Form the $n \times n$ input-output matrix, A, by placing in each column the amount of the various commodities required to produce 1 unit of a particular commodity.

2. Calculate $I - A$, where I is the $n \times n$ identity matrix.

3. Find the inverse, $(I - A)^{-1}$.

4. Multiply the inverse on the right by the demand matrix, D, to obtain $X = (I - A)^{-1}D$.

To obtain a production matrix, X, for a closed input-output model, solve the system $(I - A)X = O$.

Production matrices for actual economies are much larger than those shown in this section. An analysis of the U.S. economy in 1997 has close to 500 commodity categories. *Source: U.S. Bureau of Economic Analysis.* Such matrices require large human and computer resources for their analysis. Some of the exercises at the end of this section use actual data in which categories have been combined to simplify the work.

2.6 EXERCISES

Find the production matrix for the following input-output and demand matrices using the open model.

1. $A = \begin{bmatrix} 0.8 & 0.2 \\ 0.2 & 0.7 \end{bmatrix}$, $D = \begin{bmatrix} 2 \\ 3 \end{bmatrix}$

2. $A = \begin{bmatrix} 0.2 & 0.04 \\ 0.6 & 0.05 \end{bmatrix}$, $D = \begin{bmatrix} 3 \\ 10 \end{bmatrix}$

3. $A = \begin{bmatrix} 0.1 & 0.03 \\ 0.07 & 0.6 \end{bmatrix}$, $D = \begin{bmatrix} 5 \\ 10 \end{bmatrix}$

4. $A = \begin{bmatrix} 0.02 & 0.03 \\ 0.06 & 0.08 \end{bmatrix}$, $D = \begin{bmatrix} 100 \\ 200 \end{bmatrix}$

5. $A = \begin{bmatrix} 0.8 & 0 & 0.1 \\ 0.1 & 0.5 & 0.2 \\ 0 & 0 & 0.7 \end{bmatrix}$, $D = \begin{bmatrix} 1 \\ 6 \\ 3 \end{bmatrix}$

6. $A = \begin{bmatrix} 0.1 & 0.5 & 0 \\ 0 & 0.3 & 0.4 \\ 0.1 & 0.2 & 0.1 \end{bmatrix}$, $D = \begin{bmatrix} 10 \\ 4 \\ 2 \end{bmatrix}$

Find the ratios of products A, B, and C using a closed model.

7.
	A	B	C
A	0.3	0.1	0.8
B	0.5	0.6	0.1
C	0.2	0.3	0.1

8.
	A	B	C
A	0.3	0.2	0.3
B	0.1	0.5	0.4
C	0.6	0.3	0.3

Use a graphing calculator or computer to find the production matrix X, given the following input-output and demand matrices.

9. $A = \begin{bmatrix} 0.25 & 0.25 & 0.25 & 0.05 \\ 0.01 & 0.02 & 0.01 & 0.1 \\ 0.3 & 0.3 & 0.01 & 0.1 \\ 0.2 & 0.01 & 0.3 & 0.01 \end{bmatrix}$, $D = \begin{bmatrix} 2930 \\ 3570 \\ 2300 \\ 580 \end{bmatrix}$

10. $A = \begin{bmatrix} 0.01 & 0.2 & 0.01 & 0.2 \\ 0.5 & 0.02 & 0.03 & 0.02 \\ 0.09 & 0.05 & 0.02 & 0.03 \\ 0.3 & 0.2 & 0.2 & 0.01 \end{bmatrix}$, $D = \begin{bmatrix} 5000 \\ 1000 \\ 4000 \\ 500 \end{bmatrix}$

APPLICATIONS

Business and Economics

Input-Output Open Model In Exercises 11 and 12, refer to Example 4.

11. If the demand is changed to 925 metric tons of wheat and 1250 metric tons of oil, how many units of each commodity should be produced?

12. Change the technological matrix so that production of 1 metric ton of wheat requires 1/5 metric ton of oil (and no wheat), and production of 1 metric ton of oil requires 1/3 metric ton of wheat (and no oil). To satisfy the same demand matrix, how many units of each commodity should be produced?

Input-Output Open Model In Exercises 13–16, refer to Example 3.

13. If the demand is changed to 607 units of each commodity, how many units of each commodity should be produced?

14. Suppose 1/3 unit of manufacturing (no agriculture or transportation) is required to produce 1 unit of agriculture, 1/4 unit of transportation is required to produce 1 unit of manufacturing, and 1/2 unit of agriculture is required to produce 1 unit of transportation. How many units of each commodity should be produced to satisfy a demand of 1000 units of each commodity?

15. Suppose 1/4 unit of manufacturing and 1/2 unit of transportation are required to produce 1 unit of agriculture, 1/2 unit of agriculture and 1/4 unit of transportation to produce 1 unit of manufacturing, and 1/4 unit of agriculture and 1/4 unit of manufacturing to produce 1 unit of transportation. How many units of each commodity should be produced to satisfy a demand of 1000 units for each commodity?

16. If the input-output matrix is changed so that 1/4 unit of manufacturing and 1/2 unit of transportation are required to produce 1 unit of agriculture, 1/2 unit of agriculture and 1/4 unit of transportation are required to produce 1 unit of manufacturing, and 1/4 unit each of agriculture and manufacturing are required to produce 1 unit of transportation, find the number of units of each commodity that should be produced to satisfy a demand for 500 units of each commodity.

Input-Output Open Model

17. A primitive economy depends on two basic goods, yams and pork. Production of 1 bushel of yams requires 1/4 bushel of yams and 1/2 of a pig. To produce 1 pig requires 1/6 bushel of yams. Find the amount of each commodity that should be produced to get the following.

  a. 1 bushel of yams and 1 pig

  b. 100 bushels of yams and 70 pigs

18. A simple economy depends on three commodities: oil, corn, and coffee. Production of 1 unit of oil requires 0.2 unit of oil, 0.4 unit of corn, and no units of coffee. To produce 1 unit of corn requires 0.4 unit of oil, 0.2 unit of corn, and 0.1 unit of coffee. To produce 1 unit of coffee requires 0.2 unit of oil, 0.1 unit of corn, and 0.2 unit of coffee. Find the production required to meet a demand of 1000 units each of oil, corn, and coffee.

19. In his work *Input-Output Economics*, Leontief provides an example of a simplified economy with just three sectors: agriculture, manufacturing, and households (i.e., the sector of the economy that produces labor). It has the following input-output matrix: *Source: Input-Output Economics.*

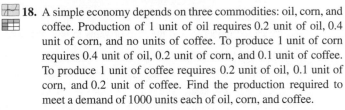

	Agriculture	Manufacturing	Households
Agriculture	0.25	0.40	0.133
Manufacturing	0.14	0.12	0.100
Households	0.80	3.60	0.133

He also gives the demand matrix

$$D = \begin{bmatrix} 35 \\ 38 \\ 40 \end{bmatrix}.$$

Find the amount of each commodity that should be produced.

20. A much-simplified version of Leontief's 42-sector analysis of the 1947 American economy has the following input-output matrix. *Source: Input-Output Economics.*

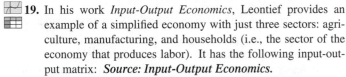

	Agriculture	Manufacturing	Households
Agriculture	0.245	0.102	0.051
Manufacturing	0.099	0.291	0.279
Households	0.433	0.372	0.011

The demand matrix (in billions of dollars) is

$$D = \begin{bmatrix} 2.88 \\ 31.45 \\ 30.91 \end{bmatrix}.$$

Find the amount of each commodity that should be produced.

21. An analysis of the 1958 Israeli economy is simplified here by grouping the economy into three sectors, with the following input-output matrix. *Source: Input-Output Economics.*

	Agriculture	Manufacturing	Energy
Agriculture	0.293	0	0
Manufacturing	0.014	0.207	0.017
Energy	0.044	0.010	0.216

The demand (in thousands of Israeli pounds) as measured by exports is

$$D = \begin{bmatrix} 138,213 \\ 17,597 \\ 1786 \end{bmatrix}.$$

Find the amount of each commodity that should be produced.

22. The 1981 Chinese economy can be simplified to three sectors: agriculture, industry and construction, and transportation and commerce. The input-output matrix is given below. *Source: Input-Output Tables of China, 1981.*

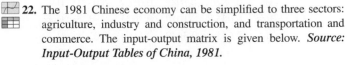

	Agriculture	Industry/ Constr.	Trans./ Commerce
Agriculture	0.158	0.156	0.009
Industry/Constr.	0.136	0.432	0.071
Trans./Commerce	0.013	0.041	0.011

The demand (in 100,000 RMB, the unit of money in China) is

$$D = \begin{bmatrix} 106,674 \\ 144,739 \\ 26,725 \end{bmatrix}.$$

  a. Find the amount of each commodity that should be produced.

  b. Interpret the economic value of an increase in demand of 1 RMB in agricultural exports.

23. Washington The 1987 economy of the state of Washington has been simplified to four sectors: natural resources, manufacturing, trade and services, and personal consumption. The input-output matrix is given below. *Source: University of Washington.*

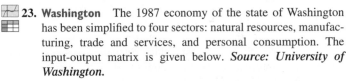

	Natural Resources	Manufacturing	Trade & Services	Personal Consumption
Natural Resources	0.1045	0.0428	0.0029	0.0031
Manufacturing	0.0826	0.1087	0.0584	0.0321
Trade & Services	0.0867	0.1019	0.2032	0.3555
Personal Consumption	0.6253	0.3448	0.6106	0.0798

Suppose the demand (in millions of dollars) is

$$D = \begin{bmatrix} 450 \\ 300 \\ 125 \\ 100 \end{bmatrix}.$$

Find the amount of each commodity that should be produced.

24. Washington In addition to solving the previous input-output model, most models of this nature also include an employment equation. For the previous model, the employment equation is added and a new system of equations is obtained as follows. *Source: University of Washington.*

$$\begin{bmatrix} x_1 \\ x_2 \\ x_3 \\ x_4 \\ N \end{bmatrix} = (I - B)^{-1}C,$$

where x_1, x_2, x_3, x_4 represent the amount, in millions of dollars, that must be produced to satisfy internal and external demands of the four sectors; N is the total workforce required for a particular set of demands; and

$$B = \begin{bmatrix} 0.1045 & 0.0428 & 0.0029 & 0.0031 & 0 \\ 0.0826 & 0.1087 & 0.0584 & 0.0321 & 0 \\ 0.0867 & 0.1019 & 0.2032 & 0.3555 & 0 \\ 0.6253 & 0.3448 & 0.6106 & 0.0798 & 0 \\ 21.6 & 6.6 & 20.2 & 0 & 0 \end{bmatrix}.$$

a. Suppose that a \$50 million change in manufacturing occurs. How will this increase in demand affect the economy?

$$\text{(Hint: Find } (I - B)^{-1}C, \text{ where } C = \begin{bmatrix} 0 \\ 50 \\ 0 \\ 0 \\ 0 \end{bmatrix}.)$$

b. Interpret the meaning of the bottom row in the matrix $(I - B)^{-1}$.

25. Community Links The use of input-output analysis can also be used to model how changes in one city can affect cities that are connected with it in some way. For example, if a large manufacturing company shuts down in one city, it is very likely that the economic welfare of all of the cities around it will suffer. Consider three Pennsylvania communities: Sharon, Farrell, and Hermitage. Due to their proximity to each other, residents of these three communities regularly spend time and money in the other communities. Suppose that we have gathered information in the form of an input-output matrix. *Source: Thayer Watkins.*

$$A = \begin{array}{c} \\ S \\ F \\ H \end{array} \begin{array}{c} \begin{array}{ccc} S & F & H \end{array} \\ \begin{bmatrix} 0.2 & 0.1 & 0.1 \\ 0.1 & 0.1 & 0 \\ 0.5 & 0.6 & 0.7 \end{bmatrix} \end{array}$$

This matrix can be thought of as the likelihood that a person from a particular community will spend money in each of the communities.

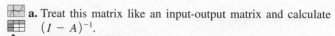

a. Treat this matrix like an input-output matrix and calculate $(I - A)^{-1}$.

b. Interpret the entries of this inverse matrix.

Input-Output Closed Model

26. Use the input-output matrix

$$\begin{array}{c} \\ \text{Yams} \\ \text{Pigs} \end{array} \begin{array}{c} \begin{array}{cc} \text{Yams} & \text{Pigs} \end{array} \\ \begin{bmatrix} \frac{1}{4} & \frac{1}{2} \\ \frac{3}{4} & \frac{1}{2} \end{bmatrix} \end{array}$$

and the closed model to find the ratio of yams to pigs produced.

27. Use the input-output matrix

$$\begin{array}{c} \\ \text{Steel} \\ \text{Coal} \end{array} \begin{array}{c} \begin{array}{cc} \text{Steel} & \text{Coal} \end{array} \\ \begin{bmatrix} \frac{3}{4} & \frac{1}{3} \\ \frac{1}{4} & \frac{2}{3} \end{bmatrix} \end{array}$$

and the closed model to find the ratio of coal to steel produced.

28. Suppose that production of 1 unit of agriculture requires 1/3 unit of agriculture, 1/3 unit of manufacturing, and 1/3 unit of transportation. To produce 1 unit of manufacturing requires 1/2 unit of agriculture, 1/4 unit of manufacturing, and 1/4 unit of transportation. To produce 1 unit of transportation requires 0 units of agriculture, 1/4 unit of manufacturing, and 3/4 unit of transportation. Find the ratio of the three commodities in the closed model.

29. Suppose that production of 1 unit of mining requires 1/5 unit of mining, 2/5 unit of manufacturing, and 2/5 unit of communication. To produce 1 unit of manufacturing requires 3/5 unit of mining, 1/5 unit of manufacturing, and 1/5 unit of communication. To produce 1 unit of communication requires 0 units of mining, 4/5 unit of manufacturing, and 1/5 unit of communication. Find the ratio of the three commodities in the closed model.

■ YOUR TURN ANSWERS

1. 749 units of agriculture, 962 units of manufacturing, and 561 units of transportation

2. (4, 2, 9)

---

2 CHAPTER REVIEW

SUMMARY

In this chapter we extended our study of linear functions to include finding solutions of systems of linear equations. Techniques such as the echelon method and the Gauss-Jordan method were developed and used to solve systems of linear equations. We introduced matrices, which are used to store mathematical information. We saw that matrices can be combined using addition, subtraction, scalar multiplication, and matrix multiplication. Two special matrices, the zero matrix and the identity matrix, were also introduced.

- The zero matrix O is a matrix whose elements are all zero.
- The identity matrix I is an $n \times n$ matrix consisting of 1's along the diagonal and 0's elsewhere.

We then developed the concept of a multiplicative inverse of a matrix and used such inverses to solve systems of equations. We concluded the chapter by introducing the Leontief input-output models, which are used to study interdependencies in an economy.

Row Operations For any augmented matrix of a system of equations, the following operations produce the augmented matrix of an equivalent system:

1. interchanging any two rows;
2. multiplying the elements of a row by a nonzero real number;
3. adding a nonzero multiple of the elements of one row to the corresponding elements of a nonzero multiple of some other row.

The Gauss-Jordan Method
1. Write each equation so that variable terms are in the same order on the left side of the equal sign and constants are on the right.
2. Write the augmented matrix that corresponds to the system.
3. Use row operations to transform the first column so that all elements except the element in the first row are zero.
4. Use row operations to transform the second column so that all elements except the element in the second row are zero.
5. Use row operations to transform the third column so that all elements except the element in the third row are zero.
6. Continue in this way, when possible, until the last row is written in the form

$$[0\,0\,0\,\cdots\,0j\,|\,k],$$

where j and k are constants with $j \neq 0$. When this is not possible, continue until every row has more zeros on the left than the previous row (except possibly for any rows of all zero at the bottom of the matrix), and the first nonzero entry in each row is the only nonzero entry in its column.
7. Multiply each row by the reciprocal of the nonzero element in that row.

Adding Matrices The sum of two $m \times n$ matrices X and Y is the $m \times n$ matrix $X + Y$ in which each element is the sum of the corresponding elements of X and Y.

Subtracting Matrices The difference of two $m \times n$ matrices X and Y is the $m \times n$ matrix $X - Y$ in which each element is found by subtracting the corresponding elements of X and Y.

Product of a Matrix and a Scalar The product of a scalar k and a matrix X is the matrix kX, each of whose elements is k times the corresponding element of X.

Product of Two Matrices Let A be an $m \times n$ matrix and let B be an $n \times k$ matrix. To find the element in the ith row and jth column of the product AB, multiply each element in the ith row of A by the corresponding element in the jth column of B, and then add these products. The product matrix AB is an $m \times k$ matrix.

Solving a System AX = B Using Matrix Inverses To solve a system of equations $AX = B$, where A is a square matrix of coefficients, X is the matrix of variables, and B is the matrix of constants, first find A^{-1}. Then, $X = A^{-1}B$.

Finding a Production Matrix
1. Form the input-output matrix, A.
2. Calculate $I - A$, where I is the $n \times n$ identity matrix.
3. Find the inverse, $(I - A)^{-1}$.
4. Multiply the inverse on the right by the demand matrix, D, to obtain $X = (I - A)^{-1}D$.

To obtain a production matrix, X, for a closed input-output model, solve the system $(I - A)X = O$.

KEY TERMS

To understand the concepts presented in this chapter, you should know the meaning and use of the following terms.
For easy reference, the section in the chapter where a word (or expression) was first used is provided.

system of equations
2.1
first-degree equation in n unknowns

unique solution
inconsistent system
dependent equations
equivalent system

echelon method
back-substitution
parameter

2.2
matrix (matrices)
element (entry)
augmented matrix

row operations	column matrix (column vector)	multiplicative inverse matrix	demand matrix
Gauss-Jordan method	**2.4**	coefficient matrix	open model
2.3	scalar	**2.6**	closed model
size	product matrix	input-output (technological)	zero matrix
square matrix	**2.5**	matrix	
row matrix (row vector)	identity matrix	production matrix	

REVIEW EXERCISES

CONCEPT CHECK

Determine whether each of the following statements is true or false, and explain why.

1. If a system of equations has three equations and four unknowns, then it could have a unique solution.

2. If $A = \begin{bmatrix} 2 & 3 \\ 1 & -1 \end{bmatrix}$ and $B = \begin{bmatrix} 3 & 4 \\ 7 & 4 \\ 1 & 0 \end{bmatrix}$, then $A + B = \begin{bmatrix} 5 & 7 \\ 8 & 3 \\ 1 & 0 \end{bmatrix}$.

3. If a system of equations has three equations and three unknowns, then it may have a unique solution, an infinite number of solutions, or no solutions.

4. The only solution to the system of equations

$$2x + 3y = 7$$
$$5x - 4y = 6$$

is $x = 2$ and $y = 1$.

5. If A is a 2×3 matrix and B is a 3×4 matrix, then $A + B$ is a 2×4 matrix.

6. If A is an $n \times k$ matrix and B is a $k \times m$ matrix, then AB is an $n \times m$ matrix.

7. If A is a 4×4 matrix and B is a 4×4 matrix, then $AB = BA$.

8. A 3×4 matrix could have an inverse.

9. It is not possible to find a matrix A such that $OA = AO = I$, where O is a 5×5 zero matrix and I is a 5×5 identity matrix.

10. When solving a system of equations by the Gauss-Jordan method, we can add a nonzero multiple of the elements of one column to the corresponding elements of some nonzero multiple of some other column.

11. Every square matrix has an inverse.

12. If A, B, and C are matrices such that $AB = C$, then $B = \dfrac{C}{A}$.

13. A system of three equations in three unknowns might have exactly five positive integer solutions.

14. If A and B are matrices such that $A = B^{-1}$, then $AB = BA$.

15. If A, B, and C are matrices such that $AB = CB$, then $A = C$.

16. The difference between an open and a closed input-output model is that in a closed model, the demand matrix D is a zero matrix.

PRACTICE AND EXPLORATIONS

17. What is true about the number of solutions to a system of m linear equations in n unknowns if $m = n$? If $m < n$? If $m > n$?

18. Suppose someone says that a more reasonable way to multiply two matrices than the method presented in the text is to multiply corresponding elements. For example, the result of

$$\begin{bmatrix} 1 & 2 \\ 3 & 4 \end{bmatrix} \cdot \begin{bmatrix} 3 & 5 \\ 7 & 11 \end{bmatrix} \quad \text{should be} \quad \begin{bmatrix} 3 & 10 \\ 21 & 44 \end{bmatrix},$$

according to this person. How would you respond?

Solve each system by the echelon method.

19. $2x - 3y = 14$
    $3x + 2y = -5$

20. $\dfrac{x}{2} + \dfrac{y}{4} = 3$
    $\dfrac{x}{4} - \dfrac{y}{2} = 4$

21. $2x - 3y + z = -5$
    $5x + 5y + 3z = 14$

22. $2x + 3y + 4z = 5$
    $3x + 4y + 5z = 6$

Solve each system by the Gauss-Jordan method.

23. $2x + 4y = -6$
    $-3x - 5y = 12$

24. $x - 4y = 10$
    $5x + 3y = 119$

25. $x - y + 3z = 13$
    $4x + y + 2z = 17$
    $3x + 2y + 2z = 1$

26. $x + 2y + 3z = 9$
    $x - 2y = 4$
    $3x + 2z = 12$

27. $3x - 6y + 9z = 12$
    $-x + 2y - 3z = -4$
    $x + y + 2z = 7$

28. $x - 2z = 5$
    $3x + 2y = 8$
    $-x + 2z = 10$

Find the size of each matrix, find the values of any variables, and identify any square, row, or column matrices.

29. $\begin{bmatrix} 2 & 3 \\ 5 & q \end{bmatrix} = \begin{bmatrix} a & b \\ c & 9 \end{bmatrix}$

30. $\begin{bmatrix} 2 & x \\ y & 6 \\ 5 & z \end{bmatrix} = \begin{bmatrix} a & -1 \\ 4 & 6 \\ p & 7 \end{bmatrix}$

31. $\begin{bmatrix} 2m & 4 & 3z & -12 \end{bmatrix} = \begin{bmatrix} 12 & k+1 & -9 & r-3 \end{bmatrix}$

32. $\begin{bmatrix} a+5 & 3b & 6 \\ 4c & 2+d & -3 \\ -1 & 4p & q-1 \end{bmatrix} = \begin{bmatrix} -7 & b+2 & 2k-3 \\ 3 & 2d-1 & 4l \\ m & 12 & 8 \end{bmatrix}$

Given the matrices

$$A = \begin{bmatrix} 4 & 10 \\ -2 & -3 \\ 6 & 9 \end{bmatrix}, \quad B = \begin{bmatrix} 2 & 3 & -2 \\ 2 & 4 & 0 \\ 0 & 1 & 2 \end{bmatrix}, \quad C = \begin{bmatrix} 5 & 0 \\ -1 & 3 \\ 4 & 7 \end{bmatrix},$$

$$D = \begin{bmatrix} 6 \\ 1 \\ 0 \end{bmatrix}, \quad E = \begin{bmatrix} 1 & 3 & -4 \end{bmatrix},$$

$$F = \begin{bmatrix} -1 & 4 \\ 3 & 7 \end{bmatrix}, \quad G = \begin{bmatrix} -2 & 0 \\ 1 & 5 \end{bmatrix},$$

find each of the following, if it exists.

33. $A + C$ **34.** $2G - 4F$ **35.** $3C + 2A$

36. $B - C$ **37.** $2A - 5C$ **38.** AG

39. AC **40.** DE **41.** ED

42. BD **43.** EC **44.** F^{-1}

45. B^{-1} **46.** $(A + C)^{-1}$

Find the inverse of each matrix that has an inverse.

47. $\begin{bmatrix} 1 & 3 \\ 2 & 7 \end{bmatrix}$ **48.** $\begin{bmatrix} -4 & 2 \\ 0 & 3 \end{bmatrix}$

49. $\begin{bmatrix} 3 & -6 \\ -4 & 8 \end{bmatrix}$ **50.** $\begin{bmatrix} 6 & 4 \\ 3 & 2 \end{bmatrix}$

51. $\begin{bmatrix} 2 & -1 & 0 \\ 1 & 0 & 1 \\ 1 & -2 & 0 \end{bmatrix}$ **52.** $\begin{bmatrix} 2 & 0 & 4 \\ 1 & -1 & 0 \\ 0 & 1 & -2 \end{bmatrix}$

53. $\begin{bmatrix} 1 & 3 & 6 \\ 4 & 0 & 9 \\ 5 & 15 & 30 \end{bmatrix}$ **54.** $\begin{bmatrix} 2 & -3 & 4 \\ 1 & 5 & 7 \\ -4 & 6 & -8 \end{bmatrix}$

Solve the matrix equation $AX = B$ for X using the given matrices.

55. $A = \begin{bmatrix} 5 & 1 \\ -2 & -2 \end{bmatrix}, \quad B = \begin{bmatrix} -8 \\ 24 \end{bmatrix}$

56. $A = \begin{bmatrix} 1 & 2 \\ 2 & 4 \end{bmatrix}, \quad B = \begin{bmatrix} 5 \\ 10 \end{bmatrix}$

57. $A = \begin{bmatrix} 1 & 0 & 2 \\ -1 & 1 & 0 \\ 3 & 0 & 4 \end{bmatrix}, \quad B = \begin{bmatrix} 8 \\ 4 \\ -6 \end{bmatrix}$

58. $A = \begin{bmatrix} 2 & 4 & 0 \\ 1 & -2 & 0 \\ 0 & 0 & 3 \end{bmatrix}, \quad B = \begin{bmatrix} 72 \\ -24 \\ 48 \end{bmatrix}$

Solve each system of equations by inverses.

59. $x + 2y = 4$
   $2x - 3y = 1$

60. $5x + 10y = 80$
   $3x - 2y = 120$

61. $x + y + z = 1$
   $2x + y = -2$
   $+ 3y + z = 2$

62. $x - 4y + 2z = -1$
   $-2x + y - 3z = -9$
   $3x + 5y - 2z = 7$

Find each production matrix, given the following input-output and demand matrices.

63. $A = \begin{bmatrix} 0.01 & 0.05 \\ 0.04 & 0.03 \end{bmatrix}, \quad D = \begin{bmatrix} 200 \\ 300 \end{bmatrix}$

64. $A = \begin{bmatrix} 0.2 & 0.1 & 0.3 \\ 0.1 & 0 & 0.2 \\ 0 & 0 & 0.4 \end{bmatrix}, \quad D = \begin{bmatrix} 500 \\ 200 \\ 100 \end{bmatrix}$

65. The following system of equations is given.

$$x + 2y + z = 7$$
$$2x - y - z = 2$$
$$3x - 3y + 2z = -5$$

a. Solve by the echelon method.

b. Solve by the Gauss-Jordan method. Compare with the echelon method.

c. Write the system as a matrix equation, $AX = B$.

d. Find the inverse of matrix A from part c.

e. Solve the system using A^{-1} from part d.

APPLICATIONS

Business and Economics

In Exercises 66–69, write a system of equations and solve.

66. Scheduling Production An office supply manufacturer makes two kinds of paper clips, standard and extra large. To make 1000 standard paper clips requires 1/4 hour on a cutting machine and 1/2 hour on a machine that shapes the clips. One thousand extra large paper clips require 1/3 hour on each machine. The manager of paper clip production has 4 hours per day available on the cutting machine and 6 hours per day on the shaping machine. How many of each kind of clip can he make?

67. Production Requirements The Waputi Indians make woven blankets, rugs, and skirts. Each blanket requires 24 hours for spinning the yarn, 4 hours for dyeing the yarn, and 15 hours for weaving. Rugs require 30, 5, and 18 hours and skirts 12, 3, and 9 hours, respectively. If there are 306, 59, and 201 hours available for spinning, dyeing, and weaving, respectively, how many of each item can be made? (*Hint:* Simplify the equations you write, if possible, before solving the system.)

68. Distribution An oil refinery in Tulsa sells 50% of its production to a Chicago distributor, 20% to a Dallas distributor, and 30% to an Atlanta distributor. Another refinery in New Orleans sells 40% of its production to the Chicago distributor, 40% to the Dallas distributor, and 20% to the Atlanta distributor. A third refinery in Ardmore sells the same distributors 30%, 40%, and 30% of its production. The three distributors received 219,000, 192,000, and 144,000 gal of oil, respectively. How many gallons of oil were produced at each of the three plants?

69. Stock Reports The New York Stock Exchange reports in daily newspapers give the dividend, price-to-earnings ratio, sales (in hundreds of shares), last price, and change in price for each company. Write the following stock reports as a 4 × 5 matrix: American Telephone & Telegraph: 1.33, 17.6, 152,000, 26.75,

+1.88; General Electric: 1.00, 20.0, 238,200, 32.36, −1.50; Sara Lee: 0.79, 25.4, 39,110, 16.51, −0.89; Walt Disney Company: 0.27, 21.2, 122,500, 28.60, +0.75.

70. Filling Orders A printer has three orders for pamphlets that require three kinds of paper, as shown in the following matrix.

		Order		
		I	II	III
	High-grade	10	5	8
Paper	Medium-grade	12	0	4
	Coated	0	10	5

The printer has on hand 3170 sheets of high-grade paper, 2360 sheets of medium-grade paper, and 1800 sheets of coated paper. All the paper must be used in preparing the order.

a. Write a 3×1 matrix for the amounts of paper on hand.

b. Write a matrix of variables to represent the number of pamphlets that must be printed in each of the three orders.

c. Write a matrix equation using the given matrix and your matrices from parts a and b.

d. Solve the equation from part c.

71. Input-Output An economy depends on two commodities, goats and cheese. It takes 2/3 of a unit of goats to produce 1 unit of cheese and 1/2 unit of cheese to produce 1 unit of goats.

a. Write the input-output matrix for this economy.

b. Find the production required to satisfy a demand of 400 units of cheese and 800 units of goats.

72. Nebraska The 1970 economy of the state of Nebraska has been condensed to six sectors: livestock, crops, food products, mining and manufacturing, households, and other. The input-output matrix is given below. *Source: University of Nebraska Lincoln.*

$$\begin{bmatrix} 0.178 & 0.018 & 0.411 & 0 & 0.005 & 0 \\ 0.143 & 0.018 & 0.088 & 0 & 0.001 & 0 \\ 0.089 & 0 & 0.035 & 0 & 0.060 & 0.003 \\ 0.001 & 0.010 & 0.012 & 0.063 & 0.007 & 0.014 \\ 0.141 & 0.252 & 0.088 & 0.089 & 0.402 & 0.124 \\ 0.188 & 0.156 & 0.103 & 0.255 & 0.008 & 0.474 \end{bmatrix}$$

a. Find the matrix $(I - A)^{-1}$ and interpret the value in row 2, column 1 of this matrix.

b. Suppose the demand (in millions of dollars) is

$$D = \begin{bmatrix} 1980 \\ 650 \\ 1750 \\ 1000 \\ 2500 \\ 3750 \end{bmatrix}.$$

Find the dollar amount of each commodity that should be produced.

Life Sciences

73. Animal Activity The activities of a grazing animal can be classified roughly into three categories: grazing, moving, and resting. Suppose horses spend 8 hours grazing, 8 moving, and 8 resting; cattle spend 10 grazing, 5 moving, and 9 resting; sheep spend 7 grazing, 10 moving, and 7 resting; and goats spend 8 grazing, 9 moving, and 7 resting. Write this information as a 4×3 matrix.

74. CAT Scans Computer Aided Tomography (CAT) scanners take X-rays of a part of the body from different directions, and put the information together to create a picture of a cross section of the body. The amount by which the energy of the X-ray decreases, measured in linear-attenuation units, tells whether the X-ray has passed through healthy tissue, tumorous tissue, or bone, based on the following table. *Source: The Mathematics Teacher.*

Type of Tissue	Linear-Attenuation Values
Healthy tissue	0.1625–0.2977
Tumorous tissue	0.2679–0.3930
Bone	0.3857–0.5108

The part of the body to be scanned is divided into cells. If an X-ray passes through more than one cell, the total linear-attenuation value is the sum of the values for the cells. For example, in the figure, let a, b, and c be the values for cells A, B, and C. The attenuation value for beam 1 is $a + b$ and for beam 2 is $a + c$.

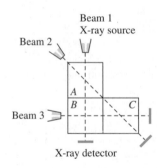

a. Find the attenuation value for beam 3.

b. Suppose that the attenuation values are 0.8, 0.55, and 0.65 for beams 1, 2, and 3, respectively. Set up and solve the system of three equations for a, b, and c. What can you conclude about cells A, B, and C?

c. Find the inverse of the coefficient matrix from part b to find a, b, and c for the following three cases, and make conclusions about cells A, B, and C for each.

Patient	Linear-Attenuation Values		
	Beam 1	Beam 2	Beam 3
X	0.54	0.40	0.52
Y	0.65	0.80	0.75
Z	0.51	0.49	0.44

75. **CAT Scans** (Refer to Exercise 74.) Four X-ray beams are aimed at four cells, as shown in the following figure. *Source: The Mathematics Teacher.*

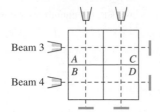

Beam 1 Beam 2

Beam 3
Beam 4

a. Suppose the attenuation values for beams 1, 2, 3, and 4 are 0.60, 0.75, 0.65, and 0.70, respectively. Do we have enough information to determine the values of a, b, c, and d? Explain.

b. Suppose we have the data from part a, as well as the following values for d. Find the values for a, b, and c, and draw conclusions about cells A, B, C, and D in each case.

(i) 0.33   (ii) 0.43

c. Two X-ray beams are added, as shown in the figure. In addition to the data in part a, we now have attenuation values for beams 5 and 6 of 0.85 and 0.50. Find the values for a, b, c, and d, and make conclusions about cells A, B, C, and D. 109

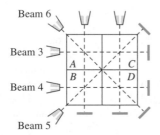

Beam 1 Beam 2

Beam 6
Beam 3
Beam 4
Beam 5

d. Six X-ray beams are not necessary because four appropriately chosen beams are sufficient. Give two examples of four beams (chosen from beams 1–6 in part c) that will give the solution. (*Note:* There are 12 possible solutions.)

e. Discuss what properties the four beams selected in part d must have in order to provide a unique solution.

76. **Hockey** In a recent study, the number of head and neck injuries among hockey players wearing full face shields and half face shields were compared. The following table provides the rates per 1000 athlete-exposures for specific injuries that caused a player wearing either shield to miss one or more events. *Source: JAMA.*

	Half Shield	Full Shield
Head and Face Injuries (excluding Concussions)	3.54	1.41
Concussions	1.53	1.57
Neck Injuries	0.34	0.29
Other	7.53	6.21

If an equal number of players in a large league wear each type of shield and the total number of athlete-exposures for the league in a season is 8000, use matrix operations to estimate the total number of injuries of each type.

Physical Sciences

77. **Roof Trusses** Linear systems occur in the design of roof trusses for new homes and buildings. The simplest type of roof truss is a triangle. The truss shown in the figure below is used to frame roofs of small buildings. If a 100-lb force is applied at the peak of the truss, then the forces or weights W_1 and W_2 exerted parallel to each rafter of the truss are determined by the following linear system of equations.

$$\frac{\sqrt{3}}{2}(W_1 + W_2) = 100$$

$$W_1 - W_2 = 0$$

Solve the system to find W_1 and W_2. *Source: Structural Analysis.*

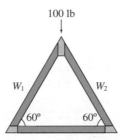

100 lb

W_1 W_2

60°   60°

78. **Roof Trusses** (Refer to Exercise 77.) Use the following system of equations to determine the force or weights W_1 and W_2 exerted on each rafter for the truss shown in the figure.

$$\frac{1}{2}W_1 + \frac{\sqrt{2}}{2}W_2 = 150$$

$$\frac{\sqrt{3}}{2}W_1 - \frac{\sqrt{2}}{2}W_2 = 0$$

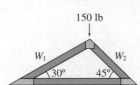

150 lb

W_1 W_2

30°   45°

79. Carbon Dioxide Determining the amount of carbon dioxide in the atmosphere is important because carbon dioxide is known to be a greenhouse gas. Carbon dioxide concentrations (in parts per million) have been measured at Mauna Loa, Hawaii, for more than 40 years. The concentrations have increased quadratically. The table lists readings for 3 years. *Source: Scripps Institution of Oceanography.*

Year	CO_2
1960	317
1980	339
2004	377

a. If the relationship between the carbon dioxide concentration C and the year t is expressed as $C = at^2 + bt + c$, where $t = 0$ corresponds to 1960, use a linear system of equations to determine the constants $a, b,$ and c.

b. Predict the year when the amount of carbon dioxide in the atmosphere will double from its 1960 level. (*Hint:* This requires solving a quadratic equation. For review on how to do this, see Section R.4.)

80. Chemistry When carbon monoxide (CO) reacts with oxygen (O_2), carbon dioxide (CO_2) is formed. This can be written as $CO + (1/2)O_2 = CO_2$ and as a matrix equation. If we form a 2×1 column matrix by letting the first element be the number of carbon atoms and the second element be the number of oxygen atoms, then CO would have the column matrix *Source: Journal of Chemical Education.*

$$\begin{bmatrix} 1 \\ 1 \end{bmatrix}.$$

Similarly, O_2 and CO_2 would have the column matrices $\begin{bmatrix} 0 \\ 2 \end{bmatrix}$ and $\begin{bmatrix} 1 \\ 2 \end{bmatrix}$, respectively.

a. Use the Gauss-Jordan method to find numbers x and y (known as *stoichiometric numbers*) that solve the system of equations

$$\begin{bmatrix} 1 \\ 1 \end{bmatrix} x + \begin{bmatrix} 0 \\ 2 \end{bmatrix} y = \begin{bmatrix} 1 \\ 2 \end{bmatrix}.$$

Compare your answers to the equation written above.

b. Repeat the process for $xCO_2 + yH_2 + zCO = H_2O$, where H_2 is hydrogen, and H_2O is water. In words, what does this mean?

General Interest

81. Students Suppose 20% of the boys and 30% of the girls in a high school like tennis, and 60% of the boys and 90% of the girls like math. If 500 students like tennis and 1500 like math, how many boys and girls are in the school? Find all possible solutions.

82. Baseball In the 2009 Major League Baseball season, slugger Ichiro Suzuki had a total of 225 hits. The number of singles he hit was 11 more than four times the combined total of doubles and home runs. The number of doubles he hit was 1 more than twice the combined total of triples and home runs. The number of singles and home runs together was 15 more than five times the combined total of doubles and triples. Find the number of singles, doubles, triples, and home runs that Suzuki hit during the season. *Source: Baseball-Reference.com.*

83. Cookies Regular Nabisco Oreo cookies are made of two chocolate cookie wafers surrounding a single layer of vanilla cream. The claim on the package states that a single serving is 34 g, which is three cookies. Nabisco Double Stuf cookies are made of the same two chocolate cookie wafers surrounding a double layer of vanilla cream. The claim on this package states that a single serving is 29 g, which is two Double Stuf cookies. If the Double Stuf cookies truly have a double layer of vanilla cream, find the weight of a single chocolate wafer and the weight of a single layer of vanilla cream.

EXTENDED APPLICATION

CONTAGION

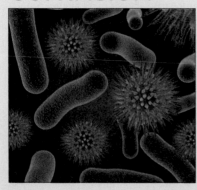

Suppose that three people have contracted a contagious disease. *Source: Finite Mathematics with Applications to Business, Life Sciences, and Social Sciences.* A second group of five people may have been in contact with the three infected persons. A third group of six people may have been in contact with the

second group. We can form a 3×5 matrix P with rows representing the first group of three and columns representing the second group of five. We enter a one in the corresponding position if a person in the first group has contact with a person in the second group. These direct contacts are called *first-order contacts*. Similarly, we form a 5×6 matrix Q representing the first-order contacts between the second and third group. For example, suppose

$$P = \begin{bmatrix} 1 & 0 & 0 & 1 & 0 \\ 0 & 0 & 1 & 1 & 0 \\ 1 & 1 & 0 & 0 & 0 \end{bmatrix} \text{ and}$$

$$Q = \begin{bmatrix} 1 & 1 & 0 & 1 & 1 & 1 \\ 0 & 0 & 0 & 0 & 1 & 0 \\ 0 & 0 & 0 & 0 & 0 & 0 \\ 0 & 1 & 0 & 1 & 0 & 0 \\ 1 & 0 & 0 & 0 & 1 & 0 \end{bmatrix}.$$

From matrix P we see that the first person in the first group had contact with the first and fourth persons in the second group. Also, none of the first group had contact with the last person in the second group.

A *second-order contact* is an indirect contact between persons in the first and third groups through some person in the second group. The product matrix PQ indicates these contacts. Verify that the second-row, fourth-column entry of PQ is 1. That is, there is one second-order contact between the second person in group one and the fourth person in group three. Let a_{ij} denote the element in the ith row and jth column of the matrix PQ. By looking at the products that form a_{24} below, we see that the common contact was with the fourth individual in group two. (The p_{ij} are entries in P, and the q_{ij} are entries in Q.)

$$a_{24} = p_{21}q_{14} + p_{22}q_{24} + p_{23}q_{34} + p_{24}q_{44} + p_{25}q_{54}$$
$$= 0 \cdot 1 + 0 \cdot 0 + 1 \cdot 0 + 1 \cdot 1 + 0 \cdot 0$$
$$= 1$$

The second person in group 1 and the fourth person in group 3 both had contact with the fourth person in group 2.

This idea could be extended to third-, fourth-, and larger-order contacts. It indicates a way to use matrices to trace the spread of a contagious disease. It could also pertain to the dispersal of ideas or anything that might pass from one individual to another.

EXERCISES

1. Find the second-order contact matrix PQ mentioned in the text.

2. How many second-order contacts were there between the second contagious person and the third person in the third group?

3. Is there anyone in the third group who has had no contacts at all with the first group?

4. The totals of the columns in PQ give the total number of second-order contacts per person, while the column totals in P and Q give the total number of first-order contacts per person. Which person(s) in the third group had the most contacts, counting first- and second-order contacts?

5. Go to the website WolframAlpha.com and enter: "multiply matrices." Study how matrix multiplication can be performed by Wolfram|Alpha. Try Exercise 1 with Wolfram|Alpha and discuss how it compares with Microsoft Excel and with your graphing calculator.

DIRECTIONS FOR GROUP PROJECT

Assume that your group (3–5 students) is trying to map the spread of a new disease. Suppose also that the information given above has been obtained from interviews with the first three people that were hospitalized with symptoms of the disease and their contacts. Using the questions above as a guide, prepare a presentation for a public meeting that describes the method of obtaining the data, the data itself, and addresses the spirit of each question. Formulate a strategy for how to handle the spread of this disease to other people. The presentation should be mathematically sound, grammatically correct, and professionally crafted. Use presentation software, such as Microsoft PowerPoint, to present your findings.

3

Linear Programming:
The Graphical Method

An oil refinery turns crude oil into many different products, including gasoline and fuel oil. Efficient management requires matching the output of each product to the demand and the available shipping capacity. In an exercise in Section 3, we explore the use of linear programming to allocate refinery production for maximum profit.

Many realistic problems involve inequalities—a factory can manufacture *no more than* 12 items on a shift, or a medical researcher must interview *at least* a hundred patients to be sure that a new treatment for a disease is better than the old treatment. *Linear inequalities* of the form $ax + by \leq c$ (or with $\geq$, $<$, or $>$ instead of $\leq$) can be used in a process called *linear programming* to *optimize* (find the maximum or minimum value of a quantity) for a given situation.

In this chapter we introduce some *linear programming* problems that can be solved by graphical methods. Then, in Chapter 4, we discuss the simplex method, a general method for solving linear programming problems with many variables.

3.1 Graphing Linear Inequalities

APPLY IT **How can a company determine the feasible number of units of each product to manufacture in order to meet all production requirements?** *We will answer this question in Example 6 by graphing a set of inequalities.*

As mentioned earlier, a linear inequality is defined as follows.

Linear Inequality
A **linear inequality** in two variables has the form

$$ax + by \leq c$$
$$ax + by < c,$$
$$ax + by \geq c,$$
or
$$ax + by > c,$$

for real numbers a, b, and c, with a and b not both 0.

EXAMPLE 1 Graphing an Inequality

Graph the linear inequality $2x - 3y \leq 12$.

SOLUTION Because of the "=" portion of $\leq$, the points of the line $2x - 3y = 12$ satisfy the linear inequality $2x - 3y \leq 12$ and are part of its graph. As in Chapter 1, find the intercepts by first letting $x = 0$ and then letting $y = 0$; use these points to get the graph of $2x - 3y = 12$ shown in Figure 1.

FOR REVIEW

Recall from Chapter 1 that one way to sketch a line is to first let $x = 0$ to find the y-intercept, then let $y = 0$ to find the x-intercept. For example, given $2x - 3y = 12$, letting $x = 0$ yields $-3y = 12$, so $y = -4$, and the corresponding point is $(0, -4)$. Letting $y = 0$ yields $2x = 12$, so $x = 6$ and the point is $(6, 0)$. Plot these two points, as in Figure 1, then use a straightedge to draw a line through them.

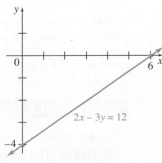

FIGURE 1

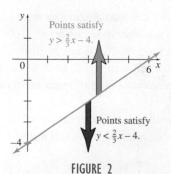

Points satisfy $y > \frac{2}{3}x - 4.$

Points satisfy $y < \frac{2}{3}x - 4.$

FIGURE 2

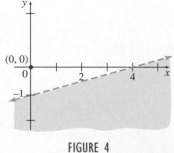

$2x - 3y \leq 12$

(0, 0)

FIGURE 3

The points on the line satisfy "$2x - 3y$ *equals* 12." To locate the points satisfying "$2x - 3y$ *is less than* or equal to 12," first solve $2x - 3y \leq 12$ for y.

$$2x - 3y \leq 12$$
$$-3y \leq -2x + 12 \qquad \text{Subtract } 2x.$$
$$y \geq \frac{2}{3}x - 4 \qquad \text{Multiply by } -\frac{1}{3}.$$

(Recall that multiplying both sides of an inequality by a negative number reverses the direction of the inequality symbol.)

As shown in Figure 2, the points *above* the line $2x - 3y = 12$ satisfy

$$y > \frac{2}{3}x - 4,$$

while those below the line satisfy

$$y < \frac{2}{3}x - 4.$$

In summary, the inequality $2x - 3y \leq 12$ is satisfied by all points *on or above* the line $2x - 3y = 12$. Indicate the points above the line by shading, as in Figure 3. The line and shaded region in Figure 3 make up the graph of the linear inequality $2x - 3y \leq 12$. ▮

CAUTION In this chapter, be sure to use a straightedge to draw lines and to plot the points with care. A sloppily drawn line could give a deceptive picture of the region being considered.

In Example 1, the line $2x - 3y = 12$, which separates the points in the solution from the points that are not in the solution, is called the **boundary**.

There is an alternative way to find the correct region to shade or to check the method shown above. Choose as a test point any point not on the boundary line. For example, in Example 1 we could choose the point $(0, 0)$, which is not on the line $2x - 3y = 12$. Substitute 0 for x and 0 for y in the given inequality.

$$2x - 3y \leq 12$$
$$2(0) - 3(0) \leq 12$$
$$0 \leq 12 \qquad \text{True}$$

Since the result $0 \leq 12$ is true, the test point $(0, 0)$ belongs on the side of the boundary where all points satisfy $2x - 3y < 12$. For this reason, we shade the side containing $(0, 0)$, as in Figure 3. Choosing a point on the other side of the line, such as $(4, -3)$, would produce a false result when the values $x = 4$ and $y = -3$ were substituted into the given inequality. In such a case, we would shade the side of the line *not including* the test point.

NOTE Many of the inequalities in this chapter are of the form $ax + by \leq c$ or $ax + by \geq c$. These are most easily graphed by changing the $\leq$ or $\geq$ to $=$ and then plotting the two intercepts by letting $x = 0$ and $y = 0$, as in Example 1. Shade the appropriate side of the line as described above. If both $a > 0$ and $b > 0$ the shaded region for $ax + by \leq c$ is *below* the line, while the shaded region for $ax + by \geq c$ is *above* the line.

EXAMPLE 2 Graphing an Inequality

Graph $x - 4y > 4$.

SOLUTION The boundary here is the line $x - 4y = 4$. Since the points on this line do not satisfy $x - 4y > 4$, the line is drawn dashed, as in Figure 4. To decide whether to shade the

(0, 0)

FIGURE 4

region above the line or the region below the line, we will choose a test point not on the boundary line. Choosing $(0, 0)$, we replace x with 0 and y with 0:

$$x - 4y > 4$$
$$0 - 4(0) > 4$$
$$0 > 4. \quad \text{False}$$

The correct half-plane is the one that does *not* contain $(0, 0)$; the region below the boundary line is shaded, as shown in Figure 4.   **TRY YOUR TURN 1**

> **CAUTION** Be careful. If the point $(0, 0)$ is on the boundary line, it cannot be used as a test point.

As the examples above suggest, the graph of a linear inequality is represented by a shaded region in the plane, perhaps including the line that is the boundary of the region. Each shaded region is an example of a **half-plane**, a region on one side of a line. For example, in Figure 5 line r divides the plane into half-planes P and Q. The points on r belong neither to P nor to Q. Line r is the boundary of each half-plane.

YOUR TURN 1

Graph $3x + 2y \leq 18$.

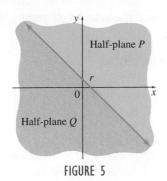

Half-plane P

r

Half-plane Q

FIGURE 5

TECHNOLOGY NOTE

Graphing calculators can shade regions on the plane. TI calculators have a DRAW menu that includes an option to shade above or below a line. For instance, to graph the inequality in Example 2, first solve the equation for y, then use your calculator to graph the line $y = (1/4)x - 1$. Then the command `Shade(-10, (1/4)X - 1)` produces Figure 6(a).

The TI-84 Plus calculator offers another way to graph the region above or below a line. Press the $Y=$ key. Note the slanted line to the left of Y_1, Y_2, and so on. Use the left arrow key to move the cursor to that position for Y_1. Press ENTER until you see the symbol ◤. This indicates that the calculator will shade below the line whose equation is entered in Y_1. (The symbol ◥ operates similarly to shade above a line.) We used this method to get the graph in Figure 6(b). For more details, see the *Graphing Calculator and Excel Spreadsheet Manual* available with this book.

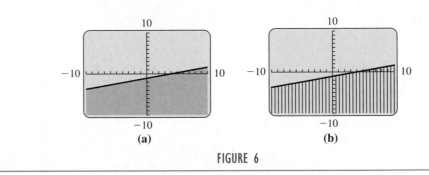

(a)   (b)

FIGURE 6

The steps in graphing a linear inequality are summarized below.

Graphing a Linear Inequality

1. Draw the graph of the boundary line. Make the line solid if the inequality involves $\leq$ or $\geq$; make the line dashed if the inequality involves $<$ or $>$.

2. Decide which half-plane to shade. Use either of the following methods.

   a. Solve the inequality for y; shade the region above the line if the inequality is of the form $y >$ or $y \geq$; shade the region below the line if the inequality is of the form $y <$ or $y \leq$.

   b. Choose any point not on the line as a test point. Shade the half-plane that includes the test point if the test point satisfies the original inequality; otherwise, shade the half-plane on the other side of the boundary line.

Systems of Inequalities

Realistic problems often involve many inequalities. For example, a manufacturing problem might produce inequalities resulting from production requirements as well as inequalities about cost requirements. A collection of at least two inequalities is called a **system of inequalities**. The solution of a system of inequalities is made up of all those points that satisfy all the inequalities of the system at the same time. To graph the solution of a system of inequalities, graph all the inequalities on the same axes and identify, by heavy shading or direction arrows, the region common to all graphs. The next example shows how this is done.

EXAMPLE 3   Graphing a System of Inequalities

Graph the system

$$y < -3x + 12$$
$$x < 2y.$$

SOLUTION   The graph of the first inequality has the line $y = -3x + 12$ as its boundary. Because of the $<$ symbol, we use a dotted line and shade *below* the line. The second inequality should first be solved for y to get $y > (1/2)x$ to see that the graph is the region *above* the dotted boundary line $y = (1/2)x$.

The heavily shaded region in Figure 7(a) shows all the points that satisfy both inequalities of the system. Since the points on the boundary lines are not in the solution, the boundary lines are dashed.

NOTE
When shading regions by hand, it may be difficult to tell what is shaded heavily and what is shaded only lightly, particularly when more than two inequalities are involved. In such cases, an alternative technique, called *reverse shading,* is to shade the region *opposite* that of the inequality. In other words, the region that is *not* wanted can be shaded. Then, when the various regions are shaded, whatever is not shaded is the desired region. We will not use this technique in this text, but you may wish to try it on your own.

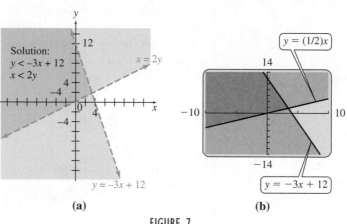

(a)   (b)

FIGURE 7

📈 **TECHNOLOGY NOTE**   A calculator graph of the system in Example 3 is shown in Figure 7(b). You can graph this system on a TI-84 Plus using `Shade(Y₂,Y₁)`.

A region consisting of the overlapping parts of two or more graphs of inequalities in a system, such as the heavily shaded region in Figure 7, is sometimes called the **region of feasible solutions** or the **feasible region**, since it is made up of all the points that satisfy (are feasible for) all inequalities of the system.

EXAMPLE 4   Graphing a Feasible Region

Graph the feasible region for the system

$$y \le -2x + 8$$
$$-2 \le x \le 1.$$

SOLUTION   The boundary line of the first inequality is $y = -2x + 8$. Because of the $\le$ symbol, we use a solid line and shade *below* the line.

YOUR TURN 2 Graph the feasible region for the system

$$3x - 4y \geq 12$$
$$x + y \geq 0.$$

The second inequality is a compound inequality, indicating $-2 \leq x$ *and* $x \leq 1$. Recall that the graph $x = -2$ is the vertical line through $(-2, 0)$ and that the graph $x = 1$ is the vertical line through $(1, 0)$. For $-2 \leq x$, we draw a vertical line and shade the region to the right. For $x \leq 1$, we draw a vertical line and shade the region to the left.

The shaded region in Figure 8 shows all the points that satisfy the system of inequalities. **TRY YOUR TURN 2**

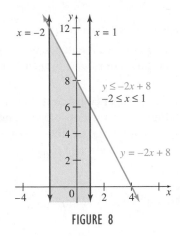

FIGURE 8

EXAMPLE 5 Graphing a Feasible Region

Graph the feasible region for the system

$$2x - 5y \leq 10$$
$$x + 2y \leq 8$$
$$x \geq 0$$
$$y \geq 0.$$

SOLUTION On the same axes, graph each inequality by graphing the boundary and choosing the appropriate half-plane. Then find the feasible region by locating the overlap of all the half-planes. This feasible region is shaded in Figure 9. The inequalities $x \geq 0$ and $y \geq 0$ restrict the feasible region to the first quadrant.

The feasible region in Example 5 is **bounded**, since the region is enclosed by boundary lines on all sides. On the other hand, the feasible regions in Examples 3 and 4 are **unbounded.**

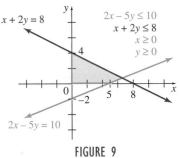

FIGURE 9

Applications
As shown in Section 3 of this chapter, many realistic problems lead to systems of linear inequalities. The next example is typical of such problems.

EXAMPLE 6 Manufacturing

Happy Ice Cream Cone Company makes cake cones and sugar cones, both of which must be processed in the mixing department and the baking department. Manufacturing one batch of cake cones requires 1 hour in the mixing department and 2 hours in the baking department, and producing one batch of sugar cones requires 2 hours in the mixing department and 1 hour in the baking department. Each department is operated for at most 12 hours per day.

APPLY IT

(a) Write a system of inequalities that expresses these restrictions.

SOLUTION Let x represent the number of batches of cake cones made and y represent the number of batches of sugar cones made. Then, make a table that summarizes the given information.

Production Requirements for Ice Cream Cones				
	Cake	Sugar		Total
Number of Units Made	x	y		
Hours in Mixing Dept.	1	2	$\leq$	12
Hours in Baking Dept.	2	1	$\leq$	12

Since the departments operate at most 12 hours per day, we put the total number of hours as ≤ 12. Putting the inequality ($\leq$ or $\geq$) next to the number in the chart may help you remember which way to write the inequality.

In the mixing department, x batches of cake cones require a total of $1 \cdot x = x$ hours, and y batches of sugar cones require $2 \cdot y = 2y$ hours. Since the mixing department can operate no more than 12 hours per day,

$$x + 2y \leq 12. \quad \text{Mixing department}$$

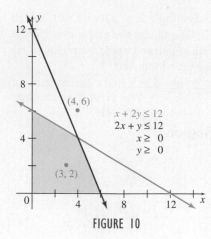

FIGURE 10

We translated "no more than" as "less than or equal to." Notice how this inequality corresponds to the row in the table for the mixing department. Similarly, the row corresponding to the baking department gives

$$2x + y \leq 12. \quad \text{Baking department}$$

Since it is not possible to produce a negative number of cake cones or sugar cones,

$$x \geq 0 \quad \text{and} \quad y \geq 0.$$

(b) Graph the feasible region.

SOLUTION The feasible region for this system of inequalities is shown in Figure 10.

(c) Using the graph from part (b), can 3 batches of cake cones and 2 batches of sugar cones be manufactured in one day? Can 4 batches of cake cones and 6 batches of sugar cones be manufactured in one day?

SOLUTION Three batches of cake cones and two batches of sugar cones correspond to the point $(3, 2)$. Since $(3, 2)$ is in the feasible region in Figure 10, it is possible to manufacture these quantities in one day. However, since $(4, 6)$ is *not* in the feasible region in Figure 10, it is *not* possible to manufacture 4 batches of cake cones and 6 batches of sugar cones in one day.

The following steps summarize the process of finding the feasible region.

1. Form a table that summarizes the information.

2. Convert the table into a set of linear inequalities.

3. Graph each linear inequality.

4. Graph the region that is common to all the regions graphed in step 3.

3.1 EXERCISES

Graph each linear inequality.

1. $x + y \leq 2$

2. $y \leq x + 1$

3. $x \geq 2 - y$

4. $y \geq x - 3$

5. $4x - y < 6$

6. $4y + x > 6$

7. $4x + y < 8$

8. $2x - y > 2$

9. $x + 3y \geq -2$

10. $2x + 3y \leq 6$

11. $x \leq 3y$

12. $2x \geq y$

13. $x + y \leq 0$

14. $3x + 2y \geq 0$

15. $y < x$

16. $y > 5x$

17. $x < 4$

18. $y > 5$

19. $y \leq -2$

20. $x \geq -4$

Graph the feasible region for each system of inequalities. Tell whether each region is bounded or unbounded.

21. $x + y \leq 1$
$x - y \geq 2$

22. $4x - y < 6$
$3x + y < 9$

23. $x + 3y \leq 6$
$2x + 4y \geq 7$

24. $-x - y < 5$
$2x - y < 4$

25. $x + y \leq 7$
$x - y \leq -4$
$4x + y \geq 0$

26. $3x - 2y \geq 6$
$x + y \leq -5$
$y \leq -6$

27. $-2 < x < 3$
$-1 \leq y \leq 5$
$2x + y < 6$

28. $1 < x < 4$
$y > 2$
$x > y$

29. $y - 2x \leq 4$
$y \geq 2 - x$
$x \geq 0$
$y \geq 0$

30. $2x + 3y \leq 12$
$2x + 3y > 3$
$3x + y < 4$
$x \geq 0$
$y \geq 0$

31. $3x + 4y > 12$
$2x - 3y < 6$
$0 \leq y \leq 2$
$x \geq 0$

32. $0 \leq x \leq 9$
$x - 2y \geq 4$
$3x + 5y \leq 30$
$y \geq 0$

Use a graphing calculator to graph the following.

33. $2x - 6y > 12$

34. $4x - 3y < 12$

35. $3x - 4y < 6$
$2x + 5y > 15$

36. $6x - 4y > 8$
$2x + 5y < 5$

37. The regions A through G in the figure can be described by the inequalities

$$x + 3y \ ? \ 6$$
$$x + y \ ? \ 3$$
$$x - 2y \ ? \ 2$$
$$x \geq 0$$
$$y \geq 0,$$

where ? can be either $\leq$ or $\geq$. For each region, tell what the ? should be in the three inequalities. For example, for region A, the ? should be $\geq$, $\leq$, and $\leq$, because region A is described by the inequalities

$$x + 3y \geq 6$$
$$x + y \leq 3$$
$$x - 2y \leq 2$$
$$x \geq 0$$
$$y \geq 0.$$

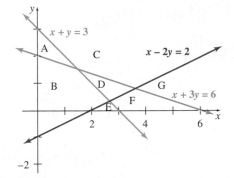

APPLICATIONS

Business and Economics

38. Production Scheduling A small pottery shop makes two kinds of planters, glazed and unglazed. The glazed type requires 1/2 hour to throw on the wheel and 1 hour in the kiln. The unglazed type takes 1 hour to throw on the wheel and 6 hours in the kiln. The wheel is available for at most 8 hours per day, and the kiln for at most 20 hours per day.

a. Complete the following table.

	Glazed	Unglazed	Total
Number Made	x	y	
Time on Wheel			
Time in Kiln			

b. Set up a system of inequalities and graph the feasible region.

c. Using your graph from part b, can 5 glazed and 2 unglazed planters be made? Can 10 glazed and 2 unglazed planters be made?

39. Time Management Carmella and Walt produce handmade shawls and afghans. They spin the yarn, dye it, and then weave it. A shawl requires 1 hour of spinning, 1 hour of dyeing, and 1 hour of weaving. An afghan needs 2 hours of spinning, 1 hour of dyeing, and 4 hours of weaving. Together, they spend at most 8 hours spinning, 6 hours dyeing, and 14 hours weaving.

a. Complete the following table.

	Shawls	Afghans	Total
Number Made	x	y	
Spinning Time			
Dyeing Time			
Weaving Time			

b. Set up a system of inequalities and graph the feasible region.

c. Using your graph from part b, can 3 shawls and 2 afghans be made? Can 4 shawls and 3 afghans be made?

For Exercises 40–45, perform the following steps.

a. Write a system of inequalities to express the conditions of the problem.

b. Graph the feasible region of the system.

40. Transportation Southwestern Oil supplies two distributors located in the Northwest. One distributor needs at least 3000 barrels of oil, and the other needs at least 5000 barrels. Southwestern can send out at most 10,000 barrels. Let $x =$ the number of barrels of oil sent to distributor 1 and $y =$ the number sent to distributor 2.

41. Finance The loan department in a bank will use at most $30 million for commercial and home loans. The bank's policy is to allocate at least four times as much money to home loans as to commercial loans. The bank's return is 6% on a home loan and 8% on a commercial loan. The manager of the loan department wants to earn a return of at least $1.6 million on these loans. Let $x =$ the amount (in millions) for home loans and $y =$ the amount (in millions) for commercial loans.

42. Transportation The California Almond Growers have at most 2400 boxes of almonds to be shipped from their plant in Sacramento to Des Moines and San Antonio. The Des Moines market needs at least 1000 boxes, while the San Antonio market must have at least 800 boxes. Let $x =$ the number of boxes to be shipped to Des Moines and $y =$ the number of boxes to be shipped to San Antonio.

43. Management The Gillette Company produces two popular battery-operated razors, the M3Power™ and the Fusion Power™. Because of demand, the number of M3Power™ razors is never more than one-half the number of Fusion Power™ razors. The factory's production cannot exceed 800 razors per day. Let $x =$ the number of M3Power™ razors and $y =$ the number of Fusion Power™ razors produced per day.

44. Production Scheduling A cement manufacturer produces at least 3.2 million barrels of cement annually. He is told by the Environmental Protection Agency (EPA) that his operation emits 2.5 lb of dust for each barrel produced. The EPA has ruled that annual emissions must be reduced to no more than 1.8 million lb. To do this, the manufacturer plans to replace the present dust collectors with two types of electronic precipitators. One type would reduce emissions to 0.5 lb per barrel and operating costs would be 16¢ per barrel. The other would reduce the dust to 0.3 lb per barrel and operating costs would be 20¢ per barrel. The manufacturer does not want to spend more than 0.8 million dollars in operating costs on the precipitators. He needs to know how many barrels he could produce with each type. Let x = the number of barrels (in millions) produced with the first type and y = the number of barrels (in millions) produced with the second type.

Life Sciences

45. Nutrition A dietician is planning a snack package of fruit and nuts. Each ounce of fruit will supply 1 unit of protein, 2 units

of carbohydrates, and 1 unit of fat. Each ounce of nuts will supply 1 unit of protein, 1 unit of carbohydrates, and 1 unit of fat. Every package must provide at least 7 units of protein, at least 10 units of carbohydrates, and no more than 9 units of fat. Let x = the ounces of fruit and y = the ounces of nuts to be used in each package.

■■■YOUR TURN ANSWERS

1.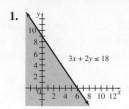

2.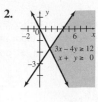

3.2 Solving Linear Programming Problems Graphically

Many mathematical models designed to solve problems in business, biology, and economics involve finding an optimum value (maximum or minimum) of a function, subject to certain restrictions. In a **linear programming** problem, we must find the maximum or minimum value of a function, called the **objective function**, and satisfy a set of restrictions, or **constraints**, given by linear inequalities. When only two variables are involved, the solution to a linear programming problem can be found by first graphing the set of constraints, then finding the feasible region as discussed in the previous section. This method is explained in the following example.

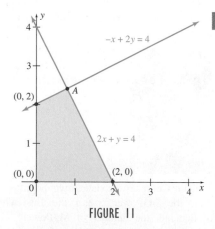

FIGURE 11

EXAMPLE 1 **Maximization**

Find the maximum value of the objective function $z = 3x + 4y$, subject to the following constraints.

$$2x + y \leq 4$$
$$-x + 2y \leq 4$$
$$x \geq 0$$
$$y \geq 0$$

SOLUTION The feasible region is graphed in Figure 11. We can find the coordinates of point A, $(4/5, 12/5)$, by solving the system

$$2x + y = 4$$
$$-x + 2y = 4.$$

Every point in the feasible region satisfies all the constraints; however, we want to find those points that produce the maximum possible value of the objective function. To see how to find this maximum value, change the graph of Figure 11 by adding lines that represent the objective function $z = 3x + 4y$ for various sample values of z. By choosing the values 0, 5, 10, and 15 for z, the objective function becomes (in turn)

$$0 = 3x + 4y, \quad 5 = 3x + 4y, \quad 10 = 3x + 4y, \quad \text{and} \quad 15 = 3x + 4y.$$

These four lines (known as **isoprofit lines**) are graphed in Figure 12. (Why are the lines parallel?) The figure shows that z cannot take on the value 15 because the graph for $z = 15$ is entirely outside the feasible region. The maximum possible value of z will be obtained from a line parallel to the others and between the lines representing the objective function when $z = 10$ and $z = 15$. The value of z will be as large as possible and all constraints will be satisfied if this line just touches the feasible region. This occurs at point A. We find that A has coordinates $(4/5, 12/5)$. (See the review in the margin.) The value of z at this point is

$$z = 3x + 4y = 3\left(\frac{4}{5}\right) + 4\left(\frac{12}{5}\right) = \frac{60}{5} = 12.$$

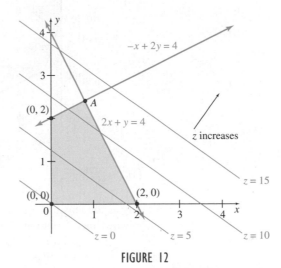

FIGURE 12

The maximum possible value of z is 12. Of all the points in the feasible region, A leads to the largest possible value of z.

A graphing calculator can be useful for finding the coordinates of intersection points such as point A. We do this by solving each equation for y, graphing each line, and then using the capability of the calculator to find the coordinates of the point of intersection.

Points such as A in Example 1 are called corner points. A **corner point** is a point in the feasible region where the boundary lines of two constraints cross. Since corner points occur where two straight lines cross, the coordinates of a corner point are the solution of a system of two linear equations. As we saw in Example 1, corner points play a key role in the solution of linear programming problems. We will make this explicit after the following example.

EXAMPLE 2 **Minimization**

Solve the following linear programming problem.

$$\text{Minimize} \quad z = 2x + 4y$$
$$\text{subject to:} \quad x + 2y \geq 10$$
$$3x + y \geq 10$$
$$x \geq 0$$
$$y \geq 0.$$

SOLUTION Figure 13 on the next page shows the feasible region and the lines that result when z in the objective function is replaced by 0, 10, 20, 40, and 50. The line representing the objective function touches the region of feasible solutions when $z = 20$. Two corner points, $(2, 4)$ and $(10, 0)$, lie on this line; both $(2, 4)$ and $(10, 0)$, as well as all the points on the boundary line between them, give the same optimum value of z. There are infinitely many equally good values of x and y that will give the same minimum value of the objective function $z = 2x + 4y$. This minimum value is 20.

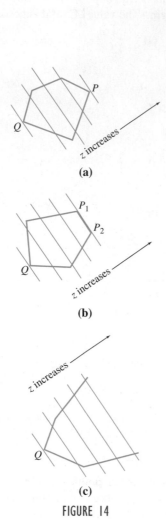

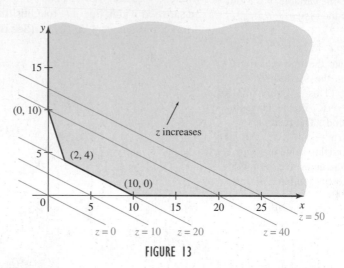

FIGURE 13

As long as the feasible region is not empty, linear programming problems with bounded regions always have solutions. On the other hand, the feasible region in Example 2 is unbounded, and no solution will *maximize* the value of the objective function because z can be made as large as you like.

Some general conclusions can be drawn from the method of solution used in Examples 1 and 2. Imagine a line sliding across a region. Figure 14 shows various feasible regions and the position of the line $ax + by = z$ for various values of z. We assume here that as the line slides from the lower left to the upper right, the value of z increases. In Figure 14(a), the objective function takes on its minimum value at corner point Q and its maximum value at P. The minimum is again at Q in part (b), but the maximum occurs at P_1 or P_2, or any point on the line segment connecting them. Finally, in part (c), the minimum value occurs at Q, but the objective function has no maximum value because the feasible region is unbounded. As long as the objective function increases as x and y increase, the objective function will have no maximum over an unbounded region.

The preceding discussion suggests the truth of the **corner point theorem**.

FIGURE 14

Corner Point Theorem

If an optimum value (either a maximum or a minimum) of the objective function exists, it will occur at one or more of the corner points of the feasible region.

This theorem simplifies the job of finding an optimum value. First, we graph the feasible region and find all corner points. Then we test each corner point in the objective function. Finally, we identify the corner point producing the optimum solution. For unbounded regions, we must decide whether the required optimum can be found (see Example 2).

With the theorem, we can solve the problem in Example 1 by first identifying the four corner points in Figure 11: $(0, 0)$, $(0, 2)$, $(4/5, 12/5)$, and $(2, 0)$. Then we substitute each of the four points into the objective function $z = 3x + 4y$ to identify the corner point that produces the maximum value of z.

Values of the Objective Function at Corner Points	
Corner Point	Value of $z = 3x + 4y$
$(0, 0)$	$3(0) + 4(0) = 0$
$(0, 2)$	$3(0) + 4(2) = 8$
$\left(\frac{4}{5}, \frac{12}{5}\right)$	$3\left(\frac{4}{5}\right) + 4\left(\frac{12}{5}\right) = 12$ Maximum
$(2, 0)$	$3(2) + 4(0) = 6$

From these results, the corner point $(4/5, 12/5)$ yields the maximum value of 12. This is the same as the result found earlier.

The following summary gives the steps to use in solving a linear programming problem by the graphical method.

NOTE
As the corner point theorem states and Example 2 illustrates, the optimal value of a linear programming problem may occur at more than one corner point. When the optimal solution occurs at two corner points, every point on the line segment between the two points is also an optimal solution.

Solving a Linear Programming Problem

1. Write the objective function and all necessary constraints.
2. Graph the feasible region.
3. Identify all corner points.
4. Find the value of the objective function at each corner point.
5. For a bounded region, the solution is given by the corner point producing the optimum value of the objective function.
6. For an unbounded region, check that a solution actually exists. If it does, it will occur at a corner point.

When asked to solve a linear programming problem, give the maximum or minimum value, as well any points where that value occurs.

EXAMPLE 3   **Maximization and Minimization**

Sketch the feasible region for the following set of constraints, and then find the maximum and minimum values of the objective function $z = x + 10y$.

$$x + 4y \geq 12$$
$$x - 2y \leq 0$$
$$2y - x \leq 6$$
$$x \leq 6$$

SOLUTION   The graph in Figure 15 shows that the feasible region is bounded. Use the corner points from the graph to find the maximum and minimum values of the objective function.

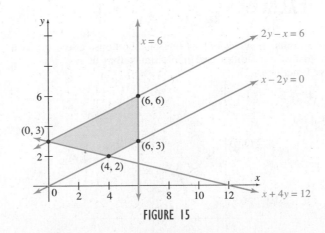

FIGURE 15

YOUR TURN 1 Find the maximum and minimum values of the objective function $z = 3x + 4y$ on the region bounded by

$$2x + \ y \geq \ 5$$
$$x + 5y \geq 16$$
$$2x + \ y \leq 14$$
$$-x + 4y \leq 20$$

Values of the Objective Function at Corner Points		
Corner Point	**Value of $z = x + 10y$**	
$(0, 3)$	$0 + 10(3) = 30$	
$(4, 2)$	$4 + 10(2) = 24$	Minimum
$(6, 3)$	$6 + 10(3) = 36$	
$(6, 6)$	$6 + 10(6) = 66$	Maximum

The minimum value of $z = x + 10y$ is 24 at the corner point $(4, 2)$. The maximum value is 66 at $(6, 6)$. **TRY YOUR TURN 1**

To verify that the minimum or maximum is correct in a linear programming problem, you might want to add the graph of the line $z = 0$ to the graph of the feasible region. For instance, in Example 3, the result of adding the line $x + 10y = 0$ is shown in Figure 16. Now imagine moving a straightedge through the feasible region parallel to this line. It appears that the first place the line touches the feasible region is at $(4, 2)$, where we found the minimum. Similarly, the last place the line touches is at $(6, 6)$, where we found the maximum. In Figure 16, these parallel lines, labeled $z = 24$ and $z = 66$, are also shown.

NOTE
The graphical method is very difficult for a problem with three variables (where the feasible region is three dimensional) and impossible with four or more variables. A method for solving such problems will be shown in the next chapter.

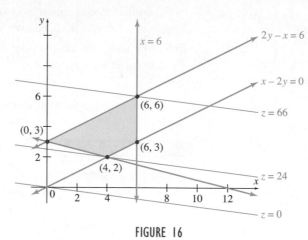

FIGURE 16

3.2   EXERCISES

The following graphs show regions of feasible solutions. Use these regions to find maximum and minimum values of the given objective functions.

1. a. $z = 3x + 2y$

  b. $z = \ x + 4y$

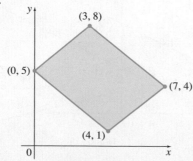

2. a. $z = x + 4y$

  b. $z = 5x + 2y$

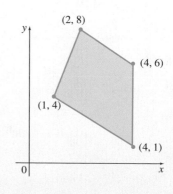

3. a. $z = 0.40x + 0.75y$

b. $z = 1.50x + 0.25y$

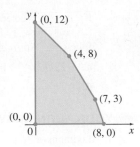

4. a. $z = 0.35x + 1.25y$

b. $z = 1.5x + 0.5y$

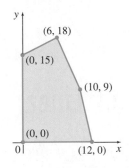

5. a. $z = 4x + 2y$

b. $z = 2x + 3y$

c. $z = 2x + 4y$

d. $z = x + 4y$

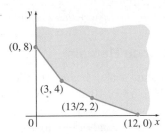

6. a. $z = 4x + y$

b. $z = 5x + 6y$

c. $z = x + 2y$

d. $z = x + 6y$

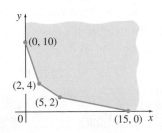

Use graphical methods to solve each linear programming problem.

7. Minimize $z = 4x + 7y$

subject to: $x - y \geq 1$

$3x + 2y \geq 18$

$x \geq 0$

$y \geq 0.$

8. Minimize $z = x + 3y$

subject to: $x + y \leq 10$

$5x + 2y \geq 20$

$-x + 2y \geq 0$

$x \geq 0$

$y \geq 0.$

9. Maximize $z = 5x + 2y$

subject to: $4x - y \leq 16$

$2x + y \geq 11$

$x \geq 3$

$y \leq 8.$

10. Maximize $z = 10x + 8y$

subject to: $2x + 3y \leq 100$

$5x + 4y \leq 200$

$x \geq 10$

$0 \leq y \leq 20.$

11. Maximize $z = 10x + 10y$

subject to: $5x + 8y \geq 200$

$25x - 10y \geq 250$

$x + y \leq 150$

$x \geq 0$

$y \geq 0.$

12. Maximize $z = 4x + 5y$

subject to: $10x - 5y \leq 100$

$20x + 10y \geq 150$

$x + y \geq 12$

$x \geq 0$

$y \geq 0.$

13. Maximize $z = 3x + 6y$

subject to: $2x - 3y \leq 12$

$x + y \leq 5$

$3x + 4y \geq 24$

$x \geq 0$

$y \geq 0.$

14. Maximize $z = 4x + 6y$

subject to: $3 \leq x + y \leq 10$

$x - y \geq 3$

$x \geq 0$

$y \geq 0.$

15. Find values of $x \geq 0$ and $y \geq 0$ that maximize $z = 10x + 12y$ subject to each set of constraints.

a. $x + y \leq 20$ **b.** $3x + y \leq 15$ **c.** $2x + 5y \geq 22$

$x + 3y \leq 24$ $x + 2y \leq 18$ $4x + 3y \leq 28$

$2x + 2y \leq 17$

16. Find values of $x \geq 0$ and $y \geq 0$ that minimize $z = 3x + 2y$ subject to each set of constraints.

a. $10x + 7y \leq 42$ **b.** $6x + 5y \geq 25$ **c.** $x + 2y \geq 10$

$4x + 10y \geq 35$ $2x + 6y \geq 15$ $2x + y \geq 12$

$x - y \leq 8$

17. You are given the following linear programming problem:*

Maximize $z = c_1x_1 + c_2x_2$

subject to: $2x_1 + x_2 \leq 11$

$-x_1 + 2x_2 \leq 2$

$x_1 \geq 0, x_2 \geq 0.$

If $c_2 > 0$, determine the range of c_1/c_2 for which $(x_1, x_2) = (4, 3)$ is an optimal solution. (Choose one of the following.) *Source: Society of Actuaries.*

a. $[-2, 1/2]$ **b.** $[-1/2, 2]$ **c.** $[-11, -1]$

d. $[1, 11]$ **e.** $[-11, 11]$

*The notation x_1 and x_2 used in this exercise is the same as x and y.

▬▬YOUR TURN ANSWERS

1. Maximum of 36 at (4,6); minimum of 15 at (1,3)

3.3 Applications of Linear Programming

APPLY IT   How many canoes and kayaks should a business purchase, given a limited budget and limited storage?

We will use linear programming to answer this question in Example 1.

EXAMPLE 1   Canoe Rentals

Andrew Crowley plans to start a new business called River Explorers, which will rent canoes and kayaks to people to travel 10 miles down the Clarion River in Cook Forest State Park. He has $45,000 to purchase new boats. He can buy the canoes for $600 each and the kayaks for $750 each. His facility can hold up to 65 boats. The canoes will rent for $25 a day, and the kayaks will rent for $30 a day. How many canoes and how many kayaks should he buy to earn the most revenue if all boats can be rented each day?

**APPLY IT**   **SOLUTION**   Let x represent the number of canoes and let y represent the number of kayaks. Summarize the given information in a table.

Rental Information				
	Canoes	Kayaks		Total
Number of Boats	x	y	$\leq$	65
Cost of Each	$600	$750	$\leq$	$45,000
Revenue	$25	$30		

The constraints, imposed by the number of boats and the cost, correspond to the rows in the table as follows.

$$x + y \leq 65$$

$$600x + 750y \leq 45{,}000$$

Dividing both sides of the second constraint by 150 gives the equivalent inequality

$$4x + 5y \leq 300.$$

Since the number of boats cannot be negative, $x \geq 0$ and $y \geq 0$. The objective function to be maximized gives the amount of revenue. If the variable z represents the total revenue, the objective function is

$$z = 25x + 30y.$$

In summary, the mathematical model for the given linear programming problem is as follows:

$$\text{Maximize} \qquad z = 25x + 30y \tag{1}$$
$$\text{subject to:} \qquad x + y \le 65 \tag{2}$$
$$4x + 5y \le 300 \tag{3}$$
$$x \ge 0 \tag{4}$$
$$y \ge 0. \tag{5}$$

Using the methods described in the previous section, graph the feasible region for the system of inequalities (2)–(5), as in Figure 17. Three of the corner points can be identified from the graph as (0, 0), (65, 0), and (0, 60). The fourth corner point, labeled Q in the figure, can be found by solving the system of equations

$$x + y = 65$$
$$4x + 5y = 300.$$

Solve this system to find that Q is the point (25, 40). Now test these four points in the objective function to determine the maximum value of z. The results are shown in the table.

YOUR TURN 1 Solve Example 1 with everything the same except the revenue for a kayak changed to $35 a day.

YOUR TURN 2 Suppose that in Example 6 in Sec. 3.1, the company earns $20 in profit for each batch of cake cones and $30 for each batch of sugar cones. How many batches of each should the company make to maximize profit?

Values of the Objective Function at Corner Points		
Corner Point	**Value of $z = 25x + 30y$**	
$(0, 0)$	$25(0) + 30(0) = 0$	
$(65, 0)$	$25(65) + 30(0) = 1625$	
$(0, 60)$	$25(0) + 30(60) = 1800$	
$(25, 40)$	$25(25) + 30(40) = 1825$	Maximum

The objective function, which represents revenue, is maximized when $x = 25$ and $y = 40$. He should buy 25 canoes and 40 kayaks for a maximum revenue of $1825 a day.

TRY YOUR TURN 1 AND 2

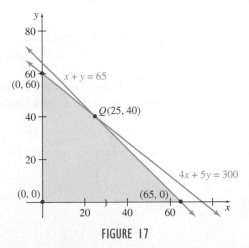

FIGURE 17

Fortunately, the answer to the linear programming problem in Example 1 is a point with integer coordinates, as the number of each type of boat must be an integer. Unfortunately, there is no guarantee that this will always happen at a corner point. When the solution to a linear programming problem is restricted to integers, it is an *integer programming* problem, which is more difficult to solve than a linear programming problem. The feasible region for an integer programming problem consists only of those points with integer coordinates that satisfy the constraints. In this text, all problems in which fractional solutions are meaningless are contrived to have integer solutions.

EXAMPLE 2 Farm Animals

A 4-H member raises only goats and pigs. She wants to raise no more than 16 animals, including no more than 10 goats. She spends $25 to raise a goat and $75 to raise a pig, and she has $900 available for this project. Each goat produces $12 in profit and each pig $40 in profit. How many goats and how many pigs should she raise to maximize total profit?

SOLUTION   First, set up a table that shows the information given in the problem.

4-H Animal Information				
	Goats	Pigs		Total
Number Raised	x	y	$\leq$	16
Goat Limit	x		$\leq$	10
Cost to Raise	$25	$75	$\leq$	$900
Profit (each)	$12	$40		

Use the table to write the necessary constraints. Since the total number of animals cannot exceed 16, the first constraint is

$$x + y \leq 16.$$

"No more than 10 goats" means

$$x \leq 10.$$

The cost to raise x goats at $25 per goat is $25x$ dollars, while the cost for y pigs at $75 each is $75y$ dollars. Since only $900 is available,

$$25x + 75y \leq 900.$$

Dividing both sides by 25 gives the equivalent inequality

$$x + 3y \leq 36.$$

The number of goats and pigs cannot be negative, so

$$x \geq 0 \quad \text{and} \quad y \geq 0.$$

The 4-H member wants to know how many goats and pigs to raise in order to produce maximum profit. Each goat yields $12 profit and each pig $40. If z represents total profit, then

$$z = 12x + 40y.$$

In summary, we have the following linear programming problem:

$$
\begin{aligned}
\text{Maximize} \quad & z = 12x + 40y \\
\text{subject to:} \quad & x + y \leq 16 \\
& x + 3y \leq 36 \\
& x \leq 10 \\
& x \geq 0 \\
& y \geq 0.
\end{aligned}
$$

A graph of the feasible region is shown in Figure 18. The corner points $(0, 12)$, $(0, 0)$, and $(10, 0)$ can be read directly from the graph. The coordinates of each of the other corner points can be found by solving a system of linear equations.

Test each corner point in the objective function to find the maximum profit.

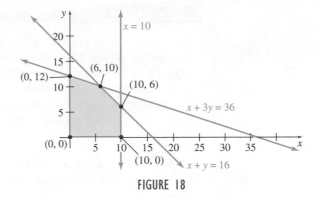

FIGURE 18

YOUR TURN 3 Solve Example 2 with everything the same except the total amount available for the project changed to $1050.

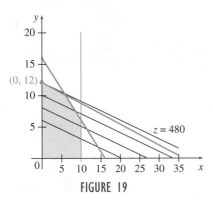

FIGURE 19

Values of the Objective Function at Corner Points		
Corner Point	**Value of $z = 12x + 40y$**	
$(0, 12)$	$12(0) + 40(12) = 480$	Maximum
$(6, 10)$	$12(6) + 40(10) = 472$	
$(10, 6)$	$12(10) + 40(6) = 360$	
$(10, 0)$	$12(10) + 40(0) = 120$	
$(0, 0)$	$12(0) + 40(0) = 0$	

The maximum of 480 occurs at $(0, 12)$. Thus, 12 pigs and no goats will produce a maximum profit of $480. **TRY YOUR TURN 3**

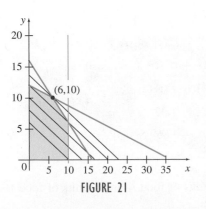

FIGURE 20

In the maximization problem in Example 2, since the profit for a single pig is $40 and the profit for a single goat is only $12, it is more profitable to raise only pigs and no goats. However, if the profit from raising pigs begins to decrease (or the profit from goats begins to increase), it will eventually be more profitable to raise both goats and pigs. In fact, if the profit from raising pigs decreases to a number below $36, then the previous solution is no longer optimal.

To see why this is true, in Figure 19 we have graphed the original objective function $(z = 12x + 40y)$ for various values of z, as we did in Example 1 of the previous section. Notice that each of these objective lines has slope $m = -12/40 = -3/10$. When $z = 480$, the line touches only one feasible point, $(0, 12)$, which is where the maximum profit occurs.

If the profit from raising pigs decreases from $40 to p, where p is a value slightly below 40, the objective function lines will have the equation $z = 12x + py$ for various values of z, and the slope of the lines becomes $m = -12/p$. Eventually, as p becomes smaller, the slope of these objective lines will be equal to the slope of the line $x + 3y = 36$ (that is, $-1/3$), corresponding to the second constraint. This occurs when $-12/p = -1/3$, or $p = 36$, as illustrated by the overlapping blue and dotted lines in Figure 20. In this case, the optimal solution occurs at every point on the line segment that joins $(0, 12)$ and $(6, 10)$.

Once the profit from raising pigs decreases to below $36, the slopes of the sample objective function lines become more negative (steeper) and the optimal solution changes, as indicated in Figure 21. As z increases, the last feasible point that the lines touch is $(6, 10)$. For profits from raising pigs that are slightly below $36, the optimal solution will occur when $x = 6$ and $y = 10$. In other words, the maximum profit will occur when she raises both goats and pigs.

FIGURE 21

EXAMPLE 3 Nutrition

Certain animals in a rescue shelter must have at least 30 g of protein and at least 20 g of fat per feeding period. These nutrients come from food A, which costs 18 cents per unit and supplies 2 g of protein and 4 g of fat; and food B, which costs 12 cents per unit and has 6 g of protein and 2 g of fat. Food B is bought under a long-term contract requiring that at least 2 units of B be used per serving. Another contract requires that the amount of food B used be no more than 3 times the amount of food A used.

(a) How much of each food must be bought to produce the minimum cost per serving?

SOLUTION Let x represent the required amount of food A and y the amount of food B. Use the given information to prepare the following table.

Rescue Animal Nutrition Information	Food A	Food B		Total
Number of Units	x	y		
Grams of Protein	2	6	$\geq$	30
Grams of Fat	4	2	$\geq$	20
Long-Term Contract		y	$\geq$	2
Cost	18¢	12¢		

Since the animals must have *at least* 30 g of protein and 20 g of fat, we use $\geq$ in the inequality. If the animals needed *at most* a certain amount of some nutrient, we would use $\leq$. The long-term contract requires that $y \geq 2$.

In addition to the information in the table, we also have the requirement that the amount of food B used be no more than 3 times the amount of food A used. We can write this as $y \leq 3x$.

The linear programming problem can be stated as follows.

$$\text{Minimize} \qquad z = 0.18x + 0.12y$$
$$\text{subject to:} \qquad 2x + 6y \geq 30 \qquad \text{Protein}$$
$$4x + 2y \geq 20 \qquad \text{Fat}$$
$$2 \leq y \leq 3x \qquad \text{Contracts}$$
$$x \geq 0.$$

$y = 3x$
(2, 6)
(3, 4)
(9, 2)
$y = 2$
$4x + 2y = 20$
$2x + 6y = 30$

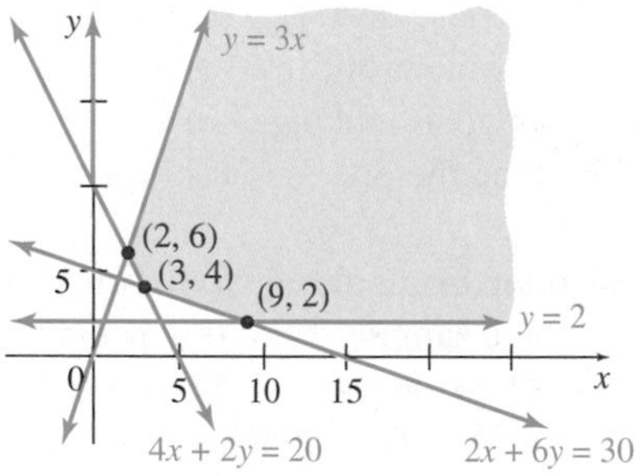

FIGURE 22

(The usual constraint $y \geq 0$ is redundant because of the constraint $y \geq 2$.) A graph of the feasible region is shown in Figure 22. The corner points are $(2, 6)$, $(3, 4)$, and $(9, 2)$. Test each corner point in the objective function to find the minimum cost.

Values of the Objective Function at Corner Points	
Corner Point	Value of $z = 0.18x + 0.12y$
$(2, 6)$	$0.18(2) + 0.12(6) = 1.08$
$(3, 4)$	$0.18(3) + 0.12(4) = 1.02$ Minimum
$(9, 2)$	$0.18(9) + 0.12(2) = 1.86$

The minimum of 1.02 occurs at $(3, 4)$. Thus, 3 units of food A and 4 units of food B will produce a minimum cost of $1.02 per serving.

(b) The rescue shelter manager notices that although the long-term contract states that at least 2 units of food B be used per serving, the solution uses 4 units of food B, which is 2 units more than the minimum amount required. Can a more economical solution be found that only uses 2 units of food B?

SOLUTION The solution found in part (a) is the most economical solution, even though it exceeds the requirement for using at least 2 units of food B. Notice from Figure 22 that the four lines representing the four constraints do not meet at a single point, so any solution in the feasible region will have to exceed at least one constraint. The rescue shelter manager might use this information to negotiate a better deal with the distributor of food B by making a guarantee to use at least 4 units of food B per serving in the future.

The notion that some constraints are not met exactly is related to the concepts of *surplus* and *slack variables*, which will be explored in the next chapter.

The feasible region in Figure 22 is an *unbounded* feasible region—the region extends indefinitely to the upper right. With this region it would not be possible to *maximize* the objective function, because the total cost of the food could always be increased by encouraging the animals to eat more.

3.3 EXERCISES

Write Exercises 1–6 as linear inequalities. Identify all variables used. (Note: Not all of the given information is used in Exercises 5 and 6.)

1. Product A requires 3 hours on a machine, while product B needs 5 hours on the same machine. The machine is available for at most 60 hours per week.

2. A cow requires a third of an acre of pasture and a sheep needs a quarter acre. A rancher wants to use at least 120 acres of pasture.

3. Jessica Corpo needs at least 1500 units of calcium supplements per day. Her calcium carbonate supplement provides 600 units, and her calcium citrate supplement supplies 250 units.

4. Pauline Wong spends 3 hours selling a small computer and 5 hours selling a larger model. She works no more than 45 hours per week.

5. Coffee costing $8 per lb is to be mixed with coffee costing $10 per lb to get at least 40 lb of a blended coffee.

6. A tank in an oil refinery holds 120 gal. The tank contains a mixture of light oil worth $1.25 per gal and heavy oil worth $0.80 per gal.

APPLICATIONS

Business and Economics

7. **Transportation** The Miers Company produces small engines for several manufacturers. The company receives orders from two assembly plants for their Top-flight engine. Plant I needs at least 45 engines, and plant II needs at least 32 engines. The company can send at most 120 engines to these two assembly plants. It costs $30 per engine to ship to plant I and $40 per engine to ship to plant II. Plant I gives Miers $20 in rebates toward its products for each engine they buy, while plant II gives similar $15 rebates. Miers estimates that they need at least $1500 in rebates to cover products they plan to buy from the two plants. How many engines should be shipped to each plant to minimize shipping costs? What is the minimum cost?

8. **Transportation** A manufacturer of refrigerators must ship at least 100 refrigerators to its two West Coast warehouses. Each warehouse holds a maximum of 100 refrigerators. Warehouse A holds 25 refrigerators already, and warehouse B has 20 on hand. It costs $12 to ship a refrigerator to warehouse A and $10 to ship one to warehouse B. Union rules require that at least 300 workers be hired. Shipping a refrigerator to warehouse A requires 4 workers, while shipping a refrigerator to warehouse B requires 2 workers. How many refrigerators should be shipped to each warehouse to minimize costs? What is the minimum cost?

9. **Insurance Premiums** A company is considering two insurance plans with the types of coverage and premiums shown in the following table.

	Policy A	Policy B
Fire/Theft	$10,000	$15,000
Liability	$180,000	$120,000
Premium	$50	$40

(For example, this means that $50 buys one unit of plan A, consisting of $10,000 fire and theft insurance and $180,000 of liability insurance.)

a. The company wants at least $300,000 fire/theft insurance and at least $3,000,000 liability insurance from these plans. How many units should be purchased from each plan to minimize the cost of the premiums? What is the minimum premium?

b. Suppose the premium for policy A is reduced to $25. Now how many units should be purchased from each plan to minimize the cost of the premiums? What is the minimum premium?

10. **Profit** The Muro Manufacturing Company makes two kinds of plasma screen television sets. It produces the Flexscan set that sells for $350 profit and the Panoramic I that sells for $500 profit. On the assembly line, the Flexscan requires 5 hours, and the Panoramic I takes 7 hours. The cabinet shop spends 1 hour on the cabinet for the Flexscan and 2 hours on the cabinet for the Panoramic I. Both sets require 4 hours for testing and packing. On a particular production run, the Muro Company has available 3600 work-hours on the assembly line, 900 work-hours in the cabinet shop, and 2600 work-hours in the testing and packing department.

a. How many sets of each type should it produce to make a maximum profit? What is the maximum profit?

b. Suppose the profit on the Flexscan goes up to $450. Now how many sets of each type should it produce to make a maximum profit? What is the maximum profit?

c. The solutions from parts a and b leave some unused time in either the assembly line, the cabinet shop, or the testing and packing department. Identify any unused time in each solution. Is it possible to have a solution that leaves no excess time? Explain.

11. **Revenue** A machine shop manufactures two types of bolts. The bolts require time on each of the three groups of machines, but the time required on each group differs, as shown in the table below.

	Type I	Type II
Machine 1	0.2 min	0.2 min
Machine 2	0.6 min	0.2 min
Machine 3	0.04 min	0.08 min

Production schedules are made up one day at a time. In a day, 300, 720, and 100 minutes are available, respectively, on these machines. Type I bolts sell for 15¢ and type II bolts for 20¢.

a. How many of each type of bolt should be manufactured per day to maximize revenue?

b. What is the maximum revenue?

c. Suppose the selling price of type I bolts began to increase. Beyond what amount would this price have to increase before a different number of each type of bolts should be produced to maximize revenue?

12. **Revenue** The manufacturing process requires that oil refineries must manufacture at least 2 gal of gasoline for every gallon of fuel oil. To meet the winter demand for fuel oil, at least 3 million gal a day must be produced. The demand for gasoline is no more than 6.4 million gal per day. It takes 15 minutes to ship each million gal of gasoline and 1 hour to ship each million gal of fuel oil out of the warehouse. No more than 4 hours and 39 minutes are available for shipping.

a. If the refinery sells gasoline for $2.50 per gal and fuel oil for $2 per gal, how much of each should be produced to maximize revenue?

b. Find the maximum revenue.

c. Suppose the price for fuel oil begins to increase. Beyond what amount would this price have to increase before a different amount of gasoline and fuel oil should be produced to maximize revenue?

13. **Revenue** A candy company has 150 kg of chocolate-covered nuts and 90 kg of chocolate-covered raisins to be sold as two different mixes. One mix will contain half nuts and half raisins and will sell for $7 per kg. The other mix will contain 3/4 nuts and 1/4 raisins and will sell for $9.50 per kg.

a. How many kilograms of each mix should the company prepare for the maximum revenue? Find the maximum revenue.

b. The company raises the price of the second mix to $11 per kg. Now how many kilograms of each mix should the company prepare for the maximum revenue? Find the maximum revenue.

14. **Profit** A small country can grow only two crops for export, coffee and cocoa. The country has 500,000 hectares of land available for the crops. Long-term contracts require that at least 100,000 hectares be devoted to coffee and at least 200,000 hectares to cocoa. Cocoa must be processed locally, and production bottlenecks limit cocoa to 270,000 hectares. Coffee requires two workers per hectare, with cocoa requiring five. No more than 1,750,000 people are available for working with these crops. Coffee produces a profit of $220 per hectare and cocoa a profit of $550 per hectare. How many hectares should the country devote to each crop in order to maximize profit? Find the maximum profit.

15. **Blending** The Mostpure Milk Company gets milk from two dairies and then blends the milk to get the desired amount of butterfat for the company's premier product. Milk from dairy I costs $2.40 per gal, and milk from dairy II costs $0.80 per gal. At most $144 is available for purchasing milk. Dairy I can supply at most 50 gal of milk averaging 3.7% butterfat. Dairy II can supply at most 80 gal of milk averaging 3.2% butterfat.

a. How much milk from each dairy should Mostpure use to get at most 100 gal of milk with the maximum total amount of butterfat? What is the maximum amount of butterfat?

b. The solution from part a leaves both dairy I and dairy II with excess capacity. Calculate the amount of additional milk each dairy could produce. Is there any way all this capacity could be used while still meeting the other constraints? Explain.

16. **Transportation** A flash drive manufacturer has 370 boxes of a particular drive in warehouse I and 290 boxes of the same drive in warehouse II. A computer store in San Jose orders 350 boxes of the drive, and another store in Memphis orders 300 boxes. The shipping costs per box to these stores from the two warehouses are shown in the following table.

		Destination	
		San Jose	Memphis
Warehouse	I	$2.50	$2.20
	II	$2.30	$2.10

How many boxes should be shipped to each city from each warehouse to minimize shipping costs? What is the minimum cost? (*Hint:* Use x, $350 - x$, y, and $300 - y$ as the variables.)

17. **Finance** A pension fund manager decides to invest a total of at most $30 million in U.S. Treasury bonds paying 4% annual interest and in mutual funds paying 8% annual interest. He plans to invest at least $5 million in bonds and at least $10 million in mutual funds. Bonds have an initial fee of $100 per million dollars, while the fee for mutual funds is $200 per million. The fund manager is allowed to spend no more than $5000 on fees. How much should be invested in each to maximize annual interest? What is the maximum annual interest?

Manufacturing (Note: Exercises 18–20 are from qualification examinations for Certified Public Accountants.) *Source: American Institute of Certified Public Accountants.* The Random Company manufactures two products, Zeta and Beta. Each product must pass through two processing operations. All materials are introduced at the start of Process No. 1. There are no work-in-process inventories. Random may produce either one product exclusively or various combinations of both products subject to the following constraints:

	Process No. 1	Process No. 2	Contribution Margin (per unit)
Hours Required to Produce One Unit:			
Zeta	1 hr	1 hr	$4.00
Beta	2 hr	3 hr	$5.25
Total Capacity (in hours per day)	1000 hr	1275 hr	

A shortage of technical labor has limited Beta production to 400 units per day. There are no constraints on the production of Zeta other than the hour constraints in the above schedule. Assume that all relationships between capacity and production are linear.

18. Given the objective to maximize total contribution margin, what is the production constraint for Process No. 1? (Choose one of the following.)

    a. Zeta + Beta ≤ 1000   **b.** Zeta + 2 Beta ≤ 1000

    c. Zeta + Beta ≥ 1000   **d.** Zeta + 2 Beta ≥ 1000

19. Given the objective to maximize total contribution margin, what is the labor constraint for production of Beta? (Choose one of the following.)

    a. Beta ≤ 400   **b.** Beta ≥ 400

    c. Beta ≤ 425   **d.** Beta ≥ 425

20. What is the objective function of the data presented? (Choose one of the following.)

    a. Zeta + 2 Beta = $9.25

    b. $4.00 Zeta + 3($5.25)Beta = Total Contribution Margin

    c. $4.00 Zeta + $5.25 Beta = Total Contribution Margin

    d. 2($4.00) Zeta + 3($5.25) Beta = Total Contribution Margin

Life Sciences

21. **Health Care** Glen Spencer takes vitamin pills. Each day he must have at least 480 IU of vitamin A, 5 mg of vitamin B_1, and 100 mg of vitamin C. He can choose between pill 1, which contains 240 IU of vitamin A, 1 mg of vitamin B_1, and 10 mg of vitamin C, and pill 2, which contains 60 IU of vitamin A, 1 mg of vitamin B_1, and 35 mg of vitamin C. Pill 1 costs 15¢, and pill 2 costs 30¢.

    a. How many of each pill should he buy in order to minimize his cost? What is the minimum cost?

    b. For the solution in part a, Glen is receiving more than he needs of at least one vitamin. Identify that vitamin, and tell how much surplus he is receiving. Is there any way he can avoid receiving that surplus while still meeting the other constraints and minimizing the cost? Explain.

22. **Predator Food Requirements** A certain predator requires at least 10 units of protein and 8 units of fat per day. One prey of species I provides 5 units of protein and 2 units of fat; one prey of species II provides 3 units of protein and 4 units of fat. Capturing and digesting each species-II prey requires 3 units of energy, and capturing and digesting each species-I prey requires 2 units of energy. How many of each prey would meet the predator's daily food requirements with the least expenditure of energy? Are the answers reasonable? How could they be interpreted?

23. **Nutrition** A dietician is planning a snack package of fruit and nuts. Each ounce of fruit will supply zero units of protein, 2 units of carbohydrates, and 1 unit of fat, and will contain 20 calories. Each ounce of nuts will supply 3 units of protein, 1 unit of carbohydrate, and 2 units of fat, and will contain 30 calories. Every package must provide at least 6 units of protein, at least 10 units of carbohydrates, and no more than 9 units of fat. Find the number of ounces of fruit and number of ounces of nuts that will meet the requirement with the least number of calories. What is the least number of calories?

24. **Health Care** Ms. Oliveras was given the following advice. She should supplement her daily diet with at least 6000 USP units of vitamin A, at least 195 mg of vitamin C, and at least 600 USP units of vitamin D. Ms. Oliveras finds that Mason's Pharmacy carries Brand X vitamin pills at 5¢ each and Brand Y vitamins at 4¢ each. Each Brand X pill contains 3000 USP units of A, 45 mg of C, and 75 USP units of D, while Brand Y pills contain 1000 USP units of A, 50 mg of C, and 200 USP units of D.

    a. What combination of vitamin pills should she buy to obtain the least possible cost? What is the least possible cost per day?

    b. For the solution in part a, Ms. Oliveras is receiving more than she needs of at least one vitamin. Identify that vitamin, and tell how much surplus she is receiving. Is there any way she can avoid receiving that surplus while still meeting the other constraints and minimizing the cost? Explain.

Social Sciences

25. **Anthropology** An anthropology article presents a hypothetical situation that could be described by a linear programming model. Suppose a population gathers plants and animals for

survival. They need at least 360 units of energy, 300 units of protein, and 8 hides during some time period. One unit of plants provides 30 units of energy, 10 units of protein, and no hides. One animal provides 20 units of energy, 25 units of protein, and 1 hide. Only 25 units of plants and 25 animals are available. It costs the population 30 hours of labor to gather one unit of a plant and 15 hours for an animal. Find how many units of plants and how many animals should be gathered to meet the requirements with a minimum number of hours of labor. *Source: Annual Review of Anthropology.*

General Interest

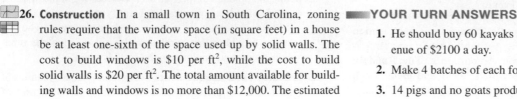

26. Construction In a small town in South Carolina, zoning rules require that the window space (in square feet) in a house be at least one-sixth of the space used up by solid walls. The cost to build windows is $10 per ft^2, while the cost to build solid walls is $20 per ft^2. The total amount available for building walls and windows is no more than $12,000. The estimated

monthly cost to heat the house is $0.32 for each square foot of windows and $0.20 for each square foot of solid walls. Find the maximum total area (windows plus walls) if no more than $160 per month is available to pay for heat.

27. Farming An agricultural advisor looks at the results of Example 2 and claims that it cannot possibly be correct. After all, the 4-H member is able to raise 16 animals, and she is only raising 12 animals. Surely she can earn more profit by raising all 16 animals. How would you respond?

▬▬YOUR TURN ANSWERS

1. He should buy 60 kayaks and no canoes for a maximum revenue of $2100 a day.

2. Make 4 batches of each for a maximum profit of $200.

3. 14 pigs and no goats produces a maximum profit of $560.

3 CHAPTER REVIEW

SUMMARY

In this chapter, we introduced linear programming, which attempts to solve maximization and minimization problems with linear constraints. Linear programming models can be used to analyze a wide range of applications from many disciplines. The corner point theorem assures us that the optimal solution to a linear program, if it exists, must occur at one or more of the corner points of the feasible region. Linear programs can be solved using the graphical method, which graphs the region described by the linear constraints and then locates the corner point corresponding to the optimal solution value. The graphical method, however, is restricted to problems with two or three variables. In the next chapter, we will study a method that does not have this restriction.

Graphing a Linear Inequality	1. Draw the graph of the boundary line. Make the line solid if the inequality involves $\leq$ or $\geq$; make the line dashed if the inequality involves $<$ or $>$.

2. Decide which half-plane to shade. Use either of the following methods.

    a. Solve the inequality for y; shade the region above the line if the inequality is of the form of $y >$ or $y \geq$; shade the region below the line if the inequality is of the form of $y <$ or $y \leq$.

    b. Choose any point not on the line as a test point. Shade the half-plane that includes the test point if the test point satisfies the original inequality; otherwise, shade the half-plane on the other side of the boundary line.

Corner Point Theorem   If an optimum value (either a maximum or a minimum) of the objective function exists, it will occur at one or more of the corner points of the feasible region.

Solving a Linear Programming Problem

1. Write the objective function and all necessary constraints.

2. Graph the feasible region.

3. Identify all corner points.

4. Find the value of the objective function at each corner point.

5. For a bounded region, the solution is given by the corner point(s) producing the optimum value of the objective function.

6. For an unbounded region, check that a solution actually exists. If it does, it will occur at one or more corner points.

KEY TERMS

To understand the concepts presented in this chapter, you should know the meaning and use of the following terms.

3.1
linear inequality
boundary
half-plane
system of inequalities

region of feasible solutions
feasible region
bounded
unbounded

3.2
linear programming
objective function
constraints

isoprofit line
corner point

REVIEW EXERCISES

CONCEPT CHECK

Determine whether each of the following statements is true or false, and explain why.

1. The graphical method can be used to solve a linear programming problem with four variables.

2. For the inequality $5x + 4y \geq 20$, the test point (3, 4) suggests that the correct half-plane to shade includes this point.

3. Let x represent the number of acres of wheat planted and y represent the number of acres of corn planted. The inequality $x \leq 2y$ implies that the number of acres of wheat planted will be at least twice the number of acres of corn planted.

4. For the variables in Exercise 3, assume that we have a total of 60 hours to plant the wheat and corn and that it takes 2 hours per acre to prepare a wheat field and 1 hour per acre to prepare a corn field. The inequality $2x + y \geq 60$ represents the constraint on the amount of time available for planting.

5. For the variables in Exercise 3, assume that we make a profit of $14 for each acre of corn and $10 for each acre of wheat. The objective function that can be used to maximize profit is $14x + 10y$.

6. The point (2, 3) is a corner point of the linear programming problem

$$\text{Maximize} \quad z = 7x + 4y$$
$$\text{subject to:} \quad 3x + 8y \leq 30$$
$$4x + 2y \leq 15$$
$$x \geq 0, y \geq 0.$$

7. The point (2, 3) is in the feasible region of the linear programming problem in Exercise 6.

8. The optimal solution to the linear programming problem in Exercise 6 occurs at point (2, 3).

9. It is possible to find a point that lies on both sides of a linear inequality.

10. Every linear programming problem with a feasible region that is not empty either has a solution or is unbounded.

11. Solutions to linear programming problems may include fractions.

12. The inequality $4^2x + 5^2y \leq 7^2$ is a linear constraint.

13. The optimal solution to a linear programming problem can occur at a point that is not a corner point.

PRACTICE AND EXPLORATIONS

14. How many constraints are we limited to in the graphical method?

Graph each linear inequality.

15. $y \geq 2x + 3$

16. $5x - 2y \leq 10$

17. $2x + 6y \leq 8$

18. $2x - 6y \geq 18$

19. $y \geq x$

20. $y \geq -2$

Graph the solution of each system of inequalities. Find all corner points and tell whether each region is bounded or unbounded.

21. $x + y \leq 6$
    $2x - y \geq 3$

22. $3x + 2y \geq 12$
    $4x - 5y \leq 20$

23. $-4 \leq x \leq 2$
    $-1 \leq y \leq 3$
    $x + y \leq 4$

24. $2 \leq x \leq 5$
    $1 \leq y \leq 7$
    $x - y \leq 3$

25. $x + 2y \leq 4$
    $5x - 6y \leq 12$
    $x \geq 0$
    $y \geq 0$

26. $x + 2y \leq 4$
    $2x - 3y \leq 6$
    $x \geq 0$
    $y \geq 0$

Use the given regions to find the maximum and minimum values of the objective function $z = 2x + 4y$.

27.

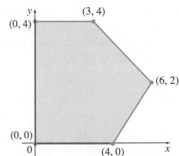

28.

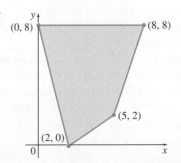

Use the graphical method to solve each linear programming problem.

29. Maximize $z = 2x + 4y$

subject to: $3x + 2y \leq 12$

$5x + y \geq 5$

$x \geq 0$

$y \geq 0.$

30. Minimize $z = 5x + 3y$

subject to: $8x + 5y \geq 40$

$4x + 10y \geq 40$

$x \geq 0$

$y \geq 0.$

31. Minimize $z = 4x + 2y$

subject to: $x + y \leq 50$

$2x + y \geq 20$

$x + 2y \geq 30$

$x \geq 0$

$y \geq 0.$

32. Maximize $z = 8x + 4y$

subject to: $3x + 12y \leq 36$

$x + y \leq 4$

$x \geq 0$

$y \geq 0.$

33. Why must the solution to a linear programming problem always occur at a corner point of the feasible region?

34. Is there necessarily a unique point in the feasible region where the maximum or minimum occurs? Why or why not?

35. It is not necessary to check all corner points in a linear programming problem. This exercise illustrates an alternative procedure, which is essentially an expansion of the ideas illustrated in Example 1 of Section 3.2.

Maximize $z = 3x + 4y$

subject to: $2x + y \leq 4$

$-x + 2y \leq 4$

$x \geq 0$

$y \geq 0.$

a. Sketch the feasible region, and add the line $z = 8$. (*Note:* 8 is chosen because the numbers work out simply, but the chosen value of z is arbitrary.)

b. Draw a line parallel to the line $z = 8$ that is as far from the origin as possible but still touches the feasible region.

c. The line you drew in part b should go through the point $(4/5, 12/5)$. Explain how you know the maximum must be located at this point.

36. Use the method described in the previous exercise to solve Exercise 32.

APPLICATIONS

Business and Economics

37. Time Management A bakery makes both cakes and cookies. Each batch of cakes requires 2 hours in the oven and 3 hours in the decorating room. Each batch of cookies needs $1\frac{1}{2}$ hours in the oven and $\frac{2}{3}$ hour in the decorating room. The oven is available no more than 15 hours per day, and the decorating room can be used no more than 13 hours per day. Set up a system of inequalities, and then graph the solution of the system.

38. Cost Analysis DeMarco's pizza shop makes two specialty pizzas, the Mighty Meaty and the Very Veggie. The Mighty Meaty is topped with 5 different meat toppings and 2 different cheeses. The Very Veggie has 6 different vegetable toppings and 4 different cheeses. The shop sells at least 4 Mighty Meaty and 6 Very Veggie pizzas every day. The cost of the toppings for each Mighty Meaty is $3, and the cost of the vegetable toppings is $2 for each Very Veggie. No more than $60 per day can be spent on these toppings. The cheese used for the Mighty Meaty is $2 per pizza, and the cheese for the Very Veggie is $4 per pizza. No more than $80 per day can be spent on cheese. Set up a system of inequalities, and then graph the solution of the system.

39. Profit Refer to Exercise 37.

a. How many batches of cakes and cookies should the bakery in Exercise 37 make in order to maximize profits if cookies produce a profit of $20 per batch and cakes produce a profit of $30 per batch?

b. How much would the profit from selling cookies have to increase before it becomes more profitable to sell only cookies?

40. Revenue How many pizzas of each kind should the pizza shop in Exercise 38 make in order to maximize revenue if the Mighty Meaty sells for $15 and the Very Veggie sells for $12?

41. Planting In Karla's garden shop, she makes two kinds of mixtures for planting. A package of gardening mixture requires 2 lb of soil, 1 lb of peat moss, and 1 lb of fertilizer. A package of potting mixture requires 1 lb of soil, 2 lb of peat moss, and 3 lb of fertilizer. She has 16 lb of soil, 11 lb of peat moss, and 15 lb of fertilizer. If a package of gardening mixture sells for $3 and a package of potting mixture for $5, how many of each should she make in order to maximize her income? What is the maximum income?

42. Construction A contractor builds boathouses in two basic models, the Atlantic and the Pacific. Each Atlantic model requires 1000 ft of framing lumber, 3000 ft^3 of concrete, and $2000 for advertising. Each Pacific model requires 2000 ft of framing lumber, 3000 ft^3 of concrete, and $3000 for advertising. Contracts call for using at least 8000 ft of framing lumber, 18,000 ft^3 of concrete, and $15,000 worth of advertising. If the construction cost for each Atlantic model is $30,000 and the construction cost for each Pacific model is $40,000, how many of each model should be built to minimize construction costs?

43. Steel A steel company produces two types of alloys. A run of type I requires 3000 lb of molybdenum and 2000 tons of iron ore pellets as well as $2000 in advertising. A run of type II requires 3000 lb of molybdenum and 1000 tons of iron ore

pellets as well as $3000 in advertising. Total costs are $15,000 on a run of type I and $6000 on a run of type II. Because of various contracts, the company must use at least 18,000 lb of molybdenum and 7000 tons of iron ore pellets and spend at least $14,000 on advertising. How much of each type should be produced to minimize costs?

Life Sciences

44. Nutrition A dietician in a hospital is to arrange a special diet containing two foods, Health Trough and Power Gunk. Each ounce of Health Trough contains 30 mg of calcium, 10 mg of iron, 10 IU of vitamin A, and 8 mg of cholesterol. Each ounce of Power Gunk contains 10 mg of calcium, 10 mg of iron, 30 IU of vitamin A, and 4 mg of cholesterol. If the minimum daily requirements are 360 mg of calcium, 160 mg of iron, and 240 IU of vitamin A, how many ounces of each food should be used to meet the minimum requirements and at the same time minimize the cholesterol intake? Also, what is the minimum cholesterol intake?

Social Sciences

45. Anthropology A simplified model of the Mountain Fur economy of central Africa has been proposed. In this model, two crops can be grown, millet and wheat, which produce 400 lb and 800 lb per acre, respectively. Millet requires 36 days to harvest one acre, while wheat requires only 8 days. There are 2 acres of land and 48 days of harvest labor available. How many acres should be devoted to each crop to maximize the pounds of grain harvested? *Source: Themes in Economic Anthropology.*

General Interest

46. Studying Jim Pringle is trying to allocate his study time this weekend. He can spend time working with either his math tutor or his accounting tutor to prepare for exams in both classes the following Monday. His math tutor charges $20 per hour, and his accounting tutor charges $40 per hour. He has $220 to spend on tutoring. Each hour that he spends working with his math tutor requires 1 aspirin and 1 hour of sleep to recover, while each hour he spends with his accounting tutor requires 1/2 aspirin and 3 hours of sleep. The maximum dosage of aspirin that he can safely take during his study time is 8 tablets, and he can only afford 15 hours of sleep this weekend. He expects that each hour with his math tutor will increase his score on the math exam by 3 points, while each hour with his accounting tutor will increase his score on the accounting exam by 5 points. How many hours should he spend with each tutor in order to maximize the number of points he will get on the two tests combined?

EXTENDED APPLICATION

SENSITIVITY ANALYSIS

In Section 3 of this chapter we used the graphical method to solve linear programming problems. A quick analysis of the strengths and weaknesses of the process of modeling real-life problems using these techniques reveals two apparent problems. First, most real-life problems require far more than two variables to adequately capture the essence of the problem. Second, our world is so dynamic that the values used to model a problem today can quickly change, leaving the current solution irrelevant or even infeasible.

In Chapter 4, we show how the simplex method can be used to solve linear programming problems with thousands or even millions of variables and constraints. In this extended application, we introduce the idea of *sensitivity analysis*, which provides a way to efficiently handle the constantly changing inputs to a linear programming problem, often without having to resolve the problem from scratch. We will limit our study to linear programs with only two variables and begin by analyzing changes in objective function coefficients.

Recall that in Example 2 of Section 3.3, a linear programming problem was developed to determine the optimal number of goats and pigs to raise for a 4-H project. The linear program was written as

$$\text{Maximize} \quad z = 12x + 40y$$
$$\text{Subject to:} \quad x + y \leq 16$$
$$x + 3y \leq 36$$
$$x \leq 10$$
$$x \geq 0, y \geq 0,$$

where x is the number of goats and y is the number of pigs.

A graph of the feasible region, with corner points illustrated, is shown in Figure 23 on the next page. We determined that a maximum profit of $480 occurs when 12 pigs and no goats are raised. This is no surprise given that the profit from a pig is $40 and the profit for a goat is just $12. In fact, we saw in that example that if the profit from raising a pig decreases to $36, then it would become profitable to raise goats, too.

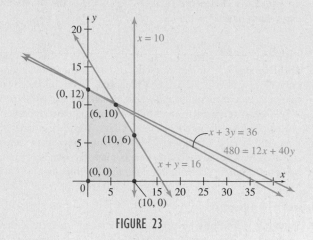

FIGURE 23

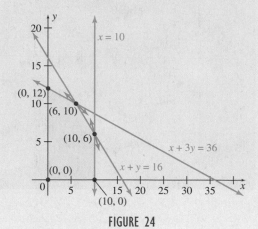

FIGURE 24

CHANGES IN OBJECTIVE FUNCTION COEFFICIENTS

Suppose that the 4-H member, whom we will call Sarah, now wonders what the profit from raising a goat would have to become so that it would become profitable for her to raise both pigs and goats.

Mathematically, we replace the $12 in the objective function with the variable $12 + c$, where c is the change in profit for raising a goat. That is, $z = (12 + c)x + 40y$. Now, for any particular profit z, we can think of this equation as a line, which we called an *isoprofit line* in Section 3.2. With $c = 0$ and $z = 480$, the isoprofit line is $480 = 12x + 40y$, which is plotted in green in Figure 23. Solving for y helps us to see that the line represented by this equation has slope $-\dfrac{12 + c}{40}$. That is,

$$z = (12 + c)x + 40y,$$

or

$$40y = -(12 + c)x + z,$$

which simplifies to

$$y = -\frac{12 + c}{40}x + \frac{z}{40}.$$

The line that represents the boundary of the constraint $x + 3y = 36$ has slope $-\frac{1}{3}$. If c begins to increase, forcing the slope of the isoprofit line to become more negative, eventually the slope of the isoprofit line corresponding to the optimal profit of $480 will have slope $-\frac{1}{3}$. If c continues to increase beyond this level, the optimal profit will occur at the corner point $(6, 10)$. That is, as long as the slope of the isoprofit line exceeds $-\frac{1}{3}$, the optimal solution will continue to be to raise 12 pigs and no goats. If, however, the slope of the isoprofit line becomes less than $-\frac{1}{3}$, then it will become profitable to raise goats as well. That is, if c increases to the point at which

$$-\frac{12 + c}{40} < -\frac{1}{3},$$

which simplifies to,

$$\frac{12 + c}{40} > \frac{1}{3},$$

or

$$c > \frac{40}{3} - 12 = \frac{4}{3},$$

the optimal solution changes from the corner point $(0, 12)$ to $(6, 10)$. Further analysis, as illustrated by Figure 24, shows that the optimal solution will remain at the corner point $(6, 10)$ as long as the slope of the isoprofit line stays between

$$-1 \le -\frac{12 + c}{40} \le -\frac{1}{3},$$

or when

$$\frac{4}{3} \le c \le 28.$$

Then the optimal solution will be to raise 6 goats and 10 pigs.

Beyond $c = $28 or a profit of $40 per goat, it will be optimal to raise 10 goats and 6 pigs. Note that this analysis can be used to completely determine the optimal solution for changes in a single cost coefficient for any two variable linear programming problem, without having to resolve the linear program.

CHANGES IN THE RIGHT HAND SIDE OF A CONSTRAINT

Sarah begins to wonder if the right-hand sides of any of the constraints might be changed so that the profit would increase. For example, the easiest thing to do would be for Sarah to increase the third constraint, corresponding to the restriction that no more than 10 goats are raised. Sarah considers the possibility of lifting this constraint. Notice, however, that if we increase the right-hand side of the third constraint, it has no affect on the optimal corner point. As you can see graphically with the red lines in Figure 25, increasing the right-hand side will not change the optimal solution. In fact, decreasing the third constraint even to zero will not change the optimal solution.

Now suppose that Sarah realizes that she has overestimated the total number of animals she can raise by 3 and wants to see how this will affect the optimal solution. The boundary of the first

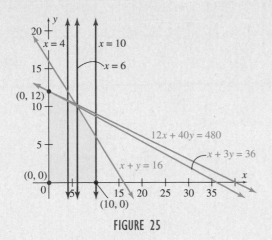

FIGURE 25

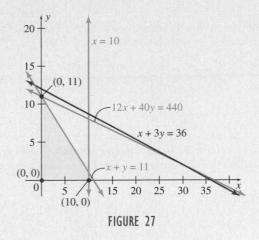

FIGURE 27

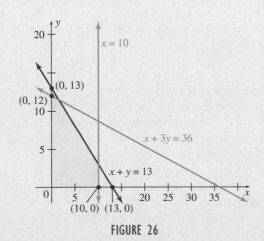

FIGURE 26

constraint can be written as the line $y = -x + 16$, which has y-intercept at the point $(0,16)$. Since Sarah has overestimated the total number of animals by three, the boundary line becomes $y = -x + 13$, which has y-intercept at the point $(0,13)$, as illustrated in Figure 26.

This change will not have an effect on the solution because raising 13 pigs is not feasible for the second constraint. That is, $(0, 13)$ is not a feasible corner point and cannot be a solution to the problem. In fact, as long as the total number of animals is at least 12 in the first constraint, the optimal solution is for Sarah to raise 12 pigs and no goats. In this case, we would say that the right-hand side of the first constraint is not sensitive to small changes. If, on the other hand, Sarah has overestimated the total number of animals she can raise and that, in fact, the total number is actually 11, then the optimal solution changes. As you can see in Figure 27, the feasible region has diminished to the point that the second constraint is now redundant and has no influence on the optimal solution. In fact, the optimal solution is to raise 11 pigs and no goats. This makes sense since the profit from raising pigs is so high that it does not become profitable to raise goats when a smaller total number of animals are raised. The same type of analysis can be performed on the right-hand side of the second constraint.

SHADOW PROFITS

Changes in the resources available in a linear programming problem usually result in a change in the value of the objective function in the optimal solution. For example, if an optimal solution uses all of the material available, then a sudden increase/decrease in the amount of that material available can change the objective function value. In the analysis above, decreasing the right-hand side of the first constraint from 16 to 12 did not change the optimal solution. In this case, we would say that the shadow profit, or the increase in the profit for changing the right-hand side of the first constraint by one unit is $0. On the other hand, if we increase the amount of money available for the project by $75, the optimal solution would change from raising 12 pigs to raising 13 pigs and thereby increase profit by $40. This implies that for each dollar increase in the amount of money available, the profit will increase by $40/75 or by about $0.53. In this case, the shadow profit would be $0.53.

CHANGES IN CONSTRAINT COEFFICIENTS

Sarah continues to explore reasons why the optimal policy is to only raise pigs. She wonders if any of the coefficients in the constraints are having an effect on the optimal solution. In the process, she realizes that it may cost more than $75 to raise a pig. In the original statement of the problem, Sarah had up to $900 to spend on her project, with each goat costing $25 to raise and each pig costing $75 to raise. This information was translated into the constraint $25x + 75y \leq 900$, which was then simplified by dividing the inequality by 25. Sarah wonders how much the $75 would have to increase before the optimal solution would include raising goats. Letting b represent the change in the cost of raising a pig, the constraint becomes $25x + (75 + b)y \leq 900$, and the corresponding boundary line can then be simplified and written as

$$y = -\frac{25}{75 + b}x + \frac{900}{75 + b}.$$

To be consistent, every dollar increase in the cost of raising a pig also forces the profit to decrease by a dollar. That is, the 40 in the objective function is replaced with $40 - b$, becoming

$$z = 12x + (40 - b)y.$$

The isoprofit lines for this objective function will each have slope of $-\dfrac{12}{40 - b}$.

When the slope of the boundary line corresponding to the altered second constraint increases to the point that it is larger than the slope of the isoprofit lines, the optimal solution will change and move to the intersection point of the first and second constraints. This occurs when

$$-\frac{25}{75 + b} > -\frac{12}{40 - b},$$

which simplifies to

$$b > \frac{100}{37} \approx 2.70.$$

In other words, as long as the cost of raising a pig is at or below \$75 + 2.70 = \$77.70, the optimal policy is to only raise pigs. When the cost exceeds this amount the optimal solution will change, and it may become profitable to raise goats, too. For example, suppose the cost of raising a pig increases to \$100. The new feasible region has corner points at (0, 0), (0, 9), (10, 0), (10, 6), and (28/3, 20/3), as indicated by Figure 28. It can be easily verified that the optimal

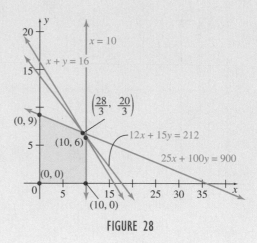

FIGURE 28

solution now occurs at (28/3, 20/3), with objective function value of $z = 12\left(\frac{28}{3}\right) + 15\left(\frac{20}{3}\right) = 212$. Of course, it is impossible to raise a fractional number of animals so some additional work is required to determine the optimal solution. Unfortunately, simply rounding the fractional solution downward to obtain $x = 9$ and $y = 6$ does not give the optimal solution. In fact, it can be shown that the optimal integer solution is $x = 10$ and $y = 6$ with an objective function value of \$210. Thus, if the cost of raising a pig is \$100 then the optimal policy is to raise 6 pigs and 10 goats.

EXERCISES

1. Use sensitivity analysis to find the optimal solution to the original 4-H problem, if the profit from raising a goat increases from \$12 to \$25.

2. Use sensitivity analysis to find the optimal solution to the original 4-H problem, if the profit from raising a pig decreases from \$40 to \$35.

3. Use sensitivity analysis to determine the smallest decrease in the cost of raising a goat, so that the optimal policy of raising 12 pigs and no goats changes. Note that changing this cost also changes the profit level for goats.

4. Katie's Landscaping Services weeds flower beds and mows lawns. Katie makes a profit of \$8 for every hour she weeds and \$25 for every hour she mows. Because Katie is also in college, she can only take on 18 jobs each week for her business. Katie has found that the flower beds in her neighborhood are very poorly maintained and that weeding a flower bed must be considered as two jobs. Each week, Katie sets aside \$72 for her brother Nate to help her and feels that a fair price to pay Nate is \$3 per hour for weeding and \$9 per hour for mowing a lawn.

a. Show that this problem can be set up as follows:

Maximize $\quad z = 8x + 25y$

Subject to: $\quad 2x + y \leq 18$

$\qquad\qquad\quad 3x + 9y \leq 72$

$\qquad\qquad\quad x \geq 0, y \geq 0$

b. Use the graphical method to solve this linear program.

c. Alter the graph from part b to determine the new optimal solution if the right hand side of the second constraint increases to 81. (*Hint:* You shouldn't have to completely resolve this problem. Simply change the intercepts of one of the constraints and use isoprofit lines to identify the corner point where the new optimal solution occurs.)

d. What is the shadow profit associated with the change from 72 to 81 in part c?

e. Using the original problem, determine the smallest increase (to the nearest penny) in the objective function coefficient for variable x that will cause the value of x to increase in the optimal solution. In this case, we would say that the cost coefficient associated with variable x is sensitive to change.

f. If the y-coefficient in the second constraint of the original problem begins to increase, forcing an equal decrease in the y-coefficient in the objective function, determine the smallest amount (to the nearest penny) that this coefficient can increase before it becomes optimal to produce positive quantities of variable x. For simplicity, assume that it is possible to have fractional amounts of each variable.

5. Go to the website WolframAlpha.com, and enter "minimize $8x + 25y$ with $2x + y < = 18$, $3x + 9y < = 72$, $x > = 0$, $y > = 0$".

Study the solution provided by Wolfram|Alpha. Notice that the solution given is the same as the solution you found in Exercise 4b. Use Wolfram|Alpha to verify your answers to Exercises 4c–4f.

DIRECTIONS FOR GROUP PROJECT

Suppose that you are the manager of a company that manufactures two types of products. Develop a simple linear program with a solution that gives how many of each type of product the company should produce. Then perform sensitivity analysis on various aspects of the model. Use presentation software to report your model and your findings to the class.

4 Linear Programming: The Simplex Method

Each type of beer has its own recipe and an associated cost per unit and brings in a specific revenue per unit. The brewery manager must meet a revenue target with minimum production costs. An exercise in Section 3 formulates the manager's goal as a linear programming problem and solves for the optimum production schedule when there are two beer varieties.

In the previous chapter we discussed solving linear programming problems by the graphical method. This method illustrates the basic ideas of linear programming, but it is practical only for problems with two variables. For problems with more than two variables, or problems with two variables and many constraints, the *simplex method* is used. This method grew out of a practical problem faced by George B. Dantzig in 1947. Dantzig was concerned with finding the least expensive way to allocate supplies for the United States Air Force.

The **simplex method** starts with the selection of one corner point (often the origin) from the feasible region. Then, in a systematic way, another corner point is found that attempts to improve the value of the objective function. Finally, an optimum solution is reached, or it can be seen that no such solution exists.

The simplex method requires a number of steps. In this chapter we divide the presentation of these steps into two parts. First, a problem is set up in Section 4.1 and the method started; then, in Section 4.2, the method is completed. Special situations are discussed in the remainder of the chapter.

4.1 Slack Variables and the Pivot

Because the simplex method is used for problems with many variables, it usually is not convenient to use letters such as x, y, z, or w as variable names. Instead, the symbols x_1 (read "x-sub-one"), x_2, x_3, and so on, are used. These variable names lend themselves easily to use on the computer.

In this section we will use the simplex method only for problems such as the following:

$$\text{Maximize} \quad z = 2x_1 - 3x_2$$
$$\text{subject to:} \quad 2x_1 + x_2 \le 10$$
$$x_1 - 3x_2 \le 5$$
$$\text{with} \quad x_1 \ge 0, \quad x_2 \ge 0.$$

This type of problem is said to be in *standard maximum form*. All constraints must be expressed in the linear form

$$a_1x_1 + a_2x_2 + a_3x_3 + \cdots + a_nx_n \le b,$$

where $x_1, x_2, x_3, \ldots, x_n$ are variables and $a_1, a_2, \ldots, a_n$ and b are constants, with $b \ge 0$.

> ## Standard Maximum Form
> A linear programming problem is in **standard maximum form** if the following conditions are satisfied.
>
> 1. The objective function is to be maximized.
> 2. All variables are nonnegative $(x_i \ge 0)$.
> 3. All remaining constraints are stated in the form
>
> $$a_1x_1 + a_2x_2 + \cdots + a_nx_n \le b \qquad \text{with } b \ge 0.$$

(Problems that do not meet all of these conditions are discussed in Sections 4.3 and 4.4.)

To use the simplex method, we start by converting the constraints, which are linear inequalities, into linear equations by adding a nonnegative variable, called a **slack variable**, to each constraint. For example, the inequality $x_1 + x_2 \leq 10$ is converted into an equation by adding the slack variable s_1 to get

$$x_1 + x_2 + s_1 = 10, \qquad \text{where } s_1 \geq 0.$$

The inequality $x_1 + x_2 \leq 10$ says that the sum $x_1 + x_2$ is less than or perhaps equal to 10. The variable s_1 "takes up any slack" and represents the amount by which $x_1 + x_2$ fails to equal 10. For example, if $x_1 + x_2$ equals 8, then s_1 is 2. If $x_1 + x_2 = 10$, then s_1 is 0.

> **CAUTION** A different slack variable must be used for each constraint.

EXAMPLE 1 Slack Variables

Restate the following linear programming problem by introducing slack variables.

$$\begin{aligned}
\text{Maximize} \quad & z = 3x_1 + 2x_2 + x_3 \\
\text{subject to:} \quad & 2x_1 + x_2 + x_3 \leq 150 \\
& 2x_1 + 2x_2 + 8x_3 \leq 200 \\
& 2x_1 + 3x_2 + x_3 \leq 320 \\
\text{with} \quad & x_1 \geq 0, \quad x_2 \geq 0, \quad x_3 \geq 0.
\end{aligned}$$

SOLUTION Rewrite the three constraints as equations by adding slack variables s_1, s_2, and s_3, one for each constraint. Then the problem can be restated as follows.

$$\begin{aligned}
\text{Maximize} \quad & z = 3x_1 + 2x_2 + x_3 \\
\text{subject to:} \quad & 2x_1 + x_2 + x_3 + s_1 \phantom{{}+ s_2 + s_3} = 150 \\
& 2x_1 + 2x_2 + 8x_3 \phantom{{}+ s_1} + s_2 \phantom{{}+ s_3} = 200 \\
& 2x_1 + 3x_2 + x_3 \phantom{{}+ s_1 + s_2} + s_3 = 320 \\
\text{with} \quad & x_1 \geq 0, \quad x_2 \geq 0, \quad x_3 \geq 0, \quad s_1 \geq 0, \quad s_2 \geq 0, \quad s_3 \geq 0.
\end{aligned}$$

In Example 4, we will take another step toward solving this linear programming problem.

Adding slack variables to the constraints converts a linear programming problem into a system of linear equations. In each of these equations, all variables should be on the left side of the equal sign and all constants on the right. All the equations in Example 1 satisfy this condition except for the objective function, $z = 3x_1 + 2x_2 + x_3$, which may be written with all variables on the left as

$$-3x_1 - 2x_2 - x_3 + z = 0.$$

Now the equations in Example 1 can be written as the following augmented matrix.

$$\begin{array}{ccccccc}
x_1 & x_2 & x_3 & s_1 & s_2 & s_3 & z \\
\end{array}$$
$$\left[\begin{array}{ccccccc|c}
2 & 1 & 1 & 1 & 0 & 0 & 0 & 150 \\
2 & 2 & 8 & 0 & 1 & 0 & 0 & 200 \\
2 & 3 & 1 & 0 & 0 & 1 & 0 & 320 \\
\hline
-3 & -2 & -1 & 0 & 0 & 0 & 1 & 0
\end{array}\right]$$
$$\text{Indicators}$$

This matrix is called the initial **simplex tableau**. The numbers in the bottom row, which are from the objective function, are called **indicators** (except for the 1 and 0 at the far right).

EXAMPLE 2 Initial Simplex Tableau

Set up the initial simplex tableau for the following problem.

A farmer has 100 acres of available land on which he wishes to plant a mixture of potatoes, corn, and cabbage. It costs him \$400 to produce an acre of potatoes, \$160 to produce an acre of corn, and $280 to produce an acre of cabbage. He has a maximum of $20,000 to spend. He makes a profit of $120 per acre of potatoes, $40 per acre of corn, and $60 per acre of cabbage. How many acres of each crop should he plant to maximize his profit?

SOLUTION   Begin by summarizing the given information as follows.

Profits and Constraints for Crops					
	Potatoes	Corn	Cabbage		Total
Number of Acres	x_1	x_2	x_3	$\leq$	100
Cost (per acre)	$400	$160	$280	$\leq$	$20,000
Profit (per acre)	$120	$40	$60		

If the number of acres allotted to each of the three crops is represented by x_1, x_2, and x_3, respectively, then the constraint pertaining to the number of acres can be expressed as

$$x_1 + x_2 + x_3 \quad \leq \quad 100 \qquad \text{Number of acres}$$

where x_1, x_2, and x_3 are all nonnegative. This constraint says that $x_1 + x_2 + x_3$ is less than or perhaps equal to 100. Use s_1 as the slack variable, giving the equation

$$x_1 + x_2 + x_3 + s_1 = 100.$$

Here s_1 represents the amount of the farmer's 100 acres that will not be used (s_1 may be 0 or any value up to 100).

The constraint pertaining to the production cost can be expressed as

$$400x_1 + 160x_2 + 280x_3 \leq 20,000, \qquad \text{Production costs}$$

or if we divide both sides by 40, as

$$10x_1 + 4x_2 + 7x_3 \leq 500.$$

This inequality can also be converted into an equation by adding a slack variable, s_2.

$$10x_1 + 4x_2 + 7x_3 + s_2 = 500$$

If we had not divided by 40, the slack variable would have represented any unused portion of the farmer's $20,000 capital. Instead, the slack variable represents 1/40 of that unused portion. (Note that s_2 may be any value from 0 to 500.)

The objective function represents the profit. The farmer wants to maximize

$$z = 120x_1 + 40x_2 + 60x_3.$$

The linear programming problem can now be stated as follows:

Maximize $\quad z = 120x_1 + 40x_2 + 60x_3$

subject to: $\quad x_1 + x_2 + x_3 + s_1 \qquad = 100$

$\qquad\qquad 10x_1 + 4x_2 + 7x_3 \qquad + s_2 = 500$

with $\qquad x_1 \geq 0, \quad x_2 \geq 0, \quad x_3 \geq 0, \quad s_1 \geq 0, \quad s_2 \geq 0.$

YOUR TURN 1 Set up the initial simplex tableau for Example 2 with the following two modifications. All the costs per acre are $100 less. Also, there are taxes of $5, $10, and $15 per acre of potatoes, corn, and cabbage, respectively, and the farmer does not want to pay more than $900 in taxes.

Rewrite the objective function as $-120x_1 - 40x_2 - 60x_3 + z = 0$, and complete the initial simplex tableau as follows.

$$\begin{array}{cccccc}
x_1 & x_2 & x_3 & s_1 & s_2 & z \\
\end{array}$$
$$\left[\begin{array}{cccccc|c}
1 & 1 & 1 & 1 & 0 & 0 & 100 \\
10 & 4 & 7 & 0 & 1 & 0 & 500 \\
\hline
-120 & -40 & -60 & 0 & 0 & 1 & 0 \\
\end{array}\right] \qquad \text{TRY YOUR TURN 1}$$

In Example 2 (which we will finish solving in the next section), the feasible region consists of the points (x_1, x_2, x_3) that satisfy the two inequalities for the number of acres and the production costs. It is difficult to draw such a region in three dimensions by hand. If there were a fourth crop, and hence a fourth dimension, plotting the region would be impossible. Nevertheless, as with the graphical method, the objective function is maximized at one of the corner points (or two corner points and all points on the line segment between the two points). With the simplex method, which we develop in this section and the next, we find that corner point using matrix methods that don't require drawing a graph of the feasible region.

After we have converted the two inequalities into equations by the addition of slack variables, the maximization problem in Example 2 consists of a system of two equations in five variables, together with the objective function. Because there are more variables than equations, the system will have an infinite number of solutions. To see this, solve the system for s_1 and s_2.

$$s_1 = 100 - x_1 - x_2 - x_3$$
$$s_2 = 500 - 10x_1 - 4x_2 - 7x_3$$

Each choice of values for x_1, x_2, and x_3 gives corresponding values for s_1 and s_2 that produce a solution of the system. But only some of these solutions are feasible.

In a feasible solution, all variables must be nonnegative. To get a unique feasible solution, we set three of the five variables equal to 0. In general, if there are m equations, then m variables can be nonzero. These m nonzero variables are called **basic variables**, and the corresponding solutions are called **basic feasible solutions**. Each basic feasible solution corresponds to a corner point. In particular, if we choose the solution with $x_1 = 0$, $x_2 = 0$, and $x_3 = 0$, then $s_1 = 100$ and $s_2 = 500$ are the basic variables. This solution, which corresponds to the corner point at the origin, is hardly optimal. It produces a profit of $0 for the farmer, since the equation that corresponds to the objective function becomes

$$-120(0) - 40(0) - 60(0) + 0s_1 + 0s_2 + z = 0.$$

In the next section we will use the simplex method to start with this solution and improve it to find the maximum possible profit.

Each step of the simplex method produces a solution that corresponds to a corner point of the region of feasible solutions. These solutions can be read directly from the matrix, as shown in the next example.

EXAMPLE 3   Basic Variables

Read a solution from the following simplex tableau.

$$
\begin{array}{cccccc}
x_1 & x_2 & x_3 & s_1 & s_2 & z \\
\end{array}
$$

$$
\left[
\begin{array}{cccccc|c}
2 & 0 & 8 & 5 & 2 & 0 & 17 \\
9 & 5 & 3 & 12 & 0 & 0 & 45 \\
\hline
-2 & 0 & -4 & 0 & 0 & 3 & 90 \\
\end{array}
\right]
$$

SOLUTION   In this solution, the variables x_2 and s_2 are basic variables. They can be identified quickly because the columns for these variables have all zeros except for one nonzero entry. All variables that are not basic variables have the value 0. This means that in the tableau just shown, x_2 and s_2 are the basic variables, while x_1, x_3, and s_1 have the value 0. The nonzero entry for x_2 is 5 in the second row. Since x_1, x_3, and s_1 are zero, the second row of the tableau represents the equation $5x_2 = 45$, so $x_2 = 9$. Similarly, from the top row, $2s_2 = 17$, so $s_2 = 17/2$. From the bottom row, $3z = 90$, so $z = 30$. The solution is thus $x_1 = 0$, $x_2 = 9$, $x_3 = 0$, $s_1 = 0$, and $s_2 = 17/2$, with $z = 30$.   ▬

Pivots   Solutions read directly from the initial simplex tableau are seldom optimal. It is necessary to proceed to other solutions (corresponding to other corner points of the feasible region) until an optimum solution is found. To get these other solutions, we use restricted

versions of the row operations from Chapter 2 to change the tableau by using one of the nonzero entries of the tableau as a **pivot**. The row operations are performed to change to 0 all entries in the column containing the pivot (except for the pivot itself, which remains unchanged). Pivoting, explained in the next example, produces a new tableau leading to another solution of the system of equations obtained from the original problem.

> **CAUTION** In this chapter, when adding a multiple of one row to a multiple of another, we will never take a negative multiple of the row being changed. For example, when changing row 2, we might use $-2R_1 + 3R_2 \rightarrow R_2$, but we will never use $2R_1 - 3R_2 \rightarrow R_2$. If you get a negative number in the rightmost column, you will know immediately that you have made an error. The reason for this restriction is that violating it turns negative numbers into positive, and vice versa. This is disastrous in the bottom row, where we will seek negative numbers when we choose our pivot column. It will also cause problems with choosing pivots, particularly in the algorithm for solving nonstandard problems in Section 4.4.

When we are performing row operations by hand, as we did in Chapter 2, we will postpone getting a 1 in each basic variable column until the final step. This will avoid fractions and decimals, which can make the operations more difficult and more prone to error. When using a graphing calculator, however, we must change the pivot to a 1 before performing row operations. The next example illustrates both of these methods.

EXAMPLE 4   Pivot

Pivot about the indicated 2 of the following initial simplex tableau.

$$
\begin{array}{ccccccc}
x_1 & x_2 & x_3 & s_1 & s_2 & s_3 & z \\
\end{array}
$$

$$
\left[
\begin{array}{ccccccc|c}
\mathbf{2} & 1 & 1 & 1 & 0 & 0 & 0 & 150 \\
1 & 2 & 8 & 0 & 1 & 0 & 0 & 200 \\
2 & 3 & 1 & 0 & 0 & 1 & 0 & 320 \\
\hline
-3 & -2 & -1 & 0 & 0 & 0 & 1 & 0 \\
\end{array}
\right]
$$

SOLUTION

**Method 1
Calculating by Hand**

Using the row operations indicated in color to get zeros in the column with the pivot, we arrive at the following tableau.

$$
\begin{array}{ccccccc}
x_1 & x_2 & x_3 & s_1 & s_2 & s_3 & z \\
\end{array}
$$

$$
\begin{array}{r}
\\
-R_1 + 2R_2 \rightarrow R_2 \\
-R_1 + R_3 \rightarrow R_3 \\
3R_1 + 2R_4 \rightarrow R_4
\end{array}
\left[
\begin{array}{ccccccc|c}
2 & 1 & 1 & 1 & 0 & 0 & 0 & 150 \\
0 & 3 & 15 & -1 & 2 & 0 & 0 & 250 \\
0 & 2 & 0 & -1 & 0 & 1 & 0 & 170 \\
0 & -1 & 1 & 3 & 0 & 0 & 2 & 450 \\
\end{array}
\right]
$$

In this simplex tableau, the variables x_1, s_2, and s_3 are basic variables. The solution is $x_1 = 75$, $x_2 = 0$, $x_3 = 0$, $s_1 = 0$, $s_2 = 125$, and $s_3 = 170$. Substituting these results into the objective function gives

$$0(75) - 1(0) + 1(0) + 3(0) + 0(125) + 0(170) + 2z = 450,$$

or $z = 225$. (This shows that the value of z can always be found using the number in the bottom row of the z column and the number in the lower right-hand corner.)

Finally, to be able to read the solution directly from the tableau, we multiply rows 1, 2, and 4 by $1/2$, getting the following tableau.

$$
\begin{array}{ccccccc}
x_1 & x_2 & x_3 & s_1 & s_2 & s_3 & z \\
\end{array}
$$

$$
\begin{array}{r}
\frac{1}{2}R_1 \rightarrow R_1 \\
\frac{1}{2}R_2 \rightarrow R_2 \\
\\
\frac{1}{2}R_4 \rightarrow R_4
\end{array}
\left[
\begin{array}{ccccccc|c}
1 & \frac{1}{2} & \frac{1}{2} & \frac{1}{2} & 0 & 0 & 0 & 75 \\
0 & \frac{3}{2} & \frac{15}{2} & -\frac{1}{2} & 1 & 0 & 0 & 125 \\
0 & 2 & 0 & -1 & 0 & 1 & 0 & 170 \\
0 & -\frac{1}{2} & \frac{1}{2} & \frac{3}{2} & 0 & 0 & 1 & 225 \\
\end{array}
\right]
$$

Method 2
Graphing Calculator

The row operations of the simplex method can also be done on a graphing calculator, as we saw in Chapter 2. Figure 1 shows the result when the tableau in this example is entered into a TI-84 Plus. The right side of the tableau is not visible but can be seen by pressing the right arrow key.

Recall that we must change the pivot to 1 before performing row operations with a graphing calculator. Figure 2 shows the result of multiplying row 1 of matrix A by $1/2$. In Figure 3 we show the same result with the decimal numbers changed to fractions.

We can now modify column 1, using the commands described in Chapter 2, to agree with the tableau under Method 1. The result is shown in Figure 4.

YOUR TURN 2 Pivot about the indicated 6 of the following initial simplex tableau, and then read a solution from the tableau.

$$\begin{array}{cccccccc}
x_1 & x_2 & x_3 & s_1 & s_2 & s_3 & z & \\
\hline
3 & \boxed{6} & 2 & 1 & 0 & 0 & 0 & 60 \\
8 & 5 & 4 & 0 & 1 & 0 & 0 & 80 \\
3 & 6 & 7 & 0 & 0 & 1 & 0 & 120 \\
\hline
-30 & -50 & -15 & 0 & 0 & 0 & 1 & 0 \\
\end{array}$$

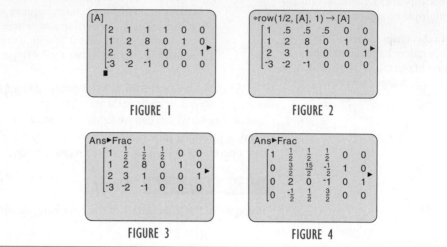

FIGURE 1   FIGURE 2

FIGURE 3   FIGURE 4

TRY YOUR TURN 2

In the simplex method, the pivoting process (without the final step of getting a 1 in each basic variable column when using Method 1) is repeated until an optimum solution is found, if one exists. In the next section we will see how to decide where to pivot to improve the value of the objective function and how to tell when an optimum solution either has been reached or does not exist.

4.1 EXERCISES

Convert each inequality into an equation by adding a slack variable.

1. $x_1 + 2x_2 \leq 6$

2. $6x_1 + 2x_2 \leq 50$

3. $2.3x_1 + 5.7x_2 + 1.8x_3 \leq 17$

4. $8x_1 + 6x_2 + 5x_3 \leq 250$

For Exercises 5–8, (a) determine the number of slack variables needed, (b) name them, and (c) use slack variables to convert each constraint into a linear equation.

5. Maximize $z = 5x_1 + 7x_2$

subject to: $2x_1 + 3x_2 \leq 15$

$4x_1 + 5x_2 \leq 35$

$x_1 + 6x_2 \leq 20$

with $x_1 \geq 0, \quad x_2 \geq 0.$

6. Maximize $z = 1.2x_1 + 3.5x_2$

subject to: $2.4x_1 + 1.5x_2 \leq 10$

$1.7x_1 + 1.9x_2 \leq 15$

with $x_1 \geq 0, \quad x_2 \geq 0.$

7. Maximize $z = 8x_1 + 3x_2 + x_3$

subject to: $7x_1 + 6x_2 + 8x_3 \leq 118$

$4x_1 + 5x_2 + 10x_3 \leq 220$

with $x_1 \geq 0, \quad x_2 \geq 0, \quad x_3 \geq 0.$

8. Maximize $z = 12x_1 + 15x_2 + 10x_3$

subject to: $2x_1 + 2x_2 + x_3 \leq 8$

$x_1 + 4x_2 + 3x_3 \leq 12$

with $x_1 \geq 0, \quad x_2 \geq 0, \quad x_3 \geq 0.$

Introduce slack variables as necessary, then write the initial simplex tableau for each linear programming problem.

9. Find $x_1 \geq 0$ and $x_2 \geq 0$ such that
$$4x_1 + 2x_2 \leq 5$$
$$x_1 + 2x_2 \leq 4$$
and $z = 7x_1 + x_2$ is maximized.

10. Find $x_1 \geq 0$ and $x_2 \geq 0$ such that
$$2x_1 + 3x_2 \leq 100$$
$$5x_1 + 4x_2 \leq 200$$
and $z = x_1 + 3x_2$ is maximized.

11. Find $x_1 \geq 0$ and $x_2 \geq 0$ such that
$$x_1 + x_2 \leq 10$$
$$5x_1 + 2x_2 \leq 20$$
$$x_1 + 2x_2 \leq 36$$
and $z = x_1 + 3x_2$ is maximized.

12. Find $x_1 \geq 0$ and $x_2 \geq 0$ such that
$$x_1 + x_2 \leq 25$$
$$4x_1 + 3x_2 \leq 48$$
and $z = 5x_1 + 3x_2$ is maximized.

13. Find $x_1 \geq 0$ and $x_2 \geq 0$ such that
$$3x_1 + x_2 \leq 12$$
$$x_1 + x_2 \leq 15$$
and $z = 2x_1 + x_2$ is maximized.

14. Find $x_1 \geq 0$ and $x_2 \geq 0$ such that
$$10x_1 + 4x_2 \leq 100$$
$$20x_1 + 10x_2 \leq 150$$
and $z = 4x_1 + 5x_2$ is maximized.

Write the solutions that can be read from each simplex tableau.

15.

x_1	x_2	x_3	s_1	s_2	z	
1	0	4	5	1	0	8
3	1	1	2	0	0	4
−2	0	2	3	0	1	28

16.

x_1	x_2	x_3	s_1	s_2	z	
1	5	0	1	2	0	6
0	2	1	2	3	0	15
0	4	0	1	−2	1	64

17.

x_1	x_2	x_3	s_1	s_2	s_3	z	
6	2	2	3	0	0	0	16
2	2	0	1	0	5	0	35
2	1	0	3	1	0	0	6
−3	−2	0	2	0	0	3	36

18.

x_1	x_2	x_3	s_1	s_2	s_3	z	
0	2	0	5	2	2	0	15
0	3	1	0	1	2	0	2
7	4	0	0	3	5	0	35
0	−4	0	0	4	3	2	40

Pivot once as indicated in each simplex tableau. Read the solution from the result.

19.

x_1	x_2	x_3	s_1	s_2	z	
1	2	4	1	0	0	56
2	**2**	1	0	1	0	40
−1	−3	−2	0	0	1	0

20.

x_1	x_2	x_3	s_1	s_2	z	
2	3	4	1	0	0	18
6	**3**	2	0	1	0	15
−1	−6	−2	0	0	1	0

21.

x_1	x_2	x_3	s_1	s_2	s_3	z	
2	2	**1**	1	0	0	0	12
1	2	3	0	1	0	0	45
3	1	1	0	0	1	0	20
−2	−1	−3	0	0	0	1	0

22.

x_1	x_2	x_3	s_1	s_2	s_3	z	
4	2	3	1	0	0	0	22
2	2	**5**	0	1	0	0	28
1	3	2	0	0	1	0	45
−3	−2	−4	0	0	0	1	0

23.

x_1	x_2	x_3	s_1	s_2	s_3	z	
2	**2**	3	1	0	0	0	500
4	1	1	0	1	0	0	300
7	2	4	0	0	1	0	700
−3	−4	−2	0	0	0	1	0

24.

x_1	x_2	x_3	x_4	s_1	s_2	s_3	z	
1	2	3	1	1	0	0	0	115
2	1	8	5	0	1	0	0	200
1	0	1	0	0	0	1	0	50
−2	−1	−1	−1	0	0	0	1	0

25. Explain the purpose of a slack variable.

26. How can you tell by looking at a linear programming problem how many slack variables will be needed?

APPLICATIONS

Set up Exercises 27–31 for solution by the simplex method. First express the linear constraints and objective function, then add slack variables to convert each constraint into a linear equation, and then set up the initial simplex tableau. The solutions of some of these problems will be completed in the exercises for the next section.

Business and Economics

27. Royalties The authors of a best-selling textbook in finite mathematics are told that, for the next edition of their book, each simple figure would cost the project $20, each figure with additions would cost $35, and each computer-drawn sketch would cost $60. They are limited to 400 figures, for which they are allowed to spend up to $2200. The number of computer-drawn sketches must be no more than the number of the other two types combined, and there must be at least twice as many simple figures as there are figures with additions. If each simple figure increases the royalties by $95, each figure with additions increases royalties by $200, and each computer-drawn figure increases royalties by $325, how many of each type of figure should be included to maximize royalties, assuming that all art costs are borne by the publisher?

28. **Manufacturing Bicycles** A manufacturer of bicycles builds racing, touring, and mountain models. The bicycles are made of both aluminum and steel. The company has available 91,800 units of steel and 42,000 units of aluminum. The racing, touring, and mountain models need 17, 27, and 34 units of steel, and 12, 21, and 15 units of aluminum, respectively. How many of each type of bicycle should be made in order to maximize profit if the company makes $8 per racing bike, $12 per touring bike, and $22 per mountain bike? What is the maximum possible profit?

29. **Production—Picnic Tables** The manager of a large park has received many complaints about the insufficient number of picnic tables available. At the end of the park season, she has surplus cash and labor resources available and decides to make as many tables as possible. She considers three possible models: redwood, stained Douglas fir, and stained white spruce (all of which last equally well). She has carpenters available for assembly work for a maximum of 90 eight-hour days, while laborers for staining work are available for no more than 60 eight-hour days. Each redwood table requires 8 hours to assemble but no staining, and it costs $159 (including all labor and materials). Each Douglas fir table requires 7 hours to assemble and 2 hours to stain, and it costs $138.85. Each white spruce table requires 8 hours to assemble and 2 hours to stain, and it costs $129.35. If no more than $15,000 is available for this project, what is the maximum number of tables which can be made, and how many of each type should be made? *Source: Karl K. Norton.*

30. **Production—Knives** The Cut-Right Company sells sets of kitchen knives. The Basic Set consists of 2 utility knives and 1 chef's knife. The Regular Set consists of 2 utility knives, 1 chef's knife, and 1 slicer. The Deluxe Set consists of 3 utility knives, 1 chef's knife, and 1 slicer. Their profit is $30 on a Basic Set, $40 on a Regular Set, and $60 on a Deluxe Set. The factory has on hand 800 utility knives, 400 chef's knives, and 200 slicers. Assuming that all sets will be sold, how many of each type should be produced in order to maximize profit? What is the maximum profit?

31. **Advertising** The Fancy Fashions, an independent, local boutique, has $8000 available each month for advertising. Newspaper ads cost $400 each, and no more than 30 can run per month. Internet banner ads cost $20 each, and no more than 60 can run per month. TV ads cost $2000 each, with a maximum of 10 available each month. Approximately 4000 women will see each newspaper ad, 3000 will see each Internet banner, and 10,000 will see each TV ad. How much of each type of advertising should be used if the store wants to maximize its ad exposure?

YOUR TURN ANSWERS

1.

x_1	x_2	x_3	s_1	s_2	s_3	z	
1	1	1	1	0	0	0	100
15	3	9	0	1	0	0	1000
1	2	3	0	0	1	0	180
−120	−40	−60	0	0	0	1	0

2.

x_1	x_2	x_3	s_1	s_2	s_3	z	
1/2	1	1/3	1/6	0	0	0	10
11/2	0	7/3	−5/6	1	0	0	30
0	0	5	−1	0	1	0	60
−5	0	5/3	25/3	0	0	1	500

$x_1 = 0, x_2 = 10, x_3 = 0, s_1 = 0, s_2 = 30, s_3 = 60, z = 500$

4.2 Maximization Problems

APPLY IT How many racing, touring, and mountain bicycles should a bicycle manufacturer make to maximize profit?

We will answer this question in Exercise 27 of this section using an algorithm called the simplex method.

In the previous section we showed how to prepare a linear programming problem for solution. First, we converted the constraints to linear equations with slack variables; then we used the coefficients of the variables from the linear equation to write an augmented matrix. Finally, we used the pivot to go from one corner point to another corner point in the region of feasible solutions.

Now we are ready to put all this together and produce an optimum value for the objective function. To see how this is done, let us complete Example 2 from Section 4.1. In this example, we were trying to determine, under certain constraints, the number of acres of potatoes (x_1), corn (x_2), and cabbage (x_3) the farmer should plant in order to optimize his profit (z). In the previous section, we set up the following simplex tableau.

$$
\begin{array}{cccccc}
x_1 & x_2 & x_3 & s_1 & s_2 & z \\
\end{array}
$$

$$
\begin{bmatrix}
1 & 1 & 1 & 1 & 0 & 0 & 100 \\
10 & 4 & 7 & 0 & 1 & 0 & 500 \\
\hline
-120 & -40 & -60 & 0 & 0 & 1 & 0
\end{bmatrix}
$$

This tableau leads to the solution $x_1 = 0$, $x_2 = 0$, $x_3 = 0$, $s_1 = 100$, and $s_2 = 500$, with s_1 and s_2 as the basic variables. These values produce a value of 0 for z. In this solution, the farmer is planting 0 acres and earning \$0 profit. We can easily see that there are other combinations of potatoes, corn, and cabbage that produce a nonzero profit, and thus we know that the farmer has better alternatives than planting nothing.

To decide which crops he should plant, we look at the original objective function representing profit,

$$z = 120x_1 + 40x_2 + 60x_3.$$

The coefficient of x_1 is the largest, which indicates that he will make the most profit per acre planting potatoes. It makes sense, then, to first try increasing x_1 to improve the profit.

To determine how much we can increase x_1, we look at the constraint equations:

$$
\begin{aligned}
x_1 + x_2 + x_3 + s_1 &= 100 \\
10x_1 + 4x_2 + 7x_3 + s_2 &= 500.
\end{aligned}
$$

Because there are two equations, only two of the five variables can be basic (and nonzero). If x_1 is nonzero in the solution, then x_1 will be a basic variable. Therefore, x_2 and x_3 will stay at 0, and the equations simplify to

$$
\begin{aligned}
x_1 + s_1 &= 100 \\
10x_1 + s_2 &= 500.
\end{aligned}
$$

Since s_1 and s_2 are both nonnegative, the first equation implies that x_1 cannot exceed 100, and the second implies that $10x_1$ cannot exceed 500, so x_1 cannot exceed 500/10, or 50. To satisfy both of these conditions, x_1 cannot exceed 50, the smaller of 50 and 100. If we let x_1 take the value of 50, then $x_1 = 50$, $x_2 = 0$, $x_3 = 0$, and $s_2 = 0$. Since $x_1 + s_1 = 100$, then $s_1 = 100 - x_1 = 100 - 50 = 50$. Therefore, s_1 is still a basic variable, while s_2 is no longer a basic variable, having been replaced in the set of basic variables by x_1. This solution gives a profit of

$$
\begin{aligned}
z &= 120x_1 + 40x_2 + 60x_3 + 0s_1 + 0s_2 \\
&= 120(50) + 40(0) + 60(0) + 0(50) + 0(0) = 6000,
\end{aligned}
$$

or \$6000, when 50 acres of potatoes are planted.

The same result could have been found from the initial simplex tableau given below. Recall that the indicators are the numbers in the bottom row in the columns labeled with real or slack variables. To use the tableau, we select the variable with the most negative indicator. (If no indicator is negative, then the value of the objective function cannot be improved.) In this example, the variable with the most negative indicator is x_1.

Basic variables

$$
\begin{array}{cccccc}
x_1 & x_2 & x_3 & s_1 & s_2 & z \\
\end{array}
$$

$$
\begin{bmatrix}
1 & 1 & 1 & 1 & 0 & 0 & 100 \\
10 & 4 & 7 & 0 & 1 & 0 & 500 \\
\hline
\mathbf{-120} & -40 & -60 & 0 & 0 & 1 & 0
\end{bmatrix}
$$

Most negative indicator

The most negative indicator identifies the variable whose value is to be made nonzero, if possible, because it indicates the variable with the largest coefficient in the objective function. To find the variable that is now basic and will become nonbasic, calculate the quotients that were found above. Do this by dividing each number from the right side of the tableau by the corresponding number from the column with the most negative indicator.

Basic variables

Quotients	x_1	x_2	x_3	s_1	s_2	z	
$100/1 = 100$	1	1	1	1	0	0	100
Smaller → $500/10 = 50$	**10**	4	7	0	1	0	500
	−120	−40	−60	0	0	1	0

Notice that we do not form a quotient for the bottom row. Of the two quotients found, the smallest is 50 (from the second row). This indicates that x_1 cannot exceed $500/10 = 50$, so 10 is the pivot. Using 10 as the pivot, perform the appropriate row operations to get zeros in the rest of the column. We will use Method 1 from Section 4.1 (calculating by hand) to perform the pivoting, but Method 2 (graphing calculator) could be used just as well. The new tableau is as follows.

Basic variables

	x_1	x_2	x_3	s_1	s_2	z	
$-\mathbf{R_2} + 10\mathbf{R_1} \rightarrow \mathbf{R_1}$	0	6	3	10	−1	0	500
	10	4	7	0	1	0	500
$12\mathbf{R_2} + \mathbf{R_3} \rightarrow \mathbf{R_3}$	0	8	24	0	12	1	6000

The solution read from this tableau is

$$x_1 = 50, \qquad x_2 = 0, \qquad x_3 = 0, \qquad s_1 = 50, \qquad s_2 = 0,$$

with $z = 6000$, the same as the result found above.

None of the indicators in the final simplex tableau are negative, which means that the value of z cannot be improved beyond \$6000. To see why, recall that the last row gives the coefficients of the objective function so that

$$0x_1 + 8x_2 + 24x_3 + 0s_1 + 12s_2 + z = 6000,$$

or
$$z = 6000 - 0x_1 - 8x_2 - 24x_3 - 0s_1 - 12s_2.$$

Since x_2, x_3, and s_2 are zero, $z = 6000$, but if any of these three variables were to increase, z would decrease.

This result suggests that the optimal solution has been found as soon as no indicators are negative. As long as an indicator is negative, the value of the objective function may be improved. If any indicators are negative, we just find a new pivot and use row operations, repeating the process until no negative indicators remain.

Once there are no longer any negative numbers in the final row, create a 1 in the columns corresponding to the basic variables and z. In the previous example, this is accomplished by dividing rows 1 and 2 by 10.

	x_1	x_2	x_3	s_1	s_2	z	
$\mathbf{R_1}/10 \rightarrow \mathbf{R_1}$	0	$\frac{6}{10}$	$\frac{3}{10}$	1	$-\frac{1}{10}$	0	50
$\mathbf{R_2}/10 \rightarrow \mathbf{R_2}$	1	$\frac{4}{10}$	$\frac{7}{10}$	0	$\frac{1}{10}$	0	50
	0	8	24	0	12	1	6000

It is now easy to read the solution from this tableau:

$$x_1 = 50, \qquad x_2 = 0, \qquad x_3 = 0, \qquad s_1 = 50, \qquad s_2 = 0,$$

with $z = 6000$.

We can finally state the solution to the problem about the farmer. The farmer will make a maximum profit of $6000 by planting 50 acres of potatoes, no acres of corn, and no acres of cabbage. The value $s_1 = 50$ indicates that of the 100 acres of land available, 50 acres should be left unplanted. It may seem strange that leaving assets unused can produce a maximum profit, but such results actually occur often.

Note that since each variable can be increased by a different amount, the most negative indicator is not always the best choice. On average, though, it has been found that the most negative indicator is the best choice.

In summary, the following steps are involved in solving a standard maximum linear programming problem by the simplex method.

Simplex Method For Standard Maximization Problems

1. Determine the objective function.

2. Write all the necessary constraints.

3. Convert each constraint into an equation by adding a slack variable in each.

4. Set up the initial simplex tableau.

5. Locate the most negative indicator. If there are two such indicators, choose the one farther to the left.

6. Form the necessary quotients to find the pivot. Disregard any quotients with 0 or a negative number in the denominator. The smallest nonnegative quotient gives the location of the pivot. If all quotients must be disregarded, no maximum solution exists. If two quotients are both equal and smallest, choose the pivot in the row nearest the top of the matrix.

7. Use row operations to change all other numbers in the pivot column to zero by adding a suitable multiple of the pivot row to a positive multiple of each row.

8. If the indicators are all positive or 0, this is the final tableau. If not, go back to Step 5 and repeat the process until a tableau with no negative indicators is obtained.

9. Read the solution from the final tableau.

In Steps 5 and 6, the choice of the column farthest to the left or the row closest to the top is arbitrary. You may choose another row or column in case of a tie, and you will get the same final answer, but your intermediate results will be different.

| CAUTION | In performing the simplex method, a negative number in the right-hand column signals that a mistake has been made. One possible error is using a negative value for c_2 in the operation $c_1R_i + c_2R_j \rightarrow R_j$.

EXAMPLE 1 Using the Simplex Method

To compare the simplex method with the graphical method, we use the simplex method to solve the problem in Example 1, Section 3.3. The graph is shown again in Figure 5. The objective function to be maximized was

$$z = 25x_1 + 30x_2. \quad \text{Revenue}$$

(Since we are using the simplex method, we use x_1 and x_2 instead of x and y as variables.) The constraints were as follows:

$$x_1 + x_2 \leq 65 \quad \text{Number}$$
$$4x_1 + 5x_2 \leq 300 \quad \text{Cost}$$

with

$$x_1 \geq 0, \quad x_2 \geq 0.$$

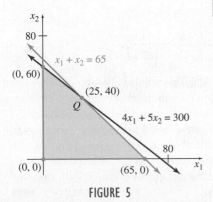

FIGURE 5

Add a slack variable to each constraint:

$$x_1 + x_2 + s_1 \qquad\quad = 65$$
$$4x_1 + 5x_2 \qquad + s_2 = 300$$

with

$$x_1 \geq 0, \ x_2 \geq 0, \ s_1 \geq 0, \ s_2 \geq 0.$$

Write the initial tableau.

$$
\begin{array}{ccccc}
x_1 & x_2 & s_1 & s_2 & z \\
\end{array}
$$
$$
\left[\begin{array}{ccccc|c}
1 & 1 & 1 & 0 & 0 & 65 \\
4 & 5 & 0 & 1 & 0 & 300 \\
\hline
-25 & -30 & 0 & 0 & 1 & 0 \\
\end{array}\right]
$$

This tableau leads to the solution $x_1 = 0$, $x_2 = 0$, $s_1 = 65$, and $s_2 = 300$, with $z = 0$, which corresponds to the origin in Figure 5. The most negative indicator is -30, which is in column 2 of row 3. The quotients of the numbers in the right-hand column and in column 2 are

$$\frac{65}{1} = 65 \qquad \text{and} \qquad \frac{300}{5} = 60.$$

The smaller quotient is 60, giving 5 as the pivot. Use row operations to get the new tableau. For clarity, we will continue to label the columns with x_1, x_2, and so on, although this is not necessary in practice.

$$
\begin{array}{ccccc}
& x_1 & x_2 & s_1 & s_2 & z \\
\end{array}
$$

$-R_2 + 5R_1 \rightarrow R_1$

$6R_2 + R_3 \rightarrow R_3$

$$
\left[\begin{array}{ccccc|c}
1 & 0 & 5 & -1 & 0 & 25 \\
4 & 5 & 0 & 1 & 0 & 300 \\
\hline
-1 & 0 & 0 & 6 & 1 & 1800 \\
\end{array}\right]
$$

The solution from this tableau is $x_1 = 0$ and $x_2 = 60$, with $z = 1800$. (From now on, we will list only the original variables when giving the solution.) This corresponds to the corner point $(0, 60)$ in Figure 5. Verify that if we instead pivoted on the column with the indicator of -25, we would arrive at the corner point $(65, 0)$, where the objective function has the value 1625, which is smaller than its value at $(0, 60)$.

Because of the indicator -1, the value of z might be improved. We compare quotients and choose the 1 in row 1, column 1, as pivot to get the final tableau.

$$
\begin{array}{ccccc}
x_1 & x_2 & s_1 & s_2 & z \\
\end{array}
$$

$-4R_1 + R_2 \rightarrow R_2$

$R_1 + R_3 \rightarrow R_3$

$$
\left[\begin{array}{ccccc|c}
1 & 0 & 5 & -1 & 0 & 25 \\
0 & 5 & -20 & 5 & 0 & 200 \\
\hline
0 & 0 & 5 & 5 & 1 & 1825 \\
\end{array}\right]
$$

There are no more negative indicators, so the optimum solution has been achieved. Create a 1 in column 2 by multiplying row 2 by $1/5$.

$$
\begin{array}{ccccc}
x_1 & x_2 & s_1 & s_2 & z \\
\end{array}
$$

$(1/5)R_2 \rightarrow R_2$

$$
\left[\begin{array}{ccccc|c}
1 & 0 & 5 & -1 & 0 & 25 \\
0 & 1 & -4 & 1 & 0 & 40 \\
\hline
0 & 0 & 5 & 5 & 1 & 1825 \\
\end{array}\right]
$$

Here the solution is $x_1 = 25$ and $x_2 = 40$, with $z = 1825$. This solution, which corresponds to the corner point $(25, 40)$ in Figure 5, is the same as the solution found earlier.

Each simplex tableau above gave a solution corresponding to one of the corner points of the feasible region. As shown in Figure 6, the first solution corresponded to the origin, with $z = 0$. By choosing the appropriate pivot, we moved systematically to a new corner point, $(0, 60)$, which improved the value of z to 1800. The next tableau took us to $(25, 40)$, producing the optimum value of $z = 1825$. There was no reason to test the last corner point, $(65, 0)$, since the optimum value z was found before that point was reached. **TRY YOUR TURN 1**

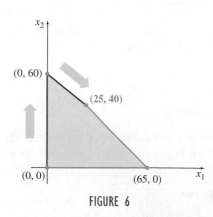

FIGURE 6

YOUR TURN 1 Use the simplex method to solve the problem in Example 2, Section 3.3.

It is good practice to verify your intermediate answers after each new tableau is calculated. You can check your answers by substituting these values for the original variables and the slack variables in the constraint equations and in the objective function.

CAUTION Never choose a zero or a negative number as the pivot. The reason for this is explained in the next example.

EXAMPLE 2 Finding the Pivot

Find the pivot for the following initial simplex tableau.

$$
\begin{array}{cccccc}
x_1 & x_2 & s_1 & s_2 & s_3 & z \\
\end{array}
$$

$$
\left[
\begin{array}{cccccc|c}
1 & -2 & 1 & 0 & 0 & 0 & 100 \\
3 & 4 & 0 & 1 & 0 & 0 & 200 \\
5 & 0 & 0 & 0 & 1 & 0 & 150 \\
\hline
-10 & -25 & 0 & 0 & 0 & 1 & 0 \\
\end{array}
\right]
$$

SOLUTION The most negative indicator is -25. To find the pivot, we find the quotients formed by the entries in the rightmost column and in the x_2 column: $100/(-2)$, $200/4$, and $150/0$. The quotients predict the value of a variable in the solution. Thus, since we want all variables to be nonnegative, we must reject a negative quotient. Furthermore, we cannot choose 0 as the pivot, because no multiple of the row with 0, when added to the other rows, will cause the other entries in the x_2 column to become 0.

The only usable quotient is $200/4 = 50$, making 4 the pivot. If all the quotients either are negative or have zero denominators, no unique optimum solution will be found. Such a situation indicates an unbounded feasible region because the variable corresponding to that column can be made as large as you like. The quotients, then, determine whether an optimum solution exists.

YOUR TURN 2 Pivot on the 4 in Example 2 and write the solution.

TRY YOUR TURN 2

CAUTION If there is a 0 in the right-hand column, do not disregard that row, unless the corresponding number in the pivot column is negative or zero. In fact, such a row gives a quotient of 0, so it will automatically have the smallest ratio. It will not cause an increase in z, but it may lead to another tableau in which z can be further increased.

TECHNOLOGY NOTE We saw earlier that graphing calculators can be used to perform row operations. A program to solve a linear programming problem with a graphing calculator is given in the *Graphing Calculator and Excel Spreadsheet Manual* available with this book.

TECHNOLOGY NOTE Spreadsheets often have a program for solving linear programming problems built in. Figure 7 on the next page shows the Solver feature of Microsoft Excel (under the Tools menu) for Example 1. (On some versions of Excel, the Solver must be installed from an outside source.) For details, see the *Graphing Calculator and Excel Spreadsheet Manual* available with this text.

In addition, Solver provides a **sensitivity analysis**, which allows us to see how much the constraints could be varied without changing the solution. Under the Solver options, make sure that "Assume Linear Model" and "Assume Non-Negative" are selected. Figure 8 on the next page shows a sensitivity analysis for Example 1. Notice that the value of the first coefficient in the objective function is 25, with an allowable increase of 5 and an allowable decrease of 1. This means that, while keeping the second coefficient at 30, the first coefficient of 25 could be increased by 5 (to 30) or decreased by 1 (to 24), and (25, 40) would still be a solution to the maximization problem. Similarly, for the second coefficient of 30, increasing it by 1.25 (to 31.25) or decreasing it by 5 (to 25) would still leave (25, 40) as a solution to the maximization problem. This would be useful to the owner who decides on the solution of (25, 40) (25 canoes and 40 kayaks) and wonders how much the objective function would have to change before the solution would no longer be optimal. The original revenue for a canoe was $25, which is the source of the first coefficient in the objective function. Assuming that everything else stays the same, the revenue could change to anything from $24 to $30, and the original decision would still be optimal.

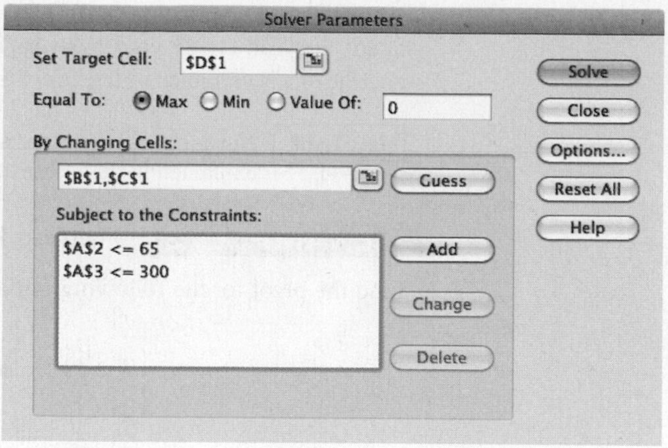

FIGURE 7

Notice, however, that any change in one of the revenues will change the total revenue in the optimal solution. For example, if the first coefficient of 25 is increased by 5 to 30, then the optimal objective value will increase by $5 \times 25 = 125$. One can perform similar changes to other parameters of the problem, but that is beyond the scope of this text.

Adjustable Cells

Cell	Name	Final Value	Reduced Cost	Objective Coefficient	Allowable Increase	Allowable Decrease
B1		25	0	25	5	1
C1		40	0	30	1.25	5

Constraints

Cell	Name	Final Value	Shadow Price	Constraint R.H. Side	Allowable Increase	Allowable Decrease
A2		65	5	65	10	5
A3		300	5	300	25	40

FIGURE 8

In many real-life problems, the number of variables and constraints may be in the hundreds, if not the thousands, in which case a computer is used to implement the simplex algorithm. Computer programs for the simplex algorithm differ in some ways from the algorithm we have shown. For example, it is not necessary for a computer to divide common factors out of inequalities to simplify the arithmetic. In fact, computer versions of the algorithm do not necessarily keep all the numbers as integers. As we saw in the previous section, dividing a row by a number may introduce decimals, which makes the arithmetic more difficult to do by hand, but creates no problem for a computer other than round-off error. Several linear programming models in actual use are presented on the website for this textbook.

If you use a graphing calculator to perform the simplex algorithm, we suggest that you review the pivoting procedure described in Method 2 of the previous section. It differs slightly from Method 1, because it converts each pivot element into a 1, but it works nicely with a calculator to keep track of the arithmetic details.

On the other hand, if you carry out the steps of the simplex method by hand, we suggest that you first eliminate fractions and decimals when setting up the initial tableau. For example, we would rewrite the constraint

$$\frac{2}{3}x_1 + \frac{5}{2}x_2 \leq 7$$

as
$$4x_1 + 15x_2 \leq 42,$$

by multiplying both sides of the equation by 6. Similarly, we would write

$$5.2x_1 + 4.4x_2 \le 8.5$$

as

$$52x_1 + 44x_2 \le 85$$

by multiplying both sides of the equation by 10. We must be cautious, however, in remembering that the value of the slack and surplus variables in the optimal solution must be adjusted by this factor to represent the original constraint.

NOTE Sometimes the simplex method cycles and returns to a previously visited solution, rather than making progress. Methods are available for handling cycling. In this text, we will avoid examples with this behavior. For more details, see Alan Sultan's *Linear Programming: An Introduction with Applications*, Academic Press, 1993. In real applications, cycling is rare and tends not to come up because of computer rounding.

4.2 EXERCISES

In Exercises 1–6, the initial tableau of a linear programming problem is given. Use the simplex method to solve each problem.

1.

x_1	x_2	x_3	s_1	s_2	z	
1	4	4	1	0	0	16
2	1	5	0	1	0	20
-3	-1	-2	0	0	1	0

2.

x_1	x_2	x_3	s_1	s_2	z	
3	3	2	1	0	0	18
2	2	3	0	1	0	16
-4	-6	-2	0	0	1	0

3.

x_1	x_2	s_1	s_2	s_3	z	
1	3	1	0	0	0	12
2	1	0	1	0	0	10
1	1	0	0	1	0	4
-2	-1	0	0	0	1	0

4.

x_1	x_2	x_3	s_1	s_2	s_3	z	
2	1	2	1	0	0	0	25
4	3	2	0	1	0	0	40
3	1	6	0	0	1	0	60
-4	-2	-3	0	0	0	1	0

5.

x_1	x_2	x_3	s_1	s_2	s_3	z	
2	2	8	1	0	0	0	40
4	-5	6	0	1	0	0	60
2	-2	6	0	0	1	0	24
-14	-10	-12	0	0	0	1	0

6.

x_1	x_2	x_3	s_1	s_2	z	
3	2	4	1	0	0	18
2	1	5	0	1	0	8
-1	-4	-2	0	0	1	0

Use the simplex method to solve each linear programming problem.

7. Maximize $z = 3x_1 + 5x_2$

subject to: $4x_1 + x_2 \le 25$

$2x_1 + 3x_2 \le 15$

with $x_1 \ge 0, \quad x_2 \ge 0.$

8. Maximize $z = 5x_1 + 2x_2$

subject to: $2x_1 + 4x_2 \le 15$

$3x_1 + x_2 \le 10$

with $x_1 \ge 0, \quad x_2 \ge 0.$

9. Maximize $z = 10x_1 + 12x_2$

subject to: $4x_1 + 2x_2 \le 20$

$5x_1 + x_2 \le 50$

$2x_1 + 2x_2 \le 24$

with $x_1 \ge 0, \quad x_2 \ge 0.$

10. Maximize $z = 1.5x_1 + 4.2x_2$

subject to: $2.8x_1 + 3.4x_2 \le 21$

$1.4x_1 + 2.2x_2 \le 11$

with $x_1 \ge 0, \quad x_2 \ge 0.$

11. Maximize $z = 8x_1 + 3x_2 + x_3$

subject to: $x_1 + 6x_2 + 8x_3 \le 118$

$x_1 + 5x_2 + 10x_3 \le 220$

with $x_1 \ge 0, \quad x_2 \ge 0, \quad x_3 \ge 0.$

12. Maximize $z = 8x_1 + 10x_2 + 7x_3$

subject to: $x_1 + 3x_2 + 2x_3 \le 10$

$x_1 + 5x_2 + x_3 \le 8$

with $x_1 \ge 0, \quad x_2 \ge 0, \quad x_3 \ge 0.$

13. Maximize $z = 10x_1 + 15x_2 + 10x_3 + 5x_4$

subject to: $x_1 + x_2 + x_3 + x_4 \le 300$

$x_1 + 2x_2 + 3x_3 + x_4 \le 360$

with $x_1 \ge 0, \quad x_2 \ge 0, \quad x_3 \ge 0, \quad x_4 \ge 0.$

14. Maximize $z = x_1 + x_2 + 4x_3 + 5x_4$

subject to: $x_1 + 2x_2 + 3x_3 + x_4 \le 115$

$2x_1 + x_2 + 8x_3 + 5x_4 \le 200$

$x_1 + x_3 \le 50$

with $x_1 \ge 0, \quad x_2 \ge 0, \quad x_3 \ge 0, \quad x_4 \ge 0.$

15. Maximize $\quad z = 4x_1 + 6x_2$

subject to: $\quad x_1 - 5x_2 \leq 25$

$4x_1 - 3x_2 \leq 12$

with $\quad x_1 \geq 0, \quad x_2 \geq 0.$

16. Maximize $\quad z = 2x_1 + 5x_2 + x_3$

subject to: $\quad x_1 - 5x_2 + 2x_3 \leq 30$

$4x_1 - 3x_2 + 6x_3 \leq 72$

with $\quad x_1 \geq 0, \quad x_2 \geq 0, \quad x_3 \geq 0.$

Use a graphing calculator, Excel, or other technology to solve the following linear programming problems.

17. Maximize $\quad z = 37x_1 + 34x_2 + 36x_3 + 30x_4 + 35x_5$

subject to: $\quad 16x_1 + 19x_2 + 23x_3 + 15x_4 + 21x_5 \leq 42{,}000$

$15x_1 + 10x_2 + 19x_3 + 23x_4 + 10x_5 \leq 25{,}000$

$9x_1 + 16x_2 + 14x_3 + 12x_4 + 11x_5 \leq 23{,}000$

$18x_1 + 20x_2 + 15x_3 + 17x_4 + 19x_5 \leq 36{,}000$

with $\quad x_1 \geq 0, \quad x_2 \geq 0, \quad x_3 \geq 0, \quad x_4 \geq 0, \quad x_5 \geq 0.$

18. Maximize $\quad z = 2.0x_1 + 1.7x_2 + 2.1x_3 + 2.4x_4 + 2.2x_5$

subject to: $\quad 12x_1 + 10x_2 + 11x_3 + 12x_4 + 13x_5 \leq 4250$

$8x_1 + 8x_2 + 7x_3 + 18x_4 + 5x_5 \leq 4130$

$9x_1 + 10x_2 + 12x_3 + 11x_4 + 8x_5 \leq 3500$

$5x_1 + 3x_2 + 4x_3 + 5x_4 + 4x_5 \leq 1600$

with $\quad x_1 \geq 0, \quad x_2 \geq 0, \quad x_3 \geq 0, \quad x_4 \geq 0, \quad x_5 \geq 0.$

19. The simplex algorithm still works if an indicator other than the most negative one is chosen. (Try it!) List the disadvantages that might occur if this is done.

20. What goes wrong if a quotient other than the smallest nonnegative quotient is chosen in the simplex algorithm? (Try it!)

21. Add lines corresponding to $z = 0$, $z = 1800$, and $z = 1825$ to Figure 6. Then explain how the increase in the objective function as we move from (0, 0) to (0, 60) to (25, 40) relates to the discussion of the Corner Point Theorem in Section 3.2.

22. Explain why the objective function can be made larger as long as there are negative numbers in the bottom row of the simplex tableau. (*Hint:* Consider an example with negative numbers in the last row, and rewrite the last row as an equation.)

APPLICATIONS

Set up and solve Exercises 23–29 by the simplex method.

Business and Economics

23. Charitable Contributions Carrie Green is working to raise money for the homeless by sending information letters and making follow-up calls to local labor organizations and church groups. She discovers that each church group requires 2 hours of letter writing and 1 hour of follow-up, while for each labor union she needs 2 hours of letter writing and 3 hours of follow-up. Carrie can raise $100 from each church group and $200 from each union local, and she has a maximum of 16 hours of letter-writing

time and a maximum of 12 hours of follow-up time available per month. Determine the most profitable mixture of groups she should contact and the most money she can raise in a month.

24. Profit The Muro Manufacturing Company makes two kinds of plasma screen television sets. It produces the Flexscan set that sells for $350 profit and the Panoramic I that sells for $500 profit. On the assembly line, the Flexscan requires 5 hours, and the Panoramic I takes 7 hours. The cabinet shop spends 1 hour on the cabinet for the Flexscan and 2 hours on the cabinet for the Panoramic I. Both sets require 4 hours for testing and packing. On a particular production run, the Muro Company has available 3600 work-hours on the assembly line, 900 work-hours in the cabinet shop, and 2600 work-hours in the testing and packing department. (See Exercise 10 in Section 3.3.)

a. How many sets of each type should it produce to make a maximum profit? What is the maximum profit?

b. Find the values of any nonzero slack variables and describe what they tell you about any unused time.

25. Poker The Texas Poker Company assembles three different poker sets. Each Royal Flush poker set contains 1000 poker chips, 4 decks of cards, 10 dice, and 2 dealer buttons. Each Deluxe Diamond poker set contains 600 poker chips, 2 decks of cards, 5 dice, and one dealer button. The Full House poker set contains 300 poker chips, 2 decks of cards, 5 dice, and one dealer button. The Texas Poker Company has 2,800,000 poker chips, 10,000 decks of cards, 25,000 dice, and 6000 dealer buttons in stock. They earn a profit of $38 for each Royal Flush poker set, $22 for each Deluxe Diamond poker set, and $12 for each Full House poker set.

a. How many of each type of poker set should they assemble to maximize profit? What is the maximum profit?

b. Find the values of any nonzero slack variables and describe what they tell you about any unused components.

26. Income A baker has 150 units of flour, 90 of sugar, and 150 of raisins. A loaf of raisin bread requires 1 unit of flour, 1 of sugar, and 2 of raisins, while a raisin cake needs 5, 2, and 1 units, respectively.

a. If raisin bread sells for $1.75 a loaf and raisin cake for $4.00 each, how many of each should be baked so that gross income is maximized?

b. What is the maximum gross income?

c. Does it require all of the available units of flour, sugar, and raisins to produce the number of loaves of raisin bread and raisin cakes that produce the maximum profit? If not, how much of each ingredient is left over? Compare any leftover to the value of the relevant slack variable.

27. APPLY IT Manufacturing Bicycles A manufacturer of bicycles builds racing, touring, and mountain models. The bicycles are made of both aluminum and steel. The company has available 91,800 units of steel and 42,000 units of aluminum. The racing, touring, and mountain models need 17, 27, and 34 units of steel, and 12, 21, and 15 units of aluminum, respectively. (See Exercise 28 in Section 4.1.)

a. How many of each type of bicycle should be made in order to maximize profit if the company makes $8 per racing bike, $12 per touring bike, and $22 per mountain bike?

b. What is the maximum possible profit?

c. Does it require all of the available units of steel and aluminum to build the bicycles that produce the maximum profit? If not, how much of each material is left over? Compare any leftover to the value of the relevant slack variable.

d. There are many unstated assumptions in the problem given above. Even if the mathematical solution is to make only one or two types of the bicycles, there may be demand for the type(s) not being made, which would create problems for the company. Discuss this and other difficulties that would arise in a real situation.

28. Production The Cut-Right Company sells sets of kitchen knives. The Basic Set consists of 2 utility knives and 1 chef's knife. The Regular Set consists of 2 utility knives, 1 chef's knife, and 1 slicer. The Deluxe Set consists of 3 utility knives, 1 chef's knife, and 1 slicer. Their profit is $30 on a Basic Set, $40 on a Regular Set, and $60 on a Deluxe Set. The factory has on hand 800 utility knives, 400 chef's knives, and 200 slicers. (See Exercise 30 in Section 4.1.)

a. Assuming that all sets will be sold, how many of each type should be made up in order to maximize profit? What is the maximum profit?

b. A consultant for the Cut-Right Company notes that more profit is made on a Regular Set of knives than on a Basic Set, yet the result from part a recommends making up 100 Basic Sets but no Regular Sets. She is puzzled how this can be the best solution. How would you respond?

29. Advertising The Fancy Fashions, an independent, local boutique, has $8000 available each month for advertising. Newspaper ads cost $400 each, and no more than 30 can run per month. Internet banner ads cost $20 each, and no more than 60 can run per month. TV ads cost $2000 each, with a maximum of 10 available each month. Approximately 4000 women will see each newspaper ad, 3000 will see each Internet banner, and 10,000 will see each TV ad. (See Exercise 31 in Section 4.1.)

a. How much of each type of advertising should be used if the store wants to maximize its ad exposure?

b. A marketing analyst is puzzled by the results of part a. More women see each TV ad than each newspaper ad or Internet banner, he reasons, so it makes no sense to use the newspaper ads and Internet banners and no TV ads. How would you respond?

⚏ 30. Profit A manufacturer makes two products, toy trucks and toy fire engines. Both are processed in four different departments, each of which has a limited capacity. The sheet metal department requires 2 hours for each truck and 3 hours for each fire engine and has a total of 24,000 hours available. The truck assembly department can handle at most 6600 trucks per week; and the fire engine assembly department assembles at most 5500 fire engines weekly. The painting department, which finishes both toys, has a maximum capacity of 10,000 per week.

a. If the profit is $8.50 for a toy truck and $12.10 for a toy fire engine, how many of each item should the company produce to maximize profit?

⚏ b. Keeping the profit for a toy truck at $8.50, use the graphical method to find the largest possible profit less than $12.10 for a toy fire engine that would result in an additional solution

besides the one found in part a. What is the additional solution? Verify this solution using the simplex method.

⚏ c. Keeping the profit for a toy truck at $8.50, use the graphical method to find the smallest possible profit greater than $12.10 for a toy fire engine that would result in an additional solution besides the one found in part a. What is the additional solution? Verify this solution using the simplex method.

Exercises 31 and 32 come from past CPA examinations. *Source: American Institute of Certified Public Accountants, Inc.* **Select the appropriate answer for each question.**

31. Profit The Ball Company manufactures three types of lamps, labeled A, B, and C. Each lamp is processed in two departments, I and II. Total available work-hours per day for departments I and II are 400 and 600, respectively. No additional labor is available. Time requirements and profit per unit for each lamp type are as follows:

	A	B	C
Work-hours in I	2	3	1
Work-hours in II	4	2	3
Profit per Unit	$5	$4	$3

The company has assigned you as the accounting member of its profit planning committee to determine the numbers of types of A, B, and C lamps that it should produce in order to maximize its total profit from the sale of lamps. The following questions relate to a linear programming model that your group has developed. (For each part, choose one of the four answers.)

a. The coefficients of the objective function would be

(**1**) 4, 2, 3.   (**2**) 2, 3, 1.

(**3**) 5, 4, 3.   (**4**) 400, 600.

b. The constraints in the model would be

(**1**) 2, 3, 1.   (**2**) 5, 4, 3.

(**3**) 4, 2, 3.   (**4**) 400, 600.

c. The constraint imposed by the available work-hours in department I could be expressed as

(**1**) $4X_1 + 2X_2 + 3X_3 \le 400$.

(**2**) $4X_1 + 2X_2 + 3X_3 \ge 400$.

(**3**) $2X_1 + 3X_2 + 1X_3 \le 400$.

(**4**) $2X_1 + 3X_2 + 1X_3 \ge 400$.

32. Profit The Golden Hawk Manufacturing Company wants to maximize the profits on products A, B, and C. The contribution margin for each product follows:

Product	Contribution Margin
A	$2
B	$5
C	$4

The production requirements and departmental capacities, by departments, are as follows:

Department	Production Requirements by Product (hours)			Departmental Capacity (total hours)
	A	*B*	*C*	
Assembling	2	3	2	30,000
Painting	1	2	2	38,000
Finishing	2	3	1	28,000

a. What is the profit-maximization formula for the Golden Hawk Company? (Choose one of the following.)

   (1) $2A + $5B + $4C = X$ (where X = profit)

   (2) $5A + 8B + 5C \leq 96,000$

   (3) $2A + $5B + $4C \leq X$

   (4) $2A + $5B + $4C = 96,000$

b. What is the constraint for the painting department of the Golden Hawk Company? (Choose one of the following.)

   (1) $1A + 2B + 2C \geq 38,000$

   (2) $2A + $5B + $4C \geq 38,000$

   (3) $1A + 2B + 2C \leq 38,000$

   (4) $2A + 3B + 2C \leq 30,000$

33. Sensitivity Analysis Using a computer spreadsheet, perform a sensitivity analysis for the objective function in Exercise 23. What are the highest and lowest possible values for the amount raised from each church group that would yield the same solution as the original problem? Answer the same question for the amount raised from each union local.

34. Sensitivity Analysis Using a computer spreadsheet, perform a sensitivity analysis for the objective function in Exercise 24. What are the highest and lowest possible values for profit on a Flexscan set that would yield the same solution as the original problem? Answer the same question for a Panoramic I set.

Set up and solve Exercises 35–40 by the simplex method.

Life Sciences

35. Calorie Expenditure Rachel Reeve, a fitness trainer, has an exercise regimen that includes running, biking, and walking. She has no more than 15 hours per week to devote to exercise, including at most 3 hours running. She wants to walk at least twice as many hours as she bikes. According to a website, a 130-pound person like Rachel will burn on average 531 calories per hour running, 472 calories per hour biking, and 354 calories per hour walking. How many hours per week should Rachel spend on each exercise to maximize the number of calories she burns? What is the maximum number of calories she will burn? (*Hint:* Write the constraint involving walking and biking in the form ≤ 0.) *Source: NutriStrategy.com.*

36. Calorie Expenditure Joe Vetere's exercise regimen includes light calisthenics, swimming, and playing the drums. He has at most 10 hours per week to devote to these activities. He wants the total time he does calisthenics and plays the drums to be at

least twice as long as he swims. His neighbors, however, will tolerate no more than 4 hours per week on the drums. According to a website, a 190-pound person like Joe will burn an average of 388 calories per hour doing calisthenics, 518 calories per hour swimming, and 345 calories per hour playing the drums. *Source: NutriStrategy.com.*

a. How many hours per week should Joe spend on each exercise to maximize the number of calories he burns? What is the maximum number of calories he will burn?

b. What conclusions can you draw about Joe's selection of activities?

37. Resource Management The average weights of the three species stocked in the lake referred to in Section 2.2, Exercise 59, are 1.62, 2.14, and 3.01 kg for species A, B, and C, respectively.

a. If the largest amounts of food that can be supplied each day are as given in Exercise 59, how should the lake be stocked to maximize the weight of the fish supported by the lake?

b. Does it require all of the available food to produce the maximum weight of fish? If not, how much of each type of food is left over?

c. Find a value for each of the average weights of the three species that would result in none of species B or C being stocked to maximize the weight of the fish supported by the lake, given the constraints in part a.

d. Find a value for each of the average weights of the three species that would result in none of species A or B being stocked to maximize the weight of the fish supported by the lake, given the constraints in part a.

38. Blending Nutrients A biologist has 500 kg of nutrient A, 600 kg of nutrient B, and 300 kg of nutrient C. These nutrients will be used to make four types of food, whose contents (in percent of nutrient per kilogram of food) and whose "growth values" are as shown in the table.

	P	**Q**	**R**	**S**
A	0	0	37.5	62.5
B	0	75	50	37.5
C	100	25	12.5	0
Growth Value	90	70	60	50

a. How many kilograms of each food should be produced in order to maximize total growth value?

b. Find the maximum growth value.

c. Does it require all of the available nutrients to produce the four types of food that maximizes the total growth value? If not, how much of each nutrient is left over?

Social Sciences

39. Politics A political party is planning a half-hour television show. The show will have at least 3 minutes of direct requests for money from viewers. Three of the party's politicians will be on the show—a senator, a congresswoman, and a governor. The senator, a party "elder statesman," demands that he be on screen at least twice as long as the governor. The total time taken by the senator and the governor must be at least twice the time taken by the congresswoman. Based on a pre-show survey, it is believed that 35, 40, and 45 (in thousands) viewers will watch the program for each minute the senator, congresswoman, and governor, respectively, are on the air. Find the time that should be allotted to each politician in order to get the maximum number of viewers. Find the maximum number of viewers.

40. Fund Raising The political party in Exercise 39 is planning its fund-raising activities for a coming election. It plans to raise money through large fund-raising parties, letters requesting funds, and dinner parties where people can meet the candidate personally. Each large fund-raising party costs $3000, each mailing costs $1000, and each dinner party costs $12,000. The party can spend up to $102,000 for these activities. From experience, the planners know that each large party will raise $200,000, each letter campaign will raise $100,000, and each dinner party will raise $600,000. They are able to carry out a total of 25 of these activities.

a. How many of each should the party plan in order to raise the maximum amount of money? What is the maximum amount?

b. Dinner parties are more expensive than letter campaigns, yet the optimum solution found in part a includes dinner parties but no letter campaigns. Explain how this is possible.

YOUR TURN ANSWERS

1. $x_1 = 0$, $x_2 = 12$, for a maximum value of $z = 480$.
2. $x_1 = 0$, $x_2 = 50$, $s_1 = 200$, $s_2 = 0$, $s_3 = 150$, $z = 1250$

4.3 Minimization Problems; Duality

APPLY IT How many units of different types of feed should a dog breeder purchase to meet the nutrient requirements of her golden retrievers at a minimum cost? *Using the method of duals, we will answer this question in Example 5.*

Minimization Problems
The definition of a problem in standard maximum form was given earlier in this chapter. Now we can define a linear programming problem in *standard minimum form*, as follows.

Standard Minimum Form

A linear programming problem is in **standard minimum form** if the following conditions are satisfied.

1. The objective function is to be minimized.
2. All variables are nonnegative.
3. All remaining constraints are stated in the form

$$a_1y_1 + a_2y_2 + \cdots + a_ny_n \geq b, \quad \text{with } b \geq 0.$$

NOTE
In this section, we require that all coefficients in the objective function be positive, so $c_1 \geq 0$, $c_2 \geq 0, \ldots, c_n \geq 0$.

The difference between maximization and minimization problems is in conditions 1 and 3: In problems stated in standard minimum form, the objective function is to be *minimized*, rather than maximized, and all constraints must have $\geq$ instead of $\leq$.

We use y_1, y_2, etc., for the variables and w for the objective function as a reminder that these are minimizing problems. Thus, $w = c_1y_1 + c_2y_2 + \cdots + c_ny_n$.

Duality
An interesting connection exists between standard maximization and standard minimization problems: any solution of a standard maximization problem produces the solution of an associated standard minimization problem, and vice versa. Each of these associated problems is called the **dual** of the other. One advantage of duals is that standard minimization problems can be solved by the simplex method discussed in the first two sections of this chapter. Let us explain the idea of a dual with an example.

EXAMPLE 1 Duality

Minimize $w = 8y_1 + 16y_2$

subject to: $y_1 + 5y_2 \geq 9$

$2y_1 + 2y_2 \geq 10$

with $y_1 \geq 0, \quad y_2 \geq 0.$

SOLUTION   Without considering slack variables just yet, write the augmented matrix of the system of inequalities and include the coefficients of the objective function (not their negatives) as the last row in the matrix.

$$
\begin{array}{c}
\hspace{5.5cm}\text{Constants} \\
\text{Objective function} \longrightarrow
\begin{bmatrix}
1 & 5 & 9 \\
2 & 2 & 10 \\
8 & 16 & 0
\end{bmatrix}
\end{array}
$$

Now look at the following matrix, which we obtain from the one above by interchanging rows and columns.

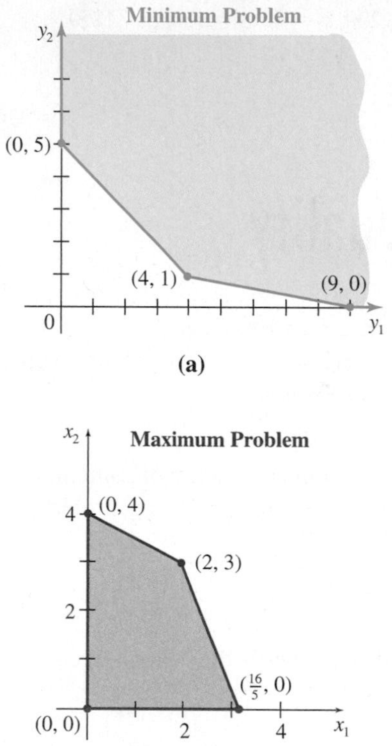

$$
\begin{array}{c}
\hspace{5.5cm}\text{Constants} \\
\text{Objective function} \longrightarrow
\begin{bmatrix}
1 & 2 & 8 \\
5 & 2 & 16 \\
9 & 10 & 0
\end{bmatrix}
\end{array}
$$

The *rows* of the first matrix (for the minimization problem) are the *columns* of the second matrix.

The entries in this second matrix could be used to write the following maximization problem in standard form (again ignoring the fact that the numbers in the last row are not negative):

Maximize $z = 9x_1 + 10x_2$

subject to: $x_1 + 2x_2 \leq 8$

$5x_1 + 2x_2 \leq 16$

with all variables nonnegative.

Figure 9(a) shows the region of feasible solutions for the minimization problem just given, while Figure 9(b) shows the region of feasible solutions for the maximization problem produced by exchanging rows and columns. The solutions of the two problems are given below.

Minimum Problem

(a)

Maximum Problem

(b)

FIGURE 9

Minimum Problem	
Corner Point	$w = 8y_1 + 16y_2$
$(0, 5)$	80
$(4, 1)$	48 Minimum
$(9, 0)$	72

The minimum is 48 when $y_1 = 4$ and $y_2 = 1$.

Maximum Problem	
Corner Point	$z = 9x_1 + 10x_2$
$(0, 0)$	0
$(0, 4)$	40
$(2, 3)$	48 Maximum
$(16/5, 0)$	28.8

The maximum is 48 when $x_1 = 2$ and $x_2 = 3$.

The two feasible regions in Figure 9 are different and the corner points are different, but the values of the objective functions are equal—both are 48. An even closer connection between the two problems is shown by using the simplex method to solve this maximization problem.

Maximization Problem

$$
\begin{array}{ccccc}
x_1 & x_2 & s_1 & s_2 & z \\
\end{array}
$$

$$
\left[\begin{array}{ccccc|c}
1 & \mathbf{2} & 1 & 0 & 0 & 8 \\
5 & 2 & 0 & 1 & 0 & 16 \\
\hline
-9 & -10 & 0 & 0 & 1 & 0
\end{array}\right]
$$

$$
\begin{array}{ccccc}
x_1 & x_2 & s_1 & s_2 & z \\
\end{array}
$$

$-R_1 + R_2 \rightarrow R_2$

$5R_1 + R_3 \rightarrow R_3$

$$
\left[\begin{array}{ccccc|c}
1 & 2 & 1 & 0 & 0 & 8 \\
\mathbf{4} & 0 & -1 & 1 & 0 & 8 \\
-4 & 0 & 5 & 0 & 1 & 40
\end{array}\right]
$$

$$
\begin{array}{ccccc}
x_1 & x_2 & s_1 & s_2 & z \\
\end{array}
$$

$-R_2 + 4R_1 \rightarrow R_1$

$R_2 + R_3 \rightarrow R_3$

$$
\left[\begin{array}{ccccc|c}
0 & 8 & 5 & -1 & 0 & 24 \\
4 & 0 & -1 & 1 & 0 & 8 \\
0 & 0 & 4 & 1 & 1 & 48
\end{array}\right]
$$

$$
\begin{array}{ccccc}
x_1 & x_2 & s_1 & s_2 & z \\
\end{array}
$$

$R_1/8 \rightarrow R_1$

$R_2/4 \rightarrow R_2$

$$
\left[\begin{array}{ccccc|c}
0 & 1 & \frac{5}{8} & -\frac{1}{8} & 0 & 3 \\
1 & 0 & -\frac{1}{4} & \frac{1}{4} & 0 & 2 \\
0 & 0 & \mathbf{4} & \mathbf{1} & 1 & 48
\end{array}\right]
$$

The maximum is 48 when
$x_1 = 2$ and $x_2 = 3$.

Notice that the solution to the *minimization problem* is found in the bottom row and slack variable columns of the final simplex tableau for the maximization problem. This result suggests that standard minimization problems can be solved by forming the dual standard maximization problem, solving it by the simplex method, and then reading the solution for the minimization problem from the bottom row of the final simplex tableau. ▬

Before using this method to actually solve a minimization problem, let us find the duals of some typical linear programming problems. The process of exchanging the rows and columns of a matrix, which is used to find the dual, is called *transposing* the matrix, and each of the two matrices is the **transpose** of the other. The transpose of an $m \times n$ matrix A, written A^T, is an $n \times m$ matrix.

EXAMPLE 2 **Transposes**

Find the transpose of each matrix.

(a) $A = \begin{bmatrix} 2 & -1 & 5 \\ 6 & 8 & 0 \\ -3 & 7 & -1 \end{bmatrix}$

SOLUTION Both matrix A and its transpose are 3×3 matrices. Write the rows of matrix A as the columns of the transpose.

$$
A^T = \begin{bmatrix} 2 & 6 & -3 \\ -1 & 8 & 7 \\ 5 & 0 & -1 \end{bmatrix}
$$

(b) $B = \begin{bmatrix} 1 & 2 & 4 & 0 \\ 2 & 1 & 7 & 6 \end{bmatrix}$

SOLUTION The matrix B is 2×4, so B^T is the 4×2 matrix

$$
B^T = \begin{bmatrix} 1 & 2 \\ 2 & 1 \\ 4 & 7 \\ 0 & 6 \end{bmatrix}.
$$

EXAMPLE 3 Duals

Write the dual of each standard linear programming problem.

(a) Maximize $\quad z = 2x_1 + 5x_2$

subject to: $\quad x_1 + x_2 \le 10$

$\qquad\qquad 2x_1 + x_2 \le 8$

with $\qquad\quad x_1 \ge 0, \quad x_2 \ge 0.$

SOLUTION Begin by writing the augmented matrix for the given problem.

$$\begin{bmatrix} 1 & 1 & | & 10 \\ 2 & 1 & | & 8 \\ 2 & 5 & | & 0 \end{bmatrix}$$

Form the transpose of the matrix as follows:

$$\begin{bmatrix} 1 & 2 & | & 2 \\ 1 & 1 & | & 5 \\ 10 & 8 & | & 0 \end{bmatrix}.$$

The dual problem is stated from this second matrix as follows (using y instead of x):

Minimize $\quad w = 10y_1 + 8y_2$

subject to: $\quad y_1 + 2y_2 \ge 2$

$\qquad\qquad y_1 + \; y_2 \ge 5$

with $\qquad\quad y_1 \ge 0, \quad y_2 \ge 0.$

(b) Minimize $\quad w = 7y_1 + 5y_2 + 8y_3$

subject to: $\quad 3y_1 + 2y_2 + y_3 \ge 10$

$\qquad\qquad 4y_1 + 5y_2 \qquad \ge 25$

with $\qquad\quad y_1 \ge 0, \quad y_2 \ge 0, \quad y_3 \ge 0.$

SOLUTION The dual problem is stated as follows.

Maximize $\quad z = 10x_1 + 25x_2$

subject to: $\quad 3x_1 + 4x_2 \le 7$

$\qquad\qquad 2x_1 + 5x_2 \le 5$

$\qquad\qquad x_1 \qquad\quad \le 8$

with $\qquad\quad x_1 \ge 0, \quad x_2 \ge 0.$ **TRY YOUR TURN 1**

YOUR TURN 1 Write the dual of the following linear programming problem.

Minimize $\quad w = 25y_1 + 12y_2 + 27y_3$

subject to: $\quad 3y_1 + 3y_2 + 4y_3 \ge 24$

$\qquad\qquad 5y_1 + \; y_2 + 3y_3 \ge 27$

with $\qquad\quad y_1 \ge 0, y_2 \ge 0, y_3 \ge 0.$

NOTE You might find it easier to set up the dual if you put the objective function *after* the constraints in the original problem, and then line up the variables, as in the following rewriting of Example 3 (b):

$$3y_1 + 2y_2 + y_3 \ge 10$$

$$4y_1 + 5y_2 \qquad \ge 25$$

Minimize $\quad 7y_1 + 5y_2 + 8y_3.$

Notice that you can read down the first column to get the coefficients of the first constraint in the dual $(3x_1 + 4x_2 \le 7)$. Reading down the second and third columns gives the coefficients of the next two constraints, and the last column gives the coefficients of the objective function.

In Example 3, all the constraints of the given standard maximization problems were $\le$ inequalities, while all those in the dual minimization problems were $\ge$ inequalities. This is generally the case; inequalities are reversed when the dual problem is stated.

NOTE
To solve a minimization problem with duals, all of the coefficients in the objective function must be positive. (To investigate what would happen without this requirement, see Exercise 18). For a method that does not have this restriction, see the next section.

The following table shows the close connection between a problem and its dual.

Duality	
Given Problem	**Dual Problem**
m variables	n variables
n constraints	m constraints
Coefficients from objective function	Constraint constants
Constraint constants	Coefficients from objective function

The next theorem, whose proof requires advanced methods, guarantees that a standard minimization problem can be solved by forming a dual standard maximization problem.

Theorem of Duality

The objective function w of a minimization linear programming problem takes on a minimum value if and only if the objective function z of the corresponding dual maximization problem takes on a maximum value. The maximum value of z equals the minimum value of w.

This method is illustrated in the following example.

EXAMPLE 4 **Duality**

Minimize $w = 3y_1 + 2y_2$

subject to: $y_1 + 3y_2 \geq 6$

$2y_1 + y_2 \geq 3$

with $y_1 \geq 0, \quad y_2 \geq 0.$

SOLUTION Use the given information to write the matrix.

$$\begin{bmatrix} 1 & 3 & | & 6 \\ 2 & 1 & | & 3 \\ \hline 3 & 2 & | & 0 \end{bmatrix}$$

Transpose to get the following matrix for the dual problem.

$$\begin{bmatrix} 1 & 2 & | & 3 \\ 3 & 1 & | & 2 \\ \hline 6 & 3 & | & 0 \end{bmatrix}$$

Write the dual problem from this matrix, as follows:

Maximize $z = 6x_1 + 3x_2$

subject to: $x_1 + 2x_2 \leq 3$

$3x_1 + x_2 \leq 2$

with $x_1 \geq 0, \quad x_2 \geq 0.$

Solve this standard maximization problem using the simplex method. Start by introducing slack variables to give the system

$$x_1 + 2x_2 + s_1 \qquad = 3$$
$$3x_1 + x_2 \qquad + s_2 \qquad = 2$$
$$-6x_1 - 3x_2 \qquad \qquad + z = 0$$

with $x_1 \geq 0, \quad x_2 \geq 0, \quad s_1 \geq 0, \quad s_2 \geq 0.$

The first tableau for this system is given below, with the pivot as indicated.

$$
\begin{array}{c}
\text{Quotients} \\
3/1 = 3 \\
2/3
\end{array}
\begin{array}{cccccc}
x_1 & x_2 & s_1 & s_2 & z & \\
\left[\begin{array}{ccccc|c}
1 & 2 & 1 & 0 & 0 & 3 \\
\mathbf{3} & 1 & 0 & 1 & 0 & 2 \\
\hline
-6 & -3 & 0 & 0 & 1 & 0
\end{array}\right]
\end{array}
$$

The simplex method gives the following as the final tableau.

$$
\begin{array}{ccccc}
x_1 & x_2 & s_1 & s_2 & z \\
\left[\begin{array}{ccccc|c}
0 & 1 & \frac{3}{5} & -\frac{1}{5} & 0 & \frac{7}{5} \\
1 & 0 & -\frac{1}{5} & \frac{2}{5} & 0 & \frac{1}{5} \\
\hline
0 & 0 & \frac{3}{5} & \frac{9}{5} & 1 & \frac{27}{5}
\end{array}\right]
\end{array}
$$

YOUR TURN 2

Minimize $w = 15y_1 + 12y_2$

subject to: $3y_1 + 5y_2 \geq 20$

$\quad\quad\quad\quad 3y_1 + \ y_2 \geq 18$

with $\quad\quad y_1 \geq 0, y_2 \geq 0.$

Since a 1 has been created in the z column, the last row of this final tableau gives the solution to the minimization problem. The minimum value of $w = 3y_1 + 2y_2$, subject to the given constraints, is $27/5$ and occurs when $y_1 = 3/5$ and $y_2 = 9/5$. The minimum value of w, $27/5$, is the same as the maximum value of z. **TRY YOUR TURN 2**

Let us summarize the steps in solving a standard minimization linear programming problem by the method of duals.

Solving a Standard Minimum Problem with Duals

1. Find the dual standard maximization problem.
2. Solve the maximization problem using the simplex method.
3. The minimum value of the objective function w is the maximum value of the objective function z.
4. The optimum solution to the minimization problem is given by the entries in the bottom row of the columns corresponding to the slack variables, so long as the entry in the z column is equal to 1.

> **CAUTION** (1) If the final entry in the z column is a value other than 1, divide the bottom row through by that value so that it will become 1. Only then can the solution of the minimization problem be found in the bottom row of the columns corresponding to the slack variables.
>
> (2) Do not simplify an inequality in the dual by dividing out a common factor. For example, if an inequality in the dual is $3x_1 + 3x_2 \leq 6$, do not simplify to $x_1 + x_2 \leq 2$ by dividing out the 3. Doing so will give an incorrect solution to the original problem.

Further Uses of the Dual
The dual is useful not only in solving minimization problems but also in seeing how small changes in one variable will affect the value of the objective function.

EXAMPLE 5   Nutrition

Suppose a dog breeder needs at least 6 units per day of nutrient A and at least 3 units of nutrient B for her golden retrievers, and that the breeder can choose between two different feeds, feed 1 and feed 2. Find the minimum cost for the breeder if each bag of feed 1 costs $3 and provides 1 unit of nutrient A and 2 units of B, while each bag of feed 2 costs $2 and provides 3 units of nutrient A and 1 of B.

APPLY IT

SOLUTION If y_1 represents the number of bags of feed 1 and y_2 represents the number of bags of feed 2, the given information leads to the following problem.

Minimize $\quad w = 3y_1 + 2y_2$

subject to: $\quad\quad y_1 + 3y_2 \geq 6 \quad$ Nutrient A

$\quad\quad\quad\quad\quad 2y_1 + \; y_2 \geq 3 \quad$ Nutrient B

with $\quad\quad\quad\quad y_1 \geq 0, \quad y_2 \geq 0.$

This standard minimization linear programming problem is the one solved in Example 4 of this section. In that example, the dual was formed and the following tableau was found.

$$
\begin{array}{ccccc}
x_1 & x_2 & s_1 & s_2 & z \\
\left[\begin{array}{ccccc|c}
0 & 1 & \frac{3}{5} & -\frac{1}{5} & 0 & \frac{7}{5} \\
1 & 0 & -\frac{1}{5} & \frac{2}{5} & 0 & \frac{1}{5} \\
0 & 0 & \frac{3}{5} & \frac{9}{5} & 1 & \frac{27}{5}
\end{array}\right]
\end{array}
$$

This final tableau shows that the breeder will obtain minimum feed costs by using $3/5$ bag of feed 1 and $9/5$ bags of feed 2 per day, for a daily cost of $27/5 = \$5.40$.

Notice that the solution to the dual (maximization) problem is

$$x_1 = \frac{1}{5} = 0.20 \quad\quad \text{and} \quad\quad x_2 = \frac{7}{5} = 1.40.$$

These represent the **shadow costs** of nutrients A and B. The nutrients don't have actual costs because you can't buy a unit of nutrient A or B; all you can buy is a bag of Feed 1 or Feed 2, each of which contains both nutrients A and B. Shadow costs, however, are a convenient way of allocating costs. Suppose that a unit of nutrient A is said to cost $\$0.20$, its shadow cost, and a unit of nutrient B is said to cost $\$1.40$. Then the minimum daily cost, which we previously found to be $\$5.40$ (providing 6 units of A and 3 units of B), can be found by the following procedure.

$$
\begin{array}{r}
(\$0.20 \text{ per unit of A}) \times (6 \text{ units of A}) = \$1.20 \\
+ \; (\$1.40 \text{ per unit of B}) \times (3 \text{ units of B}) = \$4.20 \\
\hline
\text{Minimum daily cost} = \$5.40
\end{array}
$$

Furthermore, the shadow costs allow the breeder to calculate feed costs for small changes in nutrient requirements. For example, an increase of one unit in the requirement for each nutrient would produce a total cost of $\$7.00$:

$\$5.40$	6 units of A, 3 of B
0.20	1 extra unit of A
$+ \; 1.40$	1 extra unit of B
$\$7.00$	Total cost per day

Shadow costs only give the exact answer for a limited range. Unfortunately, finding that range is somewhat complicated. In the dog feed example, we can add up to 3 units or delete up to 4 units of A, and shadow costs will give the exact answer. If, however, we add 4 units of A, shadow costs give an answer of $\$6.20$, while the true cost is $\$6.67$.

NOTE
Shadow costs become shadow profits in maximization problems. For example, see Exercises 21 and 22.

CAUTION If you wish to use shadow costs, do not simplify an inequality in the original problem by dividing out a common factor. For example, if an inequality in the original problem is $3y_1 + 3y_2 \geq 6$, do not simplify to $y_1 + y_2 \geq 2$ by dividing out the 3. Doing so will give incorrect shadow costs.

TECHNOLOGY NOTE

The Solver in Microsoft Excel provides the values of the dual variables. See the *Graphing Calculator and Excel Spreadsheet Manual* available with this book for more details.

4.3 EXERCISES

Find the transpose of each matrix.

1. $\begin{bmatrix} 1 & 2 & 3 \\ 3 & 2 & 1 \\ 1 & 10 & 0 \end{bmatrix}$

2. $\begin{bmatrix} 3 & 4 & -2 & 0 & 1 \\ 2 & 0 & 11 & 5 & 7 \end{bmatrix}$

3. $\begin{bmatrix} 4 & 5 & -3 & 15 \\ 7 & 14 & 20 & -8 \\ 5 & 0 & -2 & 23 \end{bmatrix}$

4. $\begin{bmatrix} 1 & 11 & 15 \\ 0 & 10 & -6 \\ 4 & 12 & -2 \\ 1 & -1 & 13 \\ 2 & 25 & -1 \end{bmatrix}$

State the dual problem for each linear programming problem.

5. Maximize $z = 4x_1 + 3x_2 + 2x_3$

subject to: $\quad x_1 + x_2 + x_3 \le 5$

$\qquad\qquad x_1 + x_2 \qquad \le 4$

$\qquad\qquad 2x_1 + x_2 + 3x_3 \le 15$

with $\qquad x_1 \ge 0, \quad x_2 \ge 0, \quad x_3 \ge 0.$

6. Maximize $z = 2x_1 + 7x_2 + 4x_3$

subject to: $\quad 4x_1 + 2x_2 + x_3 \le 26$

$\qquad\qquad x_1 + 7x_2 + 8x_3 \le 33$

with $\qquad x_1 \ge 0, \quad x_2 \ge 0, \quad x_3 \ge 0.$

7. Minimize $\quad w = 3y_1 + 6y_2 + 4y_3 + y_4$

subject to: $\quad y_1 + y_2 + y_3 + y_4 \ge 150$

$\qquad\qquad 2y_1 + 2y_2 + 3y_3 + 4y_4 \ge 275$

with $\qquad y_1 \ge 0, \quad y_2 \ge 0, \quad y_3 \ge 0, \quad y_4 \ge 0.$

8. Minimize $\quad w = y_1 + y_2 + 4y_3$

subject to: $\quad y_1 + 2y_2 + 3y_3 \ge 115$

$\qquad\qquad 2y_1 + y_2 + 8y_3 \ge 200$

$\qquad\qquad y_1 \qquad + y_3 \ge 50$

with $\qquad y_1 \ge 0, \quad y_2 \ge 0, \quad y_3 \ge 0.$

Use the simplex method to solve.

9. Find $y_1 \ge 0$ and $y_2 \ge 0$ such that

$$2y_1 + 3y_2 \ge 6$$

$$2y_1 + y_2 \ge 7$$

and $w = 5y_1 + 2y_2$ is minimized.

10. Find $y_1 \ge 0$ and $y_2 \ge 0$ such that

$$2y_1 + 3y_2 \ge 15$$

$$5y_1 + 6y_2 \ge 35$$

and $w = 2y_1 + 3y_2$ is minimized.

11. Find $y_1 \ge 0$ and $y_2 \ge 0$ such that

$$10y_1 + 5y_2 \ge 100$$

$$20y_1 + 10y_2 \ge 150$$

and $w = 4y_1 + 5y_2$ is minimized.

12. Minimize $\quad w = 29y_1 + 10y_2$

subject to: $\quad 3y_1 + 2y_2 \ge 2$

$\qquad\qquad 5y_1 + y_2 \ge 3$

with $\qquad y_1 \ge 0, \quad y_2 \ge 0.$

13. Minimize $\quad w = 6y_1 + 10y_2$

subject to: $\quad 3y_1 + 5y_2 \ge 15$

$\qquad\qquad 4y_1 + 7y_2 \ge 20$

with $\qquad y_1 \ge 0, \quad y_2 \ge 0.$

14. Minimize $\quad w = 3y_1 + 2y_2$

subject to: $\quad y_1 + 2y_2 \ge 10$

$\qquad\qquad y_1 + y_2 \ge 8$

$\qquad\qquad 2y_1 + y_2 \ge 12$

with $\qquad y_1 \ge 0, \quad y_2 \ge 0.$

15. Minimize $\quad w = 2y_1 + y_2 + 3y_3$

subject to: $\quad y_1 + y_2 + y_3 \ge 100$

$\qquad\qquad 2y_1 + y_2 \qquad \ge 50$

with $\qquad y_1 \ge 0, \quad y_2 \ge 0, \quad y_3 \ge 0.$

16. Minimize $\quad w = 4y_1 + 7y_2 + 9y_3$

subject to: $\quad 2y_1 + 3y_2 + 4y_3 \ge 45$

$\qquad\qquad y_1 + 5y_2 + 2y_3 \ge 40$

with $\qquad y_1 \ge 0, \quad y_2 \ge 0, \quad y_3 \ge 0.$

17. You are given the following linear programming problem (P):

Minimize $\quad z = x_1 + 2x_2$

subject to: $\quad -2x_1 + x_2 \ge 1$

$\qquad\qquad x_1 - 2x_2 \ge 1$

$\qquad\qquad x_1 \ge 0, \quad x_2 \ge 0.$

The dual of (P) is (D). Which of the statements below is true? *Source: Society of Actuaries.*

a. (P) has no feasible solution and the objective function of (D) is unbounded.

b. (D) has no feasible solution and the objective function of (P) is unbounded.

c. The objective functions of both (P) and (D) are unbounded.

d. Both (P) and (D) have optimal solutions.

e. Neither (P) nor (D) has feasible solutions.

18. Suppose the coefficient of 3 in the objective function of Example 4 is changed to -3 and all the constraints are kept the same. Explain why the objective function now has no minimum.

APPLICATIONS

Business and Economics

19. Production Costs   A brewery produces regular beer and a lower-carbohydrate "light" beer. Steady customers of the brewery buy 10 units of regular beer and 15 units of light beer monthly. While setting up the brewery to produce the beers, the management decides to produce extra beer, beyond that needed to satisfy customers. The cost per unit of regular beer is $32,000 and the cost per unit of light beer is $50,000. Every unit of regular beer brings in $120,000 in revenue, while every unit of light beer brings in $300,000 in revenue. The brewery wants at least $9,000,000 in revenue. At least 20 additional units of beer can be sold.

  a. How much of each type of beer should be made so as to minimize total production costs?

  b. Suppose the minimum revenue is increased to $9,500,000. Use shadow costs to calculate the total production costs.

20. Supply Costs   The chemistry department at a local college decides to stock at least 900 small test tubes and 600 large test tubes. It wants to buy at least 2700 test tubes to take advantage of a special price. Since the small test tubes are broken twice as often as the large, the department will order at least twice as many small tubes as large.

  a. If the small test tubes cost 18 cents each and the large ones, made of a cheaper glass, cost 15 cents each, how many of each size should be ordered to minimize cost?

  b. Suppose the minimum number of test tubes is increased to 3000. Use shadow costs to calculate the total cost in this case.

In most examples of this section, the original problem is a minimization problem and the dual is a maximization problem whose solution gives shadow costs. The reverse is true in Exercises 21 and 22. The dual here is a minimization problem whose solution can be interpreted as shadow profits.

21. Agriculture   Refer to the original information in Example 2, Section 4.1.

  a. Give the dual problem.

  b. Use the shadow profits to estimate the farmer's profit if land is cut to 90 acres but capital increases to $21,000.

  c. Suppose the farmer has 110 acres but only $19,000. Find the optimum profit and the planting strategy that will produce this profit.

22. Toy Manufacturing   A small toy manufacturing firm has 200 squares of felt, 600 oz of stuffing, and 90 ft of trim available to make two types of toys, a small bear and a monkey. The bear requires 1 square of felt and 4 oz of stuffing. The monkey requires 2 squares of felt, 3 oz of stuffing, and 1 ft of trim. The firm makes $1 profit on each bear and $1.50 profit on each monkey.

  a. Set up the linear programming problem to maximize profit.

  b. Solve the linear programming problem in part a.

  c. What is the corresponding dual problem?

  d. What is the optimal solution to the dual problem?

  e. Use the shadow profits to calculate the profit the firm will make if its supply of felt increases to 210 squares.

  f. How much profit will the firm make if its supply of stuffing is cut to 590 oz and its supply of trim is cut to 80 ft?

  g. Explain why it makes sense that the shadow profit for trim is 0.

23. Interview Time   Joan McKee has a part-time job conducting public opinion interviews. She has found that a political interview takes 45 min and a market interview takes 55 min. She needs to minimize the time she spends doing interviews to allow more time for her full-time job. Unfortunately, to keep her part-time job, she must complete at least 8 interviews each week. Also, she must earn at least $60 per week at this job; she earns $8 for each political interview and $10 for each market interview. Finally, to stay in good standing with her supervisor, she must earn at least 40 bonus points per week; she receives 6 bonus points for each political interview and 5 points for each market interview. How many of each interview should she do each week to minimize the time spent?

24. Animal Food   An animal food must provide at least 54 units of vitamins and 60 calories per serving. One gram of soybean meal provides 2.5 units of vitamins and 5 calories. One gram of meat byproducts provides 4.5 units of vitamins and 3 calories. One gram of grain provides 5 units of vitamins and 10 calories. A gram of soybean meal costs 8¢, a gram of meat byproducts 9¢, and a gram of grain 10¢.

  a. What mixture of these three ingredients will provide the required vitamins and calories at minimum cost?

  b. What is the minimum cost?

  c. There is more than one optimal basic solution to this problem. The answer found in part a depends on whether the tie in the minimum ratio rule was broken by pivoting on the second row or third row of the dual. Find the other solution.

25. Feed Costs   Refer to Example 5 in this section on minimizing the daily cost of feeds.

  a. Find a combination of feeds that will cost $7 and give 7 units of A and 4 units of B.

  b. Use the dual variables to predict the daily cost of feed if the requirements change to 5 units of A and 4 units of B. Find a combination of feeds to meet these requirements at the predicted price.

26. Pottery   Karla Harby makes three items in her pottery shop: large bowls, small bowls, and pots for plants. A large bowl requires 3 lb of clay and 6 fl oz of glaze. A small bowl requires 2 lb of clay, and 6 fl oz of glaze. A pot requires 4 lb of clay and 2 fl oz of glaze. She must use up 72 lb of old clay and 108 fl oz of old glaze; she can order more if necessary. If Karla can make a large bowl in 5 hours, a small bowl in 6 hours, and a pot in 4 hours, how many of each should she make to minimize her time? What is the minimum time?

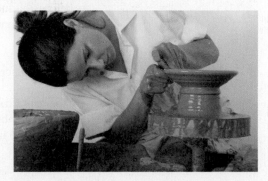

Life Sciences

27. Calorie Expenditure Francesca wants to start exercising to burn at least 1500 extra calories per week, but she does not have much spare time for exercise. According to a website, she can burn an average of 3.5 calories per minute walking, 4 calories per minute cycling, and 8 calories per minute swimming. She would like her total time walking and cycling to be at least 3 times as long as she spends swimming. She would also like to walk at least 30 minutes per week. How much time should she spend on each activity not only to meet her goals but also to minimize her total exercise time per week? What is her minimum exercise time per week? *Source: BrianMac.co.uk.*

28. Health Care Marty McDonald takes vitamin pills. Each day he must have at least 3200 IU of vitamin A, 5 mg of vitamin B_1, and 200 mg of vitamin C. He can choose between pill 1, which costs 10¢ and contains 1600 IU of A, 1 mg of B_1, and 20 mg of C; and pill 2, which costs 20¢ and contains 400 IU of A, 1 mg of B_1, and 70 mg of C. How many of each pill should he buy in order to minimize his cost?

29. Blending Nutrients A biologist must make a nutrient for her algae. The nutrient must contain the three basic elements D, E, and F, and must contain at least 10 kg of D, 12 kg of E, and 20 kg of F. The nutrient is made from three ingredients, I, II, and

III. The quantity of D, E, and F in one unit of each of the ingredients is as given in the following chart.

		Ingredient		
		I	*II*	*III*
Kilograms of	*D*	4	1	10
Elements (per	*E*	3	2	1
unit of ingredient)	*F*	0	4	5
Cost per unit (in $)		4	7	5

How many units of each ingredient are required to meet the biologist's needs at minimum cost?

■■■YOUR TURN ANSWERS

1. Maximize $z = 24x_1 + 27x_2$

   subject to: $3x_1 + 5x_2 \le 25$

                   $3x_1 + x_2 \le 12$

                   $4x_1 + 3x_2 \le 27$

   with $x_1 \ge 0, \quad x_2 \ge 0.$

2. Minimum is 187/2 when $y_1 = 35/6$ and $y_2 = 1/2.$

4.4 Nonstandard Problems

APPLY IT **How many cars should an auto manufacturer send from each of its two plants to each of two dealerships in order to minimize the cost while meeting each dealership's needs?**

We will learn techniques in this section for answering questions like the one above, which will be answered in Example 2.

So far we have used the simplex method to solve linear programming problems in standard maximum or minimum form only. Now this work is extended to include linear programming problems with mixed $\le$ and $\ge$ constraints.

For example, suppose a new constraint is added to the farmer problem in Example 2 of Section 4.1: To satisfy orders from regular buyers, the farmer must plant a total of at least 60 acres of the three crops. Notice that our solution from Section 4.2 (plant 50 acres of potatoes, no acres of corn, and no acres of cabbage) does not satisfy this constraint, which introduces the new inequality

$$x_1 + x_2 + x_3 \ge 60.$$

As before, this inequality must be rewritten as an equation in which the variables all represent nonnegative numbers. The inequality $x_1 + x_2 + x_3 \ge 60$ means that

$$x_1 + x_2 + x_3 - s_3 = 60$$

for some nonnegative variable s_3. (Remember that s_1 and s_2 are the slack variables in the problem.)

The new variable, s_3, is called a **surplus variable**. The value of this variable represents the excess number of acres (over 60) that may be planted. Since the total number of acres planted is to be no more than 100 but at least 60, the value of s_3 can vary from 0 to 40.

We must now solve the system of equations

$$
\begin{aligned}
x_1 + x_2 + x_3 + s_1 &&&= 100 \\
10x_1 + 4x_2 + 7x_3 &+ s_2 &&= 500 \\
x_1 + x_2 + x_3 &&- s_3 &= 60 \\
-120x_1 - 40x_2 - 60x_3 &&+ z &= 0,
\end{aligned}
$$

with $x_1, x_2, x_3, s_1, s_2,$ and s_3 all nonnegative.

Set up the initial simplex tableau.

$$
\begin{bmatrix}
x_1 & x_2 & x_3 & s_1 & s_2 & s_3 & z & \\
1 & 1 & 1 & 1 & 0 & 0 & 0 & 100 \\
10 & 4 & 7 & 0 & 1 & 0 & 0 & 500 \\
1 & 1 & 1 & 0 & 0 & -1 & 0 & 60 \\
\hline
-120 & -40 & -60 & 0 & 0 & 0 & 1 & 0
\end{bmatrix}
$$

This tableau gives the solution

$$
x_1 = 0, \quad x_2 = 0, \quad x_3 = 0, \quad s_1 = 100, \quad s_2 = 500, \quad s_3 = -60.
$$

But this is not a feasible solution, since s_3 is negative. This means that the third constraint is not satisfied; we have $x_1 + x_2 + x_3 = 0$, but the sum is supposed to be at least 60. All the variables in any feasible solution must be nonnegative if the solution is to correspond to a point in the feasible region.

When a negative value of a variable appears in the solution, row operations are used to transform the matrix until a solution is found in which all variables are nonnegative. Here the problem is the -1 in a column corresponding to a basic variable. If the number in that row of the right-hand column were 0, we could simply multiply this row by -1 to remove the negative from the column. But we cannot do this with a positive number in the right-hand column. Instead, we find the positive entry that is farthest to the left in the third row (the row containing the -1); namely, the 1 in row 3, column 1. We will pivot using this column. (Actually, any column with a positive entry in row 3 will do; we chose the column farthest to the left arbitrarily.*) Use quotients as before to find the pivot, which is the 10 in row 2, column 1. Then use row operations to get the following tableau.

$$
\begin{array}{c}
-R_2 + 10R_1 \rightarrow R_1 \\
\\
-R_2 + 10R_3 \rightarrow R_3 \\
12R_2 + R_4 \rightarrow R_4
\end{array}
\begin{bmatrix}
x_1 & x_2 & x_3 & s_1 & s_2 & s_3 & z & \\
0 & 6 & 3 & 10 & -1 & 0 & 0 & 500 \\
10 & 4 & 7 & 0 & 1 & 0 & 0 & 500 \\
0 & 6 & 3 & 0 & -1 & -10 & 0 & 100 \\
\hline
0 & 8 & 24 & 0 & 12 & 0 & 1 & 6000
\end{bmatrix}
$$

Notice from the s_3 column that $-10s_3 = 100$, so s_3 is still negative. Therefore, we apply the procedure again. The 6 in row 3, column 2, is the positive entry farthest to the left in row 3, and by investigating quotients, we see that it is also the pivot. This leads to the following tableau.

$$
\begin{array}{c}
-R_3 + R_1 \rightarrow R_1 \\
-2R_3 + 3R_2 \rightarrow R_2 \\
\\
-4R_3 + 3R_4 \rightarrow R_4
\end{array}
\begin{bmatrix}
x_1 & x_2 & x_3 & s_1 & s_2 & s_3 & z & \\
0 & 0 & 0 & 10 & 0 & 10 & 0 & 400 \\
30 & 0 & 15 & 0 & 5 & 20 & 0 & 1300 \\
0 & 6 & 3 & 0 & -1 & -10 & 0 & 100 \\
\hline
0 & 0 & 60 & 0 & 40 & 40 & 3 & 17{,}600
\end{bmatrix}
$$

The value of s_3 is now 0 and the solution is feasible. We now continue with the simplex method until an optimal solution is found. We check for negative indicators, but since

*We use this rule for simplicity. There are, however, more complicated methods for choosing the pivot element that require, on average, fewer pivots to find the solution.

there are none, we have merely to create a 1 in each column corresponding to a basic variable or z.

$$
\begin{array}{c}
\begin{array}{cccccccc}
& x_1 & x_2 & x_3 & s_1 & s_2 & s_3 & z
\end{array}
\end{array}
$$

$$
\begin{array}{r}
R_1/10 \to R_1 \\
R_2/30 \to R_2 \\
R_3/6 \to R_3 \\
R_4/3 \to R_4
\end{array}
\left[
\begin{array}{ccccccc|c}
0 & 0 & 0 & 1 & 0 & 1 & 0 & 40 \\
1 & 0 & \frac{1}{2} & 0 & \frac{1}{6} & \frac{2}{3} & 0 & \frac{130}{3} \\
0 & 1 & \frac{1}{2} & 0 & -\frac{1}{6} & -\frac{5}{3} & 0 & \frac{50}{3} \\
0 & 0 & 20 & 0 & \frac{40}{3} & \frac{40}{3} & 1 & \frac{17{,}600}{3}
\end{array}
\right]
$$

The solution is

$$
x_1 = \frac{130}{3} = 43\frac{1}{3}, \quad x_2 = \frac{50}{3} = 16\frac{2}{3}, \quad x_3 = 0, \quad z = \frac{17{,}600}{3} = 5866.67.
$$

NOTE
If we ever reach a point where a surplus variable still has a negative solution, but there are no positive elements left in the row, then we have no way to make the surplus variable positive, so the problem has no feasible solution.

For maximum profit with this new constraint, the farmer should plant $43\frac{1}{3}$ acres of potatoes, $16\frac{2}{3}$ acres of corn, and no cabbage. The profit will be $5866.67, less than the $6000 profit if the farmer were to plant only 50 acres of potatoes. Because of the additional constraint that at least 60 acres must be planted, the profit is reduced. Notice that $s_1 = 40$. This is the slack variable for the constraint that no more than 100 acres are available. It indicates that 40 of the 100 available acres are still unused.

The procedure we have followed is a simplified version of the **two-phase method**, which is widely used for solving problems with mixed constraints. To see the complete method, including how to handle some complications that may arise, see *Linear Programming: An Introduction with Applications* by Alan Sultan, Academic Press, 1993.

In the previous section we solved standard minimum problems using duals. If a minimizing problem has mixed $\leq$ and $\geq$ constraints, the dual method cannot be used. We solve such problems with the method presented in this section. To see how, consider the simple fact: When a number t gets smaller, then $-t$ gets larger, and vice versa. For instance, if t goes from 6 to 1 to 0 to -8, then $-t$ goes from -6 to -1 to 0 to 8. Thus, if w is the objective function of a minimizing linear programming problem, the feasible solution that produces the minimum value of w also produces the maximum value of $z = -w$, and vice versa. Therefore, to solve a minimization problem with objective function w, we need only solve the maximization problem with the same constraints and objective function $z = -w$.

In summary, the following steps are involved in solving the nonstandard problems in this section.

Solving a Nonstandard Problem

1. If necessary, convert the problem to a maximization problem.
2. Add slack variables and subtract surplus variables as needed.
3. Write the initial simplex tableau.
4. If any basic variable has a negative value, locate the nonzero number in that variable's column, and note what row it is in.
5. In the row located in Step 4, find the positive entry that is farthest to the left, and note what column it is in.
6. In the column found in Step 5, choose a pivot by investigating quotients.
7. Use row operations to change the other numbers in the pivot column to 0.
8. Continue Steps 4 through 7 until all basic variables are nonnegative. If it ever becomes impossible to continue, then the problem has no feasible solution.
9. Once a feasible solution has been found, continue to use the simplex method until the optimal solution is found.

In the next example, we use this method to solve a minimization problem with mixed constraints.

EXAMPLE 1 Minimization

Minimize $w = 3y_1 + 2y_2$

subject to: $y_1 + 3y_2 \leq 6$

               $2y_1 + y_2 \geq 3$

with $y_1 \geq 0, \quad y_2 \geq 0.$

SOLUTION   Change this to a maximization problem by letting z equal the *negative* of the objective function: $z = -w$. Then find the *maximum* value of

$$z = -w = -3y_1 - 2y_2.$$

The problem can now be stated as follows.

Maximize $z = -3y_1 - 2y_2$

subject to: $y_1 + 3y_2 \leq 6$

               $2y_1 + y_2 \geq 3$

with $y_1 \geq 0, \quad y_2 \geq 0.$

To begin, we add slack and surplus variables, and rewrite the objective function.

$$y_1 + 3y_2 + s_1 \qquad\qquad = 6$$
$$2y_1 + y_2 \qquad - s_2 \qquad = 3$$
$$3y_1 + 2y_2 \qquad\qquad + z = 0$$

Set up the initial simplex tableau.

$$
\begin{array}{ccccc}
y_1 & y_2 & s_1 & s_2 & z \\
\end{array}
$$

$$
\left[\begin{array}{ccccc|c}
1 & 3 & 1 & 0 & 0 & 6 \\
2 & 1 & 0 & -1 & 0 & 3 \\
3 & 2 & 0 & 0 & 1 & 0
\end{array}\right]
$$

The solution $y_1 = 0$, $y_2 = 0$, $s_1 = 6$, and $s_2 = -3$, is not feasible. Row operations must be used to get a feasible solution. We start with s_2, which has a -1 in row 2. The positive entry farthest to the left in row 2 is the 2 in column 1. The element in column 1 that gives the smallest quotient is 2, so it becomes the pivot. Pivoting produces the following matrix.

$$
\begin{array}{ccccc}
y_1 & y_2 & s_1 & s_2 & z \\
\end{array}
$$

$$
\begin{array}{l}
-R_2 + 2R_1 \to R_1 \\
\\
-3R_2 + 2R_3 \to R_3
\end{array}
\left[\begin{array}{ccccc|c}
0 & 5 & 2 & 1 & 0 & 9 \\
2 & 1 & 0 & -1 & 0 & 3 \\
0 & 1 & 0 & 3 & 2 & -9
\end{array}\right]
$$

Now $s_2 = 0$, so the solution is feasible. Furthermore, there are no negative indicators, so the solution is optimal. Divide row 1 by 2, row 2 by 2, and row 3 by 2 to find the final solution: $y_1 = 3/2$ and $y_2 = 0$. Since $z = -w = -9/2$, the minimum value is $w = 9/2$.

TRY YOUR TURN 1

An important application of linear programming is the problem of minimizing the cost of transporting goods. This type of problem is often referred to as a *transportation problem* or *warehouse problem*. Some problems of this type were included in the exercise sets in previous chapters. The next example is based on Exercise 52 from Section 2.2, in which the transportation costs were set equal to $10,640. We will now use the simplex method to minimize the transportation costs.

EXAMPLE 2 Transportation Problem

An auto manufacturer sends cars from two plants, I and II, to dealerships A and B located in a midwestern city. Plant I has a total of 28 cars to send, and plant II has 8. Dealer A needs 20 cars, and dealer B needs 16. Transportation costs per car based on the distance of each

NOTE
1) Slack variables are used for $\leq$ constraints, while surplus variables are used for $\geq$ constraints
2) Remember to convert from z to w as the last step in solving a minimization problem.

YOUR TURN 1

Minimize $w = 6y_1 + 4y_2$

subject to: $3y_1 + 4y_2 \geq 10$

               $9y_1 + 7y_2 \leq 18$

with $y_1 \geq 0, \quad y_2 \geq 0.$

dealership from each plant are $220 from I to A, $300 from I to B, $400 from II to A, and $180 from II to B. How many cars should be sent from each plant to each of the two dealerships to minimize transportation costs? Use the simplex method to find the solution.

APPLY IT

SOLUTION To begin, let

$$y_1 = \text{the number of cars shipped from I to A;}$$
$$y_2 = \text{the number of cars shipped from I to B;}$$
$$y_3 = \text{the number of cars shipped from II to A;}$$

and $\qquad y_4 = \text{the number of cars shipped from II to B.}$

Plant I has only 28 cars to ship, so

$$y_1 + y_2 \le 28.$$

Similarly, plant II has only 8 cars to ship, so

$$y_3 + y_4 \le 8.$$

Since dealership A needs 20 cars and dealership B needs 16 cars,

$$y_1 + y_3 \ge 20 \qquad \text{and} \qquad y_2 + y_4 \ge 16.$$

The manufacturer wants to minimize transportation costs, so the objective function is

$$w = 220y_1 + 300y_2 + 400y_3 + 180y_4.$$

Now write the problem as a system of linear equations, adding slack or surplus variables as needed, and let $z = -w$.

$$
\begin{aligned}
y_1 + y_2 && + s_1 &&&&&&&& = 28 \\
y_3 + y_4 && + s_2 &&&&&&&& = 8 \\
y_1 \qquad + y_3 &&&& - s_3 &&&&&& = 20 \\
y_2 \qquad + y_4 &&&&&& - s_4 && = 16 \\
220y_1 + 300y_2 + 400y_3 + 180y_4 &&&&&&&& + z &= 0
\end{aligned}
$$

Set up the initial simplex tableau.

$$
\begin{bmatrix}
y_1 & y_2 & y_3 & y_4 & s_1 & s_2 & s_3 & s_4 & z & \\
1 & 1 & 0 & 0 & 1 & 0 & 0 & 0 & 0 & 28 \\
0 & 0 & 1 & 1 & 0 & 1 & 0 & 0 & 0 & 8 \\
\boxed{1} & 0 & 1 & 0 & 0 & 0 & -1 & 0 & 0 & 20 \\
0 & 1 & 0 & 1 & 0 & 0 & 0 & -1 & 0 & 16 \\
\hline
220 & 300 & 400 & 180 & 0 & 0 & 0 & 0 & 1 & 0
\end{bmatrix}
$$

Because $s_3 = -20$, we choose the positive entry farthest to the left in row 3, which is the 1 in column 1. After forming the necessary quotients, we find that the 1 is also the pivot, leading to the following tableau.

$$
\begin{array}{c}
-R_3 + R_1 \to R_1 \\ \\ \\ \\ \\ -220R_3 + R_5 \to R_5
\end{array}
\begin{bmatrix}
y_1 & y_2 & y_3 & y_4 & s_1 & s_2 & s_3 & s_4 & z & \\
0 & \boxed{1} & -1 & 0 & 1 & 0 & 1 & 0 & 0 & 8 \\
0 & 0 & 1 & 1 & 0 & 1 & 0 & 0 & 0 & 8 \\
1 & 0 & 1 & 0 & 0 & 0 & -1 & 0 & 0 & 20 \\
0 & 1 & 0 & 1 & 0 & 0 & 0 & -1 & 0 & 16 \\
\hline
0 & 300 & 180 & 180 & 0 & 0 & 220 & 0 & 1 & -4400
\end{bmatrix}
$$

We still have $s_4 = -16$. Verify that the 1 in row 1, column 2, is the next pivot, leading to the following tableau.

$$\begin{bmatrix} y_1 & y_2 & y_3 & y_4 & s_1 & s_2 & s_3 & s_4 & z & \\ 0 & 1 & -1 & 0 & 1 & 0 & 1 & 0 & 0 & 8 \\ 0 & 0 & 1 & 1 & 0 & 1 & 0 & 0 & 0 & 8 \\ 1 & 0 & 1 & 0 & 0 & 0 & -1 & 0 & 0 & 20 \\ 0 & 0 & \boxed{1} & 1 & -1 & 0 & -1 & -1 & 0 & 8 \\ 0 & 0 & 480 & 180 & -300 & 0 & -80 & 0 & 1 & -6800 \end{bmatrix}$$

with the row operations $-R_1 + R_4 \rightarrow R_4$ and $-300R_1 + R_5 \rightarrow R_5$.

We still have $s_4 = -8$. Choosing column 3 to pivot, there is a tie between rows 2 and 4. Ordinarily in such cases, we arbitrarily choose the pivot in the row nearest to the top of the matrix. With surplus variables, however, we have the immediate goal of making all basic variables nonnegative. Because choosing row 4 will remove s_4 from the set of basic variables, we will choose as the pivot the 1 in column 3, row 4. The result is the following tableau.

$$\begin{bmatrix} y_1 & y_2 & y_3 & y_4 & s_1 & s_2 & s_3 & s_4 & z & \\ 0 & 1 & 0 & 1 & 0 & 0 & 0 & -1 & 0 & 16 \\ 0 & 0 & 0 & 0 & 1 & 1 & 1 & 1 & 0 & 0 \\ 1 & 0 & 0 & -1 & 1 & 0 & 0 & 1 & 0 & 12 \\ 0 & 0 & 1 & 1 & -1 & 0 & -1 & -1 & 0 & 8 \\ 0 & 0 & 0 & -300 & 180 & 0 & 400 & 480 & 1 & -10640 \end{bmatrix}$$

with row operations $R_4 + R_1 \rightarrow R_1$, $-R_4 + R_2 \rightarrow R_2$, $-R_4 + R_3 \rightarrow R_3$, $-480R_4 + R_5 \rightarrow R_5$.

We now have the feasible solution

$$y_1 = 12, \quad y_2 = 16, \quad y_3 = 8, \quad y_4 = 0, \quad s_1 = 0, \quad s_2 = 0, \quad s_3 = 0, \quad s_4 = 0,$$

with $w = 10{,}640$. But there are still negative indicators in the bottom row, so we can keep going. After two more tableaus, we find that

$$y_1 = 20, \quad y_2 = 8, \quad y_3 = 0, \quad y_4 = 8,$$

with $w = 8240$. Therefore, the manufacturer should send 20 cars from plant I to dealership A and 8 cars to dealership B. From plant II, 8 cars should be sent to dealership B and none to dealership A. The transportation cost will then be $8240, a savings of $2400 over the original stated cost of $10,640. **TRY YOUR TURN 2**

YOUR TURN 2 Finish the missing steps in Example 2 and show the final tableau.

When one or more of the constraints in a linear programming problem is an equation, rather than an inequality, we add an **artificial variable** to each equation, rather than a slack or surplus variable. The first goal of the simplex method is to eliminate any artificial variables as basic variables, since they must have a value of 0 in the solution. We do this exactly the way we changed the surplus variables from having negative values to being zero. The only difference is that once an artificial variable is made nonbasic, we must never pivot on that column again, because that would change it to a basic variable. We must never carry out any pivot that would result in an artificial variable becoming nonzero.

EXAMPLE 3 Artificial Variables

In the transportation problem discussed in Example 2, it would be more realistic for the dealerships to order exactly 20 and 16 cars, respectively. Solve the problem with these two equality constraints.

SOLUTION Using the same variables, we can state the problem as follows.

$$\begin{aligned} \text{Minimize} \quad & w = 220y_1 + 300y_2 + 400y_3 + 180y_4 \\ \text{subject to:} \quad & y_1 + y_2 \le 28 \\ & y_3 + y_4 \le 8 \\ & y_1 + y_3 = 20 \\ & y_2 + y_4 = 16 \end{aligned}$$

with all variables nonnegative.

The corresponding system of equations requires slack variables s_1 and s_2 and two artificial variables that we shall call a_1 and a_2, to remind us that they require special handling. The system

$$
\begin{aligned}
y_1 + y_2 &&&&+ s_1 &&&&&&= 28 \\
&& y_3 + y_4 &&&+ s_2 &&&&&= 8 \\
y_1 && + y_3 &&&&&+ a_1 &&&= 20 \\
&& y_2 && + y_4 &&&&&+ a_2 &= 16 \\
220y_1 + 300y_2 + 400y_3 + 180y_4 &&&&&&&&&+ z &= 0
\end{aligned}
$$

produces a tableau exactly the same as in Example 2, except that the columns labeled s_3 and s_4 in that example are now labeled a_1 and a_2, and there is initially a 1 rather than a -1 in the columns for a_1 and a_2. The first three pivots are the same as in Example 2, resulting in the following tableau.

y_1	y_2	y_3	y_4	s_1	s_2	a_1	a_2	z	
0	1	0	1	0	0	0	1	0	16
0	0	0	0	1	1	-1	-1	0	0
1	0	0	-1	1	0	0	-1	0	12
0	0	1	1	-1	0	1	1	0	8
0	0	0	-300	180	0	-400	-480	1	-10640

NOTE
Another way to handle this situation is by solving for y_3 and y_4 in terms of y_1 and y_2. Then proceed with the usual method for standard problems.

Ordinarily our next step would be to pivot on the column with the indicator of -480, but that is not allowed because it would make a_2 a basic variable. Similarly, we cannot pivot on the column with the indicator of -400. Instead, we pivot on the column with the indicator of -300. Verify that after this step and one more tableau, we reach the same solution as in Example 2. In other problems, equality constraints can result in a higher cost. ▬

CAUTION If the artificial variables cannot be made equal to zero, the problem has no feasible solution.

Applications requiring the simplex method often have constraints that have a zero on the right-hand side. For example, in Exercise 35 of Section 4.2 a person wants to walk at least twice as many hours as she bikes. This results in one of the constraints $x_1 - 2x_2 \leq 0$ or $-x_1 + 2x_2 \geq 0$. For the purposes of using the simplex method to solve problems in the standard maximum form, it is always better to write constraints in the first form, since the first constraint can be readily handled by the basic simplex method by adding a slack variable.

Several linear programming models in actual use are presented on the Web site for this textbook. These models illustrate the usefulness of linear programming. In most real applications, the number of variables is so large that these problems could not be solved without using methods (like the simplex method) that can be adapted to computers.

4.4 EXERCISES

Rewrite each system of inequalities as a system of linear equations, adding slack variables or subtracting surplus variables as necessary.

1. $2x_1 + 3x_2 \leq 8$
$x_1 + 4x_2 \geq 7$

2. $3x_1 + 7x_2 \leq 9$
$4x_1 + 5x_2 \geq 11$

3. $2x_1 + x_2 + 2x_3 \leq 50$
$x_1 + 3x_2 + x_3 \geq 35$
$x_1 + 2x_2 \geq 15$

4. $2x_1 + x_3 \leq 40$
$x_1 + x_2 \geq 18$
$x_1 + x_3 \geq 20$

Convert each problem into a maximization problem.

5. Minimize $w = 3y_1 + 4y_2 + 5y_3$
subject to: $y_1 + 2y_2 + 3y_3 \geq 9$
$y_2 + 2y_3 \geq 8$
$2y_1 + y_2 + 2y_3 \geq 6$
with $y_1 \geq 0, \quad y_2 \geq 0, \quad y_3 \geq 0.$

6. Minimize $\quad w = 8y_1 + 3y_2 + y_3$

subject to: $\quad 7y_1 + 6y_2 + 8y_3 \geq 18$

$\qquad\qquad 4y_1 + 5y_2 + 10y_3 \geq 20$

with $\qquad y_1 \geq 0, \quad y_2 \geq 0, \quad y_3 \geq 0.$

7. Minimize $\quad w = y_1 + 2y_2 + y_3 + 5y_4$

subject to: $\quad y_1 + y_2 + y_3 + y_4 \geq 50$

$\qquad\qquad 3y_1 + y_2 + 2y_3 + y_4 \geq 100$

with $\qquad y_1 \geq 0, \quad y_2 \geq 0, \quad y_3 \geq 0, \quad y_4 \geq 0.$

8. Minimize $\quad w = y_1 + y_2 + 7y_3$

subject to: $\quad 5y_1 + 2y_2 + y_3 \geq 125$

$\qquad\qquad 4y_1 + y_2 + 6y_3 \leq 75$

$\qquad\qquad 6y_1 + 8y_2 \qquad \geq 84$

with $\qquad y_1 \geq 0, \quad y_2 \geq 0, \quad y_3 \geq 0.$

Use the simplex method to solve.

9. Find $x_1 \geq 0$ and $x_2 \geq 0$ such that

$$x_1 + 2x_2 \geq 24$$

$$x_1 + x_2 \leq 40$$

and $z = 12x_1 + 10x_2$ is maximized.

10. Find $x_1 \geq 0$ and $x_2 \geq 0$ such that

$$2x_1 + x_2 \geq 20$$

$$2x_1 + 5x_2 \leq 80$$

and $z = 6x_1 + 2x_2$ is maximized.

11. Find $x_1 \geq 0$, $x_2 \geq 0$, and $x_3 \geq 0$ such that

$$x_1 + x_2 + x_3 \leq 150$$

$$x_1 + x_2 + x_3 \geq 100$$

and $z = 2x_1 + 5x_2 + 3x_3$ is maximized.

12. Find $x_1 \geq 0$, $x_2 \geq 0$, and $x_3 \geq 0$ such that

$$x_1 + x_2 + x_3 \leq 15$$

$$4x_1 + 4x_2 + 2x_3 \geq 48$$

and $z = 2x_1 + x_2 + 3x_3$ is maximized.

13. Find $x_1 \geq 0$ and $x_2 \geq 0$ such that

$$x_1 + x_2 \leq 100$$

$$2x_1 + 3x_2 \leq 75$$

$$x_1 + 4x_2 \geq 50$$

and $z = 5x_1 - 3x_2$ is maximized.

14. Find $x_1 \geq 0$ and $x_2 \geq 0$ such that

$$x_1 + 2x_2 \leq 18$$

$$x_1 + 3x_2 \geq 12$$

$$2x_1 + 2x_2 \leq 24$$

and $z = 5x_1 - 10x_2$ is maximized.

15. Find $y_1 \geq 0$, $y_2 \geq 0$, and $y_3 \geq 0$ such that

$$5y_1 + 3y_2 + 2y_3 \leq 150$$

$$5y_1 + 10y_2 + 3y_3 \geq 90$$

and $w = 10y_1 + 12y_2 + 10y_3$ is minimized.

16. Minimize $\quad w = 3y_1 + 2y_2 + 3y_3$

subject to: $\quad 2y_1 + 3y_2 + 6y_3 \leq 60$

$\qquad\qquad y_1 + 4y_2 + 5y_3 \geq 40$

with $\qquad y_1 \geq 0, \quad y_2 \geq 0, \quad y_3 \geq 0.$

Solve using artificial variables.

17. Maximize $\quad z = 3x_1 + 2x_2$

subject to: $\quad x_1 + x_2 = 50$

$\qquad\qquad 4x_1 + 2x_2 \geq 120$

$\qquad\qquad 5x_1 + 2x_2 \leq 200$

with $\qquad x_1 \geq 0, \quad x_2 \geq 0.$

18. Maximize $\quad z = 5x_1 + 7x_2$

subject to: $\quad x_1 + x_2 = 15$

$\qquad\qquad 2x_1 + 4x_2 \geq 30$

$\qquad\qquad 3x_1 + 5x_2 \geq 10$

with $\qquad x_1 \geq 0, \quad x_2 \geq 0.$

19. Minimize $\quad w = 32y_1 + 40y_2 + 48y_3$

subject to: $\quad 20y_1 + 10y_2 + 5y_3 = 200$

$\qquad\qquad 25y_1 + 40y_2 + 50y_3 \leq 500$

$\qquad\qquad 18y_1 + 24y_2 + 12y_3 \geq 300$

with $\qquad y_1 \geq 0, \quad y_2 \geq 0, \quad y_3 \geq 0.$

20. Minimize $\quad w = 15y_1 + 12y_2 + 18y_3$

subject to: $\quad y_1 + 2y_2 + 3y_3 \leq 12$

$\qquad\qquad 3y_1 + y_2 + 3y_3 \geq 18$

$\qquad\qquad y_1 + y_2 + y_3 = 10$

with $\qquad y_1 \geq 0, \quad y_2 \geq 0, \quad y_3 \geq 0.$

21. Explain how, in any linear programming problem, the value of the objective function can be found without using the number in the lower right-hand corner of the final tableau.

22. Explain why, for a maximization problem, you write the negative of the coefficients of the objective function on the bottom row, while, for a minimization problem, you write the coefficients themselves.

APPLICATIONS

Business and Economics

23. Transportation Southwestern Oil supplies two distributors in the Northwest from two outlets, S_1 and S_2. Distributor D_1 needs at least 3000 barrels of oil, and distributor D_2 needs at least 5000 barrels. The two outlets can each furnish up to 5000 barrels of oil. The costs per barrel to ship the oil are given in the table.

		Distributors	
		D_1	D_2
Outlets	S_1	$30	$20
	S_2	$25	$22

There is also a shipping tax per barrel as given in the table below. Southwestern Oil is determined to spend no more than $40,000 on shipping tax.

	D_1	D_2
S_1	$2	$6
S_2	$5	$4

a. How should the oil be supplied to minimize shipping costs?

b. Find and interpret the values of any nonzero slack or surplus variables.

24. **Transportation** Change Exercise 23 so that the two outlets each furnish exactly 5000 barrels of oil, with everything else the same. Use artificial variables to solve the problem, following the steps outlined in Example 3.

25. **Finance** A bank has set aside a maximum of $25 million for commercial and home loans. Every million dollars in commercial loans requires 2 lengthy application forms, while every million dollars in home loans requires 3 lengthy application forms. The bank cannot process more than 72 application forms at this time. The bank's policy is to loan at least four times as much for home loans as for commercial loans. Because of prior commitments, at least $10 million will be used for these two types of loans. The bank earns 10% on commercial loans and 12% on home loans. What amount of money should be allotted for each type of loan to maximize the interest income?

26. **Blending Seed** Topgrade Turf lawn seed mixture contains three types of seed: bluegrass, rye, and Bermuda. The costs per pound of the three types of seed are 16 cents, 14 cents, and 12 cents, respectively. In each batch there must be at least 25% bluegrass seed, and the amount of Bermuda must be no more than 2/3 the amount of rye. To fill current orders, the company must make at least 6000 lb of the mixture. How much of each kind of seed should be used to minimize cost?

27. **Blending Seed** Change Exercise 26 so that the company must make exactly 6000 lb of the mixture. Use artificial variables to solve the problem.

28. **Investments** Lynda Rago has decided to invest a $100,000 inheritance in government securities that earn 7% per year, municipal bonds that earn 6% per year, and mutual funds that earn an average of 10% per year. She will spend at least $40,000 on government securities, and she wants at least half the inheritance to go to bonds and mutual funds. Government securities have an initial fee of 2%, municipal bonds have an initial fee of 1%, and mutual funds have an initial fee of 3%. Lynda has $2400 available to pay initial fees. How much should be invested in each way to maximize the interest yet meet the constraints? What is the maximum interest she can earn?

29. **Transportation** The manufacturer of a popular personal computer has orders from two dealers. Dealer D_1 wants at least 32 computers, and dealer D_2 wants at least 20 computers. The manufacturer can fill the orders from either of two warehouses, W_1 or W_2. There are 25 computers on hand at W_1, and 30 at W_2. The costs (in dollars) to ship one computer to each dealer from each warehouse are given in the following table.

		Dealer	
		D_1	D_2
Warehouse	W_1	$14	$12
	W_2	$12	$10

a. How should the orders be filled to minimize shipping costs?

b. Find and interpret the values of any nonzero slack or surplus variables.

30. **Calorie Expenditure** Joe Vetere's exercise regimen includes light calisthenics, swimming, and playing the drums. He has at most 10 hours per week to devote to these activities. He wants the total time he does calisthenics and plays the drums to be at least twice as long as he swims. His neighbors, however, will tolerate no more than 4 hours per week on the drums. According to a website, a 190-pound person like Joe will burn an average of 388 calories per hour doing calisthenics, 518 calories per hour swimming, and 345 calories per hour playing the drums. In Section 4.2, Exercise 36, Joe found that he could maximize calories burned in an exercise routine that did not include playing the drums as part of his exercise plan. *Source: NutriStrategy.com.*

a. Joe really likes to play the drums and insists that his exercise plan include at least 1 hour of playing the drums per week. With this added constraint, now how many hours per week should Joe spend on each exercise to maximize the number of calories he burns? What is the maximum number of calories he will burn?

b. Without the added constraint from part a, Joe's maximum calorie expenditure was $4313\frac{1}{3}$ calories. Compare this number with the new optimal solution. What conclusions can you draw when additional constraints are placed on a problem?

c. Show how the solution from part a can be found without using the simplex method by considering the constraints and the number of calories for each activity.

31. **Blending Chemicals** Natural Brand plant food is made from three chemicals, labeled I, II, and III. In each batch of the plant food, the amounts of chemicals II and III must be in the ratio of 4 to 3. The amount of nitrogen must be at least 30 kg. The percent of nitrogen in the three chemicals is 9%, 4%, and 3%, respectively. If the three chemicals cost $1.09, $0.87, and $0.65 per kilogram, respectively, how much of each should be used to minimize the cost of producing at least 750 kg of the plant food?

32. **Blending a Soft Drink** A popular soft drink called Sugarlo, which is advertised as having a sugar content of no more than 10%, is blended from five ingredients, each of which has some sugar content. Water may also be added to dilute the mixture. The sugar content of the ingredients and their costs per gallon are given below.

	Ingredient					
	1	2	3	4	5	Water
Sugar Content (%)	0.28	0.19	0.43	0.57	0.22	0
Cost ($/gal)	0.48	0.32	0.53	0.28	0.43	0.04

At least 0.01 of the content of Sugarlo must come from ingredients 3 or 4, 0.01 must come from ingredients 2 or 5, and 0.01 from ingredients 1 or 4. How much of each ingredient should be used in preparing 15,000 gal of Sugarlo to minimize the cost?

33. Blending Gasoline A company is developing a new additive for gasoline. The additive is a mixture of three liquid ingredients, I, II, and III. For proper performance, the total amount of additive must be at least 10 oz per gal of gasoline. However, for safety reasons, the amount of additive should not exceed 15 oz per gal of gasoline. At least 1/4 oz of ingredient I must be used for every ounce of ingredient II, and at least 1 oz of ingredient III must be used for every ounce of ingredient I. If the costs of I, II, and III are $0.30, $0.09, and $0.27 per oz, respec-

tively, find the mixture of the three ingredients that produces the minimum cost of the additive. How much of the additive should be used per gal of gasoline?

YOUR TURN ANSWERS

1. Minimum is 10 when $y_1 = 0$ and $y_2 = 5/2$.

2. $$\begin{bmatrix} 0 & 1 & -1 & 0 & 0 & -1 & 0 & -1 & 0 & | & 8 \\ 0 & 0 & 0 & 0 & 1 & 1 & 1 & 1 & 0 & | & 0 \\ 1 & 0 & 1 & 0 & 0 & 0 & -1 & 0 & 0 & | & 20 \\ 0 & 0 & 1 & 1 & 0 & 1 & 0 & 0 & 0 & | & 8 \\ \hline 0 & 0 & 300 & 0 & 0 & 120 & 220 & 300 & 1 & | & -8240 \end{bmatrix}$$

4 CHAPTER REVIEW

SUMMARY

In this chapter, we introduced the simplex method, which is a procedure for solving any linear programming problem. To apply this method, we first had to write the problem as a standard maximization problem in matrix form. This form tells us an initial basic feasible solution, which the simplex method uses to determine other basic feasible solutions. Each successive iteration of the simplex method gives us a new basic feasible solution, whose objective function value is greater than or equal to the objective function value of the previous basic feasible solution. We then introduced duality, which tells us that every time we solve a linear programming problem, we are actually solving two problems—a maximization problem and a minimization problem. This has far-reaching consequences in the fields of operations research and decision sciences, including the fact that standard minimization problems can be solved by the simplex method. Finally, we extended the simplex method to solve problems that are not standard because they have inequalities going in both directions (and perhaps equalities as well).

Standard Maximum Form A linear programming problem is in standard maximum form if the following conditions are satisfied.

1. The objective function is to be maximized.
2. All variables are nonnegative.
3. All remaining constraints are stated in the form

$$a_1x_1 + a_2x_2 + \cdots + a_nx_n \le b \qquad \text{with } b \ge 0.$$

Simplex Method 1. Determine the objective function.

2. Write all the necessary constraints.
3. Convert each constraint into an equation by adding a slack variable in each.
4. Set up the initial simplex tableau.
5. Locate the most negative indicator. If there are two such indicators, choose the one farther to the left.
6. Form the necessary quotients to find the pivot. Disregard any quotients with 0 or a negative number in the denominator. The smallest nonnegative quotient gives the location of the pivot. If all quotients must be disregarded, no maximum solutions exist. If two quotients are both equal and smallest, choose the pivot in the row nearest the top of the matrix.
7. Use row operations to change all other numbers in the pivot column to zero by adding a suitable multiple of the pivot row to a positive multiple of each row.
8. If the indicators are all positive or 0, this is the final tableau. If not, go back to step 5 and repeat the process until a tableau with no negative indicators is obtained.
9. Read the solution from the final tableau.

Standard Minimum Form

A linear programming problem is in standard minimum form if the following conditions are satisfied.

1. The objective function is to be minimized.

2. All variables are nonnegative.

3. All remaining constraints are stated in the form

$$a_1 y_1 + a_2 y_2 + \cdots + a_n y_n \geq b \qquad \text{with } b \geq 0.$$

Theorem of Duality

The objective function w of a minimization linear programming problem takes on a minimum value if and only if the objective function z of the corresponding dual maximization problem takes on a maximum value. The maximum value of z equals the minimum value of w.

Solving a Standard Minimum Problem with Duals

1. Find the dual standard maximization problem.

2. Solve the maximization problem using the simplex method.

3. The minimum value of the objective function w is the maximum value of the objective function z.

4. The optimum solution is given by the entries in the bottom row of the columns corresponding to the slack variables, as long as the entry in the z column is equal to 1.

Solving a Nonstandard Problem

1. If necessary, convert the problem to a maximization problem.

2. Add slack variables and subtract surplus variables as needed.

3. Write the initial simplex tableau.

4. If any basic variable has a negative value, locate the nonzero number in that variable's column, and note what row it is in.

5. In the row located in step 4, find the positive entry that is farthest to the left, and note what column it is in.

6. In the column found in step 5, choose a pivot by investigating quotients.

7. Use row operations to change the other numbers in the pivot column to 0.

8. Continue steps 4 through 7 until all basic variables are nonnegative. If it ever becomes impossible to continue, then the problem has no feasible solution.

9. Once a feasible solution has been found, continue to use the simplex method until the optimal solution is found.

Artificial Variables

When one or more of the constraints in a linear programming problem is an equation, rather than an inequality, an artificial variable is added to each equation. Artificial variables are handled in the same way as surplus variables, except that once an artificial variable is no longer basic, never pivot on its column. If in the optimal solution an artificial variable has a positive value, then the original problem does not have a solution.

KEY TERMS

simplex method

4.1
standard maximum form
slack variable
simplex tableau
indicators

basic variable
basic feasible solution
pivot

4.2
sensitivity analysis

4.3
standard minimum form
dual
transpose
shadow costs

4.4
surplus variable
two-phase method
artificial variable

REVIEW EXERCISES

CONCEPT CHECK

Determine whether each of the following statements is true or false, and explain why.

1. The simplex method can be used to solve all linear programming problems.

2. If the feasible region of a linear programming problem is unbounded, then the objective function value is unbounded.

3. A linear programming problem in standard maximization form always has a feasible solution.

4. A linear programming problem in standard minimization form always has a feasible solution.

5. A linear programming problem in standard maximization form always has a finite optimal solution.

6. The tableau below for a linear program in standard maximization form shows that it has no finite maximum value.

$$\begin{array}{cccccc} x_1 & x_2 & s_1 & s_2 & z & \\ \left[\begin{array}{ccccc|c} -1 & 1 & 0 & 1 & 0 & 1 \\ -4 & 0 & 1 & -2 & 0 & 3 \\ \hline -1 & 0 & 0 & 2 & 1 & 4 \end{array}\right] \end{array}$$

7. One must always use the minimum quotient when choosing a pivot row.

8. If there is a 0 in the right-hand column, we can disregard it when determining the quotients used to choose the pivot row.

9. One must always pick the most negative number in the indicator row when choosing the pivot column.

10. A basic variable can be assigned a value of zero by the simplex method.

11. A slack variable of a linear programming problem in standard maximization form may become negative during the intermediate stages of the simplex method.

12. The dual of the dual of a linear programming problem is the original problem.

13. The simplex method guarantees that each iteration will yield a feasible solution whose objective function value is bigger than the objective function value of all previous solutions.

14. Standard maximization problems can only have slack variables, and standard minimization problems (not solved by the dual method) can only have surplus variables.

PRACTICE AND EXPLORATIONS

15. When is it necessary to use the simplex method rather than the graphical method?

16. What can you conclude if a surplus variable cannot be made nonnegative?

For each problem, (a) add slack variables or subtract surplus variables, and (b) set up the initial simplex tableau.

17. Maximize $z = 2x_1 + 7x_2$

subject to: $4x_1 + 6x_2 \leq 60$

$3x_1 + x_2 \leq 18$

$2x_1 + 5x_2 \leq 20$

$x_1 + x_2 \leq 15$

with $x_1 \geq 0, \quad x_2 \geq 0.$

18. Maximize $z = 25x_1 + 30x_2$

subject to: $3x_1 + 5x_2 \leq 47$

$x_1 + x_2 \leq 25$

$5x_1 + 2x_2 \leq 35$

$2x_1 + x_2 \leq 30$

with $x_1 \geq 0, \quad x_2 \geq 0.$

19. Maximize $z = 5x_1 + 8x_2 + 6x_3$

subject to: $x_1 + x_2 + x_3 \leq 90$

$2x_1 + 5x_2 + x_3 \leq 120$

$x_1 + 3x_2 \geq 80$

with $x_1 \geq 0, \quad x_2 \geq 0, \quad x_3 \geq 0.$

20. Maximize $z = 4x_1 + 6x_2 + 8x_3$

subject to: $x_1 + x_2 + 2x_3 \geq 200$

$8x_1 + 6x_3 \leq 400$

$3x_1 + 5x_2 + x_3 \leq 300$

with $x_1 \geq 0, \quad x_2 \geq 0, \quad x_3 \geq 0.$

Use the simplex method to solve each maximization linear programming problem with the given initial tableau.

21.
$$\begin{array}{cccccc} x_1 & x_2 & x_3 & s_1 & s_2 & z \\ \left[\begin{array}{cccccc|c} 4 & 5 & 2 & 1 & 0 & 0 & 18 \\ 2 & 8 & 6 & 0 & 1 & 0 & 24 \\ \hline -5 & -3 & -6 & 0 & 0 & 1 & 0 \end{array}\right] \end{array}$$

22.
$$\begin{array}{ccccc} x_1 & x_2 & s_1 & s_2 & z \\ \left[\begin{array}{ccccc|c} 2 & 7 & 1 & 0 & 0 & 14 \\ 2 & 3 & 0 & 1 & 0 & 10 \\ \hline -2 & -4 & 0 & 0 & 1 & 0 \end{array}\right] \end{array}$$

23.
$$\begin{array}{ccccccc} x_1 & x_2 & x_3 & s_1 & s_2 & s_3 & z \\ \left[\begin{array}{ccccccc|c} 1 & 2 & 2 & 1 & 0 & 0 & 0 & 50 \\ 3 & 1 & 0 & 0 & 1 & 0 & 0 & 20 \\ 1 & 0 & 2 & 0 & 0 & -1 & 0 & 15 \\ \hline -5 & -3 & -2 & 0 & 0 & 0 & 1 & 0 \end{array}\right] \end{array}$$

24.
$$\begin{array}{cccccc} x_1 & x_2 & s_1 & s_2 & s_3 & z \\ \left[\begin{array}{cccccc|c} 3 & 6 & -1 & 0 & 0 & 0 & 28 \\ 1 & 1 & 0 & 1 & 0 & 0 & 12 \\ 2 & 1 & 0 & 0 & 1 & 0 & 16 \\ \hline -1 & -2 & 0 & 0 & 0 & 1 & 0 \end{array}\right] \end{array}$$

Convert each problem into a maximization problem and then solve each problem using both the dual method and the method of Section 4.4.

25. Minimize $w = 10y_1 + 15y_2$

subject to: $y_1 + y_2 \geq 17$

$5y_1 + 8y_2 \geq 42$

with $y_1 \geq 0, \quad y_2 \geq 0.$

26. Minimize $w = 22y_1 + 44y_2 + 33y_3$

subject to: $y_1 + 2y_2 + y_3 \geq 3$

$y_1 + y_3 \geq 3$

$3y_1 + 2y_2 + 2y_3 \geq 8$

with $y_1 \geq 0, \quad y_2 \geq 0, \quad y_3 \geq 0.$

27. Minimize $w = 7y_1 + 2y_2 + 3y_3$

subject to: $y_1 + y_2 + 2y_3 \geq 48$

$y_1 + y_2 \geq 12$

$y_3 \geq 10$

$3y_1 + y_3 \geq 30$

with $y_1 \geq 0, \quad y_2 \geq 0, \quad y_3 \geq 0.$

28. Minimize $\quad w = 3y_1 + 4y_2 + y_3 + 2y_4$

subject to: $\quad 4y_1 + 6y_2 + 3y_3 + 8y_4 \geq 19$

$\quad\quad\quad\quad 13y_1 + 7y_2 + 2y_3 + 6y_4 \geq 16$

with $\quad\quad y_1 \geq 0, \quad y_2 \geq 0, \quad y_3 \geq 0, \quad y_4 \geq 0.$

Use the simplex method to solve each problem. (You may need to use artificial variables.)

29. Maximize $\quad z = 20x_1 + 30x_2$

subject to: $\quad 5x_1 + 10x_2 \leq 120$

$\quad\quad\quad\quad 10x_1 + 15x_2 \geq 200$

with $\quad\quad x_1 \geq 0, \quad x_2 \geq 0.$

30. Minimize $\quad w = 4y_1 + 2y_2$

subject to: $\quad y_1 + 3y_2 \geq 6$

$\quad\quad\quad\quad 2y_1 + 8y_2 \leq 21$

with $\quad\quad y_1 \geq 0, \quad y_2 \geq 0.$

31. Maximize $\quad z = 10x_1 + 12x_2$

subject to: $\quad 2x_1 + 2x_2 = 17$

$\quad\quad\quad\quad 2x_1 + 5x_2 \geq 22$

$\quad\quad\quad\quad 4x_1 + 3x_2 \leq 28$

with $\quad\quad x_1 \geq 0, \quad x_2 \geq 0.$

32. Minimize $\quad w = 24y_1 + 30y_2 + 36y_3$

subject to: $\quad 5y_1 + 10y_2 + 15y_3 \geq 1200$

$\quad\quad\quad\quad y_1 + \quad y_2 + \quad y_3 \leq \quad 50$

with $\quad\quad y_1 \geq 0, \quad y_2 \geq 0, \quad y_3 \geq 0.$

33. What types of problems can be solved using slack, surplus, and artificial variables?

34. What kind of problems can be solved using the method of duals?

35. In solving a linear programming problem, you are given the following initial tableau.

$$\begin{bmatrix} 4 & 2 & 3 & 1 & 0 & 0 & 9 \\ 5 & 4 & 1 & 0 & 1 & 0 & 10 \\ -6 & -7 & -5 & 0 & 0 & 1 & 0 \end{bmatrix}$$

a. What is the problem being solved?

b. If the 1 in row 1, column 4 were a -1 rather than a 1, how would it change your answer to part a?

c. After several steps of the simplex algorithm, the following tableau results.

$$\begin{bmatrix} 3 & 0 & 5 & 2 & -1 & 0 & 8 \\ 11 & 10 & 0 & -1 & 3 & 0 & 21 \\ 47 & 0 & 0 & 13 & 11 & 10 & 227 \end{bmatrix}$$

What is the solution? (List only the values of the original variables and the objective function. Do not include slack or surplus variables.)

d. What is the dual of the problem you found in part a?

e. What is the solution of the dual you found in part d? (Do not perform any steps of the simplex algorithm; just examine the tableau given in part c.)

36. In Chapter 2 we wrote a system of linear equations using matrix notation. We can do the same thing for the system of linear inequalities in this chapter.

a. Find matrices A, B, C, and X such that the maximization problem in Example 1 of Section 4.1 can be written as

$$\begin{aligned} &\text{Maximize} &&CX \\ &\text{subject to:} &&AX \leq B \\ &\text{with} &&X \geq O. \end{aligned}$$

(*Hint:* Let B and X be column matrices, and C a row matrix.)

b. Show that the dual of the problem in part a can be written as

$$\begin{aligned} &\text{Minimize} &&YB \\ &\text{subject to:} &&YA \geq C \\ &\text{with} &&Y \geq O, \end{aligned}$$

where Y is a row matrix.

c. Show that for any feasible solutions X and Y to the original and dual problems, respectively, $CX \leq YB$. (*Hint:* Multiply both sides of $AX \leq B$ by Y on the left. Then substitute for YA.)

d. For the solution X to the maximization problem and Y to the dual, it can be shown that

$$CX = YB$$

is always true. Verify this for Example 1 of Section 4.1. What is the significance of the value in CX (or YB)?

APPLICATIONS

For Exercises 37–40, **(a) select appropriate variables; (b) write the objective functions; (c) write the constraints as inequalities.**

Business and Economics

37. Production The Bronze Forge produces and ships three different hand-crafted bronze plates: a dogwood-engraved cake plate, a wheat-engraved bread plate, and a lace-engraved dinner plate. Each cake plate requires \$15 in materials, 5 hours of labor, and \$6 to ship. Each bread plate requires \$10 in materials, 4 hours of labor, and \$5 to ship. Each dinner plate requires \$8 in materials, 4 hours of labor, and \$5 to deliver. The profit on the cake plate is \$15, on the bread plate is \$12, and on the dinner plate is \$5. The company has available up to 2700 hours of labor per week. Each week, they can spend at most \$1500 on materials and \$1200 on delivery. How many of each plate should the company produce to maximize their weekly profit? What is their maximum profit?

38. Investments An investor is considering three types of investments: a high-risk venture into oil leases with a potential return of 15%, a medium-risk investment in stocks with a 9% return, and a relatively safe bond investment with a 5% return. He has \$50,000 to invest. Because of the risk, he will limit his investment in oil leases and stocks to 30% and his investment in oil leases and bonds to 50%. How much should he invest in each to maximize his return, assuming investment returns are as expected?

39. Profit The Aged Wood Winery makes two white wines, Fruity and Crystal, from two kinds of grapes and sugar. One gallon of Fruity wine requires 2 bushels of Grape A, 2 bushels of Grape B, and 2 lb of sugar, and produces a profit of $12. One gallon of Crystal wine requires 1 bushel of Grape A, 3 bushels of Grape B, and 1 lb of sugar, and produces a profit of $15. The winery has available 110 bushels of grape A, 125 bushels of grape B, and 90 lb of sugar. How much of each wine should be made to maximize profit?

40. Production Costs Cauchy Canners produces canned whole tomatoes and tomato sauce. This season, the company has available 3,000,000 kg of tomatoes for these two products. To meet the demands of regular customers, it must produce at least 80,000 kg of sauce and 800,000 kg of whole tomatoes. The cost per kilogram is $4 to produce canned whole tomatoes and $3.25 to produce tomato sauce. Labor agreements require that at least 110,000 person-hours be used. Each kilogram can of sauce requires 3 minutes for one worker, and each kilogram can of whole tomatoes requires 6 minutes for one worker. How many kilograms of tomatoes should Cauchy use for each product to minimize cost? (For simplicity, assume production of y_1 kg of canned whole tomatoes and y_2 kg of tomato sauce requires $y_1 + y_2$ kg of tomatoes.)

41. Solve Exercise 37.   **42.** Solve Exercise 38.

43. Solve Exercise 39.   **44.** Solve Exercise 40.

45. Canning Cauchy Canners produces canned corn, beans, and carrots. Demand for vegetables requires it to produce at least 1000 cases per month. Based on past sales, it should produce at least twice as many cases of corn as of beans, and at least 340 cases of carrots. It costs $10 to produce a case of corn, $15 to produce a case of beans, and $25 to produce a case of carrots.

   a. Using the method of surplus variables, find how many cases of each vegetable should be produced to minimize costs. What is the minimum cost?

   b. Using the method of duals, find how many cases of each vegetable should be produced to minimize costs. What is the minimum cost?

   c. Suppose the minimum number of cases that must be produced each month rises to 1100. Use shadow costs to calculate the total cost in this case.

46. Food Cost A store sells two brands of snacks. A package of Sun Hill costs $3 and contains 10 oz of peanuts, 4 oz of raisins, and 2 oz of rolled oats. A package of Bear Valley costs $2 and contains 2 oz of peanuts, 4 oz of raisins, and 8 oz of rolled oats. Suppose you wish to make a mixture that contains at least 20 oz of peanuts, 24 oz of raisins, and 24 oz of rolled oats.

   a. Using the method of surplus variables, find how many packages of each you should buy to minimize the cost. What is the minimum cost?

   b. Using the method of duals, find how many packages of each you should buy to minimize the cost. What is the minimum cost?

   c. Suppose the minimum amount of peanuts is increased to 28. Use shadow costs to calculate the total cost in this case.

   d. Explain why it makes sense that the shadow cost for rolled oats is 0.

Life Sciences

47. Calorie Expenditure Ginger's exercise regimen includes doing tai chi, riding a unicycle, and fencing. She has at most 10 hours per week to devote to these activities. Her fencing partner can work with her at most only 2 hours per week. She wants the total time she does tai chi to be at least twice as long as she unicycles. According to a website, a 130-pound person like Ginger will burn an average of 236 calories per hour doing tai chi, 295 calories per hour riding a unicycle, and 354 calories per hour fencing. *Source: NutriStrategy.com.*

   a. How many hours per week should Ginger spend on each activity to maximize the number of calories she burns? What is the maximum number of calories she will burn?

   b. Show how the solution from part a can be found without using the simplex method by considering the constraints and the number of calories for each activity.

EXTENDED APPLICATION

USING INTEGER PROGRAMMING IN THE STOCK-CUTTING PROBLEM

In Chapter 3 Section 3 we noted that some problems require solutions in integers because the resources to be allocated are items that can't be split into pieces, like cargo containers or airplanes. These *integer programming* problems are generally harder than the linear programming

problems we have been solving by the simplex method, but often linear programming can be combined with other techniques to solve integer problems. Even if the number of variables and constraints is small, some help from software is usually required. We will introduce integer programming with the basic but important *stock-cutting problem*. (To get a feeling for the issues involved, you may want to try the simple stock-cutting problem given in Exercise 1.)

A paper mill produces rolls of paper that are much wider than most customers require, often as wide as 200 in. The mill then cuts these wide rolls into smaller widths to fill orders for paper rolls to

Source: Northwestern University and the Special Interest Group on Cutting and Packing

be used in printing and packaging and other applications. The stock-cutting problem is the following:

> Given a list of roll widths and the number of rolls ordered for each width, decide how to cut the raw rolls that come from the paper-making machine into smaller rolls so as to fill all the orders with a minimum amount of waste.

Another way to state the problem is: What is the minimum number of raw rolls required to fill the orders? This is an integer problem because the customers have ordered whole numbers of rolls, and each roll is cut in a single piece from one of the raw rolls.

As an example, suppose the paper machine produces rolls 100 in. wide. The manufacturer offers rolls in the following six widths: 14 in., 17 in., 31 in., 33 in., 36 in., and 45 in. (We'll call these the standard widths.) The current orders to be filled are as follows:

Rolls of Paper Ordered						
Width in Inches	14	17	31	33	36	45
Number Ordered	100	123	239	121	444	87

The cutting machine can make four simultaneous cuts, so a raw roll can be cut into as many as five pieces. With luck, all five pieces might be usable for filling orders, but there will usually be unusable waste on the end, and we also might end up with more rolls of some standard width than we need. We'll consider both the end pieces that are too narrow and any unused standard-width rolls as waste, and this is the waste we want to minimize.

The first question is, what are the possible cutting patterns? We're restricted to at most five standard rolls from any given raw roll, and we'll elect to use as much as possible in each raw roll so that the waste remaining at the end will always be less than 14 in. So, for example, 14|36|45| is a possible pattern, but 14|14|14|14|14| is not, because it has too many cuts, and 45|36| is not, because more than 14 in. is left at the end. (Each vertical bar represents a cut; if the piece on the end happens to be a standard width, then we don't need a cut after it, since we've reached the end of the roll.) This is already a tricky problem, and variations of it appear in many industrial applications involving packing objects of different sizes into a fixed space (for example, packing crates into a container for shipment overseas). In the Exercises we'll ask you to write down some more possible patterns, but finding all of them is a job for a computer, and it turns out that there are exactly 33 possible cutting patterns. In Chapter 8 you'll learn some counting techniques that might help you write the program to find all possible patterns.

The next question is, what's the best we can do? We have to use an integral number of 100-in. raw rolls, and we can find the total "roll-inches" ordered by multiplying the width of each standard roll by the number ordered for this width. This computation is a natural one for the matrix notation that you have learned. If W and O are 6×1 column matrices, then the total roll inches used is $W^{\mathrm{T}}O$:

$$W = \begin{bmatrix} 14 \\ 17 \\ 31 \\ 33 \\ 36 \\ 45 \end{bmatrix} \qquad O = \begin{bmatrix} 100 \\ 123 \\ 239 \\ 121 \\ 444 \\ 87 \end{bmatrix} \qquad W^{\mathrm{T}}O = 34{,}792$$

Since each raw roll is 100 in., the best we can do is to use 348 rolls with a total width of 34,800. As a percentage of the raw material, the corresponding waste is

$$\frac{8}{34{,}800} \approx 0.02\%,$$

which represents very low waste. Of course, we'll only reach this target if we can lay out the cutting with perfect efficiency.

As we noted previously, these integer programming problems are difficult, but many mathematical analysis and spreadsheet programs have built-in optimization routines that can handle problems of modest size. We submitted this problem to one such program, giving it the lists of orders and widths and a list of the 33 allowable cutting patterns. Figure 10 shows the seven cutting patterns chosen by the minimizer software, with a graphical representation, and the total number of times each pattern was used.

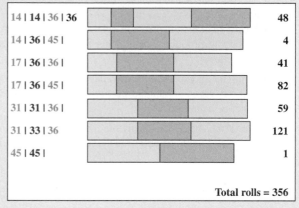

FIGURE 10

With these cutting choices we generate the following numbers of each standard width:

Solution to Stock-cutting Problem						
Width	14	17	31	33	36	45
Quantity Produced	100	123	239	121	444	88
Quantity Ordered	100	123	239	121	444	87

We figured that the minimum possible number of raw rolls was 348, so we have used only 8 more than the minimum. In the Exercises you'll figure the percentage of waste with this cutting plan.

Manufacturers of glass and sheet metal encounter a two-dimensional version of this problem: They need to cut the rectangular pieces that have been ordered from a larger rectangular piece of glass or metal, laying out the ordered sizes so as to minimize waste.

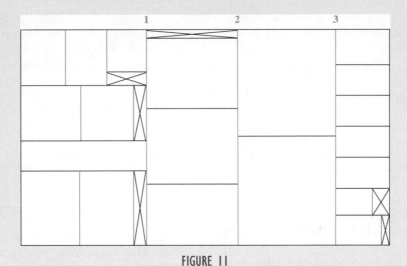

FIGURE 11

Besides the extra dimension, this problem is complicated by another constraint: The typical cutting machine can make only "guillotine cuts" that go completely across the sheet being cut, so a cutting algorithm must usually begin with a few long cuts that cut the original rectangle into strips, followed by crossways cuts that begin to create the order sizes. A typical finished cutting layout might look like Figure 11.

The first cuts would be the three vertical cuts labeled 1, 2, and 3, followed by horizontal cuts in each of the four resulting strips, then vertical cuts in these new rectangles, and so on. Areas of waste are marked with **X**. An additional complication in designing the layout is that any given stock rectangle can be oriented in two different directions (unless it's square), so the packing problem has many alternative solutions.

In three dimensions, a comparable problem is to fill a shipping container with smaller boxes (rectangular prisms) with the minimum wasted space. These packing problems are complicated geometric versions of a basic problem called the *knapsack problem*:

Given n objects with weights $w_1, w_2, \ldots, w_n$ and cash values $v_1, v_2, \ldots, v_n$ and a knapsack that can hold a weight of at most W, choose the objects that will pack in the greatest value. In the Exercises you can try a small example, but as soon as n gets large, this problem "explodes," that is, the number of possibilities becomes too large for a trial-and-error solution, even with a computer to do the bookkeeping. The development of good algorithms for cutting and packing problems is an active research specialty in the field of optimization.

EXERCISES

1. Suppose you plan to build a raised flower bed using landscape timbers, which come in 8-ft lengths. You want the bed's outer dimensions to be 6 ft by 4 ft, and you will use three layers of timbers. The timbers are 6 in. by 6 in. in cross section, so if you make the bottom and top layers with 6-ft lengths on the sides and 3-ft lengths on the ends, and the middle layer with 5-ft lengths on the sides and 4-ft lengths on the ends, you could build the bed out of the following lengths.

Plan A	
Length	Number Needed
3 ft	4
4 ft	2
5 ft	2
6 ft	4

a. What is the smallest number of timbers you can buy to build your bed? How will you lay out the cuts? How much wood will you waste?

b. If you overlap the corners in a different way, you can build the bed with this plan:

Plan B	
Length	Needed
3 ft	2
4 ft	4
5 ft	4
6 ft	2

Does plan B allow you to build the bed with fewer 8-ft timbers?

c. What is the smallest length for the uncut timbers that would allow you to build the bed with no waste?

2. For the list of standard paper roll widths given earlier, write down four more possible cutting patterns that use at most four cuts and leave less than 14 in. of waste on the end. See if you can find ones that aren't in the list of patterns returned by the optimizer.

3. Four of the 33 possible patterns use up the raw roll with no waste, that is, the widths add up to exactly 100 in. Find these four patterns.

4. For the computer solution of the cutting problem, figure out the percent of the 356 rolls used that is wasted.

185

5. In our cutting plan, we elected to use up as much as possible of each 100-in, roll with standard widths. Why might it be a better idea to allow leftover rolls that are *wider* than 14 in.?

6. The following table shows the weights of six objects and their values.

Weight	2	2.5	3	3.5	4	4.5
Value	12	11	7	13	10	11

   If your knapsack holds a maximum weight of 9, what is the highest value you can pack in?

7. Suppose that of the original 33 cutting patterns, we allow only the 7 that occur in the final solution. Let y_1 = the number of copies of the first pattern, and similarly define y_2 through y_7. Write the integer programming problem that we are trying to solve.

8. Try solving the problem from Exercise 7 with the Solver in Microsoft Excel. First try it only requiring that the variables be nonnegative, not that they be integers. Then add constraints in the constraint box that the variables be integers.

9. Compare your two answers from Exercise 8 with each other and with the answer given in Figure 10. Discuss the effect on the minimum value of the requirement that the variables take on integer values. Also discuss what these solutions tell you about uniqueness of this integer programming problem.

10. Go to the website WolframAlpha.com and enter: "Minimize $y_1 + y_2$ with $2y_1 + y_2 \geq 100$, $y_1 \geq 0$, $y_2 \geq 0$". Study the solution provided by Wolfram|Alpha. As of this writing, Wolfram|Alpha cannot solve linear programming problems with more than five variables nor does it provide a way to do integer programming. Investigate whether this is still true. If these restrictions no longer exist, try Exercise 8 with Wolfram|Alpha and discuss how it compares with Microsoft Excel.

DIRECTIONS FOR GROUP PROJECT

Suppose you and three of the students from class have met at your house to study and your father questions each of you on what you are learning in college. While this is happening, your mother is busy planning a new raised-bed garden and your sister is attempting to choose which items she will put in a backpack for a field trip. Using the data in Exercises 1 and 6, prepare a presentation for your family on the value of what you're learning in college.

5

Mathematics of Finance

Buying a car usually requires both some savings for a down payment and a loan for the balance. An exercise in Section 2 calculates the regular deposits that would be needed to save up the full purchase price, and other exercises and examples in this chapter compute the payments required to amortize a loan.

Everybody uses money. Sometimes you work for your money and other times your money works for you. For example, unless you are attending college on a full scholarship, it is very likely that you and your family have either saved money or borrowed money, or both, to pay for your education. When we borrow money, we normally have to pay interest for that privilege. When we save money, for a future purchase or retirement, we are lending money to a financial institution and we expect to earn interest on our investment. We will develop the mathematics in this chapter to understand better the principles of borrowing and saving. These ideas will then be used to compare different financial opportunities and make informed decisions.

5.1 Simple and Compound Interest

APPLY IT

If you can borrow money at 8% interest compounded annually or at 7.9% compounded monthly, which loan would cost less?

In this section we will learn how to compare different interest rates with different compounding periods. The question above will be answered in Example 8.

Simple Interest

Interest on loans of a year or less is frequently calculated as **simple interest**, a type of interest that is charged (or paid) only on the amount borrowed (or invested) and not on past interest. The amount borrowed is called the **principal**. The **rate** of interest is given as a percentage per year, expressed as a decimal. For example, $6\% = 0.06$ and $11\frac{1}{2}\% = 0.115$. The **time** the money is earning interest is calculated in years. One year's interest is calculated by multiplying the principal times the interest rate, or Pr. If the time that the money earns interest is other than one year, we multiply the interest for one year by the number of years, or Prt.

> ### Simple Interest
>
> $$I = Prt$$
>
> where
>
> P is the principal;
>
> r is the annual interest rate;
>
> t is the time in years.

EXAMPLE 1   Simple Interest

To buy furniture for a new apartment, Candace Cooney borrowed $5000 at 8% simple interest for 11 months. How much interest will she pay?

SOLUTION Use the formula $I = Prt$, with $P = 5000$, $r = 0.08$, and $t = 11/12$ (in years). The total interest she will pay is

$$I = 5000(0.08)(11/12) \approx 366.67,$$

or $366.67.

A deposit of P dollars today at a rate of interest r for t years produces interest of $I = Prt$. The interest, added to the original principal P, gives

$$P + Prt = P(1 + rt).$$

This amount is called the *future value* of P dollars at an interest rate r for time t in years. When loans are involved, the future value is often called the *maturity value* of the loan. This idea is summarized as follows.

Future or Maturity Value for Simple Interest

The **future** or **maturity value** A of P dollars at a simple interest rate r for t years is

$$A = P(1 + rt).$$

EXAMPLE 2   Maturity Values

Find the maturity value for each loan at simple interest.

(a) A loan of $2500 to be repaid in 8 months with interest of 4.3%

SOLUTION   The loan is for 8 months, or $8/12 = 2/3$ of a year. The maturity value is

$$A = P(1 + rt)$$

$$A = 2500\left[1 + 0.043\left(\frac{2}{3}\right)\right]$$

$$A \approx 2500(1 + 0.028667) = 2571.67,$$

or $2571.67. (The answer is rounded to the nearest cent, as is customary in financial problems.) Of this maturity value,

$$\$2571.67 - \$2500 = \$71.67$$

represents interest.

(b) A loan of $11,280 for 85 days at 7% interest

SOLUTION   It is common to assume 360 days in a year when working with simple interest. We shall usually make such an assumption in this book. The maturity value in this example is

$$A = 11,280\left[1 + 0.07\left(\frac{85}{360}\right)\right] \approx 11,466.43,$$

or $11,466.43.

YOUR TURN 1 Find the maturity value for a $3000 loan at 5.8% interest for 100 days.

TRY YOUR TURN 1

CAUTION   When using the formula for future value, as well as all other formulas in this chapter, we often neglect the fact that in real life, money amounts are rounded to the nearest penny. As a consequence, when the amounts are rounded, their values may differ by a few cents from the amounts given by these formulas. For instance, in Example 2(a), the interest in each monthly payment would be $2500(0.043/12) \approx$ $8.96, rounded to the nearest penny. After 8 months, the total is 8($8.96) = $71.68, which is 1¢ more than we computed in the example.

In part (b) of Example 2 we assumed 360 days in a year. Historically, to simplify calculations, it was often assumed that each year had twelve 30-day months, making a year 360 days long. Treasury bills sold by the U.S. government assume a 360-day year in calculating interest. Interest found using a 360-day year is called *ordinary interest* and interest found using a 365-day year is called *exact interest*.

The formula for future value has four variables, P, r, t, and A. We can use the formula to find any of the quantities that these variables represent, as illustrated in the next example.

EXAMPLE 3 **Simple Interest**

Theresa Cortesini wants to borrow $8000 from Christine O'Brien. She is willing to pay back $8180 in 6 months. What interest rate will she pay?

SOLUTION Use the formula for future value, with $A = 8180$, $P = 8000$, $t = 6/12 = 0.5$, and solve for r.

$$A = P(1 + rt)$$
$$8180 = 8000(1 + 0.5r)$$
$$8180 = 8000 + 4000r \qquad \text{Distributive property}$$
$$180 = 4000r \qquad \text{Subtract 8000.}$$
$$r = 0.045 \qquad \text{Divide by 4000.}$$

YOUR TURN 2 Find the interest rate if $5000 is borrowed, and $5243.75 is paid back 9 months later.

Thus, the interest rate is 4.5%. **TRY YOUR TURN 2**

When you deposit money in the bank and earn interest, it is as if the bank borrowed the money from you. Reversing the scenario in Example 3, if you put $8000 in a bank account that pays simple interest at a rate of 4.5% annually, you will have accumulated $8180 after 6 months.

Compound Interest

As mentioned earlier, simple interest is normally used for loans or investments of a year or less. For longer periods compound interest is used. With **compound interest**, interest is charged (or paid) on interest as well as on principal. For example, if $1000 is deposited at 5% interest for 1 year, at the end of the year the interest is $1000(0.05)(1) = 50. The balance in the account is $1000 + $50 = 1050. If this amount is left at 5% interest for another year, the interest is calculated on $1050 instead of the original $1000, so the amount in the account at the end of the second year is $1050 + $1050(0.05)(1) = 1102.50. Note that simple interest would produce a total amount of only

$$\$1000[1 + (0.05)(2)] = \$1100.$$

The additional $2.50 is the interest on $50 at 5% for one year.

To find a formula for compound interest, first suppose that P dollars is deposited at a rate of interest r per year. The amount on deposit at the end of the first year is found by the simple interest formula, with $t = 1$.

$$A = P(1 + r \cdot 1) = P(1 + r)$$

If the deposit earns compound interest, the interest earned during the second year is paid on the total amount on deposit at the end of the first year. Using the formula $A = P(1 + rt)$ again, with P replaced by $P(1 + r)$ and $t = 1$, gives the total amount on deposit at the end of the second year.

$$A = [P(1 + r)](1 + r \cdot 1) = P(1 + r)^2$$

In the same way, the total amount on deposit at the end of the third year is

$$P(1 + r)^3.$$

Generalizing, in t years the total amount on deposit is

$$A = P(1 + r)^t,$$

called the **compound amount**.

NOTE Compare this formula for compound interest with the formula for simple interest.

Compound interest	$A = P(1 + r)^t$
Simple interest	$A = P(1 + rt)$

The important distinction between the two formulas is that in the compound interest formula, the number of years, t, is an *exponent*, so that money grows much more rapidly when interest is compounded.

Interest can be compounded more than once per year. Common compounding periods include *semiannually* (two periods per year), *quarterly* (four periods per year), *monthly* (twelve periods per year), or *daily* (usually 365 periods per year). The *interest rate per period*, i, is found by dividing the annual interest rate, r, by the number of compounding periods, m, per year. To find the total number of compounding periods, n, we multiply the number of years, t, by the number of compounding periods per year, m. The following formula can be derived in the same way as the previous formula.

Compound Amount

$$A = P(1 + i)^n$$

where $i = \dfrac{r}{m}$ and $n = mt$,

A is the future (maturity) value;
P is the principal;
r is the annual interest rate;
m is the number of compounding periods per year;
t is the number of years;
n is the number of compounding periods;
i is the interest rate per period.

EXAMPLE 4   Compound Interest

Suppose $1000 is deposited for 6 years in an account paying 4.25% per year compounded annually.

(a) Find the compound amount.

SOLUTION   In the formula for the compound amount, $P = 1000$, $i = 0.0425/1$, and $n = 6(1) = 6$. The compound amount is

$$A = P(1 + i)^n$$
$$A = 1000(1.0425)^6.$$

Using a calculator, we get

$$A \approx \$1283.68,$$

the compound amount.

(b) Find the amount of interest earned.

SOLUTION   Subtract the initial deposit from the compound amount.

$$\text{Amount of interest} = \$1283.68 - \$1000 = \$283.68$$

EXAMPLE 5   Compound Interest

Find the amount of interest earned by a deposit of $2450 for 6.5 years at 5.25% compounded quarterly.

SOLUTION   Interest compounded quarterly is compounded 4 times a year. In 6.5 years, there are $6.5(4) = 26$ periods. Thus, $n = 26$. Interest of 5.25% per year is 5.25%/4 per quarter, so $i = 0.0525/4$. Now use the formula for compound amount.

YOUR TURN 3 Find the amount of interest earned by a deposit of $1600 for 7 years at 4.2% compounded monthly.

$$A = P(1 + i)^n$$
$$A = 2450(1 + 0.0525/4)^{26} \approx 3438.78$$

Rounded to the nearest cent, the compound amount is $3438.78, so the interest is $3438.78 - $2450 = $988.78.   **TRY YOUR TURN 3**

CAUTION As shown in Example 5, compound interest problems involve two rates—the annual rate r and the rate per compounding period i. Be sure you understand the distinction between them. When interest is compounded annually, these rates are the same. In all other cases, $i \neq r$. Similarly, there are two quantities for time: the number of years t and the number of compounding periods n. When interest is compounded annually, these variables have the same value. In all other cases, $n \neq t$.

It is interesting to compare loans at the same rate when simple or compound interest is used. Figure 1 shows the graphs of the simple interest and compound interest formulas with $P = 1000$ at an annual rate of 10% from 0 to 20 years. The future value after 15 years is shown for each graph. After 15 years of compound interest, $1000 grows to $4177.25, whereas with simple interest, it amounts to $2500.00, a difference of $1677.25.

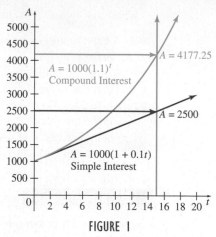

FIGURE 1

TECHNOLOGY NOTE Spreadsheets are ideal for performing financial calculations. Figure 2 shows a Microsoft Excel spreadsheet with the formulas for compound and simple interest used to create columns B and C, respectively, when $1000 is invested at an annual rate of 10%. Compare row 16 with Figure 1. For more details on the use of spreadsheets in the mathematics of finance, see the *Graphing Calculator and Excel Spreadsheet Manual* available with this book.

	A	B	C
1	period	compound	simple
2	1	1100	1100
3	2	1210	1200
4	3	1331	1300
5	4	1464.1	1400
6	5	1610.51	1500
7	6	1771.561	1600
8	7	1948.7171	1700
9	8	2143.58881	1800
10	9	2357.947691	1900
11	10	2593.74246	2000
12	11	2853.116706	2100
13	12	3138.428377	2200
14	13	3452.27124	2300
15	14	3797.498336	2400
16	15	4177.248169	2500
17	16	4594.972986	2600
18	17	5054.470285	2700
19	18	5559.917313	2800
20	19	6115.909045	2900
21	20	6727.499949	3000

FIGURE 2

We can also solve the compound amount formula for the interest rate, as in the following example.

EXAMPLE 6 Compound Interest Rate

Suppose Carol Merrigan invested $5000 in a savings account that paid quarterly interest. After 6 years the money had accumulated to $6539.96. What was the annual interest rate?

SOLUTION Because $m = 4$ and $t = 6$, the number of compounding periods is $n = 4 \times 6 = 24$. Using this value along with $P = 5000$ and $A = 6539.96$ in the formula for compound amount, we have

$$5000(1 + r/4)^{24} = 6539.96$$

$(1 + r/4)^{24} = 1.30799$	Divide both sides by 5000.
$1 + r/4 = 1.30799^{1/24} \approx 1.01125$	Take both sides to the 1/24 power.
$r/4 = 0.01125$	Subtract 1 from both sides.
$r = 0.045$	Multiply both sides by 4.

YOUR TURN 4 Find the annual interest rate if $6500 is worth $8665.69 after being invested for 8 years in an account that compounded interest monthly.

The annual interest rate was 4.5%. **TRY YOUR TURN 4**

Effective Rate
If $1 is deposited at 4% compounded quarterly, a calculator can be used to find that at the end of one year, the compound amount is $1.0406, an increase of 4.06% over the original $1. The actual increase of 4.06% in the money is somewhat higher than the stated increase of 4%. To differentiate between these two numbers, 4% is called the **nominal** or **stated rate** of interest, while 4.06% is called the *effective rate*.* To avoid confusion between stated rates and effective rates, we shall continue to use r for the stated rate and we will use r_E for the effective rate.

EXAMPLE 7 Effective Rate

Find the effective rate corresponding to a stated rate of 6% compounded semiannually.

SOLUTION Here, $i = r/m = 0.06/2 = 0.03$ for $m = 2$ periods. Use a calculator to find that $(1.03)^2 \approx 1.06090$, which shows that $1 will increase to $1.06090, an actual increase of 6.09%. The effective rate is $r_E = 6.09\%$.

Generalizing from this example, the effective rate of interest is given by the following formula.

> **Effective Rate**
>
> The **effective rate** corresponding to a stated rate of interest r compounded m times per year is
>
> $$r_E = \left(1 + \frac{r}{m}\right)^m - 1.$$

EXAMPLE 8 Effective Rate

Joe Vetere needs to borrow money. His neighborhood bank charges 8% interest compounded semiannually. A downtown bank charges 7.9% interest compounded monthly. At which bank will Joe pay the lesser amount of interest?

*When applied to consumer finance, the effective rate is called the annual percentage rate, APR, or annual percentage yield, APY.

APPLY IT

SOLUTION Compare the effective rates.

Neighborhood bank: $\quad r_E = \left(1 + \dfrac{0.08}{2}\right)^2 - 1 = 0.0816 = 8.16\%$

Downtown bank: $\quad r_E = \left(1 + \dfrac{0.079}{12}\right)^{12} - 1 \approx 0.081924 \approx 8.19\%$

YOUR TURN 5 Find the effective rate for an account that pays 2.7% compounded monthly.

The neighborhood bank has the lower effective rate, although it has a higher stated rate.

TRY YOUR TURN 5

Present Value

The formula for compound interest, $A = P(1 + i)^n$, has four variables: A, P, i, and n. Given the values of any three of these variables, the value of the fourth can be found. In particular, if A (the future amount), i, and n are known, then P can be found. Here P is the amount that should be deposited today to produce A dollars in n periods.

EXAMPLE 9 Present Value

Rachel Reeve must pay a lump sum of $6000 in 5 years. What amount deposited today at 6.2% compounded annually will amount to $6000 in 5 years?

SOLUTION Here $A = 6000$, $i = 0.062$, $n = 5$, and P is unknown. Substituting these values into the formula for the compound amount gives

$$6000 = P(1.062)^5$$

$$P = \frac{6000}{(1.062)^5} \approx 4441.49,$$

or $4441.49. If Rachel leaves $4441.49 for 5 years in an account paying 6.2% compounded annually, she will have $6000 when she needs it. To check your work, use the compound interest formula with $P = \$4441.49$, $i = 0.062$, and $n = 5$. You should get $A = \$6000.00$.

As Example 9 shows, $6000 in 5 years is approximately the same as $4441.49 today (if money can be deposited at 6.2% compounded annually). An amount that can be deposited today to yield a given sum in the future is called the *present value* of the future sum. Generalizing from Example 9, by solving $A = P(1 + i)^n$ for P, we get the following formula for present value.

> **Present Value for Compound Interest**
>
> The **present value** of A dollars compounded at an interest rate i per period for n periods is
>
> $$P = \frac{A}{(1 + i)^n} \qquad \text{or} \qquad P = A(1 + i)^{-n}.$$

EXAMPLE 10 Present Value

Find the present value of $16,000 in 9 years if money can be deposited at 6% compounded semiannually.

SOLUTION In 9 years there are $2 \cdot 9 = 18$ semiannual periods. A rate of 6% per year is 3% in each semiannual period. Apply the formula with $A = 16,000$, $i = 0.03$, and $n = 18$.

YOUR TURN 6 Find the present value of $10,000 in 7 years if money can be deposited at 4.25% compounded quarterly.

$$P = \frac{A}{(1 + i)^n} = \frac{16,000}{(1.03)^{18}} \approx 9398.31$$

A deposit of $9398.31 today, at 6% compounded semiannually, will produce a total of $16,000 in 9 years.

TRY YOUR TURN 6

We can solve the compound amount formula for n also, as the following example shows.

EXAMPLE 11 Compounding Time

Suppose the $2450 from Example 5 is deposited at 5.25% compounded quarterly until it reaches at least $10,000. How much time is required?

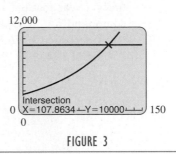 **Method 1**
Graphing Calculator

SOLUTION Graph the functions $y = 2450(1 + 0.0525/4)^x$ and $y = 10,000$ in the same window, and then find the point of intersection. As Figure 3 shows, the functions intersect at $x = 107.8634$. Note, however, that interest is only added to the account every quarter, so we must wait 108 quarters, or $108/4 = 27$ years, for the money to be worth at least $10,000.

FIGURE 3

Method 2
Using Logarithms
(Optional)

The goal is to solve the equation

$$2450(1 + 0.0525/4)^n = 10,000.$$

Divide both sides by 2450, and simplify the expression in parentheses to get

$$1.013125^n = \frac{10,000}{2450}.$$

Now take the logarithm (either base 10 or base e) of both sides to get

$$\log(1.013125^n) = \log(10,000/2450)$$
$$n \log(1.013125) = \log(10,000/2450) \qquad \text{Use logarithm property } \log x^r = r \log x.$$
$$n = \frac{\log(10,000/2450)}{\log(1.013125)} \qquad \text{Divide both sides by } \log(1.013125).$$
$$\approx 107.86.$$

As in Method 1, this means that we must wait 108 quarters, or $108/4 = 27$ years, for the money to be worth at least $10,000.

FOR REVIEW

For a review of logarithmic functions, please refer to Appendix B if you are using *Finite Mathematics*, or to Section 10.5 if you are using *Finite Mathematics and Calculus with Applications*. The only property of logarithms that is needed to find the compounding time is $\log x^r = r \log x$. Logarithms may be used in base 10, using the LOG button on a calculator, or in base e, using the LN button.

EXAMPLE 12 Price Doubling

Suppose the general level of inflation in the economy averages 8% per year. Find the number of years it would take for the overall level of prices to double.

SOLUTION To find the number of years it will take for $1 worth of goods or services to cost $2, find n in the equation

$$2 = 1(1 + 0.08)^n,$$

where $A = 2$, $P = 1$, and $i = 0.08$. This equation simplifies to

$$2 = (1.08)^n.$$

Solving this equation using either a graphing calculator or logarithms, as in Example 11, shows that $n = 9.00647$. Thus, the overall level of prices will double in about 9 years.

TRY YOUR TURN 7

YOUR TURN 7 Find the time needed for $3800 deposited at 3.5% compounded semiannually to be worth at least $7000.

You can quickly estimate how long it takes a sum of money to double, when compounded annually, by using either the **rule of 70** or the **rule of 72**. The rule of 70 (used for small rates of growth) says that for $0.001 \leq r < 0.05$, the value of $70/100r$ gives a good approximation of the doubling time. The rule of 72 (used for larger rates of growth) says that for $0.05 \leq r \leq 0.12$, the value of $72/100r$ approximates the doubling time well. In Example 12, the inflation rate is 8%, so the doubling time is approximately $72/8 = 9$ years.*

Continuous Compounding
Suppose that a bank, in order to attract more business, offers to not just compound interest every quarter, or every month, or every day, or even every hour, but constantly? This type of compound interest, in which the number of times a year that the interest is compounded becomes infinite, is known as **continuous compounding.** To see how it works, look back at Example 5, where we found that $2450, when deposited for 6.5 years at 5.25% compounded quarterly, resulted in a compound amount of $3438.78. We can find the compound amount if we compound more often by putting different values of n in the formula $A = 2450(1 + 0.0525/n)^{6.5n}$, as shown in the following table.

Compounding n Times Annually		
n	Type of Compounding	Compound Amount
4	quarterly	$3438.78
12	monthly	$3443.86
360	daily	$3446.34
8640	every hour	$3446.42

Notice that as n becomes larger, the compound amount also becomes larger, but by a smaller and smaller amount. In this example, increasing the number of compounding periods a year from 360 to 8640 only earns 8¢ more. It is shown in calculus that as n becomes infinitely large, $P(1 + r/n)^{nt}$ gets closer and closer to Pe^{rt}, where e is a very important irrational number whose approximate value is 2.718281828. To calculate interest with continuous compounding, use the e^x button on your calculator. You will learn more about the number e if you study calculus, where e plays as important a role as π does in geometry.

Continuous Compounding

If a deposit of P dollars is invested at a rate of interest r compounded continuously for t years, the compound amount is

$$A = Pe^{rt} \text{ dollars.}$$

EXAMPLE 13 Continuous Compounding

Suppose that $2450 is deposited at 5.25% compounded continuously.

(a) Find the compound amount and the interest earned after 6.5 years.

> **SOLUTION** Using the formula for continuous compounding with $P = 2450$, $r = 0.0525$, and $t = 6.5$, the compound amount is
>
> $$A = 2450e^{0.0525(6.5)} \approx 3446.43.$$

The compound amount is $3446.43, which is just a penny more than if it had been compounded hourly, or 9¢ more than daily compounding. Because it makes so little difference, continuous compounding has dropped in popularity in recent years. The interest in this case is $3446.43 - 2450 = 996.43, or $7.65 more than if it were compounded quarterly, as in Example 5.

*To see where the rule of 70 and the rule of 72 come from, see the section on Taylor Series in *Calculus with Applications* by Margaret L. Lial, Raymond N. Greenwell, and Nathan P. Ritchey, Pearson, 2012.

(b) Find the effective rate.

SOLUTION As in Example 7, the effective rate is just the amount of interest that $1 would earn in one year, or

$$e^{0.0525} - 1 \approx 0.0539,$$

or 5.39%. In general, the effective rate for interest compounded continuously at a rate r is just $e^r - 1$.

(c) Find the time required for the original $2450 to grow to $10,000.

SOLUTION Similar to the process in Example 11, we must solve the equation

$$10,000 = 2450e^{0.0525t}.$$

Divide both sides by 2450, and solve the resulting equation as in Example 11, either by taking logarithms of both sides or by using a graphing calculator to find the intersection of the graphs of $y = 2450e^{0.0525t}$ and $y = 10,000$. If you use logarithms, you can take advantage of the fact that $\ln(e^x) = x$, where $\ln x$ represents the logarithm in base e. In either case, the answer is 26.79 years.

Notice that unlike in Example 11, you don't need to wait until the next compounding period to reach this amount, because interest is being added to the account continuously.

TRY YOUR TURN 8 ◼

YOUR TURN 8 Find the interest earned on $5000 deposited at 3.8% compounded continuously for 9 years.

At this point, it seems helpful to summarize the notation and the most important formulas for simple and compound interest. We use the following variables.

P = principal or present value

A = future or maturity value

r = annual (stated or nominal) interest rate

t = number of years

m = number of compounding periods per year

i = interest rate per period $\quad i = r/m$

n = total number of compounding periods $\quad n = tm$

r_E = effective rate

Simple Interest	Compound Interest	Continuous Compounding
$A = P(1 + rt)$	$A = P(1 + i)^n$	$A = Pe^{rt}$
$P = \dfrac{A}{1 + rt}$	$P = \dfrac{A}{(1 + i)^n} = A(1 + i)^{-n}$	$P = Ae^{-rt}$
	$r_E = \left(1 + \dfrac{r}{m}\right)^m - 1$	$r_E = e^r - 1$

5.1 EXERCISES

1. What factors determine the amount of interest earned on a fixed principal?

2. In your own words, describe the *maturity value* of a loan.

3. What is meant by the *present value* of money?

4. We calculated the loan in Example 2(b) assuming 360 days in a year. Find the maturity value using 365 days in a year. Which is more advantageous to the borrower?

Find the simple interest.

5. $25,000 at 3% for 9 months

6. $4289 at 4.5% for 35 weeks

7. $1974 at 6.3% for 25 weeks

8. $6125 at 1.25% for 6 months

Find the simple interest. Assume a 360-day year.

9. $8192.17 at 3.1% for 72 days

10. $7236.15 at 4.25% for 30 days

Find the maturity value and the amount of simple interest earned.

11. $3125 at 2.85% for 7 months

12. $12,000 at 5.3% for 11 months

13. If $1500 earned simple interest of $56.25 in 6 months, what was the simple interest rate?

14. If $23,500 earned simple interest of $1057.50 in 9 months, what was the simple interest rate?

15. Explain the difference between simple interest and compound interest.

16. What is the difference between r and i?

17. What is the difference between t and n?

18. In Figure 1, one line is straight and the other is curved. Explain why this is, and which represents each type of interest.

Find the compound amount for each deposit and the amount of interest earned.

19. $1000 at 6% compounded annually for 8 years

20. $1000 at 4.5% compounded annually for 6 years

21. $470 at 5.4% compounded semiannually for 12 years

22. $15,000 at 6% compounded monthly for 10 years

23. $8500 at 8% compounded quarterly for 5 years

24. $9100 at 6.4% compounded quarterly for 9 years

Find the interest rate for each deposit and compound amount.

25. $8000 accumulating to $11,672.12, compounded quarterly for 8 years

26. $12,500 accumulating to $20,077.43, compounded quarterly for 9 years

27. $4500 accumulating to $5994.79, compounded monthly for 5 years

28. $6725 accumulating to $10,353.47, compounded monthly for 7 years

Find the effective rate corresponding to each nominal rate.

29. 4% compounded quarterly

30. 6% compounded quarterly

31. 7.25% compounded semiannually

32. 6.25% compounded semiannually

Find the present value (the amount that should be invested now to accumulate the following amount) if the money is compounded as indicated.

33. $12,820.77 at 4.8% compounded annually for 6 years

34. $36,527.13 at 5.3% compounded annually for 10 years

35. $2000 at 6% compounded semiannually for 8 years

36. $2000 at 7% compounded semiannually for 8 years

37. $8800 at 5% compounded quarterly for 5 years

38. $7500 at 5.5% compounded quarterly for 9 years

39. How do the nominal or stated interest rate and the effective interest rate differ?

40. If interest is compounded more than once per year, which rate is higher, the stated rate or the effective rate?

Using either logarithms or a graphing calculator, find the time required for each initial amount to be at least equal to the final amount.

41. $5000, deposited at 4% compounded quarterly, to reach at least $9000

42. $8000, deposited at 3% compounded quarterly, to reach at least $23,000

43. $4500, deposited at 3.6% compounded monthly, to reach at least $11,000

44. $6800, deposited at 5.4% compounded monthly, to reach at least $15,000

Find the doubling time for each of the following levels of inflation using (a) logarithms or a graphing calculator, and (b) the rule of 70 or 72, whichever is appropriate.

45. 3.3%   46. 6.25%

For each of the following amounts at the given interest rate compounded continuously, find (a) the future value after 9 years, (b) the effective rate, and (c) the time to reach $10,000.

47. $5500 at 3.1%   48. $4700 at 4.65%

APPLICATIONS

Business and Economics

49. **Loan Repayment** Tanya Kerchner borrowed $7200 from her father to buy a used car. She repaid him after 9 months, at an annual interest rate of 6.2%. Find the total amount she repaid. How much of this amount is interest?

50. **Delinquent Taxes** An accountant for a corporation forgot to pay the firm's income tax of $321,812.85 on time. The government charged a penalty based on an annual interest rate of 13.4% for the 29 days the money was late. Find the total amount (tax and penalty) that was paid. (Use a 365-day year.)

51. **Savings** A $1500 certificate of deposit held for 75 days was worth $1521.25. To the nearest tenth of a percent, what interest rate was earned? Assume a 360-day year.

52. **Bond Interest** A bond with a face value of $10,000 in 10 years can be purchased now for $5988.02. What is the simple interest rate?

53. **Stock Growth** A stock that sold for $22 at the beginning of the year was selling for $24 at the end of the year. If the stock paid a dividend of $0.50 per share, what is the simple interest rate on an investment in this stock? (*Hint:* Consider the interest to be the increase in value plus the dividend.)

 54. **Wealth** A 1997 article in *The New York Times* discussed how long it would take for Bill Gates, the world's second richest person at the time (behind the Sultan of Brunei), to become the world's first trillionaire. His birthday is October 28, 1955, and on

July 16, 1997, he was worth $42 billion. (*Note:* A trillion dollars is 1000 billion dollars.) *Source: The New York Times.*

a. Assume that Bill Gates's fortune grows at an annual rate of 58%, the historical growth rate through 1997 of Microsoft stock, which made up most of his wealth in 1997. Find the age at which he becomes a trillionaire. (*Hint:* Use the formula for interest compounded annually, $A = P(1 + i)^n$, with $P = 42$. Graph the future value as a function of n on a graphing calculator, and find where the graph crosses the line $y = 1000$.)

b. Repeat part a using 10.5% growth, the average return on all stocks since 1926. *Source: CNN.*

c. What rate of growth would be necessary for Bill Gates to become a trillionaire by the time he is eligible for Social Security on January 1, 2022, after he has turned 66?

d. *Forbes* magazine's listings of billionaires for 2006 and 2010 have given Bill Gates's worth as roughly $50.0 billion and $53.0 billion, respectively. What was the rate of growth of his wealth between 2006 and 2010? *Source: Forbes.*

55. Student Loan Upon graduation from college, Kelly was able to defer payment on his $40,000 subsidized Stafford student loan for 6 months. Since the interest will no longer be paid on his behalf, it will be added to the principal until payments begin. If the interest is 6.54% compounded monthly, what will the principal amount be when he must begin repaying his loan? *Source: SallieMae.*

56. Comparing Investments Two partners agree to invest equal amounts in their business. One will contribute $10,000 immediately. The other plans to contribute an equivalent amount in 3 years, when she expects to acquire a large sum of money. How much should she contribute at that time to match her partner's investment now, assuming an interest rate of 6% compounded semiannually?

57. Comparing Investments As the prize in a contest, you are offered $1000 now or $1210 in 5 years. If money can be invested at 6% compounded annually, which is larger?

58. Retirement Savings The pie graph below shows the percent of baby boomers aged 46–49 who said they had investments with a total value as shown in each category. *Source: The New York Times.*

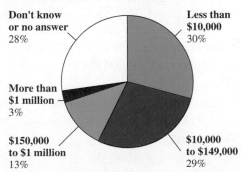

Figures add to more than 100% because of rounding.

Note that 30% have saved less than $10,000. Assume the money is invested at an average rate of 8% compounded quarterly. What will the top numbers in each category amount to in 20 years, when this age group will be ready for retirement?

Negative Interest Under certain conditions, Swiss banks pay negative interest: they charge you. (You didn't think all that secrecy was free?) Suppose a bank "pays" -2.4% interest compounded annually. Find the compound amount for a deposit of $150,000 after each period.

59. 4 years

60. 8 years

61. Investment In the New Testament, Jesus commends a widow who contributed 2 mites to the temple treasury (Mark 12: 42–44). A mite was worth roughly 1/8 of a cent. Suppose the temple invested those 2 mites at 4% interest compounded quarterly. How much would the money be worth 2000 years later?

62. Investments Eric Cobbe borrowed $5200 from his friend Frank Cronin to pay for remodeling work on his house. He repaid the loan 10 months later with simple interest at 7%. Frank then invested the proceeds in a 5-year certificate of deposit paying 6.3% compounded quarterly. How much will he have at the end of 5 years? (*Hint:* You need to use both simple and compound interest.)

63. Investments Suppose $10,000 is invested at an annual rate of 5% for 10 years. Find the future value if interest is compounded as follows.

a. Annually   **b.** Quarterly   **c.** Monthly

d. Daily (365 days)   **e.** Continuously

64. Investments In Exercise 63, notice that as the money is compounded more often, the compound amount becomes larger and larger. Is it possible to compound often enough so that the compound amount is $17,000 after 10 years? Explain.

The following exercise is from an actuarial examination. *Source: The Society of Actuaries.*

65. Savings On January 1, 2000, Jack deposited $1000 into bank X to earn interest at a rate of j per annum compounded semiannually. On January 1, 2005, he transferred his account to bank Y to earn interest at the rate of k per annum compounded quarterly. On January 1, 2008, the balance of bank Y is $1990.76. If Jack could have earned interest at the rate of k per annum compounded quarterly from January 1, 2000, through January 1, 2008, his balance would have been $2203.76. Calculate the ratio k/j.

66. Interest Rate In 1995, O. G. McClain of Houston, Texas, mailed a $100 check to a descendant of Texas independence hero Sam Houston to repay a $100 debt of McClain's great-great-grandfather, who died in 1835, to Sam Houston. A bank estimated the interest on the loan to be $420 million for the 160 years it was due. Find the interest rate the bank was using, assuming interest is compounded annually. *Source: The New York Times.*

67. Comparing CD Rates Marine Bank offered the following CD (Certificates of Deposit) rates. The rates are annual percentage yields, or effective rates, which are higher than the corresponding nominal rates. Assume quarterly compounding. Solve for r to approximate the corresponding nominal rates to the nearest hundredth. *Source: Marine Bank.*

Term	6 mo	Special! 9 mo	1 yr	2 yr	3 yr
APY%	2.50	5.10	4.25	4.50	5.25

68. Effective Rate A Web site for E*TRADE Financial claims that they have "one of the highest yields in the nation" on a 6-month CD. The stated yield was 5.46%; the actual rate was not stated. Assuming monthly compounding, find the actual rate. *Source: E*TRADE.*

69. Effective Rate On August 18, 2006, Centennial Bank of Fountain Valley, California, paid 5.5% interest, compounded monthly, on a 1-year CD, while First Source Bank of South Bend, Indiana, paid 5.63% compounded annually. What are the effective rates for the two CDs, and which bank pays a higher effective rate? *Source: Bankrate.com.*

70. Savings A department has ordered 8 new Dell computers at a cost of $2309 each. The order will not be delivered for 6 months. What amount could the department deposit in a special 6-month CD paying 4.79% compounded monthly to have enough to pay for the machines at time of delivery?

71. Buying a House Steve May wants to have $30,000 available in 5 years for a down payment on a house. He has inherited $25,000. How much of the inheritance should he invest now to accumulate $30,000, if he can get an interest rate of 5.5% compounded quarterly?

72. Rule of 70 On the day of their first grandchild's birth, a new set of grandparents invested $10,000 in a trust fund earning 4.5% compounded monthly.

  a. Use the rule of 70 to estimate how old the grandchild will be when the trust fund is worth $20,000.

  b. Use your answer to part a to determine the actual amount that will be in the trust fund at that time. How close was your estimate in part a?

Doubling Time Use the ideas from Example 12 to find the time it would take for the general level of prices in the economy to double at each average annual inflation rate.

73. 4%   **74.** 5%

75. Doubling Time The consumption of electricity has increased historically at 6% per year. If it continues to increase at this rate indefinitely, find the number of years before the electric utilities will need to double their generating capacity.

76. Doubling Time Suppose a conservation campaign coupled with higher rates causes the demand for electricity to increase at only 2% per year, as it has recently. Find the number of years before the utilities will need to double generating capacity.

77. Mitt Romney According to *The New York Times*, "During the fourteen years [Mitt Romney] ran it, Bain Capital's investments reportedly earned an annual rate of return of over 100 percent, potentially turning an initial investment of $1 million into more than $14 million by the time he left in 1998." *Source: The New York Times.*

  a. What rate of return, compounded annually, would turn $1 million into $14 million by 1998?

  b. The actual rate of return of Bain Capital during the 14 years that Romney ran it was 113%. *Source: The American.* How much would $1 million, compounded annually at this rate, be worth after 14 years?

YOUR TURN ANSWERS

1. $3048.33	**2.** 6.5%	**3.** $545.75
4. 3.6%	**5.** 2.73%	**6.** $7438.39
7. 18 years	**8.** $2038.80	

5.2 Future Value of an Annuity

APPLY IT If you deposit $1500 each year for 6 years in an account paying 8% interest compounded annually, how much will be in your account at the end of this period?

In this section and the next, we develop future value and present value formulas for such periodic payments. To develop these formulas, we must first discuss sequences.

Geometric Sequences

If a and r are nonzero real numbers, the infinite list of numbers $a, ar, ar^2, ar^3, ar^4, \ldots, ar^n, \ldots$ is called a **geometric sequence**. For example, if $a = 3$ and $r = -2$, we have the sequence

$$3, 3(-2), 3(-2)^2, 3(-2)^3, 3(-2)^4, \ldots,$$

or

$$3, -6, 12, -24, 48, \ldots.$$

In the sequence $a, ar, ar^2, ar^3, ar^4, \ldots$, the number a is called the **first term** of the sequence, ar is the **second term**, ar^2 is the **third term**, and so on. Thus, for any $n \geq 1$,

$$ar^{n-1} \text{ is the } n\text{th term of the sequence.}$$

Each term in the sequence is r times the preceding term. The number r is called the **common ratio** of the sequence.

| CAUTION | Do not confuse r, the ratio of two successive terms in a geometric series, with r, the annual interest rate. Different letters might have been helpful, but the usage in both cases is almost universal. |

EXAMPLE 1   Geometric Sequence

Find the seventh term of the geometric sequence $5, 20, 80, 320, \ldots$.

SOLUTION   The first term in the sequence is 5, so $a = 5$. The common ratio, found by dividing the second term by the first, is $r = 20/5 = 4$. We want the seventh term, so $n = 7$. Use ar^{n-1}, with $a = 5, r = 4$, and $n = 7$.

$$ar^{n-1} = (5)(4)^{7-1} = 5(4)^6 = 20{,}480 \qquad \blacksquare$$

Next, we need to find the sum S_n of the first n terms of a geometric sequence, where

$$S_n = a + ar + ar^2 + ar^3 + ar^4 + \cdots + ar^{n-1}. \tag{1}$$

If $r = 1$, then

$$\underbrace{S_n = a + a + a + a + \cdots + a}_{n \text{ terms}} = na.$$

If $r \neq 1$, multiply both sides of equation (1) by r to get

$$rS_n = ar + ar^2 + ar^3 + ar^4 + \cdots + ar^n. \tag{2}$$

Now subtract corresponding sides of equation (1) from equation (2).

$$
\begin{aligned}
rS_n &= ar + ar^2 + ar^3 + ar^4 + \cdots + ar^{n-1} + ar^n \\
-S_n &= -(a + ar + ar^2 + ar^3 + ar^4 + \cdots + ar^{n-1}) \\
\hline
rS_n - S_n &= -a + ar^n \\
S_n(r - 1) &= a(r^n - 1) \qquad \text{Factor.} \\
S_n &= \frac{a(r^n - 1)}{r - 1} \qquad \text{Divide both sides by } r - 1.
\end{aligned}
$$

This result is summarized below.

Sum of Terms

If a geometric sequence has first term a and common ratio r, then the sum S_n of the first n terms is given by

$$S_n = \frac{a(r^n - 1)}{r - 1}, \quad r \neq 1.$$

EXAMPLE 2   Sum of a Geometric Sequence

Find the sum of the first six terms of the geometric sequence $3, 12, 48, \ldots$.

SOLUTION   Here $a = 3, r = 4$, and $n = 6$. Find S_6 by the formula above.

$$
\begin{aligned}
S_6 &= \frac{3(4^6 - 1)}{4 - 1} \qquad n = 6, a = 3, r = 4. \\
&= \frac{3(4096 - 1)}{3} \\
&= 4095 \qquad \qquad \text{TRY YOUR TURN 1} \; \blacksquare
\end{aligned}
$$

YOUR TURN 1   Find the sum of the first 9 terms of the geometric series $4, 12, 36, \ldots$.

Ordinary Annuities

A sequence of equal payments made at equal periods of time is called an **annuity**. If the payments are made at the end of the time period, and if the frequency of payments is the same as the frequency of compounding, the annuity is called an **ordinary annuity**. The time between payments is the **payment period**, and the time from the beginning of the first payment period to the end of the last period is called the **term** of the annuity. The **future value of the annuity**, the final sum on deposit, is defined as the sum of the compound amounts of all the payments, compounded to the end of the term.

Two common uses of annuities are to accumulate funds for some goal or to withdraw funds from an account. For example, an annuity may be used to save money for a large purchase, such as an automobile, a college education, or a down payment on a home. An annuity also may be used to provide monthly payments for retirement. We explore these options in this and the next section.

For example, suppose $1500 is deposited at the end of each year for the next 6 years in an account paying 8% per year compounded annually. Figure 4 shows this annuity. To find the future value of the annuity, look separately at each of the $1500 payments. The first of these payments will produce a compound amount of

$$1500(1 + 0.08)^5 = 1500(1.08)^5.$$

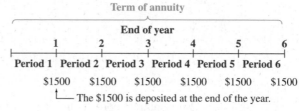

FIGURE 4

Use 5 as the exponent instead of 6, since the money is deposited at the *end* of the first year and earns interest for only 5 years. The second payment of $1500 will produce a compound amount of $1500(1.08)^4$. As shown in Figure 5, the future value of the annuity is

$$1500(1.08)^5 + 1500(1.08)^4 + 1500(1.08)^3 + 1500(1.08)^2$$
$$+ 1500(1.08)^1 + 1500.$$

(The last payment earns no interest at all.)

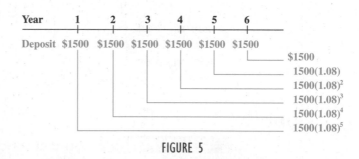

FIGURE 5

Reading this sum in reverse order, we see that it is the sum of the first six terms of a geometric sequence, with $a = 1500$, $r = 1.08$, and $n = 6$. Thus, the sum equals

$$\frac{a(r^n - 1)}{r - 1} = \frac{1500[(1.08)^6 - 1]}{1.08 - 1} \approx \$11{,}003.89.$$

To generalize this result, suppose that payments of R dollars each are deposited into an account at the end of each period for n periods, at a rate of interest i per period. The first payment of R dollars will produce a compound amount of $R(1 + i)^{n-1}$ dollars, the second payment will produce $R(1 + i)^{n-2}$ dollars, and so on; the final payment earns no interest and contributes just R dollars to the total. If S represents the future value (or sum) of the annuity, then (as shown in Figure 6),

$$S = R(1 + i)^{n-1} + R(1 + i)^{n-2} + R(1 + i)^{n-3} + \cdots + R(1 + i) + R,$$

or, written in reverse order,

$$S = R + R(1 + i)^1 + R(1 + i)^2 + \cdots + R(1 + i)^{n-1}.$$

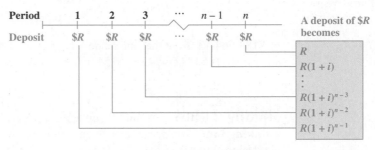

FIGURE 6

This result is the sum of the first n terms of the geometric sequence having first term R and common ratio $1 + i$. Using the formula for the sum of the first n terms of a geometric sequence,

$$S = \frac{R[(1 + i)^n - 1]}{(1 + i) - 1} = \frac{R[(1 + i)^n - 1]}{i} = R\left[\frac{(1 + i)^n - 1}{i}\right].$$

The quantity in brackets is commonly written $s_{\overline{n}|i}$ (read "s-angle-n at i"), so that

$$S = R \cdot s_{\overline{n}|i}.$$

Values of $s_{\overline{n}|i}$ can be found with a calculator.

A formula for the future value of an annuity S of n payments of R dollars each at the end of each consecutive interest period, with interest compounded at a rate i per period, follows.* Recall that this type of annuity, with payments at the *end* of each time period, is called an ordinary annuity.

Future Value of an Ordinary Annuity

$$S = R\left[\frac{(1 + i)^n - 1}{i}\right] \qquad \text{or} \qquad S = Rs_{\overline{n}|i}$$

where

S is the future value;

R is the periodic payment;

i is the interest rate per period;

n is the number of periods.

TECHNOLOGY NOTE   A calculator will be very helpful in computations with annuities. The TI-84 Plus graphing calculator has a special FINANCE menu that is designed to give any desired result after entering the basic information. If your calculator does not have this feature, many calculators can easily be programmed to evaluate the formulas introduced in this section and the next. We include these programs in the *Graphing Calculator and Excel Spreadsheet Manual* available for this text.

*We use S for the future value here, instead of A as in the compound interest formula, to help avoid confusing the two formulas.

EXAMPLE 3 Ordinary Annuity

Bethany Ward is an athlete who believes that her playing career will last 7 years. To prepare for her future, she deposits $22,000 at the end of each year for 7 years in an account paying 6% compounded annually. How much will she have on deposit after 7 years?

SOLUTION Her payments form an ordinary annuity, with $R = 22,000$, $n = 7$, and $i = 0.06$. Using the formula for future value of an annuity,

$$S = 22,000\left[\frac{(1.06)^7 - 1}{0.06}\right] \approx 184,664.43,$$

YOUR TURN 2 Find the accumulated amount after 11 years if $250 is deposited every month in an account paying 3.3% interest compounded monthly.

or $184,664.43. Note that she made 7 payments of $22,000, or $154,000. The interest that she earned is $184,664.43 - $154,000 = $30,664.43.   **TRY YOUR TURN 2**

Sinking Funds
A fund set up to receive periodic payments as in Example 3 is called a **sinking fund**. The periodic payments, together with the interest earned by the payments, are designed to produce a given sum at some time in the future. For example, a sinking fund might be set up to receive money that will be needed to pay off the principal on a loan at some future time. If the payments are all the same amount and are made at the end of a regular time period, they form an ordinary annuity.

EXAMPLE 4 Sinking Fund

Experts say that the baby boom generation (Americans born between 1946 and 1960) cannot count on a company pension or Social Security to provide a comfortable retirement, as their parents did. It is recommended that they start to save early and regularly. Nancy Hart, a baby boomer, has decided to deposit $200 each month for 20 years in an account that pays interest of 7.2% compounded monthly.

(a) How much will be in the account at the end of 20 years?

SOLUTION This savings plan is an annuity with $R = 200$, $i = 0.072/12$, and $n = 12(20)$. The future value is

$$S = 200\left[\frac{(1 + (0.072/12))^{12(20)} - 1}{0.072/12}\right] \approx 106,752.47,$$

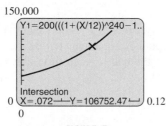

FIGURE 7

or $106,752.47. Figure 7 shows a calculator graph of the function

$$S = 200\left[\frac{(1 + (x/12))^{12(20)} - 1}{x/12}\right]$$

where r, the annual interest rate, is designated x. The value of the function at $x = 0.072$, shown at the bottom of the window, agrees with our result above.

(b) Nancy believes she needs to accumulate $130,000 in the 20-year period to have enough for retirement. What interest rate would provide that amount?

SOLUTION

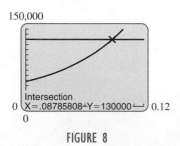

FIGURE 8

Method I
Graphing Calculator

One way to answer this question is to solve the equation for S in terms of x with $S = 130,000$. This is a difficult equation to solve. Although trial and error could be used, it would be easier to use the graphing calculator graph in Figure 7. Adding the line $y = 130,000$ to the graph and then using the capability of the calculator to find the intersection point with the curve shows the annual interest rate must be at least 8.79% to the nearest hundredth. See Figure 8.

Method 2
TVM Solver

Using the TVM Solver under the FINANCE menu on the TI-84 Plus calculator, enter 240 for N (the number of periods), 0 for PV (present value), −200 for PMT (negative because the money is being paid out), 130000 for FV (future value), and 12 for P/Y (payments per year). Put the cursor next to I% (payment) and press SOLVE. The result, shown in Figure 9, indicates that an interest rate of 8.79% is needed.

```
N=240
▪I%=8.785807706
PV=0▪
PMT=-200
FV=130000
P/Y=12
C/Y=12
PMT: END BEGIN
```

FIGURE 9

In Example 4 we used sinking fund calculations to determine the amount of money that accumulates over time through monthly payments and interest. We can also use this formula to determine the amount of money necessary to periodically invest at a given interest rate to reach a particular goal. Start with the annuity formula

$$S = R\left[\frac{(1 + i)^n - 1}{i}\right],$$

and multiply both sides by $i/[(1 + i)^n - 1]$ to derive the following formula.

Sinking Fund Payment

$$R = \frac{Si}{(1 + i)^n - 1} \quad \text{or} \quad R = \frac{S}{s_{\overline{n}|i}}$$

where

R is the periodic payment;

S is the future value;

i is the interest rate per period;

n is the number of periods.

EXAMPLE 5 **Sinking Fund Payment**

Suppose Nancy, in Example 4, cannot get the higher interest rate to produce $130,000 in 20 years. To meet that goal, she must increase her monthly payment. What payment should she make each month?

SOLUTION Nancy's goal is to accumulate $130,000 in 20 years at 7.2% compounded monthly. Therefore, the future value is $S = 130,000$, the monthly interest rate is $i = 0.072/12$, and the number of periods is $n = 12(20)$. Use the sinking fund payment formula to find the payment R.

$$R = \frac{(130,000)(0.072/12)}{(1 + (0.072/12))^{12(20)} - 1} \approx 243.5540887$$

Nancy will need payments of $243.56 each month for 20 years to accumulate at least $130,000. Notice that $243.55 is not quite enough, so round up here. Figure 10(a) shows the point of intersection of the graphs of

$$Y_1 = X\left[\frac{(1 + 0.072/12)^{12(20)} - 1}{0.072/12}\right]$$

YOUR TURN 3 Find the quarterly payment needed to produce $13,500 in 14 years at 3.75% interest compounded quarterly.

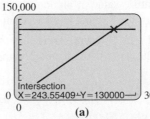

(a)

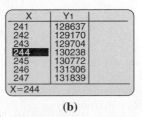

(b)

FIGURE 10

and $Y_2 = 130{,}000$. The result agrees with the answer we found analytically. The table shown in Figure 10(b) confirms that the payment should be between $243 and $244.

TRY YOUR TURN 3

TECHNOLOGY NOTE

We can also use a graphing calculator or spreadsheet to make a table of the amount in a sinking fund. In the formula for future value of an annuity, simply let n be a variable with values from 1 to the total number of payments. Figure 11(a) shows the beginning of such a table generated on a TI-84 Plus for Example 5. Figure 11(b) shows the beginning of the same table using Microsoft Excel.

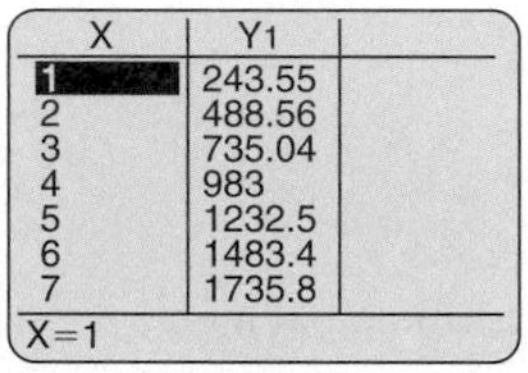

n	Amount in Fund
1	243.55
2	488.56
3	735.04
4	983.00
5	1232.45
6	1483.40
7	1735.85
8	1989.81
9	2245.30
10	2502.32
11	2760.89
12	3021.00

X	Y1
1	243.55
2	488.56
3	735.04
4	983
5	1232.5
6	1483.4
7	1735.8
X=1	

(a)   (b)

FIGURE 11

Annuities Due

The formula developed above is for *ordinary annuities*—those with payments made at the *end* of each time period. These results can be modified slightly to apply to **annuities due**—annuities in which payments are made at the *beginning* of each time period. To find the future value of an annuity due, treat each payment as if it were made at the *end* of the *preceding* period. That is, find $s_{\overline{n}|i}$ for *one additional period*; to compensate for this, subtract the amount of one payment.

Thus, the **future value of an annuity due** of n payments of R dollars each at the beginning of consecutive interest periods, with interest compounded at the rate of i per period, is

$$S = R\left[\frac{(1+i)^{n+1} - 1}{i}\right] - R \quad \text{or} \quad S = Rs_{\overline{n+1}|i} - R.$$

TECHNOLOGY NOTE

The finance feature of the TI-84 Plus can be used to find the future value of an annuity due as well as an ordinary annuity. If this feature is not built in, you may wish to program your calculator to evaluate this formula, too.

EXAMPLE 6   Future Value of an Annuity Due

Find the future value of an annuity due if payments of $500 are made at the beginning of each quarter for 7 years, in an account paying 6% compounded quarterly.

SOLUTION In 7 years, there are $n = 28$ quarterly periods. Add one period to get $n + 1 = 29$, and use the formula with $i = 0.06/4 = 0.015$.

$$S = 500\left[\frac{(1.015)^{29} - 1}{0.015}\right] - 500 \approx 17{,}499.35$$

The account will have a total of $17,499.35 after 7 years.

TRY YOUR TURN 4

YOUR TURN 4 Find the future value of an annuity due with $325 made at the beginning of each month for 5 years in an account paying 3.3% compounded monthly.

5.2 EXERCISES

Find the fifth term of each geometric sequence.

1. $a = 3$; $r = 2$ **2.** $a = 7$; $r = 5$

3. $a = -8$; $r = 3$ **4.** $a = -6$; $r = 2$

5. $a = 1$; $r = -3$ **6.** $a = 12$; $r = -2$

7. $a = 256$; $r = \dfrac{1}{4}$ **8.** $a = 729$; $r = \dfrac{1}{3}$

Find the sum of the first four terms for each geometric sequence.

9. $a = 1$; $r = 2$ **10.** $a = 4$; $r = 4$

11. $a = 5$; $r = \dfrac{1}{5}$ **12.** $a = 6$; $r = \dfrac{1}{2}$

13. $a = 128$; $r = -\dfrac{3}{2}$ **14.** $a = 64$; $r = -\dfrac{3}{4}$

15. Explain how a geometric sequence is related to an ordinary annuity.

16. Explain the difference between an ordinary annuity and an annuity due.

Find the future value of each ordinary annuity. Interest is compounded annually.

17. $R = 100$; $i = 0.06$; $n = 4$

18. $R = 1000$; $i = 0.06$; $n = 5$

19. $R = 25{,}000$; $i = 0.045$; $n = 36$

20. $R = 29{,}500$; $i = 0.058$; $n = 15$

Find the future value of each ordinary annuity, if payments are made and interest is compounded as given. Then determine how much of this value is from contributions and how much is from interest.

21. $R = 9200$; 10% interest compounded semiannually for 7 years

22. $R = 1250$; 5% interest compounded semiannually for 18 years

23. $R = 800$; 6.51% interest compounded semiannually for 12 years

24. $R = 4600$; 8.73% interest compounded quarterly for 9 years

25. $R = 12{,}000$; 4.8% interest compounded quarterly for 16 years

26. $R = 42{,}000$; 10.05% interest compounded semiannually for 12 years

27. What is meant by a sinking fund? Give an example of a sinking fund.

28. List some reasons for establishing a sinking fund.

Determine the interest rate needed to accumulate the following amounts in a sinking fund, with monthly payments as given.

29. Accumulate $56,000, monthly payments of $300 over 12 years

30. Accumulate $120,000, monthly payments of $500 over 15 years

Find the periodic payment that will amount to each given sum under the given conditions.

31. $S = \$10{,}000$; interest is 5% compounded annually; payments are made at the end of each year for 12 years.

32. $S = \$150{,}000$; interest is 6% compounded semiannually; payments are made at the end of each semiannual period for 11 years.

Find the amount of each payment to be made into a sinking fund so that enough will be present to accumulate the following amounts. Payments are made at the end of each period.

33. $8500; money earns 8% compounded annually; there are 7 annual payments.

34. $2750; money earns 5% compounded annually; there are 5 annual payments.

35. $75,000; money earns 6% compounded semiannually for $4\frac{1}{2}$ years.

36. $25,000; money earns 5.7% compounded quarterly for $3\frac{1}{2}$ years.

37. $65,000; money earns 7.5% compounded quarterly for $2\frac{1}{2}$ years.

38. $9000; money earns 4.8% compounded monthly for $2\frac{1}{2}$ years.

Find the future value of each annuity due. Assume that interest is compounded annually.

39. $R = 600$; $i = 0.06$; $n = 8$

40. $R = 1700$; $i = 0.04$; $n = 15$

41. $R = 16{,}000$; $i = 0.05$; $n = 7$

42. $R = 4000$; $i = 0.06$; $n = 11$

Find the future value of each annuity due. Then determine how much of this value is from contributions and how much is from interest.

43. Payments of $1000 made at the beginning of each semiannual period for 9 years at 8.15% compounded semiannually

44. $750 deposited at the beginning of each month for 15 years at 5.9% compounded monthly

45. $250 deposited at the beginning of each quarter for 12 years at 4.2% compounded quarterly

46. $1500 deposited at the beginning of each semiannual period for 11 years at 5.6% compounded semiannually

APPLICATIONS

Business and Economics

47. Comparing Accounts Laurie Campbell deposits $12,000 at the end of each year for 9 years in an account paying 8% interest compounded annually.

   a. Find the final amount she will have on deposit.

   b. Laurie's brother-in-law works in a bank that pays 6% compounded annually. If she deposits money in this bank instead of the one above, how much will she have in her account?

c. How much would Laurie lose over 9 years by using her brother-in-law's bank?

48. Savings Matthew Pastier is saving for a Plasma HDTV. At the end of each month he puts $100 in a savings account that pays 2.25% interest compounded monthly. How much is in the account after 2 years? How much did Matthew deposit? How much interest did he earn?

49. Savings A typical pack-a-day smoker spends about $136.50 per month on cigarettes. Suppose the smoker invests that amount each month in a savings account at 4.8% interest compounded monthly. What would the account be worth after 40 years? *Source: MSN.*

50. Retirement Planning A 45-year-old man puts $2500 in a retirement account at the end of each quarter until he reaches the age of 60, then makes no further deposits. If the account pays 6% interest compounded quarterly, how much will be in the account when the man retires at age 65?

51. Retirement Planning At the end of each quarter, a 50-year-old woman puts $3000 in a retirement account that pays 5% interest compounded quarterly. When she reaches 60, she withdraws the entire amount and places it in a mutual fund that pays 6.9% interest compounded monthly. From then on she deposits $300 in the mutual fund at the end of each month. How much is in the account when she reaches age 65?

Individual Retirement Accounts Suppose a 40-year-old person deposits $4000 per year in an Individual Retirement Account until age 65. Find the total in the account with the following assumptions of interest rates. (Assume quarterly compounding, with payments of $1000 made at the end of each quarter period.) Find the total amount of interest earned.

52. 6%   **53.** 8%   **54.** 4%   **55.** 10%

56. Savings Greg Tobin needs $10,000 in 8 years.

  a. What amount should he deposit at the end of each quarter at 8% compounded quarterly so that he will have his $10,000?

  b. Find Greg's quarterly deposit if the money is deposited at 6% compounded quarterly.

57. Buying Equipment Harv, the owner of Harv's Meats, knows that he must buy a new deboner machine in 4 years. The machine costs $12,000. In order to accumulate enough money to pay for the machine, Harv decides to deposit a sum of money at the end of each 6 months in an account paying 6% compounded semiannually. How much should each payment be?

58. Buying a Car Amanda Perdaris wants to have a $20,000 down payment when she buys a new car in 6 years. How much money must she deposit at the end of each quarter in an account paying 3.2% compounded quarterly so that she will have the down payment she desires?

59. Savings Stacy Schrank is paid on the first day of the month and $80 is automatically deducted from her pay and deposited in a savings account. If the account pays 2.5% interest compounded monthly, how much will be in the account after 3 years and 9 months?

60. Savings A father opened a savings account for his daughter on the day she was born, depositing $1000. Each year on her birthday he deposits another $1000, making the last deposit on her 21st birthday. If the account pays 5.25% interest compounded annually, how much is in the account at the end of the day on his daughter's 21st birthday? How much interest has been earned?

61. Savings Beth Dahlke deposits $2435 at the beginning of each semiannual period for 8 years in an account paying 6% compounded semiannually. She then leaves that money alone, with no further deposits, for an additional 5 years. Find the final amount on deposit after the entire 13-year period.

62. Savings David Kurzawa deposits $10,000 at the beginning of each year for 12 years in an account paying 5% compounded annually. He then puts the total amount on deposit in another account paying 6% compounded semiannually for another 9 years. Find the final amount on deposit after the entire 21-year period.

In Exercises 63 and 64, use a graphing calculator to find the value of *i* that produces the given value of *S*. (See Example 4(b).)

63. Retirement To save for retirement, Karla Harby put $300 each month into an ordinary annuity for 20 years. Interest was compounded monthly. At the end of the 20 years, the annuity was worth $147,126. What annual interest rate did she receive?

64. Rate of Return Caroline DiTullio made payments of $250 per month at the end of each month to purchase a piece of property. At the end of 30 years, she completely owned the property, which she sold for $330,000. What annual interest rate would she need to earn on an annuity for a comparable rate of return?

65. Lottery In a 1992 Virginia lottery, the jackpot was $27 million. An Australian investment firm tried to buy all possible combinations of numbers, which would have cost $7 million. In fact, the firm ran out of time and was unable to buy all combinations but ended up with the only winning ticket anyway. The firm received the jackpot in 20 equal annual payments of $1.35 million. Assume these payments meet the conditions of an ordinary annuity. *Source: The Washington Post.*

  a. Suppose the firm can invest money at 8% interest compounded annually. How many years would it take until the investors would be further ahead than if they had simply invested the $7 million at the same rate? (*Hint:* Experiment with different values of *n*, the number of years, or use a graphing calculator to plot the value of both investments as a function of the number of years.)

  b. How many years would it take in part a at an interest rate of 12%?

66. Buying Real Estate Vicki Kakounis sells some land in Nevada. She will be paid a lump sum of $60,000 in 7 years. Until then, the buyer pays 8% simple interest quarterly.

  a. Find the amount of each quarterly interest payment on the $60,000.

  b. The buyer sets up a sinking fund so that enough money will be present to pay off the $60,000. The buyer will make semiannual payments into the sinking fund; the account pays 6% compounded semiannually. Find the amount of each payment into the fund.

67. Buying Rare Stamps Phil Weaver bought a rare stamp for his collection. He agreed to pay a lump sum of $4000 after 5 years. Until then, he pays 6% simple interest semiannually on the $4000.

a. Find the amount of each semiannual interest payment.

b. Paul sets up a sinking fund so that enough money will be present to pay off the $4000. He will make annual payments into the fund. The account pays 8% compounded annually. Find the amount of each payment.

68. Down Payment A conventional loan, such as for a car or a house, is similar to an annuity but usually includes a down payment. Show that if a down payment of D dollars is made at the beginning of the loan period, the future value of all the payments, including the down payment, is

$$S = D(1 + i)^n + R\left[\frac{(1 + i)^n - 1}{i}\right].$$

■YOUR TURN ANSWERS

1. 39,364

2. $39,719.98

3. $184.41

4. $21,227.66

5.3 Present Value of an Annuity; Amortization

APPLY IT **What monthly payment will pay off a $17,000 car loan in 36 monthly payments at 6% annual interest?**
The answer to this question is given in Example 2 in this section. We shall see that it involves finding the present value of an annuity.

Suppose that at the end of each year, for the next 10 years, $500 is deposited in a savings account paying 7% interest compounded annually. This is an example of an ordinary annuity. The **present value of an annuity** is the amount that would have to be deposited in one lump sum today (at the same compound interest rate) in order to produce exactly the same balance at the end of 10 years. We can find a formula for the present value of an annuity as follows.

Suppose deposits of R dollars are made at the end of each period for n periods at interest rate i per period. Then the amount in the account after n periods is the future value of this annuity:

$$S = R \cdot s_{\overline{n}|i} = R\left[\frac{(1 + i)^n - 1}{i}\right].$$

On the other hand, if P dollars are deposited today at the same compound interest rate i, then at the end of n periods, the amount in the account is $P(1 + i)^n$. If P is the present value of the annuity, this amount must be the same as the amount S in the formula above; that is,

$$P(1 + i)^n = R\left[\frac{(1 + i)^n - 1}{i}\right].$$

To solve this equation for P, multiply both sides by $(1 + i)^{-n}$.

$$P = R(1 + i)^{-n}\left[\frac{(1 + i)^n - 1}{i}\right]$$

Use the distributive property; also recall that $(1 + i)^{-n}(1 + i)^n = 1$.

$$P = R\left[\frac{(1 + i)^{-n}(1 + i)^n - (1 + i)^{-n}}{i}\right] = R\left[\frac{1 - (1 + i)^{-n}}{i}\right]$$

The amount P is the *present value of the annuity*. The quantity in brackets is abbreviated as $a_{\overline{n}|i}$, so

$$a_{\overline{n}|i} = \frac{1 - (1 + i)^{-n}}{i}.$$

(The symbol $a_{\overline{n}|i}$ is read "*a*-angle-*n* at *i*." Compare this quantity with $s_{\overline{n}|i}$ in the previous section.) The formula for the present value of an annuity is summarized on the next page.

FOR REVIEW

Recall from Section R.6 that for any nonzero number a, $a^0 = 1$. Also, by the product rule for exponents, $a^x \cdot a^y = a^{x+y}$. In particular, if a is any nonzero number, $a^n \cdot a^{-n} = a^{n+(-n)} = a^0 = 1$.

Present Value of an Ordinary Annuity

The present value P of an annuity of n payments of R dollars each at the end of consecutive interest periods with interest compounded at a rate of interest i per period is

$$P = R\left[\frac{1 - (1 + i)^{-n}}{i}\right] \quad \text{or} \quad P = Ra_{\overline{n}|i}.$$

CAUTION Don't confuse the formula for the present value of an annuity with the one for the future value of an annuity. Notice the difference: the numerator of the fraction in the present value formula is $1 - (1 + i)^{-n}$, but in the future value formula, it is $(1 + i)^n - 1$.

TECHNOLOGY NOTE The financial feature of the TI-84 Plus calculator can be used to find the present value of an annuity by choosing that option from the menu and entering the required information. If your calculator does not have this built-in feature, it will be useful to store a program to calculate present value of an annuity in your calculator. A program is given in the *Graphing Calculator and Excel Spreadsheet Manual* available with this book.

EXAMPLE 1 Present Value of an Annuity

John Cross and Wendy Mears are both graduates of the Brisbane Institute of Technology (BIT). They both agree to contribute to the endowment fund of BIT. John says that he will give $500 at the end of each year for 9 years. Wendy prefers to give a lump sum today. What lump sum can she give that will equal the present value of John's annual gifts, if the endowment fund earns 7.5% compounded annually?

YOUR TURN 1 Find the present value of an annuity of $120 at the end of each month put into an account yielding 4.8% compounded monthly for 5 years.

SOLUTION Here, $R = 500$, $n = 9$, and $i = 0.075$, and we have

$$P = R \cdot a_{\overline{9}|0.075} = 500\left[\frac{1 - (1.075)^{-9}}{0.075}\right] \approx 3189.44.$$

Therefore, Wendy must donate a lump sum of $3189.44 today.   **TRY YOUR TURN 1**

One of the most important uses of annuities is in determining the equal monthly payments needed to pay off a loan, as illustrated in the next example.

EXAMPLE 2 Car Payments

A car costs $19,000. After a down payment of $2000, the balance will be paid off in 36 equal monthly payments with interest of 6% per year on the unpaid balance. Find the amount of each payment.

APPLY IT

SOLUTION A single lump sum payment of $17,000 today would pay off the loan. So, $17,000 is the present value of an annuity of 36 monthly payments with interest of $6\%/12 = 0.5\%$ per month. Thus, $P = 17,000$, $n = 36$, $i = 0.005$, and we must find the monthly payment R in the formula

$$P = R\left[\frac{1 - (1 + i)^{-n}}{i}\right]$$

YOUR TURN 2 Find the car payment in Example 2 if there are 48 equal monthly payments and the interest rate is 5.4%.

$$17,000 = R\left[\frac{1 - (1.005)^{-36}}{0.005}\right]$$

$$R \approx 517.17.$$

A monthly payment of $517.17 will be needed.   **TRY YOUR TURN 2**

Each payment in Example 2 includes interest on the unpaid balance, with the remainder going to reduce the loan. For example, the first payment of $517.17 includes interest of $0.005(\$17{,}000) = \$85$ and is divided as follows.

monthly   interest   to reduce
payment   due   the balance

$\$517.17 - \$85 = \$432.17$

At the end of this section, amortization schedules show that this procedure does reduce the loan to $0 after all payments are made (the final payment may be slightly different).

Amortization
A loan is **amortized** if both the principal and interest are paid by a sequence of equal periodic payments. In Example 2, a loan of $17,000 at 6% interest compounded monthly could be amortized by paying $517.17 per month for 36 months.

The periodic payment needed to amortize a loan may be found, as in Example 2, by solving the present value equation for R.

Amortization Payments
A loan of P dollars at interest rate i per period may be amortized in n equal periodic payments of R dollars made at the end of each period, where

$$R = \frac{P}{\left[\dfrac{1 - (1 + i)^{-n}}{i}\right]} = \frac{Pi}{1 - (1 + i)^{-n}} \quad \text{or} \quad R = \frac{P}{a_{\overline{n}|i}}.$$

▮ EXAMPLE 3   Home Mortgage

The Perez family buys a house for $275,000, with a down payment of $55,000. They take out a 30-year mortgage for $220,000 at an annual interest rate of 6%.

(a) Find the amount of the monthly payment needed to amortize this loan.

> **SOLUTION**   Here $P = 220{,}000$ and the monthly interest rate is $i = 0.06/12 = 0.005$.* The number of monthly payments is $12(30) = 360$. Therefore,
>
> $$R = \frac{220{,}000}{a_{\overline{360}|0.005}} = \frac{220{,}000}{\left[\dfrac{1 - (1.005)^{-360}}{0.005}\right]} = 1319.01.$$

Monthly payments of $1319.01 are required to amortize the loan.

(b) Find the total amount of interest paid when the loan is amortized over 30 years.

> **SOLUTION**   The Perez family makes 360 payments of $1319.01 each, for a total of $474,843.60. Since the amount of the loan was $220,000, the total interest paid is
>
> $$\$474{,}843.60 - \$220{,}000 = \$254{,}843.60.$$

This large amount of interest is typical of what happens with a long mortgage. A 15-year mortgage would have higher payments but would involve significantly less interest.

(c) Find the part of the first payment that is interest and the part that is applied to reducing the debt.

> **SOLUTION**   During the first month, the entire $220,000 is owed. Interest on this amount for 1 month is found by the formula for simple interest, with $r =$ annual interest rate and $t =$ time in years.
>
> $$I = Prt = 220{,}000(0.06)\frac{1}{12} = \$1100$$

---

*Mortgage rates are quoted in terms of annual interest, but it is always understood that the monthly rate is 1/12 of the annual rate and that interest is compounded monthly.

YOUR TURN 3 Find the monthly payment and total amount of interest paid in Example 3 if the mortgage is for 15 years and the interest rate is 7%.

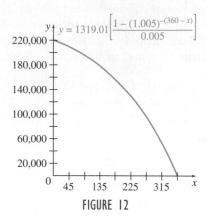

FIGURE 12

At the end of the month, a payment of $1319.01 is made; since $1100 of this is interest, a total of

$$\$1319.01 - \$1100 = \$219.01$$

is applied to the reduction of the original debt.   **TRY YOUR TURN 3**

It can be shown that the unpaid balance after x payments is given by the function

$$y = R\left[\frac{1 - (1 + i)^{-(n-x)}}{i}\right],$$

although this formula will only give an approximation if R is rounded to the nearest penny. For example, the unrounded value of R in Example 3 is 1319.011155. When this value is put into the above formula, the unpaid balance is found to be

$$y = 1319.011155\left[\frac{1 - (1.005)^{-359}}{0.005}\right] = 219,780.99,$$

while rounding R to 1319.01 in the above formula gives an approximate balance of $219,780.80. A graph of this function is shown in Figure 12.

We can find the unpaid balance after any number of payments, x, by finding the y-value that corresponds to x. For example, the remaining balance after 5 years or 60 payments is shown at the bottom of the graphing calculator screen in Figure 13(a). You may be surprised that the remaining balance on a $220,000 loan is as large as $204,719.41. This is because most of the early payments on a loan go toward interest, as we saw in Example 3(c).

By adding the graph of $y = (1/2)220,000 = 110,000$ to the figure, we can find when half the loan has been repaid. From Figure 13(b) we see that 252 payments are required. Note that only 108 payments remain at that point, which again emphasizes the fact that the earlier payments do little to reduce the loan.

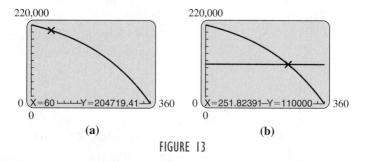

FIGURE 13

Amortization Schedules
In the preceding example, 360 payments are made to amortize a $220,000 loan. The loan balance after the first payment is reduced by only $219.01, which is much less than $(1/360)(220,000) \approx \$611.11$. Therefore, even though equal *payments* are made to amortize a loan, the loan *balance* does not decrease in equal steps. This fact is very important if a loan is paid off early.

EXAMPLE 4 Early Payment

Ami Aigen borrows $1000 for 1 year at 12% annual interest compounded monthly. Verify that her monthly loan payment is $88.8488, which is rounded to $88.85. After making three payments, she decides to pay off the remaining balance all at once. How much must she pay?

SOLUTION Since nine payments remain to be paid, they can be thought of as an annuity consisting of nine payments of $88.85 at 1% interest per period. The present value of this annuity is

$$88.8488\left[\frac{1 - (1.01)^{-9}}{0.01}\right] \approx 761.08.$$

So Ami's remaining balance, computed by this method, is $761.08.

An alternative method of figuring the balance is to consider the payments already made as an annuity of three payments. At the beginning, the present value of this annuity was

$$88.85\left[\frac{1 - (1.01)^{-3}}{0.01}\right] \approx 261.31.$$

So she still owes the difference $1000 - \$261.31 = \$738.69$. Furthermore, she owes the interest on this amount for 3 months, for a total of

$$(738.69)(1.01)^3 \approx \$761.07.$$

This balance due differs from the one obtained by the first method by 1 cent because the monthly payment and the other calculations were rounded to the nearest penny. If we had used the more accurate value of $R = 88.8488$ and not rounded any intermediate answers, both methods would have given the same value of $761.08. **TRY YOUR TURN 4**

YOUR TURN 4 Find the remaining balance in Example 4 if the balance was to be paid off after four months. Use the unrounded value for R of $88.8488.

Although most people would not quibble about a difference of 1 cent in the balance due in Example 4, the difference in other cases (larger amounts or longer terms) might be more than that. A bank or business must keep its books accurately to the nearest penny, so it must determine the balance due in such cases unambiguously and exactly. This is done by means of an **amortization schedule**, which lists how much of each payment is interest and how much goes to reduce the balance, as well as how much is owed after *each* payment.

EXAMPLE 5 Amortization Table

Determine the exact amount Ami Aigen in Example 4 owes after three monthly payments.

SOLUTION An amortization table for the loan is shown below. It is obtained as follows. The annual interest rate is 12% compounded monthly, so the interest rate per month is $12\%/12 = 1\% = 0.01$. When the first payment is made, 1 month's interest—namely $0.01(1000) = \$10$—is owed. Subtracting this from the $88.85 payment leaves $78.85 to be applied to repayment. Hence, the principal at the end of the first payment period is $1000 - \$78.85 = \$921.15$, as shown in the "payment 1" line of the chart.

When payment 2 is made, 1 month's interest on $921.15 is owed, namely $0.01(921.15) = \$9.21$. Subtracting this from the $88.85 payment leaves $79.64 to reduce the principal. Hence, the principal at the end of payment 2 is $921.15 - \$79.64 = \$841.51$. The interest portion of payment 3 is based on this amount, and the remaining lines of the table are found in a similar fashion.

The schedule shows that after three payments, she still owes $761.08, an amount that agrees with the first method in Example 4.

Amortization Table				
Payment Number	Amount of Payment	Interest for Period	Portion to Principal	Principal at End of Period
0	—	—	—	$1000.00
1	$88.85	$10.00	$78.85	$921.15
2	$88.85	$9.21	$79.64	$841.51
3	$88.85	$8.42	$80.43	$761.08
4	$88.85	$7.61	$81.24	$679.84
5	$88.85	$6.80	$82.05	$597.79
6	$88.85	$5.98	$82.87	$514.92
7	$88.85	$5.15	$83.70	$431.22
8	$88.85	$4.31	$84.54	$346.68
9	$88.85	$3.47	$85.38	$261.30
10	$88.85	$2.61	$86.24	$175.06
11	$88.85	$1.75	$87.10	$87.96
12	$88.84	$0.88	$87.96	$0.00

The amortization schedule in Example 5 is typical. In particular, note that all payments are the same except the last one. It is often necessary to adjust the amount of the final payment to account for rounding off earlier and to ensure that the final balance is exactly 0.

An amortization schedule also shows how the periodic payments are applied to interest and principal. The amount going to interest decreases with each payment, while the amount going to reduce the principal increases with each payment.

TECHNOLOGY NOTE

A graphing calculator program to produce an amortization schedule is available in the *Graphing Calculator and Excel Spreadsheet Manual* available with this book. The TI-84 Plus includes a built-in program to find the amortization payment. Spreadsheets are another useful tool for creating amortization tables. Microsoft Excel has a built-in feature for calculating monthly payments. Figure 14 shows an Excel amortization table for Example 5. For more details, see the *Graphing Calculator and Excel Spreadsheet Manual*, also available with this book.

	A	B	C	D	E	F
1	Pmt#	Payment	Interest	Principal	End Prncpl	
2	0				1000	
3	1	88.85	10.00	78.85	921.15	
4	2	88.85	9.21	79.64	841.51	
5	3	88.85	8.42	80.43	761.08	
6	4	88.85	7.61	81.24	679.84	
7	5	88.85	6.80	82.05	597.79	
8	6	88.85	5.98	82.87	514.92	
9	7	88.85	5.15	83.70	431.22	
10	8	88.85	4.31	84.54	346.68	
11	9	88.85	3.47	85.38	261.30	
12	10	88.85	2.61	86.24	175.06	
13	11	88.85	1.75	87.10	87.96	
14	12	88.85	0.88	87.97	-0.01	

FIGURE 14

EXAMPLE 6 Paying Off a Loan Early

Suppose that in Example 2, the car owner decides that she can afford to make payments of $700 rather than $517.17. How much earlier would she pay off the loan? How much interest would she save?

SOLUTION Putting $R = 700$, $P = 17,000$, and $i = 0.005$ into the formula for the present value of an annuity gives

$$17,000 = 700 \left[\frac{1 - 1.005^{-n}}{0.005} \right].$$

Multiply both sides by 0.005 and divide by 700 to get

$$\frac{85}{700} = 1 - 1.005^{-n},$$

or

$$1.005^{-n} = 1 - \frac{85}{700}.$$

Solve this using either logarithms or a graphing calculator, as in Example 11 in Section 5.1, to get $n = 25.956$. This means that 25 payments of $700, plus a final, smaller payment, would be sufficient to pay off the loan. Create an amortization table to verify that the final payment would be $669.47 (the sum of the principal after the penultimate payment plus the interest on that principal for the final month). The loan would then be paid off after 26 months, or 10 months early.

The original loan required 36 payments of $517.17, or 36(517.17) = $18,618.12, although the amount is actually $18,618.24 because the final payment was $517.29, as an amortization table would show. With larger payments, the car owner paid 25(700) + 669.47 = $18,169.47. Therefore, the car owner saved $18,618.24 − $18,169.47 = $448.77 in interest by making larger payments each month.

5.3 EXERCISES

1. Explain the difference between the present value of an annuity and the future value of an annuity. For a given annuity, which is larger? Why?

2. What does it mean to amortize a loan?

Find the present value of each ordinary annuity.

3. Payments of $890 each year for 16 years at 6% compounded annually

4. Payments of $1400 each year for 8 years at 6% compounded annually

5. Payments of $10,000 semiannually for 15 years at 5% compounded semiannually

6. Payments of $50,000 quarterly for 10 years at 4% compounded quarterly

7. Payments of $15,806 quarterly for 3 years at 6.8% compounded quarterly

8. Payments of $18,579 every 6 months for 8 years at 5.4% compounded semiannually

Find the lump sum deposited today that will yield the same total amount as payments of $10,000 at the end of each year for 15 years at each of the given interest rates.

9. 4% compounded annually

10. 6% compounded annually

Find (a) the payment necessary to amortize each loan; (b) the total payments and the total amount of interest paid based on the calculated monthly payments, and (c) the total payments and total amount of interest paid based upon an amortization table.

11. $2500; 6% compounded quarterly; 6 quarterly payments

12. $41,000; 8% compounded semiannually; 10 semiannual payments

13. $90,000; 6% compounded annually; 12 annual payments

14. $140,000; 8% compounded quarterly; 15 quarterly payments

15. $7400; 6.2% compounded semiannually; 18 semiannual payments

16. $5500; 10% compounded monthly; 24 monthly payments

Suppose that in the loans described in Exercises 13–16, the borrower paid off the loan after the time indicated below. Calculate the amount needed to pay off the loan, using either of the two methods described in Example 4.

17. After 3 years in Exercise 13

18. After 5 quarters in Exercise 14

19. After 3 years in Exercise 15

20. After 7 months in Exercise 16

Use the amortization table in Example 5 to answer the questions in Exercises 21–24.

21. How much of the fourth payment is interest?

22. How much of the eleventh payment is used to reduce the debt?

23. How much interest is paid in the first 4 months of the loan?

24. How much interest is paid in the last 4 months of the loan?

25. What sum deposited today at 5% compounded annually for 8 years will provide the same amount as $1000 deposited at the end of each year for 8 years at 6% compounded annually?

26. What lump sum deposited today at 8% compounded quarterly for 10 years will yield the same final amount as deposits of $4000 at the end of each 6-month period for 10 years at 6% compounded semiannually?

Find the monthly house payments necessary to amortize each loan. Then calculate the total payments and the total amount of interest paid.

27. $199,000 at 7.01% for 25 years

28. $175,000 at 6.24% for 30 years

29. $253,000 at 6.45% for 30 years

30. $310,000 at 5.96% for 25 years

Suppose that in the loans described in Exercises 13–16, the borrower made a larger payment, as indicated below. Calculate (a) the time needed to pay off the loan, (b) the total amount of the payments, and (c) the amount of interest saved, compared with part c of Exercises 13–16.

31. $16,000 in Exercise 13

32. $18,000 in Exercise 14

33. $850 in Exercise 15

34. $400 in Exercise 16

APPLICATIONS

Business and Economics

35. **House Payments** Calculate the monthly payment and total amount of interest paid in Example 3 with a 15-year loan, and then compare with the results of Example 3.

36. **Installment Buying** Stereo Shack sells a stereo system for $600 down and monthly payments of $30 for the next 3 years. If the interest rate is 1.25% per month on the unpaid balance, find

    a. the cost of the stereo system.

    b. the total amount of interest paid.

37. **Car Payments** Hong Le buys a car costing $14,000. He agrees to make payments at the end of each monthly period for 4 years. He pays 7% interest, compounded monthly.

    a. What is the amount of each payment?

    b. Find the total amount of interest Le will pay.

38. **Credit Card Debt** Tom Shaffer charged $8430 on his credit card to relocate for his first job. When he realized that the interest rate for the unpaid balance was 27% compounded monthly, he decided not to charge any more on that account. He wants to have this account paid off by the end of 3 years,

so he arranges to have automatic payments sent at the end of each month.

a. What monthly payment must he make to have the account paid off by the end of 3 years?

b. How much total interest will he have paid?

39. New Car In Spring 2010, some dealers offered a cash-back allowance of $2250 or 0.9% financing for 36 months on an Acura TL. *Source: cars.com.*

a. Determine the payments on an Acura TL if a buyer chooses the 0.9% financing option and needs to finance $30,000 for 36 months, compounded monthly. Find the total amount the buyer will pay for this option.

b. Determine the payments on an Acura TL if a buyer chooses the cash-back option and now needs to finance only $27,750. At the time, it was possible to get a new car loan at 6.33% for 48 months, compounded monthly. Find the total amount the buyer will pay for this option.

c. Discuss which deal is best and why.

40. New Car In Spring 2010, some dealers offered a cash-back allowance of $1500 or 1.9% financing for 36 months on a Volkswagen Tiguan. *Source: cars.com.*

a. Determine the payments on a Volkswagen Tiguan if a buyer chooses the 1.9% financing option and needs to finance $25,000 for 36 months, compounded monthly. Find the total amount the buyer will pay for this option.

b. Determine the payments on a Volkswagen Tiguan if a buyer chooses the cash-back option and now needs to finance only $23,500. At the time, it was possible to get a new car loan at 6.33% for 48 months, compounded monthly. Find the total amount the buyer will pay for this option.

c. Discuss which deal is best and why.

41. Lottery Winnings In most states, the winnings of million-dollar lottery jackpots are divided into equal payments given annually for 20 years. (In Colorado, the results are distributed over 25 years.) This means that the present value of the jackpot is worth less than the stated prize, with the actual value determined by the interest rate at which the money could be invested. *Source: The New York Times Magazine.*

a. Find the present value of a $1 million lottery jackpot distributed in equal annual payments over 20 years, using an interest rate of 5%.

b. Find the present value of a $1 million lottery jackpot distributed in equal annual payments over 20 years, using an interest rate of 9%.

c. Calculate the answer for part a using the 25-year distribution time in Colorado.

d. Calculate the answer for part b using the 25-year distribution time in Colorado.

Student Loans Student borrowers now have more options to choose from when selecting repayment plans. The standard plan repays the loan in up to 10 years with equal monthly payments.

The extended plan allows up to 25 years to repay the loan. *Source: U.S. Department of Education.* **A student borrows $55,000 at 6.80% compounded monthly.**

42. Find the monthly payment and total interest paid under the standard plan over 10 years.

43. Find the monthly payment and total interest paid under the extended plan over 25 years.

Installment Buying In Exercises 44–46, prepare an amortization schedule showing the first four payments for each loan.

44. An insurance firm pays $4000 for a new printer for its computer. It amortizes the loan for the printer in 4 annual payments at 8% compounded annually.

45. Large semitrailer trucks cost $110,000 each. Ace Trucking buys such a truck and agrees to pay for it by a loan that will be amortized with 9 semiannual payments at 8% compounded semiannually.

46. One retailer charges $1048 for a laptop computer. A firm of tax accountants buys 8 of these laptops. They make a down payment of $1200 and agree to amortize the balance with monthly payments at 6% compounded monthly for 4 years.

47. Investment In 1995, Oseola McCarty donated $150,000 to the University of Southern Mississippi to establish a scholarship fund. What is unusual about her is that the entire amount came from what she was able to save each month from her work as a washer woman, a job she began in 1916 at the age of 8, when she dropped out of school. *Sources: The New York Times.*

a. How much would Ms. McCarty have to put into her savings account at the end of every 3 months to accumulate $150,000 over 79 years? Assume she received an interest rate of 5.25% compounded quarterly.

b. Answer part a using a 2% and a 7% interest rate.

48. Loan Payments   When Nancy Hart opened her law office, she bought $14,000 worth of law books and $7200 worth of office furniture. She paid $1200 down and agreed to amortize the balance with semiannual payments for 5 years, at 8% compounded semiannually.

  a. Find the amount of each payment.

  b. Refer to the text and Figure 13. When her loan had been reduced below $5000, Nancy received a large tax refund and decided to pay off the loan. How many payments were left at this time?

49. House Payments   Ian Desrosiers buys a house for $285,000. He pays $60,000 down and takes out a mortgage at 6.5% on the balance. Find his monthly payment and the total amount of interest he will pay if the length of the mortgage is

  a. 15 years;

  b. 20 years;

  c. 25 years.

  d. Refer to the text and Figure 13. When will half the 20-year loan in part b be paid off?

50. House Payments   The Chavara family buys a house for $225,000. They pay $50,000 down and take out a 30-year mortgage on the balance. Find their monthly payment and the total amount of interest they will pay if the interest rate is

  a. 6%;

  b. 6.5%;

  c. 7%.

  d. Refer to the text and Figure 13. When will half the 7% loan in part c be paid off?

51. Refinancing a Mortgage   Fifteen years ago, the Budai family bought a home and financed $150,000 with a 30-year mortgage at 8.2%.

  a. Find their monthly payment, the total amount of their payments, and the total amount of interest they will pay over the life of this loan.

  b. The Budais made payments for 15 years. Estimate the unpaid balance using the formula

$$y = R\left[\frac{1 - (1 + i)^{-(n-x)}}{i}\right],$$

  and then calculate the total of their remaining payments.

  c. Suppose interest rates have dropped since the Budai family took out their original loan. One local bank now offers a 30-year mortgage at 6.5%. The bank fees for refinancing are $3400. If the Budais pay this fee up front and refinance the balance of their loan, find their monthly payment. Including the refinancing fee, what is the total amount of their payments? Discuss whether or not the family should refinance with this option.

  d. A different bank offers the same 6.5% rate but on a 15-year mortgage. Their fee for financing is $4500. If the Budais pay this fee up front and refinance the balance of their loan, find their monthly payment. Including the refinancing fee, what is the total amount of their payments? Discuss whether or not the family should refinance with this option.

52. Inheritance   Deborah Harden has inherited $25,000 from her grandfather's estate. She deposits the money in an account offering 6% interest compounded annually. She wants to make equal annual withdrawals from the account so that the money (principal and interest) lasts exactly 8 years.

  a. Find the amount of each withdrawal.

  b. Find the amount of each withdrawal if the money must last 12 years.

53. Charitable Trust   The trustees of a college have accepted a gift of $150,000. The donor has directed the trustees to deposit the money in an account paying 6% per year, compounded semiannually. The trustees may make equal withdrawals at the end of each 6-month period; the money must last 5 years.

  a. Find the amount of each withdrawal.

  b. Find the amount of each withdrawal if the money must last 6 years.

Amortization   Prepare an amortization schedule for each loan.

54. A loan of $37,948 with interest at 6.5% compounded annually, to be paid with equal annual payments over 10 years.

55. A loan of $4836 at 7.25% interest compounded semi-annually, to be repaid in 5 years in equal semiannual payments.

56. Perpetuity   A *perpetuity* is an annuity in which the payments go on forever. We can derive a formula for the present value of a perpetuity by taking the formula for the present value of an annuity and looking at what happens when n gets larger and larger. Explain why the present value of a perpetuity is given by

$$P = \frac{R}{i}.$$

57. Perpetuity   Using the result of Exercise 56, find the present value of perpetuities for each of the following.

  a. Payments of $1000 a year with 4% interest compounded annually

  b. Payments of $600 every 3 months with 6% interest compounded quarterly

YOUR TURN ANSWERS

1. $6389.86	**2.** $394.59
3. $1977.42, $135,935.60	**4.** $679.84

5

CHAPTER REVIEW

SUMMARY

In this chapter we introduced the mathematics of finance. We first extended simple interest calculations to compound interest, which is interest earned on interest previously earned. We then developed the mathematics associated with the following financial concepts.

- In an annuity, money continues to be deposited at regular intervals, and compound interest is earned on that money as well.
- In an ordinary annuity, payments are made at the end of each time period, and the compounding period is the same as the time between payments, which simplifies the calculations.

- An annuity due is slightly different, in that the payments are made at the beginning of each time period.
- A sinking fund is like an ordinary annuity; a fund is set up to receive periodic payments. The payments plus the compound interest will produce a desired sum by a certain date.
- The present value of an annuity is the amount that would have to be deposited today to produce the same amount as the annuity at the end of a specified time.
- An amortization table shows how a loan is paid back after a specified time. It shows the payments broken down into interest and principal.

We have presented a lot of new formulas in this chapter. By answering the following questions, you can decide which formula to use for a particular problem.

1. Is simple or compound interest involved?

   Simple interest is normally used for investments or loans of a year or less; compound interest is normally used in all other cases.

2. If simple interest is being used, what is being sought: interest amount, future value, present value, or interest rate?

3. If compound interest is being used, does it involve a lump sum (single payment) or an annuity (sequence of payments)?

   a. For a lump sum, what is being sought: present value, future value, number of periods at interest, or effective rate?

   b. For an annuity,

      i. Is it an ordinary annuity (payment at the end of each period) or an annuity due (payment at the beginning of each period)?

      ii. What is being sought: present value, future value, or payment amount?

Once you have answered these questions, choose the appropriate formula and work the problem. As a final step, consider whether the answer you get makes sense. For instance, present value should always be less than future value. The amount of interest or the payments in an annuity should be fairly small compared to the total future value.

List of Variables r is the annual interest rate.

i is the interest rate per period.

t is the number of years.

n is the number of periods.

m is the number of periods per year.

P is the principal or present value.

A is the future value of a lump sum.

S is the future value of an annuity.

R is the periodic payment in an annuity.

$$i = \frac{r}{m} \qquad n = tm$$

	Simple Interest	Compound Interest	Continuous Compounding
Interest	$I = Prt$	$I = A - P$	$I = A - P$
Future Value	$A = P(1 + rt)$	$A = P(1 + i)^n$	$A = Pe^{rt}$
Present Value	$P = \dfrac{A}{1 + rt}$	$P = \dfrac{A}{(1 + i)^n} = A(1 + i)^{-n}$	$P = Ae^{-rt}$
Effective Rate		$r_E = \left(1 + \dfrac{r}{m}\right)^m - 1$	$r_E = e^r - 1$

Ordinary Annuity Future Value $\quad S = R\left[\dfrac{(1 + i)^n - 1}{i}\right] = R \cdot s_{\overline{n}|i}$

Present Value $\quad P = R\left[\dfrac{1 - (1 + i)^{-n}}{i}\right] = R \cdot a_{\overline{n}|i}$

Annuity Due Future Value $\quad S = R\left[\dfrac{(1 + i)^{n+1} - 1}{i}\right] - R$

Sinking Fund Payment $\quad R = \dfrac{Si}{(1 + i)^n - 1} = \dfrac{S}{s_{\overline{n}|i}}$

Amortization Payments $\quad R = \dfrac{Pi}{1 - (1 + i)^{-n}} = \dfrac{P}{a_{\overline{n}|i}}$

KEY TERMS

5.1
simple interest
principal
rate
time
future value
maturity value
compound interest
compound amount

nominal (stated) rate
effective rate
present value
rule of 70
rule of 72
continuous compounding

5.2
geometric sequence
terms

common ratio
annuity
ordinary annuity
payment period
future value of an annuity
term of an annuity
future value of an ordinary
   annuity
sinking fund

annuity due
future value of an annuity due

5.3
present value of an annuity
amortize a loan
amortization schedule

REVIEW EXERCISES

CONCEPT CHECK

Determine whether each of the following statements is true or false, and explain why.

1. For a particular interest rate, compound interest is always better than simple interest.

2. The sequence 1, 2, 4, 6, 8, . . . is a geometric sequence.

3. If a geometric sequence has first term 3 and common ratio 2, then the sum of the first 5 terms is $S_5 = 93$.

4. The value of a sinking fund should decrease over time.

5. For payments made on a mortgage, the (noninterest) portion of the payment applied on the principal increases over time.

6. On a 30-year conventional home mortgage, at recent interest rates, it is common to pay more money on the interest on the loan than the actual loan itself.

7. One can use the amortization payments formula to calculate the monthly payment of a car loan.

8. The effective rate formula can be used to calculate the present value of a loan.

9. The following calculation gives the monthly payment on a $25,000 loan, compounded monthly at a rate of 5% for a period of six years:

$$25{,}000\left[\dfrac{(1 + 0.05/12)^{72} - 1}{0.05/12}\right].$$

10. The following calculation gives the present value of an annuity of $5,000 payments at the end of each year for 10 years. The fund earns 4.5% compounded annually.

$$5000\left[\dfrac{1 - (1.045)^{-10}}{0.045}\right]$$

PRACTICE AND EXPLORATION

Find the simple interest for each loan.

11. $15,903 at 6% for 8 months

12. $4902 at 5.4% for 11 months

13. $42,368 at 5.22% for 7 months

14. $3478 at 6.8% for 88 days (assume a 360-day year)

15. For a given amount of money at a given interest rate for a given time period, does simple interest or compound interest produce more interest?

Find the compound amount in each loan.

16. $2800 at 7% compounded annually for 10 years

17. $19,456.11 at 8% compounded semiannually for 7 years

18. $312.45 at 5.6% compounded semiannually for 16 years

19. $57,809.34 at 6% compounded quarterly for 5 years

Find the amount of interest earned by each deposit.

20. $3954 at 8% compounded annually for 10 years

21. $12,699.36 at 5% compounded semiannually for 7 years

22. $12,903.45 at 6.4% compounded quarterly for 29 quarters

23. $34,677.23 at 4.8% compounded monthly for 32 months

24. What is meant by the present value of an amount A?

Find the present value of each amount.

25. $42,000 in 7 years, 6% compounded monthly

26. $17,650 in 4 years, 4% compounded quarterly

27. $1347.89 in 3.5 years, 6.77% compounded semiannually

28. $2388.90 in 44 months, 5.93% compounded monthly

29. Write the first five terms of the geometric sequence with $a = 2$ and $r = 3$.

30. Write the first four terms of the geometric sequence with $a = 4$ and $r = 1/2$.

31. Find the sixth term of the geometric sequence with $a = -3$ and $r = 2$.

32. Find the fifth term of the geometric sequence with $a = -2$ and $r = -2$.

33. Find the sum of the first four terms of the geometric sequence with $a = -3$ and $r = 3$.

34. Find the sum of the first five terms of the geometric sequence with $a = 8000$ and $r = -1/2$.

35. Find $s_{\overline{30}|0.02}$.

36. Find $s_{\overline{20}|0.06}$.

37. What is meant by the future value of an annuity?

Find the future value of each annuity and the amount of interest earned.

38. $500 deposited at the end of each 6-month period for 10 years; money earns 6% compounded semiannually.

39. $1288 deposited at the end of each year for 14 years; money earns 4% compounded annually.

40. $4000 deposited at the end of each quarter for 7 years; money earns 5% compounded quarterly.

41. $233 deposited at the end of each month for 4 years; money earns 4.8% compounded monthly.

42. $672 deposited at the beginning of each quarter for 7 years; money earns 4.4% compounded quarterly.

43. $11,900 deposited at the beginning of each month for 13 months; money earns 6% compounded monthly.

44. What is the purpose of a sinking fund?

Find the amount of each payment that must be made into a sinking fund to accumulate each amount.

45. $6500; money earns 5% compounded annually for 6 years.

46. $57,000; money earns 4% compounded semiannually for $8\frac{1}{2}$ years.

47. $233,188; money earns 5.2% compounded quarterly for $7\frac{3}{4}$ years.

48. $1,056,788; money earns 7.2% compounded monthly for $4\frac{1}{2}$ years.

Find the present value of each ordinary annuity.

49. Deposits of $850 annually for 4 years at 6% compounded annually

50. Deposits of $1500 quarterly for 7 years at 5% compounded quarterly

51. Payments of $4210 semiannually for 8 years at 4.2% compounded semiannually

52. Payments of $877.34 monthly for 17 months at 6.4% compounded monthly

53. Give two examples of the types of loans that are commonly amortized.

Find the amount of the payment necessary to amortize each loan. Calculate the total interest paid.

54. $80,000; 5% compounded annually; 9 annual payments

55. $3200; 8% compounded quarterly; 12 quarterly payments

56. $32,000; 6.4% compounded quarterly; 17 quarterly payments

57. $51,607; 8% compounded monthly; 32 monthly payments

Find the monthly house payments for each mortgage. Calculate the total payments and interest.

58. $256,890 at 5.96% for 25 years

59. $177,110 at 6.68% for 30 years

A portion of an amortization table is given below for a $127,000 loan at 8.5% interest compounded monthly for 25 years.

Payment Number	Amount of Payment	Interest for Period	Portion to Principal	Principal at End of Period
1	$1022.64	$899.58	$123.06	$126,876.94
2	$1022.64	$898.71	$123.93	$126,753.01
3	$1022.64	$897.83	$124.81	$126,628.20
4	$1022.64	$896.95	$125.69	$126,502.51
5	$1022.64	$896.06	$126.58	$126,375.93
6	$1022.64	$895.16	$127.48	$126,248.45
7	$1022.64	$894.26	$128.38	$126,120.07
8	$1022.64	$893.35	$129.29	$125,990.78
9	$1022.64	$892.43	$130.21	$125,860.57
10	$1022.64	$891.51	$131.13	$125,729.44
11	$1022.64	$890.58	$132.06	$125,597.38
12	$1022.64	$889.65	$132.99	$125,464.39

Use the table to answer the following questions.

60. How much of the fifth payment is interest?

61. How much of the twelfth payment is used to reduce the debt?

62. How much interest is paid in the first 3 months of the loan?

63. How much has the debt been reduced at the end of the first year?

APPLICATIONS

Business and Economics

64. Personal Finance Jane Fleming owes $5800 to her mother. She has agreed to repay the money in 10 months at an interest rate of 5.3%. How much will she owe in 10 months? How much interest will she pay?

65. Business Financing Julie Ward needs to borrow $9820 to buy new equipment for her business. The bank charges her 6.7% simple interest for a 7-month loan. How much interest will she be charged? What amount must she pay in 7 months?

66. Business Financing An accountant loans $28,000 at simple interest to her business. The loan is at 6.5% and earns $1365 interest. Find the time of the loan in months.

67. Business Investment A developer deposits $84,720 for 7 months and earns $4055.46 in simple interest. Find the interest rate.

68. Personal Finance In 3 years Beth Rechsteiner must pay a pledge of $7500 to her college's building fund. What lump sum can she deposit today, at 5% compounded semiannually, so that she will have enough to pay the pledge?

69. Personal Finance Tom, a graduate student, is considering investing $500 now, when he is 23, or waiting until he is 40 to invest $500. How much more money will he have at the age of 65 if he invests now, given that he can earn 5% interest compounded quarterly?

70. Pensions Pension experts recommend that you start drawing at least 40% of your full pension as early as possible. Suppose you have built up a pension of $12,000-annual payments by working 10 years for a company. When you leave to accept a better job, the company gives you the option of collecting half of the full pension when you reach age 55 or the full pension at age 65. Assume an interest rate of 8% compounded annually. By age 75, how much will each plan produce? Which plan would produce the larger amount? *Source: Smart Money.*

71. Business Investment A firm of attorneys deposits $5000 of profit-sharing money at the end of each semiannual period for $7\frac{1}{2}$ years. Find the final amount in the account if the deposits earn 10% compounded semiannually. Find the amount of interest earned.

72. Business Financing A small resort must add a swimming pool to compete with a new resort built nearby. The pool will cost $28,000. The resort borrows the money and agrees to repay it with equal payments at the end of each quarter for $6\frac{1}{2}$ years at an interest rate of 8% compounded quarterly. Find the amount of each payment.

73. Business Financing The owner of Eastside Hallmark borrows $48,000 to expand the business. The money will be repaid in equal payments at the end of each year for 7 years. Interest is 6.5%. Find the amount of each payment and the total amount of interest paid.

74. Personal Finance To buy a new computer, Mark Nguyen borrows $3250 from a friend at 4.2% interest compounded annually for 4 years. Find the compound amount he must pay back at the end of the 4 years.

75. Effective Rate On May 21, 2010, Ascencia (a division of PBI Bank) paid 1.49% interest, compounded monthly, on a 1-year CD, while giantbank.com paid 1.45% compounded daily. What are the effective rates for the two CDs, and which bank paid a higher effective rate? *Source: Bankrate.com.*

76. Home Financing When the Lee family bought their home, they borrowed $315,700 at 7.5% compounded monthly for 25 years. If they make all 300 payments, repaying the loan on schedule, how much interest will they pay? (Assume the last payment is the same as the previous ones.)

77. New Car In Spring 2010, some dealers offered the following options on a Chevrolet HHR: a cash-back allowance of $4000, 0% financing for 60 months, or 3.9% financing for 72 months. *Source: cars.com.*

a. Determine the payments on a Chevrolet HHR if a buyer chooses the 0% financing option and needs to finance $16,000 for 60 months. Find the total amount of the payments.

b. Repeat part a for the 3.9% financing option for 72 months.

c. Determine the payments on a Chevrolet HHR if a buyer chooses the cash-back option and now needs to finance only $12,000. At the time, it was possible to get a new car loan at 6.33% for 48 months, compounded monthly. Find the total amount of the payments.

d. Discuss which deal is best and why.

78. New Car In Spring 2010, some dealers offered a cash-back allowance of $5000 or 0% financing for 72 months on a Chevrolet Silverado. *Source: cars.com.*

a. Determine the payments on a Chevrolet Silverado if a buyer chooses the 0% financing option and needs to finance $30,000 for 72 months. Find the total amount of the payments.

b. Determine the payments on a Chevrolet Silverado if a buyer chooses the cash-back option and now needs to finance only $25,000. At the time, it was possible to get a new car loan at 6.33% for 48 months, compounded monthly. Find the total amount of the payments.

c. Discuss which deal is best and why.

d. Find the interest rate at the bank that would make the total amount of payments for the two options equal.

79. Buying and Selling a House The Bahary family bought a house for $191,000. They paid $40,000 down and took out a 30-year mortgage for the balance at 6.5%.

a. Find their monthly payment.

b. How much of the first payment is interest?

After 180 payments, the family sells its house for $238,000. They must pay closing costs of $3700 plus 2.5% of the sale price.

c. Estimate the current mortgage balance at the time of the sale using one of the methods from Example 4 in Section 3.

d. Find the total closing costs.

e. Find the amount of money they receive from the sale after paying off the mortgage.

The following exercise is from an actuarial examination. *Source: The Society of Actuaries.*

80. Death Benefit The proceeds of a $10,000 death benefit are left on deposit with an insurance company for 7 years at an annual effective interest rate of 5%. The balance at the end of 7 years is paid to the beneficiary in 120 equal monthly payments of X, with the first payment made immediately. During the payout period, interest is credited at an annual effective interest rate of 3%. Calculate X. Choose one of the following.

a. 117   **b.** 118   **c.** 129   **d.** 135   **e.** 158

81. Investment *The New York Times* posed a scenario with two individuals, Sue and Joe, who each have $1200 a month to spend on housing and investing. Each takes out a mortgage for $140,000. Sue gets a 30-year mortgage at a rate of 6.625%. Joe gets a 15-year mortgage at a rate of 6.25%. Whatever money is left after the mortgage payment is invested in a mutual fund with a return of 10% annually. *Source: The New york Times.*

a. What annual interest rate, when compounded monthly, gives an effective annual rate of 10%?

b. What is Sue's monthly payment?

c. If Sue invests the remainder of her $1200 each month, after the payment in part b, in a mutual fund with the interest rate in part a, how much money will she have in the fund at the end of 30 years?

d. What is Joe's monthly payment?

e. You found in part d that Joe has nothing left to invest until his mortgage is paid off. If he then invests the entire $1200 monthly in a mutual fund with the interest rate in part a, how much money will he have at the end of 30 years (that is, after 15 years of paying the mortgage and 15 years of investing)?

f. Who is ahead at the end of the 30 years and by how much?

g. Discuss to what extent the difference found in part f is due to the different interest rates or to the different amounts of time.

EXTENDED APPLICATION

TIME, MONEY, AND POLYNOMIALS

A *time line* is often helpful for evaluating complex investments. For example, suppose you buy a $1000 CD at time t_0. After one year $2500 is added to the CD at t_1. By time t_2, after another year, your money has grown to $3851 with interest. What rate of interest, called *yield to maturity* (YTM), did your money earn? A time line for this situation is shown in Figure 15.

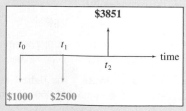

FIGURE 15

Assuming interest is compounded annually at a rate i, and using the compound interest formula, gives the following description of the YTM.

$$1000(1 + i)^2 + 2500(1 + i) = 3851$$

Source: COMAP.

To determine the yield to maturity, we must solve this equation for i. Since the quantity $1 + i$ is repeated, let $x = 1 + i$ and first solve the second-degree (quadratic) polynomial equation for x.

$$1000x^2 + 2500x - 3851 = 0$$

We can use the quadratic formula with $a = 1000$, $b = 2500$, and $c = -3851$.

$$x = \frac{-2500 \pm \sqrt{2500^2 - 4(1000)(-3851)}}{2(1000)}$$

We get $x = 1.0767$ and $x = -3.5767$. Since $x = 1 + i$, the two values for i are $0.0767 = 7.67\%$ and $-4.5767 = -457.67\%$. We reject the negative value because the final accumulation is greater than the sum of the deposits. In some applications, however, negative rates may be meaningful. By checking in the first equation, we see that the yield to maturity for the CD is 7.67%.

Now let us consider a more complex but realistic problem. Suppose Austin Caperton has contributed for 4 years to a retirement fund. He contributed $6000 at the beginning of the first year. At the beginning of the next 3 years, he contributed $5840, $4000, and $5200, respectively. At the end of the fourth year, he had $29,912.38 in his fund. The interest rate earned by the fund varied between 21% and -3%, so Caperton would like to know the YTM $= i$ for his hard-earned retirement dollars. From a time line (see Figure 16), we set up the following equation in $1 + i$ for Caperton's savings program.

$$6000(1 + i)^4 + 5840(1 + i)^3 + 4000(1 + i)^2$$
$$+ 5200(1 + i) = 29,912.38$$

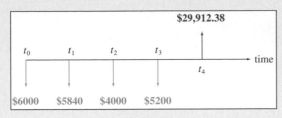

$29,912.38

t_0 t_1 t_2 t_3 time
                    t_4

$6000 $5840 $4000 $5200

FIGURE 16

Let $x = 1 + i$. We need to solve the fourth-degree polynomial equation

$$f(x) = 6000x^4 + 5840x^3 + 4000x^2 + 5200x$$
$$-29{,}912.38 = 0.$$

There is no simple way to solve a fourth-degree polynomial equation, so we will use a graphing calculator.

We expect that $0 < i < 1$, so that $1 < x < 2$. Let us calculate $f(1)$ and $f(2)$. If there is a change of sign, we will know that there is a solution to $f(x) = 0$ between 1 and 2. We find that

$$f(1) = -8872.38 \quad \text{and} \quad f(2) = 139{,}207.62.$$

Using a graphing calculator, we find that there is one positive solution to this equation, $x = 1.14$, so $i = \text{YTM} = 0.14 = 14\%$.

EXERCISES

1. Lorri Morgan received $50 on her 16th birthday, and $70 on her 17th birthday, both of which she immediately invested in the bank, with interest compounded annually. On her 18th birthday, she had $127.40 in her account. Draw a time line, set up a polynomial equation, and calculate the YTM.

2. At the beginning of the year, Blake Allvine invested $10,000 at 5% for the first year. At the beginning of the second year, he added $12,000 to the account. The total account earned 4.5% for the second year.

   a. Draw a time line for this investment.
   b. How much was in the fund at the end of the second year?
   c. Set up and solve a polynomial equation and determine the YTM. What do you notice about the YTM?

3. On January 2 each year for 3 years, Michael Bailey deposited bonuses of $1025, $2200, and $1850, respectively, in an account. He received no bonus the following year, so he made no deposit. At the end of the fourth year, there was $5864.17 in the account.

   a. Draw a time line for these investments.
   b. Write a polynomial equation in x ($x = 1 + i$) and use a graphing calculator to find the YTM for these investments.
   c. Go to the website WolframAlpha.com, and ask it to solve the polynomial from part b. Compare this method of solving the equation with using a graphing calculator.

4. Don Beville invested yearly in a fund for his children's college education. At the beginning of the first year, he invested $1000; at the beginning of the second year, $2000; at the third through the sixth, $2500 each year, and at the beginning of the seventh, he invested $5000. At the beginning of the eighth year, there was $21,259 in the fund.

   a. Draw a time line for this investment program.
   b. Write a seventh-degree polynomial equation in $1 + i$ that gives the YTM for this investment program.
   c. Use a graphing calculator to show that the YTM is less than 5.07% and greater than 5.05%.
   d. Use a graphing calculator to calculate the solution for $1 + i$ and find the YTM.
   e. Go to the website WolframAlpha.com, and ask it to solve the polynomial from part b. Compare this method of solving the equation with using a graphing calculator.

5. People often lose money on investments. Melissa Fischer invested $50 at the beginning of each of 2 years in a mutual fund, and at the end of 2 years her investment was worth $90.

   a. Draw a time line and set up a polynomial equation in $1 + i$. Solve for i.
   b. Examine each negative solution (rate of return on the investment) to see if it has a reasonable interpretation in the context of the problem. To do this, use the compound interest formula on each value of i to trace each $50 payment to maturity.

DIRECTIONS FOR GROUP PROJECT

Assume that you are in charge of a group of financial analysts and that you have been asked by the broker at your firm to develop a time line for each of the people listed in the exercises above. Prepare a report for each client that presents the YTM for each investment strategy. Make sure that you describe the methods used to determine the YTM in a manner that the average client should understand.

6 Logic

The rules of a game often include complex conditional statements, such as "if you roll doubles, you can roll again, but if you roll doubles twice in a row, you lose a turn." As exercises in this chapter illustrate, logical analysis of complex statements helps us clarify not only the rules of games but any precise use of language, from legal codes to medical diagnoses.

n 1943, Thomas Watson, head of IBM, made the now-infamous prediction, "I think there is a world market for maybe five computers." In 1977, Ken Olson, founder of Digital Equipment Corp., prophesied, "There is no reason anyone would want a computer in their home." ***Source: Microsoft Corp.*** Perhaps such predictions were so wrong because of the difficulty, in the early days of computing, of foreseeing later uses of the computer, such as for communication, shopping, entertainment, or information retrieval. It's rather amazing that such modern conveniences are made possible by a machine made of components based on mathematical logic. Even when you are using your computer for some seemingly nonmathematical activity, components, known as gates or switches, that have been developed from concepts that you will learn about in this chapter, are making it all possible.

6.1 Statements

APPLY IT   **How can a complex statement be analyzed to determine whether it is a truthful statement?**
You will be asked to answer this question in Exercise 89.

   Symbolic logic uses formal mathematics with symbols to represent statements and arguments in everyday language. Logic can help us determine whether a complex statement is true, as well as determine whether a conclusion necessarily follows from a set of assumptions.

   Many kinds of sentences occur in ordinary language, including factual statements, opinions, commands, and questions. Symbolic logic discusses statements and opinions that may be true or false, but not commands or questions.

Statements
   A **statement** is a declarative sentence that is either true or false, but not both simultaneously. For example, both of the following are statements:

> Mount McKinley is the tallest mountain in North America.

$$15 - 9 = 10.$$

The first sentence is true, while the second is false. None of the following sentences are considered statements in logic, however, because they cannot be identified as being either true or false:

> Tie your shoes.
>
> Are we having fun yet?
>
> This sentence is false.

The first sentence is a command or suggestion, while the second is a question. The third sentence is a paradox: If the sentence is true, then the sentence itself tells us that it must be false. But if what it says is false, then the sentence must be true. We avoid such paradoxes by disallowing statements that refer to themselves.

   When one or more simple statements are combined with **logical connectives** such as *and*, *or*, *not*, and *if . . . then*, the result is called a **compound statement**, while the simple statements that make up the compound statement are called **component statements**.

EXAMPLE 1 Compound Statements

Decide whether each statement is compound.

(a) George Washington was the first U.S. president, and John Adams was his vice president.

SOLUTION This statement is compound using the connective *and*. The component statements are "George Washington was the first U.S. president" and "John Adams was his vice president."

(b) If what you've told me is true, then we are in great peril.

SOLUTION This statement is also compound. The component statements "what you've told me is true" and "we are in great peril" are linked with the connective *if . . . then.*

(c) We drove across New Mexico toward the town with the curious name Truth or Consequences.

SOLUTION This statement is not compound. Even though *or* is a connective, here it is part of the name of the city. It is not connecting two statements.

(d) The money is not there.

SOLUTION Most logicians consider this statement to be compound, even though it has just one component statement, and we will do so in this book. The connective *not* is applied to the component statement "The money is there." **TRY YOUR TURN 1**

YOUR TURN 1 Is the following statement compound? "I bought Ben and Jerry's ice cream."

We will study the *and, or,* and *not* connectives in detail in this section, and return to the *if . . . then* connective in the third section of this chapter.

Negation

The **negation** of the statement "I play the guitar" is "I do not play the guitar." There are equivalent ways to say the same thing, such as "It is not true that I play the guitar." The negation of a true statement is false, and the negation of a false statement is true.

EXAMPLE 2 Negation

Give the negation of each statement.

(a) California is the most populous state in the country.

SOLUTION Form the negation using the word *not*: "California is not the most populous state in the country."

(b) It is not raining today.

SOLUTION If a statement already has the word *not*, we can remove it to form the negation: "It is raining today." **TRY YOUR TURN 2**

YOUR TURN 2 Write the negation of the following statement. "Wal-Mart is not the largest corporation in the USA."

EXAMPLE 3 Negation

Write the negation of each inequality.

(a) $x > 11$

SOLUTION The negation of "x is greater than 11" is "x is *not* greater than 11," or $x \leq 11$.

(b) $4x + 9y \leq 36$

SOLUTION The negation is $4x + 9y > 36$. **TRY YOUR TURN 3**

YOUR TURN 3 Write the negation of the following inequality. $4x + 2y < 5$

Symbols

To simplify work with logic, symbols are used. Statements are represented with letters, such as p, q, or r, while several symbols for connectives are shown in the following table. The table also gives the type of compound statement having the given connective.

Logic Symbols		
Connective	**Symbol**	**Type of Statement**
and	$\wedge$	Conjunction
or	$\vee$	Disjunction
not	$\sim$	Negation

The symbol $\sim$ represents the connective *not*. If p represents the statement "Barack Obama was president in 2010" then $\sim p$ represents "Barack Obama was not president in 2010." The statement $\sim p$ could also be translated as "It is not true that Barack Obama was president in 2010." There is usually more than one way to express a negation, and so your answer may not always agree exactly with ours. We recommend avoiding convoluted wording.

In applications, choosing meaningful letters will help you remember what the letter represents. While p may be perfectly good for representing a generic statement, a statement such as "Django is a good dog" might be better represented by the letter d.

EXAMPLE 4 Symbolic Statements

Let h represent "My backpack is heavy," and r represent "It's going to rain." Write each symbolic statement in words.

(a) $h \wedge r$

SOLUTION From the table, $\wedge$ represents *and*, so the statement represents

My backpack is heavy and it's going to rain.

(b) $\sim h \vee r$

SOLUTION The *not* applies only to the first symbol, not the entire expression:

My backpack is not heavy or it's going to rain.

(c) $\sim (h \vee r)$

SOLUTION Because of the parentheses, the *not* applies to the entire expression:

It is not the case that either my backpack is heavy or it's going to rain.

(d) $\sim (h \wedge r)$

SOLUTION Again, the *not* applies to the entire expression:

It is not the case that both my backpack is heavy and it's going to rain.

TRY YOUR TURN 4

YOUR TURN 4 Write the symbolic statement in words. $h \wedge \sim r$.

The statement in Example 4(c) is usually translated, "Neither h nor r," as in "Neither is my backpack heavy nor is it going to rain."

The negation of the negation of a statement is simply the statement itself. For example, the negation of the statement in Example 4(d) is $h \wedge r$, or "My backpack is heavy and it's going to rain." Symbolically, $\sim(\sim p)$ is equivalent to p.

We can represent the fact that the negation of a true statement is false and the negation of a false statement is true using a table known as a **truth table**, which shows all possible combinations of truth values for the component statements, as well as the corresponding truth value for the compound statement under consideration. Here is the truth table for negation.

Truth Table for the Negation *not p*

p	$\sim p$
T	F
F	T

Conjunction

The word logicians use for "*p and q*," denoted $p \wedge q$, is **conjunction**. In everyday language, *and* conveys the idea that both component statements are true. For example, the statement

My birthday is in April and yours is in May

would be true only if my birthday is indeed in April and yours is in May. If either part were false, the statement would be considered false. We represent this definition of $p \wedge q$ symbolically using the following truth table for conjunction. Notice that there are four possible combinations of truth values for p and q.

Truth Table for the Conjunction *p and q*

p	q	$p \wedge q$
T	T	T
T	F	F
F	T	F
F	F	F

Although the order of the rows in a truth table is arbitrary, you should follow the order that we use, which is not only standard but also organized and easy to remember. A different method can make it hard to compare answers, and with a disorganized method one could easily miss a case or even have duplicate cases. Notice that in the columns labeled with a single variable (p or q), the rightmost column (the q column) alternates T, F, T, F. The next column to the left (the p column) has two T's followed by two F's. When we introduce truth tables with three variables, the leftmost column will have four T's followed by four F's.

YOUR TURN 5 Let p represent "7 < 2" and q represent "4 > 3." Find the truth value of $\sim p \wedge q$.

EXAMPLE 5 Truth Value

Let p represent "5 > 3" and let q represent "6 < 0." Find the truth value of $p \wedge q$.

SOLUTION

Here p is true and q is false. Looking in the second row of the conjunction truth table shows that $p \wedge q$ is false. **TRY YOUR TURN 5**

TECHNOLOGY NOTE

Some calculators have logic functions that allow the user to test the truth or falsity of statements involving =, ≠, >, ≥, <, and ≤. On the TI-84 Plus calculator, functions from the TEST menu, illustrated in Figure 1, return a 1 when a statement is true and a 0 when a statement is false. Figure 2 shows the input of 4 > 9 and the corresponding output of 0, indicating that the statement is false.

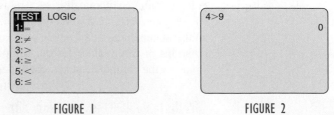

FIGURE 1

FIGURE 2

Example 5 can also be completed with the help of a TI-84 Plus calculator. Using the LOGIC menu, illustrated in Figure 3, we input 5 > 3 and 6 < 0 into the calculator. The output is zero, shown in Figure 4, which means the statement is false.

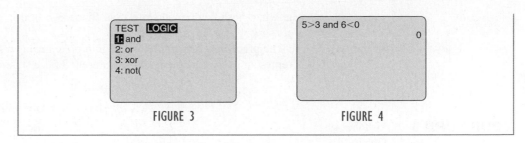

FIGURE 3   FIGURE 4

There's another word that has the same logical meaning as the word *and*, namely *but*. For example, the statement

I was not ready yesterday, but I am ready today

conveys the same logical meaning as "I was not ready yesterday, and I am ready today," with the additional idea of contrast between my status on the two days. Contrast may be relevant in normal conversation, but it makes no difference in logic.

Disjunction
In ordinary language, the word *or* can be ambiguous. For example, the statement

Those with a passport or driver's license will be admitted

means that anyone will be admitted who has a passport or a driver's license or both. On the other hand, the statement

You can have a piece of cake or a piece of fruit

probably means you cannot have both.

In logic, the word *or* has the first meaning, known as *inclusive disjunction* or just **disjunction**. It is written with the symbol $\vee$, so that $p \vee q$ means "*p* or *q* or both." It is only false when both component statements are false. The truth table for disjunction is given below.

Truth Table for the Disjunction *p or q*

p	q	$p \vee q$
T	T	T
T	F	T
F	T	T
F	F	F

NOTE
In English, the *or* is often interpreted as exclusive disjunction, not the way it is used in logic. For more on the exclusive disjunction, see Exercises 35–38 in the next section.

Compound Statements
The next two examples show how we can find the truth value of compound statements.

EXAMPLE 6 Truth Value of a Compound Statement

Suppose p is false, q is true, and r is false. What is the truth value of the compound statement $\sim p \wedge (q \vee \sim r)$?

SOLUTION Here parentheses are used to group q and $\sim r$ together. Work first inside the parentheses. Since r is false, $\sim r$ will be true. Since $\sim r$ is true and q is true and an *or* statement is true when either component is true, $q \vee \sim r$ must be true. An *and* statement is only true when both components are true. Since $\sim p$ is true and $q \vee \sim r$ is true, the statement $\sim p \wedge (q \vee \sim r)$ is true.

The preceding paragraph may be interpreted using a short-cut symbolic method, letting T represent a true statement and F represent a false statement:

$$\sim p \wedge (q \vee \sim r)$$
$$\sim F \wedge (T \vee \sim F) \qquad \text{\textit{p} is false, \textit{q} is true, \textit{r} is false.}$$
$$T \wedge (T \vee T) \qquad \text{\textasciitilde F gives T.}$$
$$T \wedge T \qquad \text{T} \vee \text{T gives T.}$$
$$T. \qquad \text{T} \wedge \text{T gives T.}$$

YOUR TURN 6 If p is false, q is true, and r is false, find the truth value of the statement: $(\sim p \wedge q) \vee r$.

The T in the final row indicates that the compound statement is true. **TRY YOUR TURN 6**

EXAMPLE 7 Mathematical Statements

Let p represent the statement "$3 > 2$," q represent "$5 < 4$," and r represent "$3 < 8$." Decide whether the following statements are *true* or *false*.

(a) $\sim p \wedge \sim q$

SOLUTION Since p is true, $\sim p$ is false. By the *and* truth table, if one part of an "and" statement is false, the entire statement is false. This makes $\sim p \wedge \sim q$ false.

(b) $\sim (p \wedge q)$

SOLUTION First, work within the parentheses. Since p is true and q is false, $p \wedge q$ is false by the *and* truth table. Next, apply the negation. The negation of a false statement is true, making $\sim (p \wedge q)$ a true statement.

(c) $(\sim p \wedge r) \vee (\sim q \wedge \sim p)$

SOLUTION Here p is true, q is false, and r is true. This makes $\sim p$ false and $\sim q$ true. By the *and* truth table, the statement $\sim p \wedge r$ is false, and the statement $\sim q \wedge \sim p$ is also false. Finally,

$$(\sim p \wedge r) \vee (\sim q \wedge \sim p)$$
$$\downarrow \qquad\qquad \downarrow$$
$$\text{F} \qquad \vee \qquad \text{F,}$$

YOUR TURN 7 Let p represent "$7 < 2$," q represent "$4 > 3$," and r represent "$2 > 8$." Find the truth value of $p \vee (\sim q \wedge r)$.

which is false by the *or* truth table. (For an alternate solution, see Example 3(b) in the next section.) **TRY YOUR TURN 7**

NOTE The expression $\sim p \wedge \sim q$ in part (a) is often expressed in English as "Neither p nor q." We saw at the end of Example 4 that the expression $\sim (p \vee q)$ can also be expressed as "Neither p nor q." As we shall see in the next section, the expressions $\sim p \wedge \sim q$ and $\sim (p \vee q)$ are equivalent.

TECHNOLOGY NOTE Figure 5 shows the results of using a T1-84 Plus calculator to solve parts (a) and (b) of Example 7.

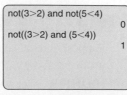

FIGURE 5

6.1 EXERCISES

Decide whether each of the following is a statement. If it is a statement, decide whether or not it is compound.

1. Montevideo is the capital of Uruguay.

2. John Jay was the first chief justice of the United States.

3. Don't feed the animals.

4. Do unto others as you would have them do unto you.

5. $2 + 2 = 5$ and $3 + 3 = 7$

6. $x < 7$ or $x > 14$

7. Got milk?

8. Is that all there is?

9. I am not a crook.

10. China does not have a population of more than 1 billion.

11. She enjoyed the comedy team of Penn and Teller.

12. The New Hampshire motto is "Live free or die."

13. If I get an A, I will celebrate.

14. If it's past 8:00, then we are late.

Write a negation for each statement.

15. My favorite flavor is chocolate.

16. This is not the time to complain.

Give a negation of each inequality.

17. $y > 12$ 18. $x < -6$

19. $q \geq 5$ 20. $r \leq 19$

21. Try to negate the sentence "The exact number of words in this sentence is 10" and see what happens. Explain the problem that arises.

22. Explain why the negation of "$r > 4$" is not "$r < 4$."

Let b represent the statement "I'm getting better" and d represent the statement "My parrot is dead." Translate each symbolic statement into words.

23. $\sim b$ 24. $\sim d$

25. $\sim b \vee d$ 26. $b \wedge \sim d$

27. $\sim(b \wedge \sim d)$ 28. $\sim(b \vee d)$

Use the concepts introduced in this section to answer Exercises 29–34.

29. If q is false, what must be the truth value of $(p \wedge \sim q) \wedge q$?

30. If q is true, what must be the truth value of $q \vee (q \wedge \sim p)$?

31. If $p \wedge q$ is true, then q must be _____.

32. If $p \vee q$ is false, and p is false, then q must be _____.

33. If $\sim(p \vee q)$ is true, what must be the truth values of each of the component statements?

34. If $\sim(p \wedge q)$ is false, what must be the truth values of the component statements?

Let p represent a false statement and let q represent a true statement. Find the truth value of each compound statement.

35. $\sim p$ 36. $\sim q$

37. $p \vee q$ 38. $p \wedge q$

39. $p \vee \sim q$ 40. $\sim p \wedge q$

41. $\sim p \vee \sim q$ 42. $p \wedge \sim q$

43. $\sim(p \wedge \sim q)$ 44. $\sim(\sim p \vee \sim q)$

45. $\sim[\sim p \wedge (\sim q \vee p)]$ 46. $\sim[(\sim p \wedge \sim q) \vee \sim q]$

47. Is the statement $3 \geq 1$ a conjunction or a disjunction? Why?

48. Why is the statement $6 \geq 2$ true? Why is $6 \geq 6$ true?

Let p represent a true statement, and q and r represent false statements. Find the truth value of each compound statement.

49. $(p \wedge r) \vee \sim q$ 50. $(q \vee \sim r) \wedge p$

51. $p \wedge (q \vee r)$ 52. $(\sim p \wedge q) \vee \sim r$

53. $\sim(p \wedge q) \wedge (r \vee \sim q)$ 54. $(\sim p \wedge \sim q) \vee (\sim r \wedge q)$

55. $\sim[(\sim p \wedge q) \vee r]$ 56. $\sim[r \vee (\sim q \wedge \sim p)]$

Let p represent the statement "$2 > 7$," let q represent the statement "$8 \leq 6$," and let r represent the statement "$19 \leq 19$". Find the truth value of each compound statement.

57. $p \wedge r$ 58. $p \vee \sim q$

59. $\sim q \vee \sim r$ 60. $\sim p \wedge \sim r$

61. $(p \wedge q) \vee r$ 62. $\sim p \vee (\sim r \vee \sim q)$

63. $(\sim r \wedge q) \vee \sim p$ 64. $\sim(p \vee \sim q) \vee \sim r$

APPLICATIONS

Business and Economics

Income Tax The following excerpts appear in a guide for preparing income tax reports. *Source: Your Income Tax 2010.*

  a. *Which filing status should you use?*

  b. *Scholarships and fellowships of a degree candidate are tax free to the extent that the grants pay for tuition and course-related fees, books, supplies, and equipment that are required for courses.*

  c. *You do not want to withhold too little from your pay and you do not want to withhold too much.*

  d. *An individual does not have to be your biological child to be a "qualifying" child.*

65. Which of these excerpts are statements?

66. Which of these excerpts are compound statements?

67. Write the negation of excerpt d.

68. Determine p and q to symbolically represent excerpt c.

Technology For Exercises 69–74, let a represent the statement "Apple Computer, Inc. developed the iPad," and j represent the statement, "Steve Jobs is the CEO of Apple Computer, Inc." Convert each compound statement into symbols.

69. Apple Computer, Inc. developed the iPad, and Steve Jobs is the CEO of Apple Computer, Inc.

70. Apple Computer, Inc. did not develop the iPad, and Steve Jobs is not the CEO of Apple Computer, Inc.

71. Apple Computer, Inc. did not develop the iPad, or Steve Jobs is the CEO of Apple Computer, Inc.

72. Apple Computer, Inc. developed the iPad, or Steve Jobs is the CEO of Apple Computer, Inc.

73. Suppose the statements that Apple Computer, Inc. developed the iPad and that Steve Jobs is CEO of Apple Computer, Inc. are both true. Which of Exercises 69–72 are true statements?

74. Suppose the statements that Apple Computer, Inc. developed the iPad and that Steve Jobs is CEO of Apple Computer, Inc. are both false. Which of Exercises 69–72 are true statements?

Life Sciences

Medicine The following excerpts appear in a home medical reference book. *Source: American College of Physicians Complete Home Medical Guide.*

a. *Can you climb one or two flights of stairs without shortness of breath or heaviness or fatigue in your legs?*

b. *Regularly doing exercises that concentrate on strengthening particular muscle groups and improving overall flexibility can help prevent back pain and keep you mobile.*

c. *If you answered yes to all of the questions above, you are reasonably fit.*

d. *These chemical compounds act as natural antidepressants, and they can help you feel more relaxed.*

e. *You may find that exercise helps you cope with stress.*

75. Which of these excerpts are statements?

76. Which of these excerpts are compound statements?

77. Write the negation of excerpt e.

Social Sciences

Law The following excerpts appear in a guide to common laws. *Source: Law for Dummies.*

a. *If you're involved in an accident, regardless of who you think is at fault, stop your vehicle and inspect the damage.*

b. *Don't be pressured into signing a contract.*

c. *You can file a civil lawsuit yourself, or your attorney can do it for you.*

d. *You can't marry unless you're at least 18 years old or unless you have the permission of your parents or guardian.*

e. *The Bill of Rights defines the fundamental rights of all Americans.*

78. Which of these excerpts are statements?

79. Which of these excepts are compound statements?

80. Write the negation of excerpt e.

81. Philosophy Read each of the following quotes from ancient philosophers. Provide an argument why these quotes may or may not be called statements. *Source: Masterworks of Philosophy.*

a. "A friend is a friend of someone."—Socrates

b. "Every art, and every science reduced to a teachable form, and in like manner every action and moral choice, aims, it is thought, at some good: for which reason a common and by no means a bad description of what the Chief Good is, 'that which all things aim at.'"—Aristotle

c. "Furthermore, Friendship helps the young to keep from error: the old, in respect of attention and such deficiencies in action as their weakness makes them liable to; and those who are in their prime, in respect of noble deeds, because they are thus more able to devise plans and carry them out."—Aristotle

82. Bible Read each of the following quotes from the biblical book Proverbs. Provide an argument why these quotes may or may not be called statements. *Source: TNIV Bible.*

a. "A gentle answer turns away wrath."—Proverbs 15:1

b. "The hot-tempered stir up dissension, but those who are patient calm a quarrel."—Proverbs 15:18

c. "When justice is done, it brings joy to the righteous but terror to evildoers."—Proverbs 21:15

d. "Do not exploit the poor because they are poor and do not crush the needy in court."—Proverbs 22:22

e. "Apply your heart to instruction and your ears to words of knowledge."—Proverbs 23:12

General Interest

Football For Exercises 83–88, let n represent the statement "New Orleans won the Super Bowl" and m represent the statement "Peyton Manning is the best quarterback." Convert each compound statement into symbols.

83. New Orleans won the Super Bowl but Peyton Manning is not the best quarterback.

84. New Orleans did not win the Super Bowl or Peyton Manning is not the best quarterback.

85. New Orleans did not win the Super Bowl or Peyton Manning is the best quarterback.

86. New Orleans did not win the Super Bowl but Peyton Manning is the best quarterback.

87. Neither did New Orleans win the Super Bowl nor is Peyton Manning the best quarterback.

88. Either New Orleans won the Super Bowl or Peyton Manning is the best quarterback, and it is not the case that both New Orleans won the Super Bowl and Peyton Manning is the best quarterback.

89. APPLY IT Suppose the statements that New Orleans won the Super Bowl and that Peyton Manning is the best quarterback are both true. Which of Exercises 83–88 are true statements?

90. Suppose the statements that New Orleans won the Super Bowl and that Peyton Manning is the best quarterback are both false. Which of Exercises 83–88 are true statements?

YOUR TURN ANSWERS

1. No

2. Wal-Mart is the largest corporation in the USA.

3. $4x + 2y \geq 5$

4. My backpack is heavy, and it's not going to rain.

5. True

6. True

7. False

6.2 Truth Tables and Equivalent Statements

APPLY IT When using a search engine on the Internet, you are asked to supply key words. How does the search engine connect these key words?
You will be asked to explore this question in Exercise 51.

In the previous section, we created truth tables for some simple logical expressions. We will now create truth tables for more complex statements, and determine for what values of the individual component statements the complex statement is true. We will continue to use the following standard format for listing the possible truth values in compound statements involving two statements.

p	q	**Compound Statement**
T	T	
T	F	
F	T	
F	F	

EXAMPLE 1 **Truth Tables**

(a) Construct a truth table for $(\sim p \wedge q) \vee \sim q$.

SOLUTION Begin by listing all possible combinations of truth values for p and q. Then find the truth values of $\sim p \wedge q$. Start by listing the truth values of $\sim p$, which are the opposite of those of p.

p	q	$\sim p$
T	T	F
T	F	F
F	T	T
F	F	T

Use only the "$\sim p$" column and the "q" column, along with the *and* truth table, to find the truth values of $\sim p \wedge q$. List them in a separate column.

p	q	$\sim p$	$\sim p \wedge q$
T	T	F	F
T	F	F	F
F	T	T	T
F	F	T	F

Next include a column for $\sim q$.

p	q	$\sim p$	$\sim p \wedge q$	$\sim q$
T	T	F	F	F
T	F	F	F	T
F	T	T	T	F
F	F	T	F	T

Finally, make a column for the entire compound statement. To find the truth values, use *or* to combine $\sim p \wedge q$ with $\sim q$.

p	q	$\sim p$	$\sim p \wedge q$	$\sim q$	$(\sim p \wedge q) \vee \sim q$
T	T	F	F	F	F
T	F	F	F	T	T
F	T	T	T	F	T
F	F	T	F	T	T

YOUR TURN 1 Construct a truth table for $p \wedge (\sim p \vee q)$. If p and q are both true, find the truth value of $p \wedge (\sim p \vee q)$.

(b) Suppose both p and q are true. Find the truth value of $(\sim p \wedge q) \vee \sim q$.

SOLUTION Look in the first row of the final truth table above, where both p and q have truth value T. Read across the row to find that the compound statement is false.

TRY YOUR TURN 1

EXAMPLE 2 Truth Table

Construct a truth table for the following statement:

I'm bringing the food, or Heather Peck's bringing the food and I'm not.

SOLUTION If we let i represent "I'm bringing the food" and h represent "Heather Peck is bringing the food," the statement can be represented symbolically as $i \vee (h \wedge \sim i)$. Proceed as shown in the following truth table.

i	h	$\sim i$	$h \wedge \sim i$	$i \vee (h \wedge \sim i)$
T	T	F	F	T
T	F	F	F	T
F	T	T	T	T
F	F	T	F	F

YOUR TURN 2 Construct a truth statement for the following statement: "I do not order pizza, or you do not make dinner and I order pizza."

Notice from the truth table above that the only circumstances under which the original statement is false is when both statements "I'm bringing the food" and "Heather Peck is bringing the food" are false.

TRY YOUR TURN 2

If a compound statement involves three component statements p, q, and r, we will use the following format in setting up the truth table.

p	q	r	**Compound Statement**
T	T	T	
T	T	F	
T	F	T	
T	F	F	
F	T	T	
F	T	F	
F	F	T	
F	F	F	

As we mentioned in the previous section, the rightmost column (the r column) alternates T, F, T, F. The next column to the left (the q column) has two T's followed by two F's, and then repeats this pattern. The leftmost column has four T's followed by four F's.

EXAMPLE 3 Truth Tables

(a) Construct a truth table for $(\sim p \wedge r) \vee (\sim q \wedge \sim p)$.

SOLUTION This statement has three component statements, p, q, and r. The truth table thus requires eight rows to list all possible combinations of truth values of p, q, and r. The final truth table is formed in much the same way as in the previous examples.

p	q	r	$\sim p$	$\sim p \wedge r$	$\sim q$	$\sim q \wedge \sim p$	$(\sim p \wedge r) \vee (\sim q \wedge \sim p)$
T	T	T	F	F	F	F	F
T	T	F	F	F	F	F	F
T	F	T	F	F	T	F	F
T	F	F	F	F	T	F	F
F	T	T	T	T	F	F	T
F	T	F	T	F	F	F	F
F	F	T	T	T	T	T	T
F	F	F	T	F	T	T	T

(b) Suppose p is true, q is false, and r is true. Find the truth value of $(\sim p \wedge r) \vee (\sim q \wedge \sim p)$.

SOLUTION By the third row of the truth table in part (a), the compound statement is false. (This is an alternate method for working part (c) of Example 7 of the previous section.)

Notice that the truth table in Example 3(a) has three component statements and eight rows. The truth tables for the conjunction and disjunction have two component statements, and each table has four rows. The truth table for the negation has one component and two rows. These are summarized in the next table. Can we use this information to determine the number of rows in a truth table with n components?

Number of Rows in a Truth Table	
Number of Statements	**Number of Rows**
1	$2 = 2^1$
2	$4 = 2^2$
3	$8 = 2^3$

One strategy for solving this type of problem is noticing a pattern and using *inductive reasoning*, or reasoning that uses particular facts to find a general rule. If n is a counting number, and a logical statement is composed of n component statements, we can use inductive reasoning to conjecture that its truth table will have 2^n rows. This can be proved using ideas in Chapter 8.

Intuitively, it's not hard to see why adding a statement doubles the number of rows. For example, if we wanted to construct a truth table with the four statements p, q, r, and s, we could start with the truth table for p, q, and r, which has eight rows. We must let s have the value of T for each of these rows, and we must also let s have the value of F for each of these rows, giving a total of 16 rows.

Number of Rows in a Truth Table

A logical statement having n component statements will have 2^n rows in its truth table.

Alternative Method for Constructing Truth Tables

After making a reasonable number of truth tables, some people prefer the shortcut method shown in Example 4, which repeats Examples 1 and 3.

EXAMPLE 4 Truth Tables

Construct the truth table for each statement.

(a) $(\sim p \wedge q) \vee \sim q$

SOLUTION Start by inserting truth values for $\sim p$ and for q.

p	q	$(\sim p \wedge q) \vee \sim q$	
T	T	F	T
T	F	F	F
F	T	T	T
F	F	T	F

Next, use the *and* truth table to obtain the truth values of $\sim p \wedge q$.

p	q	$(\sim p \wedge q) \vee \sim q$		
T	T	F	F	T
T	F	F	F	F
F	T	T	T	T
F	F	T	F	F

Now disregard the two preliminary columns of truth values for $\sim p$ and for q, and insert truth values for $\sim q$.

p	q	$(\sim p \wedge q) \vee \sim q$	
T	T	F	F
T	F	F	T
F	T	T	F
F	F	F	T

Finally, use the *or* truth table.

p	q	$(\sim p \wedge q) \vee \sim q$		
T	T	F	F	F
T	F	F	T	T
F	T	T	T	F
F	F	F	T	T

These steps can be summarized as follows.

p	q	$(\sim p \wedge q) \vee \sim q$				
T	T	F	F	T	F	F
T	F	F	F	F	T	T
F	T	T	T	T	T	F
F	F	T	F	F	T	T
		①	②	①	④	③

The circled numbers indicate the order in which the various columns of the truth table were found.

(b) $(\sim p \wedge r) \vee (\sim q \wedge \sim p)$

SOLUTION   Work as follows.

p	q	r			$(\sim p \wedge r) \vee (\sim q \wedge \sim p)$				
T	T	T	F	F T	F	F	F	F
T	T	F	F	F F	F	F	F	F
T	F	T	F	F T	F	T	F	F
T	F	F	F	F F	F	T	F	F
F	T	T	T	T T	T	F	F	T
F	T	F	T	F F	F	F	F	T
F	F	T	T	T T	T	T	T	T
F	F	F	T	F F	T	T	T	T
			①	② ①	⑤	③	④	③

Equivalent Statements

One application of truth tables is illustrated by showing that two statements are equivalent; by definition, two statements are **equivalent** if they have the same truth value in *every* possible situation. The columns of each truth table that were the last to be completed will be exactly the same for equivalent statements.

EXAMPLE 5   **Equivalent Statements**

Are the statements

$$\sim p \wedge \sim q \qquad \text{and} \qquad \sim(p \vee q)$$

equivalent?

SOLUTION   To find out, make a truth table for each statement, with the following results.

p	q	$\sim p \wedge \sim q$
T	T	F
T	F	F
F	T	F
F	F	T

p	q	$\sim(p \vee q)$
T	T	F
T	F	F
F	T	F
F	F	T

Since the truth values are the same in all cases, as shown in the columns in color, the statements $\sim p \wedge \sim q$ and $\sim(p \vee q)$ are equivalent. Equivalence is written with a three-bar symbol, $\equiv$. Using this symbol, $\sim p \wedge \sim q \equiv \sim(p \vee q)$.

In the same way, the statements $\sim p \vee \sim q$ and $\sim(p \wedge q)$ are equivalent. We call these equivalences **De Morgan's Laws.***

De Morgan's Laws

For any statements p and q,

$$\sim(p \vee q) \equiv \sim p \wedge \sim q$$

$$\sim(p \wedge q) \equiv \sim p \vee \sim q.$$

DeMorgan's Laws can be used to find the negations of certain compound statements.

---

*Augustus De Morgan (1806–1871) was born in India, where his father was serving as an English army officer. After studying mathematics at Cambridge, he prepared for a career in law because his performance on the challenging Tripos Exam in mathematics was not very good. Nevertheless, De Morgan was offered a position as chair of the department of mathematics at London University. He wrote and taught mathematics with great clarity and dedication, and he was also an excellent flutist.

EXAMPLE 6 Negation

Find the negation of each statement, applying De Morgan's Laws to simplify.

(a) It was a dark and stormy night.

SOLUTION If we first rephrase the statement as "The night was dark and the night was stormy," we can let d represent "the night was dark" and s represent "the night was stormy." The original compound statement can then be written $d \wedge s$. The negation of $d \wedge s$ is $\sim(d \wedge s)$. Using the second of De Morgan's Laws, $\sim(d \wedge s) \equiv \sim d \vee \sim s$. In words this reads:

The night was not dark or the night was not stormy.

In retrospect, this should be obvious. If it's not true that the night was dark and stormy, then either the night wasn't dark or it wasn't stormy.

(b) Either John will play the guitar or George will not sing.

SOLUTION In this book, we interpret "either . . . or" as disjunction; the word "either" just highlights that a disjunction is about to occur. Using the first of De Morgan's Laws, $\sim(j \vee \sim g) \equiv \sim j \wedge \sim(\sim g)$. Using the observation from the last section that the negation of a negation of a statement is simply the statement, we can simplify the last statement to $\sim j \wedge g$. In words this reads:

John will not play the guitar and George will sing.

(c) $(\sim p \wedge q) \vee \sim r$

SOLUTION After negating, we apply the first of De Morgan's Laws, changing $\vee$ to $\wedge$:

$$\sim[(\sim p \wedge q) \vee \sim r] \equiv \sim(\sim p \wedge q) \wedge \sim(\sim r).$$

The last term simplifies to r. Applying the second of De Morgan's Laws to the first term yields

$$\sim(\sim p \wedge q) \wedge r \equiv (p \vee \sim q) \wedge r,$$

where we have replaced $\sim(\sim p)$ with p. (From now on, we will make this simplification without mentioning it.) **TRY YOUR TURN 3**

YOUR TURN 3 Find the negation of the following statement: "You do not make dinner or I order pizza."

6.2 EXERCISES

Give the number of rows in the truth table for each compound statement.

1. $p \vee \sim r$

2. $p \wedge (r \wedge \sim s)$

3. $(\sim p \wedge q) \vee (\sim r \vee \sim s) \wedge r$

4. $[(p \vee q) \wedge (r \wedge s)] \wedge (t \vee \sim p)$

5. $[(\sim p \wedge \sim q) \wedge (\sim r \wedge s \wedge \sim t)] \wedge (\sim u \vee \sim v)$

6. $[(\sim p \wedge \sim q) \vee (\sim r \vee \sim s)] \vee [(\sim m \wedge \sim n) \wedge (u \wedge \sim v)]$

7. If the truth table for a certain compound statement has 64 rows, how many distinct component statements does it have?

8. Is it possible for the truth table of a compound statement to have exactly 48 rows? Why or why not?

Construct a truth table for each compound statement.

9. $\sim p \wedge q$

10. $\sim p \vee \sim q$

11. $\sim(p \wedge q)$

12. $p \vee \sim q$

13. $(q \vee \sim p) \vee \sim q$

14. $(p \wedge \sim q) \wedge p$

15. $\sim q \wedge (\sim p \vee q)$

16. $\sim p \vee (\sim q \wedge \sim p)$

17. $(p \vee \sim q) \wedge (p \wedge q)$

18. $(\sim p \wedge \sim q) \vee (\sim p \vee q)$

19. $(\sim p \wedge q) \wedge r$

20. $r \vee (p \wedge \sim q)$

21. $(\sim p \wedge \sim q) \vee (\sim r \vee \sim p)$

22. $(\sim r \vee \sim p) \wedge (\sim p \vee \sim q)$

23. $\sim(\sim p \wedge \sim q) \vee (\sim r \vee \sim s)$

24. $(\sim r \vee s) \wedge (\sim p \wedge q)$

Write the negation of each statement, applying De Morgan's Laws to simplify.

25. It's vacation, and I am having fun.

26. Rachel Reeve was elected president and Joanne Dill was elected treasurer.

27. Either the door was unlocked or the thief broke a window.

28. Sue brings the wrong book or she forgets the notes.

29. I'm ready to go, but Naomi Bahary isn't.

30. You can lead a horse to water, but you cannot make him drink.

31. $12 > 4$ or $8 = 9$

32. $2 + 3 = 5$ and $12 + 13 = 15$

33. Larry or Moe is out sick today.

34. You and I have gone through a lot.

35. Complete the truth table for *exclusive disjunction.* The symbol $\underline{\vee}$ represents "one or the other is true, but not both."

p	q	$p \underline{\vee} q$
T	T	
T	F	
F	T	
F	F	

Exclusive disjunction

Decide whether the following compound statements are true or false. Remember from Exercise 35 that $\underline{\vee}$ is the exclusive disjunction; that is, assume "either p or q is true, but not both."

36. $(3 + 1 = 4) \underline{\vee} (2 + 5 = 7)$

37. $(3 + 1 = 4) \underline{\vee} (2 + 5 = 9)$

38. $(3 + 1 = 7) \underline{\vee} (2 + 5 = 7)$

39. Let p represent $2\sqrt{6} - 4\sqrt{5} > -1$, q represent

$$\frac{14 - 7\sqrt{8}}{2.5 - \sqrt{5}} > -22, \text{ and } s \text{ represent } \frac{7 - \dfrac{5}{\sqrt{3}}}{\sqrt{8} - 2} < \frac{\sqrt{3}}{\sqrt{2}}.$$

Use the LOGIC menu on a graphing calculator to find the truth value of each statement.

a. $p \wedge q$ **b.** $\sim p \wedge q$

c. $\sim(p \vee q)$ **d.** $(s \wedge \sim p) \vee (\sim s \wedge q)$

APPLICATIONS

Business and Economics

40. **Income Tax** The following statement appears in a guide for preparing income tax reports. Use one of De Morgan's Laws to write the negation of this statement. *Source: Your Income Tax 2010.*

    Tips of less than $20 per month are taxable but are not subject to withholding.

41. **Warranty** The following statement appears in an iPhone warranty guide. Use one of De Morgan's Laws to negate this statement. *Source: AppleCare Protection Plan.*

Service will be performed at the location, or the store may send the Covered Equipment to an Apple repair service location to be repaired.

42. **eBay** The eBay Buyer Protection plan guarantees that a buyer will receive an item and that it will be as described, or eBay will cover the purchase price plus shipping. Let r represent "A buyer will receive an item," d represent "It will be as described," and e represent "eBay will cover the purchase price plus shipping." Write the guarantee symbolically, and then construct a truth table for the statement, putting the variables in the order r, d, and e. Under what conditions would the guarantee be false? *Source: eBay®.*

43. **Guarantees** The guarantee on a brand of vacuum cleaner reads: "You will be completely satisfied or we will refund your money without asking any questions." Let s represent "You will be completely satisfied," r represent "We will refund your money," and q represent "We will ask you questions." Write the guarantee symbolically and then construct a truth table for the statement, putting the variables in the order s, r, q. Under what conditions would the guarantee be false?

Life Sciences

44. **Medicine** The following statements appear in a home medical reference book. Define p and q so that the statements can be written symbolically. Then negate each statement. *Source: American College of Physicians Complete Home Medical Guide.*

    a. Tissue samples may be taken from almost anywhere in the body, and the procedure used depends on the site.

    b. The procedure can be carried out quickly in the doctor's office and is not painful.

    c. Fluid samples may be examined for infection, or the cells in the fluid may be separated and examined to detect other abnormalities.

Social Sciences

45. **Law** Attorneys sometimes use the phrase "and/or." This phrase corresponds to which usage of the word *or*: inclusive or exclusive?

46. **Law** The following statement appears in a guide to common laws. Define p and q so that the statement can be written symbolically. Then negate the statement. *Source: Law for Dummies.*

    You can file a complaint yourself as you would if you were using small claims court, or your attorney can file one for you.

47. **Presidential Quote** Use both of De Morgan's Laws to rewrite the negation of the following quote made by President John F. Kennedy at Vanderbilt University on March 18, 1963:

    Liberty without learning is always in peril, and learning without liberty is always in vain. *Source: Masters of Chiasmus.*

48. **Politician** Senator Pompous B. Blowhard made the following campaign promise: "I will cut taxes and eliminate the deficit, or I will not run for reelection." Let c represent "I will cut taxes," let e represent "I will eliminate the deficit," and let r represent "I will run for reelection." Write the promise symbolically and then construct a truth table for the statement, putting the variables in the order c, e, r. Under what conditions would the promise be false?

General Interest

49. Yahtzee®   The following statement appears in the instructions for the Milton Bradley game Yahtzee®. *Source: Milton Bradley Company.* Negate the statement.

> You could reroll the die again for your Large Straight or set aside the 2 Twos and roll for your Twos or for 3 of a Kind.

50. Logic Puzzles   Raymond Smullyan is one of today's foremost writers of logic puzzles. Smullyan proposed a question, based on the classic Frank Stockton short story, in which a prisoner must make a choice between two doors: Behind one is a beautiful lady, and behind the other is a hungry tiger. What if each door has a sign, and the prisoner knows that only one sign is true? The sign on Door 1 reads: In this room there is a lady and in the other room there is a tiger. The sign on Door 2 reads: In one of these rooms there is a lady and in one of these rooms there is a tiger. With this information, determine what is behind each door. *Source: The Lady or the Tiger? And Other Logic Puzzles.*

51. APPLY IT   Describe how a search engine on the Internet uses key words and logical connectives to locate information.

YOUR TURN ANSWERS

1.

p	q	$\sim p$	$\sim p \vee q$	$p \wedge (\sim p \vee q)$
T	T	F	T	T
T	F	F	F	F
F	T	T	T	F
F	F	T	T	F

true

2.

p	d	$\sim d$	$\sim d \wedge p$	$\sim p$	$\sim p \vee (\sim d \wedge p)$
T	T	F	F	F	F
T	F	T	T	F	T
F	T	F	F	T	T
F	F	T	F	T	T

3. You make dinner and I do not order pizza.

6.3   The Conditional and Circuits

APPLY IT   How can logic be used in the design of electrical circuits?
This question will be answered in this section.

Conditionals

A **conditional** statement is a compound statement that uses the connective *if . . . then*, or anything equivalent. For example:

> *If* it rains, *then* I carry my umbrella.
>
> *If* the president comes, *then* security will be tight.
>
> *If* the check doesn't arrive today, I will call to find out why.

In the last statement, the word *then* was implied but not explicitly stated; the sentence is equivalent to the statement "If the check doesn't arrive today, *then* I will call to find out why." In each of these statements, the component after the word *if* gives a condition under which the last component is true. The last component is the statement coming after the word *then* (or, in the third statement, the implied *then*). There may be other conditions under which the last component is true. In the second statement, for example, it might be true that even if the president doesn't come, security will be tight because the vice president is coming.

The conditional is written with an arrow, so that "if *p*, then *q*" is symbolized as

$$p \rightarrow q.$$

We read $p \rightarrow q$ as "*p* implies *q*" or "if *p*, then *q*." In the conditional $p \rightarrow q$, the statement *p* is the **antecedent**, while *q* is the **consequent**.

There are many equivalent forms of the conditional. For example, the statement

> Winners never quit

can be rephrased as

> If you are a winner, then you never quit.

There are other forms of the conditional that we will study in the next section.

Just as we defined conjunction and disjunction using a truth table, we will now do the same for the conditional. To see how such a table should be set up, we will analyze the following statement that might be made by a cereal company:

If you eat Wheat Crunchies, then you'll be full of energy.

Let *e* represent "You eat Wheat Crunchies" and *f* represent "You'll be full of energy." As before, there are four possible combinations of truth values for the two component statements.

	Wheat Crunchies Analysis		
Possibility	Eat Wheat Crunchies?	Full of Energy?	
1	Yes	Yes	*e* is T, *f* is T
2	Yes	No	*e* is T, *f* is F
3	No	Yes	*e* is F, *f* is T
4	No	No	*e* is F, *f* is F

Let's consider each of these possibilities.

1. If you eat Wheat Crunchies and indeed find that you are full of energy, then you must conclude that the company's claim is true, so place T in the first row of the truth table. This does not necessarily mean that eating Wheat Crunchies caused you to be full of energy; perhaps you are full of energy for some other reason unrelated to what you ate for breakfast.

2. If you eat Wheat Crunchies and are not full of energy, then the company's claim is false, so place F in the second row of the truth table.

3. If you don't eat Wheat Crunchies and yet find that you are full of energy, this doesn't invalidate the company's claim. They only promised results if you ate Wheat Crunchies; they made no claims about what would happen if you don't eat Wheat Crunchies. We will, therefore, place T in the third row of the truth table.

4. If you don't eat Wheat Crunchies and are not full of energy, you can't very well blame the company. Because they can still claim that their promise is true, we will place T in the fourth row.

The discussion above leads to the following truth table.

Truth Table for the Conditional *if p, then q*

p	q	$p \rightarrow q$
T	T	T
T	F	F
F	T	T
F	F	T

The truth table for the conditional leads to some counterintuitive conclusions, because *if . . . then* sometimes has other connotations in English, as the following examples illustrate.

EXAMPLE 1 **Truth Value**

Suppose you get a 61 on the test and you pass the course. Find the truth value of the following statement:

If you get a 70 or higher on the test, then you pass the course.

SOLUTION The first component of this conditional is false, while the second is true. According to the third line of the truth table for the conditional, this statement is true.

This result may surprise you. Even though the statement only says what happens if you get a 70 or higher on the test, perhaps you also infer from the statement that if you don't get 70 or higher, you won't pass the course. This is a common interpretation of *if . . . then*, but it is not consistent with our truth table above. This interpretation is called the *biconditional* and will be discussed in the next section.

EXAMPLE 2 Truth Value

Find the truth value of each of the following statements.

(a) If the earth is shaped like a cube, then elephants are smaller than mice.

> **SOLUTION** Both components of this conditional are false. According to the fourth line of the truth table for the conditional, this statement is true.

(b) If the earth is shaped like a cube, then George Washington was the first president of the United States.

> **SOLUTION** The first component of this conditional is false, while the second is true. According to the third line of the truth table for the conditional, this statement is true, even though it may seem like an odd statement. **TRY YOUR TURN 1**

YOUR TURN 1 Find the truth value of the following statement. "If Little Rock is the capital of Arkansas, then New York City is the capital of New York."

The following observations come from the truth table for $p \rightarrow q$.

Special Characteristics of Conditional Statements

1. $p \rightarrow q$ is false only when the antecedent (p) is *true* and the consequent (q) is *false*.
2. If the antecedent (p) is *false*, then $p \rightarrow q$ is automatically *true*.
3. If the consequent (q) is *true*, then $p \rightarrow q$ is automatically *true*.

EXAMPLE 3 Conditional Statement

Write *true* or *false* for each statement. Here T represents a true statement, and F represents a false statement.

(a) $T \rightarrow (6 = 3)$

> **SOLUTION** Since the antecedent is true, while the consequent, $6 = 3$, is false, the given statement is false by the first point mentioned above.

(b) $(5 < 2) \rightarrow F$

> **SOLUTION** The antecedent is false, so the given statement is true by the second observation.

YOUR TURN 2 Determine if the following statement is *true* or *false:* $(4 > 5) \rightarrow F$.

(c) $(3 \neq 2 + 1) \rightarrow T$

> **SOLUTION** The consequent is true, making the statement true by the third characteristic of conditional statements. **TRY YOUR TURN 2**

EXAMPLE 4 Truth Value

Given that p, q, and r are all false, find the truth value of the statement

$$(p \rightarrow \sim q) \rightarrow (\sim r \rightarrow q).$$

SOLUTION Using the shortcut method explained in Example 6 of Section 6.1, we can replace p, q, and r with F (since each is false) and proceed as before, using the negation and conditional truth tables as necessary.

$$(p \rightarrow \sim q) \rightarrow (\sim r \rightarrow q)$$
$$(F \rightarrow \sim F) \rightarrow (\sim F \rightarrow F) \qquad \text{p, q, r are false.}$$
$$(F \rightarrow T) \rightarrow (T \rightarrow F) \qquad \text{Use the negation truth table.}$$
$$T \rightarrow F \qquad\qquad \text{Use the conditional truth table.}$$
$$F$$

YOUR TURN 3 If p, q, and r are all false, find the truth value of the statement $\sim q \to (p \to r)$.

The statement $(p \to \sim q) \to (\sim r \to q)$ is false when p, q, and r are all false.

TRY YOUR TURN 3

Truth tables for compound statements involving conditionals are found using the techniques described in the previous section. The next example shows how this is done.

EXAMPLE 5 **Truth Table**

Construct a truth table for each statement.

(a) $(\sim p \to \sim q) \wedge \sim(p \vee \sim q)$

SOLUTION First insert the truth values of $\sim p$ and of $\sim q$. Then find the truth values of $\sim p \to \sim q$.

p	q	$\sim p$	$\sim q$	$\sim p \to \sim q$
T	T	F	F	T
T	F	F	T	T
F	T	T	F	F
F	F	T	T	T

Next use p and $\sim q$ to find the truth values of $p \vee \sim q$ and then $\sim(p \vee \sim q)$.

p	q	$\sim p$	$\sim q$	$\sim p \to \sim q$	$p \vee \sim q$	$\sim(p \vee \sim q)$
T	T	F	F	T	T	F
T	F	F	T	T	T	F
F	T	T	F	F	F	T
F	F	T	T	T	T	F

Finally, add the column for the statement $(\sim p \to \sim q) \wedge \sim(p \vee \sim q)$.

p	q	$\sim p$	$\sim q$	$\sim p \to \sim q$	$p \vee \sim q$	$\sim(p \vee \sim q)$	$(\sim p \to \sim q) \wedge \sim(p \vee \sim q)$
T	T	F	F	T	T	F	F
T	F	F	T	T	T	F	F
F	T	T	F	F	F	T	F
F	F	T	T	T	T	F	F

(b) $(p \to q) \to (\sim p \vee q)$

SOLUTION Go through steps similar to the ones above.

p	q	$p \to q$	$\sim p$	$\sim p \vee q$	$(p \to q) \to (\sim p \vee q)$
T	T	T	F	T	T
T	F	F	F	F	T
F	T	T	T	T	T
F	F	T	T	T	T

As the truth table in Example 5(a) shows, the statement $(\sim p \to \sim q) \wedge \sim(p \vee \sim q)$ is always false, regardless of the truth values of the components. Such a statement is called a **contradiction**. Most contradictions consist of two statements that cannot simultaneously be true that are united with a conjunction. The simplest such statement is $p \wedge \sim p$.

Notice from the truth table in Example 5(b) that the statement $(p \rightarrow q) \rightarrow (\sim p \vee q)$ is always true, regardless of the truth values of the components. Such a statement is called a **tautology**. Other examples of tautologies (as can be checked by forming truth tables) include $p \vee \sim p$, $p \rightarrow p$, $(\sim p \vee \sim q) \rightarrow \sim (q \wedge p)$, and so on. An important point here is that there are compound statements (such as $p \vee \sim p$) that are true (or false) independent of the truth values of the component parts; the truth depends on the logical structure alone. By the way, the truth tables in Example 5 also could have been found by the alternative method shown in the previous section.

Notice from the third and fifth columns of the truth table in Example 5(b) that $p \rightarrow q$ and $\sim p \vee q$ are equivalent.

> ## Writing a Conditional as an *or* Statement
> $p \rightarrow q$ is equivalent to $\sim p \vee q$.

EXAMPLE 6 Equivalent Statements

Write the following statement without using *if . . . then.*

> If this is your first time, we're glad you're here.

SOLUTION Letting f represent "This is your first time" and g represent "We're glad you're here," the conditional may be restated as $\sim f \vee g$, or in words:

> This is not your first time or we're glad you're here. **TRY YOUR TURN 4**

YOUR TURN 4 Write the following statement without using *if . . . then*: "If you do the homework, then you will pass the quiz."

Since

$$p \rightarrow q \equiv \sim p \vee q,$$

we can take the negation of both sides of this equivalence to get

$$\sim (p \rightarrow q) \equiv \sim (\sim p \vee q).$$

By applying De Morgan's Law to the right side, we have

$$\sim (p \rightarrow q) \equiv \sim (\sim p) \wedge \sim q$$
$$\equiv p \wedge \sim q.$$

This gives us an equivalent form for the negation of the conditional.

> ## Negation of $p \rightarrow q$
> The negation of $p \rightarrow q$ is $p \wedge \sim q$.

EXAMPLE 7 Negation

Write the negation of each statement.

(a) If you go to the left, I'll go to the right.

SOLUTION Let y represent "You go to the left" and i represent "I'll go to the right." Then the original statement is represented $y \rightarrow i$. As we showed earlier, the negation of this is $y \wedge \sim i$, which can be translated into words as

> You go to the left and I won't go to the right.

(b) It must be alive if it is breathing.

SOLUTION First, we must restate the given statement in *if . . . then* form:

> If it is breathing, then it must be alive.

YOUR TURN 5 Write the negation of the following statement: "If you are on time, then we will be on time."

Based on our earlier discussion, the negation is

> It is breathing and it is not alive. **TRY YOUR TURN 5**

A common error occurs when students try to write the negation of a conditional statement as another conditional statement. As seen in Example 7, the negation of a conditional statement is written as a conjunction.

Circuits One of the first nonmathematical applications of symbolic logic was seen in the master's thesis of Claude Shannon in 1937. Shannon showed how the logic developed almost a century earlier by British mathematician George Boole could be used as an aid in designing electrical **circuits**. His work was immediately taken up by the designers of computers. These computers, then in the developmental stage, could be simplified and built for less money using the ideas of Shannon.

To see how Shannon's ideas work, look at the electrical switch shown in Figure 6. We assume that current will flow through this switch when it is closed and not when it is open.

Figure 7 shows two switches connected in **series**; in such a circuit, current will flow only when both switches are closed. Note how closely a series circuit corresponds to the conjunction $p \wedge q$. We know that $p \wedge q$ is true only when both p and q are true.

A circuit corresponding to the disjunction $p \vee q$ can be found by drawing a **parallel** circuit, as in Figure 8. Here, current flows if either p or q is closed or if both p and q are closed.

The circuit in Figure 9 corresponds to the statement $(p \vee q) \wedge \sim q$, which is a compound statement involving both a conjunction and a disjunction.

The way that logic is used to simplify an electrical circuit depends on the idea of equivalent statements, from Section 6.2. Recall that two statements are equivalent if they have exactly the same truth table final column. The symbol $\equiv$ is used to indicate that the two statements are equivalent. Some of the equivalent statements that we shall need are shown in the following box.

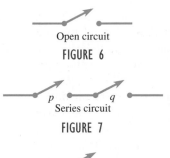

Open circuit

FIGURE 6

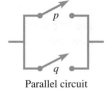

p $\qquad$ q

Series circuit

FIGURE 7

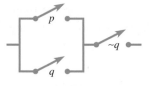

p

q

Parallel circuit

FIGURE 8

p

$\sim q$

q

FIGURE 9

Equivalent Statements

1. a. $p \vee q \equiv q \vee p$ $\qquad$ Commutative Laws
   b. $p \wedge q \equiv q \wedge p$

2. a. $p \vee (q \vee r) \equiv (p \vee q) \vee r$ $\qquad$ Associative Laws
   b. $p \wedge (q \wedge r) \equiv (p \wedge q) \wedge r$

3. a. $p \vee (q \wedge r) \equiv (p \vee q) \wedge (p \vee r)$ $\qquad$ Distributive Laws
   b. $p \wedge (q \vee r) \equiv (p \wedge q) \vee (p \wedge r)$

4. a. $\sim(p \wedge q) \equiv \sim p \vee \sim q$ $\qquad$ De Morgan's Laws
   b. $\sim(p \vee q) \equiv \sim p \wedge \sim q$

5. a. $p \vee p \equiv p$ $\qquad$ Idempotent Laws
   b. $p \wedge p \equiv p$

6. a. $(p \wedge q) \vee p \equiv p$ $\qquad$ Absorption Laws
   b. $(p \vee q) \wedge p \equiv p$

7. $\sim(\sim p) \equiv p$ $\qquad$ Double Negative

8. $p \rightarrow q \equiv \sim p \vee q$ $\qquad$ Conditional as an "or"

9. $p \rightarrow q \equiv \sim q \rightarrow \sim p$ $\qquad$ Contrapositive

If T represents any true statement and F represents any false statement, then

10. a. $p \vee T \equiv T$ $\qquad$ Identity Laws
    b. $p \wedge T \equiv p$
    c. $p \vee F \equiv p$
    d. $p \wedge F \equiv F$

11. a. $p \vee \sim p \equiv T$ $\qquad$ Negation Laws
    b. $p \wedge \sim p \equiv F$

This list may seem formidable, but if we break it down, it turns out to be not so bad.

First, notice that the first six equivalences come in pairs, in which $\lor$ is replaced with $\land$ and vice versa. This illustrates the **Principle of Duality**, which states that if a logical equivalence contains no logical operators other than $\lor$, $\land$, and $\sim$ (that is, it does not contain $\rightarrow$), then we may replace $\lor$ with $\land$ and vice versa, and replace T with F and vice versa, and the new equivalence is still true. This means that we need to keep track of just one from each pair of equivalences, although we have put both in the table for completeness.

Next, notice that equivalences 1 and 2 are just the familiar Commutative and Associative laws for addition or multiplication, which means you can rearrange the order and rearrange the parentheses in an expression involving only $\land$ or only $\lor$. You are asked to prove these using a truth table in Exercises 61–68.

Equivalence 3 is just the familiar Distributive Law of multiplication over addition. In normal multiplication, however, you *cannot* distribute $2 + (3 \times 5)$ to get $(2 + 3) \times (2 + 5)$. In other words, you cannot distribute addition over multiplication. But you can distribute $\land$ over $\lor$ and vice versa. You are asked to prove these using a truth table in Exercises 65 and 66.

De Morgan's Laws, given as equivalence 4, were discussed in the previous section.

Equivalence 5, the Idempotent Laws, may seem trivially true, so you may wonder why we bother stating them. The reason is that they are useful in simplifying circuits, as we shall see in Example 8. Equivalence 7, the Double Negative, is similar in this regard.

Equivalence 6, the Absorption Laws, may be less obvious, but they, too, are useful in simplifying circuits, as we shall see in Example 9. Their proofs are in Exercises 67 and 68.

Equivalence 8, the Conditional as an "or" statement, was discussed earlier in this section. Equivalence 9, the Contrapositive, is very important and will be discussed in the next section.

Equivalences 10 and 11, the Identity and Negation laws, should be fairly obvious. We only list them here because, just like equivalences 5 and 7, they can help us simplify circuits. Notice that the Principle of Duality applies to these equivalences.

Circuits can be used as models of compound statements, with a closed switch corresponding to T, while an open switch corresponds to F. The method for simplifying circuits is explained in the following example.

EXAMPLE 8 Circuit

Simplify the circuit in Figure 10.

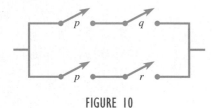

FIGURE 10

SOLUTION At the top of Figure 10, p and q are connected in series, and at the bottom, p and r are connected in series. These are interpreted as the compound statements $p \land q$ and $p \land r$, respectively. These two conjunctions are connected in parallel, as indicated by the figure treated as a whole. Therefore, we write the disjunction of the two conjunctions:

$$(p \land q) \lor (p \land r).$$

(Think of the two switches labeled "p" as being controlled by the same handle.) By the Distributive Law (equivalence statement 3),

$$(p \land q) \lor (p \land r) \equiv p \land (q \lor r),$$

which has the circuit of Figure 11. This new circuit is logically equivalent to the one in Figure 10 and yet contains only three switches instead of four—which might well lead to a large savings in manufacturing costs.

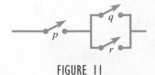

FIGURE 11

EXAMPLE 9 Circuit

Simplify the circuit in Figure 12.

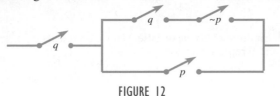

FIGURE 12

SOLUTION The diagram shows q in series with a circuit containing two parallel components, one containing q and $\sim p$, and the other containing p. Thus the circuit can be represented by the logical statement

$$q \wedge [(q \wedge \sim p) \vee p].$$

Notice that for the expression in brackets, we can use the Commutative Law to put the p first, and then we can use the Distributive Law, which leads to a series of simplifications.

$$
\begin{aligned}
q \wedge [(q \wedge \sim p) \vee p] &\equiv q \wedge [p \vee (q \wedge \sim p)] &&\text{Commutative Law} \\
&\equiv q \wedge [(p \vee q) \wedge (p \vee \sim p)] &&\text{Distributive Law} \\
&\equiv q \wedge [(p \vee q) \wedge \text{T}] &&\text{Negation Law} \\
&\equiv q \wedge (p \vee q) &&\text{Identity Law} \\
&\equiv (p \vee q) \wedge q &&\text{Commutative Law} \\
&\equiv (q \vee p) \wedge q &&\text{Commutative Law} \\
&\equiv q &&\text{Absorption Law}
\end{aligned}
$$

The circuit simplified to a single switch! We hope this convinces you of the power of logic. Notice that we used the Commutative Law three times simply to get terms into a form in which another equivalence applies. This could be avoided if we just added more equivalences to our previous list, such as the Distributive Laws in the reverse order. This would, however, make our list even longer, so we will not do that, even though it means that we need an occasional extra step in our proofs. ▪

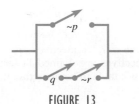

FIGURE 13

EXAMPLE 10 Circuit

Draw a circuit for $p \rightarrow (q \wedge \sim r)$.

SOLUTION By the equivalent statement Conditional as an "or," $p \rightarrow q$ is equivalent to $\sim p \vee q$. This equivalence gives $p \rightarrow (q \wedge \sim r) \equiv \sim p \vee (q \wedge \sim r)$, which has the circuit diagram in Figure 13. ▪

6.3 EXERCISES

In Exercises 1–6, decide whether each statement is true or false, and explain why.

1. If the antecedent of a conditional statement is false, the conditional statement is true.

2. If the consequent of a conditional statement is true, the conditional statement is true.

3. If q is true, then $(p \wedge q) \rightarrow q$ is true.

4. If p is true, then $\sim p \rightarrow (q \vee r)$ is true.

5. Given that $\sim p$ is true and q is false, the conditional $p \rightarrow q$ is true.

6. Given that $\sim p$ is false and q is false, the conditional $p \rightarrow q$ is true.

7. In a few sentences, explain how we determine the truth value of a conditional statement.

8. Explain why the statement "If $3 = 5$, then $4 = 6$" is true.

Tell whether each conditional is true or false. Here T represents a true statement and F represents a false statement.

9. $F \rightarrow (4 \neq 7)$

10. $(6 \geq 6) \rightarrow F$

11. $(4 = 11 - 7) \rightarrow (8 > 0)$

12. $(4^2 \neq 16) \rightarrow (4 - 4 = 8)$

Let *d* represent "She dances tonight," let *s* represent "He sings loudly," and let *e* represent "I'm leaving early." Express each compound statement in words.

13. $d \rightarrow (e \land s)$

14. $(d \land s) \rightarrow e$

15. $\sim s \rightarrow (d \lor \sim e)$

16. $(\sim d \lor \sim e) \rightarrow \sim s$

Let *d* represent "My dog ate my homework," let *f* represent "I receive a failing grade," and let *g* represent "I'll run for governor." Express each compound statement in symbols.

17. My dog ate my homework, or if I receive a failing grade, then I'll run for governor.

18. I'll run for governor, and if I receive a failing grade, then my dog did not eat my homework.

19. I'll run for governor if I don't receive a failing grade.

20. I won't receive a failing grade if my dog didn't eat my homework.

Find the truth value of each statement. Assume that *p* and *r* are false, and *q* is true.

21. $\sim r \rightarrow p$

22. $\sim q \rightarrow r$

23. $\sim p \rightarrow (q \land r)$

24. $(\sim r \lor p) \rightarrow p$

25. $\sim q \rightarrow (p \land r)$

26. $(\sim p \land \sim q) \rightarrow (p \land \sim r)$

27. $(p \rightarrow \sim q) \rightarrow (\sim p \land \sim r)$

28. $(p \rightarrow \sim q) \land (p \rightarrow r)$

29. Explain why, if we know that *p* is true, we also know that

$$[r \lor (p \lor s)] \rightarrow (p \lor q)$$

is true, even if we are not given the truth values of *q*, *r*, and *s*.

30. Construct a true statement involving a conditional, a conjunction, a disjunction, and a negation (not necessarily in that order), that consists of component statements *p*, *q*, and *r*, with all of these component statements false.

31. Using the table of equivalent statements rather than a truth table, explain why the statement $(\sim p \rightarrow \sim q) \land \sim (p \lor \sim q)$ in Example 5(a) must be a contradiction.

32. What is the minimum number of times that F must appear in the final column of a truth table for us to be assured that the statement is not a tautology?

Construct a truth table for each statement. Identify any tautologies or contradictions.

33. $\sim q \rightarrow p$

34. $p \rightarrow \sim q$

35. $(p \lor \sim p) \rightarrow (p \land \sim p)$

36. $(p \land \sim q) \land (p \rightarrow q)$

37. $(p \lor q) \rightarrow (q \lor p)$

38. $(\sim p \rightarrow \sim q) \rightarrow (p \land q)$

39. $r \rightarrow (p \land \sim q)$

40. $[(r \lor p) \land \sim q] \rightarrow p$

41. $(\sim r \rightarrow s) \lor (p \rightarrow \sim q)$

42. $(\sim p \land \sim q) \rightarrow (\sim r \rightarrow \sim s)$

Write each statement as an equivalent statement that does not use the *if . . . then* connective. Remember that $p \rightarrow q$ is equivalent to $\sim p \lor q$.

43. If your eyes are bad, your whole body will be full of darkness.

44. If you meet me halfway, this will work.

45. I'd buy that car if I had the money.

46. I would watch out if I were you.

Write the negation of each statement. Remember that the negation of $p \rightarrow q$ is $p \land \sim q$.

47. If you ask me, I will do it.

48. If you are not part of the solution, you are part of the problem.

49. If you don't love me, I won't be happy.

50. If he's my brother, then he's not heavy.

Use truth tables to decide which of the pairs of statements are equivalent.

51. $p \rightarrow q; \sim p \lor q$

52. $\sim (p \rightarrow q); p \land \sim q$

53. $p \rightarrow q; q \rightarrow p$

54. $q \rightarrow p; \sim p \rightarrow \sim q$

55. $p \rightarrow \sim q; \sim p \lor \sim q$

56. $p \rightarrow q; \sim q \rightarrow \sim p$

57. $p \land \sim q; \sim q \rightarrow \sim p$

58. $\sim p \land q; \sim p \rightarrow q$

In some approaches to logic, the only connectives are $\sim$ and $\rightarrow$, and the other logical connectives are defined in terms of these.* Verify this by using a truth table to demonstrate the following equivalences.

59. $p \land q \equiv \sim (p \rightarrow \sim q)$

60. $p \lor q \equiv \sim p \rightarrow q$

In Exercises 61–68, construct a truth table to prove each law.

61. $p \lor q \equiv q \lor p$, the Commutative Law for $\lor$

62. $p \land q \equiv q \land p$, the Commutative Law for $\land$

63. $p \lor (q \lor r) \equiv (p \lor q) \lor r$, the Associative Law for $\lor$

64. $p \land (q \land r) \equiv (p \land q) \land r$, the Associative Law for $\land$

65. $p \lor (q \land r) \equiv (p \lor q) \land (p \lor r)$, the Distributive Law for $\lor$ over $\land$

66. $p \land (q \lor r) \equiv (p \land q) \lor (p \land r)$, the Distributive Law for $\land$ over $\lor$

*For example, see Stefan Bilaniuk, A Problem Course in Mathematical Logic, http://euclid.trentu.ca/math/sb/pcml/welcome.html.

67. $(p \wedge q) \vee p \equiv p$, the first Absorption Law

68. $(p \vee q) \wedge p \equiv p$, the second Absorption Law

Write a logical statement representing each circuit. Simplify each circuit when possible.

69.

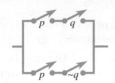

70.

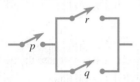

71.

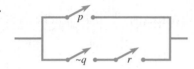

72.

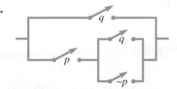

73.

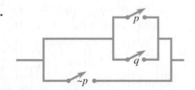

74.

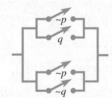

75.

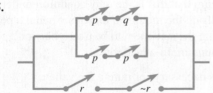

76.

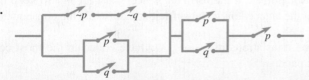

Draw circuits representing the following statements as they are given. Simplify if possible.

77. $p \wedge (q \vee \sim p)$

78. $(\sim p \wedge \sim q) \wedge \sim r$

79. $(p \vee q) \wedge (\sim p \wedge \sim q)$

80. $(\sim q \wedge \sim p) \vee (\sim p \vee q)$

81. $[(p \vee q) \wedge r] \wedge \sim p$

82. $[(\sim p \wedge \sim r) \vee \sim q] \wedge (\sim p \wedge r)$

83. $\sim q \rightarrow (\sim p \rightarrow q)$

84. $\sim p \rightarrow (\sim p \vee \sim q)$

85. $[(p \wedge q) \vee p] \wedge [(p \vee q) \wedge q]$

86. $[(p \wedge q) \vee (p \wedge q)] \vee (p \wedge r)$

87. Explain why the circuit

will always have exactly one open switch. What does this circuit simplify to?

88. Refer to Figures 10 and 11 in Example 8. Suppose the cost of the use of one switch for an hour is 3¢. By using the circuit in Figure 11 rather than the circuit in Figure 10, what is the savings for a year of 365 days, assuming that the circuit is in continuous use?

APPLICATIONS

Business and Economics

89. Income Tax The following statements appear in a guide for preparing income tax reports. Rewrite each statement with an equivalent statement using *or. Source: Your Income Tax 2010.*

  a. If you are married at the end of the year, you may file a joint return with your spouse.

  b. A bequest received by an executor from an estate is tax free if it is not compensation for services.

  c. If a course improves your current job skills but leads to qualification for a new profession, the course is not deductable.

90. Warranty A protection plan for the iPhone states: "Service will be performed at the location, or the store may send the Covered Equipment to an Apple repair service location to be repaired." *Source: AppleCare Protection Plan.*

  a. Rewrite the statement with an equivalent statement using the *if . . . then* connective.

  b. Write the negation of the statement.

91. Stocks An investor announces, "If the value of my portfolio exceeds $100,000 or the price of my stock in Ford Motor Company falls below $50 per share, then I will sell all my shares of Ford stock and I will give the proceeds to the United Way." Let v represent "The value of my portfolio exceeds $100,000," p represent "The price of my stock in Ford Motor Company falls below $50 per share," s represent "I will sell all my shares of Ford stock," and *g* represent "I will give the proceeds to the United Way."

a. Convert the statement into symbols.

b. Find the truth value of the statement if my portfolio is worth $80,000, Ford Motor stock is at $56 per share, I sell my Ford Motor stock, and I keep the proceeds.

c. Explain under what circumstances the statement will be true or false.

d. Find the negation of the statement.

Life Sciences

92. Medicine The following statement appears in a home medical reference book. Rewrite this statement replacing the *if . . . then* with an *or* statement. Then negate that statement. *Source: American College of Physicians Complete Home Medical Guide.*

If you are exercising outside, it is important to protect your skin and eyes from the sun.

Social Sciences

93. The following statements appear in a guide to common law. Write an equivalent statement using the *if . . . then* connective. *Source: Law for Dummies.*

a. You can file a civil lawsuit yourself, or your attorney can do it for you.

b. Your driver's license may come with restrictions, or restrictions may sometimes be added on later.

c. You can't marry unless you're at least 18 years old, or unless you have the permission of your parents or guardian.

YOUR TURN ANSWERS

1. False   **2.** True   **3.** True

4. You do not do the homework, or you will pass the quiz.

5. You are on time and we will not be on time.

6.4 More on the Conditional

APPLY IT   Is it possible to rewrite statements in a medical guide in one or more ways? *This question will be answered in Exercise 44.*

The conditional can be written in several ways. For example, the word *then* might not be explicitly stated, so the statement "If I'm here, then you're safe" might become "If I'm here, you're safe." Also, the consequent (the *then* component) might come before the antecedent (the *if* component), as in "You're safe if I'm here." In this section we will examine other translations of the conditional.

Alternative Forms of the Conditional
Another way to express the conditional is with the word *sufficient*. For example, the statement

If it rains in the valley, then snow is falling on the mountain can be written

Rain in the valley is sufficient for snow to fall on the mountain.

This statement claims that one condition guaranteeing snow to fall on top of the mountain is rain in the valley. The statement doesn't claim that this is the only condition under which snow falls on the mountain; perhaps snow falls on the mountain on days when it's perfectly clear in the valley. On the other hand, assuming the statement to be true, you won't see rain in the valley unless snow is falling on the mountain. In other words,

Snow falling on the mountain is necessary for rain in the valley.

To summarize this discussion, the statement "if *p*, then *q*" is sometimes stated in the form "*p* is sufficient for *q*" and sometimes in the form "*q* is necessary for *p*." To keep track of these two forms, notice that the antecedent is the sufficient part, while the consequent is the necessary part.

There are other common translations of $p \rightarrow q$. We have collected the most common ones in the box on the next page.

Common Translations of $p \rightarrow q$

The conditional $p \rightarrow q$ can be translated in any of the following ways.

If p, then q.	p is sufficient for q.
If p, q.	q is necessary for p.
p implies q.	All p's are q's.
p only if q.	q if p.
q when p.	

The translation of $p \rightarrow q$ into these various word forms does not in any way depend on the truth or falsity of $p \rightarrow q$.

EXAMPLE 1   Equivalent Statements

Write the following statement in eight different equivalent ways, using the common translations of $p \rightarrow q$ in the box above:

If you answer this survey, then you will be entered in the drawing.

SOLUTION

If you answer this survey, you will be entered in the drawing.

Answering this survey implies that you will be entered in the drawing.

You answer this survey only if you will be entered in the drawing.

You will be entered in the drawing when you answer this survey.

Answering this survey is sufficient for you to be entered in the drawing.

Being entered in the drawing is necessary for you to answer this survey.

All who answer this survey will be entered in the drawing.

You will be entered in the drawing if you answer this survey.

EXAMPLE 2   Equivalent Statements

Write each statement in the form "if p, then q."

(a) Possession of a valid identification card is necessary for admission.

  SOLUTION   If you are admitted, then you possess a valid identification card.

(b) You should use this door only if there is an emergency.

  SOLUTION   If you should use this door, then there is an emergency.

(c) All who are weary can come and rest.

  SOLUTION   If you are weary, then you can come and rest.   **TRY YOUR TURN 1**

> **YOUR TURN 1** Write each statement in the form "if p, then q."
> **(a)** Being happy is sufficient for you to clap your hands.
> **(b)** All who seek shall find.

CAUTION   Notice that "p only if q" is not the same as "p if q." The first means $p \rightarrow q$, while the second means $q \rightarrow p$.

EXAMPLE 3   Symbolic Statements

Let r represent "A triangle is a right triangle" and s represent "The sum of the squares of the two sides equals the square of the hypotenuse." Write each statement in symbols.

(a) A triangle is a right triangle if the sum of the squares of the two sides equals the square of the hypotenuse.

SOLUTION

$$s \rightarrow r$$

(b) A triangle is a right triangle only if the sum of the squares of the two sides equals the square of the hypotenuse.

SOLUTION

$$r \rightarrow s$$

Converse, Inverse, and Contrapositive
Example 3(b) gives a statement of the Pythagorean theorem. Example 3(a) gives what is called the **converse** of the Pythagorean theorem. The converse of a statement $p \rightarrow q$ is the statement $q \rightarrow p$. For example, the converse of the statement

> If today is Monday, then we have to put out the garbage

is the statement

> If we have to put out the garbage, then today is Monday.

Although both the Pythagorean theorem and its converse are true, the converse of a true statement is not necessarily true, as we shall see in Example 4.

A second statement related to the conditional is the **inverse**, in which both the antecedent and consequent are negated. The inverse of the first statement above is

> If today is not Monday, then we do not have to put out the garbage.

The inverse of a statement $p \rightarrow q$ is the statement $\sim p \rightarrow \sim q$. The inverse, like the converse, is not necessarily true, even if the original statement is true.

The third and final statement related to the conditional is the **contrapositive**, in which the antecedent and consequent are negated and interchanged. The contrapositive of the first statement above is

> If we do not have to put out the garbage, then today is not Monday.

The contrapositive of a statement $p \rightarrow q$ is the statement $\sim q \rightarrow \sim p$. We will show in a moment that the contrapositive is logically equivalent to the original statement, so that if one is true, the other is also true. First, we summarize the three statements related to the original conditional statement below.

Related Conditional Statements

Original Statement	$p \rightarrow q$	(If p, then q.)
Converse	$q \rightarrow p$	(If q, then p.)
Inverse	$\sim p \rightarrow \sim q$	(If not p, then not q.)
Contrapositive	$\sim q \rightarrow \sim p$	(If not q, then not p.)

In the following truth table, we include the conditional $p \rightarrow q$ as well as the three related forms. Notice that the original conditional and the contrapositive are equivalent; that is, whenever one is true, so is the other. Notice also that the converse and the inverse are equivalent to each other. Finally, notice that the original conditional is not equivalent to the converse or the inverse; for example, in line 2 of the table, where p is true and q is false, the original conditional and the contrapositive are both false, but the inverse and converse are true.

		Original	Converse	Inverse	Contrapositive
			Equivalent		
				Equivalent	
p	q	$p \rightarrow q$	$q \rightarrow p$	$\sim p \rightarrow \sim q$	$\sim q \rightarrow \sim p$
T	T	T	T	T	T
T	F	F	T	T	F
F	T	T	F	F	T
F	F	T	T	T	T

This discussion is summarized in the following sentence.

Equivalences

The original statement and the contrapositive are equivalent, and the converse and the inverse are equivalent.

EXAMPLE 4   Related Conditional Statements

Consider the following statement:

If Cauchy is a cat, then Cauchy is a mammal.

Write each of the following.

(a) The converse

SOLUTION Let c represent "Cauchy is a cat" and m represent "Cauchy is a mammal." Then the original statement is $c \rightarrow m$, and the converse is $m \rightarrow c$, or

If Cauchy is a mammal, then Cauchy is a cat.

Notice that in this case the original statement is true, while the converse is false, because Cauchy might be a mammal that is not a cat, such as a horse.

(b) The inverse

SOLUTION The inverse of $c \rightarrow m$ is $\sim c \rightarrow \sim m$, or

If Cauchy is not a cat, then Cauchy is not a mammal.

The inverse, like the converse, is false in this case.

(c) The contrapositive

SOLUTION The contrapositive of $c \rightarrow m$ is $\sim m \rightarrow \sim c$, or

If Cauchy is not a mammal, then Cauchy is not a cat.

In this case the contrapositive is true, as it must be if the original statement is true. ▪

YOUR TURN 2 Write the converse, inverse, and contrapositive of the following statement: "If I get another ticket, then I lose my license." Which is equivalent to the original statement?

Biconditionals   As we saw earlier, both the Pythagorean theorem and its converse are true. That is, letting r represent "A triangle is a right triangle" and s represent "The sum of the squares of the two sides equals the square of the hypotenuse," both $s \rightarrow r$ and $r \rightarrow s$ are true. In words, we say s *if and only if* r (often abbreviated s iff r). (Recall that *only if* is the opposite of *if . . . then*.) This is called the **biconditional** and is written $s \leftrightarrow r$. You can think of the biconditional as being defined by the equivalence

$$p \leftrightarrow q \equiv (p \rightarrow q) \wedge (q \rightarrow p)$$

or by the following truth table.

Truth Table for the Biconditional *p if and only if q*

p	q	$p \leftrightarrow q$
T	T	T
T	F	F
F	T	F
F	F	T

EXAMPLE 5 **Biconditional Statements**

Tell whether each statement is true or false.

(a) George Washington is the first president of the United States if and only if John Adams is the second president of the United States.

SOLUTION Since both component statements are true, the first row of the truth table tells us that this biconditional is true.

(b) Alaska is one of the original 13 states if and only if kangaroos can fly.

SOLUTION Since both component statements are false, the last row of the truth table tells us that this biconditional is true.

(c) $2 + 2 = 4$ if and only if $7 > 10$.

SOLUTION The first component statement is true, but the second is false. The second row of the truth table tells us that this biconditional is false. **TRY YOUR TURN 3**

YOUR TURN 3 Determine if the following statement is *true* or *false*. "New York City is the capital of the United States if and only if Paris is the capital of France."

Notice in the truth table and the previous example that when p and q have the same truth value, $p \leftrightarrow q$ is true; when p and q have different truth values, $p \leftrightarrow q$ is false.

In this and the previous two sections, truth tables have been derived for several important types of compound statements. The summary that follows describes how these truth tables may be remembered.

Summary of Basic Truth Tables

1. $\sim p$, the **negation** of p, has truth value opposite of p.
2. $p \wedge q$, the **conjunction**, is true only when both p and q are true.
3. $p \vee q$, the **disjunction**, is false only when both p and q are false.
4. $p \rightarrow q$, the **conditional**, is false only when p is true and q is false.
5. $p \leftrightarrow q$, the **biconditional**, is true only when p and q have the same truth value.

6.4 EXERCISES

For each given statement, write (a) the converse, (b) the inverse, and (c) the contrapositive in *if . . . then* form. In some of the exercises, it may be helpful to restate the statement in *if . . . then* form.

1. If the exit is ahead, then I don't see it.

2. If I finish reading this novel, then I'll write a review.

3. If I knew you were coming, I'd have cleaned the house.

4. If I'm the bottom, you're the top.

5. Mathematicians wear pocket protectors.

6. Beggars can't be choosers.

7. $p \rightarrow \sim q$

8. $\sim q \rightarrow \sim p$

9. $p \rightarrow (q \vee r)$ (*Hint:* Use one of De Morgan's Laws as necessary.)

10. $(r \vee \sim q) \rightarrow p$ (*Hint:* Use one of De Morgan's Laws as necessary.)

11. Discuss the equivalences that exist among the direct conditional statement, the converse, the inverse, and the contrapositive.

12. State the contrapositive of "If the square of a natural number is even, then the natural number is even." The two statements must have the same truth value. Use several examples and inductive reasoning to decide whether both are true or both are false.

Write each statement in the form "if p . . . then q."

13. Your signature implies that you accept the conditions.

14. His tardiness implies that he doesn't care.

15. You can take this course pass/fail only if you have prior permission.

16. You can purchase this stock only if you have $1000.

17. You can skate on the pond when the temperature is below 10°.

18. The party will be stopped when more than 200 people attend.

19. Eating 10 hot dogs is sufficient to make someone sick.

20. Two hours in the desert sun is sufficient to give the typical person a sunburn.

21. A valid passport is necessary for travel to France.

22. Support from the party bosses is necessary to get the nomination.

23. For a number to have a real square root, it is necessary that it be nonnegative.

24. For a number to have a real square root, it is sufficient that it be nonnegative.

25. All brides are beautiful.

26. All passengers for Hempstead must change trains at Jamaica station.

27. A number is divisible by 3 if the sum of its digits is divisible by 3.

28. A number is even if its last digit is even.

29. One of the following statements is not equivalent to all the others. Which one is it?

a. r only if s.

b. r implies s.

c. If r, then s.

d. r is necessary for s.

30. Use the statement "Being 65 years old is sufficient for being eligible for Medicare" to explain why "p is sufficient for q" is equivalent to "if p, then q."

31. Use the statement "Being over 21 is necessary for entering this club" to explain why "p is necessary for q" is equivalent to "if q, then p."

32. Explain why the statement "Elephants can fly if and only if Africa is the smallest continent" is true.

Identify each statement as *true* or *false*.

33. $5 = 9 - 4$ if and only if $8 + 2 = 10$.

34. $3 + 1 \neq 6$ if and only if $8 \neq 8$.

35. $8 + 7 \neq 15$ if and only if $3 \times 5 \neq 9$.

36. $6 \times 2 = 14$ if and only if $9 + 7 \neq 16$.

37. China is in Asia if and only if Mexico is in Europe.

38. The moon is made of green cheese if and only if Hawaii is one of the United States.

Construct a truth table for each statement.

39. $(\sim p \wedge q) \leftrightarrow (p \rightarrow q)$

40. $(p \leftrightarrow \sim q) \leftrightarrow (\sim p \vee q)$

APPLICATIONS

Business and Economics

41. Income Tax The following excerpts appear in a guide for preparing income tax reports. Write each of the statements in *if . . . then* form. *Source: Your Income Tax 2010.*

a. All employer contributions must be reported on Form 8889.

b. Certain tax benefits may be claimed by married persons only if they file jointly.

c. A child is not a qualifying child if he or she provides over half of his or her own support.

42. Income Tax The following excerpt appears in a guide for preparing income tax reports. Write the statement in *if . . . then* form. Find the contrapositive of this statement. *Source: Your Income Tax 2010.*

A refund of your state or local income tax is not taxable if you did not previously claim the tax as an itemized deduction in a prior year.

43. Credit Cards The following statement appeared in a card member agreement for a Chase Visa Card. Write the converse, inverse, and contrapositive. Which statements are equivalent? *Source: JP Morgan Chase & Co.*

If your account is in default, we may close your account without notice.

Life Sciences

44. APPLY IT The following statement is from a home medical guide. Write the statement in four different ways, using the common translations of $p \rightarrow q$. (See Example 1.) *Source: American College of Physicians Complete Home Medical Guide.*

When you sleep well, you wake up feeling refreshed and alert.

45. Polar Bears The following statement is with regard to polar bear cubs. *Source: National Geographic.*

If there are triplets, the most persistent stands to gain an extra meal and it may eat at the expense of another.

a. Use symbols to write this statement.

b. Write the contrapositive of this statement.

Social Sciences

46. Law The following statements appear in a guide to common laws. Write the statement in *if . . . then* form. **Source: Law for Dummies.**

a. When you buy a car, you must register it right away.

b. All states have financial responsibility laws.

c. Your insurer pays the legitimate claims of the injured party and defends you in court if you're sued.

47. Law The following statements appear in a guide to common laws. Write the converse, inverse, and contrapositive of the statements. Which statements are equivalent? **Source: Law for Dummies.**

a. If you are married, then you can't get married again.

b. If you pay for your purchase with a credit card, you are protected by the Fair Credit Billing Act.

c. If you hit a parked car, you're expected to make a reasonable effort to locate the owner.

48. Philosophy Aristotle once said, "If liberty and equality, as is thought by some, are chiefly to be found in democracy, they will be best attained when all persons alike share in the government to the utmost." Write the contrapositive of this statement. **Source: Bartlett's Familiar Quotations.**

49. Political Development It has been argued that political development in Western Europe will increase if and only if social assimilation is increasing. **Source: International Organizational.**

a. Express this statement symbolically, and construct a truth table for it.

b. The author of the article quoted above says that it is true that political development in Western Europe is increasing, but it is false that social assimilation is increasing. What can then be said about the original statement?

50. Libya Referring to Libya's offer to pay $2.7 billion in compensation for the families of those killed in the 1988 crash of Pan Am flight 103, a White House official said, "This is a necessary step, but it is not sufficient [for the United States to drop sanctions against Libya]." Letting *d* represent "the United States drops sanctions against Libya" and *l* represent "Libya offers to pay compensation," write the White House official's statement as a statement in symbolic logic. **Source: The New York Times.**

51. Education It has been argued that "a high level of education . . . comes close to being a necessary [condition for democracy]." For the purpose of this exercise, consider "comes close to being" as meaning "is," and assume the statement refers to a country. Write the statement in *if . . . then* form, and then write the converse, inverse, and contrapositive of the statement. Which one of these is equivalent to the original statement? **Source: The American Political Science Review.**

52. Political Alliances According to political scientist Howard Rosenthal, "the presence of [a Modéré] incumbent can be regarded as a necessary condition for a R.P.F. alliance." Write the statement in *if . . . then* form, and then write the converse, inverse, and contrapositive of the statement. Which one of these is equivalent to the original statement? **Source: The American Political Science Review.***

*The R.P.F. and Modéré are the names of political parties in France.

53. Test of Reasoning A test devised by psychologist Peter Wason is designed to test how people reason. **Source: Science News.** As an example of this test, volunteers are given the rule, "If a card has a D on one side, then it must have a 3 on the other side." Volunteers view four cards displaying D, F, 3, and 7, respectively. They are told that each card has a letter on one side and a number on the other. Which cards do they need to turn over to determine if the rule has been violated? In Wason's experiments, fewer than one-fourth of the participants gave the correct answer.

54. Test of Reasoning In another example of a Wason test (see previous exercise), volunteers are given the rule, "If an employee works on the weekend, then that person gets a day off during the week." Volunteers are given four cards displaying "worked on the weekend," "did not work on the weekend," "did get a day off," and "did not get a day off." Volunteers were told that one side of the card tells whether an employee worked on the weekend, and the other side tells whether an employee got a day off. Which cards must be turned over to determine if the rule has been violated? In a set of experiments, volunteers told to take the perspective of the employees tended to give the correct answer, while volunteers told to take the perspective of employers tended to turn over the second and third card.

General Interest

55. Sayings Rewrite each of the following statements as a conditional in *if . . . then* form. Then write two statements that are equivalent to the *if . . . then* statements. (*Hint:* Write the contrapositive of the statement and rewrite the conditional statement using *or.*)

a. Nothing ventured, nothing gained.

b. The best things in life are free.

c. Every cloud has a silver lining.

56. Sayings Think of some wise sayings that have been around for a long time, and state them in *if . . . then* form.

57. Games Statements similar to the ones below appear in the instructions for various Milton Bradley games. Rewrite each statement in *if . . . then* form and then write an equivalent statement using *or.* **Source: Milton Bradley Company.**

a. You can score in this box only if the dice show any sequence of four numbers.

b. When two or more words are formed in the same play, each is scored.

c. All words labeled as a part of speech are permitted.

▬▬**YOUR TURN ANSWERS**

1. a. If you are happy, then you clap your hands.

  b. If you seek, then you shall find.

2. *Converse:* If I lose my license, then I get another ticket. *Inverse:* If I didn't get another ticket, then I didn't lose my license. *Contrapositive:* If I didn't lose my license, then I didn't get another ticket. The contrapositive is equivalent.

3. False.

6.5 Analyzing Arguments and Proofs

APPLY IT If I could be in two places at the same time, then I'd come to your game. I did not come to your game, so can I conclude that I can't be in two places at the same time?

This question will be analyzed using truth tables in Example 2 of this section.

In this section, we will analyze and construct logical arguments, or proofs, that can be used to determine whether a given set of statements produces a sensible conclusion. A logical argument is made up of **premises** and **conclusions**. The premises are statements that we accept for the sake of the argument. It is not the purpose of logic to determine whether or not the premises are actually true. In logic, we suppose the premises to be true and then ask what statements follow using the laws of logic. The statements that follow are the conclusions. The argument is considered *valid* if the conclusions must be true when the premises are true.

Valid and Invalid Arguments

An argument is **valid** if the fact that all the premises are true forces the conclusion to be true. An argument that is not valid is **invalid**, or a **fallacy**.

It is very important to note that *valid* and *true* are not the same—an argument can be valid even though the conclusion is false. (See the discussion after Example 3.)

We will begin by using truth tables to determine whether certain types of arguments are valid or invalid. We will then use the results from these examples to demonstrate a more powerful method of proving that an argument is valid or invalid. As our first example, consider the following argument.

> If it's after midnight, then I must go to sleep.
>
> It's after midnight.
> _____
> I must go to sleep.

Here we use the common method of placing one premise over another, with the conclusion below a line. Alternatively, we could indicate that the last line is a conclusion using "therefore," as in "Therefore, I must go to sleep."

To test the validity of this argument, we begin by identifying the component statements found in the argument. We will use the generic variables p and q here, rather than meaningful variable names, so the argument will have a generic form.

> p represents "It's after midnight";
>
> q represents "I must go to sleep."

Now we write the two premises and the conclusion in symbols:

> Premise 1: $p \rightarrow q$
> Premise 2: p
> _____
> Conclusion: q .

To decide if this argument is valid, we must determine whether the conjunction of both premises implies the conclusion for all possible cases of truth values for p and q. Therefore, write the conjunction of the premises as the antecedent of a conditional statement, and the conclusion as the consequent.

$$[(p \rightarrow q) \quad \wedge \quad p] \quad \rightarrow \quad q$$

Premise   and   premise   implies conclusion.

Finally, construct the truth table for the conditional statement, as shown on the next page.

p	q	$p \to q$	$(p \to q) \wedge p$	$[(p \to q) \wedge p] \to q$
T	T	T	T	T
T	F	F	F	T
F	T	T	F	T
F	F	T	F	T

Since the final column, shown in color, indicates that the conditional statement that represents the argument is true for all possible truth values of p and q, the statement is a tautology. Thus, the argument is valid.

The pattern of the argument in the preceding example,

$$\begin{array}{c} p \to q \\ \underline{p } \\ q \end{array} \, ,$$

is a common one and is called **Modus Ponens**, or the *law of detachment*.

In summary, to test the validity of an argument using a truth table, go through the steps in the box that follows.

Testing the Validity of an Argument with a Truth Table

1. Assign a letter to represent each component statement in the argument.

2. Express each premise and the conclusion symbolically.

3. Form the symbolic statement of the entire argument by writing the *conjunction* of *all* the premises as the antecedent of a conditional statement and the conclusion of the argument as the consequent.

4. Complete the truth table for the conditional statement formed in step 3. If it is a tautology, then the argument is valid; otherwise, it is invalid.

EXAMPLE 1 Determining Validity

Determine whether the argument is *valid* or *invalid*.

> If I win the lottery, then I'll buy a new house.
> I bought a new house.
> _____
> I won the lottery.

SOLUTION Let p represent "I win the lottery" and let q represent "I'll buy a new house." Using these symbols, the argument can be written in the form

$$\begin{array}{c} p \to q \\ \underline{q } \\ p \end{array} \, .$$

To test for validity, construct a truth table for the statement

$$[(p \to q) \wedge q] \to p.$$

p	q	$p \to q$	$(p \to q) \wedge q$	$[(p \to q) \wedge q] \to p$
T	T	T	T	T
T	F	F	F	T
F	T	T	T	F
F	F	T	F	T

The third row of the final column of the truth table shows F, and this is enough to conclude that the argument is invalid. Even if the premises are true, the conclusion that I won the lottery is not necessarily true. Perhaps I bought a new house with money I inherited. ∎

If a conditional and its converse were logically equivalent, then an argument of the type found in Example 1 would be valid. Since a conditional and its converse are *not* equivalent, the argument is an example of what is sometimes called the **Fallacy of the Converse**.

EXAMPLE 2   Determining Validity

Determine whether the argument is *valid* or *invalid*.

> If I could be in two places at the same time, I'd come to your game.
>
> I did not come to your game.
> _____
>
> I cannot be in two places at the same time.

APPLY IT

SOLUTION   If p represents "I could be in two places at the same time" and q represents "I'd come to your game," the argument becomes

$$p \rightarrow q$$
$$\underline{\sim q}$$
$$\sim p \quad .$$

The symbolic statement of the entire argument is

$$[(p \rightarrow q) \wedge \sim q] \rightarrow \sim p.$$

The truth table for this argument, shown below, indicates a tautology, and the argument is valid.

p	q	$p \rightarrow q$	$\sim q$	$(p \rightarrow q) \wedge \sim q$	$\sim p$	$[(p \rightarrow q) \wedge \sim q] \rightarrow \sim p$
T	T	T	F	F	F	T
T	F	F	T	F	F	T
F	T	T	F	F	T	T
F	F	T	T	T	T	T

The pattern of reasoning of this example is called **Modus Tollens**, or the *law of contraposition*.

With reasoning similar to that used to name the fallacy of the converse, the fallacy

$$p \rightarrow q$$
$$\underline{\sim p}$$
$$\sim q$$

is often called the **Fallacy of the Inverse**. An example of such a fallacy is "If it rains, I get wet. It doesn't rain. Therefore, I don't get wet."

EXAMPLE 3   Determining Validity

Determine whether the argument is *valid* or *invalid*.

> I'll win this race or I'll eat my hat.
>
> I didn't win this race.
> _____
>
> I'll eat my hat.

SOLUTION   Let p represent "I'll win this race" and let q represent "I'll eat my hat." Using these symbols, the argument can be written in the form

$$p \vee q$$
$$\underline{\sim p}$$
$$q \quad .$$

Set up a truth table for

$$[(p \vee q) \wedge \sim p] \rightarrow q.$$

p	q	$p \vee q$	$\sim p$	$(p \vee q) \wedge \sim p$	$[(p \vee q) \wedge \sim p] \to q$
T	T	T	F	F	T
T	F	T	F	F	T
F	T	T	T	T	T
F	F	F	T	F	T

The statement is a tautology and the argument is valid. Any argument of this form is valid by the law of **Disjunctive Syllogism**.

Suppose you notice a few months later that I didn't win the race or eat my hat. You object that the conclusion of the argument in Example 3 is false. Did that make the argument invalid? No. The problem is that I lied about the first premise, namely, that I would win the race or eat my hat. A valid argument only guarantees that *if* all the premises are true, then the conclusion must also be true. Whether the premises of an argument are actually true is a separate issue. I might have a perfectly valid argument, yet you disagree with my conclusion because you disagree with one or more of my premises.

EXAMPLE 4 Determining Validity

Determine whether the following argument is *valid* or *invalid*.

If you make your debt payments late, you will damage your credit record.

If you damage your credit record, you will have difficulty getting a loan.

If you make your debt payments late, you will have difficulty getting a loan.

SOLUTION Let p represent "you make your debt payments late," let q represent "you will damage your credit record," and let r represent "you will have difficulty getting a loan." The argument takes on the general form

$$p \to q$$
$$q \to r$$
$$p \to r.$$

Make a truth table for the following statement:

$$[(p \to q) \wedge (q \to r)] \to (p \to r).$$

It will require eight rows.

p	q	r	$p \to q$	$q \to r$	$p \to r$	$(p \to q) \wedge (q \to r)$	$[(p \to q) \wedge (q \to r)] \to (p \to r)$
T	T	T	T	T	T	T	T
T	T	F	T	F	F	F	T
T	F	T	F	T	T	F	T
T	F	F	F	T	F	F	T
F	T	T	T	T	T	T	T
F	T	F	T	F	T	F	T
F	F	T	T	T	T	T	T
F	F	F	T	T	T	T	T

This argument is valid since the final statement is a tautology. The pattern of argument shown in this example is called **Reasoning by Transitivity**, or the *law of hypothetical syllogism*.

A summary of the valid and invalid forms of argument presented so far is given on the following page.

Valid Argument Forms

Modus Ponens	Modus Tollens	Disjunctive Syllogism	Reasoning by Transitivity
$p \rightarrow q$	$p \rightarrow q$	$p \lor q$	$p \rightarrow q$
p	$\sim q$	$\sim p$	$q \rightarrow r$
q	$\sim p$	q	$p \rightarrow r$

Invalid Argument Forms (Fallacies)

Fallacy of the Converse	Fallacy of the Inverse
$p \rightarrow q$	$p \rightarrow q$
q	$\sim p$
p	$\sim q$

When an argument contains three or more premises, it will be necessary to determine the truth values of the conjunction of all of them. Remember that if *at least one* premise in a conjunction of several premises is false, then the entire conjunction is false. This will be used in the next example.

EXAMPLE 5   Determining Validity

Determine whether the following argument is *valid* or *invalid*.

> If Ed reaches the semifinals, then Liz will be happy.
>
> If Liz is not happy, then Roxanne will bake her a cake.
>
> Roxanne did not bake Liz a cake.
> _____
> Therefore, Ed reaches the semifinals.

SOLUTION   Let e represent "Ed reaches the semifinals," l represent "Liz is happy," and r represent "Roxanne bakes Liz a cake." The symbolic form of the argument is as follows.

1. $e \rightarrow l$ Premise

2. $\sim l \rightarrow r$ Premise

3. $\underline{\sim r}$ Premise

   e Conclusion

Our strategy will be to apply the valid argument forms either to the premises or to conclusions that have already been reached. Notice, for example, that we can apply Modus Tollens to lines 2 and 3 to reach $\sim(\sim l) \equiv l$. We write this as follows, giving the statements and the argument form used.

4. l 2, 3, Modus Tollens

From statements 1 and 4, can we conclude e? Only if we use the Fallacy of the Converse! Thus this argument appears to be invalid.

Someone might object that the argument is actually valid by a clever proof that we haven't discovered. We could refute this objection by creating a truth table, as we did in Examples 1 through 4. A less tedious approach is to find a truth value for all the components e, l, and r that make the premises true and the conclusion false. If we can do that, we know for certain that the conclusion does not follow from the premises. Write the premises and the conclusion with the desired truth value next to each.

1. $e \rightarrow l$ T
2. $\sim l \rightarrow r$ T
3. $\underline{\sim r}$ T
     e F

YOUR TURN 1 Determine whether the following argument is *valid* or *invalid*. You watch television tonight or you write your paper tonight. If you write your paper tonight, then you will get a good grade. You get a good grade. Therefore, you did not watch television tonight.

In statement 3, to make $\sim r$ true, we know that r must be false. From the conclusion, e must be false. Statement 1 is then automatically true. To make statement 2 true, simply make l true, so $\sim l$ is false. In summary, the following assignment of truth values make all the premises true, yet the conclusion is false.

$$e = \text{"Ed reaches the semifinals"} = \text{F}$$
$$l = \text{"Liz is happy"} = \text{T}$$
$$r = \text{"Roxanne bakes Liz a cake"} = \text{F} \quad \textbf{TRY YOUR TURN 1}$$

Let's summarize what we have seen.

- To show that an argument is valid, prove it using the four valid argument forms discussed in this section. We can also use any of the laws given in the Equivalent Statement box in Section 6.3.

- To show that an argument is invalid, give an assignment of truth values that makes the premises true and the conclusion false.

It's not always clear which of these two strategies you should try first, but if one doesn't work, try the other. An argument may be valid, but it may not be provable using only the rules learned so far. In the next section we will study additional rules for proving arguments. When all else fails, you can always create a truth table, but if the argument has many variables, this could be very tedious.

EXAMPLE 6   Determining Validity

Lewis Carroll* gave humorous logic puzzles in his book *Symbolic Logic*. In each puzzle, he presented several premises, and the reader was to find valid conclusions. Here is one of his puzzles. What is the valid conclusion? *Source: The Complete Works of Lewis Carroll.*

> Babies are illogical.
>
> Nobody is despised who can manage a crocodile.
>
> Illogical persons are despised.

SOLUTION   First, write each premise in *if . . . then* form.

> If you are a baby, then you are illogical.
>
> If you can manage a crocodile, then you are not despised.
>
> If you are illogical, then you are despised.

Let b represent "you are a baby," i represent "you are illogical," m represent "you can manage a crocodile," and d represent "you are despised." The statements can then be written symbolically as

1. $b \rightarrow i$ Premise
2. $m \rightarrow \sim d$ Premise
3. $i \rightarrow d$ Premise

Notice that we can combine the first and third statements using reasoning by transitivity.

4. $b \rightarrow d$ 1, 3, Transitivity

*Lewis Carroll is the pseudonym for Charles Dodgson (1832–1898), mathematician and author of *Alice in Wonderland.*

How can we combine this with statement 2, which has not yet been used? Both statements 2 and 4 have d at the end, but in statement 2 the d is negated. Use the contrapositive to get statement 2 in a more useful form.

5. $d \rightarrow \sim m$ 2, Contrapositive

Now we can use transitivity.

6. $b \rightarrow \sim m$ 4, 5, Transitivity

In words, the conclusion is "If you are a baby, then you cannot manage a crocodile," or, as Lewis Carroll put it, "Babies cannot manage crocodiles." ▬▬

Often there are different ways to do a proof. In the previous example, you might first apply contrapositive to statement 2, and then combine the result with statement 3 using transitivity, and finally combine that result with statement 1 using transitivity. In the exercises in this textbook, your proofs might look different from those in the back of the book but still be correct.

▮ **EXAMPLE 7**   **Determining Validity**

Determine whether the following argument is *valid* or *invalid*.

>   If tomorrow is Saturday and sunny, then it is a beach day.
>
>   Tomorrow is Saturday.
>
>   Tomorrow is not a beach day.
>
>   Therefore, tomorrow must not be sunny.

SOLUTION   Let t represent "tomorrow is Saturday," s represent "tomorrow is sunny," and b represent "tomorrow is a beach day." We give a proof that this argument is valid.

1. $(t \wedge s) \rightarrow b$ Premise

2. t Premise

3. $\sim b$ Premise

4. $\sim(t \wedge s)$ 1, 3, Modus Tollens

5. $\sim t \vee \sim s$ 4, De Morgan's Law

6. $\sim s$ 2, 5, Disjunctive Syllogism

YOUR TURN 2 Determine whether the following argument is *valid* or *invalid.* You will not put money in the parking meter or you will not buy a cup of coffee. If you do not put money in the parking meter, you will get a ticket. You did not get a ticket. Therefore, you did not buy a cup of coffee. ▬▬

How did we figure out this proof? We started by looking for a rule of logic that combines two of the three premises. Notice that statement 3 has the negation of the right side of statement 1; this is a clue that Modus Tollens might be helpful. Once we get statement 4, using one of De Morgan's Laws is an obvious choice, as it should be whenever we see a negation over $\wedge$ or $\vee$. Statement 2 still hasn't been used at this point, but notice that it is the negation of one part of the $\vee$ in statement 5, leading us to try Disjunctive Syllogism.   **TRY YOUR TURN 2** ▬▬

▮ **EXAMPLE 8**   **Determining Validity**

Determine whether the following argument is *valid* or *invalid*.

>   If your teeth are white, you use Extreme Bright toothpaste.
>
>   If your teeth are white, you will be more attractive.
>
>   Therefore, if you use Extreme Bright toothpaste, you will be more attractive.

SOLUTION   Let w represent "your teeth are white," t represent "you use Extreme Bright toothpaste," and a represent "you will be more attractive."

In symbolic form, this argument can be written as follows.

$$w \rightarrow t$$
$$\underline{w \rightarrow a}$$
$$t \rightarrow a$$

This looks like a misguided attempt at using reasoning by transitivity and so appears invalid. To show this, we need to find a way to make the premises true and the conclusion false. The only way the conclusion, $t \rightarrow a$, can be false is if t is true and a is false. Both premises will automatically be true if we make w false. Thus the argument is invalid. The following assignment of truth values make all the premises true and the conclusion false.

$w =$ "your teeth are white" $=$ F

$t =$ "you use Extreme Bright toothpaste" $=$ T

$a =$ "you will be more attractive" $=$ F

6.5 EXERCISES

Each of the following arguments is either valid by one of the forms of valid arguments discussed in this section or is a fallacy by one of the forms of invalid arguments discussed. (See the summary boxes.) Decide whether the argument is *valid* or *invalid*, and give the form that applies.

1. If she weighs the same as a duck, she's made of wood.

   If she's made of wood, she's a witch.

   If she weighs the same as a duck, she's a witch.

2. If passing is out of the question, there's no point in going to class.

   If there's no point in going to class, then attending college makes no sense.

   If passing is out of the question, then attending college makes no sense.

3. If I had the money, I'd go on vacation.

   I have the money.

   I go on vacation.

4. If I were a rabbit, I'd hop away.

   I am a rabbit.

   I hop away.

5. If you want to make trouble, the door is that way.

   The door is that way.

   You want to make trouble.

6. If you finish the test, you can leave early.

   You can leave early.

   You finish the test.

7. If Andrew Crowley plays, the opponent gets shut out.

   The opponent does not get shut out.

   Andrew Crowley does not play.

8. If you want to follow along, we're on p. 315.

   We're not on p. 315.

   You don't want to follow along.

9. "If we evolved a race of Isaac Newtons, that would not be progress." (quote from Aldous Huxley)

   We have not evolved a race of Isaac Newtons.

   That is progress.

10. "If I have seen farther than others, it is because I stood on the shoulders of giants." (quote from Sir Isaac Newton)

    I have not seen farther than others.

    I have not stood on the shoulders of giants.

11. Something is rotten in the state of Denmark, or my name isn't Hamlet.

    My name is Hamlet.

    Something is rotten in the state of Denmark.

12. "We shall conquer together or we shall die together." (quote from Winston Churchill)

    We shall not die together.

    We shall conquer together.

Determine whether each argument is valid or invalid. If it is valid, give a proof. If it is invalid, give an assignment of truth values to the variables that makes the premises true and the conclusion false.

13. $p \vee q$
    $\underline{p}$
    $\sim q$

14. $p \vee \sim q$
    $\underline{p}$
    $\sim q$

15. $p \rightarrow q$
    $\underline{q \rightarrow p}$
    $p \wedge q$

16. $\sim p \rightarrow q$
    $\underline{p}$
    $\sim q$

17. $\sim p \rightarrow \sim q$
    $\underline{q}$
    p

18. $p \rightarrow \sim q$
    $\underline{q}$
    $\sim p$

19. $p \rightarrow q$
$\sim q$
$\underline{\sim p \rightarrow r}$
r

20. $p \vee q$
$\sim p$
$\underline{r \rightarrow \sim q}$
$\sim r$

21. $p \rightarrow q$
$q \rightarrow r$
$\underline{\sim r}$
$\sim p$

22. $p \rightarrow q$
$\underline{r \rightarrow \sim q}$
$p \rightarrow \sim r$

23. $p \rightarrow q$
$q \rightarrow \sim r$
$\underline{p}$
$r \vee s$
s

24. $p \rightarrow q$
$\sim p \rightarrow r$
$s \rightarrow \sim q$
$\underline{\sim r \rightarrow \sim s}$

Use a truth table, similar to those in Examples 1–4, to prove each rule of logic. The rules in Exercises 25–27 are known as *simplification*, *amplification*, and *conjunction*, respectively.

25. $\underline{p \wedge q}$
p

26. $\underline{p}$
$p \vee q$

27. p
$\underline{q}$
$p \wedge q$

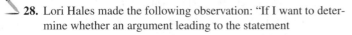

28. Lori Hales made the following observation: "If I want to determine whether an argument leading to the statement

$$[(p \rightarrow q) \wedge \sim q] \rightarrow \sim p$$

is valid, I only need to consider the lines of the truth table that lead to T for the column headed $(p \rightarrow q) \wedge \sim q$." Lori was very perceptive. Can you explain why her observation was correct?

APPLICATIONS

For Exercises 29–37, determine whether each of the following arguments is *valid* or *invalid*. If it is valid, give a proof. If it is invalid, give an assignment of truth values to the variables that makes the premises true and the conclusion false.

Business and Economics

29. Investment If Alex invests in AT&T, Sophia will invest in Sprint Nextel. Victor will invest in Verizon or Alex invests in AT&T. Victor will not invest in Verizon. Therefore, Sophia will invest in Sprint Nextel.

30. Credit If you make your debt payments late, you will damage your credit record. If you damage your credit record, you will have trouble getting a loan. You do not have trouble getting a loan. Therefore, you do not make your debt payments late.

31. Stock Market It is a bearish market. If prices are rising, then it is not a bearish market. If prices are not rising, then the investor will sell stocks. Therefore, the investor will not sell stocks.

Life Sciences

32. Classification If the animal is a reptile, then it belongs to the chordate phylum. The animal either belongs to the echinoderm phylum or the chordate phylum. It does not belong to the echinoderm phylum. Therefore, it is not a reptile.

33. Classification The animal is a spider or it is an insect. If it is a spider, then it has eight legs and it has two main body parts. It does not have eight legs or it does not have two main body parts. Therefore, it is an insect.

Social Sciences

34. Politics If Boehme runs for senator, then Hoffman will run for governor. If Tobin runs for congressman, then Hoffman will not run for governor. Therefore, if Tobin runs for congressman, then Boehme will not run for senator.

General Interest

35. Marriage If I am married to you, then we are one. If we are one, then you are not really a part of me. We are one or you are really a part of me. Therefore, if I am married to you, then you are really a part of me.

36. Love If I were your husband and you were my wife, then I'd never stop loving you. I've stopped loving you. Therefore, I am not your husband or you are not my wife.

37. Baseball The Yankees will be in the World Series or the Phillies won't be there. If the Phillies are not in the World Series, the National League cannot win. In fact, the National League wins. Therefore, the Yankees are in the World Series.

38. Time Suppose that you ask someone for the time and you get the following response:

"If I tell you the time, then we'll start chatting. If we start chatting, then you'll want to meet me at a truck stop. If we meet at a truck stop, then we'll discuss my family. If we discuss my family, then you'll find out that my daughter is available for marriage. If you find out that she is available for marriage, then you'll want to marry her. If you want to marry her, then my life will be miserable since I don't want my daughter married to some fool who can't afford a $20 watch."

Use Reasoning by Transitivity to draw a valid conclusion.

Lewis Carroll Exercises 39–44 are from problems in Lewis Carroll's book *Symbolic Logic*. Write each premise in symbols, and then give a conclusion that uses all the premises and yields a valid argument. *Source: The Complete Works of Lewis Carroll.*

39. Let d be "it is a duck," p be "it is my poultry," o be "one is an officer," and w be "one waltzes."

a. No ducks waltz.

b. No officers ever decline to waltz.

c. All my poultry are ducks.

d. Give a conclusion that yields a valid argument.

40. Let l be "one is able to do logic," j be "one is fit to serve on a jury," s be "one is sane," and y be "he is your son."

a. Everyone who is sane can do logic.

b. No lunatics are fit to serve on a jury.

c. None of your sons can do logic.

d. Give a conclusion that yields a valid argument.

41. Let h be "one is honest," p be "one is a pawnbroker," b be "one is a promise breaker," t be "one is trustworthy," c be "one is very communicative," and w be "one is a wine drinker."

  a. Promise breakers are untrustworthy.

  b. Wine drinkers are very communicative.

  c. A man who keeps his promise is honest.

  d. No teetotalers are pawnbrokers. (*Hint:* Assume "teetotaler" is the opposite of "wine drinker.")

  e. One can always trust a very communicative person.

  f. Give a conclusion that yields a valid argument.

42. Let g be "it is a guinea pig," i be "it is hopelessly ignorant of music," s be "it keeps silent while the *Moonlight Sonata* is being played," and a be "it appreciates Beethoven."

  a. Nobody who really appreciates Beethoven fails to keep silent while the *Moonlight Sonata* is being played.

  b. Guinea pigs are hopelessly ignorant of music.

  c. No one who is hopelessly ignorant of music ever keeps silent while the *Moonlight Sonata* is being played.

  d. Give a conclusion that yields a valid argument.

43. Let s be "it begins with 'Dear Sir'," c be "it is crossed," d be "it is dated," f be "it is filed," i be "it is in black ink," t be "it is in the third person," r be "I can read it," p be "it is on blue paper," o be "it is on one sheet," and b be "it is written by Brown."

  a. All the dated letters in this room are written on blue paper.

  b. None of them are in black ink, except those that are written in the third person.

  c. I have not filed any of them that I can read.

  d. None of them that are written on one sheet are undated.

  e. All of them that are not crossed are in black ink.

  f. All of them written by Brown begin with "Dear Sir."

  g. All of them written on blue paper are filed.

  h. None of them written on more than one sheet are crossed.

  i. None of them that begin with "Dear Sir" are written in the third person.

  j. Give a conclusion that yields a valid argument.

44. Let p be "he is going to a party," b be "he brushes his hair," s be "he has self-command," l be "he looks fascinating," o be "he is an opium eater," t be "he is tidy," and w be "he wears white kid gloves."

  a. No one who is going to a party ever fails to brush his hair.

  b. No one looks fascinating if he is untidy.

  c. Opium eaters have no self-command.

  d. Everyone who has brushed his hair looks fascinating.

  e. No one wears white kid gloves unless he is going to a party. (*Hint:* "not a unless b" $\equiv a \rightarrow b$.)

  f. A man is always untidy if he has no self-command.

  g. Give a conclusion that yields a valid argument.

YOUR TURN ANSWERS

  1. Invalid;

  $t =$ "you watch television tonight" $=$ T,

  $p =$ "you write your paper tonight" $=$ F,

  $g =$ "you get a good grade" $=$ T.

  2. Valid

1. $\sim m \vee \sim c$	Premise	
2. $\sim m \rightarrow t$	Premise	
3. $\sim t$	Premise	
4. m	2,3, Modus Tollens	
5. $\sim c$	1,4 Disjunctive Syllogism	

6.6 Analyzing Arguments with Quantifiers

APPLY IT   If some U.S. presidents won the popular vote and George W. Bush is the U.S. president, did he win the popular vote?
This question will be analyzed in Example 6 of this section.

There is a subtle but important point that we have ignored in some of the logic puzzles of the previous section. Consider the statement "Babies are illogical" in Example 6 in that section. We reinterpreted this statement as "If you are a baby, then you are illogical." Letting b represent "you are a baby" and i represent "you are illogical," we wrote this statement as $b \rightarrow i$. This was perfectly adequate for solving that puzzle. But now suppose we want to add two more premises: "Madeleine is a baby" and "Noa is a baby." It would seem that two valid conclusions are: "Madeleine is illogical" and "Noa is illogical." But how do we write these statements symbolically? We can't very well have b represent both "Madeleine is a baby" and "Noa is a baby." We might try letting these two statements be represented by m and n, but then it's not clear how we can combine m or n with $b \rightarrow i$.

Logicians solve this problem with the use of **quantifiers**. The words *all*, *each*, *every*, and *no(ne)* are called **universal quantifiers**, while words and phrases such as *some, there exists*, and *(for) at least one* are called **existential quantifiers**. Quantifiers are used extensively in mathematics to indicate *how many* cases of a particular situation exist.

In the previous example, we could let $b(x)$ represent "x is a baby," where x could represent any baby. Then $b(m)$ could represent "Madeleine is a baby" and $b(n)$ could represent "Noa is a baby." We want to say that $b(x) \rightarrow i(x)$ *for all x.* Logicians use the symbol $\forall$ to represent "for all." We could then represent the statement "Babies are illogical" as

$$\forall x \, [b(x) \rightarrow i(x)].$$

Suppose, instead, that we don't want to claim that all babies are illogical, but that some babies are illogical, or, equivalently, that there exists someone who is a baby and who is illogical. Logicians use the symbol $\exists$ to represent "there exists." We could then write the statement "Some babies are illogical" as

$$\exists x \, [b(x) \wedge i(x)].$$

NOTE
The important point to note here is that when a statement contains the word "all," a universal quantifier ($\forall$) is usually called for, and when it contains the word "some," an existential quantifier ($\exists$) is usually called for.

Negation of Quantifiers

We must be careful in forming the negation of a statement involving quantifiers. Suppose we wish to say that the statement "Babies are illogical" is false. This is *not* the same as saying "Babies are logical." Maybe some babies are illogical and some are logical. Let $b(x)$ represent "x is a baby" and $i(x)$ represent "x is illogical." To deny the claim that for all x, the statement $b(x) \rightarrow i(x)$ is true, is equivalent to saying that for some x, the statement $b(x) \rightarrow i(x)$ is false. Thus,

$$\sim\{\forall x \, [b(x) \rightarrow i(x)]\} \equiv \exists x \, \{\sim[b(x) \rightarrow i(x)]\}.$$

Recall from Section 6.3 that the right side of this last statement is equivalent to $\exists x \, \{[b(x) \wedge \sim i(x)]\}$. Therefore,

$$\sim\{\forall x \, [b(x) \rightarrow i(x)]\} \equiv \exists x \, \{[b(x) \wedge \sim i(x)]\}.$$

This last statement makes intuitive sense; it says that if it's false that all babies are illogical, then there exists someone who is a baby and who is not illogical. In other words, some babies are logical.

In a similar way, suppose we wish to say that the statement "Some babies are illogical" is false. To deny the existence of an x such that $b(x) \wedge i(x)$ is the same as saying that for all x, the statement $b(x) \wedge i(x)$ is false. Thus,

$$\sim\{\exists x \, [b(x) \wedge i(x)]\} \equiv \forall x \, \{\sim[b(x) \wedge i(x)]\}.$$

Recall from Section 6.3 that $\sim(p \rightarrow q) \equiv p \wedge \sim q$, so negating both sides of this equivalence and replacing q with $\sim q$ implies $\sim(p \wedge q) \equiv p \rightarrow \sim q$. We can use this fact to rewrite the previous equivalence as

$$\sim\{\exists x \, [b(x) \wedge i(x)]\} \equiv \forall x \, \{[b(x) \rightarrow \sim i(x)]\}.$$

This equivalence makes intuitive sense. It says that if it's false that some babies are illogical, then all babies are not illogical, which can be phrased more clearly as "All babies are logical."

Let's summarize the above discussion.

Negations of Quantified Statements

Statement	Symbolic	Negation	Symbolic for Negation
$s(x)$ is true for all x.	$\forall x \, [s(x)]$	$s(x)$ is false for some x.	$\exists x \, [\sim s(x)]$
$s(x)$ is true for some x.	$\exists x \, [s(x)]$	$s(x)$ is false for all x.	$\forall x \, [\sim s(x)]$

Essentially, negation changes a $\forall$ into a $\exists$ and vice versa.

EXAMPLE 1 Negation

Write each statement symbolically. Then write the negation symbolically, and translate the negation back into words.

(a) Some Texans eat quiche.

SOLUTION Let $t(x)$ represent "x is a Texan" and $q(x)$ represent "x eats quiche." The statement can be written as

$$\exists x[t(x) \wedge q(x)].$$

Its negation is

$$\forall x\{\sim[t(x) \wedge q(x)]\},$$

which is equivalent to

$$\forall x[t(x) \rightarrow \sim q(x)].$$

In words, "All Texans don't eat quiche," which can be expressed more clearly as "No Texan eats quiche."

(b) Some Texans do not eat quiche.

SOLUTION Using the notation from part (a), the statement can be written as

$$\exists x[t(x) \wedge \sim q(x)].$$

Its negation is

$$\forall x\{\sim[t(x) \wedge \sim q(x)]\},$$

which is equivalent to

$$\forall x[t(x) \rightarrow q(x)].$$

In words, "All Texans eat quiche."

(c) No Texans eat quiche.

SOLUTION This is the same as saying "All Texans do not eat quiche," which can be written as

$$\forall x [t(x) \rightarrow \sim q(x)].$$

Its negation is

$$\exists x\{ \sim[t(x) \rightarrow \sim q(x)]\},$$

which is equivalent to

$$\exists x [t(x) \wedge q(x)].$$

In words, "Some Texans eat quiche." Notice that the statements in parts (a) and (c) are negations of each other. **TRY YOUR TURN 1**

YOUR TURN 1 Write each statement symbolically. Then write the negation symbolically, and translate the negation back into words.

a. All college students study.
b. Some professors are not organized.

Just as we had rules for analyzing arguments without quantifiers in the last section, there are also rules for analyzing arguments with quantifiers. Instead of presenting such rules, we will instead demonstrate a visual technique based on **Euler diagrams**, illustrated in the following examples. Euler (pronounced "oiler") diagrams are named after the great Swiss mathematician Leonhard Euler (1707–1783).

EXAMPLE 2 Determining Validity

Represent the following argument symbolically. Is the argument valid?

> All elephants have wrinkles.
> Babar is an elephant.
> _____
> Babar has wrinkles.

SOLUTION If we let $e(x)$ represent "x is an elephant," $w(x)$ represent "x has wrinkles," and b represent "Babar," we could represent the argument symbolically as follows.

$$\forall x \, [e(x) \rightarrow w(x)]$$
$$\underline{e(b)}$$
$$w(b)$$

Notice that this argument resembles Modus Ponens from the previous section, but with a quantifier, so we suspect that the argument is valid. To verify our suspicion, we draw regions to represent the first premise. One is the region for "elephants." Since all elephants have wrinkles, the region for "elephants" goes inside the region for "have wrinkles," as in Figure 14(a).

The second premise, "Babar is an elephant," suggests that "Babar" should go inside the region representing "elephants." Let b represent "Babar." Figure 14(b) shows that "Babar" is also inside the region for "have wrinkles." Therefore, if both premises are true, the conclusion that "Babar has wrinkles" must also be true. Euler diagrams illustrate that this argument is valid. **TRY YOUR TURN 2**

YOUR TURN 2 Represent the following argument symbolically. Is the argument valid?
All insects are arthropods.
A bee is an insect.
A bee is an arthropod.

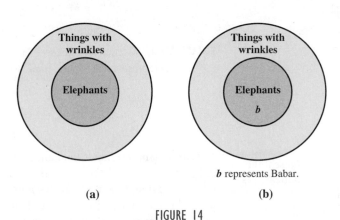

b represents Babar.

(a)   (b)

FIGURE 14

EXAMPLE 3 Determining Validity

Represent the following argument symbolically. Is the argument valid?

All frogs are slimy.

Kermit is not slimy.

Kermit is not a frog.

SOLUTION If we let $f(x)$ represent "x is a frog," $s(x)$ represent "x is slimy," and k represent "Kermit," we could represent the argument symbolically as follows.

$$\forall x \, [f(x) \rightarrow s(x)]$$
$$\underline{\sim s(k)}$$
$$\sim f(k)$$

This argument resembles Modus Tollens with a quantifier, so it is probably valid. We will use an Euler diagram to verify this.

In Figure 15(a) on the next page, the region for "frogs" is drawn entirely inside the region for "slimy." Since "Kermit is *not* slimy," place a k for "Kermit" *outside* the region for "slimy." (See Figure 15(b).) Placing the k outside the region for "slimy" automatically places it outside the region for "frogs." Thus, if the first two premises are true, the conclusion that Kermit is not a frog must also be true, so the argument is valid. **TRY YOUR TURN 3**

YOUR TURN 3 Represent the following argument symbolically. Is the argument valid?
All birds have wings.
Rover does not have wings.
Rover is not a bird.

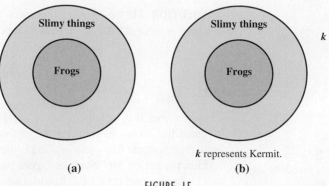

k represents Kermit.

(a)   (b)

FIGURE 15

EXAMPLE 4   **Determining Validity**

Represent the following argument symbolically. Is the argument valid?

> All well-run businesses generate profits.
>
> Monsters, Inc., generates profits.
>
> Monsters, Inc., is a well-run business.

SOLUTION If we let $w(x)$ represent "*x* is a well-run business," $p(x)$ represent "*x* generates profits," and *m* represent "Monsters, Inc.," we could represent the argument symbolically as follows.

$$\forall x \, [w(x) \to p(x)]$$
$$p(m)$$
$$\overline{}$$
$$w(m)$$

YOUR TURN 4   Represent the following argument symbolically. Is the argument valid?
Every man has his price.
Sam has his price.
Sam is a man.

This argument resembles the Fallacy of the Converse with a quantifier. We will use an Euler diagram to verify that the argument is invalid.

The region for "well-run business" goes entirely inside the region for "generates a profit." (See Figure 16.) It is not clear where to put the *m* for "Monsters, Inc." It must go inside the region for "generates profits," but it could go inside or outside the region "well-run business." Even if the premises are true, the conclusion may or may not be true, so the argument is invalid.   **TRY YOUR TURN 4**

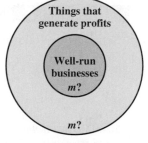

FIGURE 16

As we mentioned in the previous section, the validity of an argument is not the same as the truth of its conclusion. The argument in Example 4 was invalid, but the conclusion "Monsters, Inc." may or may not be true. We cannot make a valid conclusion.

EXAMPLE 5   **Determining Validity**

Represent the following argument symbolically. Is the argument valid?

> All squirrels eat nuts.
>
> All those who eat nuts are healthy.
>
> All who are healthy avoid cigarettes.
>
> All squirrels avoid cigarettes.

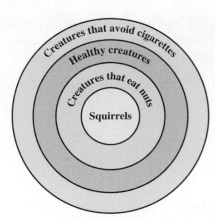

FIGURE 17

SOLUTION If we let $s(x)$ represent "x is a squirrel," $e(x)$ represent "x eats nuts," $h(x)$ represent "x is healthy," and $a(x)$ represent "x avoids cigarettes," we could represent the argument symbolically as follows.

$$\forall x\,[s(x) \rightarrow e(x)]$$
$$\forall x\,[e(x) \rightarrow h(x)]$$
$$\underline{\forall x\,[h(x) \rightarrow a(x)]}$$
$$\forall x\,[s(x) \rightarrow a(x)]$$

This argument should remind you of Reasoning by Transitivity. We will use the Euler diagram in Figure 17 to verify that the argument is valid. If each premise is true, then the conclusion must be true because the region for "squirrels" lies completely within the region for "avoid cigarettes." Thus, the argument is valid.

Our last example in this section illustrates an argument with the word "some," which means that an existential quantifier is needed.

EXAMPLE 6 Determining Validity

Represent the following argument symbolically. Is the argument valid?

Some U.S. presidents have won the popular vote.

George W. Bush is president of the United States.

George W. Bush won the popular vote.

APPLY IT

SOLUTION If we let $p(x)$ represent "x is a U.S. president," $v(x)$ represent "x won the popular vote," and w represent "George W. Bush," we could represent the argument symbolically as follows.

$$\exists x\,[p(x) \land v(x)]$$
$$\underline{p(w)}$$
$$v(w)$$

This argument doesn't resemble any of those in the previous section. An Euler diagram might help us see whether this argument is valid or not. The first premise is sketched in Figure 18(a). We have indicated that some U.S. presidents have won the popular vote by putting an x in the region that belongs to both the set of presidential candidates who have won the popular vote and the set of U.S. presidents. There are two possibilities for w, as shown in Figure 18(b).

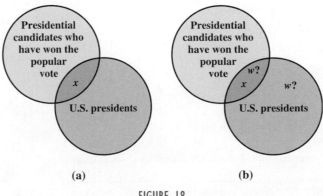

(a)   (b)

FIGURE 18

YOUR TURN 5 Represent the following argument symbolically. Is the argument valid?
Some vegetarians eat eggs.
Sarah is a vegetarian.
Sarah eats eggs.

One possibility is that Bush won the popular vote; the other is that Bush did not win the popular vote. Since the truth of the premises does not force the conclusion to be true, the argument is invalid.   **TRY YOUR TURN 5**

This argument is not valid regardless of whether George W. Bush won the popular vote. In fact, in 2000 George W. Bush was elected president even though he received 50,456,002 votes, compared with 50,999,897 for Al Gore. Bush did win the popular vote in 2004. *Source: Federal Election Commission.*

6.6 EXERCISES

For Exercises 1–6, (a) write the statement symbolically, (b) write the negative of the statement in part (a), and (c) translate your answer from part (b) into words.

1. Some books are bestsellers.

2. Every dog has his day.

3. No CEO sleeps well at night.

4. There's no place like Alaska.

5. All the leaves are brown.

6. Some days are better than other days.

In Exercises 7–20, (a) represent the argument symbolically, and (b) use an Euler diagram to determine if the argument is valid.

7. Graduates want to find good jobs.
   Theresa Cortesini is a graduate.
   _____
   Theresa Cortesini wants to find a good job.

8. All sophomores have earned at least 60 credits.
   Lucy Banister is a sophomore.
   _____
   Lucy Banister has earned at least 60 credits.

9. All professors are covered with chalk dust.
   Otis Taylor is covered with chalk dust.
   _____
   Otis Taylor is a professor.

10. All dinosaurs are extinct.
    The dodo is extinct.
    _____
    The dodo is a dinosaur.

11. All accountants use spreadsheets.
    Nancy Hart does not use spreadsheets.
    _____
    Nancy Hart is not an accountant.

12. All fish have gills.
    Whales do not have gills.
    _____
    Whales are not fish.

13. Some people who are turned down for a mortgage have a second income.
    All people who are turned down for a loan need a mortgage broker.
    _____
    Some people with a second income need a mortgage broker.

14. Some residents of Minnesota don't like snow.
    All skiers like snow.
    _____
    Some residents of Minnesota are not skiers.

15. Some who wander are lost.
    Marty McDonald wanders.
    _____
    Marty McDonald is lost.

16. Some old houses have root cellars.
    My house has a root cellar.
    _____
    My house is old.

17. Some psychologists are university professors.
    Some psychologists have a private practice.
    _____
    Some university professors have a private practice.

18. Someone who is responsible must pay for this.
    Neil Hunnewell is responsible.
    _____
    Neil Hunnewell must pay for this.

19. Everybody is either a saint or a sinner.
    Some people are not saints.
    _____
    Some people are sinners.

20. If you're here for the first time, we want to make you feel welcome.
    If you're here after being gone for a while, we want to make you feel welcome.
    Some people are here for the first time or are here after being gone for a while.
    _____
    There are some people here whom we want to make feel welcome.

21. Refer to Example 4. If the second premise and the conclusion were interchanged, would the argument then be valid?

22. Refer to Example 5. Give a different conclusion than the one given there, so that the argument is still valid.

Construct a valid argument based on the Euler diagram shown.

23.

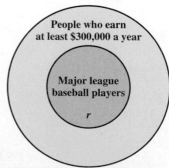

r represents Ryan Howard.

24.

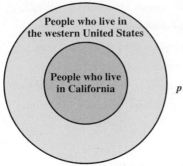

p represents Phyllis Crittenden.

As mentioned in the text, an argument can have a true conclusion yet be invalid. In these exercises, each argument has a true conclusion. Identify each argument as valid or invalid.

25. All houses have roofs.

All roofs have nails.

All houses have nails.

26. All platypuses have bills.

All ducks have bills.

A platypus is not a duck.

27. All mammals have fur.

All tigers have fur.

All tigers are mammals.

28. All mammals have fur.

All tigers are mammals.

All tigers have fur.

29. California is adjacent to Arizona.

Arizona is adjacent to Nevada.

California is adjacent to Nevada.

30. Seattle is northwest of Boise.

Seattle is northwest of Salt Lake City.

Boise is northwest of Salt Lake City.

31. A rectangle has four sides.

A square has four sides.

A square is a rectangle.

32. No integer is irrational.

The number π is irrational.

The number π is not an integer.

33. Explain the difference between the following statements:

All students did not pass the test.
Not all students passed the test.

34. Write the following statement using *every:* There is no one here who has not done that at one time or another.

APPLICATIONS

Business and Economics

35. Advertising Incorrect use of quantifiers often is heard in everyday language. Suppose you hear that a local electronics

chain is having a 30% off sale, and the radio advertisement states, "All items are not available in all stores." Do you think that, literally translated, the ad really means what it says? What do you think really is meant? Explain your answer.

36. Portfolios Repeat Exercise 35 for the following: "All people don't have the time to devote to maintaining their financial portfolios properly."

Life Sciences

37. Schizophrenia The following diagram shows the relationship between people with schizophrenia, people with a mental disorder, and people who live in California.

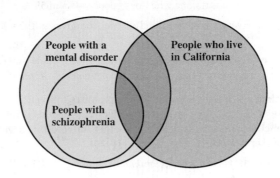

Which of the following conclusions are valid?

a. All those with schizophrenia have a mental disorder.

b. All those with schizophrenia live in California.

c. Some of those with a mental disorder live in California.

d. Some of those who live in California do not have schizophrenia.

e. All those with a mental disorder have schizophrenia.

Social Sciences

Constitution Each of the following exercises gives a passage from the U.S. Constitution, followed by another statement. (a) Translate the passage from the Constitution into a statement in symbolic logic. (b) Find a valid conclusion using as premises the passage from the Constitution and the statement that follows it. (c) Illustrate the argument with an Euler diagram.

38. All legislative powers herein granted shall be vested in a Congress of the United States . . . (Article 1, Section 1)

The power to collect taxes is a legislative power herein granted.

39. No person shall be a Representative who shall not have attained to the age of twenty-five years, and been seven years a citizen of the United States, and who shall not, when elected, be an inhabitant of that State in which he shall be chosen. (Article 1, Section 2)

John Boehner is a Representative.

40. No bill of attainder or ex post facto law shall be passed. (Article 1, Section 9)

The law forbidding members of the Communist Party to serve as an officer or as an employee of a labor union was a bill of attainder. *Source: United States V. Brown.*

41. No State shall enter into any treaty, alliance, or confederation. (Article 1, Section 10)

Texas is a state.

General Interest

42. Bible Write the negation of each of the following quotes from the Bible. *Source: NIV Bible.*

    a. "Everyone who hears about this will laugh with me."—Genesis 21:6

    b. "Someone came to destroy your lord the king."—1 Samuel 26:15

    c. "There is no one who does good."—Psalm 53:3

    d. "Everyone is the friend of one who gives gifts."—Proverbs 19:6

    e. "Everyone who quotes proverbs will quote this proverb about you: 'Like mother, like daughter.'"—Ezekiel 16:44

Animals In Exercises 43–48, the premises marked *A*, *B*, and *C* are followed by several possible conclusions. Take each conclusion in turn, and check whether the resulting argument is *valid* or *invalid.*

    A. *All kittens are cute animals.*

    B. *All cute animals are admired by animal lovers.*

    C. *Some dangerous animals are admired by animal lovers.*

43. Some kittens are dangerous animals.

44. Some cute animals are dangerous.

45. Some dangerous animals are cute.

46. Kittens are not dangerous animals.

47. All kittens are admired by animal lovers.

48. Some things admired by animal lovers are dangerous animals.

YOUR TURN ANSWERS

1. a. $\forall x\,[c(x) \rightarrow s(x)]$; negation: $\exists x\,[c(x) \wedge \sim s(x)]$, "Some college students do not study."

    b. $\exists x\,[p(x) \wedge \sim o(x)]$; negation: $\forall x\,[p(x) \rightarrow o(x)]$, "All professors are organized."

2. $\forall x\,[i(x) \rightarrow a(x)]$

$$\frac{i(b)}{a(b)}$$

Valid.

3. $\forall x\,[b(x) \rightarrow w(x)]$

$$\frac{\sim w(r)}{\sim b(r)}$$

Valid.

4. $\forall x\,[m(x) \rightarrow p(x)]$

$$\frac{p(s)}{m(s)}$$

Invalid.

5. $\exists x\,[v(x) \wedge e(x)]$

$$\frac{v(s)}{e(s)}$$

Invalid.

6 CHAPTER REVIEW

SUMMARY

In this chapter we introduced symbolic logic, which uses letters to represent statements, and symbols for words such as

- *and* (conjunction, denoted by $\wedge$),
- *or* (disjunction, denoted by $\vee$),
- *not* (negation, denoted by $\sim$),
- *if . . . then* (conditional, denoted by $\rightarrow$), and
- *if and only if* (biconditional, denoted by $\leftrightarrow$).

Statements are declarative sentences that are either true or false, but not both simultaneously. Using logical connectives, two or more statements can be combined to form a compound statement. Truth values of various compound statements were explored using truth tables. We saw that two logical statements are equivalent (denoted as $\equiv$) if they have the same truth value. We used symbolic logic to design circuits, and then used logical equivalences to simplify the circuits. We saw that the contrapositive, a statement related to the conditional, is equivalent to the original conditional statement, but that two other related statements, the inverse and the converse, are not equivalent to the original conditional statement. We next explored how to prove valid arguments and give counterexamples to invalid arguments. Finally, we discussed the quantifiers *for all* (denoted by $\forall$) and *there exists* (denoted by $\exists$), and we used Euler diagrams to determine the validity of an argument involving quantifiers.

Truth Tables for Logical Operators

p	q	$p \wedge q$	$p \vee q$	$p \rightarrow q$	$p \leftrightarrow q$
T	T	T	T	T	T
T	F	F	T	F	F
F	T	F	T	T	F
F	F	F	F	T	T

Writing a Conditional as an or Statement	$p \to q \equiv \sim p \vee q$
Negation of a Conditional Statement	$\sim(p \to q) \equiv p \wedge \sim q$

Equivalent Statements

1a. $p \vee q \equiv q \vee p$ Commutative Laws
  b. $p \wedge q \equiv q \wedge p$
2a. $p \vee (q \vee r) \equiv (p \vee q) \vee r$ Associative Laws
  b. $p \wedge (q \wedge r) \equiv (p \wedge q) \wedge r$
3a. $p \vee (q \wedge r) \equiv (p \vee q) \wedge (p \vee r)$ Distributive Laws
  b. $p \wedge (q \vee r) \equiv (p \wedge q) \vee (p \wedge r)$
4a. $\sim(p \wedge q) \equiv \sim p \vee \sim q$ De Morgan's Laws
  b. $\sim(p \vee q) \equiv \sim p \wedge \sim q$
5a. $p \vee p \equiv p$ Idempotent Laws
  b. $p \wedge p \equiv p$
6a. $(p \wedge q) \vee p \equiv p$ Absorption Laws
  b. $(p \vee q) \wedge p \equiv p$
7. $\sim(\sim p) \equiv p$ Double Negative
8. $p \to q \equiv \sim p \vee q$ Conditional as an "or"
9. $p \to q \equiv \sim q \to \sim p$ Contrapositive

If T represents any true statement and F represents any false statement, then

10a. $p \vee T \equiv T$ Identity Laws
  b. $p \wedge T \equiv p$
  c. $p \vee F \equiv p$
  d. $p \wedge F \equiv F$
11a. $p \vee \sim p \equiv T$ Negation Laws
  b. $p \wedge \sim p \equiv F$

Common Translations of $p \to q$

If p, then q. p is sufficient for q.
If p, q. q is necessary for p.
p implies q. All p's are q's.
p only if q. q if p.
q when p.

Related Conditional Statements

The contrapositive is equivalent to the original statement. However, the converse and inverse are not equivalent to the original statement, although they are equivalent to each other.

Original Statement $p \to q$
Converse $q \to p$
Inverse $\sim p \to \sim q$
Contrapositive $\sim q \to \sim p$

Valid Argument Forms

Modus Ponens	Modus Tollens	Disjunctive Syllogism	Reasoning by Transitivity
$p \to q$	$p \to q$	$p \vee q$	$p \to q$
p	$\sim q$	$\sim p$	$q \to r$
q	$\sim p$	q	$p \to r$

Invalid Argument Forms (Fallacies)

Fallacy of the Converse	Fallacy of the Inverse
$p \to q$	$p \to q$
q	$\sim p$
p	$\sim q$

KEY TERMS

6.1
statement
logical connective
compound statement
component statement
negation
truth table
conjunction
disjunction

6.2
equivalent
De Morgan's Laws

6.3
conditional
antecedent
consequent
contradiction
tautology
circuit
series
parallel
Principle of Duality

6.4
converse

inverse
contrapositive
biconditional

6.5
premise
conclusion
valid
invalid
fallacy
Modus Ponens
Fallacy of the Converse
Modus Tollens

Fallacy of the Inverse
Disjunctive Syllogism
Reasoning by Transitivity

6.6
quantifier
universal quantifier
existential quantifier
Euler diagram

REVIEW EXERCISES

CONCEPT CHECK

Determine whether each of the following statements is true or false, and explain why.

1. A compound statement is a negation, a conjunction, a disjunction, a conditional, or a biconditional.

2. A truth table with 5 variables has 10 rows.

3. A truth table can have an odd number of rows.

4. Using one of De Morgan's Laws, the negation of a disjunction may be written as a conditional.

5. The negation of a conditional statement is a disjunction.

6. Elements in a circuit that are in parallel are connected in a logic statement with an *or*.

7. A tautology might be false.

8. A statement might be true even though its inverse is false.

9. The conclusion of a valid argument must be true.

10. The conclusion of a fallacy must be false.

11. Euler diagrams can be used to determine whether an argument with quantifiers is valid or invalid.

12. The negation of a statement with the universal quantifier involves the existential quantifier.

PRACTICE AND EXPLORATIONS

Write the negation of each statement.

13. If she doesn't pay me, I won't have enough cash.

14. We played the Titans and the Titans won.

Let *l* represent "He loses the election" and let *w* represent "He wins the hearts of the voters." Write each statement in symbols.

15. He loses the election, but he wins the hearts of the voters.

16. If he wins the hearts of the voters, then he doesn't lose the election.

17. He loses the election only if he doesn't win the hearts of the voters.

18. He loses the election if and only if he doesn't win the hearts of the voters.

Using the same statements as for Exercises 15–18, write each mathematical statement in words.

19. $\sim l \wedge w$

20. $\sim(l \vee \sim w)$

Assume that *p* is true and that *q* and *r* are false. Find the truth value of each statement.

21. $\sim q \wedge \sim r$

22. $r \vee (p \wedge \sim q)$

23. $r \to (s \vee r)$ (The truth value of the statement *s* is unknown.)

24. $p \leftrightarrow (p \to q)$

25. Explain in your own words why, if *p* is a statement, the biconditional $p \leftrightarrow \sim p$ must be false.

26. State the necessary conditions for

    a. a conditional statement to be false;

    b. a conjunction to be true;

    c. a disjunction to be false.

Construct a truth table for each statement. Is the statement a tautology?

27. $p \wedge (\sim p \vee q)$

28. $\sim(p \wedge q) \to (\sim p \vee \sim q)$

Write each conditional statement in *if . . . then* form.

29. All mathematicians are loveable.

30. You can have dessert only if you eat your vegetables.

31. Having at least as many equations as unknowns is necessary for a system to have a unique solution.

32. Having a feasible region is sufficient for a linear programming problem to have a minimum.

For each statement, write (a) the converse, (b) the inverse, and (c) the contrapositive.

33. If the proposed regulations have been approved, then we need to change the way we do business.

34. $(p \vee q) \to \sim r$ (Use one of De Morgan's Laws to simplify.)

Write a logical statement representing each circuit. Simplify each circuit when possible.

35.

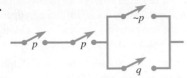

36.

Draw the circuit representing each statement as it is given. Simplify if possible.

37. $(p \wedge q) \vee (p \wedge p)$

38. $p \wedge (p \vee q)$

For Exercises 39 and 40, consider the exclusive disjunction introduced in Exercises 35–38 of Section 6.2. In exclusive disjunction, denoted by $p \veebar q$, either p or q is true, but not both.

39. Use a truth table to show that
$p \veebar q \equiv (p \vee q) \wedge \sim(p \wedge q)$.

40. Use a truth table to show that
$p \veebar q \equiv \sim[(p \vee q) \rightarrow (p \wedge q)]$.

41. Consider the statement "If this year is 2010, then $1 + 1 = 3$."

  a. Is the above statement true?

  b. Was the above statement true in 2010?

   c. Discuss how a statement such as the one above changes its truth value over time and how this agrees or disagrees with your intuitive notion of whether a conditional statement is true or false.

42. a. Consider the statement "If Shakespeare didn't write *Hamlet*, then someone else did." Explain why this statement is true.

  b. Consider the statement "If Shakespeare hadn't written *Hamlet*, then someone else would have." Explain why this statement should be false in any reasonable logic.

  c. If we let s represent "Shakespeare wrote *Hamlet*" and e represent "someone else wrote *Hamlet*," then the logic of this chapter leads us to represent either the statement of part (a) or part (b) as $\sim s \rightarrow e$. Discuss the issue of how this statement could be true in part (a) but false in part (b). *Source: Formal Logic: Its Scope and Limit.*

Each of the following arguments is either valid by one of the forms of valid arguments discussed in this chapter or is a fallacy by one of the forms of invalid arguments discussed. Decide whether the argument is valid or invalid, and give the form that applies.

43. If you're late one more time, you'll be docked.
You're late one more time.
$\overline{\text{You'll be docked.}}$

44. If the company makes a profit, its stock goes up.
The company doesn't make a profit.
$\overline{\text{Its stock doesn't go up.}}$

45. The instructor is late or my watch is wrong.
My watch is not wrong.
$\overline{\text{The instructor is late.}}$

46. If the parent is loving, then the child will be happy.
If the child is happy, then the teacher can teach.
$\overline{\text{If the parent is loving, then the teacher can teach.}}$

47. If you play that song one more time, I'm going nuts.
I'm going nuts.
$\overline{\text{You play that song one more time.}}$

48. If it's after five, the store is closed.
The store is not closed.
$\overline{\text{It's not after five.}}$

Determine whether each argument is *valid* or *invalid*. If it is valid, give a proof. If it is invalid, give an assignment of truth values to the variables that makes the premises true and the conclusion false.

49. If we hire a new person, then we'll spend more on training. If we rewrite the manual, then we won't spend more on training. We rewrite the manual. Therefore, we don't hire a new person.

50. It is not true that Smith or Jones received enough votes to qualify. But if Smith got enough votes to qualify, then the election was rigged. Therefore, the election was not rigged.

51. $\sim p \rightarrow \sim q$
$\dfrac{q \rightarrow p}{p \vee q}$

52. $p \rightarrow q$
$\dfrac{r \rightarrow \sim q}{p \rightarrow \sim r}$

In Exercises 53 and 54, (a) write the statement symbolically, (b) write the negation of the statement in part (a), and (c) translate your answer from part (b) into words.

53. All dogs have a license.

54. Some cars have manual transmissions.

In Exercises 55 and 56, (a) represent the argument symbolically, and (b) use an Euler diagram to determine whether the argument is valid.

55. All members of that fraternity do well academically.
Jordan Enzor is a member of that fraternity.
$\overline{\text{Jordan Enzor does well academically.}}$

56. Some members of that fraternity don't do well academically.
Bruce Collin does well academically.
$\overline{\text{Bruce Collin is not a member of that fraternity.}}$

57. Construct a truth table for $p \rightarrow (q \rightarrow r)$ and $(p \rightarrow q) \rightarrow r$. Are these two statements equivalent?

58. a. Convert the statement $p \rightarrow (q \rightarrow r)$ into an equivalent statement with an "or," but without the conditional.

b. Convert the statement $(p \rightarrow q) \rightarrow r$ into an equivalent statement with an "or," but without the conditional.

59. a. Construct a truth table for the statement $(p \wedge \sim p) \rightarrow q$.

b. Based on the truth table for part a, explain why any conclusion may be reached from a contradictory premise.

APPLICATIONS

Business and Economics

Income Tax The following excerpts appear in a guide for preparing income tax reports. *Source: Your Income Tax 2010.*

a. *The tax law provides several tax benefits for people attending school.*

b. *If you use the Tax Table, you do not have to compute your tax mathematically.*

c. *You may not deduct the expenses of seeking your first job.*

d. *Make sure the names used when you and your spouse file your joint return match the names you have provided to the Social Security Administration.*

60. Which of these excerpts are statements?

61. Which of these excerpts are compound statements?

62. Write the negation of statement c.

63. Write statement b as an equivalent statement using the connective *or*.

64. Write the contrapositive of statement b.

Life Sciences

65. Medicine The following statements appear in a home medical reference book. Write each conditional statement in *if . . . then* form. *Source: American College of Physicians Complete Home Medical Guide.*

a. When you exercise regularly, your heart becomes stronger and more efficient.

b. All teenagers need to be aware of the risks of drinking and driving.

c. You may need extra immunizations if you are visiting a country that has a high incidence of infectious diseases.

d. Food is essential (necessary) for good health.

Social Sciences

66. Law The following statements appear in a guide to common laws. Write the negation of each statement. *Source: Law For Dummies.*

a. If you can't afford to hire an attorney, the judge will arrange for you to have legal representation.

b. Your attorney may discover it in the course of working on your case, or the other side may unearth it.

c. You don't have a Constitutional right to make a phone call from jail, but you may be allowed to make one call.

67. Democratization According to one sociologist, "If, for example, democratization always occurs in the company of widening splits within ruling oligarchies (that is, such splits are active candidates for necessary conditions of democratization), a valid causal story will most likely connect democratization with such splits." Let w represent "widening splits occur within ruling oligarchies," d represent "democratization occurs," and v represent "a valid causal story connects democratization with splits." Represent the sociologist's statement as a statement in logic. *Source: Sociological Theory.*

68. Philosophy In an article in *Skeptical Inquirer*, Ralph Estling said, "Positivists such as [Stephen] Hawking tell us that reality cannot be dealt with . . . [and that] we can only deal with what we can measure in some way. . . . This seems sensible . . . until we watch as our physicists slowly slide down the slippery slopes . . . [and] tell us that only the measurable is real." *Source: Skeptical Inquirer.*

a. Draw an Euler diagram with three circles showing real things, things we can measure, and things we can deal with. Your diagram should simultaneously illustrate the following statements:

  i. Real things are not things we can deal with.

  ii. Only things we can measure are things we can deal with.

  iii. Only things we can measure are real things.

b. Using the Euler diagram from part a, explain why things that we can measure and deal with are not real, assuming the previous statements are true.

c. Philosopher Timothy Chambers suggests adding the plausible premise, "All things we can measure are things we can deal with." Explain why this premise, added to the previous three, implies that nothing is real. *Source: The Mathematics Teacher.*

General Interest

Scrabble The following excerpts can be found on the box to the game, Scrabble. *Source: Milton Bradley Company.*

a. *Turn all the letter tiles facedown at the side of the board or pour them into the bag or other container, and shuffle.*

b. *Any word may be challenged before the next player starts a turn.*

c. *If the word challenged is acceptable, the challenger loses his or her next turn.*

d. *When the game ends, each player's score is reduced by the sum of his or her unplayed letters.*

69. Which of these excerpts are statements?

70. Which of these excerpts are compound statements?

Lewis Carroll The following exercises are from problems by Lewis Carroll. Write each premise in symbols, and then give a conclusion that uses all the premises and yields a valid argument. *Source: The Complete Works of Lewis Carroll.*

71. Let s be "the puppy lies still," g be "the puppy is grateful to be lent a skipping rope," l be "the puppy is lame," and w be "the puppy cares to do worsted work."

a. Puppies that will not lie still are always grateful for the loan of a skipping rope.

b. A lame puppy would not say "thank you" if you offered to lend it a skipping rope.

c. None but lame puppies ever care to do worsted work.

d. Give a conclusion that yields a valid argument.

72. Let *o* be "the bird is an ostrich," *h* be "the bird is at least 9 feet high," *a* be "the bird is in this aviary," *m* be "the bird belongs to me," and *p* be "the bird lives on mince pies."

  a. No birds, except ostriches, are 9 feet high. (*Hint:* Interpret as: If a bird is at least 9 feet high, it is an ostrich.)

  b. There are no birds in this aviary that belong to any one but me.

  c. No ostrich lives on mince pies.

  d. I have no birds less than 9 feet high.

  e. Give a conclusion that yields a valid argument.

73. Let *f* be "the kitten loves fish," *t* be "the kitten is teachable," *a* be "the kitten has a tail," *g* be "the kitten will play with a gorilla," *w* be "the kitten has whiskers," and *e* be "the kitten has green eyes."

  a. No kitten that loves fish is unteachable.

  b. No kitten without a tail will play with a gorilla.

  c. Kittens with whiskers always love fish.

  d. No teachable kitten has green eyes.

  e. No kittens have tails unless they have whiskers. (*Hint:* "($\sim a$) unless *b*" $\equiv a \rightarrow b$)

  f. Give a conclusion that yields a valid argument.

74. Let *u* be "the writer understands human nature," *c* be "the writer is clever," *p* be "the writer is a true poet," *r* be "the writer can stir the hearts of men," *s* be "the writer is Shakespeare," and *h* be "the writer wrote *Hamlet*."

  a. All writers who understand human nature are clever.

  b. No one is a true poet unless he can stir the hearts of men. (*Hint:* "($\sim a$) unless *b*" $\equiv a \rightarrow b$)

  c. Shakespeare wrote *Hamlet*.

  d. No writer who does not understand human nature can stir the hearts of men.

  e. None but a true poet could have written *Hamlet*.

  f. Give a conclusion that yields a valid argument.

EXTENDED APPLICATION

LOGIC PUZZLES

Some people find that logic puzzles, which appear in periodicals such as *World-Class Logic Problems* (Penny Press) and *Logic Puzzles* (Dell), provide hours of enjoyment. They are based on deductive reasoning, and players answer questions based on clues given. The following explanation on solving such problems appeared in an issue of *World-Class Logic Problems*.

HOW TO SOLVE LOGIC PROBLEMS

Solving logic problems is entertaining and challenging. All the information you need to solve a logic problem is given in the introduction and clues, and in illustrations, when provided. If you've never solved a logic problem before, our sample should help you get started. Fill in the Sample Solving Chart in Figure 19 as you follow our explanation. We use a "•" to signify "Yes" and an "×" to signify "No."

Five couples were married last week, each on a different weekday. From the information provided, determine the woman (one is Cathy) and man (one is Paul) who make up each couple, as well as the day on which each couple was married.

1. Anne was married on Monday, but not to Wally.

2. Stan's wedding was on Wednesday. Rob was married on Friday, but not to Ida.

3. Vern (who married Fran) was married the day after Eve.

Anne was married Monday (1), so put a "•" at the intersection of Anne and Monday. Put "×" in all the other days in Anne's row and all the other names in the Monday column. (Whenever you establish a relationship, as we did here, be sure to place "×" at the intersections of all relationships that become impossible as a result.)

SAMPLE SOLVING CHART:

FIGURE 19

Anne wasn't married to Wally (1), so put an "×" at the intersection of Anne and Wally. Stan's wedding was Wednesday (2), so put a "•" at the intersection of Stan and Wednesday (don't forget the "×"s). Stan didn't marry Anne, who was married Monday, so put an "×" at the intersection of Anne and Stan. Rob was married Friday, but not to Ida (2), so put a "•" at the intersection of Rob and Friday, and "×" at the intersections of Rob and Ida and Ida and Friday. Rob also didn't marry Anne, who was married Monday, so put an "×" at the intersection of Anne and Rob. Now your chart should look like Figure 20 on the next page.

Vern married Fran (3), so put a "•" at the intersection of Vern and Fran. This leaves Anne's only possible husband as Paul, so put a "•" at the intersection of Anne and Paul and Paul and Monday. Vern and Fran's wedding was the day after Eve's (3), which wasn't

FIGURE 20

Monday [Anne], so Vern's wasn't Tuesday. It must have been Thursday [see chart], so Eve's was Wednesday (3). Put "•" at the intersections of Vern and Thursday, Fran and Thursday, and Eve and Wednesday. Now your chart should look like Figure 21.

FIGURE 21

The chart shows that Cathy was married Friday, Ida was married Tuesday, and Wally was married Tuesday. Ida married Wally, and Cathy's wedding was Friday, so she married Rob. After this information is filled in, Eve could only have married Stan. You've completed the puzzle, and your chart should now look like Figure 22

FIGURE 22

In summary: Anne and Paul, Monday; Cathy and Rob, Friday; Eve and Stan, Wednesday; Fran and Vern, Thursday; Ida and Wally, Tuesday.

280

In some problems, it may be necessary to make a logical guess based on facts you've established. When you do, always look for clues or other facts that disprove it. If you find that your guess is incorrect, eliminate it as a possibility.

EXERCISES

1. *Water, Water, Everywhere* After an invigorating workout, five fitness-conscious friends know that nothing is more refreshing than a tall, cool glass of mineral water! Each person (including Annie) has a different, favorite form of daily exercise (one likes to rollerblade), and each drinks a different brand of mineral water (one is Crystal Spring). From the information provided, determine the type of exercise and brand of water each person prefers. *Source: World-Class Logic Problems Special.*

   a. The one who bicycles in pursuit of fitness drinks Bevé.

   b. Tim enjoys aerobicizing every morning before work. Ben is neither the one who drinks Sparkling Creek nor the one who imbibes Bevé.

   c. Page (who is neither the one who jogs nor the one who walks to keep in shape) drinks Purity. Meg drinks Mountain Clear, but not after jogging.

2. *Let's Get Physical* The Anytown Community Center, in conjunction with the Board of Education's adult-outreach program, has scheduled a week-long series of lectures this fall on topics in physics. The goals are to increase awareness of the physical sciences and to attract renowned scientists (including Dr. Denton) to the community. Each of the five lectures will be held on a different weekday, and each will feature a different physicist lecturing on a different topic (one is magnetism). So far, the community has shown great interest in their upcoming physical training! From the information provided, determine the physicist who will speak on each weekday and the topic of his or her lecture. *Source: World-Class Logic Problems Special.*

   a. Dr. Hoo, who is from Yale, will not be lecturing on Thursday. Dr. Zhivago's lecture will be exactly three days after the lecture on chaos theory.

   b. If Dr. Jay is lecturing on Thursday, then the person giving the kinetic-energy lecture will appear on Tuesday; otherwise, Dr. Jay will speak on Tuesday, and kinetic energy will be the topic of Monday's lecture.

c. Dr. Know (who is not giving the lecture on quantum mechanics) is not the Harvard physicist who will speak on Monday. The photonics lecture will not be given on either Wednesday or Thursday.

		PHYSICIST					TOPIC				
		DR. DENTON	DR. HOO	DR. JAY	DR. KNOW	DR. ZHIVAGO	CHAOS THEORY	KINETIC ENERGY	MAGNETISM	PHOTONICS	QUANTUM MECHANICS
WEEKDAY	MONDAY										
	TUESDAY										
	WEDNESDAY										
	THURSDAY										
	FRIDAY										
TOPIC	CHAOS THEORY										
	KINETIC ENERGY										
	MAGNETISM										
	PHOTONICS										
	QUANTUM MECHANICS										

3. ***What's in Store?*** I had a day off from work yesterday, so I figured it was the perfect time to do some shopping. I hit the road shortly after breakfast and visited five stores (one was Bullseye). At each shop, I had intended to buy a different one of five items (a pair of andirons, a Crock-pot, pruning shears, a pair of sneakers, or a toaster oven). Unfortunately, no store had the item I was looking for in stock. The trips weren't a total loss, however, as I purchased a different item (a CD, a fondue pot, a garden gnome, spark plugs, or a winter coat) that had caught my eye in each store. Despite my failure to acquire any of the things I had sought, there were a couple of positive outcomes. I now have some nifty new things that I know I'll enjoy, and I have a shopping list written and ready to go for my next day off! From the information provided, determine the order in which I visited the five stores, as well as the item I sought and the item I bought at each store. ***Source: World-Class Logic Problems Special.***

a. I went to the store where I bought a CD (which isn't where I sought pruning shears) immediately after I visited PJ Nickle but immediately before I went to the shop where I intended to buy a Crock-pot.

b. I didn't purchase the fondue pot at Costington's. I went into one store intending to buy a Crock-pot, but came out with a garden gnome instead. I didn't go to PJ Nickle for a pair of andirons.

c. The store at which I sought a toaster oven (which wasn't the third one I visited) isn't the place where I eventually bought a winter coat. Neither Lacy's nor S-Mart was the fourth shop I visited.

d. I went to Costington's immediately after I visited the shop where I sought pruning shears (which wasn't S-Mart) but immediately before I went to the store where I purchased a set of spark plugs.

4. ***High Five*** Otis Lifter is the elevator operator at Schwarzenbach Tower, downtown Brownsville's tallest building. Since he works in such a towering edifice, Otis gets a chance to chat with his passengers on the way to their destinations. Five people who always have a friendly word for Otis work on the Schwarzenbach's top floors. Each person works on a different floor, which is home to a different company (one is the Watershed Co.). Each company is in a different business (one is a real-estate agency). Otis is content with his job, but he'll be the first to tell you that, like any profession, it has its ups and downs! From the information provided, can you determine the floor (41st through 45th) to which Otis took each person, as well as the name of his or her company and the type of business it conducts? ***Source: World-Class Logic Problems Special.***

a. Edwina's company is exactly 1 floor above Nelson & Leopold but exactly 1 floor below the accounting firm. Brierwood Ltd. is on the 44th floor.

b. Zed's company is exactly 1 floor above Ogden's. Keith's business is on the 45th floor. Trish works at the public-relations firm (which is exactly 2 floors above Glyptic).

c. Glyptic and the Thebes Group are the literary agency and the Web-design firm, in some order. The Web-design firm is not on the 41st floor.

		STORE					ITEM SOUGHT					ITEM BOUGHT				
		BULLSEYE	COSTINGTON'S	LACY'S	PJ NICKLE	S-MART	ANDIRONS	CROCK-POT	PRUNING SHEARS	SNEAKERS	TOASTER OVEN	CD	FONDUE POT	GARDEN GNOME	SPARK PLUGS	WINTER COAT
ORDER	FIRST															
	SECOND															
	THIRD															
	FOURTH															
	FIFTH															
ITEM BOUGHT	CD															
	FONDUE POT															
	GARDEN GNOME															
	SPARK PLUGS															
	WINTER COAT															
ITEM SOUGHT	ANDIRONS															
	CROCK-POT															
	PRUNING SHEARS															
	SNEAKERS															
	TOASTER OVEN															

		FLOOR					COMPANY					BUSINESS				
		41st	42nd	43rd	44th	45th	BRIERWOOD LTD.	GLYPTIC	NELSON & LEOPOLD	THEBES GROUP	WATERSHED CO.	ACCOUNTING	LITERARY AGENCY	PUBLIC RELATIONS	REAL ESTATE	WEB DESIGN
PERSON	EDWINA															
	KEITH															
	OGDEN															
	TRISH															
	ZED															
BUSINESS	ACCOUNTING															
	LITERARY AGENCY															
	PUBLIC RELATIONS															
	REAL ESTATE															
	WEB DESIGN															
COMPANY	BRIERWOOD LTD.															
	GLYPTIC															
	NELSON & LEOPOLD															
	THEBES GROUP															
	WATERSHED CO.															

5. *First Ratings* At long last, Macrocosm Industries has released Q Sphere, its new video-game console. To fully demonstrate the Q Sphere's capabilities, each of Macrocosm's five Q Sphere games (including Idle Hands) is a different genre. Anxious to be the first publication to spotlight this new gaming system, *All Game* magazine featured reviews of the Q Sphere games in its latest issue. Each game was played extensively by a different *All Game* staff reviewer (including Chadwick) and given a different rating (from lowest to highest, "don't bother," "just okay," "pretty cool," "almost perfect," or "totally awesome"). In the end, though, true video-game aficionados will want to try all of the Q Sphere games for themselves, despite the ratings! From the information provided, determine the genre of the game reviewed by each *All Game* staff member (identified by first and last names—one surname is Ploof), as well as the rating given to each game. *Source: World-Class Logic Problems Special.*

- **a.** At least one game was given a lower rating than the sports game (which is called Pitching Duel). The person surnamed Corley reviewed the action game.

- **b.** The puzzle game's rating was "just okay," which was higher than the rating Darren Castles gave.

- **c.** King of the Road is the racing game. The person surnamed Munoz (who isn't Milton) gave the "totally awesome" rating.

- **d.** The person surnamed Gilligan reviewed the simulation game, which isn't Hypnotic Trace (which was rated "almost perfect").

- **e.** Alise gave one game a "pretty cool" rating. Kourtney spent many hours playing Fiji in order to write her review.

		LAST NAME					GAME					GENRE					RATING				
		CASTLES	CORLEY	GILLIGAN	MUNOZ	PLOOF	FIJI	HYPNOTIC TRACE	IDLE HANDS	KING OF THE ROAD	PITCHING DUEL	ACTION	PUZZLE	RACING	SIMULATION	SPORTS	DON'T BOTHER	JUST OKAY	PRETTY COOL	ALMOST PERFECT	TOTALLY AWESOME
FIRST NAME	ALISE																				
	CHADWICK																				
	DARREN																				
	KOURTNEY																				
	MILTON																				
RATING	DON'T BOTHER																				
	JUST OKAY																				
	PRETTY COOL																				
	ALMOST PERFECT																				
	TOTALLY AWESOME																				
GENRE	ACTION																				
	PUZZLE																				
	RACING																				
	SIMULATION																				
	SPORTS																				
GAME	FIJI																				
	HYPNOTIC																				
	IDLE HANDS																				
	KING OF THE ROAD																				
	PITCHING DUEL																				

DIRECTIONS FOR GROUP PROJECT

Construct your own logic puzzle.

7

Sets and Probability

The study of probability begins with counting. An exercise in Section 2 of this chapter counts trucks carrying different combinations of early, late, and extra late peaches from the orchard to canning facilities. You'll see trees in another context in Section 5, where we use branching tree diagrams to calculate conditional probabilities.

Our lives are bombarded by seemingly chance events — the chance of rain, the risk of an accident, the possibility the stock market increases, the likelihood of winning the lottery — all whose particular outcome may appear to be quite random. The field of probability attempts to quantify the likelihood of chance events and helps us to prepare for this uncertainty. In short, probability helps us to better understand the world in which we live. In this chapter and the next, we introduce the basic ideas of probability theory and give a sampling of some of its uses. Since the language of sets and set operations is used in the study of probability, we begin there.

7.1 Sets

APPLY IT **In how many ways can two candidates win the 50 states plus the District of Columbia in a U.S. presidential election?**
Using knowledge of sets, we will answer this question in Exercise 69.

Think of a **set** as a well-defined collection of objects in which it is possible to determine if a given object is included in the collection. A set of coins might include one of each type of coin now put out by the U.S. government. Another set might be made up of all the students in your English class. By contrast, a collection of young adults does not constitute a set unless the designation "young adult" is clearly defined. For example, this set might be defined as those aged 18 to 29.

In mathematics, sets are often made up of numbers. The set consisting of the numbers 3, 4, and 5 is written

$$\{3, 4, 5\},$$

with set braces, $\{\ \}$, enclosing the numbers belonging to the set. The numbers 3, 4, and 5 are called the **elements** or **members** of this set. To show that 4 is an element of the set $\{3, 4, 5\}$, we use the symbol $\in$ and write

$$4 \in \{3, 4, 5\},$$

read "4 is an element of the set containing 3, 4, and 5." Also, $5 \in \{3, 4, 5\}$.

To show that 8 is *not* an element of this set, place a slash through the symbol:

$$8 \notin \{3, 4, 5\}.$$

Sets often are named with capital letters, so that if

$$B = \{5, 6, 7\},$$

then, for example, $6 \in B$ and $10 \notin B$.

It is possible to have a set with no elements. Some examples are the set of counting numbers less than one, the set of foreign-born presidents of the United States, and the set of men more than 10 feet tall. A set with no elements is called the **empty set** (or **null set**) and is written $\emptyset$.

CAUTION   Be careful to distinguish between the symbols 0, $\emptyset$, $\{0\}$, and $\{\emptyset\}$. The symbol 0 represents a *number*; $\emptyset$ represents a *set* with 0 elements; $\{0\}$ represents a set with one element, 0; and $\{\emptyset\}$ represents a set with one element, $\emptyset$.

We use the symbol $n(A)$ to indicate the *number* of elements in a finite set A. For example, if $A = \{a, b, c, d, e\}$, then $n(A) = 5$. Using this notation, we can write the information in the previous Caution as $n(\emptyset) = 0$ and $n(\{0\}) = n(\{\emptyset\}) = 1$.

Two sets are *equal* if they contain the same elements. The sets $\{5, 6, 7\}$, $\{7, 6, 5\}$, and $\{6, 5, 7\}$ all contain exactly the same elements and are equal. In symbols,

$$\{5, 6, 7\} = \{7, 6, 5\} = \{6, 5, 7\}.$$

This means that the ordering of the elements in a set is unimportant. Note that each element of the set is only listed once. Sets that do not contain exactly the same elements are *not equal*. For example, the sets $\{5, 6, 7\}$ and $\{7, 8, 9\}$ do not contain exactly the same elements and, thus, are not equal. To indicate that these sets are not equal, we write

$$\{5, 6, 7\} \neq \{7, 8, 9\}.$$

Sometimes we are interested in a common property of the elements in a set, rather than a list of the elements. This common property can be expressed by using **set-builder notation**, for example,

$$\{x \mid x \text{ has property } P\}$$

(read "the set of all elements x such that x has property P") represents the set of all elements x having some stated property P.

EXAMPLE 1 Sets

Write the elements belonging to each set.

(a) $\{x \mid x \text{ is a natural number less than 5}\}$

SOLUTION The natural numbers less than 5 make up the set $\{1, 2, 3, 4\}$.

(b) $\{x \mid x \text{ is a state that borders Florida}\}$

SOLUTION The states that border Florida make up the set $\{\text{Alabama, Georgia}\}$.

TRY YOUR TURN 1

YOUR TURN 1 Write the elements belonging to the set. $\{x \mid x \text{ is a state whose name begins with the letter O}\}$.

The **universal set** for a particular discussion is a set that includes all the objects being discussed. In elementary school arithmetic, for instance, the set of whole numbers might be the universal set, while in a college algebra class the universal set might be the set of real numbers. The universal set will be specified when necessary, or it will be clearly understandable from the context of the problem.

Subsets
Sometimes every element of one set also belongs to another set. For example, if

$$A = \{3, 4, 5, 6\}$$

and

$$B = \{2, 3, 4, 5, 6, 7, 8\},$$

then every element of A is also an element of B. This is an example of the following definition.

Subset

Set A is a **subset** of set B (written $A \subseteq B$) if every element of A is also an element of B. Set A is a *proper subset* (written $A \subset B$) if $A \subseteq B$ and $A \neq B$.

To indicate that A is *not* a subset of B, we write $A \not\subseteq B$.

EXAMPLE 2 Sets

Decide whether the following statements are *true* or *false*.

(a) $\{3, 4, 5, 6\} = \{4, 6, 3, 5\}$

SOLUTION Both sets contain exactly the same elements, so the sets are equal and the given statement is true. (The fact that the elements are listed in a different order does not matter.)

(b) $\{5, 6, 9, 12\} \subseteq \{5, 6, 7, 8, 9, 10, 11\}$

SOLUTION The first set is not a subset of the second because it contains an element, 12, that does not belong to the second set. Therefore, the statement is false.

TRY YOUR TURN 2

The empty set, $\emptyset$, by default, is a subset of every set A since it is impossible to find an element of the empty set (it has no elements) that is not an element of the set A. Similarly, A is a subset of itself, since every element of A is also an element of the set A.

Subset Properties
For any set A,

$$\emptyset \subseteq A \quad \text{and} \quad A \subseteq A.$$

EXAMPLE 3 Subsets

List all possible subsets for each set.

(a) $\{7, 8\}$

SOLUTION There are 4 subsets of $\{7, 8\}$:

$$\emptyset, \quad \{7\}, \quad \{8\}, \quad \text{and} \quad \{7, 8\}.$$

(b) $\{a, b, c\}$

SOLUTION There are 8 subsets of $\{a, b, c\}$:

$$\emptyset, \quad \{a\}, \quad \{b\}, \quad \{c\}, \quad \{a, b\}, \quad \{a, c\}, \quad \{b, c\}, \quad \text{and} \quad \{a, b, c\}.$$

A good way to find the subsets of $\{7, 8\}$ and the subsets of $\{a, b, c\}$ in Example 3 is to use a **tree diagram**—a systematic way of listing all the subsets of a given set. Figure 1 shows tree diagrams for finding the subsets of $\{7, 8\}$ and $\{a, b, c\}$.

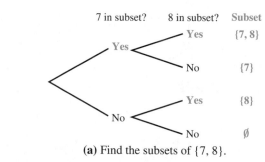

(a) Find the subsets of $\{7, 8\}$.

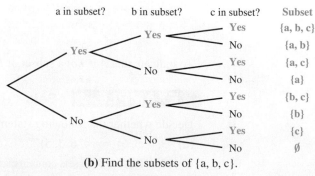

(b) Find the subsets of $\{a, b, c\}$.

FIGURE 1

As Figure 1 shows, there are two possibilities for each element (either it's in the subset or it's not), so a set with 2 elements has $2 \cdot 2 = 2^2 = 4$ subsets, and a set with 3 elements has $2^3 = 8$ subsets. This idea can be extended to a set with any finite number of elements, which leads to the following conclusion.

Number of Subsets

A set of k distinct elements has 2^k subsets.

In other words, if $n(A) = k$, then $n(\text{the set of all subsets of } A) = 2^k$.

EXAMPLE 4 Subsets

Find the number of subsets for each set.

(a) $\{3, 4, 5, 6, 7\}$

SOLUTION This set has 5 elements; thus, it has 2^5 or 32 subsets.

(b) $\{x \mid x \text{ is a day of the week}\}$

SOLUTION This set has 7 elements and therefore has $2^7 = 128$ subsets.

(c) $\emptyset$

SOLUTION Since the empty set has 0 elements, it has $2^0 = 1$ subset—itself.

TRY YOUR TURN 3

YOUR TURN 3 Find the number of subsets for the set $\{x \mid x \text{ is a season of the year}\}$.

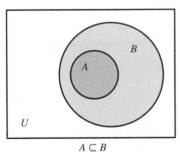

$A \subseteq B$

FIGURE 2

Figure 2 shows a set A that is a subset of set B. The rectangle and everything inside it represents the universal set, U. Such diagrams, called **Venn diagrams**—after the English logician John Venn (1834–1923), who invented them in 1876—are used to help illustrate relationships among sets. Venn diagrams are very similar to Euler diagrams, described in Section 6.6. Euler diagrams are used in logic to denote variables having a certain property or not, while Venn diagrams are used in the context of sets to denote something being an element of a set or not.

Set Operations
It is possible to form new sets by combining or manipulating one or more existing sets. Given a set A and a universal set U, the set of all elements of U that do *not* belong to A is called the *complement* of set A. For example, if set A is the set of all the female students in a class, and U is the set of all students in the class, then the complement of A would be the set of all male students in the class. The complement of set A is written A', read "A-prime."

Complement of a Set

Let A be any set, with U representing the universal set. Then the **complement** of A, colored pink in the figure, is

$$A' = \{x \mid x \notin A \text{ and } x \in U\}.$$

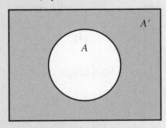

(Recall that the rectangle represents the universal set U.)

EXAMPLE 5 Set Operations

Let $U = \{1, 2, 3, 4, 5, 6, 7, 8, 9, 10, 11\}$, $A = \{1, 2, 4, 5, 7\}$, and $B = \{2, 4, 5, 7, 9, 11\}$. Find each set.

(a) A'

SOLUTION Set A' contains the elements of U that are not in A.

$$A' = \{3, 6, 8, 9, 10, 11\}$$

(b) $B' = \{1, 3, 6, 8, 10\}$

(c) $\emptyset' = U$ and $U' = \emptyset$

(d) $(A')' = A$

Given two sets A and B, the set of all elements belonging to *both* set A and set B is called the *intersection* of the two sets, written $A \cap B$. For example, the elements that belong to both set $A = \{1, 2, 4, 5, 7\}$ and set $B = \{2, 4, 5, 7, 9, 11\}$ are 2, 4, 5, and 7, so that

$$A \cap B = \{1, 2, 4, 5, 7\} \cap \{2, 4, 5, 7, 9, 11\} = \{2, 4, 5, 7\}.$$

Intersection of Two Sets

The **intersection** of sets A and B, shown in green in the figure, is

$$A \cap B = \{x \mid x \in A \text{ and } x \in B\}.$$

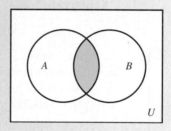

EXAMPLE 6 Set Operations

Let $A = \{3, 6, 9\}$, $B = \{2, 4, 6, 8\}$, and the universal set $U = \{0, 1, 2, \ldots, 10\}$. Find each set.

(a) $A \cap B$

SOLUTION

$$A \cap B = \{3, 6, 9\} \cap \{2, 4, 6, 8\} = \{6\}$$

(b) $A \cap B'$

SOLUTION

$$A \cap B' = \{3, 6, 9\} \cap \{0, 1, 3, 5, 7, 9, 10\} = \{3, 9\}$$

YOUR TURN 4 For the sets in Example 6, find $A' \cap B$.

TRY YOUR TURN 4

Two sets that have no elements in common are called *disjoint sets*. For example, there are no elements common to both $\{50, 51, 54\}$ and $\{52, 53, 55, 56\}$, so these two sets are disjoint, and

$$\{50, 51, 54\} \cap \{52, 53, 55, 56\} = \emptyset.$$

This result can be generalized as follows.

Disjoint Sets

For any sets A and B, if A and B are **disjoint sets**, then $A \cap B = \emptyset$.

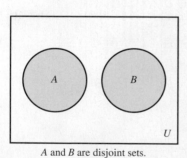

A and B are disjoint sets.

FIGURE 3

Figure 3 shows a pair of disjoint sets.

The set of all elements belonging to set A, to set B, or to both sets is called the *union* of the two sets, written $A \cup B$. For example,

$$\{1, 3, 5\} \cup \{3, 5, 7, 9\} = \{1, 3, 5, 7, 9\}.$$

Union of Two Sets

The **union** of sets A and B, shown in blue in the figure, is

$$A \cup B = \{x \mid x \in A \text{ or } x \in B \text{ (or both)}\}.$$

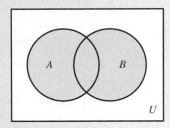

EXAMPLE 7   **Union of Sets**

Let $A = \{1, 3, 5, 7, 9, 11\}$, $B = \{3, 6, 9, 12\}$, $C = \{1, 2, 3, 4, 5\}$, and the universal set $U = \{0, 1, 2, \ldots, 12\}$. Find each set.

(a) $A \cup B$

SOLUTION   Begin by listing the elements of the first set, $\{1, 3, 5, 7, 9, 11\}$. Then include any elements from the second set *that are not already listed*. Doing this gives

$$A \cup B = \{1, 3, 5, 7, 9, 11\} \cup \{3, 6, 9, 12\} = \{1, 3, 5, 7, 9, 11, 6, 12\}$$
$$= \{1, 3, 5, 6, 7, 9, 11, 12\}.$$

(b) $(A \cup B) \cap C'$

SOLUTION   Begin with the expression in parentheses, which we calculated in part (a), and then intersect this with C'.

$$(A \cup B) \cap C' = \{1, 3, 5, 6, 7, 9, 11, 12\} \cap \{0, 6, 7, 8, 9, 10, 11, 12\}$$
$$= \{6, 7, 9, 11, 12\} \qquad \textbf{TRY YOUR TURN 5}$$

YOUR TURN 5   For the sets in Example 7, find $A \cup (B \cap C')$.

NOTE
1. As Example 7 shows, when forming sets, do not list the same element more than once. In our final answer, we listed the elements in numerical order to make it easier to see what elements are in the set, but the set is the same, regardless of the order of the elements.
2. As shown in the definitions, an element is in the *intersection* of sets A and B if it is in A *and* B. On the other hand, an element is in the *union* of sets A and B if it is in A *or* B (or both).
3. In mathematics, "A or B" implies A or B, or both. Since "or" includes the possibility of both, we will usually omit the words "or both".

EXAMPLE 8   **Stocks**

The following table gives the 52-week low and high prices, the closing price, and the change from the previous day for six stocks in the Standard & Poor's 100 on February 19, 2010. *Source: New York Stock Exchange.*

New York Stock Exchange				
Stock	Low	High	Close	Change
AT&T	21.44	28.73	25.10	−0.55
Coca-Cola	37.44	59.45	55.72	−0.34
FedEx	34.02	92.59	81.76	+2.07
McDonald's	50.44	65.75	64.74	+0.40
PepsiCo	45.39	64.48	62.66	+0.21
Walt Disney	15.14	32.75	31.23	+0.68

Let the universal set U consist of the six stocks listed in the table. Let A contain all stocks with a high price greater than $50, B all stocks with a closing price between $60 and $70, C all stocks with a positive price change, and D all stocks with a low price less than $40. Find the following.

a. A'

SOLUTION Set A' contains all the listed stocks that are not in set A, or those with a high price less than or equal to $50, so

$$A' = \{\text{AT\&T, Disney}\}.$$

b. $A \cap C$

SOLUTION The intersection of A and C will contain those stocks that are in both sets A and C, or those with a high price greater than $50 *and* a positive price change.

$$A \cap C = \{\text{FedEx, McDonald's, PepsiCo}\}$$

c. $D \cup B$

SOLUTION The union of D and B contains all stocks with a low price less than $40 *or* a closing price between $60 and $70.

$$D \cup B = \{\text{AT\&T, Coca-Cola, FedEx, McDonald's, PepsiCo, Disney}\} = U$$

■

EXAMPLE 9 Employment

A department store classifies credit applicants by gender, marital status, and employment status. Let the universal set be the set of all applicants, M be the set of male applicants, S be the set of single applicants, and E be the set of employed applicants. Describe each set in words.

a. $M \cap E$

SOLUTION The set $M \cap E$ includes all applicants who are both male *and* employed; that is, employed male applicants.

b. $M' \cup S$

SOLUTION This set includes all applicants who are female (not male) *or* single. *All* female applicants and *all* single applicants are in this set.

c. $M' \cap S'$

SOLUTION These applicants are female *and* married (not single); thus, $M' \cap S'$ is the set of all married female applicants.

d. $M \cup E'$

SOLUTION $M \cup E'$ is the set of applicants that are male *or* unemployed. The set includes *all* male applicants and *all* unemployed applicants. ■

7.1 EXERCISES

In Exercises 1–10, write true or false for each statement.

1. $3 \in \{2, 5, 7, 9, 10\}$ **2.** $6 \in \{-2, 6, 9, 5\}$

3. $9 \notin \{2, 1, 5, 8\}$ **4.** $3 \notin \{7, 6, 5, 4\}$

5. $\{2, 5, 8, 9\} = \{2, 5, 9, 8\}$

6. $\{3, 7, 12, 14\} = \{3, 7, 12, 14, 0\}$

7. {all whole numbers greater than 7 and less than 10} $= \{8, 9\}$

8. $\{x \,|\, x$ is an odd integer; $6 \le x \le 18\} = \{7, 9, 11, 15, 17\}$

9. $0 \in \emptyset$ **10.** $\emptyset \in \{\emptyset\}$

Let $A = \{2, 4, 6, 10, 12\}$, $B = \{2, 4, 8, 10\}$, $C = \{4, 8, 12\}$, $D = \{2, 10\}$, $E = \{6\}$, and $U = \{2, 4, 6, 8, 10, 12, 14\}$. Insert $\subseteq$ or $\not\subseteq$ to make the statement true.

11. A ___ U **12.** E ___ A **13.** A ___ E

14. B ___ C **15.** $\emptyset$ ___ A **16.** $\{0, 2\}$ ___ D

17. D ___ B **18.** A ___ C

19. Repeat Exercises 11–18 except insert $\subset$ or $\not\subset$ to make the statement true.

20. What is set-builder notation? Give an example.

Insert a number in each blank to make the statement true, using the sets for Exercises 11–18.

21. There are exactly ___ subsets of A.

22. There are exactly ___ subsets of B.

23. There are exactly ___ subsets of C.

24. There are exactly ___ subsets of D.

Insert $\cap$ or $\cup$ to make each statement true.

25. $\{5, 7, 9, 19\}$ ___ $\{7, 9, 11, 15\} = \{7, 9\}$

26. $\{8, 11, 15\}$ ___ $\{8, 11, 19, 20\} = \{8, 11\}$

27. $\{2, 1, 7\}$ ___ $\{1, 5, 9\} = \{1, 2, 5, 7, 9\}$

28. $\{6, 12, 14, 16\}$ ___ $\{6, 14, 19\} = \{6, 12, 14, 16, 19\}$

29. $\{3, 5, 9, 10\}$ ___ $\emptyset = \emptyset$

30. $\{3, 5, 9, 10\}$ ___ $\emptyset = \{3, 5, 9, 10\}$

31. $\{1, 2, 4\}$ ___ $\{1, 2, 4\} = \{1, 2, 4\}$

32. $\{0, 10\}$ ___ $\{10, 0\} = \{0, 10\}$

33. Describe the intersection and union of sets. How do they differ?

34. Is it possible for two nonempty sets to have the same intersection and union? If so, give an example.

Let $U = \{1, 2, 3, 4, 5, 6, 7, 8, 9\}$, $X = \{2, 4, 6, 8\}$, $Y = \{2, 3, 4, 5, 6\}$, and $Z = \{1, 2, 3, 8, 9\}$. List the members of each set, using set braces.

35. $X \cap Y$

36. $X \cup Y$

37. X'

38. Y'

39. $X' \cap Y'$

40. $X' \cap Z$

41. $Y \cap (X \cup Z)$

42. $X' \cap (Y' \cup Z)$

43. $(X \cap Y') \cup (Z' \cap Y')$ 44. $(X \cap Y) \cup (X' \cap Z)$

45. In Example 6, what set do you get when you calculate $(A \cap B) \cup (A \cap B')$?

46. Explain in words why $(A \cap B) \cup (A \cap B') = A$.

Let $U = \{$all students in this school$\}$, $M = \{$all students taking this course$\}$, $N = \{$all students taking accounting$\}$, and $P = \{$all students taking zoology$\}$. Describe each set in words.

47. M'

48. $M \cup N$

49. $N \cap P$

50. $N' \cap P'$

51. Refer to the sets listed for Exercises 11–18. Which pairs of sets are disjoint?

52. Refer to the sets listed for Exercises 35–44. Which pairs are disjoint?

Refer to Example 8 in the text. Describe each set in Exercises 53–56 in words; then list the elements of each set.

53. B'

54. $A \cap B$

55. $(A \cap B)'$

56. $(C \cup D)'$

57. Let $A = \{1, 2, 3, \{3\}, \{1, 4, 7\}\}$. Answer each of the following as *true* or *false*.

   a. $1 \in A$ b. $\{3\} \in A$ c. $\{2\} \in A$ d. $4 \in A$

   e. $\{\{3\}\} \subset A$ f. $\{1, 4, 7\} \in A$ g. $\{1, 4, 7\} \subseteq A$

58. Let $B = \{a, b, c, \{d\}, \{e, f\}\}$. Answer each of the following as *true* or *false*.

   a. $a \in B$ b. $\{b, c, d\} \subset B$ c. $\{d\} \in B$

   d. $\{d\} \subseteq B$ e. $\{e, f\} \in B$ f. $\{a, \{e, f\}\} \subset B$

   g. $\{e, f\} \subset B$

APPLICATIONS

Business and Economics

Mutual Funds The tables in the next column show five of the largest holdings of four major mutual funds on January 31, 2010. *Sources: fidelity.com, franklintempleton.com, janus.com, vanguard.com.*

Vanguard 500	Fidelity New Millennium Fund
Exxon Mobil	Pfizer
Apple	Cisco Systems
General Electric	Wal-Mart Stores
IBM	Apple
JPMorgan Chase & Co.	JPMorgan Chase & Co.

Janus Perkins Large Cap Value	Templeton Large Cap Value Fund
Exxon Mobil	IBM
General Electric	General Electric
Wal-Mart Stores	Hewlett-Packard
AT&T	Home Depot
JPMorgan Chase & Co.	Aflac

Let U be the smallest possible set that includes all the corporations listed, and V, F, J, and T be the set of top holdings for each mutual fund, respectively. Find each set:

59. $V \cap J$

60. $V \cap (F \cup T)$

61. $(J \cup F)'$

62. $J' \cap T'$

63. **Sales Calls** Suppose that Kendra Gallegos has appointments with 9 potential customers. Kendra will be ecstatic if all 9 of these potential customers decide to make a purchase from her. Of course, in sales there are no guarantees. How many different sets of customers may place an order with Kendra? (*Hint:* Each set of customers is a subset of the original set of 9 customers.)

Life Sciences

Health The following table shows some symptoms of an overactive thyroid and an underactive thyroid. *Source: The Merck Manual of Diagnosis and Therapy.*

Underactive Thyroid	Overactive Thyroid
Sleepiness, s	Insomnia, i
Dry hands, d	Moist hands, m
Intolerance of cold, c	Intolerance of heat, h
Goiter, g	Goiter, g

Let U be the smallest possible set that includes all the symptoms listed, N be the set of symptoms for an underactive thyroid, and O be the set of symptoms for an overactive thyroid. Find each set.

64. O'

65. N'

66. $N \cap O$

67. $N \cup O$

68. $N \cap O'$

Social Sciences

69. **APPLY IT Electoral College** U.S. presidential elections are decided by the Electoral College, in which each of the 50

states, plus the District of Columbia, gives all of its votes to a candidate.* Ignoring the number of votes each state has in the Electoral College, but including all possible combinations of states that could be won by either candidate, how many outcomes are possible in the Electoral College if there are two candidates? (*Hint:* The states that can be won by a candidate form a subset of all the states.)

General Interest

70. Musicians A concert featured a cellist, a flutist, a harpist, and a vocalist. Throughout the concert, different subsets of the four musicians performed together, with at least two musicians playing each piece. How many subsets of at least two are possible?

Television Cable Services The following table lists some of the most popular cable television networks. Use this information for Exercises 71–76. *Source: The New York Times 2010 Almanac.*

Network	Subscribers (in millions)	Launch	Content
The Discovery Channel	98.0	1985	Nonfiction, nature, science
TNT	98.0	1988	Movies, sports, original programming
USA Network	97.5	1980	Sports, family entertainment
The Learning Channel (TLC)	97.3	1980	Original programming, family entertainment
TBS Superstation	97.3	1976	Movies, sports, original programming

List the elements of the following sets. For exercises 74–76, describe each set in words.

71. *F*, the set of networks that were launched before 1985.

72. *G*, the set of networks that feature sports.

73. *H*, the set of networks that have more than 97.6 million viewers.

*The exceptions are Maine and Nebraska, which allocate their electoral college votes according to the winner in each congressional district.

74. $F \cap H$ **75.** $G \cup H$ **76.** G'

77. Games In David Gale's game of Subset Takeaway, the object is for each player, at his or her turn, to pick a nonempty proper subset of a given set subject to the condition that no subset chosen earlier by either player can be a subset of the newly chosen set. The winner is the last person who can make a legal move. Consider the set $A = \{1, 2, 3\}$. Suppose Joe and Dorothy are playing the game and Dorothy goes first. If she chooses the proper subset $\{1\}$, then Joe cannot choose any subset that includes the element 1. Joe can, however, choose $\{2\}$ or $\{3\}$ or $\{2, 3\}$. Develop a strategy for Joe so that he can always win the game if Dorothy goes first. *Source: Scientific American.*

States In the following list of states, let $A = \{$states whose name contains the letter $e\}$, let $B = \{$states with a population of more than 4,000,000$\}$, and $C = \{$states with an area greater than 40,000 square miles$\}$. *Source: The New York Times 2010 Almanac.*

State	Population (1000s)	Area (sq. mi.)
Alabama	4662	52,419
Alaska	686	663,267
Colorado	4939	104,094
Florida	18,328	65,755
Hawaii	1288	10,931
Indiana	6377	36,418
Kentucky	4269	40,409
Maine	1316	35,385
Nebraska	1783	77,354
New Jersey	8683	8721

78. a. Describe in words the set $A \cup (B \cap C)'$.

  b. List all elements in the set $A \cup (B \cap C)'$.

79. a. Describe in words the set $(A \cup B)' \cap C$.

  b. List all elements in the set $(A \cup B)' \cap C$.

▮ YOUR TURN ANSWERS

1. $\{$Ohio, Oklahoma, Oregon$\}$

2. True

3. $2^4 = 16$ subsets

4. $\{2, 4, 8\}$

5. $\{1, 3, 5, 6, 7, 9, 11, 12\}$

7.2 Applications of Venn Diagrams

APPLY IT The responses to a survey of 100 households show that 76 have a DVD player, 21 have a Blue Ray player, and 12 have both. How many have neither a DVD player nor Blue Ray player?
In Example 3 we show how a Venn diagram can be used to sort out this information to answer the question.

Venn diagrams were used in the previous section to illustrate set union and intersection. The rectangular region of a Venn diagram represents the universal set U. Including only a single set A inside the universal set, as in Figure 4, divides U into two regions. Region 1

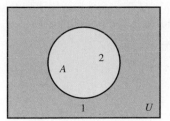

One set leads to 2 regions
(numbering is arbitrary).

FIGURE 4

represents those elements of U outside set A (that is, the elements in A'), and region 2 represents those elements belonging to set A. (The numbering of these regions is arbitrary.)

The Venn diagram in Figure 5(a) shows two sets inside U. These two sets divide the universal set into four regions. As labeled in Figure 5(a), region 1 represents the set whose elements are outside both set A and set B. Region 2 shows the set whose elements belong to A and not to B. Region 3 represents the set whose elements belong to both A and B. Which set is represented by region 4? (Again, the labeling is arbitrary.)

Two other situations can arise when representing two sets by Venn diagrams. If it is known that $A \cap B = \emptyset$, then the Venn diagram is drawn as in Figure 5(b). If it is known that $A \subseteq B$, then the Venn diagram is drawn as in Figure 5(c). For the material presented throughout this chapter we will only refer to Venn diagrams like the one in Figure 5(a), and note that some of the regions of the Venn diagram may be equal to the empty (or null) set.

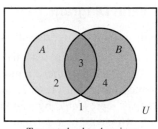

Two sets lead to 4 regions
(numbering is arbitrary).

(a)

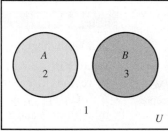

Two sets lead to 3 regions
(numbering is arbitrary).

(b)

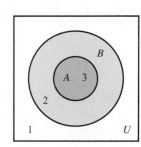

Two sets lead to 3 regions
(numbering is arbitrary).

(c)

FIGURE 5

EXAMPLE 1 **Venn Diagrams**

Draw Venn diagrams similar to Figure 5(a) and shade the regions representing each set.

(a) $A' \cap B$

SOLUTION Set A' contains all the elements outside set A. As labeled in Figure 5(a), A' is represented by regions 1 and 4. Set B is represented by regions 3 and 4. The intersection of sets A' and B, the set $A' \cap B$, is given by the region common to both sets. The result is the set represented by region 4, which is blue in Figure 6. When looking for the intersection, remember to choose the area that is in one region *and* the other region.

In addition to the fact that region 4 in Figure 6 is $A' \cap B$, notice that region 1 is $A' \cap B'$, region 2 is $A \cap B'$, and region 3 is $A \cap B$.

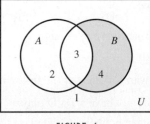

FIGURE 6

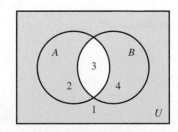

FIGURE 7

(b) $A' \cup B'$

SOLUTION Again, set A' is represented by regions 1 and 4, and set B' by regions 1 and 2. To find $A' \cup B'$, identify the region that represents the set of all elements in A', B', or both. The result, which is blue in Figure 7, includes regions 1, 2, and 4. When looking for the union, remember to choose the area that is in one region *or* the other region (or both). **TRY YOUR TURN 1**

YOUR TURN 1 Draw a Venn diagram and shade the region representing $A \cup B'$.

Venn diagrams also can be drawn with three sets inside U. These three sets divide the universal set into eight regions, which can be numbered (arbitrarily) as in Figure 8.

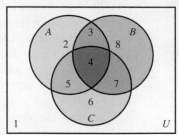

Three sets lead to 8 regions.
FIGURE 8

EXAMPLE 2 Venn Diagram

In a Venn diagram, shade the region that represents $A' \cup (B \cap C')$.

YOUR TURN 2 Draw a Venn diagram and shade the region representing $A' \cap (B \cup C)$.

SOLUTION First find $B \cap C'$. Set B is represented by regions 3, 4, 7, and 8, and set C' by regions 1, 2, 3, and 8. The overlap of these regions (regions 3 and 8) represents the set $B \cap C'$. Set A' is represented by regions 1, 6, 7, and 8. The union of the set represented by regions 3 and 8 and the set represented by regions 1, 6, 7, and 8 is the set represented by regions 1, 3, 6, 7, and 8, which are blue in Figure 9.   **TRY YOUR TURN 2**

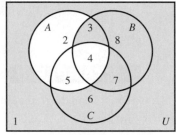

FIGURE 9

Applications
Venn diagrams can be used to analyze many applications, as illustrated in the following examples.

EXAMPLE 3 Entertainment Technology

A researcher collecting data on 100 households finds that

   76 have a DVD player;

   21 have a Blue Ray player; and

   12 have both.

The researcher wants to answer the following questions.

(a) How many do not have a DVD player?

(b) How many have neither a DVD player nor a Blue Ray player?

(c) How many have a Blue Ray player but not a DVD player?

APPLY IT

SOLUTION A Venn diagram like the one in Figure 10 will help sort out this information. In Figure 10(a), we put the number 12 in the region common to both a DVD player and a Blue Ray player, because 12 households have both. Of the 21 with a Blue Ray player, $21 - 12 = 9$ have no DVD player, so in Figure 10(b) we put 9 in the region for a Blue Ray player but no DVD player. Similarly, $76 - 12 = 64$ households have a DVD player but not a Blue Ray player, so we put 64 in that region. Finally, the diagram shows that $100 - 64 - 12 - 9 = 15$ households have neither a DVD player nor a Blue Ray player. Now we can answer the questions:

(a) $15 + 9 = 24$ do not have a DVD player.

(b) 15 have neither.

(c) 9 have a Blue Ray player but not a DVD player.

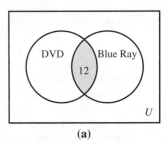
(a)

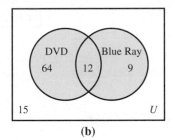
(b)

FIGURE 10

EXAMPLE 4 Magazines

A survey of 77 freshman business students at a large university produced the following results.

25 of the students read *Bloomberg Businessweek*;

19 read *The Wall Street Journal*;

27 do not read *Fortune*;

11 read *Bloomberg Businessweek* but not *The Wall Street Journal*;

11 read *The Wall Street Journal* and *Fortune*;

13 read *Bloomberg Businessweek* and *Fortune*;

9 read all three.

Use this information to answer the following questions.

(a) How many students read none of the publications?

(b) How many read only *Fortune*?

(c) How many read *Bloomberg Businessweek* and *The Wall Street Journal*, but not *Fortune*?

SOLUTION Since 9 students read all three publications, begin by placing 9 in the area that belongs to all three regions, as shown in Figure 11. Of the 13 students who read *Bloomberg Businessweek* and *Fortune*, 9 also read *The Wall Street Journal*. Therefore, only $13 - 9 = 4$ read just *Bloomberg Businessweek* and *Fortune*. Place the number 4 in the area of Figure 11 common only to *Bloomberg Businessweek* and *Fortune* readers.

In the same way, place $11 - 9 = 2$ in the region common only to *Fortune* and *The Wall Street Journal*. Of the 11 students who read *Bloomberg Businessweek* but not *The Wall Street Journal*, 4 read *Fortune*, so place $11 - 4 = 7$ in the region for those who read only *Bloomberg Businessweek*.

The data show that 25 students read *Bloomberg Businessweek*. However, $7 + 4 + 9 = 20$ readers have already been placed in the region representing *Bloomberg Businessweek*. The balance of this region will contain only $25 - 20 = 5$ students. These students read *Bloomberg Businessweek* and *The Wall Street Journal* but not *Fortune*. In the same way, $19 - (5 + 9 + 2) = 3$ students read only *The Wall Street Journal*.

Using the fact that 27 of the 77 students do not read *Fortune*, we know that 50 do read *Fortune*. We already have $4 + 9 + 2 = 15$ students in the region representing *Fortune*, leaving $50 - 15 = 35$ who read only *Fortune*.

A total of $7 + 4 + 35 + 5 + 9 + 2 + 3 = 65$ students are placed in the three circles in Figure 11. Since 77 students were surveyed, $77 - 65 = 12$ students read none of the three publications, and 12 is placed outside all three regions.

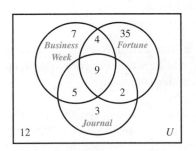

FIGURE 11

YOUR TURN 3 One hundred students were asked which fast food restaurants they had visited in the past month. The results are as follows:

47 ate at McDonald's;

46 ate at Taco Bell;

44 ate at Wendy's;

17 ate at McDonald's and Taco Bell;

19 ate at Taco Bell and Wendy's;

22 ate at Wendy's and McDonald's;

13 ate at all three.

Determine how many ate only at Taco Bell.

Now Figure 11 can be used to answer the questions asked above.

(a) There are 12 students who read none of the three publications.

(b) There are 35 students who read only *Fortune*.

(c) The overlap of the regions representing readers of *Bloomberg Businessweek* and *The Wall Street Journal* shows that 5 students read *Bloomberg Businessweek* and *The Wall Street Journal* but not *Fortune*. **TRY YOUR TURN 3** ▨

CAUTION A common error in solving problems of this type is to make a circle represent one set and another circle represent its complement. In Example 4, with one circle representing those who read *Bloomberg Businessweek*, we did not draw another for those who do not read *Bloomberg Businessweek*. An additional circle is not only unnecessary (because those not in one set are automatically in the other) but very confusing, because the region outside or inside both circles must be empty. Similarly, if a problem involves men and women, do not draw one circle for men and another for women. Draw one circle; if you label it "women," for example, then men are automatically those outside the circle.

EXAMPLE 5 Utility Maintenance

Jeff Friedman is a section chief for an electric utility company. The employees in his section cut down trees, climb poles, and splice wire. Friedman reported the following information to the management of the utility.

"Of the 100 employees in my section,

45 can cut trees;

50 can climb poles;

57 can splice wire;

22 can climb poles but can't cut trees;

20 can climb poles and splice wire;

25 can cut trees and splice wire;

14 can cut trees and splice wire but can't climb poles;

 9 can't do any of the three (management trainees)."

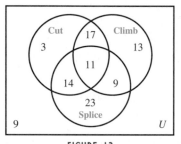

FIGURE 12

The data supplied by Friedman lead to the numbers shown in Figure 12. Add the numbers from all of the regions to get the total number of employees:

$$9 + 3 + 14 + 23 + 11 + 9 + 17 + 13 = 99.$$

Friedman claimed to have 100 employees, but his data indicated only 99. Management decided that Friedman didn't qualify as a section chief, and he was reassigned as a night-shift meter reader in Guam. (*Moral:* He should have taken this course.) ▬

In all the examples above, we started with a piece of information specifying the relationship with all the categories. This is usually the best way to begin solving problems of this type.

As we saw in the previous section, we use the symbol $n(A)$ to indicate the *number* of elements in a finite set A. The following statement about the number of elements in the union of two sets will be used later in our study of probability.

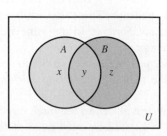

FIGURE 13

Union Rule for Sets

$$n(A \cup B) = n(A) + n(B) - n(A \cap B)$$

To prove this statement, let $x + y$ represent $n(A)$, y represent $n(A \cap B)$, and $y + z$ represent $n(B)$, as shown in Figure 13. Then

$$n(A \cup B) = x + y + z,$$
$$n(A) + n(B) - n(A \cap B) = (x + y) + (y + z) - y = x + y + z,$$

so

$$n(A \cup B) = n(A) + n(B) - n(A \cap B).$$

EXAMPLE 6 School Activities

A group of 10 students meet to plan a school function. All are majoring in accounting or economics or both. Five of the students are economics majors and 7 are majors in accounting. How many major in both subjects?

SOLUTION Let A represent the set of accounting majors and B represent the set of economics majors. Use the union rule, with $n(A) = 5$, $n(B) = 7$, and $n(A \cup B) = 10$. Find $n(A \cap B)$.

$$n(A \cup B) = n(A) + n(B) - n(A \cap B)$$
$$10 = 5 + 7 - n(A \cap B),$$

so $n(A \cap B) = 5 + 7 - 10 = 2.$ **TRY YOUR TURN 4**

YOUR TURN 4 A group of students are sitting in the lounge. All are texting or listening to music or both. Eleven are listening to music and 15 are texting. Eight are doing both. How many students are in the lounge?

When A and B are disjoint, then $n(A \cap B) = 0$, so the union rule simplifies to $n(A \cup B) = n(A) + n(B)$.

CAUTION The rule $n(A \cup B) = n(A) + n(B)$ is only valid when A and B are disjoint. When A and B are *not* disjoint, use the rule $n(A \cup B) = n(A) + n(B) - n(A \cap B)$.

EXAMPLE 7 Endangered Species

The following table gives the number of threatened and endangered animal species in the world as of January, 2010. *Source: U.S. Fish and Wildlife Service.*

Endangered and Threatened Species			
	Endangered (E)	Threatened (T)	Totals
Amphibians and reptiles (A)	101	52	153
Arachnids and insects (I)	63	10	73
Birds (B)	256	23	279
Clams, crustaceans, corals and snails (C)	108	24	132
Fishes (F)	85	66	151
Mammals (M)	325	35	360
Totals	938	210	1148

Using the letters given in the table to denote each set, find the number of species in each of the following sets.

(a) $E \cap B$

SOLUTION The set $E \cap B$ consists of all species that are endangered *and* are birds. From the table, we see that there are 256 such species.

(b) $E \cup B$

SOLUTION The set $E \cup B$ consists of all species that are endangered *or* are birds. We include all 938 endangered species, plus the 23 bird species who are threatened but not endangered, for a total of 961. Alternatively, we could use the formula $n(E \cup B) = n(E) + n(B) - n(E \cap B) = 938 + 279 - 256 = 961$.

(c) $(F \cup M) \cap T'$

SOLUTION Begin with the set $F \cup M$, which is all species that are fish or mammals. This consists of the four categories with 85, 66, 325, and 35 species. Of this set, take those that are *not* threatened, for a total of $85 + 325 = 410$ species. This is the number of species of fish and mammals that are not threatened.

EXAMPLE 8 **Online Activities**

Suppose that a group of 150 students have done at least one of these activities online: purchasing an item, banking, and watching a video. In addition,

90 students have made an online purchase;

50 students have banked online;

70 students have watched an online video;

15 students have made an online purchase and watched an online video;

12 have banked online and watched an online video; and

10 students have done all three.

How many students have made an online purchase and banked online?

SOLUTION Let P represent the set of students who have made a purchase online, B the set who have banked online, and V the set who have watched a video online. Since 10 students did all three activities, begin by placing 10 in the area that belongs to all three regions, as shown in Figure 14. Of the 15 students who belong in sets P and V, 10 also belong to set B. Thus, only $15 - 10 = 5$ students were in the area of Figure 14 common only to sets P and V. Likewise, there are $12 - 10 = 2$ students who belong to only sets B and V. Since there are already $5 + 10 + 2 = 17$ students in set V, there are $70 - 17 = 53$ students in only set V.

We cannot use the information about set P, since there are two regions in P for which we have no information. Similarly, we cannot use the information about set B. In such cases, we label a region with the variable x. Here we place x in the region common only to P and B, as shown in Figure 14.

Of the 90 students in set P, the number who are only in set P must be $90 - x - 10 - 5 = 75 - x$, and this expression is placed in the appropriate region in Figure 14. Similarly, the number who are in only set B is $50 - x - 10 - 2 = 38 - x$. Notice that because all 150 students participated in at least one activity, there are no elements in the region outside the three circles.

Now that the diagram is filled out, we can determine the value of x by recalling that the total number of students was 150. Thus,

$$(75 - x) + 5 + x + 10 + (38 - x) + 2 + 53 = 150.$$

Simplifying, we have $183 - x = 150$, implying that $x = 33$. The number of students who made an online purchase and banked online is

$$33 + 10 = 43.$$

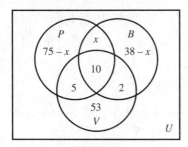

FIGURE 14

7.2 EXERCISES

Sketch a Venn diagram like the one in the figure, and use shading to show each set.

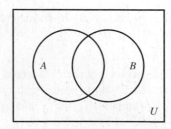

1. $B \cap A'$

2. $A \cup B'$

3. $A' \cup B$

4. $A' \cap B'$

5. $B' \cup (A' \cap B')$

6. $(A \cap B) \cup B'$

7. U'

8. $\emptyset'$

9. Three sets divide the universal set into at most _____ regions.

10. What does the notation $n(A)$ represent?

Sketch a Venn diagram like the one shown, and use shading to show each set.

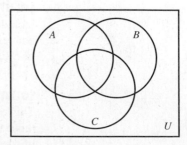

11. $(A \cap B) \cap C$

12. $(A \cap C') \cup B$

13. $A \cap (B \cup C')$

14. $A' \cap (B \cap C)$

15. $(A' \cap B') \cap C'$

16. $(A \cap B') \cap C$

17. $(A \cap B') \cup C'$

18. $A' \cap (B' \cup C)$

19. $(A \cup B') \cap C$

20. $A' \cup (B' \cap C')$

Use the union rule to answer the following questions.

21. If $n(A) = 5$, $n(B) = 12$, and $n(A \cap B) = 4$, what is $n(A \cup B)$?

22. If $n(A) = 15$, $n(B) = 30$, and $n(A \cup B) = 33$, what is $n(A \cap B)$?

23. Suppose $n(B) = 9$, $n(A \cap B) = 5$, and $n(A \cup B) = 22$. What is $n(A)$?

24. Suppose $n(A \cap B) = 5$, $n(A \cup B) = 38$, and $n(A) = 13$. What is $n(B)$?

Draw a Venn diagram and use the given information to fill in the number of elements for each region.

25. $n(U) = 41, n(A) = 16, n(A \cap B) = 12, n(B') = 20$

26. $n(A) = 28, n(B) = 12, n(A \cup B) = 32, n(A') = 19$

27. $n(A \cup B) = 24, n(A \cap B) = 6, n(A) = 11$, $n(A' \cup B') = 25$

28. $n(A') = 31, n(B) = 25, n(A' \cup B') = 46, n(A \cap B) = 12$

29. $n(A) = 28, n(B) = 34, n(C) = 25, n(A \cap B) = 14$, $n(B \cap C) = 15, n(A \cap C) = 11, n(A \cap B \cap C) = 9$, $n(U) = 59$

30. $n(A) = 54, n(A \cap B) = 22, n(A \cup B) = 85$, $n(A \cap B \cap C) = 4, n(A \cap C) = 15, n(B \cap C) = 16$, $n(C) = 44, n(B') = 63$

31. $n(A \cap B) = 6, n(A \cap B \cap C) = 4, n(A \cap C) = 7$, $n(B \cap C) = 4, n(A \cap C') = 11, n(B \cap C') = 8$, $n(C) = 15, n(A' \cap B' \cap C') = 5$

32. $n(A) = 13, n(A \cap B \cap C) = 4, n(A \cap C) = 6$, $n(A \cap B') = 6, n(B \cap C) = 6, n(B \cap C') = 11$, $n(B \cup C) = 22, n(A' \cap B' \cap C') = 5$

In Exercises 33–36, show that the statement is true by drawing Venn diagrams and shading the regions representing the sets on each side of the equals sign.*

33. $(A \cup B)' = A' \cap B'$

34. $(A \cap B)' = A' \cup B'$

35. $A \cap (B \cup C) = (A \cap B) \cup (A \cap C)$

36. $A \cup (B \cap C) = (A \cup B) \cap (A \cup C)$

37. Use the union rule of sets to prove that $n(A \cup B \cup C) = n(A) + n(B) + n(C) - n(A \cap B) - n(A \cap C) - n(B \cap C) + n(A \cap B \cap C)$. (*Hint:* Write $A \cup B \cup C$ as $A \cup (B \cup C)$ and use the formula from Exercise 35.)

*The statements in Exercises 33 and 34 are known as De Morgan's Laws. They are named for the English mathematician Augustus De Morgan (1806–1871). They are analogous to De Morgan's Laws for logic seen in the previous chapter.

Business and Economics

Use Venn diagrams to answer the following questions.

38. Cooking Preferences Jeff Friedman, of Example 5 in the text, was again reassigned, this time to the home economics department of the electric utility. He interviewed 140 people in a suburban shopping center to discover some of their cooking habits. He obtained the following results:

58 use microwave ovens;
63 use electric ranges;
58 use gas ranges;
19 use microwave ovens and electric ranges;
17 use microwave ovens and gas ranges;
4 use both gas and electric ranges;
1 uses all three;
2 use none of the three.

Should he be reassigned one more time? Why or why not?

39. Harvesting Fruit Toward the middle of the harvesting season, peaches for canning come in three types, early, late, and extra late, depending on the expected date of ripening. During a certain week, the following data were recorded at a fruit delivery station:

34 trucks went out carrying early peaches;
61 carried late peaches;
50 carried extra late;
25 carried early and late;
30 carried late and extra late;
8 carried early and extra late;
6 carried all three;
9 carried only figs (no peaches at all).

a. How many trucks carried only late variety peaches?

b. How many carried only extra late?

c. How many carried only one type of peach?

d. How many trucks (in all) went out during the week?

40. Cola Consumption Market research showed that the adult residents of a certain small town in Georgia fit the following categories of cola consumption. (We assume here that no one drinks both regular cola and diet cola.)

Age	Drink Regular Cola (R)	Drink Diet Cola (D)	Drink No Cola (N)	Totals
21–25 (Y)	40	15	15	70
26–35 (M)	30	30	20	80
Over 35 (O)	10	50	10	70
Totals	80	95	45	220

Using the letters given in the table, find the number of people in each set.

a. $Y \cap R$

b. $M \cap D$

c. $M \cup (D \cap Y)$

d. $Y' \cap (D \cup N)$

e. $O' \cup N$

f. $M' \cap (R' \cap N')$

g. Describe the set $M \cup (D \cap Y)$ in words.

41. Investment Habits The following table shows the results of a survey taken by a bank in a medium-sized town in Tennessee. The survey asked questions about the investment habits of bank customers. (We assume here that no one invests in more than one of type of investment.)

Age	Stocks (S)	Bonds (B)	Savings Accounts (A)	Totals
18–29 (Y)	6	2	15	23
30–49 (M)	14	5	14	33
50 or over (O)	32	20	12	64
Totals	52	27	41	120

Using the letters given in the table, find the number of people in each set.

a. $Y \cap B$ **b.** $M \cup A$ **c.** $Y \cap (S \cup B)$

d. $O' \cup (S \cup A)$ **e.** $(M' \cup O') \cap B$

f. Describe the set $Y \cap (S \cup B)$ in words.

42. Investment Survey Most mathematics professors love to invest their hard-earned money. A recent survey of 150 math professors revealed that

111 invested in stocks;
98 invested in bonds;
100 invested in certificates of deposit;
80 invested in stocks and bonds;
83 invested in bonds and certificates of deposit;
85 invested in stocks and certificates of deposit;
9 did not invest in any of the three.

How many mathematics professors invested in stocks and bonds and certificates of deposit?

Life Sciences

43. Genetics After a genetics experiment on 50 pea plants, the number of plants having certain characteristics was tallied, with the following results.

22 were tall;
25 had green peas;
39 had smooth peas;
9 were tall and had green peas;
20 had green peas and smooth peas;
6 had all three characteristics;
4 had none of the characteristics.

a. Find the number of plants that were tall and had smooth peas.

b. How many plants were tall and had peas that were neither smooth nor green?

c. How many plants were not tall but had peas that were smooth and green?

44. Blood Antigens Human blood can contain the A antigen, the B antigen, both the A and B antigens, or neither antigen. A third antigen, called the Rh antigen, is important in human reproduction, and again may or may not be present in an individual. Blood is called type A-positive if the individual has the A and Rh but not the B antigen. A person having only the A and B antigens is said to have type AB-negative blood. A person having only the Rh antigen has type O-positive blood. Other blood types are defined in a similar manner. Identify the blood types of the individuals in regions (a)–(h) below.

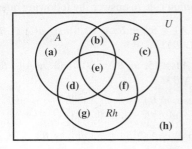

45. Blood Antigens (Use the diagram from Exercise 44.) In a certain hospital, the following data were recorded.

25 patients had the A antigen;
8 had the A and not the B antigen;
27 had the B antigen;
22 had the B and Rh antigens;
30 had the Rh antigen;
12 had none of the antigens;
16 had the A and Rh antigens;
15 had all three antigens.

How many patients

a. were represented? **b.** had exactly one antigen?

c. had exactly two antigens? **d.** had O-positive blood?

e. had AB-positive blood? **f.** had B-negative blood?

g. had O-negative blood? **h.** had A-positive blood?

46. Mortality The table lists the number of deaths in the United States during 2006 according to race and gender. Use this information and the letters given to find the number of people in each set. *Source: National Vital Statistics Reports.*

	White (W)	Black (B)	American Indian (I)	Asian or Pacific Islander (A)
Female (F)	1,055,221	141,369	6407	21,325
Male (M)	1,022,328	148,602	7630	23,382

a. F **b.** $F \cap (I \cup A)$

c. $M \cup B$ **d.** $W' \cup I' \cup A'$

e. In words, describe the set in part b.

47. Hockey The table lists the number of head and neck injuries for 319 ice hockey players wearing either a full shield or half shield in the Canadian Inter-University Athletics Union during one season. Using the letters given in the table, find the number of injuries in each set. *Source: JAMA.*

	Half Shield (H)	Full Shield (F)
Head and Face Injuries (A)	95	34
Concussions (B)	41	38
Neck Injuries (C)	9	7
Other Injuries (D)	202	150

a. $A \cap F$
b. $C \cap (H \cup F)$
c. $D \cup F$
d. $B' \cap C'$

Social Sciences

48. Military The number of female military personnel in 2009 is given in the following table. Use this information and the letters given to find the number of female military personnel in each set. *Source: Department of Defense.*

	Army (A)	Air Force (B)	Navy (C)	Marines (D)	Totals
Officers (O)	14,322	12,097	7884	1202	35,505
Enlisted (E)	59,401	51,965	42,225	11,749	165,340
Cadets & Midshipmen (M)	688	922	920	0	2530
Totals	74,411	64,984	51,029	12,951	203,375

a. $A \cup B$
b. $E \cup (C \cup D)$
c. $O' \cap M'$

U.S. Population The projected U.S. population in 2020 (in millions) by age and race or ethnicity is given in the following table. Use this information in Exercises 49–54. *Source: U.S. Bureau of the Census.*

	Non-Hispanic White (A)	Hispanic (B)	Black (C)	Asian (D)	American Indian (E)
Under 45 (F)	110.6	37.6	30.2	13.1	2.2
45–64 (G)	55.3	10.3	9.9	4.3	0.6
65 and over (H)	41.4	4.7	5.0	2.2	0.3
Totals	207.3	52.6	45.1	19.6	3.1

Using the letters given in the table, find the number of people in each set.

49. $A \cap F$
50. $G \cup B$
51. $G \cup (C \cap H)$
52. $F \cap (B \cup H)$
53. $H \cup D$
54. $G' \cap (A' \cap C')$

Marital Status The following table gives the population breakdown (in thousands) of the U.S. population in 2008 based on marital status and race or ethnic origin. *Source: New York Times 2010 Almanac.*

	White (W)	Black (B)	Hispanic (H)	Asian or Pacific Islander (A)
Never married (N)	54,205	13,547	12,021	3518
Married (M)	106,517	9577	16,111	6741
Widowed (I)	11,968	1740	1068	507
Divorced/ separated (D)	23,046	4590	3477	665

Find the number of people in each set. Describe each set in words.

55. $N \cap (B \cup H)$
56. $(M \cup I) \cap A$
57. $(D \cup W) \cap A'$
58. $M' \cap (B \cup A)$

General Interest

59. Chinese New Year A survey of people attending a Lunar New Year celebration in Chinatown yielded the following results:

> 120 were women;
> 150 spoke Cantonese;
> 170 lit firecrackers;
> 108 of the men spoke Cantonese;
> 100 of the men did not light firecrackers;
> 18 of the non-Cantonese-speaking women lit firecrackers;
> 78 non-Cantonese-speaking men did not light firecrackers;
> 30 of the women who spoke Cantonese lit firecrackers.

a. How many attended?

b. How many of those who attended did not speak Cantonese?

c. How many women did not light firecrackers?

d. How many of those who lit firecrackers were Cantonese-speaking men?

60. Native American Ceremonies At a pow-wow in Arizona, 75 Native American families from all over the Southwest came to participate in the ceremonies. A coordinator of the pow-wow took a survey and found that

> 15 families brought food, costumes, and crafts;
> 25 families brought food and crafts;
> 42 families brought food;
> 35 families brought crafts;
> 14 families brought crafts but not costumes;
> 10 families brought none of the three items;
> 18 families brought costumes but not crafts.

a. How many families brought costumes and food?

b. How many families brought costumes?

c. How many families brought food, but not costumes?

d. How many families did not bring crafts?

e. How many families brought food or costumes?

61. Poultry Analysis A chicken farmer surveyed his flock with the following results. The farmer had

    9 fat red roosters;
    13 thin brown hens;
    15 red roosters;
    11 thin red chickens (hens and roosters);
    17 red hens;
    56 fat chickens (hens and roosters);
    41 roosters;
    48 hens.

Assume all chickens are thin or fat, red or brown, and hens (female) or roosters (male). How many chickens were

a. in the flock?

b. red?

c. fat roosters?

d. fat hens?

e. thin and brown?

f. red and fat?

YOUR TURN ANSWERS

1. 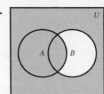 2. **3.** 23 **4.** 18

7.3 Introduction to Probability

APPLY IT What is the probability that a randomly selected person in the United States is Hispanic or Black?
After introducing probability, we will answer this question in Exercise 58.

If you go to a supermarket and buy 5 pounds of peaches at 99 cents per pound, you can easily find the *exact* price of your purchase: $4.95. On the other hand, the produce manager of the market is faced with the problem of ordering peaches. The manager may have a good estimate of the number of pounds of peaches that will be sold during the day, but it is impossible to predict the *exact* amount. The number of pounds that customers will purchase during a day is *random*: the quantity cannot be predicted exactly. A great many problems that come up in applications of mathematics involve random phenomena—those for which exact prediction is impossible. The best that we can do is determine the *probability* of the possible outcomes.

Sample Spaces In probability, an **experiment** is an activity or occurrence with an observable result. Each repetition of an experiment is called a **trial**. The possible results of each trial are called **outcomes**. The set of all possible outcomes for an experiment is the **sample space** for that experiment. A sample space for the experiment of tossing a coin is made up of the outcomes heads (*h*) and tails (*t*). If *S* represents this sample space, then

$$S = \{h, t\}.$$

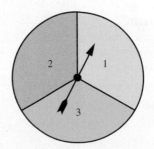

FIGURE 15

EXAMPLE 1 Sample Spaces

Give the sample space for each experiment.

(a) A spinner like the one in Figure 15 is spun.

SOLUTION The three outcomes are 1, 2, or 3, so the sample space is

$$\{1, 2, 3\}.$$

(b) For the purposes of a public opinion poll, respondents are classified as young, middle-aged, or senior, and as male or female.

SOLUTION A sample space for this poll could be written as a set of ordered pairs:

$$\{(\text{young, male}), (\text{young, female}), (\text{middle-aged, male}),$$
$$(\text{middle-aged, female}), (\text{senior, male}), (\text{senior, female})\}.$$

(c) An experiment consists of studying the numbers of boys and girls in families with exactly 3 children. Let b represent *boy* and g represent *girl*.

SOLUTION A three-child family can have 3 boys, written bbb, 3 girls, ggg, or various combinations, such as bgg. A sample space with four outcomes (not equally likely) is

$$S_1 = \{3 \text{ boys}, 2 \text{ boys and } 1 \text{ girl}, 1 \text{ boy and } 2 \text{ girls}, 3 \text{ girls}\}.$$

Notice that a family with 3 boys or 3 girls can occur in just one way, but a family of 2 boys and 1 girl or 1 boy and 2 girls can occur in more than one way. If the *order* of the births is considered, so that bgg is different from gbg or ggb, for example, another sample space is

$$S_2 = \{bbb, bbg, bgb, gbb, bgg, gbg, ggb, ggg\}.$$

YOUR TURN 1 Two coins are tossed, and a head or a tail is recorded for each coin. Give a sample space where each outcome is equally likely.

The second sample space, S_2, has equally likely outcomes if we assume that boys and girls are equally likely. This assumption, while not quite true, is approximately true, so we will use it throughout this book. The outcomes in S_1 are not equally likely, since there is more than one way to get a family with 2 boys and 1 girl (bbg, bgb, or gbb) or a family with 2 girls and 1 boy (ggb, gbg, or bgg), but only one way to get 3 boys (bbb) or 3 girls (ggg). **TRY YOUR TURN 1**

CAUTION An experiment may have more than one sample space, as shown in Example 1(c). The most convenient sample spaces have equally likely outcomes, but it is not always possible to choose such a sample space.

Events

An **event** is a subset of a sample space. If the sample space for tossing a coin is $S = \{h, t\}$, then one event is $E = \{h\}$, which represents the outcome "heads."

An ordinary die is a cube whose six different faces show the following numbers of dots: 1, 2, 3, 4, 5, and 6. If the die is fair (not "loaded" to favor certain faces over others), then any one of the faces is equally likely to come up when the die is rolled. The sample space for the experiment of rolling a single fair die is $S = \{1, 2, 3, 4, 5, 6\}$. Some possible events are listed below.

The die shows an even number: $E_1 = \{2, 4, 6\}$.

The die shows a 1: $E_2 = \{1\}$.

The die shows a number less than 5: $E_3 = \{1, 2, 3, 4\}$.

The die shows a multiple of 3: $E_4 = \{3, 6\}$.

Using the notation introduced earlier in this chapter, notice that $n(S) = 6, n(E_1) = 3$, $n(E_2) = 1, n(E_3) = 4$, and $n(E_4) = 2$.

EXAMPLE 2 Events

For the sample space S_2 in Example 1(c), write the following events.

(a) Event H: the family has exactly two girls

SOLUTION Families with three children can have exactly two girls with either bgg, gbg, or ggb, so event H is

$$H = \{bgg, gbg, ggb\}.$$

(b) Event K: the three children are the same sex

SOLUTION Two outcomes satisfy this condition: all boys or all girls.

$$K = \{bbb, ggg\}$$

YOUR TURN 2 Two coins are tossed, and a head or a tail is recorded for each coin. Write the event E: the coins show exactly one head.

(c) Event J: the family has three girls

SOLUTION Only ggg satisfies this condition, so

$$J = \{ggg\}.$$ **TRY YOUR TURN 2**

In Example 2(c), event J had only one possible outcome, ggg. Such an event, with only one possible outcome, is a **simple event**. If event E equals the sample space S, then E is called a **certain event**. If event $E = \emptyset$, then E is called an **impossible event**.

EXAMPLE 3 Events

Suppose a coin is flipped until both a head and a tail appear, or until the coin has been flipped four times, whichever comes first. Write each of the following events in set notation.

(a) The coin is flipped exactly three times.

SOLUTION This means that the first two flips of the coin did not include both a head and a tail, so they must both be heads or both be tails. Because the third flip is the last one, it must show the side of the coin not yet seen. Thus the event is

$$\{hht, tth\}.$$

(b) The coin is flipped at least three times.

SOLUTION In addition to the outcomes listed in part (a), there is also the possibility that the coin is flipped four times, which only happens when the first three flips are all heads or all tails. Thus the event is

$$\{hht, tth, hhhh, hhht, tttt, ttth\}.$$

(c) The coin is flipped at least two times.

SOLUTION This event consists of the entire sample space:

$$S = \{ht, th, hht, tth, hhhh, hhht, tttt, ttth\}.$$

This is an example of a certain event.

(d) The coin is flipped fewer than two times.

SOLUTION The coin cannot be flipped fewer than two times under the rules described, so the event is the empty set $\emptyset$. This is an example of an impossible event.

Since events are sets, we can use set operations to find unions, intersections, and complements of events. A summary of the set operations for events is given below.

Set Operations for Events

Let E and F be events for a sample space S.

$E \cap F$ occurs when both E **and** F occur;

$E \cup F$ occurs when E **or** F **or both** occur;

E' occurs when E does **not** occur.

EXAMPLE 4 Minimum-Wage Workers

A study of workers earning the minimum wage grouped such workers into various categories, which can be interpreted as events when a worker is selected at random. *Source: Economic Policy Institute.* Consider the following events:

E: worker is under 20;

F: worker is white;

G: worker is female.

Describe the following events in words.

(a) E'

SOLUTION E' is the event that the worker is 20 or over.

(b) $F \cap G'$

SOLUTION $F \cap G'$ is the event that the worker is white and not a female, that is, the worker is a white male.

(c) $E \cup G$

SOLUTION $E \cup G$ is the event that the worker is under 20 or is female. Note that this event includes all workers under 20, both male and female, and all female workers of any age.
TRY YOUR TURN 3

YOUR TURN 3 Describe the following event in words: $E' \cap F'$.

Two events that cannot both occur at the same time, such as rolling an even number and an odd number with a single roll of a die, are called *mutually exclusive events*.

Mutually Exclusive Events
Events E and F are **mutually exclusive events** if $E \cap F = \emptyset$.

Any event E and its complement E' are mutually exclusive. By definition, mutually exclusive events are disjoint sets.

$E \cap G = \emptyset$

FIGURE 16

EXAMPLE 5 Mutually Exclusive Events

Let $S = \{1, 2, 3, 4, 5, 6\}$, the sample space for tossing a single die. Let $E = \{4, 5, 6\}$, and let $G = \{1, 2\}$. Then E and G are mutually exclusive events since they have no outcomes in common: $E \cap G = \emptyset$. See Figure 16.

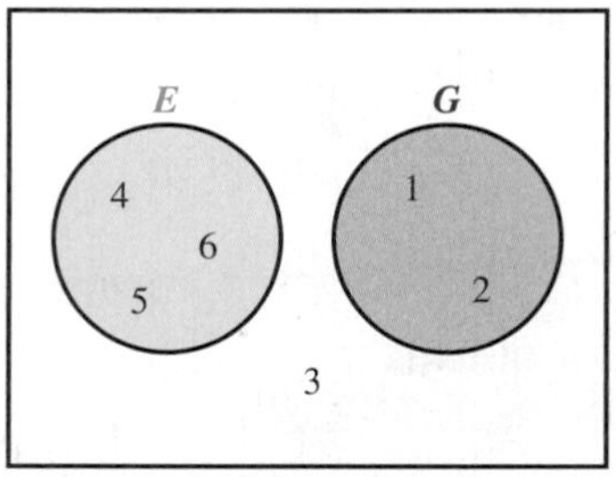

Probability
For sample spaces with *equally likely* outcomes, the probability of an event is defined as follows.

Basic Probability Principle
Let S be a sample space of equally likely outcomes, and let event E be a subset of S. Then the **probability** that event E occurs is

$$P(E) = \frac{n(E)}{n(S)}.$$

By this definition, the probability of an event is a number that indicates the relative likelihood of the event.

CAUTION The basic probability principle only applies when the outcomes are equally likely.

EXAMPLE 6 Basic Probabilities

Suppose a single fair die is rolled. Use the sample space $S = \{1, 2, 3, 4, 5, 6\}$ and give the probability of each event.

(a) E: the die shows an even number

SOLUTION Here, $E = \{2, 4, 6\}$, a set with three elements. Since S contains six elements,

$$P(E) = \frac{3}{6} = \frac{1}{2}.$$

(b) F: the die shows a number less than 10

SOLUTION Event F is a certain event, with

$$F = \{1, 2, 3, 4, 5, 6\},$$

so that

$$P(F) = \frac{6}{6} = 1.$$

YOUR TURN 4 In Example 6, find the probability of event H: the die shows a number less than 5.

(c) G: the die shows an 8

SOLUTION This event is impossible, so

$$P(G) = 0.$$

TRY YOUR TURN 4

A standard deck of 52 cards has four suits: hearts (♥), clubs (♣), diamonds (♦), and spades (♠), with 13 cards in each suit. The hearts and diamonds are red, and the spades and clubs are black. Each suit has an ace (A), a king (K), a queen (Q), a jack (J), and cards numbered from 2 to 10. The jack, queen, and king are called *face cards* and for many purposes can be thought of as having values 11, 12, and 13, respectively. The ace can be thought of as the low card (value 1) or the high card (value 14). See Figure 17. We will refer to this standard deck of cards often in our discussion of probability.

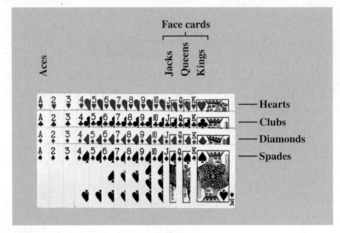

FIGURE 17

EXAMPLE 7 Playing Cards

If a single playing card is drawn at random from a standard 52-card deck, find the probability of each event.

(a) Drawing an ace

SOLUTION There are 4 aces in the deck. The event "drawing an ace" is

{heart ace, diamond ace, club ace, spade ace}.

Therefore,

$$P(\text{ace}) = \frac{4}{52} = \frac{1}{13}.$$

(b) Drawing a face card

SOLUTION Since there are 12 face cards (three in each of the four suits),

$$P(\text{face card}) = \frac{12}{52} = \frac{3}{13}.$$

(c) Drawing a spade

SOLUTION The deck contains 13 spades, so

$$P(\text{spade}) = \frac{13}{52} = \frac{1}{4}.$$

(d) Drawing a spade or a heart

SOLUTION Besides the 13 spades, the deck contains 13 hearts, so

$$P(\text{spade or heart}) = \frac{26}{52} = \frac{1}{2}.$$ **TRY YOUR TURN 5**

YOUR TURN 5 Find the probability of drawing a jack or a king.

In the preceding examples, the probability of each event was a number between 0 and 1. The same thing is true in general. Any event E is a subset of the sample space S, so $0 \le n(E) \le n(S)$. Since $P(E) = n(E)/n(S)$, it follows that $0 \le P(E) \le 1$. Note that a certain event has probability 1 and an impossible event has probability 0, as seen in Example 6.

For any event E, $0 \le P(E) \le 1$.

Empirical Probability

In many real-life problems, it is not possible to establish exact probabilities for events. Instead, useful approximations are often found by drawing on past experience. The next example shows one approach to such **empirical probabilities**.

EXAMPLE 8 Injuries

The following table lists the estimated number of injuries in the United States associated with recreation equipment. *Source: National Safety Council.*

Recreation Equipment Injuries	
Equipment	**Number of Injuries**
Bicycles	515,871
Skateboards	143,682
Trampolines	107,345
Playground climbing equipment	77,845
Swings or swing sets	59,144

Find the probability that a randomly selected person whose injury is associated with recreation equipment was hurt on a trampoline.

SOLUTION We first find the total number of injuries. Verify that the amounts in the table sum to 903,887. The probability is then found by dividing the number of people injured on trampolines by the total number of people injured. Thus,

$$P(\text{Trampolines}) = \frac{107,345}{903,887} \approx 0.1188.$$

7.3 EXERCISES

1. What is meant by a "fair" coin or die?

2. What is the sample space for an experiment?

Write sample spaces for the experiments in Exercises 3–10.

3. A month of the year is chosen for a wedding.

4. A day in April is selected for a bicycle race.

5. A student is asked how many points she earned on a recent 80-point test.

6. A person is asked the number of hours (to the nearest hour) he watched television yesterday.

7. The management of an oil company must decide whether to go ahead with a new oil shale plant or to cancel it.

8. A record is kept each day for three days about whether a particular stock goes up or down.

9. A coin is tossed, and a die is rolled.

10. A box contains five balls, numbered 1, 2, 3, 4, and 5. A ball is drawn at random, the number on it recorded, and the ball replaced. The box is shaken, a second ball is drawn, and its number is recorded.

11. Define an event.

12. What is a simple event?

For the experiments in Exercises 13–18, write out the sample space *S*, choosing an *S* with equally likely outcomes, if possible. Then give the value of *n(S)* and tell whether the outcomes in *S* are equally likely. Finally, write the indicated events in set notation.

13. A committee of 2 people is selected at random from 5 executives: Alam, Bartolini, Chinn, Dickson, and Ellsberg.

   a. Chinn is on the committee.

   b. Dickson and Ellsberg are not both on the committee.

   c. Both Alam and Chinn are on the committee.

14. Five states are being considered as the location for three new high-energy physics laboratories: California (CA), Colorado (CO), New Jersey (NJ), New York (NY), and Utah (UT). Three states will be chosen at random. Write elements of the sample space in the form (CA, CO, NJ).

   a. All three states border an ocean.

   b. Exactly two of the three states border an ocean.

   c. Exactly one of the three states is west of the Mississippi River.

15. Slips of paper marked with the numbers 1, 2, 3, 4, and 5 are placed in a box. After being mixed, two slips are drawn simultaneously.

   a. Both slips are marked with even numbers.

   b. One slip is marked with an odd number and the other is marked with an even number.

   c. Both slips are marked with the same number.

16. An unprepared student takes a three-question, true/false quiz in which he guesses the answers to all three questions, so each answer is equally likely to be correct or wrong.

   a. The student gets three answers wrong.

   b. The student gets exactly two answers correct.

   c. The student gets only the first answer correct.

17. A coin is flipped until two heads appear, up to a maximum of four flips. (If three tails are flipped, the coin is still tossed a fourth time to complete the experiment).

   a. The coin is tossed four times.

   b. Exactly two heads are tossed.

   c. No heads are tossed.

18. One jar contains four balls, labeled 1, 2, 3, and 4. A second jar contains five balls, labeled 1, 2, 3, 4, and 5. An experiment consists of taking one ball from the first jar, and then taking a ball from the second jar.

   a. The number on the first ball is even.

   b. The number on the second ball is even.

   c. The sum of the numbers on the two balls is 5.

   d. The sum of the numbers on the two balls is 1.

A single fair die is rolled. Find the probabilities of each event.

19. Getting a 2

20. Getting an odd number

21. Getting a number less than 5

22. Getting a number greater than 2

23. Getting a 3 or a 4

24. Getting any number except 3

A card is drawn from a well-shuffled deck of 52 cards. Find the probability of drawing the following.

25. A 9   26. A black card

27. A black 9   28. A heart

29. The 9 of hearts   30. A face card

31. A 2 or a queen   32. A black 7 or a red 8

33. A red card or a 10   34. A spade or a king

A jar contains 3 white, 4 orange, 5 yellow, and 8 black marbles. If a marble is drawn at random, find the probability that it is the following.

35. White   36. Orange

37. Yellow   38. Black

39. Not black   40. Orange or yellow

Which of Exercises 41–48 are examples of empirical probability?

41. The probability of heads on 5 consecutive tosses of a coin

42. The probability that a freshman entering college will graduate with a degree

43. The probability that a person is allergic to penicillin

44. The probability of drawing an ace from a standard deck of 52 cards

45. The probability that a person will get lung cancer from smoking cigarettes

46. A weather forecast that predicts a 70% chance of rain tomorrow

47. A gambler's claim that on a roll of a fair die, $P(\text{even}) = 1/2$

48. A surgeon's prediction that a patient has a 90% chance of a full recovery

49. The student sitting next to you in class concludes that the probability of the ceiling falling down on both of you before class ends is 1/2, because there are two possible outcomes—the ceiling will fall or not fall. What is wrong with this reasoning?

50. The following puzzler was given on the *Car Talk* radio program. *Source: Car Talk, Feb 24, 2001.*

   "Three different numbers are chosen at random, and one is written on each of three slips of paper. The slips are then placed face down on the table. The objective is to choose the slip upon which is written the largest number. Here are the rules: You can turn over any slip of paper and look at the amount written on it. If for any reason you think this is the largest, you're done; you keep it. Otherwise you discard it and turn over a second slip. Again, if you think this is the one with the biggest number, you keep that one and the game is over. If you don't, you discard that one too. . . . The chance of getting the highest number is one in three. Or is it? Is there a strategy by which you can improve the odds?"

The answer to the puzzler is that you can indeed improve the probability of getting the highest number by the following strategy. Pick one of the slips of paper, and after looking at the number, throw it away. Then pick a second slip; if it has a larger number than the first slip, stop. If not, pick the third slip. Find the probability of winning with this strategy.*

APPLICATIONS

Business and Economics

51. Survey of Workers The management of a firm wishes to check on the opinions of its assembly line workers. Before the workers are interviewed, they are divided into various categories. Define events E, F, and G as follows.

E: worker is female
F: worker has worked less than 5 years
G: worker contributes to a voluntary retirement plan

Describe each event in words.

a. E' **b.** $E \cap F$ **c.** $E \cup G'$

d. F' **e.** $F \cup G$ **f.** $F' \cap G'$

52. Research Funding In 2008, funding for research and development in the United States totaled $397.6 billion dollars. Support came from various sources, as shown in the following table. *Source: National Science Foundation.*

Source	Amount (in billions of dollars)
Federal government	103.7
State and local government	3.5
Industry	267.8
Academic institutions	10.6
Other	12.0

Find the probability that funds for a particular project came from each source.

a. Federal government

b. Industry

c. Academic institutions

53. Investment Survey Exercise 42 of the previous section presented a survey of 150 mathematics professors. Use the information given in that exercise to find each probability.

a. A randomly chosen professor invested in stocks and bonds and certificates of deposit.

b. A randomly chosen professor invested in only bonds.

*This is a special case of the famous Googol problem. For more details, see "Recognizing the Maximum of a Sequence" by John P. Gilbert and Frederick Mosteller, *Journal of the American Statistical Association*, Vol. 61, No. 313, March 1966, pp. 35–73.

54. Labor Force The 2008 and the 2018 (projected) civilian labor forces by age are given in the following table. *Source: U.S. Department of Labor.*

Age (in years)	2008 (in millions)	2018 (in millions)
16 to 24	22.0	21.1
25 to 54	104.4	105.9
55 and over	27.9	39.8
Total	154.3	166.8

a. In 2008, find the probability that a member of the civilian labor force is age 55 or older.

b. In 2018, find the probability that a member of the civilian labor force is age 55 or over.

c. What do these projections imply about the future civilian labor force?

Life Sciences

55. Medical Survey For a medical experiment, people are classified as to whether they smoke, have a family history of heart disease, or are overweight. Define events E, F, and G as follows.

E: person smokes
F: person has a family history of heart disease
G: person is overweight

Describe each event in words.

a. G' **b.** $F \cap G$ **c.** $E \cup G'$

56. Medical Survey Refer to Exercise 55. Describe each event in words.

a. $E \cup F$ **b.** $E' \cap F$ **c.** $F' \cup G'$

57. Causes of Death There were 2,424,059 U.S. deaths in 2007. They are listed according to cause in the following table. If a randomly selected person died in 2007, use this information to find the following probabilities. *Source: Centers for Disease Control and Prevention.*

Cause	Number of Deaths
Heart Disease	615,651
Cancer	560,187
Cerebrovascular disease	133,990
Chronic lower respiratory disease	129,311
Accidents	117,075
Alzheimer's disease	74,944
Diabetes mellitus	70,905
Influenza and pneumonia	52,847
All other causes	669,149

a. The probability that the cause of death was heart disease

b. The probability that the cause of death was cancer or heart disease

c. The probability that the cause of death was not an accident and was not diabetes mellitus

Social Sciences

58. APPLY IT U.S. Population The projected U.S. population (in thousands) by race in 2020 and 2050 is given in the table. *Source: Bureau of the Census.*

Race	2020	2050
White	207,393	207,901
Hispanic	52,652	96,508
Black	41,538	53,555
Asian and Pacific Islander	18,557	32,432
Other	2602	3535

Find the probability that a randomly selected person in the given year is of the race specified.

a. Hispanic in 2020   **b.** Hispanic in 2050

c. Black in 2020   **d.** Black in 2050

59. Congressional Service The following table gives the number of years of service of senators in the 111th Congress of the United States of America. Find the probability that a randomly selected senator of the 111th Congress served 20–29 years when Congress convened. *Source: Roll Call.*

Years of Service	Number of Senators
0–9	48
10–19	27
20–29	17
30–39	6
40 or more	2

60. Civil War Estimates of the Union Army's strength and losses for the battle of Gettysburg are given in the following table, where *strength* is the number of soldiers immediately preceding the battle and *loss* indicates a soldier who was killed, wounded, captured, or missing. *Source: Regimental Strengths and Losses of Gettysburg.*

Unit	Strength	Loss
I Corps (Reynolds)	12,222	6059
II Corps (Hancock)	11,347	4369
III Corps (Sickles)	10,675	4211
V Corps (Sykes)	10,907	2187
VI Corps (Sedgwick)	13,596	242
XI Corps (Howard)	9188	3801
XII Corps (Slocum)	9788	1082
Cavalry (Pleasonton)	11,851	610
Artillery (Tyler)	2376	242
Total	91,950	22,803

a. Find the probability that a randomly selected union soldier was from the XI Corps.

b. Find the probability that a soldier was lost in the battle.

c. Find the probability that a I Corps soldier was lost in the battle.

d. Which group had the highest probability of not being lost in the battle?

e. Which group had the highest probability of loss?

f. Explain why these probabilities vary.

61. Civil War Estimates of the Confederate Army's strength and losses for the battle of Gettysburg are given in the following table, where *strength* is the number of soldiers immediately preceding the battle and *loss* indicates a soldier who was killed, wounded, captured, or missing. *Source: Regimental Strengths and Losses at Gettysburg.*

Unit	Strength	Loss
I Corps (Longstreet)	20,706	7661
II Corps (Ewell)	20,666	6603
III Corps (Hill)	22,083	8007
Cavalry (Stuart)	6621	286
Total	70,076	22,557

a. Find the probability that a randomly selected confederate soldier was from the III Corps.

b. Find the probability that a confederate soldier was lost in the battle.

c. Find the probability that a I Corps soldier was lost in the battle.

d. Which group had the highest probability of not being lost in the battle?

e. Which group had the highest probability of loss?

General Interest

62. Native American Ceremonies Exercise 60 of the previous section presented a survey of families participating in a pow-wow in Arizona. Use the information given in that exercise to find each probability.

a. A randomly chosen family brought costumes and food.

b. A randomly chosen family brought crafts, but neither food nor costumes.

c. A randomly chosen family brought food or costumes.

63. Chinese New Year Exercise 59 of the previous section presented a survey of people attending a Lunar New Year celebration in Chinatown. Use the information given in that exercise to find each of the following probabilities.

a. A randomly chosen attendee speaks Cantonese.

b. A randomly chosen attendee does not speak Cantonese.

c. A randomly chosen attendee was a woman that did not light a firecracker.

YOUR TURN ANSWERS

1. $S = \{HH, HT, TH, TT\}$ **2.** $E = \{HT, TH\}$

3. $E' \cap F'$ is the event that the worker is 20 or over and is not white.

4. $P(H) = \dfrac{4}{6} = \dfrac{2}{3}.$ **5.** $P(\text{jack or a king}) = 8/52 = 2/13.$

7.4 Basic Concepts of Probability

APPLY IT **What is the probability that a dollar of advertising in the United States is spent on magazines or newspapers?**
We will determine this probability in Example 8. But first we need to develop additional rules for calculating probability, beginning with the probability of a union of two events.

The Union Rule
To determine the probability of the union of two events E and F in a sample space S, use the union rule for sets,

$$n(E \cup F) = n(E) + n(F) - n(E \cap F),$$

which was proved in Section 7.2. Assuming that the events in the sample space S are equally likely, divide both sides by $n(S)$, so that

$$\frac{n(E \cup F)}{n(S)} = \frac{n(E)}{n(S)} + \frac{n(F)}{n(S)} - \frac{n(E \cap F)}{n(S)}$$

$$P(E \cup F) = P(E) + P(F) - P(E \cap F).$$

Although our derivation is valid only for sample spaces with equally likely events, the result is valid for any events E and F from any sample space, and is called the **union rule for probability**.

> **Union Rule for Probability**
> For any events E and F from a sample space S,
> $$P(E \cup F) = P(E) + P(F) - P(E \cap F).$$

EXAMPLE 1 Probabilities with Playing Cards

If a single card is drawn from an ordinary deck of cards, find the probability that it will be a red or a face card.

SOLUTION Let R represent the event "red card" and F the event "face card." There are 26 red cards in the deck, so $P(R) = 26/52$. There are 12 face cards in the deck, so $P(F) = 12/52$. Since there are 6 red face cards in the deck, $P(R \cap F) = 6/52$. By the union rule, the probability of the card being red or a face card is

$$P(R \cup F) = P(R) + P(F) - P(R \cap F)$$

$$= \frac{26}{52} + \frac{12}{52} - \frac{6}{52} = \frac{32}{52} = \frac{8}{13}.$$

YOUR TURN 1 In Example 1, find the probability of an ace or a club.

TRY YOUR TURN 1

EXAMPLE 2 Probabilities with Dice

Suppose two fair dice are rolled. Find each probability.

(a) The first die shows a 2, or the sum of the results is 6 or 7.

SOLUTION The sample space for the throw of two dice is shown in Figure 18 on the next page, where 1-1 represents the event "the first die shows a 1 and the second die shows a 1," 1-2 represents "the first die shows a 1 and the second die shows a 2," and so on. Let A represent the event "the first die shows a 2," and B represent the event "the sum of the results is 6 or 7." These events are indicated in Figure 18. From the diagram, event A has 6 elements, B has 11 elements, the intersection of A and B has 2 elements, and the sample space has 36 elements. Thus,

$$P(A) = \frac{6}{36}, \quad P(B) = \frac{11}{36}, \quad \text{and} \quad P(A \cap B) = \frac{2}{36}.$$

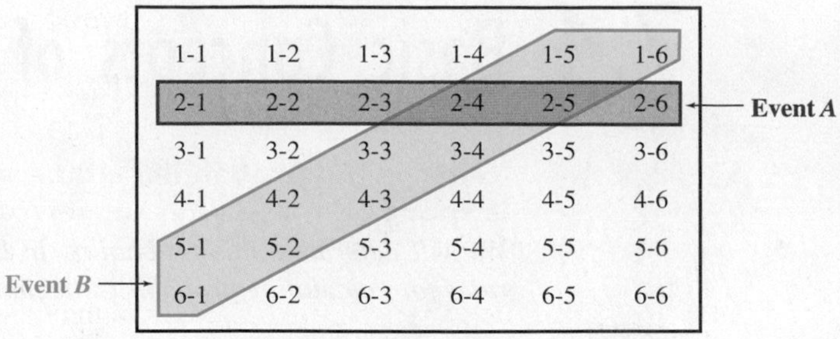

FIGURE 18

By the union rule,

$$P(A \cup B) = P(A) + P(B) - P(A \cap B)$$

$$P(A \cup B) = \frac{6}{36} + \frac{11}{36} - \frac{2}{36} = \frac{15}{36} = \frac{5}{12}.$$

(b) The sum of the results is 11, or the second die shows a 5.

SOLUTION $P(\text{sum is } 11) = 2/36$, $P(\text{second die shows a } 5) = 6/36$, and $P(\text{sum is } 11 \text{ and second die shows a } 5) = 1/36$, so

$$P(\text{sum is } 11 \text{ or second die shows a } 5) = \frac{2}{36} + \frac{6}{36} - \frac{1}{36} = \frac{7}{36}.$$

YOUR TURN 2 In Example 2, find the probability that the sum is 8, or both die show the same number.

TRY YOUR TURN 2

CAUTION You may wonder why we did not use $S = \{2, 3, 4, 5, \ldots, 12\}$ as the sample space in Example 2. Remember, we prefer to use a sample space with equally likely outcomes. The outcomes in set S above are not equally likely—a sum of 2 can occur in just one way, a sum of 3 in two ways, a sum of 4 in three ways, and so on, as shown in Figure 18.

If events E and F are mutually exclusive, then $E \cap F = \emptyset$ by definition; hence, $P(E \cap F) = 0$. In this case the union rule simplifies to $P(E \cup F) = P(E) + P(F)$.

CAUTION The rule $P(E \cup F) = P(E) + P(F)$ is only valid when E and F are mutually exclusive. When E and F are *not* mutually exclusive, use the rule $P(E \cup F) = P(E) + P(F) - P(E \cap F)$.

The Complement Rule
By the definition of E', for any event E from a sample space S,

$$E \cup E' = S \quad \text{and} \quad E \cap E' = \emptyset.$$

Since $E \cap E' = \emptyset$, events E and E' are mutually exclusive, so that

$$P(E \cup E') = P(E) + P(E').$$

However, $E \cup E' = S$, the sample space, and $P(S) = 1$. Thus

$$P(E \cup E') = P(E) + P(E') = 1.$$

Rearranging these terms gives the following useful rule for complements.

Complement Rule

$$P(E) = 1 - P(E') \quad \text{and} \quad P(E') = 1 - P(E).$$

EXAMPLE 3 Complement Rule

If a fair die is rolled, what is the probability that any number but 5 will come up?

SOLUTION If E is the event that 5 comes up, then E' is the event that any number but 5 comes up. Since $P(E) = 1/6$, we have $P(E') = 1 - 1/6 = 5/6$.

EXAMPLE 4 Complement Rule

If two fair dice are rolled, find the probability that the sum of the numbers rolled is greater than 3. Refer to Figure 18.

SOLUTION To calculate this probability directly, we must find the probabilities that the sum is 4, 5, 6, 7, 8, 9, 10, 11, or 12 and then add them. It is much simpler to first find the probability of the complement, the event that the sum is less than or equal to 3.

$$P(\text{sum} \le 3) = P(\text{sum is } 2) + P(\text{sum is } 3)$$

$$= \frac{1}{36} + \frac{2}{36}$$

$$= \frac{3}{36} = \frac{1}{12}$$

YOUR TURN 3 Find the probability that when two fair dice are rolled, the sum is less than 11.

Now use the fact that $P(E) = 1 - P(E')$ to get

$$P(\text{sum} > 3) = 1 - P(\text{sum} \le 3)$$

$$= 1 - \frac{1}{12} = \frac{11}{12}.$$ **TRY YOUR TURN 3**

Odds
Sometimes probability statements are given in terms of **odds**, a comparison of $P(E)$ with $P(E')$. For example, suppose $P(E) = 4/5$. Then $P(E') = 1 - 4/5 = 1/5$. These probabilities predict that E will occur 4 out of 5 times and E' will occur 1 out of 5 times. Then we say the *odds in favor* of E are 4 to 1.

> ### Odds
> The **odds in favor** of an event E are defined as the ratio of $P(E)$ to $P(E')$, or
>
> $$\frac{P(E)}{P(E')}, \text{ where } P(E') \ne 0.$$

EXAMPLE 5 Odds in Favor of Rain

Suppose the weather forecaster says that the probability of rain tomorrow is 1/3. Find the odds in favor of rain tomorrow.

SOLUTION Let E be the event "rain tomorrow." Then E' is the event "no rain tomorrow." Since $P(E) = 1/3$, $P(E') = 2/3$. By the definition of odds, the odds in favor of rain are

$$\frac{1/3}{2/3} = \frac{1}{2}, \qquad \text{written} \qquad 1 \text{ to } 2, \text{ or } 1:2.$$

YOUR TURN 4 If the probability of snow tomorrow is 3/10, find the odds in favor of snow tomorrow.

On the other hand, the odds that it will *not* rain, or the *odds against* rain, are

$$\frac{2/3}{1/3} = \frac{2}{1}, \qquad \text{written} \qquad 2 \text{ to } 1, \text{ or } 2:1. \quad \textbf{TRY YOUR TURN 4}$$

If the odds in favor of an event are, say, 3 to 5, then the probability of the event is 3/8, while the probability of the complement of the event is 5/8. (Odds of 3 to 5 indicate

3 outcomes in favor of the event out of a total of 8 possible outcomes.) This example suggests the following generalization.

> If the odds favoring event E are m to n, then
> $$P(E) = \frac{m}{m+n} \quad \text{and} \quad P(E') = \frac{n}{m+n}.$$

EXAMPLE 6 Package Delivery

The odds that a particular package will be delivered on time are 25 to 2.

(a) Find the probability that the package will be delivered on time.

SOLUTION Odds of 25 to 2 show 25 favorable chances out of $25 + 2 = 27$ chances altogether:

$$P(\text{package will be delivered on time}) = \frac{25}{25+2} = \frac{25}{27}.$$

(b) Find the odds against the package being delivered on time.

SOLUTION Using the complement rule, there is a $2/27$ chance that the package will not be delivered on time. So the odds against the package being delivered on time are

$$\frac{P(\text{package will not be delivered on time})}{P(\text{package will be delivered on time})} = \frac{2/27}{25/27} = \frac{2}{25},$$

or $2 : 25$.

YOUR TURN 5 If the odds that a package will be delivered on time are 17 to 3, find the probability that the package will not be delivered on time.

TRY YOUR TURN 5

EXAMPLE 7 Odds in Horse Racing

If the odds in favor of a particular horse's winning a race are 5 to 7, what is the probability that the horse will win the race?

SOLUTION The odds indicate chances of 5 out of 12 ($5 + 7 = 12$) that the horse will win, so

$$P(\text{winning}) = \frac{5}{12}.$$

Race tracks generally give odds *against* a horse winning. In this case, the track would give the odds as 7 to 5. Of course, race tracks, casinos, and other gambling establishments need to give odds that are more favorable to the house than those representing the actual probabilities, because they need to make a profit.

YOUR TURN 6 If the odds against a particular horse winning a race are 7 to 3, what is the probability the horse will win the race?

TRY YOUR TURN 6

Probability Distribution
A table listing each possible outcome of an experiment and its corresponding probability is called a **probability distribution**. The assignment of probabilities may be done in any reasonable way (on an empirical basis, as in the next example, or by theoretical reasoning, as in Section 7.3), provided that it satisfies the following conditions.

Properties of Probability
Let S be a sample space consisting of n distinct outcomes, $s_1, s_2, \ldots, s_n$. An acceptable probability assignment consists of assigning to each outcome s_i a number p_i (the probability of s_i) according to these rules.

1. The probability of each outcome is a number between 0 and 1.

$$0 \le p_1 \le 1, \quad 0 \le p_2 \le 1, \ldots, \quad 0 \le p_n \le 1$$

2. The sum of the probabilities of all possible outcomes is 1.

$$p_1 + p_2 + p_3 + \cdots + p_n = 1$$

EXAMPLE 8   Advertising Volume

The following table lists U.S. advertising volume in millions of dollars by medium in 2008. *Source: Advertising Age.*

Advertising Volume	
Medium	**Expenditures**
Television	65,137
Magazines	28,580
Newspapers	25,059
Internet	9727
Radio	9501
Outdoor	3964

(a) Construct a probability distribution for the probability that a dollar of advertising is spent on each medium.

SOLUTION We first find the total spent and then divide the amount spent on each medium by the total. Verify that the amounts in the table sum to 141,968. The probability that a dollar is spent on newspapers, for example, is $P(\text{newspapers}) = 25,059/141,968 \approx 0.1765$. Similarly, we could divide each amount by 141,968, with the results (rounded to four decimal places) shown in the following table.

Advertising Volume	
Medium	**Probabilities**
Television	0.4588
Magazines	0.2013
Newspapers	0.1765
Internet	0.0685
Radio	0.0669
Outdoor	0.0279

Verify that this distribution satisfies the conditions for probability. Each probability is between 0 and 1, so the first condition holds. The probabilities in this table sum to 0.9999. In theory, to satisfy the second condition, they should total 1.0000, but this does not always occur when the individual numbers are rounded.

(b) Find the probability that a dollar of advertising in the United States is spent on newspapers or magazines.

APPLY IT

SOLUTION The categories in the table are mutually exclusive simple events. Thus, to find the probability that an advertising dollar is spent on newspapers or magazines, we use the union rule to calculate

$$P(\text{newspapers or magazines}) = 0.1765 + 0.2013 = 0.3778.$$

We could get this same result by summing the amount spent on newspapers and magazines, and dividing the total by 141,968.

Thus, more than a third of all advertising dollars are spent on these two media, a figure that should be of interest to both advertisers and the owners of the various media. ▬

Probability distributions are discussed further in the next chapter.

EXAMPLE 9   Clothing

Susan is a college student who receives heavy sweaters from her aunt at the first sign of cold weather. Susan has determined that the probability that a sweater is the wrong size is 0.47, the probability that it is a loud color is 0.59, and the probability that it is both the wrong size and a loud color is 0.31.

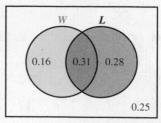

FIGURE 19

(a) Find the probability that the sweater is the correct size and not a loud color.

SOLUTION Let W represent the event "wrong size," and L represent "loud color." Place the given information on a Venn diagram, starting with 0.31 in the intersection of the regions W and L (see Figure 19). As stated earlier, event W has probability 0.47. Since 0.31 has already been placed inside the intersection of W and L,

$$0.47 - 0.31 = 0.16$$

goes inside region W, but outside the intersection of W and L, that is, in the region $W \cap L'$. In the same way,

$$0.59 - 0.31 = 0.28$$

goes inside the region for L, and outside the overlap, that is, in the region $L \cap W'$.

Using regions W and L, the event we want is $W' \cap L'$. From the Venn diagram in Figure 19, the labeled regions have a total probability of

$$0.16 + 0.31 + 0.28 = 0.75.$$

Since the entire region of the Venn diagram must have probability 1, the region outside W and L, or $W' \cap L'$, has probability

$$1 - 0.75 = 0.25.$$

The probability is 0.25 that the sweater is the correct size and not a loud color.

(b) Find the probability that the sweater is the correct size or is not loud.

SOLUTION The corresponding region, $W' \cup L'$, has probability

$$0.25 + 0.16 + 0.28 = 0.69.$$

7.4 EXERCISES

 1. Define mutually exclusive events in your own words.

 2. Explain the union rule for mutually exclusive events.

Decide whether the events in Exercises 3–8 are mutually exclusive.

3. Owning a dog and owning an MP3 player

4. Being a business major and being from Texas

5. Being retired and being 70 years old

6. Being a teenager and being 70 years old

7. Being one of the ten tallest people in the United States and being under 4 feet tall

8. Being male and being a nurse

Two dice are rolled. Find the probabilities of rolling the given sums.

9. a. 2   **b.** 4   **c.** 5   **d.** 6

10. a. 8   **b.** 9   **c.** 10   **d.** 13

11. a. 9 or more   **b.** Less than 7

c. Between 5 and 8 (exclusive)

12. a. Not more than 5   **b.** Not less than 8

c. Between 3 and 7 (exclusive)

Two dice are rolled. Find the probabilities of the following events.

13. The first die is 3 or the sum is 8.

14. The second die is 5 or the sum is 10.

One card is drawn from an ordinary deck of 52 cards. Find the probabilities of drawing the following cards.

15. a. A 9 or 10

b. A red card or a 3

c. A 9 or a black 10

d. A heart or a black card

e. A face card or a diamond

16. a. Less than a 4 (count aces as ones)

b. A diamond or a 7

c. A black card or an ace

d. A heart or a jack

e. A red card or a face card

Kristi Perez invites 13 relatives to a party: her mother, 2 aunts, 3 uncles, 2 brothers, 1 male cousin, and 4 female cousins. If the

chances of any one guest arriving first are equally likely, find the probabilities that the first guest to arrive is as follows.

17. a. A brother or an uncle **b.** A brother or a cousin

   c. A brother or her mother

18. a. An uncle or a cousin **b.** A male or a cousin

   c. A female or a cousin

The numbers 1, 2, 3, 4, and 5 are written on slips of paper, and 2 slips are drawn at random one at a time without replacement. Find the probabilities in Exercises 19 and 20.

19. a. The sum of the numbers is 9.

   b. The sum of the numbers is 5 or less.

   c. The first number is 2 or the sum is 6.

20. a. Both numbers are even.

   b. One of the numbers is even or greater than 3.

   c. The sum is 5 or the second number is 2.

Use Venn diagrams to work Exercises 21 and 22.

21. Suppose $P(E) = 0.26$, $P(F) = 0.41$, and $P(E \cap F) = 0.16$. Find the following.

   a. $P(E \cup F)$ **b.** $P(E' \cap F)$

   c. $P(E \cap F')$ **d.** $P(E' \cup F')$

22. Let $P(Z) = 0.42$, $P(Y) = 0.35$, and $P(Z \cup Y) = 0.59$. Find each probability.

   a. $P(Z' \cap Y')$ **b.** $P(Z' \cup Y')$

   c. $P(Z' \cup Y)$ **d.** $P(Z \cap Y')$

23. Three unusual dice, A, B, and C, are constructed such that die A has the numbers 3, 3, 4, 4, 8, 8; die B has the numbers 1, 1, 5, 5, 9, 9; and die C has the numbers 2, 2, 6, 6, 7, 7.

   a. If dice A and B are rolled, find the probability that B beats A, that is, the number that appears on die B is greater than the number that appears on die A.

   b. If dice B and C are rolled, find the probability that C beats B.

   c. If dice A and C are rolled, find the probability that A beats C.

   d. Which die is better? Explain.

24. In the "Ask Marilyn" column of *Parade* magazine, a reader wrote about the following game: You and I each roll a die. If your die is higher than mine, you win. Otherwise, I win. The reader thought that the probability that each player wins is 1/2. Is this correct? If not, what is the probability that each player wins? *Source: Parade magazine.*

25. Define what is meant by odds.

26. On page 134 of Roger Staubach's autobiography, *First Down, Lifetime to Go*, Staubach makes the following statement regarding his experience in Vietnam: *Source: First Down, Lifetime to Go.*

   "Odds against a direct hit are very low but when your life is in danger, you don't worry too much about the odds."

   Is this wording consistent with our definition of odds, for and against? How could it have been said so as to be technically correct?

A single fair die is rolled. Find the odds in favor of getting the results in Exercises 27–30.

27. 3 **28.** 4, 5, or 6

29. 2, 3, 4, or 5 **30.** Some number less than 6

31. A marble is drawn from a box containing 3 yellow, 4 white, and 11 blue marbles. Find the odds in favor of drawing the following.

   a. A yellow marble **b.** A blue marble

   c. A white marble **d.** Not drawing a white marble

32. Two dice are rolled. Find the odds of rolling the following. (Refer to Figure 18.)

   a. A sum of 3 **b.** A sum of 7 or 11

   c. A sum less than 5 **d.** Not a sum of 6

33. What is a probability distribution?

34. What conditions must hold for a probability distribution to be acceptable?

An experiment is conducted for which the sample space is $S = \{s_1, s_2, s_3, s_4, s_5\}$. Which of the probability assignments in Exercises 35–40 are possible for this experiment? If an assignment is not possible, tell why.

35.

Outcomes	s_1	s_2	s_3	s_4	s_5
Probabilities	0.09	0.32	0.21	0.25	0.13

36.

Outcomes	s_1	s_2	s_3	s_4	s_5
Probabilities	0.92	0.03	0	0.02	0.03

37.

Outcomes	s_1	s_2	s_3	s_4	s_5
Probabilities	1/3	1/4	1/6	1/8	1/10

38.

Outcomes	s_1	s_2	s_3	s_4	s_5
Probabilities	1/5	1/3	1/4	1/5	1/10

39.

Outcomes	s_1	s_2	s_3	s_4	s_5
Probabilities	0.64	−0.08	0.30	0.12	0.02

40.

Outcomes	s_1	s_2	s_3	s_4	s_5
Probabilities	0.05	0.35	0.5	0.2	−0.3

One way to solve a probability problem is to repeat the experiment many times, keeping track of the results. Then the probability can be approximated using the basic definition of the probability of an event E: $P(E) = n(E)/n(S)$, where E occurs $n(E)$ times out of $n(S)$ trials of an experiment. This is called the Monte Carlo method of finding probabilities. If physically repeating the experiment is too tedious, it may be simulated using a random-number generator, available on most computers and scientific or graphing calculators. To simulate a coin toss or the roll of a die on the TI-84 Plus, change the setting to fixed decimal mode with 0 digits displayed, and enter `rand` or `rand*6+.5`, respectively. **For a coin toss, interpret 0 as a head and 1 as a tail. In either case, the** ENTER **key can be pressed repeatedly to perform multiple simulations.**

41. Suppose two dice are rolled. Use the Monte Carlo method with at least 50 repetitions to approximate the following probabilities. Compare with the results of Exercise 11.

    a. P(the sum is 9 or more)

    b. P(the sum is less than 7)

42. Suppose two dice are rolled. Use the Monte Carlo method with at least 50 repetitions to approximate the following probabilities. Compare with the results of Exercise 12.

    a. P(the sum is not more than 5)

    b. P(the sum is not less than 8)

43. Suppose three dice are rolled. Use the Monte Carlo method with at least 100 repetitions to approximate the following probabilities.

    a. P(the sum is 5 or less)

    b. P(neither a 1 nor a 6 is rolled)

44. Suppose a coin is tossed 5 times. Use the Monte Carlo method with at least 50 repetitions to approximate the following probabilities.

    a. P(exactly 4 heads)

    b. P(2 heads and 3 tails)

45. The following description of the classic "Linda problem" appeared in the *New Yorker:* "In this experiment, subjects are told, 'Linda is thirty-one years old, single, outspoken, and very bright. She majored in philosophy. As a student, she was deeply concerned with issues of discrimination and social justice and also participated in antinuclear demonstrations.' They are then asked to rank the probability of several possible descriptions of Linda today. Two of them are 'bank teller' and 'bank teller and active in the feminist movement.'" Many people rank the second event as more likely. Explain why this violates basic concepts of probability. *Source: New Yorker.*

46. You are given $P(A \cup B) = 0.7$ and $P(A \cup B') = 0.9$. Determine $P(A)$. Choose one of the following. *Source: Society of Actuaries.*

    a. 0.2   **b.** 0.3   **c.** 0.4   **d.** 0.6   **e.** 0.8

APPLICATIONS

Business and Economics

47. **Defective Merchandise** Suppose that 8% of a certain batch of calculators have a defective case, and that 11% have defective batteries. Also, 3% have both a defective case and defective batteries. A calculator is selected from the batch at random. Find the probability that the calculator has a good case and good batteries.

48. **Profit** The probability that a company will make a profit this year is 0.74.

    a. Find the probability that the company will not make a profit this year.

    b. Find the odds against the company making a profit.

49. **Credit Charges** The table shows the probabilities of a person accumulating specific amounts of credit card charges over a 12-month period. Find the probabilities that a person's total charges during the period are the following.

    a. $500 or more   **b.** Less than $1000

    c. $500 to $2999   **d.** $3000 or more

Charges	Probability
Under $100	0.21
$100–$499	0.17
$500–$999	0.16
$1000–$1999	0.15
$2000–$2999	0.12
$3000–$4999	0.08
$5000–$9999	0.07
$10,000 or more	0.04

50. **Employment** The table shows the projected probabilities of a worker employed by different occupational groups in 2018. *Source: U.S. Department of Labor.*

Occupation	Probability
Management and business	0.1047
Professional	0.2182
Service	0.2024
Sales	0.1015
Office and administrative support	0.1560
Farming, fishing, forestry	0.0061
Construction	0.0531
Production	0.0585
Other	0.0995

If a worker in 2018 is selected at random, find the following.

a. The probability that the worker is in sales or service.

b. The probability that the worker is not in construction.

c. The odds in favor of the worker being in production.

51. **Labor Force** The following table gives the 2018 projected civilian labor force probability distribution by age and gender. *Source: U.S. Department of Labor.*

Age	Male	Female	Total
16–24	0.066	0.061	0.127
25–54	0.343	0.291	0.634
55 and over	0.122	0.117	0.239
Total	0.531	0.469	1.000

Find the probability that a randomly selected worker is the following.

a. Female and 16 to 24 years old

b. 16 to 54 years old

c. Male or 25 to 54 years old

d. Female or 16 to 24 years old

Life Sciences

52. Body Types A study on body types gave the following results: 45% were short; 25% were short and overweight; and 24% were tall and not overweight. Find the probabilities that a person is the following.

a. Overweight

b. Short, but not overweight

c. Tall and overweight

53. Color Blindness Color blindness is an inherited characteristic that is more common in males than in females. If M represents male and C represents red-green color blindness, we use the relative frequencies of the incidences of males and red-green color blindness as probabilities to get

$$P(C) = 0.039, P(M \cap C) = 0.035, P(M \cup C) = 0.491.$$

Source: Parsons' Diseases of the Eye.

Find the following probabilities.

a. $P(C')$ **b.** $P(M)$ **c.** $P(M')$

d. $P(M' \cap C')$ **e.** $P(C \cap M')$ **f.** $P(C \cup M')$

54. Genetics Gregor Mendel, an Austrian monk, was the first to use probability in the study of genetics. In an effort to understand the mechanism of character transmittal from one generation to the next in plants, he counted the number of occurrences of various characteristics. Mendel found that the flower color in certain pea plants obeyed this scheme:

Pure red crossed with pure white produces red.

From its parents, the red offspring received genes for both red (R) and white (W), but in this case red is *dominant* and white *recessive*, so the offspring exhibits the color red. However, the offspring still carries both genes, and when two such offspring are crossed, several things can happen in the third generation. The table below, which is called a *Punnet square*, shows the equally likely outcomes.

		Second Parent	
		R	**W**
First Parent	**R**	RR	RW
	W	WR	WW

Use the fact that red is dominant over white to find the following. Assume that there are an equal number of red and white genes in the population.

a. P(a flower is red)

b. P(a flower is white)

55. Genetics Mendel found no dominance in snapdragons, with one red gene and one white gene producing pink-flowered offspring. These second-generation pinks, however, still carry one red and one white gene, and when they are crossed, the next generation still yields the Punnet square from Exercise 54. Find each probability.

a. P(red)

b. P(pink)

c. P(white)

(Mendel verified these probability ratios experimentally and did the same for many characteristics other than flower color. His work, published in 1866, was not recognized until 1890.)

56. Genetics In most animals and plants, it is very unusual for the number of main parts of the organism (such as arms, legs, toes, or flower petals) to vary from generation to generation. Some species, however, have *meristic variability,* in which the number of certain body parts varies from generation to generation. One researcher studied the front feet of certain guinea pigs and produced the following probabilities. *Source: Genetics.*

$$P(\text{only four toes, all perfect}) = 0.77$$
$$P(\text{one imperfect toe and four good ones}) = 0.13$$
$$P(\text{exactly five good toes}) = 0.10$$

Find the probability of each event.

a. No more than four good toes

b. Five toes, whether perfect or not

57. Doctor Visit The probability that a visit to a primary care physician's (PCP) office results in neither lab work nor referral to a specialist is 35%. Of those coming to a PCP's office, 30% are referred to specialists and 40% require lab work. Determine the probability that a visit to a PCP's office results in both lab work and referral to a specialist. Choose one of the following. (*Hint:* Use the union rule for probability.) *Source: Society of Actuaries.*

a. 0.05 **b.** 0.12 **c.** 0.18

d. 0.25 **e.** 0.35

58. Shoulder Injuries Among a large group of patients recovering from shoulder injuries, it is found that 22% visit both a physical therapist and a chiropractor, whereas 12% visit neither of these. The probability that a patient visits a chiropractor exceeds by 0.14 the probability that a patient visits a physical therapist. Determine the probability that a randomly chosen member of this group visits a physical therapist. Choose one of the following. (*Hint:* Use the union rule for probability, and let $x = P$(patient visits a physical therapist).) *Source: Society of Actuaries.*

a. 0.26 **b.** 0.38 **c.** 0.40

d. 0.48 **e.** 0.62

59. Health Plan An insurer offers a health plan to the employees of a large company. As part of this plan, the individual employees may choose exactly two of the supplementary coverages A, B, and C, or they may choose no supplementary coverage. The proportions of the company's employees that choose coverages A, B, and C are 1/4, 1/3, and 5/12, respectively. Determine the probability that a randomly chosen employee will choose no supplementary coverage. Choose one of the following. (*Hint:* Draw a Venn diagram with three sets, and let $x = P(A \cap B)$. Use the fact that 4 of the 8 regions in the Venn diagram have a probability of 0.) *Source: Society of Actuaries.*

a. 0 **b.** 47/144 **c.** 1/2

d. 97/144 **e.** 7/9

Social Sciences

60. Presidential Candidates In 2002, *The New York Times* columnist William Safire gave the following odds against various prominent Democrats receiving their party's presidential nomination in 2004.

> Al Gore: 2 to 1
> Tom Daschle: 4 to 1
> John Kerry: 4 to 1
> Chris Dodd: 4 to 1
> Joe Lieberman: 5 to 1
> Joe Biden: 5 to 1
> Pat Leahy: 6 to 1
> Russell Feingold: 8 to 1
> John Edwards: 9 to 1
> Dick Gephardt: 15 to 1

John Allen Paulos observed that there is something wrong with those odds. Translate these odds into probabilities of winning the nomination, and then explain why these are not possible. *Sources: The New York Times and ABC News.*

61. Earnings The following data were gathered for 130 adult U.S. workers: 55 were women; 3 women earned more than $40,000; and 62 men earned $40,000 or less. Find the probability that an individual is

a. a woman earning $40,000 or less;

b. a man earning more than $40,000;

c. a man or is earning more than $40,000;

d. a woman or is earning $40,000 or less.

62. Expenditures for Music A survey of 100 people about their music expenditures gave the following information: 38 bought rock music; 20 were teenagers who bought rock music; and 26 were teenagers. Find the probabilities that a person is

a. a teenager who buys nonrock music;

b. someone who buys rock music or is a teenager;

c. not a teenager;

d. not a teenager, but a buyer of rock music.

63. Refugees In a refugee camp in southern Mexico, it was found that 90% of the refugees came to escape political oppression, 80% came to escape abject poverty, and 70% came to escape both. What is the probability that a refugee in the camp was not poor nor seeking political asylum?

64. Community Activities At the first meeting of a committee to plan a local Lunar New Year celebration, the persons attending are 3 Chinese men, 4 Chinese women, 3 Vietnamese women, 2 Vietnamese men, 4 Korean women, and 2 Korean men. A chairperson is selected at random. Find the probabilities that the chairperson is the following.

a. Chinese

b. Korean or a woman

c. A man or Vietnamese

d. Chinese or Vietnamese

e. Korean and a woman

65. Elections If the odds that a given candidate will win an election are 3 to 2, what is the probability that the candidate will lose?

66. Military There were 203,375 female military personnel in 2009 in various ranks and military branches, as listed in the table. *Source: Department of Defense.*

	Army (A)	Air Force (B)	Navy (C)	Marines (D)
Officers (O)	14,322	12,097	7884	1202
Enlisted (E)	59,401	51,965	42,225	11,749
Cadets & Midshipmen (M)	688	922	920	0

a. Convert the numbers in the table to probabilities.

b. Find the probability that a randomly selected woman is in the Army.

c. Find the probability that a randomly selected woman is an officer in the Navy or Marine Corps.

d. $P(A \cup B)$

e. $P(E \cup (C \cup D))$

67. Perceptions of Threat Research has been carried out to measure the amount of intolerance that citizens of Russia have for left-wing Communists and right-wing Fascists, as indicated in the table below. Note that the numbers are given as percents and each row sums to 100 (except for rounding). *Source: Political Research Quarterly.*

Russia	None at All	Don't Know	Not Very Much	Somewhat	Extremely
Left-Wing Communists	47.8	6.7	31.0	10.5	4.1
Right-Wing Fascists	3.0	3.2	7.1	27.1	59.5

a. Find the probability that a randomly chosen citizen of Russia would be somewhat or extremely intolerant of right-wing Fascists.

b. Find the probability that a randomly chosen citizen of Russia would be completely tolerant of left-wing Communists.

c. Compare your answers to parts a and b and provide possible reasons for these numbers.

68. Perceptions of Threat Research has been carried out to measure the amount of intolerance that U.S. citizens have for left-wing Communists and right-wing Fascists, as indicated in the table. Note that the numbers are given as percents and each row sums to 100 (except for rounding). *Source: Political Research Quarterly.*

United States	None at All	Don't Know	Not Very Much	Somewhat	Extremely
Left-Wing Communists	13.0	2.7	33.0	34.2	17.1
Right-Wing Fascists	10.1	3.3	20.7	43.1	22.9

a. Find the probability that a randomly chosen U.S. citizen would have at least some intolerance of right-wing Fascists.

b. Find the probability that a randomly chosen U.S. citizen would have at least some intolerance of left-wing Communists.

c. Compare your answers to parts a and b and provide possible reasons for these numbers.

d. Compare these answers to the answers to Exercise 67.

General Interest

69. Olympics In recent winter Olympics, each part of the women's figure skating program has 12 judges, but the scores of only 9 of the judges are randomly selected for the final results. As we will see in the next chapter, there are 220 possible ways for the 9 judges whose scores are counted to be selected. *The New York Times* examined those 220 possibilities for the short program in the 2006 Olympics, based on the published scores of the judges, and listed what the results would have been for each, as shown below. *Source: The New York Times.*

a. The winner of the short program was Sasha Cohen. For a random combination of 9 judges, what is the probability of that outcome?

b. The second place finisher in the short program was Irina Slutskaya. For a random combination of 9 judges, what is the probability of that outcome?

c. The third place finisher in the short program was Shizuka Arakawa. For a random combination of 9 judges, what is the probability of that outcome? Do not include outcomes that include a tie.

70. Book of Odds The following table gives the probabilities that a particular event will occur. Convert each probability to the odds in favor of the event. *Source: The Book of Odds.*

Event	Probability for the Event
An NFL pass will be intercepted.	0.03
A U.S. president owned a dog during his term in office.	0.65
A woman owns a pair of high heels.	0.61
An adult smokes.	0.21
A flight is cancelled.	0.02

71. Book of Odds The following table gives the odds that a particular event will occur. Convert each odd to the probability that the event will occur. *Source: The Book of Odds.*

Event	Odds for the Event
A powerball entry will win the jackpot.	1 to 195,199,999
An adult will be struck by lightning during a year.	1 to 835,499
An adult will file for personal bankruptcy during a year.	1 to 157.6
A person collects stamps.	1 to 59.32

YOUR TURN ANSWERS

1. $16/52 = 4/13$ **2.** $10/36 = 5/18$ **3.** $33/36 = 11/12$
4. 3 to 7 or 3:7 **5.** $3/20$ **6.** $3/10$

Outcome	1. Slutskaya 2. Cohen 3. Arakawa	1. Slutskaya 2. Arakawa 3. Cohen	1. Slutskaya 2, 3. Arakawa and Cohen tied	1. Cohen 2. Slutskaya 3. Arakawa	1. Cohen 2. Arakawa 3. Slutskaya
Number of Possible Judging Combinations	92	33	3	67	25

7.5 Conditional Probability; Independent Events

What is the probability that a broker who uses research picks stocks that go up?

The manager for a brokerage firm has noticed that some of the firm's stockbrokers have selected stocks based on the firm's research, while other brokers tend to follow their own instincts. To see whether the research department performs better than the brokers' instincts, the manager surveyed 100 brokers, with results as shown in the following table.

| | Results of Stockbroker Survey | | |
	Picked Stocks That Went Up (A)	Didn't Pick Stocks That Went Up (A')	Totals
Used Research (B)	30	15	45
Didn't Use Research (B')	30	25	55
Totals	60	40	100

Letting A represent the event "picked stocks that went up," and letting B represent the event "used research," we can find the following probabilities.

$$P(A) = \frac{60}{100} = 0.6 \qquad P(A') = \frac{40}{100} = 0.4$$

$$P(B) = \frac{45}{100} = 0.45 \qquad P(B') = \frac{55}{100} = 0.55$$

APPLY IT To answer the question asked at the beginning of this section, suppose we want to find the probability that a broker using research will pick stocks that go up. From the table, of the 45 brokers who use research, 30 picked stocks that went up, with

$$P(\text{broker who uses research picks stocks that go up}) = \frac{30}{45} \approx 0.6667.$$

This is a different number than the probability that a broker picks stocks that go up, 0.6, since we have additional information (the broker uses research) that has *reduced the sample space*. In other words, we found the probability that a broker picks stocks that go up, A, given the additional information that the broker uses research, B. This is called the *conditional probability* of event A, given that event B has occurred. It is written $P(A \mid B)$ and read as "the probability of A given B." In this example,

$$P(A|B) = \frac{30}{45}.$$

To generalize this result, assume that E and F are two events for a particular experiment and that all events in the sample space S are equally likely. We want to find $P(E \mid F)$, the probability that E occurs given F has occurred. Since we assume that F has occurred, reduce the sample space to F: look only at the elements inside F. See Figure 20. Of those $n(F)$ elements, there are $n(E \cap F)$ elements where E also occurs. This makes

$$P(E|F) = \frac{n(E \cap F)}{n(F)}.$$

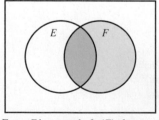

Event F has a total of $n(F)$ elements.

FIGURE 20

This equation can also be written as the quotient of two probabilities. Divide numerator and denominator by $n(S)$ to get

$$P(E|F) = \frac{n(E \cap F)/n(S)}{n(F)/n(S)} = \frac{P(E \cap F)}{P(F)}.$$

This last result motivates the definition of conditional probability.

> ### Conditional Probability
> The **conditional probability** of event E given event F, written $P(E|F)$, is
> $$P(E|F) = \frac{P(E \cap F)}{P(F)}, \quad \text{where } P(F) \neq 0.$$

Although the definition of conditional probability was motivated by an example with equally likely outcomes, it is valid in all cases. However, for *equally likely outcomes*, conditional probability can be found by directly applying the definition, or by first reducing the sample space to event F, and then finding the number of outcomes in F that are also in event E. Thus,

$$P(E|F) = \frac{n(E \cap F)}{n(F)}.$$

In the preceding example, the conditional probability could have also been found using the definition of conditional probability:

$$P(A|B) = \frac{P(A \cap B)}{P(B)} = \frac{30/100}{45/100} = \frac{30}{45} = \frac{2}{3}.$$

EXAMPLE 1 Stocks

Use the information given in the chart at the beginning of this section to find the following probabilities.

(a) $P(B|A)$

SOLUTION This represents the probability that the broker used research, given that the broker picked stocks that went up. Reduce the sample space to A. Then find $n(A \cap B)$ and $n(A)$.

$$P(B|A) = \frac{P(B \cap A)}{P(A)} = \frac{n(A \cap B)}{n(A)} = \frac{30}{60} = \frac{1}{2}$$

If a broker picked stocks that went up, then the probability is $1/2$ that the broker used research.

(b) $P(A'|B)$

SOLUTION In words, this is the probability that a broker picks stocks that do not go up, even though he used research.

$$P(A'|B) = \frac{n(A' \cap B)}{n(B)} = \frac{15}{45} = \frac{1}{3}$$

YOUR TURN 1 In Example 1, find $P(A|B')$.

(c) $P(B'|A')$

SOLUTION Here, we want the probability that a broker who picked stocks that did not go up did not use research.

$$P(B'|A') = \frac{n(B' \cap A')}{n(A')} = \frac{25}{40} = \frac{5}{8} \qquad \text{TRY YOUR TURN 1}$$

Venn diagrams are useful for illustrating problems in conditional probability. A Venn diagram for Example 1, in which the probabilities are used to indicate the number in the set defined by each region, is shown in Figure 21. In the diagram, $P(B|A)$ is found by reducing the sample space to just set A. Then $P(B|A)$ is the ratio of the number in that part of set B that is also in A to the number in set A, or $0.3/0.6 = 0.5$.

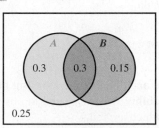

$P(A) = 0.3 + 0.3 = 0.6$

FIGURE 21

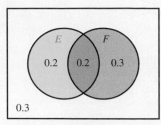

FIGURE 22

YOUR TURN 2 Given $P(E) = 0.56$, $P(F) = 0.64$, and $P(E \cup F) = 0.80$, find $P(E \mid F)$.

EXAMPLE 2 Conditional Probabilities

Given $P(E) = 0.4$, $P(F) = 0.5$, and $P(E \cup F) = 0.7$, find $P(E|F)$.

SOLUTION Find $P(E \cap F)$ first. By the union rule,

$$P(E \cup F) = P(E) + P(F) - P(E \cap F)$$
$$0.7 = 0.4 + 0.5 - P(E \cap F)$$
$$P(E \cap F) = 0.2.$$

$P(E|F)$ is the ratio of the probability of that part of E that is in F to the probability of F, or

$$P(E|F) = \frac{P(E \cap F)}{P(F)} = \frac{0.2}{0.5} = \frac{2}{5}. \quad \text{TRY YOUR TURN 2}$$

The Venn diagram in Figure 22 illustrates Example 2.

EXAMPLE 3 Tossing Coins

Two fair coins were tossed, and it is known that at least one was a head. Find the probability that both were heads.

SOLUTION At first glance, the answer may appear to be $1/2$, but this is not the case. The sample space has four equally likely outcomes, $S = \{hh, ht, th, tt\}$. Because of the condition that at least one coin was a head, the sample space is reduced to $\{hh, ht, th\}$. Since only one outcome in this reduced sample space is 2 heads,

$$P(2 \text{ heads} \mid \text{at least 1 head}) = \frac{1}{3}.$$

Alternatively, we could use the conditional probability definition. Define two events:

$$E_1 = \text{at least 1 head} = \{hh, ht, th\}$$

and

$$E_2 = 2 \text{ heads} = \{hh\}.$$

YOUR TURN 3 In Example 3, find the probability of exactly one head, given that there is at least one tail.

Since there are four equally likely outcomes, $P(E_1) = 3/4$ and $P(E_1 \cap E_2) = 1/4$. Therefore,

$$P(E_2|E_1) = \frac{P(E_2 \cap E_1)}{P(E_1)} = \frac{1/4}{3/4} = \frac{1}{3}. \quad \text{TRY YOUR TURN 3}$$

EXAMPLE 4 Playing Cards

Two cards are drawn from a standard deck, one after another without replacement. Find the probability that the second card is red, given that the first card is red.

SOLUTION According to the conditional probability formula,

$$P(\text{second card is red} \mid \text{first card is red})$$
$$= \frac{P(\text{second card is red and the first card is red})}{P(\text{first card is red})}.$$

We will soon see how to compute probabilities such as the one in the numerator. But there is a much simpler way to calculate this conditional probability. We only need to observe that with one red card gone, there are 51 cards left, 25 of which are red, so

$$P(\text{second card is red}|\text{first card is red}) = \frac{25}{51}.$$

It is important not to confuse $P(A|B)$ with $P(B|A)$. For example, in a criminal trial, a prosecutor may point out to the jury that the probability of the defendant's DNA profile matching that of a sample taken at the scene of the crime, given that the defendant is innocent, $P(D|I)$, is very small. What the jury must decide, however, is the probability that the defendant is innocent, given that the defendant's DNA profile matches the sample, $P(I|D)$. Confusing the two is an error sometimes called "the prosecutor's fallacy," and the 1990 conviction of a rape suspect in England was overturned by a panel of judges, who ordered a retrial, because the fallacy made the original trial unfair. *Source: New Scientist.*

In the next section, we will see how to compute $P(A|B)$ when we know $P(B|A)$.

Product Rule

If $P(E) \neq 0$ and $P(F) \neq 0$, then the definition of conditional probability shows that

$$P(E|F) = \frac{P(E \cap F)}{P(F)} \quad \text{and} \quad P(F|E) = \frac{P(F \cap E)}{P(E)}.$$

Using the fact that $P(E \cap F) = P(F \cap E)$, and solving each of these equations for $P(E \cap F)$, we obtain the following rule.

Product Rule of Probability

If E and F are events, then $P(E \cap F)$ may be found by either of these formulas.

$$P(E \cap F) = P(F) \cdot P(E|F) \quad \text{or} \quad P(E \cap F) = P(E) \cdot P(F|E)$$

The product rule gives a method for finding the probability that events E and F both occur, as illustrated by the next few examples.

EXAMPLE 5   Business Majors

In a class with 2/5 women and 3/5 men, 25% of the women are business majors. Find the probability that a student chosen from the class at random is a female business major.

SOLUTION   Let B and W represent the events "business major" and "woman," respectively. We want to find $P(B \cap W)$. By the product rule,

$$P(B \cap W) = P(W) \cdot P(B|W).$$

Using the given information, $P(W) = 2/5 = 0.4$ and $P(B|W) = 0.25$. Thus,

$$P(B \cap W) = 0.4(0.25) = 0.10. \quad \text{TRY YOUR TURN 4}$$

YOUR TURN 4   At a local college, 4/5 of the students live on campus. Of those who live on campus, 25% have cars on campus. Find the probability that a student lives on campus and has a car.

The next examples show how a tree diagram is used with the product rule to find the probability of a sequence of events.

EXAMPLE 6   Advertising

A company needs to hire a new director of advertising. It has decided to try to hire either person A or B, who are assistant advertising directors for its major competitor. To decide between A and B, the company does research on the campaigns managed by either A or B (no campaign is managed by both) and finds that A is in charge of twice as many advertising campaigns as B. Also, A's campaigns have satisfactory results 3 out of 4 times, while B's campaigns have satisfactory results only 2 out of 5 times. Suppose one of the competitor's advertising campaigns (managed by A or B) is selected randomly.

We can represent this situation using a tree diagram as follows. Let A denote the event "Person A manages the job" and B the event "person B manages the job." Notice that A and

B are complementary events. Since A does twice as many jobs as B, we have $P(A) = 2/3$ and $P(B) = 1/3$, as noted on the first-stage branches of the tree in Figure 23.

Let *S* be the event "satisfactory results" and *U* the event "unsatisfactory results." When A manages the job, the probability of satisfactory results is 3/4 and of unsatisfactory results 1/4 as noted on the second-stage branches. Similarly, the probabilities when B manages the job are noted on the remaining second-stage branches. The composite branches labeled 1 to 4 represent the four mutually exclusive possibilities for the managing and outcome of the selected campaign.

(a) Find the probability that A is in charge of the selected campaign and that it produces satisfactory results.

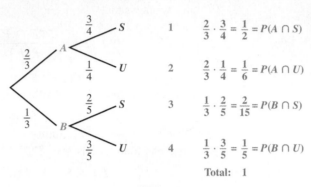

Executive	Campaign	Branch	Probability

$\frac{3}{4}$ S   1 $\frac{2}{3} \cdot \frac{3}{4} = \frac{1}{2} = P(A \cap S)$

$\frac{1}{4}$ U   2 $\frac{2}{3} \cdot \frac{1}{4} = \frac{1}{6} = P(A \cap U)$

$\frac{2}{5}$ S   3 $\frac{1}{3} \cdot \frac{2}{5} = \frac{2}{15} = P(B \cap S)$

$\frac{3}{5}$ U   4 $\frac{1}{3} \cdot \frac{3}{5} = \frac{1}{5} = P(B \cap U)$

Total: 1

FIGURE 23

SOLUTION We are asked to find $P(A \cap S)$. We know that when A does the job, the probability of success is 3/4, that is, $P(S|A) = 3/4$. Hence, by the product rule,

$$P(A \cap S) = P(A) \cdot P(S|A) = \frac{2}{3} \cdot \frac{3}{4} = \frac{1}{2}.$$

The event $A \cap S$ is represented by branch 1 of the tree, and, as we have just seen, its probability is the product of the probabilities of the pieces that make up that branch.

(b) Find the probability that B runs the campaign and that it produces satisfactory results.

SOLUTION We must find $P(B \cap S)$. The event is represented by branch 3 of the tree, and, as before, its probability is the product of the probabilities of the pieces of that branch:

$$P(B \cap S) = P(B) \cdot P(S|B) = \frac{1}{3} \cdot \frac{2}{5} = \frac{2}{15}.$$

(c) What is the probability that the selected campaign is satisfactory?

SOLUTION The event *S* is the union of the mutually exclusive events $A \cap S$ and $B \cap S$, which are represented by branches 1 and 3 of the diagram. By the union rule,

$$P(S) = P(A \cap S) + P(B \cap S) = \frac{1}{2} + \frac{2}{15} = \frac{19}{30}.$$

Thus, the probability of an event that appears on several branches is the sum of the probabilities of each of these branches.

(d) What is the probability that the selected campaign is unsatisfactory?

SOLUTION $P(U)$ can be read from branches 2 and 4 of the tree.

$$P(U) = \frac{1}{6} + \frac{1}{5} = \frac{11}{30}$$

YOUR TURN 5 In Example 6, what is the probability that A is in charge of the selected campaign and that it produces unsatisfactory results?

Alternatively, since *U* is the complement of *S*,

$$P(U) = 1 - P(S) = 1 - \frac{19}{30} = \frac{11}{30}.$$ **TRY YOUR TURN 5**

EXAMPLE 7 **Environmental Inspections**

The Environmental Protection Agency is considering inspecting 6 plants for environmental compliance: 3 in Chicago, 2 in Los Angeles, and 1 in New York. Due to a lack of inspectors, they decide to inspect two plants selected at random, one this month and one next month, with each plant equally likely to be selected, but no plant selected twice. What is the probability that 1 Chicago plant and 1 Los Angeles plant are selected?

SOLUTION   A tree diagram showing the various possible outcomes is given in Figure 24. In this diagram, the events of inspecting a plant in Chicago, Los Angeles, and New York are represented by C, LA, and NY, respectively. For the first inspection, $P(\text{C first}) = 3/6 = 1/2$ because 3 of the 6 plants are in Chicago, and all plants are equally likely to be selected. Likewise, $P(\text{LA first}) = 1/3$ and $P(\text{NY first}) = 1/6$.

For the second inspection, we first note that one plant has been inspected and, therefore, removed from the list, leaving 5 plants. For example, $P(\text{LA second} \mid \text{C first}) = 2/5$ since 2 of the 5 remaining plants are in Los Angeles. The remaining second inspection probabilities are calculated in the same manner.

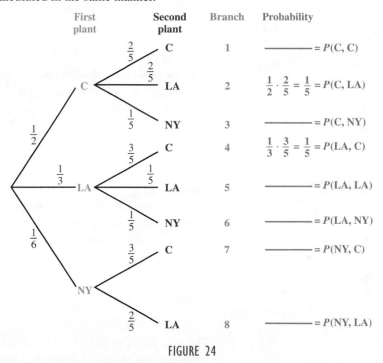

FIGURE 24

We want to find the probability of selecting exactly 1 Chicago plant and 1 Los Angeles plant. This event can occur in two ways: inspecting Chicago this month and Los Angeles next month (branch 2 of the tree diagram), or inspecting Los Angeles this month and Chicago next month (branch 4). For branch 2,

$$P(\text{C first}) \cdot P(\text{LA second} \mid \text{C first}) = \frac{1}{2} \cdot \frac{2}{5} = \frac{1}{5}.$$

For branch 4, where Los Angeles is inspected first,

$$P(\text{LA first}) \cdot P(\text{C second} \mid \text{LA first}) = \frac{1}{3} \cdot \frac{3}{5} = \frac{1}{5}.$$

Since the two events are mutually exclusive, the final probability is the sum of these two probabilities.

YOUR TURN 6   In Example 7, what is the probability that 1 New York plant and 1 Chicago plant are selected?

$$P(1 \text{ C}, 1 \text{ LA}) = P(\text{C first}) \cdot P(\text{LA second} \mid \text{C first})$$
$$+ P(\text{LA first}) \cdot P(\text{C second} \mid \text{LA first})$$
$$= \frac{2}{5}$$

TRY YOUR TURN 6

FOR REVIEW

You may wish to refer to the picture of a deck of cards shown in Figure 17 (Section 7.3) and the description accompanying it.

The product rule is often used with *stochastic processes*, which are mathematical models that evolve over time in a probabilistic manner. For example, selecting factories at random for inspection is such a process, in which the probabilities change with each successive selection.

EXAMPLE 8 Playing Cards

Two cards are drawn from a standard deck, one after another without replacement.

(a) Find the probability that the first card is a heart and the second card is red.

SOLUTION Start with the tree diagram in Figure 25. On the first draw, since there are 13 hearts among the 52 cards, the probability of drawing a heart is $13/52 = 1/4$. On the second draw, since a (red) heart has been drawn already, there are 25 red cards in the remaining 51 cards. Thus, the probability of drawing a red card on the second draw, given that the first is a heart, is $25/51$. By the product rule of probability,

$$P(\text{heart first and red second})$$
$$= P(\text{heart first}) \cdot P(\text{red second}|\text{heart first})$$
$$= \frac{1}{4} \cdot \frac{25}{51} = \frac{25}{204} \approx 0.123.$$

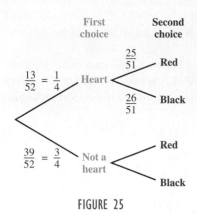

FIGURE 25

(b) Find the probability that the second card is red.

SOLUTION To solve this, we need to fill out the bottom branch of the tree diagram in Figure 25. Unfortunately, if the first card is not a heart, it is not clear how to find the probability that the second card is red, because it depends upon whether the first card is red or black. One way to solve this problem would be to divide the bottom branch into two separate branches: diamond and black card (club or spade).

There is a simpler way, however, since we don't care whether or not the first card is a heart, as we did in part (a). Instead, we'll consider whether the first card is red or black and then do the same for the second card. The result, with the corresponding probabilities, is in Figure 26. The probability that the second card is red is found by multiplying the probabilities along the two branches and adding.

$$P(\text{red second}) = \frac{1}{2} \cdot \frac{25}{51} + \frac{1}{2} \cdot \frac{26}{51}$$
$$= \frac{1}{2}$$

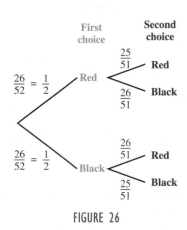

FIGURE 26

The probability is $1/2$, exactly the same as the probability that any card is red. If we know nothing about the first card, there is no reason for the probability of the second card to be anything other than $1/2$.

Independent Events
Suppose, in Example 8(a), that we draw the two cards *with* replacement rather than without replacement (that is, we put the first card back before drawing the second card). If the first card is a heart, then the probability of drawing a red card on the second draw is $26/52$, rather than $25/51$, because there are still 52 cards in the deck, 26 of them red. In this case, $P(\text{red second}|\text{heart first})$ is the same as $P(\text{red second})$. The value of the second card is not affected by the value of the first card. We say that the event that the second card is red is *independent* of the event that the first card is a heart since the knowledge of the first card does not influence what happens to the second card. On the other hand, when we draw without replacement, the events that the first card is a heart and that the second card is red are *dependent* events. The fact that the first card is a heart means there is one less red card in the deck, influencing the probability that the second card is red.

As another example, consider tossing a fair coin twice. If the first toss shows heads, the probability that the next toss is heads is still $1/2$. Coin tosses are independent events, since the outcome of one toss does not influence the outcome of the next toss. Similarly, rolls of a fair die are independent events.

On the other hand, the events "the milk is old" and "the milk is sour" are dependent events; if the milk is old, there is an increased chance that it is sour. Also, in the example at the beginning of this section, the events A (broker picked stocks that went up) and B (broker used research) are dependent events, because information about the use of research affected the probability of picking stocks that go up. That is, $P(A|B)$ is different from $P(A)$.

If events E and F are independent, then the knowledge that E has occurred gives no (probability) information about the occurrence or nonoccurrence of event F. That is, $P(F)$ is exactly the same as $P(F|E)$, or

$$P(F|E) = P(F).$$

This, in fact, is the formal definition of independent events.

Independent Events

Events E and F are **independent events** if

$$P(F|E) = P(F) \qquad \text{or} \qquad P(E|F) = P(E).$$

If the events are not independent, they are **dependent events**.

When E and F are independent events, then $P(F|E) = P(F)$ and the product rule becomes

$$P(E \cap F) = P(E) \cdot P(F|E) = P(E) \cdot P(F).$$

Conversely, if this equation holds, then it follows that $P(F) = P(F|E)$. Consequently, we have this useful fact:

Product Rule for Independent Events

Events E and F are independent events if and only if

$$P(E \cap F) = P(E) \cdot P(F).$$

EXAMPLE 9 Calculator

A calculator requires a keystroke assembly and a logic circuit. Assume that 99% of the keystroke assemblies are satisfactory and 97% of the logic circuits are satisfactory. Find the probability that a finished calculator will be satisfactory.

SOLUTION If the failure of a keystroke assembly and the failure of a logic circuit are independent events, then

$P(\text{satisfactory calculator})$

$= P(\text{satisfactory keystroke assembly}) \cdot P(\text{satisfactory logic circuit})$

$= (0.99)(0.97) \approx 0.96.$

(The probability of a defective calculator is $1 - 0.96 = 0.04$.)   **TRY YOUR TURN 7**

YOUR TURN 7 The probability that you roll a five on a single die is 1/6. Find the probability you roll two five's in a row.

CAUTION It is common for students to confuse the ideas of *mutually exclusive* events and *independent* events. Events E and F are mutually exclusive if $E \cap F = \emptyset$. For example, if a family has exactly one child, the only possible outcomes are $B = \{\text{boy}\}$ and $G = \{\text{girl}\}$. These two events are mutually exclusive. The events are *not* independent, however, since $P(G|B) = 0$ (if a family with only one child has a boy, the probability it has a girl is then 0). Since $P(G|B) \neq P(G)$, the events are not independent.

Of all the families with exactly two children, the events $G_1 = \{\text{first child is a girl}\}$ and $G_2 = \{\text{second child is a girl}\}$ are independent, since $P(G_2|G_1)$ equals $P(G_2)$. However, G_1 and G_2 are not mutually exclusive, since $G_1 \cap G_2 = \{\text{both children are girls}\} \neq \emptyset$.

To show that two events E and F are independent, show that $P(F|E) = P(F)$ or that $P(E|F) = P(E)$ or that $P(E \cap F) = P(E) \cdot P(F)$. Another way is to observe that knowledge of one outcome does not influence the probability of the other outcome, as we did for coin tosses.

NOTE In some cases, it may not be apparent from the physical description of the problem whether two events are independent or not. For example, it is not obvious whether the event that a baseball player gets a hit tomorrow is independent of the event that he got a hit today. In such cases, it is necessary to calculate whether $P(F|E) = P(F)$, or, equivalently, whether $P(E \cap F) = P(E) \cdot P(F)$.

EXAMPLE 10 Snow in Manhattan

On a typical January day in Manhattan the probability of snow is 0.10, the probability of a traffic jam is 0.80, and the probability of snow or a traffic jam (or both) is 0.82. Are the event "it snows" and the event "a traffic jam occurs" independent?

SOLUTION Let S represent the event "it snows" and T represent the event "a traffic jam occurs." We must determine whether

$$P(T|S) = P(T) \qquad \text{or} \qquad P(S|T) = P(S).$$

We know $P(S) = 0.10$, $P(T) = 0.8$, and $P(S \cup T) = 0.82$. We can use the union rule (or a Venn diagram) to find $P(S \cap T) = 0.08$, $P(T|S) = 0.8$, and $P(S|T) = 0.1$. Since

$$P(T|S) = P(T) = 0.8 \qquad \text{and} \qquad P(S|T) = P(S) = 0.1,$$

the events "it snows" and "a traffic jam occurs" are independent.   **TRY YOUR TURN 8**

YOUR TURN 8 The probability that you do your math homework is 0.8, the probability that you do your history assignment is 0.7, and the probability of you doing your math homework or your history assignment is 0.9. Are the events "do your math homework" and "do your history assignment" independent?

Although we showed $P(T|S) = P(T)$ and $P(S|T) = P(S)$ in Example 10, only one of these results is needed to establish independence. It is also important to note that independence of events does not necessarily follow intuition; it is established from the mathematical definition of independence.

7.5 EXERCISES

If a single fair die is rolled, find the probabilities of the following results.

1. A 2, given that the number rolled was odd

2. A 4, given that the number rolled was even

3. An even number, given that the number rolled was 6

4. An odd number, given that the number rolled was 6.

If two fair dice are rolled, find the probabilities of the following results.

5. A sum of 8, given that the sum is greater than 7

6. A sum of 6, given that the roll was a "double" (two identical numbers)

7. A double, given that the sum was 9

8. A double, given that the sum was 8.

If two cards are drawn without replacement from an ordinary deck, find the probabilities of the following results.

9. The second is a heart, given that the first is a heart.

10. The second is black, given that the first is a spade.

11. The second is a face card, given that the first is a jack.

12. The second is an ace, given that the first is not an ace.

13. A jack and a 10 are drawn.

14. An ace and a 4 are drawn.

15. Two black cards are drawn.

16. Two hearts are drawn.

17. In your own words, explain how to find the conditional probability $P(E|F)$.

18. In your own words, define independent events.

Decide whether each of the following pairs of events are dependent or independent.

19. A red and a green die are rolled. A is the event that the red die comes up even, and B is the event that the green die comes up even.

20. C is the event that it rains more than 10 days in Chicago next June, and D is the event that it rains more than 15 days.

21. E is the event that a resident of Texas lives in Dallas, and F is the event that a resident of Texas lives in either Dallas or Houston.

22. A coin is flipped. G is the event that today is Tuesday, and H is the event that the coin comes up heads.

In the previous section, we described an experiment in which the numbers 1, 2, 3, 4, and 5 are written on slips of paper, and 2 slips are drawn at random one at a time without replacement. Find each probability in Exercises 23 and 24.

23. The probability that the first number is 3, given the following.

   a. The sum is 7.

   b. The sum is 8.

24. The probability that the sum is 8, given the following.

   a. The first number is 5.

   b. The first number is 4.

25. Suppose two dice are rolled. Let A be the event that the sum of the two dice is 7. Find an event B related to numbers on the dice such that A and B are

   a. independent;

   b. dependent.

26. Your friend asks you to explain how the product rule for independent events differs from the product rule for dependent events. How would you respond?

27. Another friend asks you to explain how to tell whether two events are dependent or independent. How would you reply? (Use your own words.)

28. A student reasons that the probability in Example 3 of both coins being heads is just the probability that the other coin is a head, that is, $1/2$. Explain why this reasoning is wrong.

29. Let A and B be independent events with $P(A) = \dfrac{1}{4}$ and $P(B) = \dfrac{1}{5}$. Find $P(A \cap B)$ and $P(A \cup B)$.

30. If A and B are events such that $P(A) = 0.5$ and $P(A \cup B) = 0.7$, find $P(B)$ when

   a. A and B are mutually exclusive;

   b. A and B are independent.

31. The following problem, submitted by Daniel Hahn of Blairstown, Iowa, appeared in the "Ask Marilyn" column of *Parade* magazine. *Source: Parade magazine.*

   "You discover two booths at a carnival. Each is tended by an honest man with a pair of covered coin shakers. In each shaker is a single coin, and you are allowed to bet upon the chance that both coins in that booth's shakers are heads after the man in the booth shakes them, does an inspection, and can tell you that at least one of the shakers contains a head. The difference is that the man in the first booth always looks inside both of his shakers, whereas the man in the second booth looks inside only one of the shakers. Where will you stand the best chance?"

32. The following question was posed in *Chance News* by Craig Fox and Yoval Rotenstrich. You are playing a game in which a fair coin is flipped and a fair die is rolled. You win a prize if both the coin comes up heads and a 6 is rolled on the die. Now suppose the coin is tossed and the die is rolled, but you are not allowed to see either result. You are told, however, that either the head or the 6 occurred. You are then offered the chance to cancel the game and play a new game in which a die is rolled (there is no coin), and you win a prize if a 6 is rolled. *Source: Chance News.*

   a. Is it to your advantage to switch to the new game, or to stick with the original game? Answer this question by calculating your probability of winning in each case.

   b. Many people erroneously think that it's better to stick with the original game. Discuss why this answer might seem intuitive, but why it is wrong.

33. Suppose a male defendant in a court trial has a mustache, beard, tattoo, and an earring. Suppose, also, that an eyewitness has identified the perpetrator as someone with these characteristics. If the respective probabilities for the male population in this region are 0.35, 0.30, 0.10, and 0.05, is it fair to multiply these probabilities together to conclude that the probability that a person having these characteristics is 0.000525, or 21 in 40,000, and thus decide that the defendant must be guilty?

34. In a two-child family, if we assume that the probabilities of a male child and a female child are each 0.5, are the events *all children are the same sex* and *at most one male* independent? Are they independent for a three-child family?

35. Laura Johnson, a game show contestant, could win one of two prizes: a shiny new Porsche or a shiny new penny. Laura is given two boxes of marbles. The first box has 50 pink marbles in it and the second box has 50 blue marbles in it. The game show host will pick someone from the audience to be blindfolded and then draw a marble from one of the two boxes. If a pink marble is drawn, she wins the Porsche. Otherwise, Laura wins the penny. Can Laura increase her chances of winning by redistributing some of the marbles from one box to the other? Explain. *Source: Car Talk.*

APPLICATIONS

Business and Economics

Banking The Midtown Bank has found that most customers at the tellers' windows either cash a check or make a deposit. The following table indicates the transactions for one teller for one day.

	Cash Check	No Check	Totals
Make Deposit	60	20	80
No Deposit	30	10	40
Totals	90	30	120

Letting C represent "cashing a check" and D represent "making a deposit," express each probability in words and find its value.

36. $P(C|D)$ **37.** $P(D'|C)$ **38.** $P(C'|D')$

39. $P(C'|D)$ **40.** $P[(C \cap D)']$

41. Airline Delays In February 2010, the major U.S. airline with the fewest delays was United Airlines, for which 77.3% of their flights arrived on time. Assume that the event that a given flight arrives on time is independent of the event that another flight arrives on time. *Source: U.S. Department of Transportation.*

   a. Katie O'Connor plans to take four separate flights for her publisher next month on United Airlines. Assuming that the airline has the same on-time performance as in February 2010, what is the probability that all four flights arrive on time?

   b. Discuss how realistic it is to assume that the on-time arrivals of the different flights are independent.

42. Backup Computers Corporations where a computer is essential to day-to-day operations, such as banks, often have a second backup computer in case the main computer fails. Suppose there is a 0.003 chance that the main computer will fail in a given time period and a 0.005 chance that the backup computer will fail while the main computer is being repaired. Assume these failures represent independent events, and find the fraction of the time that the corporation can assume it will have computer service. How realistic is our assumption of independence?

43. ATM Transactions   Among users of automated teller machines (ATMs), 92% use ATMs to withdraw cash, and 32% use them to check their account balance. Suppose that 96% use ATMs to either withdraw cash or check their account balance (or both). Given a woman who uses an ATM to check her account balance, what is the probability that she also uses an ATM to get cash? *Source: Chicago Tribune.*

Quality Control   A bicycle factory runs two assembly lines, A and B. If 95% of line A's products pass inspection, while only 85% of line B's products pass inspection, and 60% of the factory's bikes come off assembly line A (the rest off B), find the probabilities that one of the factory's bikes did not pass inspection and came off the following.

44. Assembly line A   **45.** Assembly line B

46. Find the probability that one of the factory's bikes did not pass inspection.

Life Sciences

47. Genetics   Both of a certain pea plant's parents had a gene for red and a gene for white flowers. (See Exercise 54 in Section 7.4.) If the offspring has red flowers, find the probability that it combined a gene for red and a gene for white (rather than 2 for red).

48. Medical Experiment   A medical experiment showed that the probability that a new medicine is effective is 0.75, the probability that a patient will have a certain side effect is 0.4, and the probability that both events occur is 0.3. Decide whether these events are dependent or independent.

Genetics   Assuming that boy and girl babies are equally likely, fill in the remaining probabilities on the tree diagram below and use that information to find the probability that a family with three children has all girls, given the following.

49. The first is a girl.

50. The third is a girl.

51. The second is a girl.

52. At least 2 are girls.

53. At least 1 is a girl.

Color Blindness   The following table shows frequencies for red-green color blindness, where M represents "person is male" and C represents "person is color-blind." Use this table to find the following probabilities. (See Exercise 53, Section 7.4.)

	M	*M'*	Totals
C	0.035	0.004	0.039
C'	0.452	0.509	0.961
Totals	0.487	0.513	1.000

54. $P(M)$ **55.** $P(C)$

56. $P(M \cap C)$ **57.** $P(M \cup C)$

58. $P(M|C)$ **59.** $P(C|M)$

60. $P(M'|C)$

61. Are the events C and M, described above, dependent? What does this mean?

62. Color Blindness   A scientist wishes to determine whether there is a relationship between color blindness (C) and deafness (D).

   a. Suppose the scientist found the probabilities listed in the table. What should the findings be? (See Exercises 54–61.)

   b. Explain what your answer tells us about color blindness and deafness.

	D	*D'*	Totals
C	0.0008	0.0392	0.0400
C'	0.0192	0.9408	0.9600
Totals	0.0200	0.9800	1.0000

63. Overweight   According to a recent report, 68.3% of men and 64.1% of women in the United States were overweight. Given that 49.3% of Americans are men and 50.7% are women, find the probability that a randomly selected American fits the following description. *Source: JAMA.*

   a. An overweight man

   b. Overweight

   c. Are the events "male" and "overweight" independent?

Hockey   The table below lists the number of head and neck injuries for 319 ice hockey players' exposures wearing either a full shield or half shield in the Canadian Inter-University Athletics Union. *Source: JAMA.*

For a randomly selected injury, find each probability.

64. $P(A)$ **65.** $P(C|F)$

66. $P(A|H)$ **67.** $P(B'|H')$

	Half Shield (*H*)	Full Shield (*F*)	Totals
Head and Face Injuries (*A*)	95	34	129
Concussions (*B*)	41	38	79
Neck Injuries (*C*)	9	7	16
Other Injuries (*D*)	202	150	352
Totals	347	229	576

68. Are the events *A* and *H* independent events?*

69. Blood Pressure A doctor is studying the relationship between blood pressure and heartbeat abnormalities in her patients. She tests a random sample of her patients and notes their blood pressures (high, low, or normal) and their heartbeats (regular or irregular). She finds that:

(i) 14% have high blood pressure.

(ii) 22% have low blood pressure.

(iii) 15% have an irregular heartbeat.

(iv) Of those with an irregular heartbeat, one-third have high blood pressure.

(v) Of those with normal blood pressure, one-eighth have an irregular heartbeat.

What portion of the patients selected have a regular heartbeat and low blood pressure? Choose one of the following. (*Hint:* Make a table similar to the one for Exercises 54–61.) *Source: Society of Actuaries.*

a. 2% b. 5% c. 8% d. 9% e. 20%

70. Breast Cancer To explain why the chance of a woman getting breast cancer in the next year goes up each year, while the chance of a woman getting breast cancer in her lifetime goes down, Ruma Falk made the following analogy. Suppose you are looking for a letter that you may have lost. You have 8 drawers in your desk. There is a probability of 0.1 that the letter is in any one of the 8 drawers and a probability of 0.2 that the letter is not in any of the drawers. *Source: Chance News.*

a. What is the probability that the letter is in drawer 1?

b. Given that the letter is not in drawer 1, what is the probability that the letter is in drawer 2?

c. Given that the letter is not in drawer 1 or 2, what is the probability that the letter is in drawer 3?

d. Given that the letter is not in drawers 1–7, what is the probability that the letter is in drawer 8?

e. Based on your answers to parts a–d, what is happening to the probability that the letter is in the next drawer?

f. What is the probability that the letter is in some drawer?

g. Given that the letter is not in drawer 1, what is the probability that the letter is in some drawer?

h. Given that the letter is not in drawer 1 or 2, what is the probability that the letter is in some drawer?

i. Given that the letter is not in drawers 1–7, what is the probability that the letter is in some drawer?

j. Based on your answers to parts f–i, what is happening to the probability that the letter is in some drawer?

71. Twins A 1920 study of 17,798 pairs of twins found that 5844 consisted of two males, 5612 consisted of two females, and 6342 consisted of a male and a female. Of course, all of the

mixed-gender pairs were not identical twins. The same-gender pairs may or may not have been identical twins. The goal here is to use the data to estimate *p*, the probability that a pair of twins is identical.

a. Use the data to find the proportion of twins who were male.

b. Denoting your answer from part a by $P(B)$, show that the probability that a pair of twins is male is

$$pP(B) + (1 - p)(P(B))^2.$$

(*Hint:* Draw a tree diagram. The first set of branches should be identical and not identical. Then note that if one member of a pair of identical twins is male, the other must also be male.)

c. Using your answers from parts a and b, plus the fact that 5844 of the 17,798 twin pairs were male, find an estimate for the value of *p*.

d. Find an expression, similar to the one in part b, for the probability that a pair of twins is female.

e. Using your answers from parts a and d, plus the fact that 5612 of the 17,798 twin pairs were female, find an estimate for the value of *p*.

f. Find an expression, similar to those in parts b and d, for the probability that a pair of twins consists of one male and one female.

g. Using your answers from parts a and f, plus the fact that 6342 of the 17,798 twin pairs consisted of one male and one female, find an estimate for the value of *p*.

Social Sciences

72. Working Women A survey has shown that 52% of the women in a certain community work outside the home. Of these women, 64% are married, while 86% of the women who do not work outside the home are married. Find the probabilities that a woman in that community can be categorized as follows.

a. Married

b. A single woman working outside the home

73. Cigarette Smokers The following table gives a recent estimate (in millions) of the smoking status among persons 25 years of age and over and their highest level of education. *Source: National Health Interview Survey.*

Education	Current Smoker	Former Smoker	Non-Smoker	Total
Less than a high school diploma	7.90	6.66	14.12	28.68
High school diploma or GED	14.38	13.09	25.70	53.17
Some college	12.41	13.55	28.65	54.61
Bachelor's degree or higher	4.97	12.87	38.34	56.18
Total	39.66	46.17	106.81	192.64

a. Find the probability that a person is a current smoker.

b. Find the probability that a person has less than a high school diploma.

*We are assuming here and in other exercises that the events consist entirely of the numbers given in the table. If the numbers are interpreted as a sample of all people fitting the description of the events, then testing for independence is more complicated, requiring a technique from statistics known as a *contingency table*.

c. Find the probability that a person is a current smoker and has less than a high school diploma.

d. Find the probability that a person is a current smoker, given that the person has less than a high school diploma.

e. Are the events "current smoker" and "less than a high school diploma" independent events?

Physical Sciences

74. Rain Forecasts In a letter to the journal *Nature*, Robert A. J. Matthews gives the following table of outcomes of forecast and weather over 1000 1-hour walks, based on the United Kingdom's Meteorological office's 83% accuracy in 24-hour forecasts. *Source: Nature.*

	Rain	No Rain	Totals
Forecast of Rain	66	156	222
Forecast of No Rain	14	764	778
Totals	80	920	1000

a. Verify that the probability that the forecast called for rain, given that there was rain, is indeed 83%. Also verify that the probability that the forecast called for no rain, given that there was no rain, is also 83%.

b. Calculate the probability that there was rain, given that the forecast called for rain.

c. Calculate the probability that there was no rain, given that the forecast called for no rain.

d. Observe that your answer to part c is higher than 83% and that your answer to part b is much lower. Discuss which figure best describes the accuracy of the weather forecast in recommending whether or not you should carry an umbrella.

75. Earthquakes There are seven geologic faults (and possibly more) capable of generating a magnitude 6.7 earthquake in the region around San Francisco. Their probabilities of rupturing by the year 2032 are 27%, 21%, 11%, 10%, 4%, 3%, and 3%. *Source: Science News.*

a. Calculate the probability that at least one of these faults erupts by the year 2032, assuming that these are independent events.

b. Scientists forecast a 62% chance of an earthquake with magnitude at least 6.7 in the region around San Francisco by the year 2032. Compare this with your answer from part a. Consider the realism of the assumption of independence. Also consider the role of roundoff. For example, the probability of 10% for one of the faults is presumably rounded to the nearest percent, with the actual probability between 9.5% and 10.5%.

76. Reliability The probability that a key component of a space rocket will fail is 0.03.

a. How many such components must be used as backups to ensure that the probability of at least one of the components working is 0.999999 or more?

b. Is it reasonable to assume independence here?

General Interest

77. Titanic The table at the bottom of the page lists the number of passengers who were on the Titanic and the number of passengers who survived, according to class of ticket. Use this information to determine the following (round answers to four decimal places). *Source: Mathematics Teacher.*

a. What is the probability that a randomly selected passenger was second class?

b. What is the overall probability of surviving?

c. What is the probability of a first-class passenger surviving?

d. What is the probability of a child who was also in the third class surviving?

e. Given that the survivor is from first class, what is the probability that she was a woman?

f. Given that a male has survived, what is the probability that he was in third class?

g. Are the events third-class survival and male survival independent events? What does this imply?

78. Real Estate A real estate agent trying to sell you an attractive beachfront house claims that it will not collapse unless it is subjected simultaneously to extremely high winds and extremely high waves. According to weather service records, there is a 0.001 probability of extremely high winds, and the same for extremely high waves. The real estate agent claims, therefore, that the probability of both occurring is $(0.001)(0.001) = 0.000001$. What is wrong with the agent's reasoning?

79. Age and Loans Suppose 20% of the population are 65 or over, 26% of those 65 or over have loans, and 53% of those under 65 have loans. Find the probabilities that a person fits into the following categories.

a. 65 or over and has a loan

b. Has a loan

80. Women Joggers In a certain area, 15% of the population are joggers and 40% of the joggers are women. If 55% of those who do not jog are women, find the probabilities that

	Children		Women		Men		Totals	
	On	Survived	On	Survived	On	Survived	On	Survived
First Class	6	6	144	140	175	57	325	203
Second Class	24	24	165	76	168	14	357	114
Third Class	79	27	93	80	462	75	634	182
Totals	109	57	402	296	805	146	1316	499

an individual from that community fits the following descriptions.

a. A woman jogger

b. A man who is not a jogger

c. A woman

d. Are the events that a person is a woman and a person is a jogger independent? Explain.

81. Diet Soft Drinks Two-thirds of the population are on a diet at least occasionally. Of this group, 4/5 drink diet soft drinks, while 1/2 of the rest of the (nondieting) population drink diet soft drinks. Find the probabilities that a person fits into the following categories.

a. Drinks diet soft drinks

b. Diets, but does not drink diet soft drinks

82. Driver's License Test The Motor Vehicle Department has found that the probability of a person passing the test for a driver's license on the first try is 0.75. The probability that an individual who fails on the first test will pass on the second try is 0.80, and the probability that an individual who fails the first and second tests will pass the third time is 0.70. Find the probabilities that an individual will do the following.

a. Fail both the first and second tests

b. Fail three times in a row

c. Require at least two tries

83. Ballooning A pair of mathematicians in a hot air balloon were told that there are four independent burners, any one of which is sufficient to keep the balloon aloft. If the probability of any one burner failing during a flight is 0.001, what is the probability that the balloon will crash due to all four burners failing?

84. Speeding Tickets A smooth-talking young man has a 1/3 probability of talking a policeman out of giving him a speeding ticket. The probability that he is stopped for speeding during a given weekend is 1/2. Find the probabilities of the events in parts a and b.

a. He will receive no speeding tickets on a given weekend.

b. He will receive no speeding tickets on 3 consecutive weekends.

c. We have assumed that what happens on the second or third weekend is the same as what happened on the first weekend. Is this realistic? Will driving habits remain the same after getting a ticket?

85. Luxury Cars In one area, 4% of the population drive luxury cars. However, 17% of the CPAs drive luxury cars. Are the events "person drives a luxury car" and "person is a CPA" independent?

86. Studying A teacher has found that the probability that a student studies for a test is 0.60, the probability that a student gets a good grade on a test is 0.70, and the probability that both occur is 0.52.

a. Are these events independent?

b. Given that a student studies, find the probability that the student gets a good grade.

c. Given that a student gets a good grade, find the probability that the student studied.

87. Basketball A basketball player is fouled and now faces a one-and-one free throw situation. She shoots the first free throw. If she misses it, she scores 0 points. If she makes the first free throw, she gets to shoot a second free throw. If she misses the second free throw, she scores only one point for the first shot. If she makes the second free throw, she scores two points (one for each made shot). *Source: Mathematics Teacher.*

a. If her free-throwing percentage for this season is 60%, calculate the probability that she scores 0 points, 1 point, or 2 points.

b. Determine the free-throwing percentage necessary for the probability of scoring 0 points to be the same as the probability of scoring 2 points.* (*Hint:* Let p be the free-throwing percentage, and then solve $p^2 = 1 - p$ for p. Use only the positive value of p.)

88. Basketball The same player from Exercise 87 is now shooting two free throws (that is, she gets to shoot a second free throw whether she makes or misses the first shot.) Assume her free-throwing percentage is still 60%. Find the probability she scores 0 points, 1 point, or 2 points. *Source: Mathematics Teacher.*

89. Football A football coach whose team is 14 points behind needs two touchdowns to win. Each touchdown is worth 6 points. After a touchdown, the coach can choose either a 1-point kick, which is almost certain to succeed, or a 2-point conversion, which is roughly half as likely to succeed. After the first touchdown, the coach must decide whether to go for 1 or 2 points. If the 2-point conversion is successful, the almost certain 1-point kick after the second touchdown will win the game. If the 2-point conversion fails, the team can try another 2-point conversion after the second touchdown to tie. Some coaches, however, prefer to go for the almost certain 1-point kick after the first touchdown, hoping that the momentum will help them get a 2-point conversion after the second touchdown and win the game. They fear that an unsuccessful 2-point conversion after the first touchdown will discourage the team, which can then at best tie. *Source: The Mathematics Teacher.*

a. Draw a tree diagram for the 1-point kick after the first touchdown and the 2-point conversion after the second touchdown. Letting the probability of success for the 1-point kick and the 2-point conversion be k and r, respectively, show that

$$P(\text{win}) = kr,$$
$$P(\text{tie}) = r(1 - k), \quad \text{and}$$
$$P(\text{lose}) = 1 - r.$$

b. Consider the case of trying for a 2-point conversion after the first touchdown. If it succeeds, try a 1-point kick after the second touchdown. If the 2-point conversion fails, try

---

* The solution is the reciprocal of the number $\frac{1 + \sqrt{5}}{2}$, known as the golden ratio or the divine proportion. This number has great significance in architecture, science, and mathematics.

another one after the second touchdown. Draw a tree diagram and use it to show that

$$P(\text{win}) = kr,$$
$$P(\text{tie}) = r(2 - k - r), \quad \text{and}$$
$$P(\text{lose}) = (1 - r)^2.$$

c. What can you say about the probability of winning under each strategy?

d. Given that $r < 1$, which strategy has a smaller probability of losing? What does this tell you about the value of the two strategies?

YOUR TURN ANSWERS

1. $30/55 = 6/11$ **2.** 0.625   **3.** 2/3

4. 1/5   **5.** 1/6   **6.** 1/5

7. 1/36   **8.** No

7.6 Bayes' Theorem

APPLY IT   **What is the probability that a particular defective item was produced by a new machine operator?**
This question will be answered in Example 2 using Bayes' theorem, discussed in this section.

Suppose the probability that an applicant is hired, *given the applicant is qualified*, is known. The manager might also be interested in the probability that the applicant is qualified, *given the applicant was hired*. More generally, if $P(E|F)$ is known for two events E and F, then $P(F|E)$ can be found using a tree diagram. Since $P(E|F)$ is known, the first outcome is either F or F'. Then for each of these outcomes, either E or E' occurs, as shown in Figure 27.

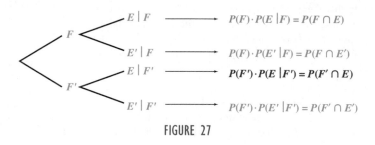

FIGURE 27

The four cases have the probabilities shown on the right. Notice $P(E \cap F)$ is the first case and $P(E)$ is the sum of the first and third cases in the tree diagram. By the definition of conditional probability,

$$P(F|E) = \frac{P(F \cap E)}{P(E)} = \frac{P(F) \cdot P(E|F)}{P(F) \cdot P(E|F) + P(F') \cdot P(E|F')}.$$

This result is a special case of Bayes' theorem, which is generalized later in this section.

Bayes' Theorem (Special Case)

$$P(F|E) = \frac{P(F) \cdot P(E|F)}{P(F) \cdot P(E|F) + P(F') \cdot P(E|F')}$$

EXAMPLE 1   **Worker Errors**

For a fixed length of time, the probability of a worker error on a certain production line is 0.1, the probability that an accident will occur when there is a worker error is 0.3, and the probability that an accident will occur when there is no worker error is 0.2. Find the probability of a worker error if there is an accident.

SOLUTION Let E represent the event of an accident, and let F represent the event of worker error. From the information given,

$$P(F) = 0.1, \qquad P(E|F) = 0.3, \qquad \text{and} \qquad P(E|F') = 0.2.$$

These probabilities are shown on the tree diagram in Figure 28.

Find $P(F|E)$ by dividing the probability that both E and F occur, given by branch 1, by the probability that E occurs, given by the sum of branches 1 and 3.

$$P(F|E) = \frac{P(F) \cdot P(E|F)}{P(F) \cdot P(E|F) + P(F') \cdot P(E|F')}$$

$$= \frac{(0.1)(0.3)}{(0.1)(0.3) + (0.9)(0.2)} = \frac{0.03}{0.21} = \frac{1}{7} \approx 0.1429$$

TRY YOUR TURN 1

YOUR TURN 1 The probability that a student will pass a math exam is 0.8 if he or she attends the review session and 0.65 if he or she does not attend the review session. Sixty percent of the students attend the review session. What is the probability that, given a student passed, the student attended the review session?

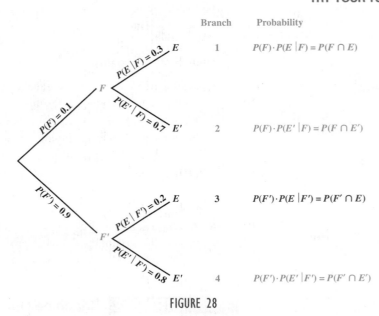

FIGURE 28

The special case of Bayes' theorem can be generalized to more than two events with the tree diagram in Figure 29. This diagram shows the paths that can produce an event E. We assume that the events $F_1, F_2, \ldots, F_n$ are mutually exclusive events (that is, disjoint events) whose union is the sample space, and that E is an event that has occurred. See Figure 30.

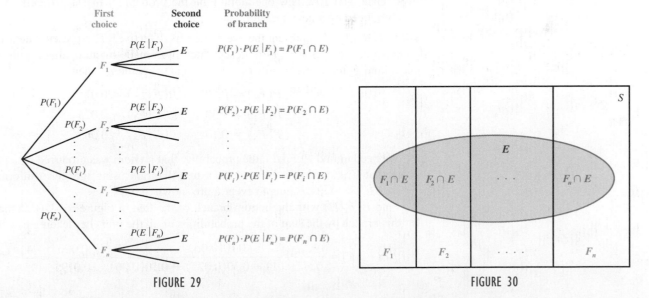

FIGURE 29   FIGURE 30

The probability $P(F_i|E)$, where $1 \le i \le n$, can be found by dividing the probability for the branch containing $P(E|F_i)$ by the sum of the probabilities of all the branches producing event E.

Bayes' Theorem

$$P(F_i|E) = \frac{P(F_i) \cdot P(E|F_i)}{P(F_1) \cdot P(E|F_1) + P(F_2) \cdot P(E|F_2) + \cdots + P(F_n) \cdot P(E|F_n)}$$

This result is known as **Bayes' theorem**, after the Reverend Thomas Bayes (1702–1761), whose paper on probability was published about three years after his death.

The statement of Bayes' theorem can be daunting. Actually, it is easier to remember the formula by thinking of the tree diagram that produced it. Go through the following steps.

Using Bayes' Theorem

1. Start a tree diagram with branches representing $F_1, F_2, \ldots, F_n$. Label each branch with its corresponding probability.

2. From the end of each of these branches, draw a branch for event E. Label this branch with the probability of getting to it, $P(E|F_i)$.

3. You now have n different paths that result in event E. Next to each path, put its probability—the product of the probabilities that the first branch occurs, $P(F_i)$, and that the second branch occurs, $P(E|F_i)$; that is, the product $P(F_i) \cdot P(E|F_i)$, which equals $P(F_i \cap E)$.

4. $P(F_i|E)$ is found by dividing the probability of the branch for F_i by the sum of the probabilities of all the branches producing event E.

EXAMPLE 2   Machine Operators

Based on past experience, a company knows that an experienced machine operator (one or more years of experience) will produce a defective item 1% of the time. Operators with some experience (up to one year) have a 2.5% defect rate, and new operators have a 6% defect rate. At any one time, the company has 60% experienced operators, 30% with some experience, and 10% new operators. Find the probability that a particular defective item was produced by a new operator.

APPLY IT

SOLUTION   Let E represent the event "item is defective," F_1 represent "item was made by an experienced operator," F_2 represent "item was made by an operator with some experience," and F_3 represent "item was made by a new operator." Then

$$P(F_1) = 0.60 \qquad P(E|F_1) = 0.01$$
$$P(F_2) = 0.30 \qquad P(E|F_2) = 0.025$$
$$P(F_3) = 0.10 \qquad P(E|F_3) = 0.06.$$

We need to find $P(F_3|E)$, the probability that an item was produced by a new operator, given that it is defective. First, draw a tree diagram using the given information, as in Figure 31. The steps leading to event E are shown in red.

Find $P(F_3|E)$ with the bottom branch of the tree in Figure 31: Divide the probability for this branch by the sum of the probabilities of all the branches leading to E, or

$$P(F_3|E) = \frac{0.10(0.06)}{0.60(0.01) + 0.30(0.025) + 0.10(0.06)} = \frac{0.006}{0.0195} = \frac{4}{13} \approx 0.3077.$$

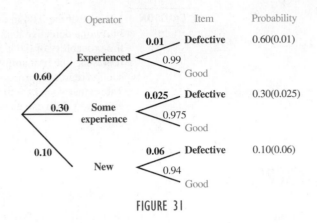

FIGURE 31

YOUR TURN 2   The English
department at a small college has
found that 12% of freshmen test into
English I, 68% test into English II,
and 20% test into English III.
Eighty percent of students in English I will seek help from the writing
center, 40% of those in English II,
and 11% of those in English III.
Given that a student received help
from the writing center, find the
probability that the student is in
English I.

In a similar way, the probability that the defective item was produced by an operator with some experience is

$$P(F_2|E) = \frac{0.30(0.025)}{0.60(0.01) + 0.30(0.025) + 0.10(0.06)} = \frac{0.0075}{0.0195} = \frac{5}{13} \approx 0.3846.$$

Finally, the probability that the defective item was produced by an experienced operator is $P(F_1|E) = 4/13 \approx 0.3077$. Check that $P(F_1|E) + P(F_2|E) + P(F_3|E) = 1$ (that is, the defective item was made by *someone*).   **TRY YOUR TURN 2**

EXAMPLE 3   Manufacturing

A manufacturer buys items from six different suppliers. The fraction of the total number of items obtained from each supplier, along with the probability that an item purchased from that supplier is defective, are shown in the following table.

Manufacturing Supplies		
Supplier	Fraction of Total Supplied	Probability of Defect
1	0.05	0.04
2	0.12	0.02
3	0.16	0.07
4	0.23	0.01
5	0.35	0.03
6	0.09	0.05

Find the probability that a defective item came from supplier 5.

SOLUTION   Let F_1 be the event that an item came from supplier 1, with $F_2, F_3, F_4, F_5,$ and F_6 defined in a similar manner. Let E be the event that an item is defective. We want to find $P(F_5|E)$. Use the probabilities in the table above to prepare a tree diagram, or work with the rows of the table to get

$$P(F_5|E) = \frac{(0.35)(0.03)}{(0.05)(0.04) + (0.12)(0.02) + (0.16)(0.07) + (0.23)(0.01) + (0.35)(0.03) + (0.09)(0.05)}$$

$$= \frac{0.0105}{0.0329} \approx 0.319.$$

There is about a 32% chance that a defective item came from supplier 5. Even though supplier 5 has only 3% defectives, his probability of being "guilty" is relatively high, about 32%, because of the large fraction supplied by 5.

| CAUTION | Notice that the 0.04 in the upper right of the previous table represents the probability of a defective item *given* that the item came from supplier 1. In contrast, the probability of 0.035 in the table for Exercises 54–61 of the previous section represents the probability that a person is color-blind *and* male. The tables in this section represent probability in a different way than those of the previous section. Tables that you encounter outside of this course might represent probability in either way. You can usually tell what is intended by the context, but be careful! |

7.6 EXERCISES

For two events M and N, $P(M) = 0.4$, $P(N|M) = 0.3$, and $P(N|M') = 0.4$. Find the following.

1. $P(M|N)$
2. $P(M'|N)$

For mutually exclusive events R_1, R_2, and R_3, we have $P(R_1) = 0.15$, $P(R_2) = 0.55$, and $P(R_3) = 0.30$. Also, $P(Q|R_1) = 0.40$, $P(Q|R_2) = 0.20$, and $P(Q|R_3) = 0.70$. Find the following.

3. $P(R_1|Q)$
4. $P(R_2|Q)$
5. $P(R_3|Q)$
6. $P(R_1'|Q)$

Suppose you have three jars with the following contents: 2 black balls and 1 white ball in the first, 1 black ball and 2 white balls in the second, and 1 black ball and 1 white ball in the third. One jar is to be selected, and then 1 ball is to be drawn from the selected jar. If the probabilities of selecting the first, second, or third jar are 1/2, 1/3, and 1/6, respectively, find the probabilities that if a white ball is drawn, it came from the following jars.

7. The second jar
8. The third jar

APPLICATIONS

Business and Economics

9. Employment Test A manufacturing firm finds that 70% of its new hires turn out to be good workers and 30% become poor workers. All current workers are given a reasoning test. Of the good workers, 85% pass it; 35% of the poor workers pass it. Assume that these figures will hold true in the future. If the company makes the test part of its hiring procedure and only hires people who meet the previous requirements and pass the test, what percent of the new hires will turn out to be good workers?

Job Qualifications Of all the people applying for a certain job, 75% are qualified and 25% are not. The personnel manager claims that she approves qualified people 85% of the time; she approves an unqualified person 20% of the time. Find each probability.

10. A person is qualified if he or she was approved by the manager.

11. A person is unqualified if he or she was approved by the manager.

Quality Control A building contractor buys 70% of his cement from supplier A and 30% from supplier B. A total of 90% of the bags from A arrive undamaged, while 95% of the bags from B arrive undamaged. Give the probabilities that a damaged bag is from the following sources.

12. Supplier A
13. Supplier B

Appliance Reliability Companies A, B, and C produce 15%, 40%, and 45%, respectively, of the major appliances sold in a certain area. In that area, 1% of the company A appliances, $1\frac{1}{2}$% of the company B appliances, and 2% of the company C appliances need service within the first year. Suppose a defective appliance is chosen at random; find the probabilities that it was manufactured by the following companies.

14. Company A
15. Company B

Television Advertising On a given weekend in the fall, a tire company can buy television advertising time for a college football game, a baseball game, or a professional football game. If the company sponsors the college football game, there is a 70% chance of a high rating, a 50% chance if they sponsor a baseball game, and a 60% chance if they sponsor a professional football game. The probabilities of the company sponsoring these various games are 0.5, 0.2, and 0.3, respectively. Suppose the company does get a high rating; find the probabilities that it sponsored the following.

16. A college football game
17. A professional football game

18. Auto Insurance An auto insurance company insures drivers of all ages. An actuary compiled the following statistics on the company's insured drivers:

Age of Driver	Probability of Accident	Portion of Company's Insured Drivers
16–20	0.06	0.08
21–30	0.03	0.15
31–65	0.02	0.49
66–99	0.04	0.28

A randomly selected driver that the company insures has an accident. Calculate the probability that the driver was age 16–20. Choose one of the following. *Source: Society of Actuaries.*

a. 0.13 **b.** 0.16 **c.** 0.19 **d.** 0.23 **e.** 0.40

19. Life Insurance An insurance company issues life insurance policies in three separate categories: standard, preferred, and ultra-preferred. Of the company's policyholders, 50% are standard, 40% are preferred, and 10% are ultra-preferred. Each standard policyholder has probability 0.010 of dying in the next year, each preferred policyholder has probability 0.005 of dying in the next year, and each ultra-preferred policyholder has probability 0.001 of dying in the next year. A policyholder dies in the next year. What is the probability that the deceased policyholder was ultra-preferred? Choose one of the following. *Source: Society of Actuaries.*

a. 0.0001   b. 0.0010   c. 0.0071   d. 0.0141   e. 0.2817

20. Automobile Collisions An actuary studied the likelihood that different types of drivers would be involved in at least one collision during any one-year period. The results of the study are presented below.

Type of Driver	Percentage of All Drivers	Probability of at Least One Collision
Teen	8%	0.15
Young Adult	16%	0.08
Midlife	45%	0.04
Senior	31%	0.05
Total	100%	

Given that a driver has been involved in at least one collision in the past year, what is the probability that the driver is a young adult driver? Choose one of the following. *Source: Society of Actuaries.*

a. 0.06   b. 0.16   c. 0.19   d. 0.22   e. 0.25

21. Shipping Errors The following information pertains to three shipping terminals operated by Krag Corp. *Source: CPA Examination.*

Terminal	Percentage of Cargo Handled	Percentage of Error
Land	50	2
Air	40	4
Sea	10	14

Krag's internal auditor randomly selects one set of shipping documents, ascertaining that the set selected contains an error. Which of the following gives the probability that the error occurred in the Land Terminal?

a. 0.02   b. 0.10

c. 0.25   d. 0.50

22. Mortgage Defaults A bank finds that the relationship between mortgage defaults and the size of the down payment is given by the following table.

Down Payment	Number of Mortgages with This Down Payment	Probability of Default
5%	1260	0.06
10%	700	0.04
20%	560	0.02
25%	280	0.01

a. If a default occurs, what is the probability that it is on a mortgage with a 5% down payment?

b. What is the probability that a mortgage that is paid to maturity has a 10% down payment?

Life Sciences

23. Colorectal Cancer Researchers found that only one out of 24 physicians could give the correct answer to the following problem: "The probability of colorectal cancer can be given as 0.3%. If a person has colorectal cancer, the probability that the hemoccult test is positive is 50%. If a person does not have colorectal cancer, the probability that he still tests positive is 3%. What is the probability that a person who tests positive actually has colorectal cancer?" What is the correct answer? *Source: Science.*

24. Hepatitis Blood Test The probability that a person with certain symptoms has hepatitis is 0.8. The blood test used to confirm this diagnosis gives positive results for 90% of people with the disease and 5% of those without the disease. What is the probability that an individual who has the symptoms and who reacts positively to the test actually has hepatitis?

25. Sensitivity and Specificity The *sensitivity* of a medical test is defined as the probability that a test will be positive given that a person has a disease, written $P(T^+|D^+)$. The *specificity* of a test is defined as the probability that a test will be negative given that the person does not have the disease, written $P(T^-|D^-)$. For example, the sensitivity and specificity for breast cancer during a mammography exam are approximately 79.6% and 90.2%, respectively. *Source: National Cancer Institute.*

a. It is estimated that 0.5% of U.S. women under the age 40 have breast cancer. Find the probability that a woman under 40 who tests positive during a mammography exam actually has breast cancer.

b. Given that a woman under 40 tests negative during a mammography exam, find the probability that she does not have breast cancer.

c. According to the National Cancer Institute, failure to diagnose breast cancer is the most common cause of medical malpractice litigation. Given a woman under 40 tests negative for breast cancer, find the probability that she does have breast cancer.

d. It is estimated that 1.5% of U.S. women over the age of 50 have breast cancer. Find the probability that a woman over 50 who tests positive actually has breast cancer.

26. Test for HIV Clinical studies have demonstrated that rapid HIV tests have a sensitivity (probability that a test will be positive given that a person has the disease) of approximately 99.9% and a specificity (probability that a test will be negative given that the person does not have the disease) of approximately 99.8%. *Source: Centers for Disease Control and Prevention*

a. In some HIV clinics, the prevalence of HIV was high, about 5%. Find the probability a person actually has HIV, given that the test came back positive from one of these clinics.

b. In other clinics, like a family planning clinic, the prevalence of HIV was low, about 0.1%. Find the probability that a person actually has HIV, given that the test came back positive from one of these clinics.

c. The answers in parts a and b are significantly different. Explain why.

27. Smokers A health study tracked a group of persons for five years. At the beginning of the study, 20% were classified as heavy smokers, 30% as light smokers, and 50% as nonsmokers. Results of the study showed that light smokers were twice as likely as nonsmokers to die during the five-year study but only half as likely as heavy smokers. A randomly selected participant from the study died over the five-year period. Calculate the probability that the participant was a heavy smoker. Choose one of the following. (*Hint:* Let $x = P$(a nonsmoker dies).) *Source: Society of Actuaries.*

a. 0.20   **b.** 0.25   **c.** 0.35   **d.** 0.42   **e.** 0.57

28. Emergency Room Upon arrival at a hospital's emergency room, patients are categorized according to their condition as critical, serious, or stable. In the past year:

  (i) 10% of the emergency room patients were critical;

  (ii) 30% of the emergency room patients were serious;

  (iii) the rest of the emergency room patients were stable;

  (iv) 40% of the critical patients died;

  (v) 10% of the serious patients died; and

  (vi) 1% of the stable patients died.

Given that a patient survived, what is the probability that the patient was categorized as serious upon arrival? Choose one of the following. *Source: Society of Actuaries.*

a. 0.06   **b.** 0.29   **c.** 0.30   **d.** 0.39   **e.** 0.64

29. Blood Test A blood test indicates the presence of a particular disease 95% of the time when the disease is actually present. The same test indicates the presence of the disease 0.5% of the time when the disease is not present. One percent of the population actually has the disease. Calculate the probability that a person has the disease, given that the test indicates the presence of the disease. Choose one of the following. *Source: Society of Actuaries.*

a. 0.324   **b.** 0.657   **c.** 0.945   **d.** 0.950   **e.** 0.995

30. Circulation The probability that a randomly chosen male has a circulation problem is 0.25. Males who have a circulation problem are twice as likely to be smokers as those who do not have a circulation problem. What is the conditional probability that a male has a circulation problem, given that he is a smoker? Choose one of the following. *Source: Society of Actuaries.*

a. 1/4   **b.** 1/3   **c.** 2/5   **d.** 1/2   **e.** 2/3

Social Sciences

31. Alcohol Abstinence The Harvard School of Public Health completed a study on alcohol consumption on college campuses in 2001. They concluded that 20.7% of women attending all-women colleges abstained from alcohol, compared to 18.6% of women attending coeducational colleges. Approximately 4.7% of women college students attend all-women schools. *Source: Harvard School of Public Health.*

a. What is the probability that a randomly selected female student abstains from alcohol?

b. If a randomly selected female student abstains from alcohol, what is the probability she attends a coeducational college?

32. Murder During the murder trial of O. J. Simpson, Alan Dershowitz, an advisor to the defense team, stated on television that only about 0.1% of men who batter their wives actually murder them. Statistician I. J. Good observed that even if, given that a husband is a batterer, the probability he is guilty of murdering his wife is 0.001, what we really want to know is the probability that the husband is guilty, given that the wife was murdered. Good estimates the probability of a battered wife being murdered, given that her husband is not guilty, as 0.001. The probability that she is murdered if her husband is guilty is 1, of course. Using these numbers and Dershowitz's 0.001 probability of the husband being guilty, find the probability that the husband is guilty, given that the wife was murdered. *Source: Nature.*

Never-Married Adults by Age Group The following tables give the proportion of men and of women 18 and older in each age group in 2009, as well as the proportion in each group who have never been married. *Source: U.S Census Bureau.*

Age	Men Proportion of Population	Proportion Never Married
18–24	0.132	0.901
25–34	0.186	0.488
35–44	0.186	0.204
45–64	0.348	0.118
65 or over	0.148	0.044

Age	Women Proportion of Population	Proportion Never Married
18–24	0.121	0.825
25–34	0.172	0.366
35–44	0.178	0.147
45–64	0.345	0.092
65 or over	0.184	0.040

33. Find the probability that a randomly selected man who has never married is between 35 and 44 years old (inclusive).

34. Find the probability that a randomly selected woman who has been married is between 18 and 24 (inclusive).

35. Find the probability that a randomly selected woman who has never been married is between 45 and 64 (inclusive).

36. Seat Belt Effectiveness A 2009 federal study showed that 63.8% of occupants involved in a fatal car crash wore seat belts. Of those in a fatal crash who wore seat belts, 2% were ejected from the vehicle. For those not wearing seat belts, 36% were ejected from the vehicle. *Source: National Highway Traffic Safety Administration.*

  a. Find the probability that a randomly selected person in a fatal car crash who was ejected from the vehicle was wearing a seat belt.

  b. Find the probability that a randomly selected person in a fatal car crash who was not ejected from the vehicle was not wearing a seat belt.

Smokers by Age Group The following table gives the proportion of U.S. adults in each age group in 2008, as well as the proportion in each group who smoke. *Source: Centers for Disease Control and Prevention.*

Age	Proportion of Population	Proportion that Smoke
18 – 44 years	0.49	0.23
45 – 64 years	0.34	0.22
64 – 74 years	0.09	0.12
75 years and over	0.08	0.06

37. Find the probability that a randomly selected adult who smokes is between 18 and 44 years of age (inclusive).

38. Find the probability that a randomly selected adult who does not smoke is between 45 and 64 years of age (inclusive).

General Interest

39. Terrorists John Allen Paulos has pointed out a problem with massive, untargeted wiretaps. To illustrate the problem, he supposes that one out of every million Americans has terrorist ties. Furthermore, he supposes that the terrorist profile is 99% accurate, so that if a person has terrorist ties, the profile will pick them up 99% of the time, and if the person does not have terrorist ties, the profile will accidentally pick them up only 1% of the time. Given that the profile has picked up a person, what is the probability that the person actually has terrorist ties? Discuss how your answer affects your opinion on domestic wiretapping. *Source: Who's Counting.*

40. Three Prisoners The famous "problem of three prisoners" is as follows.

Three men, A, B, and C, were in jail. A knew that one of them was to be set free and the other two were to be executed. But he didn't know who was the one to be spared. To the jailer who did know, A said, "Since two out of the three will be executed, it is certain that either B or C will be, at least. You will give me no information about my own chances if you give me the name of one man, B or C, who is going to be executed." Accepting this argument after some thinking, the jailer said "B will be executed." Thereupon A felt happier because now either he or C would go free, so his chance had increased from 1/3 to 1/2. *Source: Cognition.*

  a. Assume that initially each of the prisoners is equally likely to be set free. Assume also that if both B and C are to be executed, the jailer is equally likely to name either B or C. Show that A is wrong, and that his probability of being freed, given that the jailer says B will be executed, is still 1/3.

  b. Now assume that initially the probabilities of A, B, and C being freed are 1/4, 1/4, and 1/2, respectively. As in part a, assume also that if both B and C are to be executed, the jailer is equally likely to name either B or C. Now show that A's probability of being freed, given that the jailer says B will be executed, actually drops to 1/5. Discuss the reasonableness of this answer, and why this result might violate someone's intuition.

▄▄▄ **YOUR TURN ANSWERS**

  1. 0.6486   **2.** 0.2462

7 CHAPTER REVIEW

SUMMARY

We began this chapter by introducing sets, which are collections of objects. We introduced the following set operations:

- complement (A' is the set of elements not in A),
- intersection ($A \cap B$ is the set of elements belonging to both set A and set B), and
- union ($A \cup B$ is the set of elements belonging to either set A or set B or both).

We used tree diagrams and Venn diagrams to define and study concepts in set operations as well as in probability. We introduced the following terms:

- experiment (an activity or occurrence with an observable result),

- trial (a repetition of an experiment),
- outcome (a result of a trial),
- sample space (the set of all possible outcomes for an experiment), and
- event (a subset of a sample space).

We investigated how to compute various probabilities and we explored some of the properties of probability. In particular, we studied the following concepts:

- empirical probability (based on how frequently an event actually occurred),
- conditional probability (in which some other event is assumed to have occurred),

- odds (an alternative way of expressing probability),
- independent events (in which the occurrence of one event does not affect the probability of another), and
- Bayes' theorem (used to calculate certain types of conditional probability).

Throughout the chapter, many applications of probability were introduced and analyzed. In the next two chapters, we will employ these techniques to further our study into the fields of probability and statistics.

Sets Summary

Number of Subsets A set of k distinct elements has 2^k subsets.

Disjoint Sets If sets A and B are disjoint, then

$$A \cap B = \emptyset \quad \text{and} \quad n(A \cap B) = 0.$$

Union Rule for Sets For any sets A and B,

$$n(A \cup B) = n(A) + n(B) - n(A \cap B).$$

Probability Summary

Basic Probability Principle Let S be a sample space of equally likely outcomes, and let event E be a subset of S. Then the probability that event E occurs is

$$P(E) = \frac{n(E)}{n(S)}.$$

Mutually Exclusive Events If E and F are mutually exclusive events,

$$E \cap F = \emptyset \quad \text{and} \quad P(E \cap F) = 0.$$

Union Rule For any events E and F from a sample space S,

$$P(E \cup F) = P(E) + P(F) - P(E \cap F).$$

Complement Rule $P(E) = 1 - P(E') \quad \text{and} \quad P(E') = 1 - P(E)$

Odds The odds in favor of event E are $\dfrac{P(E)}{P(E')}$, where $P(E') \neq 0$.

If the odds favoring event E are m to n, then

$$P(E) = \frac{m}{m + n} \quad \text{and} \quad P(E') = \frac{n}{m + n}.$$

Properties of Probability 1. For any event E in sample space S, $0 \leq P(E) \leq 1$.

2. The sum of the probabilities of all possible distinct outcomes is 1.

Conditional Probability The conditional probability of event E, given that event F has occurred, is

$$P(E|F) = \frac{P(E \cap F)}{P(F)}, \quad \text{where } P(F) \neq 0.$$

For equally likely outcomes, conditional probability is found by reducing the sample space to event F; then

$$P(E|F) = \frac{n(E \cap F)}{n(F)}.$$

Product Rule of Probability If E and F are events, then $P(E \cap F)$ may be found by either of these formulas.

$$P(E \cap F) = P(F) \cdot P(E|F) \quad \text{or} \quad P(E \cap F) = P(E) \cdot P(F|E)$$

Independent Events If E and F are independent events,

$$P(E|F) = P(E), \quad P(F|E) = P(F), \quad \text{and} \quad P(E \cap F) = P(E) \cdot P(F).$$

Bayes' Theorem $P(F_i|E) = \dfrac{P(F_i) \cdot P(E|F_i)}{P(F_1) \cdot P(E|F_1) + P(F_2) \cdot P(E|F_2) + \cdots + P(F_n) \cdot P(E|F_n)}$

KEY TERMS

7.1
set
element (member)
empty set (or null set)
set-builder notation
universal set
subset
tree diagram
Venn diagram
complement

intersection
disjoint sets
union

7.2
union rule for sets

7.3
experiment
trial
outcome

sample space
event
simple event
certain event
impossible event
mutually exclusive events
probability
empirical probability

7.4
union rule for probability

odds
probability distribution

7.5
conditional probability
product rule
independent events
dependent events

7.6
Bayes' theorem

REVIEW EXERCISES

CONCEPT CHECK

Determine whether each of the following statements is true or false, and explain why.

1. A set is a subset of itself.

2. A set has more subsets than it has elements.

3. The union of two sets always has more elements than either set.

4. The intersection of two sets always has fewer elements than either set.

5. The number of elements in the union of two sets can be found by adding the number of elements in each set.

6. The probability of an event is always at least 0 and no larger than 1.

7. The probability of the union of two events can be found by adding the probability of each event.

8. The probability of drawing the Queen of Hearts from a deck of cards is an example of empirical probability.

9. If two events are mutually exclusive, then they are independent.

10. The probability of two independent events can be found by multiplying the probabilities of each event.

11. The probability of an event E given an event F is the same as the probability of F given E.

12. Bayes' theorem can be useful for calculating conditional probability.

PRACTICE AND EXPLORATIONS

Write true or false for each statement.

13. $9 \in \{8, 4, -3, -9, 6\}$
14. $4 \notin \{3, 9, 7\}$
15. $2 \notin \{0, 1, 2, 3, 4\}$
16. $0 \in \{0, 1, 2, 3, 4\}$
17. $\{3, 4, 5\} \subseteq \{2, 3, 4, 5, 6\}$
18. $\{1, 2, 5, 8\} \subseteq \{1, 2, 5, 10, 11\}$
19. $\{3, 6, 9, 10\} \subseteq \{3, 9, 11, 13\}$
20. $\emptyset \subseteq \{1\}$
21. $\{2, 8\} \not\subseteq \{2, 4, 6, 8\}$
22. $0 \subseteq \emptyset$

In Exercises 23–32, let $U = \{a, b, c, d, e, f, g, h\}$, $K = \{c, d, e, f, h\}$, and $R = \{a, c, d, g\}$. Find the following.

23. The number of subsets of K

24. The number of subsets of R

25. K'
26. R'
27. $K \cap R$
28. $K \cup R$
29. $(K \cap R)'$
30. $(K \cup R)'$
31. $\emptyset'$
32. U'

In Exercises 33–38, let

$U = \{$all employees of the K. O. Brown Company$\}$;
$A = \{$employees in the accounting department$\}$;
$B = \{$employees in the sales department$\}$;
$C = \{$female employees$\}$;
$D = \{$employees with an MBA degree$\}$.

Describe each set in words.

33. $A \cap C$
34. $B \cap D$
35. $A \cup D$
36. $A' \cap D$
37. $B' \cap C'$
38. $(B \cup C)'$

Draw a Venn diagram and shade each set.

39. $A \cup B'$
40. $A' \cap B$
41. $(A \cap B) \cup C$
42. $(A \cup B)' \cap C$

Write the sample space S for each experiment, choosing an S with equally likely outcomes, if possible.

43. Rolling a die

44. Drawing a card from a deck containing only the 13 spades

45. Measuring the weight of a person to the nearest half pound (the scale will not measure more than 300 lb)

46. Tossing a coin 4 times

A jar contains 5 balls labeled 3, 5, 7, 9, and 11, respectively, while a second jar contains 4 red and 2 green balls. An experiment consists of pulling 1 ball from each jar, in turn. In Exercises 47–50, write each set using set notation.

47. The sample space

48. Event E: the number on the first ball is greater than 5

49. Event F: the second ball is green

50. Are the outcomes in the sample space in Exercise 47 equally likely?

In Exercises 51–58, find the probability of each event when a single card is drawn from an ordinary deck.

51. A heart

52. A red queen

53. A face card or a heart

54. Black or a face card

55. Red, given that it is a queen

56. A jack, given that it is a face card

57. A face card, given that it is a king

58. A king, given that it is not a face card

59. Describe what is meant by disjoint sets.

60. Describe what is meant by mutually exclusive events.

61. How are disjoint sets and mutually exclusive events related?

62. Define independent events.

63. Are independent events always mutually exclusive? Are they ever mutually exclusive?

64. An uproar has raged since September 1990 over the answer to a puzzle published in *Parade* magazine, a supplement of the Sunday newspaper. In the "Ask Marilyn" column, Marilyn vos Savant answered the following question: "Suppose you're on a game show, and you're given the choice of three doors. Behind one door is a car; behind the others, goats. You pick a door, say number 1, and the host, who knows what's behind the other doors, opens another door, say number 3, which has a goat. He then says to you, 'Do you want to pick door number 2?' Is it to your advantage to take the switch?"

Ms. vos Savant estimates that she has since received some 10,000 letters; most of them, including many from mathematicians and statisticians, disagreed with her answer. Her answer has been debated by both professionals and amateurs and tested in classes at all levels from grade school to graduate school. But by performing the experiment repeatedly, it can be shown that vos Savant's answer was correct. Find the probabilities of getting the car if you switch or do not switch, and then answer the question yourself. (*Hint:* Consider the sample space.) *Source: Parade Magazine.*

Find the odds in favor of a card drawn from an ordinary deck being the following.

65. A club

66. A black jack

67. A red face card or a queen

68. An ace or a club

Find the probabilities of getting the following sums when two fair dice are rolled.

69. 8

70. 0

71. At least 10

72. No more than 5

73. An odd number greater than 8

74. 12, given that the sum is greater than 10

75. 7, given that at least one die shows a 4

76. At least 9, given that at least one die shows a 5

77. Suppose $P(E) = 0.51$, $P(F) = 0.37$, and $P(E \cap F) = 0.22$. Find the following.

a. $P(E \cup F)$ **b.** $P(E \cap F')$ **c.** $P(E' \cup F)$

d. $P(E' \cap F')$

78. An urn contains 10 balls: 4 red and 6 blue. A second urn contains 16 red balls and an unknown number of blue balls. A single ball is drawn from each urn. The probability that both balls are the same color is 0.44. Calculate the number of blue balls in the second urn. Choose one of the following. *Source: Society of Actuaries.*

a. 4   **b.** 20   **c.** 24   **d.** 44   **e.** 64

79. Box A contains 5 red balls and 1 black ball; box B contains 2 red balls and 3 black balls. A box is chosen, and a ball is selected from it. The probability of choosing box A is 3/8. If the selected ball is black, what is the probability that it came from box A?

80. Find the probability that the ball in Exercise 79 came from box B, given that it is red.

APPLICATIONS

Business and Economics

Appliance Repairs Of the appliance repair shops listed in the phone book, 80% are competent and 20% are not. A competent shop can repair an appliance correctly 95% of the time; an incompetent shop can repair an appliance correctly 55% of the time. Suppose an appliance was repaired correctly. Find the probabilities that it was repaired by the following.

81. A competent shop

82. An incompetent shop

Suppose an appliance was repaired incorrectly. Find the probabilities that it was repaired by the following.

83. A competent shop

84. An incompetent shop

85. Find the probability that an appliance brought to a shop chosen at random is repaired correctly.

86. Are the events that a repair shop is competent and that the repair is done correctly independent? Explain.

87. Sales A company sells printers and copiers. Let E be the event "a customer buys a printer," and let F be the event "a customer buys a copier." Write the following using $\cap$, $\cup$, or ' as necessary.

a. A customer buys neither machine.

b. A customer buys at least one of the machines.

88. Defective Items A sample shipment of five hair dryers is chosen at random. The probability of exactly 0, 1, 2, 3, 4, or 5 hair dryers being defective is given in the following table.

Number Defective	0	1	2	3	4	5
Probability	0.34	0.26	0.18	0.12	0.07	0.03

Find the probabilities that the following numbers of hair dryers are defective.

a. No more than 3   **b.** At least 3

89. Defective Items A manufacturer buys items from four different suppliers. The fraction of the total number of items that is obtained from each supplier, along with the probability that an item purchased from that supplier is defective, is shown in the table below.

Supplier	Fraction of Total Supplied	Probability of Defective
1	0.17	0.01
2	0.39	0.02
3	0.35	0.05
4	0.09	0.03

a. Find the probability that a randomly selected item is defective.

b. Find the probability that a defective item came from supplier 4.

c. Find the probability that a defective item came from supplier 2.

d. Are the events that an item came from supplier 4 and that the item is defective independent? Explain.

90. Car Buyers The table shows the results of a survey of buyers of a certain model of car.

Car Type	Satisfied	Not Satisfied	Totals
New	300	100	
Used	450		600
Totals		250	

a. Complete the table.

b. How many buyers were surveyed?

c. How many bought a new car and were satisfied?

d. How many were not satisfied?

e. How many bought used cars?

f. How many of those who were not satisfied had purchased a used car?

g. Rewrite the event stated in part f using the expression "given that."

h. Find the probability of the outcome in parts f and g.

i. Find the probability that a used-car buyer is not satisfied.

j. You should have different answers in parts h and i. Explain why.

k. Are the events that a car is new and that the customer is satisfied independent? Explain.

91. Auto Insurance An insurance company examines its pool of auto insurance customers and gathers the following information:

(i) All customers insure at least one car.

(ii) 70% of the customers insure more than one car.

(iii) 20% of the customers insure a sports car.

(iv) Of those customers who insure more than one car, 15% insure a sports car.

Calculate the probability that a randomly selected customer insures exactly one car and that car is not a sports car. Choose one of the following. (*Hint:* Draw a tree diagram, and let x be the probability that a customer who insures exactly one car insures a sports car.) *Source: Society of Actuaries.*

a. 0.13 **b.** 0.21 **c.** 0.24 **d.** 0.25 **e.** 0.30

92. Auto Insurance An auto insurance company has 10,000 policyholders. Each policyholder is classified as:

(i) young or old;

(ii) male or female; and

(iii) married or single.

Of these policyholders, 3000 are young, 4600 are male, and 7000 are married. The policyholders can also be classified as 1320 young males, 3010 married males, and 1400 young married persons. Finally, 600 of the policyholders are young married males. How many of the company's policyholders are young, female, and single? Choose one of the following. *Source: Society of Actuaries.*

a. 280 **b.** 423 **c.** 486 **d.** 880 **e.** 896

93. Auto Insurance An actuary studying the insurance preferences of automobile owners makes the following conclusions:

(i) An automobile owner is twice as likely to purchase collision coverage as disability coverage.

(ii) The event that an automobile owner purchases collision coverage is independent of the event that he or she purchases disability coverage.

(iii) The probability that an automobile owner purchases both collision and disability coverages is 0.15.

What is the probability that an automobile owner purchases neither collision nor disability coverage? Choose one of the following. *Source: Society of Actuaries.*

a. 0.18 **b.** 0.33 **c.** 0.48 **d.** 0.67 **e.** 0.82

94. Insurance An insurance company estimates that 40% of policyholders who have only an auto policy will renew next year and 60% of policyholders who have only a homeowners policy will renew next year. The company estimates that 80% of policyholders who have both an auto and a homeowners policy will renew at least one of these policies next year. Company records show that 65% of policyholders have an auto policy, 50% of policyholders have a homeowners policy, and 15% of policyholders have both an auto and a homeowners policy. Using the company's estimates, calculate the percentage of policyholders that will renew at least one policy next year. Choose one of the following. *Source: Society of Actuaries.*

a. 20 **b.** 29 **c.** 41 **d.** 53 **e.** 70

Life Sciences

95. Sickle Cell Anemia The square on the next page shows the four possible (equally likely) combinations when both parents are carriers of the sickle cell anemia trait. Each carrier parent has normal cells (N) and trait cells (T).

a. Complete the table.

b. If the disease occurs only when two trait cells combine, find the probability that a child born to these parents will have sickle cell anemia.

c. The child will carry the trait but not have the disease if a normal cell combines with a trait cell. Find this probability.

d. Find the probability that the child is neither a carrier nor has the disease.

	Second Parent	
	N_2	T_2
First Parent $\quad N_1$		$N_1 T_2$
$\qquad\qquad\qquad T_1$		

96. Blood Antigens In Exercise 44 of Section 7.2, we described the eight types of human blood. The percentage of the population having each type is as follows:

O^+: 38%; O^-: 8%; A^+: 32%; A^-: 7%;
B^+: 9%; B^-: 2%; AB^+: 3%; AB^-: 1%.

When a person receives a blood transfusion, it is important that the blood be compatible, which means that it introduces no new antigens into the recipient's blood. The following diagram helps illustrate what blood types are compatible. *Source: The Mathematics Teacher.*

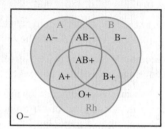

The universal blood type is O^-, since it has none of the additional antigens. The circles labeled A, B, and Rh contain blood types with the A antigen, B antigen, and Rh antigen, respectively. A person with O^- blood can only be transfused with O^- blood, because any other type would introduce a new antigen. Thus the probability that blood from a random donor is compatible is just 8%. A person with AB^+ blood already has all antigens, so the probability that blood from a random donor is compatible is 100%. Find the probability that blood from a random donor is compatible with a person with each blood type.

a. O^+ **b.** A^+ **c.** B^+

d. A^- **e.** B^- **f.** AB^-

97. Risk Factors An actuary is studying the prevalence of three health risk factors, denoted by A, B, and C, within a population of women. For each of the three factors, the probability is 0.1 that a woman in the population has only this risk factor (and no others). For any two of the three factors, the probability is 0.12 that she has exactly these two risk factors (but not the other). The probability that a woman has all three risk factors, given that she has A and B, is 1/3. What is the probability that a woman has none of the three risk factors, given that she does not have risk factor A? Choose one of the following. *Source: Society of Actuaries.*

a. 0.280   **b.** 0.311   **c.** 0.467   **d.** 0.484   **e.** 0.700

Social Sciences

98. Elections In the 2008 presidential election, over 130 million people voted, of which 46.3% were male and 53.7% were female. Of the male voters, 49% voted for Barack Obama and 48% voted for John McCain. Of the female voters, 56% voted for Obama and 43% voted for McCain. *Source: U.S. Census and The New York Times.*

a. Find the percentage of voters who voted for Obama.

b. Find the probability that a randomly selected voter for Obama was male.

c. Find the probability that a randomly selected voter for Obama was female.

99. Television Viewing Habits A telephone survey of television viewers revealed the following information:

20 watch situation comedies;
19 watch game shows;
27 watch movies;
19 watch movies but not game shows;
15 watch situation comedies but not game shows;
10 watch both situation comedies and movies;
3 watch all three;
7 watch none of these.

a. How many viewers were interviewed?

b. How many viewers watch comedies and movies but not game shows?

c. How many viewers watch only movies?

d. How many viewers do not watch movies?

100. Randomized Response Method for Getting Honest Answers to Sensitive Questions There are many personal questions that most people would rather not answer. In fact, when a person is asked such a question, a common response is to provide a false answer. In 1965, Stanley Warner developed a method to assure that the identity of an individual who answers a question remains anonymous, thus assuring an honest answer to the question. For this method, instead of one sensitive question being asked to a person, there are two questions, one sensitive and one nonsensitive, and which question the person answers depends on a randomized procedure. The interviewer, who doesn't know which question a person is answering, simply records the answer given as either "Yes" or "No." In this way, there is no way of knowing to which question the person answered "Yes" or "No." Then, using conditional probability, we can estimate the percentage of people who answer "Yes" to the sensitive question. For example, suppose that the two questions are

    A: Does your birth year end in an odd digit? (Nonsensitive)
    B: Have you ever intentionally cheated on an examination? (Sensitive)

We already know that $P(\text{"Yes"} \mid \text{Question } A) = 1/2$, since half the years are even and half are odd. The answer we seek is $P(\text{"Yes"} \mid \text{Question } B)$. In this experiment, a student is asked to flip a coin and answer question A if the coin comes up heads and otherwise answer question B. Note that the interviewer does not know the outcome of the coin flip or which question is being answered. The percentage of students answering "Yes" is used to approximate $P(\text{"Yes"})$, which is then used to estimate the percentage of students who have cheated on an examination. *Source: Journal of the American Statistical Association.*

a. Use the fact that the event "Yes" is the union of the event "Yes and Question A" with the event "Yes and Question B" to prove that

$$P(\text{"Yes"} \mid \text{Question } B)$$
$$= \frac{P(\text{"Yes"}) - P(\text{"Yes"} \mid \text{Question } A) \cdot P(\text{Question } A)}{P(\text{Question } B)}.$$

b. If this technique is tried on 100 subjects and 60 answered "Yes," what is the approximated probability that a person randomly selected from the group has intentionally cheated on an examination?

101. Police Lineup To illustrate the difficulties with eyewitness identifications from police lineups, John Allen Paulos considers a "lineup" of three pennies, in which we know that two are fair (innocent) and the third (the culprit) has a 75% probability of landing heads. The probability of picking the culprit by chance is, of course, 1/3. Suppose we observe three heads in a row on one of the pennies. If we then guess that this penny is the culprit, what is the probability that we're right? *Source: Who's Counting.*

102. SIDS On July 15, 2005, a panel in England ruled that Roy Meadow, a renowned expert on child abuse and co-founder of London's Royal College of Paediatrics and Child Health, should be erased from the register of physicians in Britain for his faulty statistics at the trial of Sally Clark, who was convicted of murdering her first two babies. Meadow testified at the trial that the probability of a baby dying of sudden death syndrome (SIDS) is 1/8543. He then calculated that the probability of two babies in a family dying of SIDS is $(1/8543)^2 \approx 1/73,000,000$. With such a small probability of both babies dying of SIDS, he concluded that the babies were instead murdered. What assumption did Meadow make in doing this calculation? Discuss reasons why this assumption may be invalid. (*Note:* Clark spent three years in prison before her conviction was reversed.) *Source: Science.*

Physical Sciences

103. Earthquake It has been reported that government scientists have predicted that the odds for a major earthquake occurring in the San Francisco Bay area during the next 30 years are 9 to 1. What is the probability that a major earthquake will occur during the next 30 years in San Francisco? *Source: The San Francisco Chronicle.*

General Interest

104. Making a First Down A first down is desirable in football—it guarantees four more plays by the team making it, assuming no score or turnover occurs in the plays. After getting a first down, a team can get another by advancing the ball at least 10 yards. During the four plays given by a first down, a team's position will be indicated by a phrase such as "third and 4," which means that the team has already had two of its four plays, and that 4 more yards are needed to get 10 yards necessary for another first down. An article in a management journal offers the following results for 189 games for a particular National Football League season. "Trials" represents the number of times a team tried to make a first down, given that it was currently playing either a third or a fourth down. Here, *n* represents the number of yards still needed for a first down. *Source: Management Science.*

n	Trials	Successes	Probability of Making First Down with n Yards to Go
1	543	388	
2	327	186	
3	356	146	
4	302	97	
5	336	91	

a. Complete the table.

b. Why is the sum of the answers in the table not equal to 1?

105. States Of the 50 United States, the following is true:

24 are west of the Mississippi River (western states);*

22 had populations less than 3.6 million in the 2010 census (small states);

26 begin with the letters A through M (early states);

9 are large late (beginning with the letters N through Z) eastern states;

14 are small western states;

11 are small early states;

7 are small early western states.

a. How many western states had populations more than 3.6 million in the 2010 census and begin with the letters N through Z?

b. How many states east of the Mississippi had populations more than 3.6 million in the 2010 census?

106. Music Country-western songs often emphasize three basic themes: love, prison, and trucks. A survey of the local country-western radio station produced the following data:

12 songs were about a truckdriver who was in love while in prison;

13 were about a prisoner in love;

28 were about a person in love;

18 were about a truckdriver in love;

33 were about people not in prison;

18 were about prisoners;

15 were about truckdrivers who were in prison;

16 were about truckdrivers who were not in prison.

a. How many songs were surveyed?

Find the number of songs about

b. truckdrivers;

c. prisoners who are not truckdrivers or in love;

d. prisoners who are in love but are not truckdrivers;

e. prisoners who are not truckdrivers;

f. people not in love.

107. Gambling The following puzzle was featured on the Puzzler part of the radio program *Car Talk* on February 23, 2002. A con man puts three cards in a bag; one card is green on both sides, one is red on both sides, and the third is green on one side and red on the other. He lets you pick one card out of the bag and put it on a table, so you can see that a red side is face up, but neither of you can see the other side. He offers to bet you even money that the other side is also red. In other words, if you bet $1, you lose if the other side is red but get back $2 if the other side is green. Is this a good bet? What is the probability that the other side is red? *Source: Car Talk.*

108. Missiles In his novel *Debt of Honor*, Tom Clancy writes the following:

"There were ten target points—missile silos, the intelligence data said, and it pleased the Colonel [Zacharias] to be eliminating the hateful things, even though the price of that was the lives of other

*We count here states such as Minnesota, which has more than half of its area to the west of the Mississippi.

men. There were only three of them [bombers], and his bomber, like the others, carried only eight weapons [smart bombs]. The total number of weapons carried for the mission was only twenty-four, with two designated for each silo, and Zacharias's last four for the last target. Two bombs each. Every bomb had a 95% probability of hitting within four meters of the aim point, pretty good numbers really, except that this sort of mission had precisely no margin for error. Even the paper probability was less than half a percent chance of a double miss, but that number times ten targets meant a 5% chance that [at least] one missile would survive, and that could not be tolerated." *Source: Debt of Honor.*

Determine whether the calculations in this quote are correct by the following steps.

a. Given that each bomb had a 95% probability of hitting the missile silo on which it was dropped, and that two bombs were dropped on each silo, what is the probability of a double miss?

b. What is the probability that a specific silo was destroyed (that is, that at least one bomb of the two bombs struck the silo)?

c. What is the probability that all ten silos were destroyed?

d. What is the probability that at least one silo survived? Does this agree with the quote?

e. What assumptions need to be made for the calculations in parts a through d to be valid? Discuss whether these assumptions seem reasonable.

109. Viewing Habits A survey of a group's viewing habits over the last year revealed the following information:

  (i) 28% watched gymnastics;

  (ii) 29% watched baseball;

  (iii) 19% watched soccer;

  (iv) 14% watched gymnastics and baseball;

  (v) 12% watched baseball and soccer;

  (vi) 10% watched gymnastics and soccer;

  (vii) 8% watched all three sports.

Calculate the percentage of the group that watched none of the three sports during the last year. Choose one of the following. *Source: Society of Actuaries.*

a. 24   **b.** 36   **c.** 41   **d.** 52   **e.** 60

EXTENDED APPLICATION

MEDICAL DIAGNOSIS

When a patient is examined, information (typically incomplete) is obtained about his or her state of health. Probability theory provides a mathematical model appropriate for this situation, as well as a procedure for quantitatively interpreting such partial information to arrive at a reasonable diagnosis. *Source: Some Mathematical Models in Biology.*

To develop a model, we list the states of health that can be distinguished in such a way that the patient can be in one and only one state at the time of the examination. For each state of health H, we associate a number, $P(H)$, between 0 and 1 such that the sum of all these numbers is 1. This number $P(H)$ represents the probability, before examination, that a patient is in the state of health H, and $P(H)$ may be chosen subjectively from medical experience, using any information available prior to the examination. The probability may be most conveniently established from clinical records; that is, a mean probability is established for patients in general, although the number would vary from patient to patient. Of course, the more information that is brought to bear in establishing $P(H)$, the better the diagnosis.

For example, limiting the discussion to the condition of a patient's heart, suppose there are exactly three states of health, with probabilities as follows.

	State of Health, H	$P(H)$
H_1	Patient has a normal heart	0.8
H_2	Patient has minor heart irregularities	0.15
H_3	Patient has a severe heart condition	0.05

Having selected $P(H)$, the information from the examination is processed. First, the results of the examination must be classified. The examination itself consists of observing the state of a number of characteristics of the patient. Let us assume that the examination for a heart condition consists of a stethoscope examination and a cardiogram. The outcome of such an examination, C, might be one of the following:

C_1 = stethoscope shows normal heart
    and cardiogram shows normal heart;

C_2 = stethoscope shows normal heart
    and cardiogram shows minor irregularities;

and so on.

It remains to assess for each state of health H the conditional probability $P(C|H)$ of each examination outcome C using only the knowledge that a patient is in a given state of health. (This may be based on the medical knowledge and clinical experience of the doctor.) The conditional probabilities $P(C|H)$ will not vary from patient to patient (although they should be reviewed periodically), so that they may be built into a diagnostic system.

Suppose the result of the examination is C_1. Let us assume the following probabilities:

$$P(C_1|H_1) = 0.9,$$
$$P(C_1|H_2) = 0.4,$$
$$P(C_1|H_3) = 0.1.$$

Now, for a given patient, the appropriate probability associated with each state of health H, after examination, is $P(H|C)$, where C is the outcome of the examination. This can be calculated by using Bayes' theorem. For example, to find $P(H_1|C_1)$—that is, the probability that

the patient has a normal heart given that the examination showed a normal stethoscope examination and a normal cardiogram—we use Bayes' theorem as follows:

$P(H_1|C_1)$

$$= \frac{P(C_1|H_1)P(H_1)}{P(C_1|H_1)P(H_1) + P(C_1|H_2)P(H_2) + P(C_1|H_3)P(H_3)}$$

$$= \frac{(0.9)(0.8)}{(0.9)(0.8) + (0.4)(0.15) + (0.1)(0.05)} \approx 0.92.$$

Hence, the probability is about 0.92 that the patient has a normal heart on the basis of the examination results. This means that in 8 out of 100 patients, some abnormality will be present and not be detected by the stethoscope or the cardiogram.

EXERCISES

1. Find $P(H_2|C_1)$.

2. Assuming the following probabilities, find $P(H_1|C_2)$.

$$P(C_2|H_1) = 0.2 \qquad P(C_2|H_2) = 0.8 \qquad P(C_2|H_3) = 0.3$$

3. Assuming the probabilities of Exercise 2, find $P(H_3|C_2)$.

DIRECTIONS FOR GROUP PROJECT

Find an article on medical decision making from a medical journal and develop a doctor–patient scenario for that particular decision. Then create a role-playing activity where the doctor and nurse present the various options and the mathematics associated with making such a decision to a patient. Make sure to present the mathematics in a manner that the average patient might understand. (Hint: Many leading medical journals include articles on medical decision making. One particular journal that certainly includes such research is Medical Decision Making.)

8 Counting Principles; Further Probability Topics

If you have 31 ice cream flavors available, how many different three-scoop cones can you make? The answer, which is surprisingly large, involves counting permutations or combinations, the subject of the first two sections in this chapter. The counting formulas we will develop have important applications in probability theory.

f we flip two coins and record heads or tails for each coin, we can list the four possible outcomes in the sample space as $S = \{hh, ht, th, tt\}$. But what if we flip 10 coins? As we shall see, there are over 1000 possible outcomes, which makes a list of the sample space impractical. Fortunately, there are methods for counting the outcomes in a set without actually listing them. We introduce these methods in the first two sections, and then we use this approach in the third section to find probabilities. In the fourth section, we introduce binomial experiments, which consist of repeated independent trials with only two possible outcomes, and develop a formula for binomial probabilities. The final section continues the discussion of probability distributions that we began in Chapter 7.

8.1 The Multiplication Principle; Permutations

APPLY IT **In how many ways can seven panelists be seated in a row of seven chairs?** *Before we answer this question in Example 8, we will begin with a simpler example.*

Suppose a breakfast consists of one pastry, for which you can choose a bagel, a muffin, or a donut, and one beverage, for which you can choose coffee or juice. How many different breakfasts can you have? For each of the 3 pastries there are 2 different beverages, or a total of $3 \cdot 2 = 6$ different breakfasts, as shown in Figure 1. This example illustrates a general principle of counting, called the **multiplication principle.**

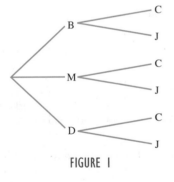

FIGURE 1

Multiplication Principle

Suppose n choices must be made, with

$$m_1 \text{ ways to make choice 1,}$$
$$m_2 \text{ ways to make choice 2,}$$

and so on, with

$$m_n \text{ ways to make choice } n.$$

Then there are

$$m_1 \cdot m_2 \cdot \cdots \cdot m_n$$

different ways to make the entire sequence of choices.

EXAMPLE 1 **Combination Lock**

A certain combination lock can be set to open to any 3-letter sequence.

(a) How many sequences are possible?

SOLUTION Since there are 26 letters in the alphabet, there are 26 choices for each of the 3 letters. By the multiplication principle, there are $26 \cdot 26 \cdot 26 = 17{,}576$ different sequences.

(b) How many sequences are possible if no letter is repeated?

SOLUTION There are 26 choices for the first letter. It cannot be used again, so there are 25 choices for the second letter and then 24 choices for the third letter. Consequently, the number of such sequences is $26 \cdot 25 \cdot 24 = 15{,}600$. **TRY YOUR TURN 1**

YOUR TURN 1 A combination lock can be set to open to any 4-digit sequence. How many sequences are possible? How many sequences are possible if no digit is repeated?

EXAMPLE 2   Morse Code

Morse code uses a sequence of dots and dashes to represent letters and words. How many sequences are possible with at most 3 symbols?

SOLUTION   "At most 3" means "1 or 2 or 3" here. Each symbol may be either a dot or a dash. Thus the following number of sequences are possible in each case.

Morse Code	
Number of Symbols	**Number of Sequences**
1	**2**
2	$2 \cdot 2 = 4$
3	$2 \cdot 2 \cdot 2 = 8$

Altogether, $2 + 4 + 8 = 14$ different sequences are possible.

EXAMPLE 3   I Ching

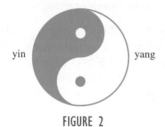

yin   yang

FIGURE 2

An ancient Chinese philosophical work known as the *I Ching (Book of Changes)* is often used as an oracle from which people can seek and obtain advice. The philosophy describes the duality of the universe in terms of two primary forces: *yin* (passive, dark, receptive) and *yang* (active, light, creative). Figure 2 shows the traditional symbol for yin and yang. The yin energy can also be represented by a broken line (– –) and the yang by a solid line (—). These lines are written on top of one another in groups of three, known as *trigrams.* For example, the trigram $\equiv\!\equiv$ is called *Tui,* the Joyous, and has the image of a lake.

(a) How many trigrams are there altogether?

   SOLUTION   Think of choosing between the 2 types of lines for each of the 3 positions in the trigram. There will be 2 choices for each position, so there are $2 \cdot 2 \cdot 2 = 8$ different trigrams.

(b) The trigrams are grouped together, one on top of the other, in pairs known as *hexagrams.* Each hexagram represents one aspect of the *I Ching* philosophy. How many hexagrams are there?

   SOLUTION   For each position in the hexagram there are 8 possible trigrams, giving $8 \cdot 8 = 64$ hexagrams.

EXAMPLE 4   Books

A teacher has 5 different books that he wishes to arrange side by side. How many different arrangements are possible?

SOLUTION   Five choices will be made, one for each space that will hold a book. Any of the 5 books could be chosen for the first space. There are 4 choices for the second space, since 1 book has already been placed in the first space; there are 3 choices for the third space, and so on. By the multiplication principle, the number of different possible arrangements is $5 \cdot 4 \cdot 3 \cdot 2 \cdot 1 = 120$.   **TRY YOUR TURN 2**

YOUR TURN 2   A teacher is lining up 8 students for a spelling bee. How many different line-ups are possible?

─FOR REVIEW─────

The natural numbers, also referred to as the positive integers, are the numbers 1, 2, 3, 4, etc.

The use of the multiplication principle often leads to products such as $5 \cdot 4 \cdot 3 \cdot 2 \cdot 1$, the product of all the natural numbers from 5 down to 1. If n is a natural number, the symbol $n!$ (read "*n factorial*") denotes the product of all the natural numbers from n down to 1. If $n = 1$, this formula is understood to give $1! = 1$.

Factorial Notation

For any natural number n,

$$n! = n(n-1)(n-2) \cdots (3)(2)(1).$$

Also, by definition,

$$0! = 1.$$

With this symbol, the product $5 \cdot 4 \cdot 3 \cdot 2 \cdot 1$ can be written as $5!$. Also, $3! = 3 \cdot 2 \cdot 1 = 6$. The definition of $n!$ could be used to show that $n[(n-1)]! = n!$ for all natural numbers $n \geq 2$. It is helpful if this result also holds for $n = 1$. This can happen only if $0!$ equals 1, as defined above.

As n gets large, $n!$ grows very quickly. For example, $5! = 120$, while $10! = 3,628,800$. Most calculators can be used to determine $n!$ for small values of n. A calculator with a 10-digit display and scientific notation capability will usually give the exact value of $n!$ for $n \leq 13$, and approximate values of $n!$ for $14 \leq n \leq 69$. The value of $70!$ is approximately 1.198×10^{100}, which is too large for most calculators. To get a sense of how large $70!$ is, suppose a computer counted the numbers from 1 to $70!$ at a rate of 1 billion numbers per second. If the computer started when the universe began, by now it would have calculated only a tiny fraction of the total.

TECHNOLOGY NOTE | On many graphing calculators, the factorial of a number is accessible through a menu. On the TI-84 Plus, for example, this menu is found by pressing the MATH key, selecting PRB (for probability), and then selecting the !.

YOUR TURN 3 A teacher wishes to place 5 out of 8 different books on her shelf. How many arrangements of 5 books are possible?

EXAMPLE 5 Books

Suppose the teacher in Example 4 wishes to place only 3 of the 5 books on his desk. How many arrangements of 3 books are possible?

SOLUTION The teacher again has 5 ways to fill the first space, 4 ways to fill the second space, and 3 ways to fill the third. Since he wants to use only 3 books, only 3 spaces can be filled (3 events) instead of 5, for $5 \cdot 4 \cdot 3 = 60$ arrangements. **TRY YOUR TURN 3**

Permutations

The answer 60 in Example 5 is called the number of *permutations* of 5 things taken 3 at a time. A **permutation** of r (where $r \geq 1$) elements from a set of n elements is any specific ordering or arrangement, *without repetition*, of the r elements. Each rearrangement of the r elements is a different permutation. The number of permutations of n things taken r at a time (with $r \leq n$) is written $P(n, r)$.* Based on the work in Example 5,

$$P(5, 3) = 5 \cdot 4 \cdot 3 = 60.$$

Factorial notation can be used to express this product as follows.

$$5 \cdot 4 \cdot 3 = 5 \cdot 4 \cdot 3 \cdot \frac{2 \cdot 1}{2 \cdot 1} = \frac{5 \cdot 4 \cdot 3 \cdot 2 \cdot 1}{2 \cdot 1} = \frac{5!}{2!} = \frac{5!}{(5-3)!}$$

This example illustrates the general rule of permutations, which can be stated as follows.

Permutations

If $P(n, r)$ (where $r \leq n$) is the number of permutations of n elements taken r at a time, then

$$P(n, r) = \frac{n!}{(n-r)!}.$$

*An alternate notation for $P(n, r)$ is $_nP_r$.

CAUTION | The letter P here represents *permutations*, not *probability*. In probability notation, the quantity in parentheses describes an *event*. In permutations notation, the quantity in parentheses always comprises *two numbers*.

The proof of the permutations rule follows the discussion in Example 5. There are n ways to choose the first of the r elements, $n - 1$ ways to choose the second, and $n - r + 1$ ways to choose the rth element, so that

$$P(n, r) = n(n - 1)(n - 2) \cdots (n - r + 1).$$

Now multiply on the right by $(n - r)!/(n - r)!$.

$$P(n, r) = n(n - 1)(n - 2) \cdots (n - r + 1) \cdot \frac{(n - r)!}{(n - r)!}$$

$$= \frac{n(n - 1)(n - 2) \cdots (n - r + 1)(n - r)!}{(n - r)!}$$

$$= \frac{n!}{(n - r)!}$$

Because we defined 0! equal to 1, the formula for permutations gives the special case

$$P(n, n) = \frac{n!}{(n - n)!} = \frac{n!}{0!} = \frac{n!}{1} = n!.$$

This result also follows from the multiplication principle, because $P(n, n)$ gives the number of permutations of n objects, and there are n choices for the first object, $n - 1$ for the second, and so on, down to just 1 choice for the last object. Example 4 illustrated this idea.

> $P(n, n)$
>
> The number of permutations of a set with n elements is $n!$; that is $P(n, n) = n!$.

EXAMPLE 6 Permutations of Letters

Find the following.

(a) The number of permutations of the letters A, B, and C.

SOLUTION By the formula $P(n, n) = n!$, with $n = 3$,

$$P(3, 3) = 3! = 3 \cdot 2 \cdot 1 = 6.$$

The 6 *permutations* (or *arrangements*) are ABC, ACB, BAC, BCA, CAB, CBA.

(b) The number of permutations if just 2 of the letters A, B, and C are to be used

SOLUTION Find $P(3, 2)$.

$$P(3, 2) = \frac{3!}{(3 - 2)!} = \frac{3!}{1!} = 3! = 6$$

This result is exactly the same answer as in part (a). This is because, in the case of $P(3, 3)$, after the first 2 choices are made, the third is already determined, as shown in the table below.

YOUR TURN 4 Find the number of permutations of the letters L, M, N, O, P, and Q, if just three of the letters are to be used.

Permutations of Three Letters						
First Two Letters	AB	AC	BA	BC	CA	CB
Third Letter	C	B	C	A	B	A

TRY YOUR TURN 4

To find $P(n, r)$, we can use either the permutations formula or direct application of the multiplication principle, as the following example shows.

EXAMPLE 7   Politics

In a recent election, eight candidates sought the Democratic nomination for president. In how many ways could voters rank their first, second, and third choices?

Method 1
Calculating By Hand

SOLUTION   This is the same as finding the number of permutations of 8 elements taken 3 at a time. Since there are 3 choices to be made, the multiplication principle gives $P(8, 3) = 8 \cdot 7 \cdot 6 = 336$. Alternatively, use the permutations formula to get

$$P(8, 3) = \frac{8!}{(8 - 3)!} = \frac{8!}{5!} = \frac{8 \cdot 7 \cdot 6 \cdot 5 \cdot 4 \cdot 3 \cdot 2 \cdot 1}{5 \cdot 4 \cdot 3 \cdot 2 \cdot 1} = 8 \cdot 7 \cdot 6 = 336.$$

Method 2
Graphing Calculator

Graphing calculators have the capacity to compute permutations. For example, on a TI-84 Plus, $P(8, 3)$ can be calculated by inputting 8 followed by `nPr` (found in the `MATH-PRB` menu), and a 3 yielding 336, as shown in Figure 3.

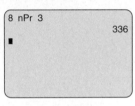

FIGURE 3

Method 3
Spreadsheet

Spreadsheets can also compute permutations. For example, in Microsoft Excel, $P(8, 3)$ can be calculated by inputting 8 and 3 in cells, say, A1 and B1, and then typing "`=FACT(A1)/ FACT(A1-B1)`" in cell C1 or, for that matter, any other cell.

CAUTION   When calculating the number of permutations with the formula, do not try to cancel unlike factorials. For example,

$$\frac{8!}{4!} \neq 2! = 2 \cdot 1 = 2.$$

$$\frac{8!}{4!} = \frac{8 \cdot 7 \cdot 6 \cdot 5 \cdot 4 \cdot 3 \cdot 2 \cdot 1}{4 \cdot 3 \cdot 2 \cdot 1} = 8 \cdot 7 \cdot 6 \cdot 5 = 1680.$$

Always write out the factors first, then cancel where appropriate.

EXAMPLE 8   Television Panel

A televised talk show will include 4 women and 3 men as panelists.

(a) In how many ways can the panelists be seated in a row of 7 chairs?

APPLY IT

   SOLUTION   Find $P(7, 7)$, the total number of ways to seat 7 panelists in 7 chairs.

$$P(7, 7) = 7! = 7 \cdot 6 \cdot 5 \cdot 4 \cdot 3 \cdot 2 \cdot 1 = 5040$$

There are 5040 ways to seat the 7 panelists.

(b) In how many ways can the panelists be seated if the men and women are to be alternated?

   SOLUTION   Use the multiplication principle. In order to alternate men and women, a woman must be seated in the first chair (since there are 4 women and only 3 men), any of the men next, and so on. Thus there are 4 ways to fill the first seat, 3 ways to fill the second seat, 3 ways to fill the third seat (with any of the 3 remaining women), and so on. This gives

$$\underset{W_1 \quad M_1 \quad W_2 \quad M_2 \quad W_3 \quad M_3 \quad W_4}{4 \cdot 3 \cdot 3 \cdot 2 \cdot 2 \cdot 1 \cdot 1} = 144$$

ways to seat the panelists.

(c) In how many ways can the panelists be seated if the men must sit together, and the women must also sit together?

SOLUTION Use the multiplication principle. We first must decide how to arrange the two groups (men and women). There are 2! ways of doing this. Next, there are 4! ways of arranging the women and 3! ways of arranging the men, for a total of

$$2! \, 4! \, 3! = 2 \cdot 24 \cdot 6 = 288$$

ways.

(d) In how many ways can one woman and one man from the panel be selected?

SOLUTION There are 4 ways to pick the woman and 3 ways to pick the man, for a total of

$$4 \cdot 3 = 12$$

ways.

TRY YOUR TURN 5

YOUR TURN 5 Two freshmen, 2 sophomores, 2 juniors, and 3 seniors are on a panel. In how many ways can the panelists be seated if each class must sit together?

If the n objects in a permutation are not all distinguishable—that is, if there are n_1 of type 1, n_2 of type 2, and so on for r different types, then the number of **distinguishable permutations** is

$$\frac{n!}{n_1! \, n_2! \cdots n_r!}.$$

For example, suppose we want to find the number of permutations of the numbers 1, 1, 4, 4, 4. We cannot distinguish between the two 1's or among the three 4's, so using 5! would give too many distinguishable arrangements. Since the two 1's are indistinguishable and account for 2! of the permutations, we divide 5! by 2!. Similarly, we also divide by 3! to account for the three indistinguishable 4's. This gives

$$\frac{5!}{2! \, 3!} = 10$$

permutations.

EXAMPLE 9 Mississippi

In how many ways can the letters in the word *Mississippi* be arranged?

SOLUTION This word contains 1 m, 4 i's, 4 s's, and 2 p's. To use the formula, let $n = 11$, $n_1 = 1$, $n_2 = 4$, $n_3 = 4$, and $n_4 = 2$ to get

$$\frac{11!}{1! \, 4! \, 4! \, 2!} = 34{,}650$$

arrangements.

TRY YOUR TURN 6

YOUR TURN 6 In how many ways can the letters in the word *Tennessee* be arranged?

NOTE If Example 9 had asked for the number of ways that the letters in a word with 11 *different* letters could be arranged, the answer would be 11! = 39,916,800.

EXAMPLE 10 Yogurt

A student buys 3 cherry yogurts, 2 raspberry yogurts, and 2 blueberry yogurts. She puts them in her dormitory refrigerator to eat one a day for the next week. Assuming yogurts of

YOUR TURN 7 A student has 4 pairs of identical blue socks, 5 pairs of identical brown socks, 3 pairs of identical black socks, and 2 pairs of identical white socks. In how many ways can he select socks to wear for the next two weeks?

the same flavor are indistinguishable, in how many ways can she select yogurts to eat for the next week?

SOLUTION This problem is again one of distinguishable permutations. The 7 yogurts can be selected in 7! ways, but since the 3 cherry, 2 raspberry, and 2 blueberry yogurts are indistinguishable, the total number of distinguishable orders in which the yogurts can be selected is

$$\frac{7!}{3!\,2!\,2!} = 210.$$

TRY YOUR TURN 7

8.1 EXERCISES

In Exercises 1–12, evaluate the factorial or permutation.

1. 6!
2. 7!
3. 15!
4. 16!
5. $P(13, 2)$
6. $P(12, 3)$
7. $P(38, 17)$
8. $P(33, 19)$
9. $P(n, 0)$
10. $P(n, n)$
11. $P(n, 1)$
12. $P(n, n - 1)$

13. How many different types of homes are available if a builder offers a choice of 6 basic plans, 3 roof styles, and 2 exterior finishes?

14. A menu offers a choice of 3 salads, 8 main dishes, and 7 desserts. How many different meals consisting of one salad, one main dish, and one dessert are possible?

15. A couple has narrowed down the choice of a name for their new baby to 4 first names and 5 middle names. How many different first- and middle-name arrangements are possible?

16. In a club with 16 members, how many ways can a slate of 3 officers consisting of president, vice-president, and secretary/treasurer be chosen?

17. Define *permutation* in your own words.

18. Explain the difference between *distinguishable* and *indistinguishable* permutations.

19. In Example 6, there are six 3-letter permutations of the letters A, B, and C. How many 3-letter subsets (unordered groups of letters) are there?

20. In Example 6, how many unordered 2-letter subsets of the letters A, B, and C are there?

21. Find the number of distinguishable permutations of the letters in each word.
    a. initial   b. little   c. decreed

22. A printer has 5 A's, 4 B's, 2 C's, and 2 D's. How many different "words" are possible that use all these letters? (A "word" does not have to have any meaning here.)

23. Wing has different books to arrange on a shelf: 4 blue, 3 green, and 2 red.
    a. In how many ways can the books be arranged on a shelf?
    b. If books of the same color are to be grouped together, how many arrangements are possible?

c. In how many distinguishable ways can the books be arranged if books of the same color are identical but need not be grouped together?

d. In how many ways can you select 3 books, one of each color, if the order in which the books are selected does not matter?

e. In how many ways can you select 3 books, one of each color, if the order in which the books are selected matters?

24. A child has a set of differently shaped plastic objects. There are 3 pyramids, 4 cubes, and 7 spheres.

a. In how many ways can she arrange the objects in a row if each is a different color?

b. How many arrangements are possible if objects of the same shape must be grouped together and each object is a different color?

c. In how many distinguishable ways can the objects be arranged in a row if objects of the same shape are also the same color but need not be grouped together?

d. In how many ways can you select 3 objects, one of each shape, if the order in which the objects are selected does not matter and each object is a different color?

e. In how many ways can you select 3 objects, one of each shape, if the order in which the objects are selected matters and each object is a different color?

25. If you already knew the value of 9!, how could you find the value of 10! quickly?

26. Given that 450! is approximately equal to $1.7333687 \times 10^{1000}$ (to 8 digits of accuracy), find 451! to 7 digits of accuracy.

27. When calculating $n!$, the number of ending zeros in the answer can be determined prior to calculating the actual number by finding the number of times 5 can be factored from $n!$. For example, 7! only has one 5 occurring in its calculation, and so there is only one ending zero in 5040. The number 10! has two 5's (one from the 5 and one from the 10) and so there must be two ending zeros in the answer 3,628,800. Use this idea to determine the number of zeros that occur in the following factorials, and then explain why this works.
    a. 13!   b. 27!   c. 75!

28. Because of the view screen, calculators only show a fixed number of digits, often 10 digits. Thus, an approximation of a

number will be shown by only including the 10 largest place values of the number. Using the ideas from the previous exercise, determine if the following numbers are correct or if they are incorrect by checking if they have the correct number of ending zeros. (*Note:* Just because a number has the correct number of zeros does not imply that it is correct.)

a. $12! = 479,001,610$

b. $23! = 25,852,016,740,000,000,000,000$

c. $15! = 1,307,643,680,000$

d. $14! = 87,178,291,200$

29. Some students find it puzzling that $0! = 1$, and think that $0!$ should equal 0. If this were true, what would be the value of $P(4, 4)$ using the permutations formula?

APPLICATIONS

Business and Economics

30. **Automobile Manufacturing** An automobile manufacturer produces 8 models, each available in 7 different exterior colors, with 4 different upholstery fabrics and 5 interior colors. How many varieties of automobile are available?

31. **Marketing** In a recent marketing campaign, Olive Garden Italian Restaurant® offered a "Never-Ending Pasta Bowl." The customer could order an array of pasta dishes, selecting from 7 types of pasta and 6 types of sauce, including 2 with meat.

    a. If the customer selects one pasta type and one sauce type, how many different "pasta bowls" can a customer order?

    b. How many different "pasta bowls" can a customer order without meat?

32. **Investments** Natalie Graham's financial advisor has given her a list of 9 potential investments and has asked her to select and rank her favorite five. In how many different ways can she do this?

33. **Scheduling** A local television station has eleven slots for commercials during a special broadcast. Six restaurants and 5 stores have bought slots for the broadcast.

    a. In how many ways can the commercials be arranged?

    b. In how many ways can the commercials be arranged so that the restaurants are grouped together and the stores are grouped together?

    c. In how many ways can the commercials be arranged so that the restaurant and store commercials are alternating?

Life Sciences

34. **Drug Sequencing** Twelve drugs have been found to be effective in the treatment of a disease. It is believed that the sequence in which the drugs are administered is important in the effectiveness of the treatment. In how many different sequences can 5 of the 12 drugs be administered?

35. **Insect Classification** A biologist is attempting to classify 52,000 species of insects by assigning 3 initials to each species. Is it possible to classify all the species in this way? If not, how many initials should be used?

36. **Science Conference** At an annual college science conference, student presentations are scheduled one after another in the afternoon session. This year, 5 students are presenting in

biology, 5 students are presenting in chemistry, and 2 students are presenting in physics.

    a. In how many ways can the presentations be scheduled?

    b. In how many ways can the presentations be scheduled so that each subject is grouped together?

    c. In how many ways can the presentations be scheduled if the conference must begin and end with a physics presentation?

Social Sciences

37. **Social Science Experiment** In an experiment on social interaction, 6 people will sit in 6 seats in a row. In how many ways can this be done?

38. **Election Ballots** In an election with 3 candidates for one office and 6 candidates for another office, how many different ballots may be printed?

General Interest

39. **Baseball Teams** A baseball team has 19 players. How many 9-player batting orders are possible?

40. **Union Elections** A chapter of union Local 715 has 35 members. In how many different ways can the chapter select a president, a vice-president, a treasurer, and a secretary?

41. **Programming Music** A concert to raise money for an economics prize is to consist of 5 works: 2 overtures, 2 sonatas, and a piano concerto.

    a. In how many ways can the program be arranged?

    b. In how many ways can the program be arranged if an overture must come first?

42. **Programming Music** A zydeco band from Louisiana will play 5 traditional and 3 original Cajun compositions at a concert. In how many ways can they arrange the program if

    a. they begin with a traditional piece?

    b. an original piece will be played last?

43. **Radio Station Call Letters** How many different 4-letter radio station call letters can be made if

    a. the first letter must be K or W and no letter may be repeated?

    b. repeats are allowed, but the first letter is K or W?

    c. the first letter is K or W, there are no repeats, and the last letter is R?

44. **Telephone Numbers** How many 7-digit telephone numbers are possible if the first digit cannot be zero and

    a. only odd digits may be used?

    b. the telephone number must be a multiple of 10 (that is, it must end in zero)?

    c. the telephone number must be a multiple of 100?

    d. the first 3 digits are 481?

    e. no repetitions are allowed?

Telephone Area Codes Several years ago, the United States began running out of telephone numbers. Telephone companies introduced new area codes as numbers were used up, and eventually almost all area codes were used up.

45. a. Until recently, all area codes had a 0 or 1 as the middle digit, and the first digit could not be 0 or 1. How many area

codes are there with this arrangement? How many telephone numbers does the current 7-digit sequence permit per area code? (The 3-digit sequence that follows the area code cannot start with 0 or 1. Assume there are no other restrictions.)

b. The actual number of area codes under the previous system was 152. Explain the discrepancy between this number and your answer to part a.

46. The shortage of area codes was avoided by removing the restriction on the second digit. (This resulted in problems for some older equipment, which used the second digit to determine that a long-distance call was being made.) How many area codes are available under the new system?

47. License Plates   For many years, the state of California used 3 letters followed by 3 digits on its automobile license plates.

  a. How many different license plates are possible with this arrangement?

  b. When the state ran out of new numbers, the order was reversed to 3 digits followed by 3 letters. How many new license plate numbers were then possible?

  c. Several years ago, the numbers described in b were also used up. The state then issued plates with 1 letter followed by 3 digits and then 3 letters. How many new license plate numbers will this provide?

48. Social Security Numbers   A social security number has 9 digits. How many social security numbers are there? The U.S. population in 2010 was about 309 million. Is it possible for every U.S. resident to have a unique social security number? (Assume no restrictions.)

49. Postal Zip Codes   The U.S. Postal Service currently uses 5-digit zip codes in most areas. How many zip codes are possible if there are no restrictions on the digits used? How many would be possible if the first number could not be 0?

50. Postal Zip Codes   The U.S. Postal Service is encouraging the use of 9-digit zip codes in some areas, adding 4 digits after the usual 5-digit code. How many such zip codes are possible with no restrictions?

51. Games   The game of Sets uses a special deck of cards. Each card has either one, two, or three identical shapes, all of the same color and style. There are three possible shapes: squiggle, diamond, and oval. There are three possible colors: green, purple, and red. There are three possible styles: solid, shaded, or outline. The deck consists of all possible combinations of shape, color, style, and number of shapes. How many cards are in the deck? *Source: Sets.*

52. Games   In the game of Scattergories, the players take 12 turns. In each turn, a 20-sided die is rolled; each side has a letter. The players must then fill in 12 categories (e.g., vegetable, city, etc.) with a word beginning with the letter rolled. Considering that a game consists of 12 rolls of the 20-sided die, and that rolling the same side more than once is allowed, how many possible games are there? *Source: Milton Bradley.*

53. Games   The game of Twenty Questions consists of asking 20 questions to determine a person, place, or thing that the other person is thinking of. The first question, which is always "Is it an animal, vegetable, or mineral?" has three possible answers. All the other questions must be answered "Yes" or "No." How many possible objects can be distinguished in this game, assuming that all 20 questions are asked? Are 20 questions enough?

54. Traveling Salesman   In the famous Traveling Salesman Problem, a salesman starts in any one of a set of cities, visits every city in the set once, and returns to the starting city. He would like to complete this circuit with the shortest possible distance.

  a. Suppose the salesman has 10 cities to visit. Given that it does not matter what city he starts in, how many different circuits can he take?

  b. The salesman decides to check all the different paths in part a to see which is shortest, but realizes that a circuit has the same distance whichever direction it is traveled. How many different circuits must he check?

  c. Suppose the salesman has 70 cities to visit. Would it be feasible to have a computer check all the different circuits? Explain your reasoning.

55. Circular Permutations   Circular permutations arise in applications involving arrangements around a closed loop, as in the previous exercise. Here are two examples.

  a. A ferris wheel has 20 seats. How many ways can 20 students arrange themselves on the ferris wheel if each student takes a different seat? We consider two arrangements to be identical if they differ only by rotations of the wheel.

  b. A necklace is to be strung with 15 beads, each of a different color. In how many ways can the beads be arranged? We consider two arrangements to be identical if they differ only by rotations of the necklace or by flipping the necklace over. (*Hint:* If every arrangement is counted twice, the correct number of arrangements can be found by dividing by 2.)

YOUR TURN ANSWERS

1. 10,000; 5040   **2.** 40,320   **3.** 6720   **4.** 120
5. 1152   **6.** 3780   **7.** 2,522,520

8.2 Combinations

APPLY IT   **In how many ways can a manager select 4 employees for promotion from 12 eligible employees?**
As we shall see in Example 5, permutations alone cannot be used to answer this question, but combinations will provide the answer.

In the previous section, we saw that there are 60 ways that a teacher can arrange 3 of 5 different books on his desk. That is, there are 60 permutations of 5 books taken 3 at a time.

Suppose now that the teacher does not wish to arrange the books on his desk but rather wishes to choose, without regard to order, any 3 of the 5 books for a book sale to raise money for his school. In how many ways can this be done?

At first glance, we might say 60 again, but this is incorrect. The number 60 counts all possible *arrangements* of 3 books chosen from 5. The following 6 arrangements, however, would all lead to the same set of 3 books being given to the book sale.

mystery-biography-textbook	biography-textbook-mystery
mystery-textbook-biography	textbook-biography-mystery
biography-mystery-textbook	textbook-mystery-biography

The list shows 6 different *arrangements* of 3 books, but only one *subset* of 3 books. A subset of items listed *without regard to order* is called a **combination**. The number of combinations of 5 things taken 3 at a time is written $C(5, 3)$.* Since they are subsets, combinations are *not ordered*.

To evaluate $C(5, 3)$, start with the $5 \cdot 4 \cdot 3$ *permutations* of 5 things taken 3 at a time. Since combinations are not ordered, find the number of combinations by dividing the number of permutations by the number of ways each group of 3 can be ordered; that is, divide by 3!.

$$C(5, 3) = \frac{5 \cdot 4 \cdot 3}{3!} = \frac{5 \cdot 4 \cdot 3}{3 \cdot 2 \cdot 1} = 10$$

There are 10 ways that the teacher can choose 3 books for the book sale.

Generalizing this discussion gives the following formula for the number of combinations of n elements taken r at a time:

$$C(n, r) = \frac{P(n, r)}{r!}.$$

Another version of this formula is found as follows.

$$C(n, r) = \frac{P(n, r)}{r!}$$

$$= \frac{n!}{(n - r)!} \cdot \frac{1}{r!} \qquad P(n, r) = \frac{n!}{(n - r)!}.$$

$$= \frac{n!}{(n - r)! \, r!}$$

The steps above lead to the following result.

Combinations

If $C(n, r)$, denotes the number of combinations of n elements taken r at a time, where $r \leq n$, then

$$C(n, r) = \frac{n!}{(n - r)! \, r!}.$$

EXAMPLE 1 Committees

How many committees of 3 people can be formed from a group of 8 people?

Method 1
Calculating By Hand

SOLUTION A committee is an unordered group, so use the combinations formula for $C(8, 3)$.

$$C(8, 3) = \frac{8!}{5!3!} = \frac{8 \cdot 7 \cdot 6 \cdot 5 \cdot 4 \cdot 3 \cdot 2 \cdot 1}{5 \cdot 4 \cdot 3 \cdot 2 \cdot 1 \cdot 3 \cdot 2 \cdot 1} = \frac{8 \cdot 7 \cdot 6}{3 \cdot 2 \cdot 1} = 56$$

*Other common notations for $C(n, r)$ are $_nC_r$, C_r^n, and $\binom{n}{r}$.

Method 2
Graphing Calculator

Graphing calculators have the capacity to compute combinations. For example, on a TI-84 Plus, $C(8, 3)$ can be calculated by inputting 8 followed by `nCr` (found in the `MATH-PRB` menu) and a 3 yielding 56, as shown in Figure 4.

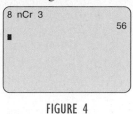

```
8 nCr 3
                    56
■
```

FIGURE 4

Method 3
Spreadsheet

YOUR TURN 1 How many committees of 4 people can be formed from a group of 10 people?

Spreadsheets can also compute combinations. For example, in Microsoft Excel, $C(8, 3)$ can be calculated by inputting 8 and 3 in cells, say, A1 and B1, and then typing "`=FACT(A1)/(FACT(A1-B1)*FACT(B1))`" in cell C1 or, for that matter, any other cell. The command "`= MULTINOMIAL(5, 3)`" also gives this answer.

TRY YOUR TURN 1

Example 1 shows an alternative way to compute $C(n, r)$. Take r or $n - r$, whichever is smaller. Write the factorial of this number in the denominator. In the numerator, write out a sufficient number of factors of $n!$ so there is one factor in the numerator for each factor in the denominator. For example, to calculate $C(8, 3)$ or $C(8, 5)$ write

$$\frac{8 \cdot 7 \cdot 6}{3 \cdot 2 \cdot 1} = 56.$$

The factors that are omitted (written in color in Example 1) cancel out of the numerator and denominator, so need not be included.

Notice from the previous discussion that $C(8, 3) = C(8, 5)$. (See Exercise 25 for a generalization of this idea.) One interpretation of this fact is that the number of ways to form a committee of 3 people chosen from a group of 8 is the same as the number of ways to choose the 5 people who are not on the committee.

Notice that this is *not* true with permutations: $P(8, 3) \neq P(8, 5)$.

EXAMPLE 2 Lawyers

Three lawyers are to be selected from a group of 30 to work on a special project.

(a) In how many different ways can the lawyers be selected?

SOLUTION Here we wish to know the number of 3-element combinations that can be formed from a set of 30 elements. (We want combinations, not permutations, since order within the group of 3 doesn't matter.)

$$C(30, 3) = \frac{30!}{27!3!} = \frac{30 \cdot 29 \cdot 28 \cdot 27!}{27! \cdot 3 \cdot 2 \cdot 1} \qquad 30! = 30 \cdot 29 \cdot 28 \cdot 27!$$
$$= \frac{30 \cdot 29 \cdot 28}{3 \cdot 2 \cdot 1}$$
$$= 4060$$

There are 4060 ways to select the project group.

(b) In how many ways can the group of 3 be selected if a certain lawyer must work on the project?

SOLUTION Since 1 lawyer already has been selected for the project, the problem is reduced to selecting 2 more from the remaining 29 lawyers.

$$C(29, 2) = \frac{29!}{27! \, 2!} = \frac{29 \cdot 28 \cdot 27!}{27! \cdot 2 \cdot 1} = \frac{29 \cdot 28}{2 \cdot 1} = 29 \cdot 14 = 406$$

In this case, the project group can be selected in 406 ways.

(c) In how many ways can a nonempty group of at most 3 lawyers be selected from these 30 lawyers?

SOLUTION Here, by "at most 3" we mean "1 or 2 or 3." (The number 0 is excluded because the group is nonempty.) Find the number of ways for each case.

Case	Number of Ways
1	$C(30, 1) = \dfrac{30!}{29! \, 1!} = \dfrac{30 \cdot 29!}{29! \, (1)} = 30$
2	$C(30, 2) = \dfrac{30!}{28! \, 2!} = \dfrac{30 \cdot 29 \cdot 28!}{28! \cdot 2 \cdot 1} = 435$
3	$C(30, 3) = \dfrac{30!}{27! \, 3!} = \dfrac{30 \cdot 29 \cdot 28 \cdot 27!}{27! \cdot 3 \cdot 2 \cdot 1} = 4060$

YOUR TURN 2 From a class of 15 students, a group of 3 or 4 students will be selected to work on a special project. In how many ways can a group of 3 or 4 students be selected?

The total number of ways to select at most 3 lawyers will be the sum

$$30 + 435 + 4060 = 4525.$$ **TRY YOUR TURN 2**

EXAMPLE 3 Sales

A salesman has 10 accounts in a certain city.

(a) In how many ways can he select 3 accounts to call on?

SOLUTION Within a selection of 3 accounts, the arrangement of the calls is not important, so there are

$$C(10, 3) = \frac{10!}{7! \, 3!} = \frac{10 \cdot 9 \cdot 8}{3 \cdot 2 \cdot 1} = 120$$

ways he can make a selection of 3 accounts.

(b) In how many ways can he select at least 8 of the 10 accounts to use in preparing a report?

SOLUTION "At least 8" means "8 or more," which is "8 or 9 or 10." First find the number of ways to choose in each case.

FOR REVIEW

Notice in Example 3 that to calculate the number of ways to select 8 or 9 or 10 accounts, we added the three numbers found. The union rule for sets from Chapter 7 says that when A and B are disjoint sets, the number of elements in A or B is the number of elements in A plus the number in B.

Case	Number of Ways
8	$C(10, 8) = \dfrac{10!}{2! \, 8!} = \dfrac{10 \cdot 9}{2 \cdot 1} = 45$
9	$C(10, 9) = \dfrac{10!}{1! \, 9!} = \dfrac{10}{1} = 10$
10	$C(10, 10) = \dfrac{10!}{0! \, 10!} = 1$

He can select at least 8 of the 10 accounts in $45 + 10 + 1 = 56$ ways.

CAUTION When we are making a first decision *and* a second decision, we *multiply* to find the total number of ways. When we are making a decision in which the first choice *or* the second choice are valid choices, we *add* to find the total number of ways.

The formulas for permutations and combinations given in this section and in the previous section will be very useful in solving probability problems in the next section. Any difficulty in using these formulas usually comes from being unable to differentiate between them. Both permutations and combinations give the number of ways to choose r objects from a set of n objects. The differences between permutations and combinations are outlined in the following table.

Permutations	Combinations
Different orderings or arrangements of the r objects are different permutations.	Each choice or subset of r objects gives one combination. Order within the group of r objects does not matter.
$$P(n, r) = \frac{n!}{(n-r)!}$$	$$C(n, r) = \frac{n!}{(n-r)!\, r!}$$
Clue words: arrangement, schedule, order	Clue words: group, committee, set, sample
Order matters!	Order does not matter!

In the next examples, concentrate on recognizing which formula should be applied.

EXAMPLE 4 **Permutations and Combinations**

For each problem, tell whether permutations or combinations should be used to solve the problem.

(a) How many 4-digit code numbers are possible if no digits are repeated?

SOLUTION Since changing the order of the 4 digits results in a different code, use permutations.

(b) A sample of 3 light bulbs is randomly selected from a batch of 15. How many different samples are possible?

SOLUTION The order in which the 3 light bulbs are selected is not important. The sample is unchanged if the items are rearranged, so combinations should be used.

(c) In a baseball conference with 8 teams, how many games must be played so that each team plays every other team exactly once?

SOLUTION Selection of 2 teams for a game is an *unordered* subset of 2 from the set of 8 teams. Use combinations again.

(d) In how many ways can 4 patients be assigned to 6 different hospital rooms so that each patient has a private room?

YOUR TURN 3 Solve the problems in Example 4.

SOLUTION The room assignments are an *ordered* selection of 4 rooms from the 6 rooms. Exchanging the rooms of any 2 patients within a selection of 4 rooms gives a different assignment, so permutations should be used. **TRY YOUR TURN 3**

EXAMPLE 5 **Promotions**

A manager must select 4 employees for promotion; 12 employees are eligible.

(a) In how many ways can the 4 be chosen?

APPLY IT

SOLUTION Since there is no reason to differentiate among the 4 who are selected, use combinations.

$$C(12, 4) = \frac{12!}{8!\, 4!} = 495$$

YOUR TURN 4 In how many ways can a committee of 3 be chosen from a group of 20 students? In how many ways can three officers (president, treasurer, and secretary) be selected from a group of 20 students?

(b) In how many ways can 4 employees be chosen (from 12) to be placed in 4 different jobs?

SOLUTION In this case, once a group of 4 is selected, they can be assigned in many different ways (or arrangements) to the 4 jobs. Therefore, this problem requires permutations.

$$P(12, 4) = \frac{12!}{8!} = 11,880$$

TRY YOUR TURN 4

FOR REVIEW

Example 6 involves a standard deck of 52 playing cards, as shown in Figure 17 in Chapter 7. Recall the discussion that accompanies the photograph.

EXAMPLE 6 Playing Cards

Five cards are dealt from a standard 52-card deck.

(a) How many such hands have only face cards?

SOLUTION The face cards are the king, queen, and jack of each suit. Since there are 4 suits, there are 12 face cards. The arrangement of the 5 cards is not important, so use combinations to get

$$C(12, 5) = \frac{12!}{7! \, 5!} = 792.$$

(b) How many such hands have a full house of aces and eights (3 aces and 2 eights)?

SOLUTION The arrangement of the 3 aces or the 2 eights does not matter, so we use combinations. There are $C(4, 3)$ ways to get 3 aces from the four aces in the deck, and $C(4, 2)$ ways to get 2 eights. By the multiplication principle we get

$$C(4, 3) \cdot C(4, 2) = 4 \cdot 6 = 24.$$

(c) How many such hands have exactly 2 hearts?

SOLUTION There are 13 hearts in the deck, so the 2 hearts will be selected from those 13 cards. The other 3 cards must come from the remaining 39 cards that are not hearts. Use combinations and the multiplication principle to get

$$C(\mathbf{13}, \mathbf{2}) \cdot C(\mathbf{39}, \mathbf{3}) = 78 \cdot 9139 = 712{,}842.$$

Notice that the two numbers in red in the combinations add up to 52, the total number of cards, and the two numbers in blue add up to 5, the number of cards in a hand.

(d) How many such hands have cards of a single suit?

SOLUTION Since the arrangement of the 5 cards is not important, use combinations. The total number of ways that 5 cards of a particular suit of 13 cards can occur is $C(13, 5)$. There are four different suits, so the multiplication principle gives

$$4 \cdot C(13, 5) = 4 \cdot 1287 = 5148$$

ways to deal 5 cards of the same suit. **TRY YOUR TURN 5**

YOUR TURN 5 How many five card hands have exactly 2 aces?

As Example 6 shows, often both combinations and the multiplication principle must be used in the same problem.

EXAMPLE 7 Soup

To illustrate the differences between permutations and combinations in another way, suppose 2 cans of soup are to be selected from 4 cans on a shelf: noodle (N), bean (B), mushroom (M), and tomato (T). As shown in Figure 5(a), there are 12 ways to select 2 cans from the 4 cans if the order matters (if noodle first and bean second is considered different from bean, then

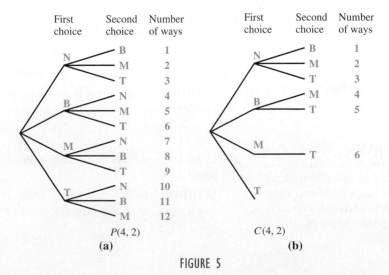

FIGURE 5

noodle, for example). On the other hand, if order is unimportant, then there are 6 ways to choose 2 cans of soup from the 4, as illustrated in Figure 5(b).

CAUTION It should be stressed that not all counting problems lend themselves to either permutations or combinations. Whenever a tree diagram or the multiplication principle can be used directly, it's often best to use it.

8.2 EXERCISES

1. Define combinations in your own words.

2. Explain the difference between a permutation and a combination.

Evaluate each combination.

3. $C(8, 3)$

4. $C(12, 5)$

5. $C(44, 20)$

6. $C(40, 18)$

7. $C(n, 0)$

8. $C(n, n)$

9. $C(n, 1)$

10. $C(n, n - 1)$

11. In how many ways can a hand of 6 clubs be chosen from an ordinary deck?

12. In how many ways can a hand of 6 red cards be chosen from an ordinary deck?

13. Five cards are marked with the numbers 1, 2, 3, 4, and 5, then shuffled, and 2 cards are drawn.

a. How many different 2-card combinations are possible?

b. How many 2-card hands contain a number less than 3?

14. An economics club has 31 members.

a. If a committee of 4 is to be selected, in how many ways can the selection be made?

b. In how many ways can a committee of at least 1 and at most 3 be selected?

15. Use a tree diagram for the following.

a. Find the number of ways 2 letters can be chosen from the set {L, M, N} if order is important and repetition is allowed.

b. Reconsider part a if no repeats are allowed.

c. Find the number of combinations of 3 elements taken 2 at a time. Does this answer differ from part a or b?

16. Repeat Exercise 15 using the set {L, M, N, P}.

In Exercises 17–24, decide whether each exercise involves permutations or combinations, and then solve the problem.

17. In a club with 9 male and 11 female members, how many 5-member committees can be chosen that have

a. all men?

b. all women?

c. 3 men and 2 women?

18. In Exercise 17, how many committees can be selected that have

a. at least 4 women?

b. no more than 2 men?

19. In a game of musical chairs, 12 children will sit in 11 chairs arranged in a row (one will be left out). In how many ways can

this happen, if we count rearrangements of the children in the chairs as different outcomes?

20. A group of 3 students is to be selected from a group of 14 students to take part in a class in cell biology.

a. In how many ways can this be done?

b. In how many ways can the group who will not take part be chosen?

21. Marbles are being drawn without replacement from a bag containing 16 marbles.

a. How many samples of 2 marbles can be drawn?

b. How many samples of 4 marbles can be drawn?

c. If the bag contains 3 yellow, 4 white, and 9 blue marbles, how many samples of 2 marbles can be drawn in which both marbles are blue?

22. There are 7 rotten apples in a crate of 26 apples.

a. How many samples of 3 apples can be drawn from the crate?

b. How many samples of 3 could be drawn in which all 3 are rotten?

c. How many samples of 3 could be drawn in which there are two good apples and one rotten one?

23. A bag contains 5 black, 1 red, and 3 yellow jelly beans; you take 3 at random. How many samples are possible in which the jelly beans are

a. all black?

b. all red?

c. all yellow?

d. 2 black and 1 red?

e. 2 black and 1 yellow?

f. 2 yellow and 1 black?

g. 2 red and 1 yellow?

24. In how many ways can 5 out of 9 plants be arranged in a row on a windowsill?

25. Show that $C(n, r) = C(n, n - r)$.

26. Padlocks with digit dials are often referred to as "combination locks." According to the mathematical definition of combination, is this an accurate description? Explain.

27. The following problem was posed on National Public Radio's *Weekend Edition*: In how many points can 6 circles intersect? *Source: National Public Radio.*

a. Find the answer for 6 circles.

b. Find the general answer for *n* circles.

28. How many different dominoes can be formed from the numbers 0...6? (*Hint:* A domino may have the same number of dots on both halves of it or it may have a different number of dots on each half.)

APPLICATIONS

Business and Economics

29. **Work Assignments** From a group of 8 newly hired office assistants, 3 are selected. Each of these 3 assistants will be assigned to a different manager. In how many ways can they be selected and assigned?

30. **Assembly Line Sampling** Five items are to be randomly selected from the first 50 items on an assembly line to determine the defect rate. How many different samples of 5 items can be chosen?

31. **Sales Schedules** A salesperson has the names of 6 prospects.

    a. In how many ways can she arrange her schedule if she calls on all 6?

    b. In how many ways can she arrange her schedule if she can call on only 4 of the 6?

32. **Worker Grievances** A group of 9 workers decides to send a delegation of 3 to their supervisor to discuss their grievances.

    a. How many delegations are possible?

    b. If it is decided that a particular worker must be in the delegation, how many different delegations are possible?

    c. If there are 4 women and 5 men in the group, how many delegations would include at least 1 woman?

33. **Hamburger Variety** Hamburger Hut sells regular hamburgers as well as a larger burger. Either type can include cheese, relish, lettuce, tomato, mustard, or catsup.

    a. How many different hamburgers can be ordered with exactly three extras?

    b. How many different regular hamburgers can be ordered with exactly three extras?

    c. How many different regular hamburgers can be ordered with at least five extras?

34. **Ice Cream Flavors** Baskin-Robbins advertises that it has 31 flavors of ice cream.

    a. How many different double-scoop cones can be made? Assume that the order of the scoops matters.

    b. How many different triple-scoop cones can be made?

    c. How many different double-scoop cones can be made if order doesn't matter?

Life Sciences

35. **Research Participants** From a group of 16 smokers and 22 nonsmokers, a researcher wants to randomly select 8 smokers and 8 nonsmokers for a study. In how many ways can the study group be selected?

36. **Plant Hardiness** In an experiment on plant hardiness, a researcher gathers 6 wheat plants, 3 barley plants, and 2 rye plants. She wishes to select 4 plants at random.

    a. In how many ways can this be done?

    b. In how many ways can this be done if exactly 2 wheat plants must be included?

Social Sciences

37. **Legislative Committee** A legislative committee consists of 5 Democrats and 4 Republicans. A delegation of 3 is to be selected to visit a small Pacific island republic.

    a. How many different delegations are possible?

    b. How many delegations would have all Democrats?

    c. How many delegations would have 2 Democrats and 1 Republican?

    d. How many delegations would include at least 1 Republican?

38. **Political Committee** From 10 names on a ballot, 4 will be elected to a political party committee. In how many ways can the committee of 4 be formed if each person will have a different responsibility, and different assignments of responsibility are considered different committees?

39. **Judges** When Paul Martinek, publisher of *Lawyers Weekly USA*, was a guest on the television news program *The O'Reilly Factor*, he discussed a decision by a three-judge panel, chosen at random from judges in the Ninth Circuit in California. The judges had ruled that the mandatory recitation of the Pledge of Allegience is unconstitutional because of the phrase "under God." According to Martinek, "Because there are 45 judges in the Ninth Circuit, there are 3000 different combinations of three-judge panels." Is this true? If not, what is the correct number? *Source: Mathematics Teacher.*

General Interest

40. **Bridge** How many different 13-card bridge hands can be selected from an ordinary deck?

41. **Poker** Five cards are chosen from an ordinary deck to form a hand in poker. In how many ways is it possible to get the following results?

    a. 4 queens   **b.** No face card

    c. Exactly 2 face cards   **d.** At least 2 face cards

    e. 1 heart, 2 diamonds, and 2 clubs

42. **Poker** In poker, a flush consists of 5 cards with the same suit, such as 5 diamonds.

    a. Find the number of ways of getting a flush consisting of cards with values from 5 to 10 by listing all the possibilities.

    b. Find the number of ways of getting a flush consisting of cards with values from 5 to 10 by using combinations.

43. **Baseball** If a baseball coach has 5 good hitters and 4 poor hitters on the bench and chooses 3 players at random, in how many ways can he choose at least 2 good hitters?

44. **Softball** The coach of the Morton Valley Softball Team has 6 good hitters and 8 poor hitters. He chooses 3 hitters at random.

    a. In how many ways can he choose 2 good hitters and 1 poor hitter?

    b. In how many ways can he choose 3 good hitters?

    c. In how many ways can he choose at least 2 good hitters?

45. **Flower Selection** Five orchids from a collection of 20 are to be selected for a flower show.

    a. In how many ways can this be done?

b. In how many ways can the 5 be selected if 2 special plants from the 20 must be included?

46. Lottery A state lottery game requires that you pick 6 different numbers from 1 to 99. If you pick all 6 winning numbers, you win the jackpot.

a. How many ways are there to choose 6 numbers if order is not important?

b. How many ways are there to choose 6 numbers if order matters?

47. Lottery In Exercise 46, if you pick 5 of the 6 numbers correctly, you win $250,000. In how many ways can you pick exactly 5 of the 6 winning numbers without regard to order?

48. Committees Suppose that out of 19 members of a club, two committees are to be formed. A nominating committee is to consist of 7 members, and a public relations committee is to consist of 5 members. No one can be on both committees.

a. Calculate the number of ways that the two committees can be formed, assuming that the nominating committee is formed first.

b. Calculate the number of ways that the two committees can be formed, assuming that the public relations committee is formed first. Verify that this answer is the same as that of part a.

c. Suppose the 7 members of the nominating committee wear red T-shirts, the 5 members of the public relations committee wear yellow T-shirts, and the remaining members of the club wear white T-shirts. A photographer lines up the members of the club to take a picture, but the picture is so blurry that people wearing the same color T-shirt are indistinguishable. In how many distinguishable ways can the club members line up? Explain why this answer is the same as the answers to parts a and b.

49. Committee A small department of 5 people decides to form a hiring committee. The only restriction on the size of the committee is that it must have at least 2 members.

a. Calculate the number of different committees possible by adding up the number of committees of different sizes.

b. Calculate the number of different committees possible by taking the total number of subsets of the 5 members and subtracting the number of committees that are invalid because they have too few members.

50. License Plates Officials from a particular state are considering a new type of license plate consisting of three letters followed by three numbers. If the letters cannot be repeated and must be in alphabetical order, calculate the number of possible distinct license plates.

51. Passwords A certain website requires users to log on using a security password.

a. If passwords must consist of six letters, followed by a single digit, determine the total number of possible distinct passwords.

b. If passwords must consist of six non-repetitive letters, followed by a single digit, determine the total number of possible distinct passwords.

52. Pizza Varieties A television commercial for Little Caesars pizza announced that with the purchase of two pizzas, one could receive free any combination of up to five toppings on each pizza. The commercial shows a young child waiting in line at Little Caesars who calculates that there are 1,048,576 possibilities for the toppings on the two pizzas. *Source: The Mathematics Teacher.*

a. Verify the child's calculation. Use the fact that Little Caesars has 11 toppings to choose from. Assume that the order of the two pizzas matters; that is, if the first pizza has combination 1 and the second pizza has combination 2, that is different from combination 2 on the first pizza and combination 1 on the second.

b. In a letter to *The Mathematics Teacher*, Joseph F. Heiser argued that the two combinations described in part a should be counted as the same, so the child has actually overcounted. Give the number of possibilities if the order of the two pizzas doesn't matter.

53. Pizza In an ad for Pizza Hut, Jessica Simpson explains to the Muppets that there are more than 6 million possibilities for their 4forAll Pizza. Griffin Weber and Glenn Weber wrote an article explaining that the number of possibilities is far more than 6 million, as described below. *Source: The College Mathematics Journal.*

a. Each pizza can have up to 3 toppings, out of 17 possible choices, or can be one of four specialty pizzas. Calculate the number of different pizzas possible.

b. Out of the total possible pizzas calculated in the first part of this exercise, a 4forAll Pizza consists of four pizzas in a box. Keeping in mind that the four pizzas could all be different, or there could be two or three different pizzas in the box, or all four pizzas could be the same, calculate the total number of 4forAll Pizzas possible.

c. The article considers another way of counting the number in part b. Suppose that only 8 pizzas were available, and they were listed in a row with lines separating each type, as in the following diagram:

$$A|B|C|D|E|F|G|H.$$

A person orders 4 pizzas by placing 4 X's in the desired places on the diagram, after which the letters can be ignored. For example, an order for 2 of A, 1 of C, and 1 of G would look like the following diagram.

$$XX||X||||X|$$

The number of ways this can be done is then the number of ways of arranging 11 objects, 4 of which are X and the other 7 of which are vertical lines, or

$$C(11, 4) = 330.$$

Use similar reasoning to verify the answer to part b.

54. Cereal The Post Corporation once introduced the cereal, *Create a Crunch™*, in which the consumers could combine ingredients to create their own unique cereal. Each box contained 8 packets of food goods. There were four types of cereal: Frosted Alpha Bits®, Cocoa Pebbles®, Fruity Pebbles®, and

Honey Comb®. Also included in the box were four "Add-Ins": granola, blue rice cereal, marshmallows, and sprinkles.

a. What is the total number of breakfasts that could be made if a breakfast is defined as any one or more cereals or add-ins?

b. If Sara Ouellette chose to mix one type of cereal with one add-in, how many different breakfasts could she make?

c. If Kristen Schmitt chose to mix two types of cereal with three add-ins, how many different breakfasts could she make?

d. If Matthew Pastier chose to mix at least one type of cereal with at least one type of add-in, how many breakfasts could he make?

e. If Inga Moffitt's favorite cereal is Fruity Pebbles®, how many different cereals could she make if each of her mixtures must include this cereal?

55. Appetizers Applebee's restaurant recently advertised "Ultimate Trios," where the customer was able to pick any three trio-sized appetizers from a list of nine options. The ad claimed that there were over 200 combinations.

a. If each customer's selection must be a different item, how many meal combinations are possible?

b. If each customer can select the same item two or even three times in each trio, how many different trios are possible?

c. Using the answers to parts a and b, discuss that restaurant's claim.

d. Two of the trio items, the Buffalo chicken wings and the boneless Buffalo wings, each have five different sauce options. This implies that there are actually 17 different trio choices that are available to a customer. In this scenario, how many different trio meal combinations are possible? (Assume that each of the three selected items is different.)

e. How many different trios are possible if 2 of the items are different flavored boneless Buffalo wings? (Assume that the third item is not a boneless Buffalo wing.)

56. Football Writer Gregg Easterbrook, discussing ESPN's unsuccessful attempt to predict the winners for the six National Football League (NFL) divisions and the six wild-card slots, claimed that there were 180 different ways to make this forecast. Reader

Milton Eisner wrote in to tell him that the actual number is much larger. To make the calculation, note that at the time the NFL consisted of two conferences, each of which consisted of three divisions. Five of the divisions had five teams, while the other had six. There was one winner from each of the six divisions, plus three wild-card slots from each of the two conferences. How many ways could the six division winners and six wild-card slots have been chosen? *Source: Slate Magazine.*

57. Music In the opera *Amahl and the Night Visitors*, the shepherds sing a chorus involving 18 different names, a challenge for singers trying to remember the names in the correct order. (Two of the three authors of this textbook have sung this chorus in public.)

a. In how many ways can the names be arranged?

b. Not all the arrangements of names in part a could be sung, because 10 of the names have 3 syllables, 4 have 2 syllables, and 4 have 4 syllables. Of the 6 lines in the chorus, 4 lines consist of a 3-syllable name repeated, followed by a 2-syllable and then a 4-syllable name (e.g., Emily, Emily, Michael, Bartholomew), and 2 lines consist of a 3-syllable name repeated, followed by two more 3-syllable names (e.g., Josephine, Josephine, Angela, Jeremy). No names are repeated except where we've indicated. (If you think this is confusing, you should try memorizing the chorus.) How many arrangements of the names could fit this pattern?

58. Olympics In recent Winter Olympics, there were 12 judges for each part of the women's figure skating program, but the scores of only 9 of the judges were randomly selected for the final results. *Source: The New York Times.*

a. In how many ways can the 9 judges whose scores are counted be selected?

b. Women's figure skating consists of a short program and a long program, with different judges for each part. How many different sets of judges' scores are possible for the entire event?

YOUR TURN ANSWERS

1. 210 **2.** 1820 **3. a.** 5040 **b.** 455 **c.** 28 **d.** 360
4. 1140; 6840 **5.** 103,776

8.3 Probability Applications of Counting Principles

APPLY IT
If 3 engines are tested from a shipping container packed with 12 diesel engines, 2 of which are defective, what is the probability that at least 1 of the defective engines will be found (in which case the container will not be shipped)?

This problem, which is solved in Example 3, could theoretically be solved with a tree diagram, but it would require a tree with a large number of branches. Many of the probability problems involving *dependent* events that were solved earlier by using tree diagrams can also be solved by using permutations or combinations. Permutations and combinations are especially helpful when the numbers involved are large.

To compare the method of using permutations or combinations with the method of tree diagrams used in Section 7.5, the first example repeats Example 7 from that section.

EXAMPLE 1   Environmental Inspections

The Environmental Protection Agency is considering inspecting 6 plants for environmental compliance: 3 in Chicago, 2 in Los Angeles, and 1 in New York. Due to a lack of inspectors, they decide to inspect 2 plants selected at random, 1 this month and 1 next month, with each plant equally likely to be selected, but no plant is selected twice. What is the probability that 1 Chicago plant and 1 Los Angeles plant are selected?

SOLUTION

Method 1
Multiplication Principle

To find the probability, we use the probability fraction, $P(E) = n(E)/n(S)$, where E is the event that 1 Chicago plant and 1 Los Angeles plant is selected and S is the sample space. We will use the multiplication principle since the plants are selected one at a time, and the first is inspected this month while the second is inspected next month.

To calculate the numerator, we find the number of elements in E. There are two ways to select a Chicago plant and a Los Angeles plant: Chicago first and Los Angeles second, or Los Angeles first and Chicago second. The Chicago plant can be selected from the 3 Chicago plants in $C(3, 1)$ ways, and the Los Angeles plant can be selected from the 2 Los Angeles plants in $C(2, 1)$ ways. Using the multiplication principle, we can select a Chicago plant then a Los Angeles plant in $C(3, 1) \cdot C(2, 1)$ ways, and a Los Angeles plant then a Chicago plant in $C(2, 1) \cdot C(3, 1)$ ways. By the union rule, we can select one Chicago plant and one Los Angeles plant in

$$C(3, 1) \cdot C(2, 1) + C(2, 1) \cdot C(3, 1) \text{ ways,}$$

giving the numerator of the probability fraction.

For the denominator, we calculate the number of elements in the sample space. There are 6 ways to select the first plant and 5 ways to select the second, for a total of $6 \cdot 5$ ways. The required probability is

$$P(1 \text{ C and } 1 \text{ LA}) = \frac{C(3, 1) \cdot C(2, 1) + C(2, 1) \cdot C(3, 1)}{6 \cdot 5}$$

$$= \frac{3 \cdot 2 + 2 \cdot 3}{6 \cdot 5} = \frac{12}{30} = \frac{2}{5}.$$

This agrees with the answer found in Example 7 of Section 7.5.

FOR REVIEW

The use of combinations to solve probability problems depends on the basic probability principle introduced in Section 7.3 and repeated here:

Let S be a sample space with equally likely outcomes, and let event E be a subset of S. Then the probability that event E occurs, written $P(E)$, is

$$P(E) = \frac{n(E)}{n(S)},$$

where $n(E)$ and $n(S)$ represent the number of elements in sets E and S.

Method 2
Combinations

This example can be solved more simply by observing that the probability that 1 Chicago plant and 1 Los Angeles plant are selected should not depend upon the order in which the plants are selected, so we may use combinations. The numerator is simply the number of ways of selecting 1 Chicago plant out of 3 Chicago plants and 1 Los Angeles plant out of 2 Los Angeles plants. The denominator is just the number of ways of selecting 2 plants out of 6. Then

$$P(1 \text{ C and } 1 \text{ LA}) = \frac{C(3, 1) \cdot C(2, 1)}{C(6, 2)} = \frac{6}{15} = \frac{2}{5}.$$

This helps explain why combinations tend to be used more often than permutations in probability. Even if order matters in the original problem, it is sometimes possible to ignore order and use combinations. Be careful to do this only when the final result does not depend on the order of events. Order often does matter. (If you don't believe this, try getting dressed tomorrow morning and then taking your shower.)

Method 3
Tree Diagram

In Section 7.5, we found this probability using the tree diagram shown in Figure 6 on the next page. Two of the branches correspond to drawing 1 Chicago plant and 1 Los Angeles plant. The probability for each branch is calculated by multiplying the probabilities along

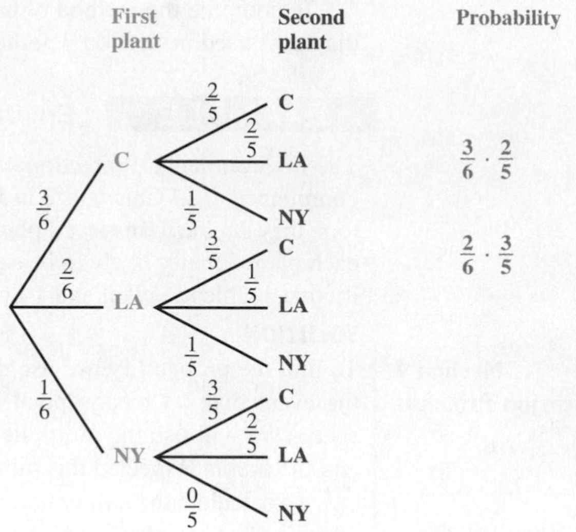

	First plant	Second plant	Probability

FIGURE 6

YOUR TURN 1 In Example 1, what is the probability that 1 New York plant and 1 Chicago plant are selected?

the branch, as we did in the previous chapter. The resulting probabilities for the two branches are then added, giving the result

$$P(1 \text{ C and } 1 \text{ LA}) = \frac{3}{6} \cdot \frac{2}{5} + \frac{2}{6} \cdot \frac{3}{5} = \frac{2}{5}.$$

TRY YOUR TURN 1

CAUTION The problems in the first two sections of this chapter asked how many ways a certain operation can be done. The problems in this section ask what is the probability that a certain event occurs; the solution involves answering questions about how many ways the event and the operation can be done.

- If a problem asks how many ways something can be done, the answer must be a nonnegative integer.
- If a problem asks for a probability, the answer must be a number between 0 and 1.

EXAMPLE 2 Nursing

From a group of 22 nurses, 4 are to be selected to present a list of grievances to management.

(a) In how many ways can this be done?

SOLUTION Four nurses from a group of 22 can be selected in $C(22, 4)$ ways. (Use combinations, since the group of 4 is an unordered set.)

$$C(22, 4) = \frac{22!}{18! \, 4!} = \frac{(22)(21)(20)(19)}{(4)(3)(2)(1)} = 7315$$

There are 7315 ways to choose 4 people from 22.

(b) One of the nurses is Lori Hales. Find the probability that Hales will be among the 4 selected.

SOLUTION The probability that Hales will be selected is given by $n(E)/n(S)$, where E is the event that the chosen group includes Hales, and S is the sample space for the experiment of choosing a group of 4. There is only $C(1, 1) = 1$ way to choose Hales. The number of ways that the other 3 nurses can be chosen from the remaining 21 nurses is

$$C(21, 3) = \frac{21!}{18! \, 3!} = 1330.$$

The probability that Hales will be one of the 4 chosen is

$$P(\text{Hales is chosen}) = \frac{n(E)}{n(S)} = \frac{C(\mathbf{1},\mathbf{1}) \cdot C(\mathbf{21},\mathbf{3})}{C(\mathbf{22},\mathbf{4})} = \frac{1330}{7315} \approx 0.1818.$$

Notice that the two numbers in red in the numerator, 1 and 21, add up to the number in red in the denominator, 22. This indicates that the 22 nurses have been split into two groups, one of size 1 (Hales) and the other of size 21 (the other nurses). Similarly, the green numbers indicate that the 4 nurses chosen consist of two groups of size 1 (Hales) and size 3 (the other nurses chosen).

YOUR TURN 2 If 8 of the 22 nurses are men in Example 2, what is the probability that exactly 2 men are among the 4 nurses selected?

(c) Find the probability that Hales will not be selected.

SOLUTION The probability that Hales will not be chosen is $1 - 0.1818 = 0.8182$.

TRY YOUR TURN 2

EXAMPLE 3 Diesel Engines

When shipping diesel engines abroad, it is common to pack 12 engines in one container. Suppose that a company has received complaints from its customers that many of the engines arrive in nonworking condition. To help solve this problem, the company decides to make a spot check of containers after loading. The company will test 3 engines from a container at random; if any of the 3 are nonworking, the container will not be shipped until each engine in it is checked. Suppose a given container has 2 nonworking engines. Find the probability that the container will not be shipped.

APPLY IT

SOLUTION The container will not be shipped if the sample of 3 engines contains 1 or 2 defective engines. If $P(1 \text{ defective})$ represents the probability of exactly 1 defective engine in the sample, then

$$P(\text{not shipping}) = P(1 \text{ defective}) + P(2 \text{ defective}).$$

There are $C(12, 3)$ ways to choose the 3 engines for testing:

$$C(12, 3) = \frac{12!}{9! \, 3!} = 220.$$

There are $C(2, 1)$ ways of choosing 1 defective engine from the 2 in the container, and for each of these ways, there are $C(10, 2)$ ways of choosing 2 good engines from among the 10 in the container. By the multiplication principle, there are

$$C(2, 1) \cdot C(10, 2) = \frac{2!}{1! \, 1!} \cdot \frac{10!}{8! \, 2!} = 2 \cdot 45 = 90$$

ways of choosing a sample of 3 engines containing 1 defective engine with

$$P(1 \text{ defective}) = \frac{90}{220} = \frac{9}{22}.$$

There are $C(2, 2)$ ways of choosing 2 defective engines from the 2 defective engines in the container, and $C(10, 1)$ ways of choosing 1 good engine from among the 10 good engines, for

$$C(2, 2) \cdot C(10, 1) = 1 \cdot 10 = 10$$

ways of choosing a sample of 3 engines containing 2 defective engines. Finally,

$$P(2 \text{ defective}) = \frac{10}{220} = \frac{1}{22}$$

and

$$P(\text{not shipping}) = P(1 \text{ defective}) + P(2 \text{ defective})$$

$$= \frac{9}{22} + \frac{1}{22} = \frac{10}{22} \approx 0.4545.$$

YOUR TURN 3 Suppose the container in Example 3 has 4 nonworking engines. Find the probability that the container will not be shipped.

—FOR REVIEW—

Recall from Section 7.4 that if E and E' are complements, then $P(E') = 1 - P(E)$. In Example 3, the event "0 defective in the sample" is the complement of the event "1 or 2 defective in the sample," since there are only 0 or 1 or 2 defective engines possible in the sample of 3 engines.

Notice that the probability is $1 - 0.4545 = 0.5455$ that the container will be shipped, even though it has 2 defective engines. The management must decide whether this probability is acceptable; if not, it may be necessary to test more than 3 engines from a container.

TRY YOUR TURN 3

Observe that in Example 3, the complement of finding 1 or 2 defective engines is finding 0 defective engines. Then instead of finding the sum $P(1 \text{ defective}) + P(2 \text{ defective})$, the result in Example 3 could be found as $1 - P(0 \text{ defective})$.

$$P(\text{not shipping}) = 1 - P(0 \text{ defective in sample})$$

$$= 1 - \frac{C(2, 0) \cdot C(10, 3)}{C(12, 3)}$$

$$= 1 - \frac{1(120)}{220}$$

$$= 1 - \frac{120}{220} = \frac{100}{220} \approx 0.4545$$

EXAMPLE 4 Poker

In a common form of the card game *poker*, a hand of 5 cards is dealt to each player from a deck of 52 cards. There are a total of

$$C(52, 5) = \frac{52!}{47! \, 5!} = 2,598,960$$

such hands possible. Find the probability of getting each of the following hands.

(a) A hand containing only hearts, called a *heart flush*

SOLUTION There are 13 hearts in a deck, with

$$C(13, 5) = \frac{13!}{8! \, 5!} = \frac{(13)(12)(11)(10)(9)}{(5)(4)(3)(2)(1)} = 1287$$

different hands containing only hearts. The probability of a heart flush is

$$P(\text{heart flush}) = \frac{C(13, 5) \cdot C(39, 0)}{C(52, 5)} = \frac{1287}{2,598,960} \approx 0.0004952.$$

You don't really need the $C(39, 0)$, since this just equals 1, but it might help to remind you that you are choosing none of the 39 cards that remain after the hearts are removed.

(b) A flush of any suit (5 cards of the same suit)

SOLUTION There are 4 suits in a deck, so

$$P(\text{flush}) = 4 \cdot P(\text{heart flush}) = 4 \cdot 0.0004952 \approx 0.001981.$$

(c) A full house of aces and eights (3 aces and 2 eights)

SOLUTION There are $C(4, 3)$ ways to choose 3 aces from among the 4 in the deck, and $C(4, 2)$ ways to choose 2 eights.

$$P(3 \text{ aces, 2 eights}) = \frac{C(4, 3) \cdot C(4, 2) \cdot C(44, 0)}{C(52, 5)} = \frac{4 \cdot 6 \cdot 1}{2,598,960} \approx 0.000009234$$

(d) Any full house (3 cards of one value, 2 of another)

SOLUTION

Method 1
Standard Procedure

The 13 values in a deck give 13 choices for the first value. As in part (c), there are $C(4, 3)$ ways to choose the 3 cards from among the 4 cards that have that value. This leaves 12 choices for the second value (order *is* important here, since a full house of 3 aces and 2 eights is not the same as a full house of 3 eights and 2 aces). From the 4 cards that have the second value, there are $C(4, 2)$ ways to choose 2. The probability of any full house is then

$$P(\text{full house}) = \frac{13 \cdot C(4, 3) \cdot 12 \cdot C(4, 2)}{2,598,960} \approx 0.001441.$$

Method 2
Alternative Procedure

As an alternative way of counting the numerator, first count the number of different values in the hand.* Since there are 13 values from which to choose, and we need 2 different values (one for the set of 3 cards and one for the set of 2), there are $C(13, 2)$ ways to choose the values. Next, of the two values chosen, select the value for which there are 3 cards, which can be done $C(2, 1)$ ways. This automatically determines that the other value is the one for which there are 2 cards. Next, choose the suits for each value. For the value with 3 cards, there are $C(4, 3)$ values of the suits, and for the value with 2 cards, there are $C(4, 2)$ values. Putting this all together,

$$P(\text{full house}) = \frac{C(13, 2) \cdot C(2, 1) \cdot C(4, 3) \cdot C(4, 2)}{2,598,960} \approx 0.001441.$$

YOUR TURN 4 In Example 4, what is the probability of a hand containing two pairs, one of aces and the other of kings? (This hand contains 2 aces, 2 kings, and a fifth card that is neither an ace nor a king.)

TRY YOUR TURN 4

EXAMPLE 5   Letters

Each of the letters w, y, o, m, i, n, and g is placed on a separate slip of paper. A slip is pulled out, and its letter is recorded in the order in which the slip was drawn. This is done four times.

(a) If the slip is not replaced after the letter is recorded, find the probability that the word "wing" is formed.

SOLUTION The sample space contains all possible arrangements of the seven letters, taken four at a time. Since order matters, use *permutations* to find the number of arrangements in the sample space.

$$P(7, 4) = \frac{7!}{3!} = 7 \cdot 6 \cdot 5 \cdot 4 = 840$$

Since there is only one way that the word "wing" can be formed, the required probability is $1/840 \approx 0.001190$.

YOUR TURN 5 Find the probability that the word "now" is formed if 3 slips are chosen without replacement in Example 5. Find the probability that the word "now" is formed if 3 slips are chosen with replacement.

(b) If the slip is replaced after the letter is recorded, find the probability that the word "wing" is formed.

SOLUTION Since the letters can be repeated, there are 7 possible outcomes for each draw of the slip. To calculate the number of arrangements in the sample space, use the *multiplication principle*. The number of arrangements in the sample space is $7^4 = 2401$, and the required probability is $1/2401 \approx 0.0004165$. **TRY YOUR TURN 5**

EXAMPLE 6   Birthdays

Suppose a group of n people is in a room. Find the probability that at least 2 of the people have the same birthday.

SOLUTION "Same birthday" refers to the month and the day, not necessarily the same year. Also, ignore leap years, and assume that each day in the year is equally likely as a birthday. To see how to proceed, we first look at the case in which $n = 5$ and find the probability that *no 2 people* from among 5 people have the same birthday. There are 365 different birthdays possible for the first of the 5 people, 364 for the second (so that the people have different birthdays), 363 for the third, and so on. The number of ways the

*We learned this approach from Professor Peter Grassi of Hofstra University.

5 people can have different birthdays is thus the number of permutations of 365 days taken 5 at a time or

$$P(365, 5) = 365 \cdot 364 \cdot 363 \cdot 362 \cdot 361.$$

The number of ways that 5 people can have the same birthday or different birthdays is

$$365 \cdot 365 \cdot 365 \cdot 365 \cdot 365 = (365)^5.$$

Finally, the *probability* that none of the 5 people have the same birthday is

$$\frac{P(365, 5)}{365^5} = \frac{365 \cdot 364 \cdot 363 \cdot 362 \cdot 361}{365 \cdot 365 \cdot 365 \cdot 365 \cdot 365} \approx 0.9729.$$

The probability that at least 2 of the 5 people *do* have the same birthday is $1 - 0.9729 = 0.0271$.

Now this result can be extended to more than 5 people. Generalizing, the probability that no 2 people among n people have the same birthday is

$$\frac{P(365, n)}{365^n}.$$

The probability that at least 2 of the n people *do* have the same birthday is

$$1 - \frac{P(365, n)}{365^n}.$$

The following table shows this probability for various values of n.

Number of People, n	Probability That Two Have the Same Birthday
5	0.0271
10	0.1169
15	0.2529
20	0.4114
22	0.4757
23	0.5073
25	0.5687
30	0.7063
35	0.8144
40	0.8912
50	0.9704
366	1

The probability that 2 people among 23 have the same birthday is 0.5073, a little more than half. Many people are surprised at this result; it seems that a larger number of people should be required.

TECHNOLOGY NOTE

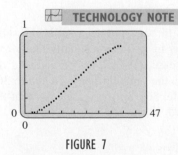

FIGURE 7

Using a graphing calculator, we can graph the probability formula in the previous example as a function of n, but care must be taken that the graphing calculator evaluates the function at integer points. Figure 7 was produced on a TI-84 Plus by letting $Y_1 = 1 - (365 \text{ nPr } X)/365 \wedge X$ on $0 \le x \le 47$. (This domain ensures integer values for x.) Notice that the graph does not extend past $x = 39$. This is because $P(365, n)$ and 365^n are too large for the calculator when $n \ge 40$.

An alternative way of doing the calculations that does not run into such large numbers is based on the concept of conditional probability. The probability that the first person's birthday does not match any so far is 365/365. The probability that the second person's birthday does not match the first's is 364/365. The probability that the third person's birthday does not match the first's or the second's is 363/365. By the product rule of probability, the probability that none of the first 3 people have matching birthdays is

$$\frac{365}{365} \cdot \frac{364}{365} \cdot \frac{363}{365}.$$

Similarly, the probability that no two people in a group of 40 have the same birthday is

$$\frac{365}{365} \cdot \frac{364}{365} \cdot \frac{363}{365} \cdots \frac{326}{365}.$$

This probability can be calculated (and then subtracted from 1 to get the probability we seek) without overflowing the calculator by multiplying each fraction times the next, rather then trying to compute the entire numerator and the entire denominator. The calculations are somewhat tedious to do by hand but can be programmed on a graphing calculator or computer.

As we saw in Examples 1 and 4(d), probability can sometimes be calculated in more than one way. We now look at one more example of this.

EXAMPLE 7 Fruit

Ray and Nate are arranging a row of fruit at random on a table. They have 5 apples, 6 oranges, and 7 lemons. What is the probability that all fruit of the same kind are together?

SOLUTION

Method 1
Distinguishable Permutations

Ray can't tell individual pieces of fruit of the same kind apart. All apples look the same to him, as do all oranges and all lemons. So in the denominator of the probability, he calculates the number of distinguishable ways to arrange the 18 pieces of fruit, given that all apples are indistinguishable, as are all oranges and all lemons.

$$\frac{18!}{5!\,6!\,7!} = 14{,}702{,}688$$

As for the numerator, the only choice is how to arrange the 3 kinds of fruit, for which there are $3! = 6$ ways. Thus

$$P(\text{all fruit of the same kind are together}) = \frac{6}{14{,}702{,}688} \approx 4.081 \times 10^{-7}.$$

Method 2
Permutations

Nate has better eyesight than Ray and can tell the individual pieces of fruit apart. So in the denominator of the probability, he calculates the number of ways to arrange the 18 pieces of fruit, which is

$$18! \approx 6.4024 \times 10^{15}.$$

For the numerator, he first must choose how to arrange the 3 kinds of fruit, for which there are $3!$ ways. Then there are $5!$ ways to arrange the apples, $6!$ ways to arrange the oranges, and $7!$ ways to arrange the lemons, for a total number of possibilities of

$$3!\,5!\,6!\,7! = 2{,}612{,}736{,}000.$$

Therefore,

YOUR TURN 6 If Ray and Nate arrange 2 kiwis, 3 apricots, 4 pineapples, and 5 coconuts in a row at random, what is the probability that all fruit of the same kind are together?

$$P(\text{all fruit of the same kind are together}) = \frac{2{,}612{,}736{,}000}{6.4024 \times 10^{15}} \approx 4.081 \times 10^{-7}.$$

The results for Method 1 and Method 2 are the same. The probability does not depend on whether a person can distinguish individual pieces of the same kind of fruit.

TRY YOUR TURN 6

8.3 EXERCISES

A basket contains 7 red apples and 4 yellow apples. A sample of 3 apples is drawn. Find the probabilities that the sample contains the following.

1. All red apples

2. All yellow apples

3. 2 yellow and 1 red apple

4. More red than yellow apples

In a club with 9 male and 11 female members, a 5-member committee will be randomly chosen. Find the probability that the committee contains the following.

5. All men

6. All women

7. 3 men and 2 women

8. 2 men and 3 women

9. At least 4 women

10. No more than 2 men

Two cards are drawn at random from an ordinary deck of 52 cards.

11. How many 2-card hands are possible?

Find the probability that the 2-card hand described above contains the following.

12. 2 aces
13. At least 1 ace
14. All spades
15. 2 cards of the same suit
16. Only face cards
17. No face cards
18. No card higher than 8 (count ace as 1)

Twenty-six slips of paper are each marked with a different letter of the alphabet and placed in a basket. A slip is pulled out, its letter recorded (in the order in which the slip was drawn), and the slip is replaced. This is done 5 times. Find the probabilities that the following "words" are formed.

19. Chuck

20. A word that starts with "p"

21. A word with no repetition of letters

22. A word that contains no "x," "y," or "z"

23. Discuss the relative merits of using tree diagrams versus combinations to solve probability problems. When would each approach be most appropriate?

24. Several examples in this section used the rule $P(E') = 1 - P(E)$. Explain the advantage (especially in Example 6) of using this rule.

For Exercises 25–28, refer to Example 6 in this section.

25. A total of 43 men have served as president through 2010.* Set up the probability that, if 43 men were selected at random, at least 2 have the same birthday.†

26. Set up the probability that at least 2 of the 100 U.S. senators have the same birthday.

27. What is the probability that at least 2 of the 435 members of the House of Representatives have the same birthday?

28. Argue that the probability that in a group of n people *exactly one pair* have the same birthday is

$$C(n, 2) \cdot \frac{P(365, n - 1)}{365^n}.$$

29. After studying all night for a final exam, a bleary-eyed student randomly grabs 2 socks from a drawer containing 9 black, 6 brown, and 2 blue socks, all mixed together. What is the probability that she grabs a matched pair?

30. Three crows, 4 blue jays, and 5 starlings sit in a random order on a section of telephone wire. Find the probability that birds of a feather flock together, that is, that all birds of the same type are sitting together.

31. If the letters l, i, t, t, l, and e are chosen at random, what is the probability that they spell the word "little"?

32. If the letters M, i, s, s, i, s, s, i, p, p, and i are chosen at random, what is the probability that they spell the word "Mississippi"?

33. An elevator has 4 passengers and stops at 7 floors. It is equally likely that a person will get off at any one of the 7 floors. Find the probability that at least 2 passengers leave at the same floor. (*Hint:* Compare this with the birthday problem.)

34. On National Public Radio, the *Weekend Edition* program posed the following probability problem: Given a certain number of balls, of which some are blue, pick 5 at random. The probability that all 5 are blue is 1/2. Determine the original number of balls and decide how many were blue. *Source: Weekend Edition.*

35. A reader wrote to the "Ask Marilyn" column in *Parade* magazine, "You have six envelopes to pick from. Two-thirds (that is, four) are empty. One-third (that is, two) contain a $100 bill. You're told to choose 2 envelopes at random. Which is more likely: (1) that you'll get at least one $100 bill, or (2) that you'll get no $100 bill at all?" Find the two probabilities. *Source: Parade Magazine.*

APPLICATIONS

Business and Economics

Quality Control A shipment of 11 printers contains 2 that are defective. Find the probability that a sample of the following sizes, drawn from the 11, will not contain a defective printer.

36. 1   37. 2   38. 3   39. 4

---

*Although Obama is the 44th President, the 22nd and 24th Presidents were the same man: Grover Cleveland.
†In fact, James Polk and Warren Harding were both born on November 2.

Refer to Example 3. The managers feel that the probability of 0.5455 that a container will be shipped even though it contains 2 defective engines is too high. They decide to increase the sample size chosen. Find the probabilities that a container will be shipped even though it contains 2 defective engines, if the sample size is increased to the following.

40. 4   **41.** 5

42. Sales Presentations Heidi Shadix and Boyd Shepherd are among 9 representatives making presentations at the annual sales meeting. The presentations are randomly ordered. Find the probability that Heidi is the first presenter and Boyd is the last presenter.

43. Sales Schedule Dan LaChapelle has the name of 6 prospects, including a customer in Scottsdale. He randomly arranges his schedule to call on only 4 of the 6 prospects. Find the probability that the customer from Scottsdale is not called upon.

Social Sciences

44. Election Ballots Five names are put on a ballot in a randomly selected order. What is the probability that they are not in alphabetical order?

45. Native American Council At the first meeting of a committee to plan a Northern California pow-wow, there were 3 women and 3 men from the Miwok tribe, 2 men and 3 women from the Hoopa tribe, and 4 women and 5 men from the Pomo tribe. If the ceremony subcouncil consists of 5 people and is randomly selected, find the probabilities that the subcouncil contains the following:

a. 3 men and 2 women;

b. exactly 3 Miwoks and 2 Pomos;

c. 2 Miwoks, 2 Hoopas, and a Pomo;

d. 2 Miwoks, 2 Hoopas, and 2 Pomos;

e. more women than men;

f. exactly 3 Hoopas;

g. at least 2 Pomos.

46. Education A school in Bangkok requires that students take an entrance examination. After the examination, there is a drawing in which 5 students are randomly selected from each group of 40 for automatic acceptance into the school, regardless of their performance on the examination. The drawing consists of placing 35 red and 5 green pieces of paper into a box. Each student picks a piece of paper from the box and then does not return the piece of paper to the box. The 5 lucky students who pick the green pieces are automatically accepted into the school. *Source: Mathematics Teacher.*

a. What is the probability that the first person wins automatic acceptance?

b. What is the probability that the last person wins automatic acceptance?

c. If the students are chosen by the order of their seating, does this give the student who goes first a better chance of winning than the second, third, . . . person? (*Hint:* Imagine that the 40 pieces of paper have been mixed up and laid in a row so that the first student picks the first piece of paper, the second student picks the second piece of paper, and so on.)

47. Education At a conference promoting excellence in education for African Americans in Detroit, special-edition books were selected to be given away in contests. There were 9 books written by Langston Hughes, 5 books by James Baldwin, and 7 books by Toni Morrison. The judge of one contest selected 6 books at random for prizes. Find the probabilities that the selection consisted of the following.

a. 3 Hughes and 3 Morrison books

b. Exactly 4 Baldwin books

c. 2 Hughes, 3 Baldwin, and 1 Morrison book

d. At least 4 Hughes books

e. Exactly 4 books written by males (Morrison is female)

f. No more than 2 books written by Baldwin

General Interest

Poker Find the probabilities of the following hands at poker. Assume aces are either high or low.

48. Royal flush (5 highest cards of a single suit)

49. Straight flush (5 in a row in a single suit, but not a royal flush)

50. Four of a kind (4 cards of the same value)

51. Straight (5 cards in a row, not all of the same suit), with ace either high or low

52. Three of a kind (3 cards of one value, with the other cards of two different values)

53. Two pairs (2 cards of one value, 2 of another value, and 1 of a third value)

54. One pair (2 cards of one value, with the other cards of three different values)

Bridge A bridge hand is made up of 13 cards from a deck of 52. Find the probabilities that a hand chosen at random contains the following.

55. Only hearts

56. At least 3 aces

57. Exactly 2 aces and exactly 2 kings

58. 6 of one suit, 4 of another, and 3 of another

Texas Hold'Em In a version of poker called Texas Hold'Em, each player has 2 cards, and by the end of the round an additional 5 cards are on the table, shared by all the players. (For more about poker hands, see Example 4 and Exercises 48–54.) Each player's hand consists of the best 5 cards out of the 7 cards available to that player. For example, if a player holds 2 kings, and on the table there is one king and 4 cards with 4 other values, then the player has three of a kind. It's possible that the player has an even better hand. Perhaps five of the seven cards are of the same suit, making a flush, or the other 4 cards are queen, jack, 10, and 9, so the player has a straight. For Exercises 59–64, calculate the probability of each hand in Texas Hold'Em, but for simplicity, ignore the possibility that the cards might form an even better hand.

59. One pair (2 cards of one value, with the other cards of five different values)

60. Two pairs (2 cards of one value, 2 of another, with the other cards of three different values)

61. Three of a kind (3 cards of one value, with the other cards of four different values)

62. Four of a kind (4 cards of one value, with the other cards of three different values)

63. Flush (at least 5 cards of the same suit)

64. Full house (3 cards of one value and two of another. Careful: The two unused cards could be a pair of another value or two different cards of other values. Also, the 7 cards could consist of 3 cards of one value, 3 of second value, and one card of a third value. We won't consider the case of 3 cards of one value and 4 of another, because even though this forms a full house, it also forms four of a kind, which is better.)

65. Suppose you are playing Texas Hold'Em and you've just received your two cards. You observe that they are both hearts. You get excited, because if at least 3 of the 5 cards that are on the table are hearts, you'll have a flush, which means that you'll likely win the round. Given that your two cards are hearts, and you know nothing of any other cards, what is the probability that you'll have a flush by the time all 5 cards are on the table?

66. Lottery In the previous section, we found the number of ways to pick 6 different numbers from 1 to 99 in a state lottery. Assuming order is unimportant, what is the probability of picking all 6 numbers correctly to win the big prize?

67. Lottery In Exercise 66, what is the probability of picking exactly 5 of the 6 numbers correctly?

68. Lottery An article in *The New York Times* discussing the odds of winning the lottery stated, "And who cares if a game-theory professor once calculated the odds of winning as equal to a poker player's chance of drawing four royal flushes in a row, all in spades—then getting up from the card table and meeting four strangers, all with the same birthday?" Calculate this probability. Does this probability seem comparable to the odds of winning the lottery? (Ignore February 29 as a birthday, and assume that all four strangers have the same birthday as each other, not necessarily the same as the poker player.) *Source: The New York Times Magazine.*

69. Lottery A reader wrote to the "Ask Marilyn" column in *Parade* magazine, "A dozen glazed doughnuts are riding on the answer to this question: Are the odds of winning in a lotto drawing higher when picking 6 numbers out of 49 or when picking 5 numbers out of 52?" Calculate each probability to answer the question. *Source: Parade Magazine.*

70. Lottery On October 22, 2005, the Powerball Lottery had a record jackpot of $340 million, which was a record at the time. To enter the lottery, 5 numbers are picked between 1 and 55, plus a bonus number between 1 and 42. All 6 numbers must be correct to win the jackpot.

a. What is the probability of winning the jackpot with a single ticket?

b. In an article for the *Minneapolis Star Tribune*, mathematician Douglas Arnold was quoted as saying, "If you were to select a group of Powerball numbers every minute for 138 years, you would have about a 50 percent chance of picking the winning Powerball ticket." Calculate the actual probability, using an estimate of 365.25 for the number of days in the year. (Arnold later told *Chance News* that this was an "off-the-top-of-my-head calculation" made when a reporter called.) *Sources: Minneapolis Star Tribune and Chance News.*

71. Canadian Lottery In June 2004, Canada introduced a change in its lottery that violated the usual convention that the smaller the probability of an event, the bigger the prize. In this lottery, participants have to guess six numbers from 1 to 49. Six numbers between 1 and 49 are then drawn at random, plus a seventh "bonus number." *Source: Chance.*

a. A fifth prize of $10 goes to those who correctly guess exactly three of the six numbers, but do not guess the bonus number. Find the probability of winning fifth prize.

b. A sixth prize of $5 goes to those who correctly guess exactly two of the six numbers plus the bonus number. Find the probability of winning sixth prize, and compare this with the probability of winning fifth prize.

72. Barbie A controversy arose in 1992 over the Teen Talk Barbie doll, each of which was programmed with four sayings randomly picked from a set of 270 sayings. The controversy was over the saying, "Math class is tough," which some felt gave a negative message toward girls doing well in math. In an interview with *Science*, a spokeswoman for Mattel, the makers of Barbie, said that "There's a less than 1% chance you're going to get a doll that says math class is tough." Is this figure correct? If not, give the correct figure. *Source: Science.*

73. Football During the 1988 college football season, the Big Eight Conference ended the season in a "perfect progression," as shown in the following table. *Source: The American Statistician.*

Won	Lost	Team
7	0	Nebraska (NU)
6	1	Oklahoma (OU)
5	2	Oklahoma State (OSU)
4	3	Colorado (CU)
3	4	Iowa State (ISU)
2	5	Missouri (MU)
1	6	Kansas (KU)
0	7	Kansas State (KSU)

Someone wondered what the probability of such an outcome might be.

a. How many games do the 8 teams play?

b. Assuming no ties, how many different outcomes are there for all the games together?

c. In how many ways could the 8 teams end in a perfect progression?

d. Assuming that each team had an equally likely probability of winning each game, find the probability of a perfect progression with 8 teams.

e. Find a general expression for the probability of a perfect progression in an n-team league with the same assumptions.

74. Unluckiest Fan During the 2009 season, the Washington Nationals baseball team won 59 games and lost 103 games. Season ticket holder Stephen Krupin reported in an interview that he watched the team lose all 19 games that he attended that season. The interviewer speculated that this must be a record for bad luck. *Source: Chance News.*

a. Based on the full 2009 season record, calculate the probability that a person would attend 19 Washington Nationals games and the Nationals would lose all 19 games.

b. However, Mr. Krupin only attended home games. The Nationals had 33 wins and 48 losses at home in 2009. Calculate the probability that a person would attend 19 Washington Nationals home games and the Nationals would lose all 19 games.

75. Bingo Bingo has become popular in the United States, and it is an efficient way for many organizations to raise money. The bingo card has 5 rows and 5 columns of numbers from 1 to 75, with the center given as a free cell. Balls showing one of the 75 numbers are picked at random from a container. If the drawn number appears on a player's card, then the player covers the number. In general, the winner is the person who first has a card with an entire row, column, or diagonal covered. *Source: Mathematics Teacher.*

a. Find the probability that a person will win bingo after just four numbers are called.

b. An L occurs when the first column and the bottom row are both covered. Find the probability that an L will occur in the fewest number of calls.

c. An X-out occurs when both diagonals are covered. Find the probability that an X-out occurs in the fewest number of calls.

d. If bingo cards are constructed so that column one has 5 of the numbers from 1 to 15, column two has 5 of the numbers from 16 to 30, column three has 4 of the numbers from 31 to 45, column four has 5 of the numbers from 46 to 60, and column five has 5 of the numbers from 61 to 75, how many different bingo cards could be constructed? (*Hint:* Order matters!)

76. Suppose a box contains 3 red and 3 blue balls. A ball is selected at random and removed, without observing its color. The box now either contains 3 red and 2 blue balls or 2 red and 3 blue balls. *Source: 40 Puzzles and Problems in Probability and Mathematical Statistics.*

a. Nate removes a ball at random from the box, observes its color, and puts the ball back. He performs this experiment a total of 6 times, and each time the ball is blue. What is the probability that a red ball was initially removed from the box? (*Hint:* Use Bayes' Theorem.)

b. Ray removes a ball at random from the box, observes its color, and puts the ball back. He performs this experiment a total of 80 times. Out of these, the ball was blue 44 times and red 36 times. What is the probability that a red ball was initially removed from the box?

c. Many people intuitively think that Nate's experiment gives more convincing evidence than Ray's experiment that a red ball was removed. Explain why this is wrong.

YOUR TURN ANSWERS

1. 1/5 **2.** 0.3483 **3.** 0.7455 **4.** 0.0006095
5. 1/210; 1/343 **6.** 9.514×10^{-6}

8.4 Binomial Probability

APPLY IT **What is the probability that 3 out of 6 randomly selected college students attend more than one institution during their college career?**
We will calculate this probability in Example 2.

The question above involves an experiment that is repeated 6 times. Many probability problems are concerned with experiments in which an event is repeated many times. Other examples include finding the probability of getting 7 heads in 8 tosses of a coin, of hitting a target 6 times out of 6, and of finding 1 defective item in a sample of 15 items. Probability problems of this kind are called **Bernoulli trials** problems, or **Bernoulli processes**, named after the Swiss mathematician Jakob Bernoulli (1654–1705), who is well known for his work in probability theory. In each case, some outcome is designated a success and any other outcome is considered a failure. This labeling is arbitrary and does not necessarily have anything to do with real success or failure. Thus, if the probability of a success in a single trial is p, the probability of failure will be $1 - p$. A Bernoulli trials problem, or **binomial experiment**, must satisfy the following conditions.

Binomial Experiment

1. The same experiment is repeated a fixed number of times.
2. There are only two possible outcomes, success and failure.
3. The repeated trials are independent, so that the probability of success remains the same for each trial.

EXAMPLE 1 Sleep

The chance that an American falls asleep with the TV on at least three nights a week is 1/4. Suppose a researcher selects 5 Americans at random and is interested in the probability that all 5 are "TV sleepers." *Source: Harper's Magazine.*

FOR REVIEW

Recall that if *A* and *B* are independent events,

$$P(A \text{ and } B) = P(A)P(B).$$

SOLUTION Here the experiment, selecting a person, is repeated 5 times. If selecting a TV sleeper is labeled a success, then getting a "non-TV sleeper" is labeled a failure. The 5 trials are almost independent. There is a very slight dependence; if, for example, the first person selected is a TV sleeper, then there is one less TV sleeper to choose from when we select the next person (assuming we never select the same person twice). When selecting a small sample out of a large population, however, the probability changes negligibly, so researchers consider such trials to be independent. Thus, the probability that all 5 in our sample are sleepers is

$$\frac{1}{4} \cdot \frac{1}{4} \cdot \frac{1}{4} \cdot \frac{1}{4} \cdot \frac{1}{4} = \left(\frac{1}{4}\right)^5 \approx 0.0009766.$$

Now suppose the problem in Example 1 is changed to that of finding the probability that exactly 4 of the 5 people in the sample are TV sleepers. This outcome can occur in more than one way, as shown below, where *s* represents a success (a TV sleeper) and *f* represents a failure (a non-TV sleeper).

outcome 1:	*s* *s* *s* *s*	***f***		
outcome 2:	*s* *s* *s*	***f***	*s*	
outcome 3:	*s* *s*	***f***	*s* *s*	
outcome 4:	*s*	***f***	*s* *s* *s*	
outcome 5:	***f***	*s* *s* *s* *s*		

Keep in mind that since the probability of success is 1/4, the probability of failure is $1 - 1/4 = 3/4$. The probability, then, of each of these 5 outcomes is

$$\left(\frac{1}{4}\right)^4 \left(\frac{3}{4}\right).$$

Since the 5 outcomes represent mutually exclusive events, add the 5 identical probabilities, which is equivalent to multiplying the above probability by 5. The result is

$$P(4 \text{ of the 5 people are TV sleepers}) = 5\left(\frac{1}{4}\right)^4 \left(\frac{3}{4}\right) = \frac{15}{4^5} \approx 0.01465.$$

In the same way, we can compute the probability of selecting 3 TV sleepers in our sample of 5. The probability of any one way of achieving 3 successes and 2 failures will be

$$\left(\frac{1}{4}\right)^3 \left(\frac{3}{4}\right)^2.$$

Rather than list all the ways of achieving 3 successes out of 5 trials, we will count this number using combinations. The number of ways to select 3 elements out of a set of 5 is $C(5, 3) = 5!/(2! \, 3!) = 10$, giving

$$P(3 \text{ of the 5 people are TV sleepers}) = 10\left(\frac{1}{4}\right)^3 \left(\frac{3}{4}\right)^2 = \frac{90}{4^5} \approx 0.08789.$$

A similar argument works in the general case.

Binomial Probability

If *p* is the probability of success in a single trial of a binomial experiment, the probability of *x* successes and $n - x$ failures in *n* independent repeated trials of the experiment, known as **binomial probability**, is

$$P(x \text{ successes in } n \text{ trials}) = C(n, x) \cdot p^x \cdot (1 - p)^{n-x}.$$

EXAMPLE 2 College Students

A recent survey found that 59% of college students attend more than one institution during their college career. Suppose a sample of 6 students is chosen. Assuming that each student's college attendance pattern is independent of the others, find the probability of each of the following. *Source: The New York Times.*

(a) Exactly 3 of the 6 students attend more than one institution.

APPLY IT

SOLUTION Think of the 6 students chosen as 6 independent trials. A success occurs if the student attends more than one institution. Then this is a binomial experiment with $n = 6$ and $p = P(\text{attend more than one institution}) = 0.59$. To find the probability that exactly 3 students attend more than one institution, let $x = 3$ and use the binomial probability formula.

$$
\begin{aligned}
P(\text{exactly 3 of 6 students}) &= C(6, 3)(0.59)^3(1 - 0.59)^{6-3} \\
&= 20(0.59)^3(0.41)^3 \\
&= 20(0.2054)(0.06892) \\
&\approx 0.2831
\end{aligned}
$$

(b) None of the 6 students attend more than one institution.

YOUR TURN 1 Find the probability that exactly 2 of the 6 students in Example 2 attend more than 1 institution.

SOLUTION Let $x = 0$.

$$
\begin{aligned}
P(\text{exactly 0 of 6 students}) &= C(6, 0)(0.59)^0(1 - 0.59)^6 \\
&= 1(1)(0.41)^6 \approx 0.00475 \quad \textbf{TRY YOUR TURN 1}
\end{aligned}
$$

EXAMPLE 3 Coin Toss

Find the probability of getting exactly 7 heads in 8 tosses of a fair coin.

YOUR TURN 2 In Example 3, find the probability of getting exactly 4 heads in 8 tosses of a fair coin.

SOLUTION The probability of success (getting a head in a single toss) is $1/2$. The probability of a failure (getting a tail) is $1 - 1/2 = 1/2$. Thus,

$$
P(7 \text{ heads in 8 tosses}) = C(8, 7)\left(\frac{1}{2}\right)^7\left(\frac{1}{2}\right)^1 = 8\left(\frac{1}{2}\right)^8 = 0.03125.
$$

TRY YOUR TURN 2

EXAMPLE 4 Defective Items

Assuming that selection of items for a sample can be treated as independent trials, and that the probability that any 1 item is defective is 0.01, find the following.

(a) The probability of 1 defective item in a random sample of 15 items from a production line

SOLUTION Here, a "success" is a defective item. Since selecting each item for the sample is assumed to be an independent trial, the binomial probability formula applies. The probability of success (a defective item) is 0.01, while the probability of failure (an acceptable item) is 0.99. This makes

$$
\begin{aligned}
P(1 \text{ defective in 15 items}) &= C(15, 1)(0.01)^1(0.99)^{14} \\
&= 15(0.01)(0.99)^{14} \\
&\approx 0.1303.
\end{aligned}
$$

(b) The probability of at most 1 defective item in a random sample of 15 items from a production line

SOLUTION "At most 1" means 0 defective items or 1 defective item. Since 0 defective items is equivalent to 15 acceptable items,

$$
P(0 \text{ defective}) = (0.99)^{15} \approx 0.8601.
$$

YOUR TURN 3 In Example 4, find the probability of 2 or 3 defective items in a random sample of 15 items from a production line.

Use the union rule, noting that 0 defective and 1 defective are mutually exclusive events, to get

$$P(\text{at most 1 defective}) = P(0 \text{ defective}) + P(1 \text{ defective})$$
$$\approx 0.8601 + 0.1303$$
$$= 0.9904. \qquad \text{TRY YOUR TURN 3}$$

EXAMPLE 5 Checkout Scanners

The Federal Trade Commission (FTC) monitors pricing accuracy to ensure that consumers are charged the correct price at the checkout. According to the FTC, 29% of stores that use checkout scanners do not accurately charge customers. *Source: Federal Trade Commission.*

(a) If you shop at 3 stores that use checkout scanners, what is the probability that you will be incorrectly charged in at least one store?

SOLUTION We can treat this as a binomial experiment, letting $n = 3$ and $p = 0.29$. We need to find the probability of "at least 1" incorrect charge, which means 1 or 2 or 3 incorrect charges. To make our calculation simpler, we will use the complement. We will find the probability of being charged incorrectly in none of the 3 stores, that is, $P(0 \text{ incorrect charges})$, and then find $1 - P(0 \text{ incorrect charges})$.

$$P(0 \text{ incorrect charges}) = C(3,0)(0.29)^0(0.71)^3$$
$$= 1(1)(0.357911) \approx 0.3579$$
$$P(\text{at least one}) = 1 - P(0 \text{ incorrect charges})$$
$$\approx 1 - 0.3579 = 0.6421$$

(b) If you shop at 3 stores that use checkout scanners, what is the probability that you will be incorrectly charged in at most one store?

YOUR TURN 4 In Example 5, if you shop at 4 stores that use checkout scanners, find the probability that you will be charged incorrectly in at least one store.

SOLUTION "At most one" means 0 or 1, so

$$P(0 \text{ or } 1) = P(0) + P(1)$$
$$= C(3,0)(0.29)^0(0.71)^3 + C(3,1)(0.29)^1(0.71)^2$$
$$= 1(1)(0.357911) + 3(0.29)(0.5041) \approx 0.7965. \qquad \text{TRY YOUR TURN 4}$$

The triangular array of numbers shown below is called **Pascal's triangle** in honor of the French mathematician Blaise Pascal (1623–1662), who was one of the first to use it extensively. The triangle was known long before Pascal's time and appears in Chinese and Islamic manuscripts from the eleventh century.

Pascal's Triangle
```
              1
           1     1
        1     2     1
     1     3     3     1
  1     4     6     4     1
1     5    10    10     5     1
```

The array provides a quick way to find binomial probabilities. The nth row of the triangle, where $n = 0, 1, 2, 3, \ldots$, gives the coefficients $C(n, r)$ for $r = 0, 1, 2, 3, \ldots, n$. For example, for $n = 4$, $1 = C(4, 0)$, $4 = C(4, 1)$, $6 = C(4, 2)$, and so on. Each number in the triangle is the sum of the two numbers directly above it. For example, in the row for

$n = 4$, 1 is the sum of 1, the only number above it, 4 is the sum of 1 and 3, 6 is the sum of 3 and 3, and so on. Adding in this way gives the sixth row:

$$1 \quad 6 \quad 15 \quad 20 \quad 15 \quad 6 \quad 1.$$

Notice that Pascal's triangle tells us, for example, that $C(4, 1) + C(4, 2) = C(5, 2)$ (that is, $4 + 6 = 10$). Using the combinations formula, it can be shown that, in general, $C(n, r) + C(n, r + 1) = C(n + 1, r + 1)$. This is left as an exercise.

EXAMPLE 6   Pascal's Triangle

Use Pascal's triangle to find the probability in Example 5 that if you shop at 6 stores that use checkout scanners, at least 3 will charge you incorrectly

SOLUTION   The probability of success is 0.29. Since at least 3 means 3, 4, 5, or 6,

$$P(\text{at least } 3) = P(3) + P(4) + P(5) + P(6)$$
$$= C(6, 3)(0.29)^3(0.71)^3 + C(6, 4)(0.29)^4(0.71)^2$$
$$+ C(6, 5)(0.29)^5(0.71)^1 + C(6, 6)(0.29)^6(0.71)^0.$$

Use the sixth row of Pascal's triangle for the combinations to get

$$P(\text{at least } 3) = 20(0.29)^3(0.71)^3 + 15(0.29)^4(0.71)^2$$
$$+ 6(0.29)^5(0.71)^1 + 1(0.29)^6(0.71)^0$$
$$= 0.1746 + 0.0535 + 0.0087 + 0.0006$$
$$= 0.2374.$$

YOUR TURN 5 In Example 5, find the probability that if you shop at 6 stores with checkout scanners, at most 3 stores will charge you incorrectly.

TRY YOUR TURN 5

EXAMPLE 7   Independent Jury

If each member of a 9-person jury acts independently of each other and makes the correct determination of guilt or innocence with probability 0.65, find the probability that the majority of jurors will reach a correct verdict. *Source: Frontiers in Economics.*

Method 1
Calculating By Hand

SOLUTION   Since the jurors in this particular situation act independently, we can treat this as a binomial experiment. Thus, the probability that the majority of the jurors will reach the correct verdict is given by

$$P(\text{at least } 5) = C(9, 5)(0.65)^5(0.35)^4 + C(9, 6)(0.65)^6(0.35)^3$$
$$+ C(9, 7)(0.65)^7(0.35)^2 + C(9, 8)(0.65)^8(0.35)^1 + C(9, 9)(0.65)^9$$
$$= 0.2194 + 0.2716 + 0.2162 + 0.1004 + 0.0207$$
$$= 0.8283.$$

Method 2
Graphing Calculator

Some graphing calculators provide binomial probabilities. On a TI-84 Plus, for example, the command `binompdf(9, .65, 5)`, found in the DISTR menu, gives 0.21939, which is the probability that $x = 5$. Alternatively, the command `binomcdf(9, .65, 4)` gives 0.17172 as the probability that 4 or fewer jurors will make the correct decision. Subtract 0.17172 from 1 to get 0.82828 as the probability that the majority of the jurors will make the correct decision. This value rounds to 0.8283, which is in agreement with Method 1. Often, graphing calculators are more accurate than calculations by hand due to the accumulation of rounding errors when doing successive calculations by hand.

Method 3
Spreadsheet

Some spreadsheets also provide binomial probabilities. In Microsoft Excel, for example, the command "=BINOMDIST(5, 9, .65, 0)" gives 0.21939, which is the probability that $x = 5$. Alternatively, the command "=BINOMDIST (4, 9, .65, 1)" gives 0.17172 as the probability that 4 or fewer jurors will make the correct decision. Subtract 0.17172 from 1 to get 0.82828 as the probability that the majority of the jurors will make the correct decision. This value agrees with the value found in Methods 1 and 2.

8.4 EXERCISES

Suppose that a family has 5 children. Also, suppose that the probability of having a girl is 1/2. Find the probabilities that the family has the following children.

1. Exactly 2 girls and 3 boys

2. Exactly 3 girls and 2 boys

3. No girls

4. No boys

5. At least 4 girls

6. At least 3 boys

7. No more than 3 boys

8. No more than 4 girls

A die is rolled 12 times. Find the probabilities of rolling the following.

9. Exactly 12 ones

10. Exactly 6 ones

11. Exactly 1 one

12. Exactly 2 ones

13. No more than 3 ones

14. No more than 1 one

A certain unfair coin lands on heads 1/4 of the time. This coin is tossed 6 times. Find the probabilities of getting the following.

15. All heads

16. Exactly 3 heads

17. No more than 3 heads

18. At least 3 heads

19. How do you identify a probability problem that involves a binomial experiment?

20. How is Pascal's triangle used to find probabilities?

21. Using the definition of combination in Section 8.2, prove that

$$C(n, r) + C(n, r + 1) = C(n + 1, r + 1)$$

(This is the formula underlying Pascal's triangle.)

In Exercises 22 and 23, argue that the use of binomial probabilities is not applicable and, thus, the probabilities that are computed are not correct.

22. In England, a woman was found guilty of smothering her two infant children. Much of the Crown's case against the lady was based on the testimony from a pediatrician who indicated that the chances of two crib deaths occurring in both siblings was only about 1 in 73 million. This number was calculated by assuming that the probability of a single crib death is 1 in 8543 and the probability of two crib deaths is 1 in 8543^2 (i.e., binomial). (See Chapter 7 Review Exercise 102.) *Source: Science.*

23. A contemporary radio station in Boston has a contest in which a caller is asked his or her date of birth. If the caller's date of birth, including the day, month, and year of birth, matches a predetermined date, the caller wins $1 million. Assuming that there were 36,525 days in the twentieth century and the contest was run 51 times on consecutive days, the probability that the grand prize will be won is

$$1 - \left(1 - \frac{1}{36,525}\right)^{51} \approx 0.0014.$$

Source: Chance News.

APPLICATIONS

Business and Economics

Management The survey discussed in Example 5 also found that customers are charged incorrectly for 1 out of every 30 items, on average. Suppose a customer purchases 15 items. Find the following probabilities.

24. A customer is charged incorrectly on 3 items.

25. A customer is not charged incorrectly for any item.

26. A customer is charged incorrectly on at least one item.

27. A customer is charged incorrectly on at least 2 items.

28. A customer is charged incorrectly on at most 2 items.

Credit Cards A survey of consumer finance found that 25.4% of credit-card-holding families hardly ever pay off the balance. Suppose a random sample of 20 credit-card-holding families is taken. Find the probabilities of each of the following results. *Source: Statistical Abstract of the United States.*

29. Exactly 6 families hardly ever pay off the balance.

30. Exactly 9 families hardly ever pay off the balance.

31. At least 4 families hardly ever pay off the balance.

32. At most 5 families hardly ever pay off the balance.

Personnel Screening A company gives prospective workers a 6-question, multiple-choice test. Each question has 5 possible answers, so that there is a 1/5 or 20% chance of answering a question correctly just by guessing. Find the probabilities of getting the following results by chance.

33. Exactly 2 correct answers

34. No correct answers

35. At least 4 correct answers

36. No more than 3 correct answers

Quality Control A factory tests a random sample of 20 transistors for defects. The probability that a particular transistor will be defective has been established by past experience as 0.05.

37. What is the probability that there are no defective transistors in the sample?

38. What is the probability that the number of defective transistors in the sample is at most 2?

39. Quality Control The probability that a certain machine turns out a defective item is 0.05. Find the probabilities that in a run of 75 items, the following results are obtained.

a. Exactly 5 defective items

b. No defective items

c. At least 1 defective item

40. Survey Results A company is taking a survey to find out whether people like its product. Its last survey indicated that 70% of the population like the product. Based on that, in a sample of 58 people, find the probabilities of the following.

a. All 58 like the product.

b. From 28 to 30 (inclusive) like the product.

41. Pecans Pecan producers blow air through the pecans so that the lighter ones are blown out. The lighter-weight pecans are generally bad and the heavier ones tend to be better. These "blow outs" and "good nuts" are often sold to tourists along the highway. Suppose 60% of the "blow outs" are good, and 80% of the "good nuts" are good. *Source: Irvin R. Hentzel.*

a. What is the probability that if you crack and check 20 "good nuts" you will find 8 bad ones?

b. What is the probability that if you crack and check 20 "blow outs" you will find 8 bad ones?

c. If we assume that 70% of the roadside stands sell "good nuts," and that out of 20 nuts we find 8 that are bad, what is the probability that the nuts are "blow outs"?

42. Hurricane Insurance A company prices its hurricane insurance using the following assumptions:

   (i) In any calendar year, there can be at most one hurricane.

   (ii) In any calendar year, the probability of a hurricane is 0.05.

   (iii) The number of hurricanes in any calendar year is independent of the number of hurricanes in any other calendar year.

Using the company's assumptions, calculate the probability that there are fewer than 3 hurricanes in a 20-year period. Choose one of the following. *Source: Society of Actuaries.*

a. 0.06   **b.** 0.19   **c.** 0.38   **d.** 0.62   **e.** 0.92

Life Sciences

Breast Cancer A recent study found that 85% of breast-cancer cases are detectable by mammogram. Suppose a random sample of 15 women with breast cancer are given mammograms. Find the probability of each of the following results, assuming that detection in the cases is independent. *Source: Harper's Index.*

43. All of the cases are detectable.

44. None of the cases are detectable.

45. Not all cases are detectable.

46. More than half of the cases are detectable.

Births of Twins The probability that a birth will result in twins is 0.012. Assuming independence (perhaps not a valid assumption), what are the probabilities that out of 100 births in a hospital, there will be the following numbers of sets of twins?

47. Exactly 2 sets of twins

48. At most 2 sets of twins

49. Effects of Radiation The probability of a mutation of a given gene under a dose of 1 roentgen of radiation is approximately 2.5×10^{-7}. What is the probability that in 10,000 genes, at least 1 mutation occurs?

50. Flu Inoculations A flu vaccine has a probability of 80% of preventing a person who is inoculated from getting the flu. A county health office inoculates 83 people. Find the probabilities of the following.

a. Exactly 10 of the people inoculated get the flu.

b. No more than 4 of the people inoculated get the flu.

c. None of the people inoculated get the flu.

51. Color Blindness The probability that a male will be color-blind is 0.042. Find the probabilities that in a group of 53 men, the following will be true.

a. Exactly 5 are color-blind.

b. No more than 5 are color-blind.

c. At least 1 is color-blind.

52. Pharmacology In placebo-controlled trials of Pravachol®, a drug that is prescribed to lower cholesterol, 7.3% of the patients who were taking the drug experienced nausea/ vomiting, whereas 7.1% of the patients who were taking the placebo experienced nausea/vomiting. *Source: Bristol-Myers Squibb Company.*

a. If 100 patients who are taking Pravachol® are selected, what is the probability that 10 or more will experience nausea/ vomiting?

b. If a second group of 100 patients receives a placebo, what is the probability that 10 or more will experience nausea/vomiting?

c. Since 7.3% is larger than 7.1%, do you believe that the Pravachol® causes more people to experience nausea/ vomiting than a placebo? Explain.

53. Genetic Fingerprinting The use of DNA has become an integral part of many court cases. When DNA is extracted from cells and body fluids, genetic information is represented by bands of information, which look similar to a bar code at a grocery store. It is generally accepted that in unrelated people, the probability of a particular band matching is 1 in 4. *Source: University of Exeter.*

a. If 5 bands are compared in unrelated people, what is the probability that all 5 of the bands match? (Express your answer in terms of "1 chance in ?".)

b. If 20 bands are compared in unrelated people, what is the probability that all 20 of the bands match? (Express your answer in terms of "1 chance in ?".)

c. If 20 bands are compared in unrelated people, what is the probability that 16 or more bands match? (Express your answer in terms of "1 chance in ?".)

d. If you were deciding paternity and there were 16 matches out of 20 bands compared, would you believe that the person being tested was the father? Explain.

54. Salmonella According to *The Salt Lake Tribune*, the Coffee Garden in Salt Lake City ran into trouble because of their four-egg quiche:

"A Salt Lake County Health Department inspector paid a visit recently and pointed out that research by the Food and Drug Administration indicates that one in four eggs carries *Salmonella* bacterium, so restaurants should never use more than three eggs when preparing quiche.

The manager on duty wondered aloud if simply throwing out three eggs from each dozen and using the remaining nine in four-egg quiches would serve the same purpose.

The inspector wasn't sure, but she said she would research it." *Source: The Salt Lake Tribune.*

a. Assuming that one in four eggs carries *Salmonella*, and that the event that any one egg is infected is independent of whether any other egg is infected, find the probability that at least one of the eggs in a four-egg quiche carries *Salmonella*.

b. Repeat part a for a three-egg quiche.

c. Discuss whether the assumption of independence is justified.

d. Discuss whether the inspector's reasoning makes sense.

55. Herbal Remedies According to Dr. Peter A.G.M. De Smet of the Netherlands, "If an herb caused an adverse reaction in 1 in 1,000 users, a traditional healer would have to treat 4,800 patients with that herb (i.e., one new patient every single working day for more than 18 years) to have a 95 percent chance of observing the reaction in more than one user." Verify this calculation by finding the probability of observing more than one reaction in 4800 patients, given that 1 in 1000 has a reaction. *Source: The New England Journal of Medicine.*

56. Vaccines A hospital receives 1/5 of its flu vaccine shipments from Company X and the remainder of its shipments from other companies. Each shipment contains a very large number of vaccine vials. For Company X's shipments, 10% of the vials are ineffective. For every other company, 2% of the vials are ineffective. The hospital tests 30 randomly selected vials from a shipment and finds that one vial is ineffective. What is the probability that this shipment came from Company X? Choose one of the following. (*Hint:* Find the probability that one out of 30 vials is ineffective, given that the shipment came from Company X and that the shipment came from other companies. Then use Bayes' theorem.) *Source: Society of Actuaries.*

a. 0.10 **b.** 0.14 **c.** 0.37 **d.** 0.63 **e.** 0.86

57. Health Study A study is being conducted in which the health of two independent groups of ten policyholders is being monitored over a one-year period of time. Individual participants in the study drop out before the end of the study with probability 0.2 (independently of the other participants). What is the probability that at least 9 participants complete the study in one of the two groups, but not in both groups? Choose one of the following. *Source: Society of Actuaries.*

a. 0.096 **b.** 0.192 **c.** 0.235 **d.** 0.376 **e.** 0.469

Social Sciences

58. Women Working A recent study found that 60% of working mothers would prefer to work part-time if money were not a concern. Find the probability that if 10 working mothers are selected at random, that at least 3 of them would prefer to work part-time. *Source: Pew Research Center.*

Volunteering A recent survey found that 83% of first-year college students were involved in volunteer work at least occasionally. Suppose a random sample of 12 college students is taken. Find the probabilities of each of the following results. *Source: The New York Times.*

59. Exactly 7 students volunteered at least occasionally.

60. Exactly 9 students volunteered at least occasionally.

61. At least 9 students volunteered at least occasionally.

62. At most 9 students volunteered at least occasionally.

63. Minority Enrollment According to the U.S. Department of Education, 32.2% of all students enrolled in degree-granting institutions (those that grant associate's or higher degrees) belong to minorities. Find the probabilities of the following results in a random sample of 10 students enrolled in degree-granting institutions. *Source: National Center for Education Statistics.*

a. Exactly 2 belong to a minority.

b. Three or fewer belong to a minority.

c. Exactly 5 do not belong to a minority.

d. Six or more do not belong to a minority.

64. Cheating According to a poll conducted by *U.S. News and World Report*, 84% of college students believe they need to cheat to get ahead in the world today. *Source: U.S. News and World Report.*

a. Do the results of this poll indicate that 84% of all college students cheat? Explain.

b. If this result is accurate and 100 college students are asked if they believe that cheating is necessary to get ahead in the world, what is the probability that 90 or more of the students will answer affirmatively to the question?

65. Education A study by Cleveland Clinic tracked a cohort of very-low-birth-weight infants for twenty years. The results of the study indicated that 74% of the very-low-birth-weight babies graduated from high school during this time period. The study also reported that 83% of the comparison group of normal-birth-weight babies graduated from high school during the same period. *Source: The New England Journal of Medicine.*

a. If 40 very-low-birth-weight babies were tracked through high school, what is the probability that at least 30 will graduate from high school by age 20?

b. If 40 babies from the comparison group were tracked through high school, what is the probability that at least 30 will graduate from high school by age 20?

66. War Dead A newspaper article questioned whether soldiers and marines from some states bear greater risks in Afghanistan and Iraq than those from others. Out of 644,066 troops deployed as of the time of the article, 1174 had been killed, for a probability of being killed of $p = 1174/644{,}066$. Assume the deaths are independent.* *Source: Valley News.*

a. Vermont had 9 deaths out of 1613 troops deployed. Find the probability of at least this many deaths.

*For further statistical analysis, see http://www.dartmouth.edu/~chance/ForWiki/GregComments.pdf.

b. Massachusetts had 28 deaths out of 7146 troops deployed. Find the probability of at least this many deaths.

c. Florida had 54 deaths out of 62,572 troops deployed. Find the probability of at most this many deaths.

d. Discuss why the assumption of independence may be questionable.

67. Sports In many sports championships, such as the World Series in baseball and the Stanley Cup final series in hockey, the winner is the first team to win four games. For this exercise, assume that each game is independent of the others, with a constant probability p that one specified team (say, the National League team) wins.

a. Find the probability that the series lasts for four, five, six, and seven games when $p = 0.5$. (*Hint:* Suppose the National

League wins the series, so they must win the last game. Consider how the previous games might come out. Then consider the probability that the American League wins.)

b. Morrison and Schmittlein have found that the Stanley Cup finals can be described by letting $p = 0.73$ be the probability that the better team wins each game. Find the probability that the series lasts for four, five, six, and seven games. *Source: Chance.*

c. Some have argued that the assumption of independence does not apply. Discuss this issue. *Source: Mathematics Magazine.*

YOUR TURN ANSWERS

1. 0.1475   **2.** 0.2734   **3.** 0.009617
4. 0.7459   **5.** 0.9372

8.5 Probability Distributions; Expected Value

APPLY IT **What is the expected payback for someone who buys one ticket in a raffle?** *In Example 3, we will calculate the expected payback or expected value of this raffle.*

We shall see that the *expected value* of a probability distribution is a type of average. Probability distributions were introduced briefly in Chapter 7 on Sets and Probability. Now we take a more complete look at probability distributions. A probability distribution depends on the idea of a *random variable*, so we begin with that.

Random Variables

When researchers carry out an experiment it is necessary to quantify the possible outcomes of the experiment. This process will enable the researcher to recognize individual outcomes and analyze the data. The most common way to keep track of the individual outcomes of the experiment is to assign a numerical value to each of the different possible outcomes of the experiment. For example, if a coin is tossed 2 times, the possible outcomes are: *hh*, *ht*, *th*, and *tt*. For each of these possible outcomes, we could record the number of heads. Then the outcome, which we will label x, is one of the numbers 0, 1, or 2. Of course, we could have used other numbers, like 00, 01, 10, and 11, to indicate these same outcomes, but the values of 0, 1, and 2 are simpler and provide an immediate description of the exact outcome of the experiment. Notice that using this random variable also gives us a way to readily know how many tails occurred in the experiment. Thus, in some sense, the values of x are random, so x is called a **random variable.**

Random Variable
A **random variable** is a function that assigns a real number to each outcome of an experiment.

Probability Distribution

A table that lists the possible values of a random variable, together with the corresponding probabilities, is called a **probability distribution**. The sum of the probabilities in a probability distribution must always equal 1. (The sum in some distributions may vary slightly from 1 because of rounding.)

EXAMPLE 1 Computer Monitors

A shipment of 12 computer monitors contains 3 broken monitors. A shipping manager checks a sample of four monitors to see if any are broken. Give the probability distribution for the number of broken monitors that the shipping manager finds.

SOLUTION Let x represent the random variable "number of broken monitors found by the manager." Since there are 3 broken monitors, the possible values of x are 0, 1, 2, and 3. We can calculate the probability of each x using the methods of Section 8.3. There are 3 broken monitors, and 9 unbroken monitors, so the number of ways of choosing 0 broken monitors (which implies 4 unbroken monitors) is $C(3, 0) \cdot C(9, 4)$. The number of ways of choosing a sample of 4 monitors is $C(12, 4)$. Therefore, the probability of choosing 0 broken monitors is

$$P(0) = \frac{C(3, 0) \cdot C(9, 4)}{C(12, 4)} = \frac{1\left(\dfrac{9 \cdot 8 \cdot 7 \cdot 6}{4 \cdot 3 \cdot 2 \cdot 1}\right)}{\left(\dfrac{12 \cdot 11 \cdot 10 \cdot 9}{4 \cdot 3 \cdot 2 \cdot 1}\right)} = \frac{126}{495} = \frac{14}{55}.$$

Similarly, the probability of choosing 1 broken monitor is

$$P(1) = \frac{C(3, 1) \cdot C(9, 3)}{C(12, 4)} = \frac{3 \cdot 84}{495} = \frac{252}{495} = \frac{28}{55}.$$

The probability of choosing 2 broken monitors is

$$P(2) = \frac{C(3, 2) \cdot C(9, 2)}{C(12, 4)} = \frac{3 \cdot 36}{495} = \frac{108}{495} = \frac{12}{55}.$$

The probability of choosing 3 broken monitors is

$$P(3) = \frac{C(3, 3) \cdot C(9, 1)}{C(12, 4)} = \frac{1 \cdot 9}{495} = \frac{9}{495} = \frac{1}{55}.$$

YOUR TURN 1 Suppose the inspector in Example 1 chose only two monitors to inspect. Find the probability distribution for the number of broken monitors.

The results can be put in a table, called a probability distribution.

Probability Distribution of Broken Monitors in Sample			
x	0	1	2	3
$P(x)$	14/55	28/55	12/55	1/55

TRY YOUR TURN 1

Instead of writing the probability distribution as a table, we could write the same information as a set of ordered pairs:

$$\{(0, 14/55), (1, 28/55), (2, 12/55), (3, 1/55)\}.$$

There is just one probability for each value of the random variable. Thus, a probability distribution defines a function, called a **probability distribution function**, or simply a **probability function**. We shall use the terms "probability distribution" and "probability function" interchangeably.

The information in a probability distribution is often displayed graphically as a special kind of bar graph called a **histogram**. The bars of a histogram all have the same width, usually 1. (The widths might be different from 1 when the values of the random variable are not consecutive integers.) The heights of the bars are determined by the probabilities. A histogram for the data in Example 1 is given in Figure 8. A histogram shows important characteristics of a distribution that may not be readily apparent in tabular form, such as the relative sizes of the probabilities and any symmetry in the distribution.

The area of the bar above $x = 0$ in Figure 8 is the product of 1 and 14/55, or $1 \cdot 14/55 = 14/55$. Since each bar has a width of 1, its area is equal to the probability that corresponds to that value of x. The probability that a particular value will occur is thus given by the area of the appropriate bar of the graph. For example, the probability that one or more monitors is broken is the sum of the areas for $x = 1$, $x = 2$, and $x = 3$. This area, shown in pink in Figure 9, corresponds to 41/55 of the total area, since

$$P(x \geq 1) = P(x = 1) + P(x = 2) + P(x = 3)$$
$$= 28/55 + 12/55 + 1/55 = 41/55.$$

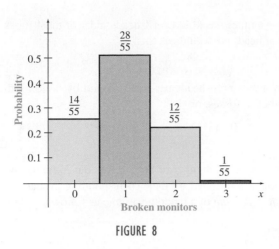

FIGURE 8

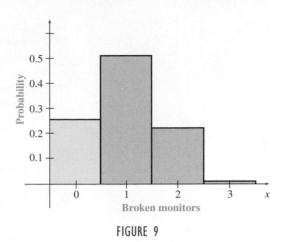

FIGURE 9

EXAMPLE 2 Probability Distributions

(a) Give the probability distribution for the number of heads showing when two coins are tossed.

SOLUTION Let x represent the random variable "number of heads." Then x can take on the values 0, 1, or 2. Now find the probability of each outcome. To find the probability of 0, 1, or 2 heads, we can either use binomial probability, or notice that there are 4 outcomes in the sample space: $\{hh, ht, th, tt\}$. The results are shown in the table with Figure 10.

Probability Distribution of Heads		
x	0	1	2
$P(x)$	1/4	1/2	1/4

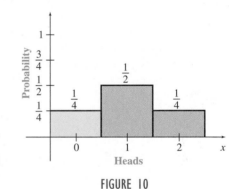

FIGURE 10

YOUR TURN 2 Find the probability distribution and draw a histogram for the number of tails showing when three coins are tossed.

(b) Draw a histogram for the distribution in the table. Find the probability that at least one coin comes up heads.

SOLUTION The histogram is shown in Figure 10. The portion in pink represents

$$P(x \geq 1) = P(x = 1) + P(x = 2)$$

$$= \frac{3}{4}.$$

TRY YOUR TURN 2

Expected Value

In working with probability distributions, it is useful to have a concept of the typical or average value that the random variable takes on. In Example 2, for instance, it seems reasonable that, on the average, one head shows when two coins are tossed. This does not tell what will happen the next time we toss two coins; we may get two heads, or we may get none. If we tossed two coins many times, however, we would expect that, in the long run, we would average about one head for each toss of two coins.

A way to solve such problems in general is to imagine flipping two coins 4 times. Based on the probability distribution in Example 2, we would expect that 1 of the 4 times we would get 0 heads, 2 of the 4 times we would get 1 head, and 1 of the 4 times we would get 2 heads. The total number of heads we would get, then, is

$$0 \cdot 1 + 1 \cdot 2 + 2 \cdot 1 = 4.$$

The expected numbers of heads per toss is found by dividing the total number of heads by the total number of tosses, or

$$\frac{0 \cdot 1 + 1 \cdot 2 + 2 \cdot 1}{4} = 0 \cdot \frac{1}{4} + 1 \cdot \frac{1}{2} + 2 \cdot \frac{1}{4} = 1.$$

Notice that the expected number of heads turns out to be the sum of the three values of the random variable x multiplied by their corresponding probabilities. We can use this idea to define the *expected value* of a random variable as follows.

Expected Value
Suppose the random variable x can take on the n values $x_1, x_2, x_3, \ldots, x_n$. Also, suppose the probabilities that these values occur are, respectively, $p_1, p_2, p_3, \ldots, p_n$. Then the **expected value** of the random variable is

$$E(x) = x_1 p_1 + x_2 p_2 + x_3 p_3 + \cdots + x_n p_n.$$

EXAMPLE 3 Computer Monitors

In Example 1, find the expected number of broken monitors that the shipping manager finds.
SOLUTION Using the values in the first table in this section and the definition of expected value, we find that

$$E(x) = 0 \cdot \frac{14}{55} + 1 \cdot \frac{28}{55} + 2 \cdot \frac{12}{55} + 3 \cdot \frac{1}{55} = 1.$$

On average, the shipping manager will find 1 broken monitor in the sample of 4. On reflection, this seems natural; 3 of the 12 monitors, or 1/4 of the total, are broken. We should expect, then, that 1/4 of the sample of 4 monitors are broken.

Physically, the expected value of a probability distribution represents a balance point. If we think of the histogram in Figure 8 as a series of weights with magnitudes represented by the heights of the bars, then the system would balance if supported at the point corresponding to the expected value.

EXAMPLE 4 Symphony Orchestra

Suppose a local symphony decides to raise money by raffling an HD television worth $400, a dinner for two worth $80, and 2 CDs worth $20 each. A total of 2000 tickets are sold at $1 each. Find the expected payback for a person who buys one ticket in the raffle.

APPLY IT

Method 1
Direct Calculation

SOLUTION
Here the random variable represents the possible amounts of payback, where payback = amount won − cost of ticket. The payback of the person winning the television is $400 (amount won) − $1 (cost of ticket) = $399. The payback for each losing ticket is $0 − $1 = −$1.

The paybacks of the various prizes, as well as their respective probabilities, are shown in the table on the next page. The probability of winning $19 is 2/2000 because there are 2 prizes worth $20. We have not reduced the fractions in order to keep all the denominators equal. Because there are 4 winning tickets, there are 1996 losing tickets, so the probability of winning −$1 is 1996/2000.

Probability Distribution of Prize Winnings				
x	$399	$79	$19	−$1
$P(x)$	1/2000	1/2000	2/2000	1996/2000

The expected payback for a person buying one ticket is

$$399\left(\frac{1}{2000}\right) + 79\left(\frac{1}{2000}\right) + 19\left(\frac{2}{2000}\right) + (-1)\left(\frac{1996}{2000}\right) = -\frac{1480}{2000}$$
$$= -0.74.$$

On average, a person buying one ticket in the raffle will lose $0.74, or 74¢.

It is not possible to lose 74¢ in this raffle: either you lose $1, or you win a prize worth $400, $80, or $20, minus the $1 you pay to play. But if you bought tickets in many such raffles over a long period of time, you would lose 74¢ per ticket on average. It is important to note that the expected value of a random variable may be a number that can never occur in any one trial of the experiment.

Method 2
Alternate Procedure

An alternative way to compute expected value in this and other examples is to calculate the expected amount won and then subtract the cost of the ticket afterward. The amount won is either $400 (with probability 1/2000), $80 (with probability 1/2000), $20 (with probability 2/2000), or $0 (with probability 1996/2000). The expected payback for a person buying one ticket is then

$$400\left(\frac{1}{2000}\right) + 80\left(\frac{1}{2000}\right) + 20\left(\frac{2}{2000}\right) + 0\left(\frac{1996}{2000}\right) - 1 = -\frac{1480}{2000}$$
$$= -0.74.$$

TRY YOUR TURN 3

YOUR TURN 3 Suppose that there is a $5 raffle with prizes worth $1000, $500, and $250. Suppose 1000 tickets are sold and you purchase a ticket. Find your expected payback for this raffle.

EXAMPLE 5 Friendly Wager

Each day Donna and Mary toss a coin to see who buys coffee ($1.20 a cup). One tosses and the other calls the outcome. If the person who calls the outcome is correct, the other buys the coffee; otherwise the caller pays. Find Donna's expected payback.

SOLUTION   Assume that an honest coin is used, that Mary tosses the coin, and that Donna calls the outcome. The possible results and corresponding probabilities are shown below.

Possible Results				
Result of Toss	Heads	Heads	Tails	Tails
Call	Heads	Tails	Heads	Tails
Caller Wins?	Yes	No	No	Yes
Probability	1/4	1/4	1/4	1/4

Donna wins a $1.20 cup of coffee whenever the results and calls match, and she loses a $1.20 cup when there is no match. Her expected payback is

$$(1.20)\left(\frac{1}{4}\right) + (-1.20)\left(\frac{1}{4}\right) + (-1.20)\left(\frac{1}{4}\right) + (1.20)\left(\frac{1}{4}\right) = 0.$$

This implies that, over the long run, Donna neither wins nor loses.

A game with an expected value of 0 (such as the one in Example 5) is called a **fair game**. Casinos do not offer fair games. If they did, they would win (on average) $0, and have a hard time paying the help! Casino games have expected winnings for the house that vary from 1.5 cents per dollar to 60 cents per dollar. Exercises 47–52 at the end of the section ask you to find the expected payback for certain games of chance.

The idea of expected value can be very useful in decision making, as shown by the next example.

EXAMPLE 6   Life Insurance

At age 50, you receive a letter from Mutual of Mauritania Insurance Company. According to the letter, you must tell the company immediately which of the following two options you will choose: take $20,000 at age 60 (if you are alive, $0 otherwise) or $30,000 at age 70 (again, if you are alive, $0 otherwise). Based *only* on the idea of expected value, which should you choose?

SOLUTION   Life insurance companies have constructed elaborate tables showing the probability of a person living a given number of years into the future. From a recent such table, the probability of living from age 50 to 60 is 0.88, while the probability of living from age 50 to 70 is 0.64. The expected values of the two options are given below.

$$\text{First option: } (20{,}000)(0.88) + (0)(0.12) = 17{,}600$$
$$\text{Second option: } (30{,}000)(0.64) + (0)(0.36) = 19{,}200$$

Based strictly on expected values, choose the second option.　■

EXAMPLE 7   Bachelor's Degrees

According to the National Center for Education Statistics, 78.7% of those earning bachelor's degrees in education in the United States in 2006–2007 were female. Suppose 5 holders of bachelor's degrees in education from 2006 to 2007 are picked at random. *Source: National Center for Education Statistics.*

(a) Find the probability distribution for the number that are female.

SOLUTION   We first note that each of the 5 people in the sample is either female (with probability 0.787) or male (with probability 0.213). As in the previous section, we may assume that the probability for each member of the sample is independent of that of any other. Such a situation is described by binomial probability with $n = 5$ and $p = 0.787$, for which we use the binomial probability formula

$$P(x \text{ successes in } n \text{ trials}) = C(n, x) \cdot p^x \cdot (1 - p)^{n-x},$$

where x is the number of females in the sample. For example, the probability of 0 females is

$$P(x = 0) = C(5, 0)(0.787)^0(0.213)^5 \approx 0.0004.$$

Similarly, we could calculate the probability that x is any value from 0 to 5, resulting in the probability distribution below (with all probabilities rounded to four places).

Probability Distribution of Female Education Graduate						
x	0	1	2	3	4	5
$P(x)$	0.0004	0.0081	0.0599	0.2211	0.4086	0.3019

(b) Find the expected number of females in the sample of 5 people.

SOLUTION   Using the formula for expected value, we have

$$E(x) = 0(0.0004) + 1(0.0081) + 2(0.0599) + 3(0.2211)$$
$$+ 4(0.4086) + 5(0.3019) = 3.935.$$

On average, 3.935 of the people in the sample of 5 will be female.

TRY YOUR TURN 4 ■

YOUR TURN 4   In the same survey quoted in Example 7, 81.6% of those earning bachelor's degrees in engineering were male. Suppose 5 holders of bachelor's degrees in engineering were picked at random, find the expected number of male engineers.

There is another way to get the answer in part (b) of the previous example. Because 78.7% of those earning bachelor's degrees in education in the United States in 2006–2007 are female, it is reasonable to expect 78.7% of our sample to be female. Thus, 78.7% of 5 is $5(0.787) = 3.935$. Notice that what we have done is to multiply n by p. It can be shown that this method always gives the expected value for binomial probability.

Expected Value for Binomial Probability

For binomial probability, $E(x) = np$. In other words, the expected number of successes is the number of trials times the probability of success in each trial.

EXAMPLE 8   Female Children

Suppose a family has 3 children.

(a) Find the probability distribution for the number of girls.

SOLUTION   Assuming girls and boys are equally likely, the probability distribution is binomial with $n = 3$ and $p = 1/2$. Letting x be the number of girls in the formula for binomial probability, we find, for example,

$$P(x = 0) = C(3, 0)\left(\frac{1}{2}\right)^0\left(\frac{1}{2}\right)^3 = \frac{1}{8}.$$

The other values are found similarly, and the results are shown in the following table.

Probability Distribution of Number of Girls			
x	0	1	2	3
$P(x)$	1/8	3/8	3/8	1/8

We can verify this by noticing that in the sample space S of all 3-child families, there are eight equally likely outcomes: $S = \{ggg, ggb, gbg, gbb, bgg, bgb, bbg, bbb\}$. One of the outcomes has 0 girls, three have 1 girl, three have 2 girls, and one has 3 girls.

(b) Find the expected number of girls in a 3-child family using the distribution from part (a).

SOLUTION   Using the formula for expected value, we have

$$\text{Expected number of girls} = 0\left(\frac{1}{8}\right) + 1\left(\frac{3}{8}\right) + 2\left(\frac{3}{8}\right) + 3\left(\frac{1}{8}\right)$$

$$= \frac{12}{8} = 1.5.$$

On average, a 3-child family will have 1.5 girls. This result agrees with our intuition that, on average, half the children born will be girls.

(c) Find the expected number of girls in a 3-child family using the formula for expected value for binomial probability.

SOLUTION   Using the formula $E(x) = np$ with $n = 3$ and $p = 1/2$, we have

YOUR TURN 5   Find the expected number of girls in a family of a dozen children.

$$\text{Expected number of girls} = 3\left(\frac{1}{2}\right) = 1.5.$$

This agrees with our answer from part (b), as it must.   **TRY YOUR TURN 5**

8.5   EXERCISES

For each experiment described below, let x determine a random variable, and use your knowledge of probability to prepare a probability distribution.

1. Four coins are tossed, and the number of heads is noted.

2. Two dice are rolled, and the total number of points is recorded.

3. Three cards are drawn from a deck. The number of aces is counted.

4. Two balls are drawn from a bag in which there are 4 white balls and 2 black balls. The number of black balls is counted.

Draw a histogram for the following, and shade the region that gives the indicated probability.

5. Exercise 1; $P(x \le 2)$

6. Exercise 2; $P(x \ge 11)$

7. Exercise 3; $P(\text{at least one ace})$

8. Exercise 4; $P(\text{at least one black ball})$

Find the expected value for each random variable.

9.

x	2	3	4	5
$P(x)$	0.1	0.4	0.3	0.2

10.

y	4	6	8	10
$P(y)$	0.4	0.4	0.05	0.15

11.

z	9	12	15	18	21
$P(z)$	0.14	0.22	0.38	0.19	0.07

12.

x	30	32	36	38	44
$P(x)$	0.31	0.29	0.26	0.09	0.05

Find the expected value for the random variable x having the probability function shown in each graph.

13.

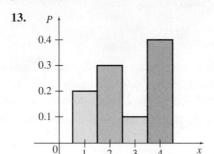

14.

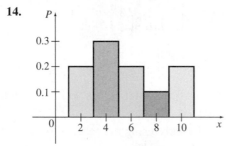

15.

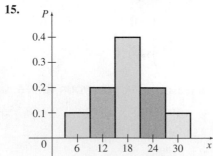

16.

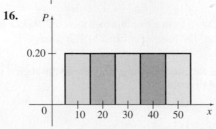

17. For the game in Example 5, find Mary's expected payback. Is it a fair game?

18. Suppose one day Mary brings a 2-headed coin and uses it to toss for the coffee. Since Mary tosses, Donna calls.

   a. Is this still a fair game?

   b. What is Donna's expected payback if she calls heads?

   c. What is Donna's expected payback if she calls tails?

Solve each exercise. Many of these exercises require the use of combinations.

19. Suppose 3 marbles are drawn from a bag containing 3 yellow and 4 white marbles.

   a. Draw a histogram for the number of yellow marbles in the sample.

   b. What is the expected number of yellow marbles in the sample?

20. Suppose 5 apples in a barrel of 25 apples are known to be rotten.

   a. Draw a histogram for the number of rotten apples in a sample of 2 apples.

   b. What is the expected number of rotten apples in a sample of 2 apples?

21. Suppose a die is rolled 4 times.

   a. Find the probability distribution for the number of times 1 is rolled.

   b. What is the expected number of times 1 is rolled?

22. A delegation of 3 is selected from a city council made up of 5 liberals and 6 conservatives.

   a. What is the expected number of liberals in the delegation?

   b. What is the expected number of conservatives in the delegation?

23. From a group of 3 women and 5 men, a delegation of 2 is selected. Find the expected number of women in the delegation.

24. In a club with 20 senior and 10 junior members, what is the expected number of junior members on a 4-member committee?

25. If 2 cards are drawn at one time from a deck of 52 cards, what is the expected number of diamonds?

26. Suppose someone offers to pay you $5 if you draw 2 diamonds in the game in Exercise 25. He says that you should pay 50 cents for the chance to play. Is this a fair game?

27. Your friend missed class the day probability distributions were discussed. How would you explain probability distribution to him?

28. Explain what expected value means in your own words.

29. Four slips of paper numbered 2, 3, 4, and 5 are in a hat. You draw a slip, note the result, and then draw a second slip and note the result (without replacing the first).

   a. Find the probability distribution for the sum of the two slips.

   b. Draw a histogram for the probability distribution in part a.

   c. Find the odds that the sum is even.

   d. Find the expected value of the sum.

APPLICATIONS

Business and Economics

30. Complaints A local used-car dealer gets complaints about his cars as shown in the table below. Find the expected number of complaints per day.

Number of Complaints per Day	0	1	2	3	4	5	6
Probability	0.02	0.06	0.16	0.25	0.32	0.13	0.06

31. Payout on Insurance Policies An insurance company has written 100 policies for $100,000, 500 policies for $50,000, and 1000 policies for $10,000 for people of age 20. If experience shows that the probability that a person will die at age 20 is 0.0012, how much can the company expect to pay out during the year the policies were written?

32. Device Failure An insurance policy on an electrical device pays a benefit of $4000 if the device fails during the first year. The amount of the benefit decreases by $1000 each successive year until it reaches 0. If the device has not failed by the beginning of any given year, the probability of failure during that year is 0.4. What is the expected benefit under this policy? Choose one of the following. *Source: Society of Actuaries.*

  a. $2234   **b.** $2400   **c.** $2500   **d.** $2667   **e.** $2694

33. Pecans Refer to Exercise 41 in Section 8.4. Suppose that 60% of the pecan "blow outs" are good, and 80% of the "good nuts" are good.

  a. If you purchase 50 pecans, what is the expected number of good nuts you will find if you purchase "blow outs"?

  b. If you purchase 50 pecans, what is the expected number of bad nuts you will find if you have purchased "good nuts"?

34. Rating Sales Accounts Levi Strauss and Company uses expected value to help its salespeople rate their accounts. For each account, a salesperson estimates potential additional volume and the probability of getting it. The product of these figures gives the expected value of the potential, which is added to the existing volume. The totals are then classified as A, B, or C, as follows: $40,000 or below, class C; from $40,000 up to and including $55,000, class B; above $55,000, class A. Complete the table below for one salesperson. *Source: James McDonald.*

35. Tour Bus A tour operator has a bus that can accommodate 20 tourists. The operator knows that tourists may not show up, so he sells 21 tickets. The probability that an individual tourist will not show up is 0.02, independent of all other tourists. Each ticket costs $50, and is non-refundable if a tourist fails to show up. If a tourist shows up and a seat is not available, the tour operator has to pay $100 (ticket cost + $50 penalty) to the tourist. What is the expected revenue of the tour operator? Choose one of the following. *Source: Society of Actuaries.*

  a. $935   **b.** $950   **c.** $967   **d.** $976   **e.** $985

Life Sciences

36. Animal Offspring In a certain animal species, the probability that a healthy adult female will have no offspring in a given year is 0.29, while the probabilities of 1, 2, 3, or 4 offspring are, respectively, 0.23, 0.18, 0.16, and 0.14. Find the expected number of offspring.

37. Ear Infections Otitis media, or middle ear infection, is initially treated with an antibiotic. Researchers have compared two antibiotics, amoxicillin and cefaclor, for their cost effectiveness. Amoxicillin is inexpensive, safe, and effective. Cefaclor is also safe. However, it is considerably more expensive and it is generally more effective. Use the tree diagram below (where the costs are estimated as the total cost of medication, office visit, ear check, and hours of lost work) to answer the following. *Source: Journal of Pediatric Infectious Disease.*

  a. Find the expected cost of using each antibiotic to treat a middle ear infection.

  b. To minimize the total expected cost, which antibiotic should be chosen?

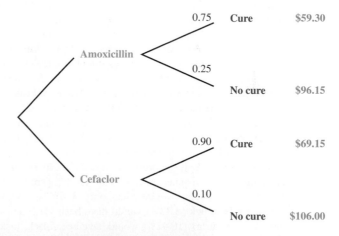

Account Number	Existing Volume	Potential Additional Volume	Probability of Getting It	Expected Value of Potential	Existing Volume + Expected Value of Potential	Class
1	$15,000	$10,000	0.25	$2500	$17,500	C
2	$40,000	$0	—	—	$40,000	C
3	$20,000	$10,000	0.20			
4	$50,000	$10,000	0.10			
5	$5000	$50,000	0.50			
6	$0	$100,000	0.60			
7	$30,000	$20,000	0.80			

38. Hospitalization Insurance An insurance policy pays an individual $100 per day for up to 3 days of hospitalization and $25 per day for each day of hospitalization thereafter. The number of days of hospitalization, X, is a discrete random variable with probability function

$$P(X = k) = \begin{cases} \dfrac{6 - k}{15} & \text{for } k = 1, 2, 3, 4, 5 \\ 0 & \text{otherwise.} \end{cases}$$

Calculate the expected payment for hospitalization under this policy. Choose one of the following. *Source: Society of Actuaries.*

a. $85   **b.** $163   **c.** $168   **d.** $213   **e.** $255

Social Sciences

39. Education Recall from Exercise 65 in Section 8.4 that a study from Cleveland Clinic reported that 74% of very-low-birth-weight babies graduate from high school by age 20. If 250 very-low-birth-weight babies are followed through high school, how many would you expect to graduate from high school? *Source: The New England Journal of Medicine.*

40. Cheating Recall from Exercise 64 in Section 8.4 that a poll conducted by *U.S. News and World Report* reported that 84% of college students believe they need to cheat to get ahead in the world today. If 500 college students were surveyed, how many would you expect to say that they need to cheat to get ahead in the world today? *Source: U.S. News and World Report.*

41. Samuel Alito When Supreme Court Justice Samuel Alito was on the U.S. Court of Appeals for the 3rd Circuit, he dissented in the successful appeal of a first-degree murder case. The prosecution used its peremptory challenges to eliminate all African Americans from the jury, as it had in three other first-degree murder trials in the same county that year. According to a majority of the judges, "An amateur with a pocket calculator can calculate the number of blacks that would have served had the State used its strikes in a racially proportionate manner. In the four capital cases there was a total of 82 potential jurors on the venires who were not removed for cause, of whom eight, or 9.76%, were black. If the prosecution had used its peremptory challenges in a manner proportional to the percentage of blacks in the overall venire, then only 3 of the 34 jurors peremptorily struck (8.82%) would have been black and 5 of the 48 actual jurors (10.42%) would have been black. Instead, none of the 48 jurors were black. Admittedly, there was no statistical analysis of these figures presented by either side in the post-conviction proceeding. But is it really necessary to have a sophisticated analysis by a statistician to conclude that there is little chance of randomly selecting four consecutive all white juries?" *Source: findlaw.com.*

a. Using binomial probability, calculate the probability that no African Americans would be selected out of 48 jurors if the percentage African American is 9.76%.

b. Binomial probability is not entirely accurate in this case, because the jurors were selected without replacement, so the selections were not independent. Recalculate the probability in part a using combinations.

c. In his dissent, Judge Alito wrote, "Statistics can be very revealing—and also terribly misleading in the hands of 'an amateur with a pocket calculator.' . . . Although only about 10% of the population is left-handed, left-handers have won five of the last six presidential elections. Our 'amateur with a calculator' would conclude that 'there is little chance of randomly selecting' left-handers in five out of six presidential elections. But does it follow that the voters cast their ballots based on whether a candidate was right- or left-handed?" Given the figures quoted by Judge Alito, what is the probability that at least 5 out of the last 6 presidents elected would be left-handed?

d. The majority of the judges, in disagreeing with Judge Alito, said, "The dissent has overlooked the obvious fact that there is no provision in the Constitution that protects persons from discrimination based on whether they are right-handed or left handed." Furthermore, according to *Chance News,* only 2 of the last 6 men elected president were left-handed. What is the probability that at least 2 out of the last 6 presidents elected would be left-handed? *Source: Chance News.*

Physical Sciences

42. Seeding Storms One of the few methods that can be used in an attempt to cut the severity of a hurricane is to *seed* the storm. In this process, silver iodide crystals are dropped into the storm. Unfortunately, silver iodide crystals sometimes cause the storm to *increase* its speed. Wind speeds may also increase or decrease even with no seeding. Use the table below to answer the following. *Source: Science.*

a. Find the expected amount of damage under each option, "seed" and "do not seed."

b. To minimize total expected damage, what option should be chosen?

	Change in wind speed	Probability	Property damage (millions of dollars)
Seed	+32%	0.038	335.8
	+16%	0.143	191.1
	0	0.392	100.0
	−16%	0.255	46.7
	−34%	0.172	16.3
Do not seed	+32%	0.054	335.8
	+16%	0.206	191.1
	0	0.480	100.0
	−16%	0.206	46.7
	+34%	0.054	16.3

General Interest

43. Cats Kimberly Workman has four cats: Riley, Abby, Beastie, and Sylvester. Each cat has a 30% probability of climbing into the chair in which Kimberly is sitting, independent of how many cats are already in the chair with Kimberly.

a. Find the probability distribution for the number of cats in the chair with Kimberly.

b. Find the expected number of cats in the chair with Kimberly using the probability distribution in part a.

c. Find the expected number of cats in the chair with Kimberly using the formula for expected value of the binomial distribution.

44. Postal Service Mr. Statistics (a feature in *Fortune* magazine) investigated the claim of the U.S. Postal Service that 83% of first class mail in New York City arrives by the next day. (The figure is 87% nationwide.) He mailed a letter to himself on 10 consecutive days; only 4 were delivered by the next day. *Source: Fortune.*

a. Find the probability distribution for the number of letters delivered by the next day if the overall probability of next-day delivery is 83%.

b. Using your answer to part a, find the probability that 4 or fewer out of 10 letters would be delivered by the next day.

c. Based on your answer to part b, do you think it is likely that the 83% figure is accurate? Explain.

d. Find the number of letters out of 10 that you would expect to be delivered by the next day if the 83% figure is accurate.

45. Raffle A raffle offers a first prize of $400 and 3 second prizes of $80 each. One ticket costs $2, and 500 tickets are sold. Find the expected payback for a person who buys 1 ticket. Is this a fair game?

46. Raffle A raffle offers a first prize of $1000, 2 second prizes of $300 each, and 20 third prizes of $10 each. If 10,000 tickets are sold at 50¢ each, find the expected payback for a person buying 1 ticket. Is this a fair game?

Find the expected payback for the games of chance described in Exercises 47–52.

47. Lottery A state lottery requires you to choose 4 cards from an ordinary deck: 1 heart, 1 club, 1 diamond, and 1 spade in that order from the 13 cards in each suit. If all four choices are selected by the lottery, you win $5000. It costs $1 to play.

48. Lottery If exactly 3 of the 4 choices in Exercise 47 are selected, the player wins $200. (Ignore the possibility that all 4 choices are selected. It still costs $1 to play.)

49. Roulette In one form of roulette, you bet $1 on "even." If 1 of the 18 even numbers comes up, you get your dollar back, plus another one. If 1 of the 20 noneven (18 odd, 0, and 00) numbers comes up, you lose your dollar.

50. Roulette In another form of roulette, there are only 19 non-even numbers (no 00).

51. Numbers *Numbers* is a game in which you bet $1 on any three-digit number from 000 to 999. If your number comes up, you get $500.

52. Keno In one form of the game *Keno*, the house has a pot containing 80 balls, each marked with a different number from 1 to 80.

You buy a ticket for $1 and mark one of the 80 numbers on it. The house then selects 20 numbers at random. If your number is among the 20, you get $3.20 (for a net winning of $2.20).

53. Contests A magazine distributor offers a first prize of $100,000, two second prizes of $40,000 each, and two third prizes of $10,000 each. A total of 2,000,000 entries are received in the contest. Find the expected payback if you submit one entry to the contest. If it would cost you $1 in time, paper, and stamps to enter, would it be worth it?

54. Contests A contest at a fast-food restaurant offered the following cash prizes and probabilities of winning on one visit. Suppose you spend $1 to buy a bus pass that lets you go to 25 different restaurants in the chain and pick up entry forms. Find your expected value.

Prize	Probability
$100,000	1/176,402,500
$25,000	1/39,200,556
$5000	1/17,640,250
$1000	1/1,568,022
$100	1/282,244
$5	1/7056
$1	1/588

55. The Hog Game In the hog game, each player states the number of dice that he or she would like to roll. The player then rolls that many dice. If a 1 comes up on any die, the player's score is 0. Otherwise, the player's score is the sum of the numbers rolled. *Source: Mathematics Teacher.*

a. Find the expected value of the player's score when the player rolls one die.

b. Find the expected value of the player's score when the player rolls two dice.

c. Verify that the expected nonzero score of a single die is 4, so that if a player rolls n dice that do not result in a score of 0, the expected score is $4n$.

d. Verify that if a player rolls n dice, there are 5^n possible ways to get a nonzero score, and 6^n possible ways to roll the dice. Explain why the expected value, E, of the player's score when the player rolls n dice is then

$$E = \frac{5^n(4n)}{6^n}.$$

56. Football After a team scores a touchdown, it can either attempt to kick an extra point or attempt a two-point conversion. During the 2005 NFL season, two-point conversions were successful 45% of the time and the extra-point kicks were successful 96% of the time. *Source: NFL.*

a. Calculate the expected value of each strategy.

b. Which strategy, over the long run, will maximize the number of points scored?

c. Using this information, should a team always only use one strategy? Explain.

57. Baseball The 2009 National League batting champion was Hanley Ramirez, with an average of 0.342. This can be interpreted as a probability of 0.342 of getting a hit whenever he bats. Assume that each time at bat is an independent event. Suppose he goes to bat four times in a game. *Source: base-ball-reference.com.*

a. Find the probability distribution for the number of hits.

b. What is the expected number of hits that Hanley Ramirez gets in a game?

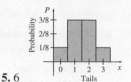

YOUR TURN ANSWERS

1.

x	0	1	2
P(x)	6/11	9/22	1/22

2.

x	0	1	2	3
P(x)	1/8	3/8	3/8	1/8

3. −$3.25   **4.** 4.08   **5.** 6

8 CHAPTER REVIEW

SUMMARY

In this chapter we continued our study of probability by introducing some elementary principles of counting. Our primary tool is the multiplication principle:

If n choices must be made, with m_1 ways to make choice 1, and for each of these ways, m_2 ways to make choice 2, and so on, with m_n ways to make choice n, then there are $m_1 \cdot m_2 \cdot \ldots \cdot m_n$ ways to make the entire sequence of choices.

We learned two counting ideas to efficiently count the number of ways we can select a number of objects without replacement:

- permutations (when order matters), and
- combinations (when order doesn't matter).

We also considered distinguishable permutations, in which some of the objects are indistinguishable. All of these concepts were then used to calculate the numerator and denominator of various probabilities. We next explored binomial probability, in which the following conditions were satisfied:

- the same experiment is repeated a fixed number of times (n),
- there are only two possible outcomes (success and failure), and

- the trials are independent, so the probability of success remains constant (p).

We showed how to quickly calculate an entire set of combinations for binomial probability using Pascal's triangle. Finally, we introduced the following terms regarding probability distributions:

- random variable (a function assigning a real number to each outcome of an experiment),
- probability distribution (the possible values of a random variable, along with the corresponding probabilities),
- histogram (a bar graph displaying a probability distribution), and
- expected value (the average value of a random variable that we would expect in the long run).

In the next chapter, we will see how probability forms the basis of the field known as statistics.

Factorial Notation $n! = n(n-1)(n-2)\cdots(3)(2)(1)$
$$0! = 1$$

Permutations $P(n, r) = \dfrac{n!}{(n-r)!}$

Distinguishable Permutations   If there are n_1 objects of type 1, n_2 of type 2, and so on for r different types, then the number of distinguishable permutations is

$$\frac{n!}{n_1! \, n_2! \cdots n_r!}.$$

Combinations $C(n, r) = \dfrac{n!}{(n-r)! \, r!}$

Binomial Probability $P(x) = C(n, x) \, p^x (1-p)^{n-x}$

Expected Value $E(x) = x_1 p_1 + x_2 p_2 + x_3 p_3 + \cdots + x_n p_n$

For binomial probability, $E(x) = np$.

KEY TERMS

8.1
multiplication principle
factorial notation
permutations
distinguishable permutations

8.2
combinations

8.4
Bernoulli trials
binomial experiment

binomial probability
Pascal's triangle

8.5
random variable
probability distribution

probability function
histogram
expected value
fair game

REVIEW EXERCISES

CONCEPT CHECK

Determine whether each of the following statements is true or false, and explain why.

1. Permutations provide a way of counting possibilities when order matters.

2. Combinations provide a way of counting possibilities when order doesn't matter.

3. The number of distinguishable permutations of n objects, when r are indistinguishable and the remaining $n - r$ are also indistinguishable, is the same as the number of combinations of r objects chosen from n.

4. Calculating the numerator or the denominator of a probability can involve either permutations or combinations.

5. The probability of at least 2 occurrences of an event is equal to the probability of 1 or fewer occurrences.

6. The probability of at least two people in a group having the same birthday is found by subtracting the probability of the complement of the event from 1.

7. The trials in binomial probability must be independent.

8. Binomial probability can be used when each trial has three possible outcomes.

9. A random variable can have negative values.

10. The expected value of a random variable must equal one of the values that the random variable can have.

11. The probabilities in a probability distribution must add up to 1.

12. A fair game can have an expected value that is greater than 0.

PRACTICE AND EXPLORATIONS

13. In how many ways can 6 shuttle vans line up at the airport?

14. How many variations in first-, second-, and third-place finishes are possible in a 100-yd dash with 6 runners?

15. In how many ways can a sample of 3 oranges be taken from a bag of a dozen oranges?

16. In how many ways can a committee of 4 be selected from a club with 10 members?

17. If 2 of the 12 oranges in Exercise 15 are rotten, in how many ways can the sample of 3 include

    a. exactly 1 rotten orange?

    b. exactly 2 rotten oranges?

    c. no rotten oranges?

    d. at most 2 rotten oranges?

18. If 6 of the 10 club members in Exercise 16 are males, in how many ways can the sample of 4 include

    a. exactly 3 males?

    b. no males?

    c. at least 2 males?

19. Five different pictures will be arranged in a row on a wall.

    a. In how many ways can this be done?

    b. In how many ways can this be done if a certain one must be first?

20. In how many ways can the 5 pictures in Exercise 19 be arranged if 2 are landscapes and 3 are puppies and if

    a. like types must be kept together?

    b. landscapes and puppies must be alternated?

21. In a Chinese restaurant the menu lists 8 items in column A and 6 items in column B.

    a. To order a dinner, the diner is told to select 3 items from column A and 2 from column B. How many dinners are possible?

    b. How many dinners are possible if the diner can select up to 3 from column A and up to 2 from column B? Assume at least one item must be included from either A or B.

22. A representative is to be selected from each of 3 departments in a small college. There are 7 people in the first department, 5 in the second department, and 4 in the third department.

    a. How many different groups of 3 representatives are possible?

    b. How many groups are possible if any number (at least 1) up to 3 representatives can form a group? (Each department is still restricted to at most one representative.)

23. Explain under what circumstances a permutation should be used in a probability problem, and under what circumstances a combination should be used.

24. Discuss under what circumstances the binomial probability formula should be used in a probability problem.

A basket contains 4 black, 2 blue, and 7 green balls. A sample of 3 balls is drawn. Find the probabilities that the sample contains the following.

25. All black balls

26. All blue balls

27. 2 black balls and 1 green ball

28. Exactly 2 black balls

29. Exactly 1 blue ball

30. 2 green balls and 1 blue ball

Suppose a family plans 6 children, and the probability that a particular child is a girl is 1/2. Find the probabilities that the 6-child family has the following children.

31. Exactly 3 girls

32. All girls

33. At least 4 girls

34. No more than 2 boys

Suppose 2 cards are drawn without replacement from an ordinary deck of 52. Find the probabilities of the following results.

35. Both cards are red.

36. Both cards are spades.

37. At least 1 card is a spade.

38. One is a face card and the other is not.

39. At least one is a face card.

40. At most one is a queen.

In Exercises 41 and 42, (a) give a probability distribution, (b) sketch its histogram, and (c) find the expected value.

41. A coin is tossed 3 times and the number of heads is recorded.

42. A pair of dice is rolled and the sum of the results for each roll is recorded.

In Exercises 43 and 44, give the probability that corresponds to the shaded region of each histogram.

43.

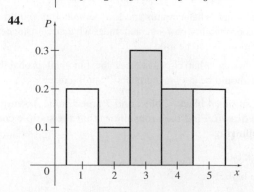

44.

45. You pay $6 to play in a game where you will roll a die, with payoffs as follows: $8 for a 6, $7 for a 5, and $4 for any other results. What are your expected winnings? Is the game fair?

46. Find the expected number of girls in a family of 5 children.

47. Three cards are drawn from a standard deck of 52 cards.

    a. What is the expected number of aces?

    b. What is the expected number of clubs?

48. Suppose someone offers to pay you $100 if you draw 3 cards from a standard deck of 52 cards and all the cards are clubs. What should you pay for the chance to win if it is a fair game?

49. Six students will decide which of them are on a committee by flipping a coin. Each student flips the coin and is on the committee if he or she gets a head. What is the probability that someone is on the committee, but not all 6 students?

50. Find the probability that at most 2 students from Exercise 49 are on the committee.

51. In this exercise we study the connection between sets (from Chapter 7) and combinations (from Chapter 8).

    a. Given a set with n elements, what is the number of subsets of size 0? of size 1? of size 2? of size n?

    b. Using your answer from part a, give an expression for the total number of subsets of a set with n elements.

    c. Using your answer from part b and a result from Chapter 7, explain why the following equation must be true:

$$C(n, 0) + C(n, 1) + C(n, 2) + \cdots + C(n, n) = 2^n.$$

    d. Verify the equation in part c for $n = 4$ and $n = 5$.

    e. Explain what the equation in part c tells you about Pascal's triangle.

In the following exercise, find the digit (0 through 9) that belongs in each box. This exercise is from the 1990 University Entrance Center Examination, given in Japan to all applicants for public universities. *Source: Japanese University Entrance Examination Problems in Mathematics.*

52. The numbers 1 through 9 are written individually on nine cards. Choose three cards from the nine, letting x, y, and z denote the numbers of the cards arranged in increasing order.

    a. There are $\square\square$ such x, y, and z combinations.

    b. The probability of having x, y, and z all even is $\dfrac{\square}{\square\square}$.

    c. The probability of having x, y, and z be consecutive numbers is $\dfrac{\square}{\square\square}$.

    d. The probability of having $x = 4$ is $\dfrac{\square}{\square\square}$.

    e. Possible values of x range from $\square$ to $\square$. If k is an integer such that $\square \le k \le \square$, the probability that $x = k$ is $\dfrac{(\square - k)(\square - k)}{\square\square\square}$. The expected value of x is $\dfrac{\square}{\square}$.

APPLICATIONS

Business and Economics

53. Music Players A popular music player is manufactured in 3 different sizes, 8 different colors, and with or without a camera. How many different varieties of this music player are available?

54. Job Qualifications Of the 12 people applying for an entry level position, 9 are qualified and 3 are not. The personnel manager will hire 4 of the applicants.

   a. In how many different ways can she hire the 4 applicants if the jobs are considered the same?

   b. In how many ways can she hire 4 applicants that are qualified?

   c. In how many ways can she hire at most 1 unqualified applicant?

   d. In how many ways can she hire the 4 applicants if the jobs are not the same?

Identity Theft According to a survey by Javelin Strategy and Research, 1 out of 6 adults in Arizona were victims of identity theft. Suppose that 12 adults are randomly selected from Arizona. Find the probabilities of each of the following results. *Source: The New York Times.*

55. None of the adults were victims of identity theft.

56. All of the adults were victims of identity theft.

57. Exactly 10 of the adults were victims of identity theft.

58. Exactly 2 of the adults were victims of identity theft.

59. At least 2 of the adults were victims of identity theft.

60. At most 3 of the adults were victims of identity theft.

61. Find the expected number of victims of identity theft in a sample of 12 adults in Arizona.

62. Land Development A developer can buy a piece of property that will produce a profit of $26,000 with probability 0.7, or a loss of $9000 with probability 0.3. What is the expected profit?

63. Insurance Claims An insurance company determines that N, the number of claims received in a week, is a random variable with $P(N = n) = 1/2^{n+1}$, where $n \geq 0$. The company also determines that the number of claims received in a given week is independent of the number of claims received in any other week. Determine the probability that exactly seven claims will be received during a given two-week period. Choose one of the following. *Source: Society of Actuaries.*

   a. 1/256   **b.** 1/128   **c.** 7/512   **d.** 1/64   **e.** 1/32

64. Injury Claims The number of injury claims per month is modeled by a random variable N with

$$P(N = n) = \frac{1}{(n + 1)(n + 2)}, \quad \text{where } n \geq 0.$$

Determine the probability of at least one claim during a particular month, given that there have been at most four claims during that month. Choose one of the following. *Source: Society of Actuaries.*

   a. 1/3   **b.** 2/5   **c.** 1/2   **d.** 3/5   **e.** 5/6

65. Product Success A company is considering the introduction of a new product that is believed to have probability 0.5 of being successful and probability 0.5 of being unsuccessful. Successful products pass quality control 80% of the time. Unsuccessful products pass quality control 25% of the time. If the product is successful, the net profit to the company will be $40 million; if unsuccessful, the net loss will be $15 million. Determine the expected net profit if the product passes quality control. Choose one of the following. *Source: Society of Actuaries.*

   a. $23 million   **b.** $24 million   **c.** $25 million
   d. $26 million   **e.** $27 million

66. Sampling Fruit A merchant buys boxes of fruit from a grower and sells them. Each box of fruit is either Good or Bad. A Good box contains 80% excellent fruit and will earn $200 profit on the retail market. A Bad box contains 30% excellent fruit and will produce a loss of $1000. The a priori probability of receiving a Good box of fruit is 0.9. Before the merchant decides to put the box on the market, he can sample one piece of fruit to test whether it is excellent. Based on that sample, he has the option of rejecting the box without paying for it. Determine the expected value of the right to sample. Choose one of the following. (*Hint:* The a priori probability is the probability before sampling a piece of fruit. If the merchant samples the fruit, what are the probabilities of accepting a Good box, accepting a Bad box, and not accepting the box? What are these probabilities if he does not sample the fruit?) *Source: Society of Actuaries.*

   a. 0   **b.** $16   **c.** $34   **d.** $72   **e.** $80

67. Overbooking Flights The March 1982 issue of *Mathematics Teacher* included "Overbooking Airline Flights," an article by Joe Dan Austin. In this article, Austin developed a model for the expected income for an airline flight. With appropriate assumptions, the probability that exactly x of n people with reservations show up at the airport to buy a ticket is given by the binomial probability formula. Assume the following: 6 reservations have been accepted for 3 seats, $p = 0.6$ is the probability that a person with a reservation will show up, a ticket costs $400, and the airline must pay $400 to anyone with a reservation who does not get a ticket. Complete the following table.

Number Who Show Up (x)	0	1	2	3	4	5	6
Airline's Income							
P(x)							

   a. Use the table to find $E(I)$, the expected airline income from the 3 seats.

   b. Find $E(I)$ for $n = 3$, $n = 4$, and $n = 5$. Compare these answers with $E(I)$ for $n = 6$. For these values of n, how many reservations should the airline book for the 3 seats in order to maximize the expected revenue?

Life Sciences

68. Pharmacology In placebo-controlled trials of Prozac®, a drug that is prescribed to fight depression, 23% of the patients who were taking the drug experienced nausea, whereas 10% of the patients who were taking the placebo experienced nausea. *Source: The New England Journal of Medicine.*

a. If 50 patients who are taking Prozac® are selected, what is the probability that 10 or more will experience nausea?

b. Of the 50 patients in part a, what is the expected number of patients who will experience nausea?

c. If a second group of 50 patients receives a placebo, what is the probability that 10 or fewer will experience nausea?

d. If a patient from a study of 1000 people, who are equally divided into two groups (those taking a placebo and those taking Prozac®), is experiencing nausea, what is the probability that he/she is taking Prozac®?

e. Since 0.23 is more than twice as large as 0.10, do you think that people who take Prozac® are more likely to experience nausea than those who take a placebo? Explain.

Social Sciences

69. Education In Exercise 46 of Section 8.3, we saw that a school in Bangkok requires that students take an entrance examination. After the examination, 5 students are randomly drawn from each group of 40 for automatic acceptance into the school regardless of their performance on the examination. The drawing consists of placing 35 red and 5 green pieces of paper into a box. If the lottery is changed so that each student picks a piece of paper from the box and then returns the piece of paper to the box, find the probability that exactly 5 of the 40 students will choose a green piece of paper. *Source: Mathematics Teacher.*

General Interest

In Exercises 70–73, (a) give a probability distribution, (b) sketch its histogram, and (c) find the expected value.

70. Candy According to officials of Mars, the makers of M&M Plain Chocolate Candies, 20% of the candies in each bag are orange. Four candies are selected from a bag and the number of orange candies is recorded. *Source: Mars, Inc.*

71. Women Athletes Since 1992, the Big 10 collegiate sports conference has been committed to increasing opportunities for women in sports. In the 2007–2008 season, 48% of the conference athletes were women. Suppose 5 athletes are picked at random from Big 10 universities. The number of women is recorded. *Source: Big Ten Conference.*

72. Race In the mathematics honors society at a college, 2 of the 8 members are African American. Three members are selected at random to be interviewed by the student newspaper, and the number of African Americans is noted.

73. Homework In a small class of 10 students, 3 did not do their homework. The professor selects half of the class to present solutions to homework problems on the board, and records how many of those selected did not do their homework.

74. Lottery A lottery has a first prize of $5000, two second prizes of $1000 each, and two $100 third prizes. A total of 10,000 tickets is sold, at $1 each. Find the expected payback of a person buying 1 ticket.

75. Contests At one time, game boards for a United Airlines contest could be obtained by sending a self-addressed, stamped envelope to a certain address. The prize was a ticket for any city to which United flies. Assume that the value of the ticket was $2000 (we might as well go first-class), and that the probability that a particular game board would win was 1/8000. If the stamps to enter the contest cost 44¢ and envelopes cost 4¢ each, find the expected payback for a person ordering 1 game board. (Notice that 2 stamps and envelopes were required to enter.)

76. Lottery On June 23, 2003, an interesting thing happened in the Pennsylvania Lottery's Big 4, in which a four-digit number from 0000 to 9999 is chosen twice a day. On this day, the number 3199 was chosen both times. *Source: Pennsylvania Lottery.*

a. What is the probability of the same number being chosen twice in one day?

b. What is the probability of the number 3199 being chosen twice in one day?

77. Lottery In the Pennsylvania Lottery's Daily Number (Evening) game, a three-digit number between 000 and 999 is chosen each day. The favorite number among players is 000, which on February 13, 2010, was the winning number for the 13th time since 1980. *Source: Pennsylvania Lottery.*

a. Find the number of times that 000 would be expected to win in 30 years of play. (Assume that the game is played 365 days a year.)

b. In 2003, a Daily Number (Midday) drawing was added. The number 000 has only occurred once. Find the number of times that 000 would be expected to win in 7 years of play. (Assume that the game is played 365 days a year.)

78. Lottery New York has a lottery game called Quick Draw, in which the player can pick anywhere from 1 up to 10 numbers from 1 to 80. The computer then picks 20 numbers, and how much you win is based on how many of your numbers match the computer's. For simplicity, we will only consider the two cases in which you pick 4 or 5 numbers. The payoffs for each dollar that you bet are given in the table below.

	How Many Numbers Match the Computer's Numbers					
	0	**1**	**2**	**3**	**4**	**5**
You Pick 4	0	0	1	5	55	
You Pick 5	0	0	0	2	20	300

a. According to the Quick Draw playing card, the "Overall Chances of Winning" when you pick 4 are "1:3.86," while the chances when you pick 5 are "1:10.34." Verify these figures.

b. Find the expected value when you pick 4 and when you pick 5, betting $1 each time.

c. Based on your results from parts a and b, are you better off picking 4 numbers or picking 5? Explain your reasoning.

79. Murphy's Law Robert Matthews wrote an article about Murphy's Law, which says that if something can go wrong, it will. He considers Murphy's Law of Odd Socks, which says that if an odd sock can be created it will be, in a drawer of 10 loose pairs of socks. *Source: Sunday Telegraph.*

a. Find the probability of getting a matching pair when the following numbers of socks are selected at random from the drawer.

  i. 5 socks   **ii.** 6 socks

b. Matthews says that it is necessary to rummage through 30% of the socks to get a matching pair. Using your answers from part a, explain precisely what he means by that.

c. Matthews claims that if you lose 6 socks at random from the drawer, then it is 100 times more likely that you will be left with the worst possible outcome—6 odd socks— than with a drawer free of odd socks. Verify this calculation by finding the probability that you will be left with 6 odd socks and the probability that you will have a drawer free of odd socks.

80. Baseball The number of runs scored in 16,456 half-innings of the 1986 National League Baseball season was analyzed by Hal Stern. Use the table to answer the following questions. *Source: Chance News.*

a. What is the probability that a given team scored 5 or more runs in any given half-inning during the 1986 season?

b. What is the probability that a given team scored fewer than 2 runs in any given half-inning of the 1986 season?

c. What is the expected number of runs that a team scored during any given half-inning of the 1986 season? Interpret this number.

Runs	Frequency	Probability
0	12,087	0.7345
1	2451	0.1489
2	1075	0.0653
3	504	0.0306
4	225	0.0137
5	66	0.0040
6	29	0.0018
7	12	0.0007
8	5	0.0003
9	2	0.0001

81. St. Petersburg Paradox Suppose you play a gambling game in which you flip a coin until you get a head. If you get a head on the first toss, you win $2. You win $4 if the first head occurs on the second toss, $8 if it occurs on the the third toss, and so forth, with a prize of 2^n if the first head occurs on the nth toss. Show that the expected value of this game is infinite. Explain why this is a paradox.*

82. Pit The card game of Pit was introduced by Parker Brothers in 1904 and is still popular. In the version owned by one of the authors of this book, there are 10 suits of 9 identical cards, plus the Bull and the Bear card, for a total of 92 cards. (Newer versions of the game have only 8 suits of cards.) For this problem, assume that all 92 cards are used, and you are dealt 9 cards.

a. What is the probability that you have one card from each of 9 different suits, but neither the Bull nor the Bear?

b. What is the probability that you have a pair of cards from one suit and one card from each of 7 other suits, but neither the Bull nor the Bear?

c. What is the probability that you have two pairs of cards from two different suits and one card from each of 5 other suits, but neither the Bull nor the Bear?

*Many articles have been written in an attempt to explain this paradox, first posed by the Swiss mathematician Daniel Bernoulli when he lived in St. Petersburg. For example, see Székely, Gábor and Donald St. P. Richards, "The St. Petersburg Paradox and the Crash of High-Tech Stocks in 2000," *The American Statistician*, Vol. 58, No. 3, Aug. 2004, pp. 225–231.

EXTENDED APPLICATION

OPTIMAL INVENTORY FOR A SERVICE TRUCK

For many different items it is difficult or impossible to take the item to a central repair facility when service is required. Washing machines, large television sets, office copiers, and computers are only a few examples of such items. Service for items of this type is commonly performed by sending a repair person to the item, with the person driving to the location in a truck containing various parts that might be required in repairing the item. Ideally, the truck should contain all the parts that might be required. However, most parts would be needed only infrequently, so that inventory costs for the parts would be high.

An optimum policy for deciding on which parts to stock on a truck would require that the probability of not being able to repair an item without a trip back to the warehouse for needed parts be as low as possible, consistent with minimum inventory costs. An analysis similar to the one below was developed at the Xerox Corporation. *Source: Management Science.*

To set up a mathematical model for deciding on the optimum truck-stocking policy, let us assume that a broken machine might require one of 5 different parts (we could assume any number of different parts—we use 5 to simplify the notation). Suppose also

Example 1

Suppose that for a particular item, only 3 possible parts might need to be replaced. By studying past records of failures of the item, and finding necessary inventory costs, suppose that the following values have been found.

p_1	p_2	p_3		H_1	H_2	H_3
0.09	0.24	0.17		$15	$40	$9

Suppose $N = 3$ and L is $54. Then, as an example,

$$C(M_1) = H_1 + NL[1 - (1 - p_2)(1 - p_3)]$$
$$= 15 + 3(54)[1 - (1 - 0.24)(1 - 0.17)]$$
$$= 15 + 3(54)[1 - (0.76)(0.83)]$$
$$\approx 15 + 59.81 = 74.81.$$

Thus, if policy M_1 is followed (carrying only part 1 on the truck), the expected cost per repair person per time period is $74.81. Also,

$$C(M_{23}) = H_2 + H_3 + NL[1 - (1 - p_1)]$$
$$= 40 + 9 + 3(54)(0.09) = 63.58,$$

so that M_{23} is a better policy than M_1. By finding the expected values for all other possible policies (see the exercises), the optimum policy may be chosen.

EXERCISES

1. Refer to the example and find the following.
   a. $C(M_0)$ b. $C(M_2)$ c. $C(M_3)$ d. $C(M_{12})$
   e. $C(M_{13})$ f. $C(M_{123})$

2. Which policy leads to the lowest expected cost?

3. In the example, $p_1 + p_2 + p_3 = 0.09 + 0.24 + 0.17 = 0.50$. Why is it not necessary that the probabilities add up to 1?

4. Suppose an item to be repaired might need one of n different parts. How many different policies would then need to be evaluated?

DIRECTIONS FOR GROUP PROJECT

Suppose you and three others are employed as service repair persons and that you have some disagreement with your supervisor as to the quantity and type of parts to have on hand for your service calls. Use the answers to Exercises 1–4 to prepare a report with a recommendation to your boss on optimal inventory. Make sure that you describe each concept well since your boss is not mathematically minded.

that the probability that a particular machine requires part 1 is p_1; that it requires part 2 is p_2; and so on. Assume also that failures of different part types are independent, and that at most one part of each type is used on a given job.

Suppose that, on the average, a repair person makes N service calls per time period. If the repair person is unable to make a repair because at least one of the parts is unavailable, there is a penalty cost, L, corresponding to wasted time for the repair person, an extra trip to the parts depot, customer unhappiness, and so on. For each of the parts carried on the truck, an average inventory cost is incurred. Let H_i be the average inventory cost for part i, where $1 \leq i \leq 5$.

Let M_1 represent a policy of carrying only part 1 on the repair truck, M_{24} represent a policy of carrying only parts 2 and 4, with M_{12345} and M_0 representing policies of carrying all parts and no parts, respectively.

For policy M_{35}, carrying parts 3 and 5 only, the expected cost per time period per repair person, written $C(M_{35})$, is

$$C(M_{35}) = (H_3 + H_5) + NL[1 - (1 - p_1)(1 - p_2)(1 - p_4)].$$

(The expression in brackets represents the probability of needing at least one of the parts not carried, 1, 2, or 4 here.) As further examples,

$$C(M_{125}) = (H_1 + H_2 + H_5) + NL[1 - (1 - p_3)(1 - p_4)],$$

while

$$C(M_{12345}) = (H_1 + H_2 + H_3 + H_4 + H_5) + NL[1 - 1]$$
$$= H_1 + H_2 + H_3 + H_4 + H_5,$$

and

$$C(M_0) = NL[1 - (1 - p_1)(1 - p_2)(1 - p_3)(1 - p_4)(1 - p_5)].$$

To find the best policy, evaluate $C(M_0)$, $C(M_1)$,..., $C(M_{12345})$, and choose the smallest result. (A general solution method is in the *Management Science* paper.)

9 Statistics

To understand the economics of large-scale farming, analysts look at historical data on the farming industry. In an exercise in Section 1 you will calculate basic descriptive statistics for U.S. wheat prices and production levels over a recent decade. Later sections in this chapter develop more sophisticated techniques for extracting useful information from this kind of data.

S tatistics is a branch of mathematics that deals with the collection and summarization of data. Methods of statistical analysis make it possible for us to draw conclusions about a population based on data from a sample of the population. Statistical models have become increasingly useful in manufacturing, government, agriculture, medicine, the social sciences, and all types of research. In this chapter we give a brief introduction to some of the key topics from statistical theory.

9.1 Frequency Distributions; Measures of Central Tendency

APPLY IT

How can the results of a survey of business executives on the number of college credits in management needed by a business major best be organized to provide useful information?
Frequency distributions will provide an answer to this question in Example 1.

Often, a researcher wishes to learn something about a characteristic of a population, but because the population is very large or mobile, it is not possible to examine all of its elements. Instead, a limited sample drawn from the population is studied to determine the characteristics of the population. For these inferences to be correct, the sample chosen must be a **random sample**. Random samples are representative of the population because they are chosen so that every element of the population is equally likely to be selected. A hand dealt from a well-shuffled deck of cards is a random sample.

A random sample can be difficult to obtain in real life. For example, suppose you want to take a random sample of voters in your congressional district to see which candidate they prefer in the next election. If you do a telephone survey, you have a random sample of people who are at home to answer the telephone, underrepresenting those who work a lot of hours and are rarely home to answer the phone, or those who have an unlisted number, or those who cannot afford a telephone, or those who only have a cell phone, or those who refuse to answer telephone surveys. Such people may have a different opinion than those you interview.

A famous example of an inaccurate poll was made by the *Literary Digest* in 1936. Their survey indicated that Alfred Landon would win the presidential election; in fact, Franklin Roosevelt won with 62% of the popular vote. The *Digest's* major error was mailing their surveys to a sample of those listed in telephone directories. During the Depression, many poor people did not have telephones, and the poor voted overwhelmingly for Roosevelt. Modern pollsters use sophisticated techniques to ensure that their sample is as random as possible.

Once a sample has been chosen and all data of interest are collected, the data must be organized so that conclusions may be more easily drawn. One method of organization is to group the data into intervals; equal intervals are usually chosen.

EXAMPLE 1   Business Executives

A survey asked a random sample of 30 business executives for their recommendations as to the number of college credits in management that a business major should have.

The results are shown below. Group the data into intervals and find the frequency of each interval.

3	25	22	16	0	9	14	8	34	21
15	12	9	3	8	15	20	12	28	19
17	16	23	19	12	14	29	13	24	18

APPLY IT

SOLUTION   The highest number in the list is 34 and the lowest is 0; one convenient way to group the data is in intervals of size 5, starting with 0–4 and ending with 30–34. This gives an interval for each number in the list and results in seven equal intervals of a convenient size. Too many intervals of smaller size would not simplify the data enough, while too few intervals of larger size would conceal information that the data might provide. A rule of thumb is to use from 6 to 15 intervals.

First tally the number of college credits falling into each interval. Then total the tallies in each interval as in the following table. This table is an example of a **grouped frequency distribution**.

Grouped Frequency Distribution		
College Credits	**Tally**	**Frequency**
0–4	|||	3
5–9	||||	4
10–14	⊮|	6
15–19	⊮ |||	8
20–24	⊮	5
25–29	|||	3
30–34			1
		Total: 30

The frequency distribution in Example 1 shows information about the data that might not have been noticed before. For example, the interval with the largest number of recommended credits is 15–19, and 19 executives (more than half) recommended between 10 and 24 credits, inclusive. Also, the frequency in each interval increases rather evenly (up to 8) and then decreases at about the same pace. However, some information has been lost; for example, we no longer know how many executives recommended 12 credits.

The information in a grouped frequency distribution can be displayed in a histogram similar to the histograms for probability distributions in Chapter 8. The intervals determine the widths of the bars; if equal intervals are used, all the bars have the same width. The heights of the bars are determined by the frequencies.

A **frequency polygon** is another form of graph that illustrates a grouped frequency distribution. The polygon is formed by joining consecutive midpoints of the tops of the histogram bars with straight line segments. The midpoints of the first and last bars are joined to endpoints on the horizontal axis where the next midpoint would appear.

NOTE
In this section, the heights of the histogram bars give the frequencies. The histograms in Chapter 8 were for probability distributions, and so the heights gave the probabilities.

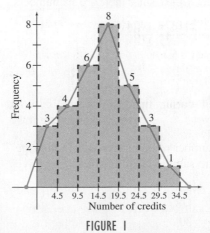

FIGURE 1

EXAMPLE 2   Frequency Distributions

A grouped frequency distribution of college credits was found in Example 1. Draw a histogram and a frequency polygon for this distribution.

SOLUTION   First draw a histogram, shown in red in Figure 1. To get a frequency polygon, connect consecutive midpoints of the tops of the bars. The frequency polygon is shown in blue.

Many graphing calculators have the capability of drawing a histogram. Figure 2 shows the data of Example 1 drawn on a TI-84 Plus. See the *Graphing Calculator and Excel Spreadsheet Manual* available with this text.

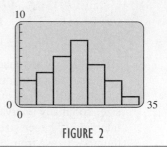

FIGURE 2

Mean

The average value of a probability distribution is the expected value of the distribution. Three measures of central tendency, or "averages," are used with frequency distributions: the mean, the median, and the mode. The most important of these is the mean, which is similar to the expected value of a probability distribution. The **arithmetic mean** (the **mean**) of a set of numbers is the sum of the numbers, divided by the total number of numbers. Recall from Section 1.3 that we can write the sum of n numbers $x_1, x_2, x_3, \ldots, x_n$ in a compact way using *summation notation*:

$$x_1 + x_2 + x_3 + \cdots + x_n = \Sigma x.$$

The symbol $\overline{x}$ (read *x*-bar) is used to represent the mean of a sample.

> ### Mean
> The **mean** of the n numbers $x_1, x_2, x_3, \ldots, x_n$ is
> $$\overline{x} = \frac{\Sigma x}{n}.$$

Bankruptcies	
Year	Petitions Filed
2004	1597
2005	2078
2006	618
2007	851
2008	1118
2009	1474

YOUR TURN 1 Find the mean of the following data: 12, 17, 21, 25, 27, 38, 49.

EXAMPLE 3 Bankruptcy

The table to the left lists the number of bankruptcy petitions (in thousands) filed in the United States in the years 2004–2009. Find the mean number of bankruptcy petitions filed annually during this period. *Source: American Bankruptcy Institute.*

SOLUTION Let $x_1 = 1597$, $x_2 = 2078$, and so on. Here, $n = 6$, since there are six numbers.

$$\overline{x} = \frac{1597 + 2078 + 618 + 851 + 1118 + 1474}{6}$$

$$\overline{x} = \frac{7736}{6} \approx 1289$$

The mean number of bankruptcy petitions filed during the given years is about 1,289,000. **TRY YOUR TURN 1**

As another example, the mean response for the number of college credits in management that a business major should have, based on the sample of 30 business executives described in Example 1, is

$$\overline{x} = \frac{(3 + 25 + 22 + \cdots + 18)}{30} = \frac{478}{30} = 15.93.$$

EXAMPLE 4   Mean for Frequency Distributions

Find the mean for the data shown in the following frequency distribution.

	Calculations for Frequency Distribution	
Value	Frequency	Value × Frequency
30	6	$30 \cdot 6 = 180$
32	9	$32 \cdot 9 = 288$
33	7	$33 \cdot 7 = 231$
37	12	$37 \cdot 12 = 444$
42	6	$42 \cdot 6 = 252$
	Total: 40	Total: 1395

SOLUTION   The value 30 appears six times, 32 nine times, and so on. To find the mean, first multiply 30 by 6, 32 by 9, and so on.

A new column, "Value × Frequency," has been added to the frequency distribution. Adding the products from this column gives a total of 1395. The total from the frequency column is 40, so $n = 40$. The mean is

$$\bar{x} = \frac{1395}{40} = 34.875.$$

The mean of grouped data is found in a similar way. For grouped data, intervals are used, rather than single values. To calculate the mean, it is assumed that all these values are located at the midpoint of the interval. The letter x is used to represent the midpoints and f represents the frequencies, as shown in the next example.

EXAMPLE 5   Business Executives

Listed below is the grouped frequency distribution for the 30 business executives described in Example 1. Find the mean from the grouped frequency distribution.

	Grouped Frequency Distribution for College Credits		
Interval	Midpoint, x	Frequency, f	Product, xf
0–4	2	3	6
5–9	7	4	28
10–14	12	6	72
15–19	17	8	136
20–24	22	5	110
25–29	27	3	81
30–34	32	1	32
		Total: 30	Total: 465

SOLUTION   A column for the midpoint of each interval has been added. The numbers in this column are found by adding the endpoints of each interval and dividing by 2. For the interval 0–4, the midpoint is $(0 + 4)/2 = 2$. The numbers in the product column on the right are found by multiplying each frequency by its corresponding midpoint. Finally, we divide the total of the product column by the total of the frequency column to get

$$\bar{x} = \frac{465}{30} = 15.5.$$

Notice that this mean is slightly different from the earlier mean of 15.93. The reason for this difference is that we have acted as if each piece of data is at the midpoint, which is not

YOUR TURN 2 Find the mean of the following grouped frequency.

Interval	Frequency
0–6	2
7–13	4
14–20	7
21–27	10
28–34	3
35–41	1

true here, and is not true in most cases. Information is always lost when the data are grouped. It is more accurate to use the original data, rather than the grouped frequency, when calculating the mean, but the original data might not be available. Furthermore, the mean based upon the grouped data is typically not too far from the mean based upon the original data, and there may be situations in which the extra accuracy is not worth the extra effort. **TRY YOUR TURN 2**

The formula for the mean of a grouped frequency distribution is given below.

Mean of a Grouped Distribution

The mean of a distribution, where x represents the midpoints, f the frequencies, and $n = \Sigma f$, is

$$\bar{x} = \frac{\Sigma xf}{n}.$$

NOTE

1. The midpoint of the intervals in a grouped frequency distribution may be values that the data cannot take on. For example, if we grouped the data for the 30 business executives into the intervals 0–5, 6–11, 12–17, 18–23, 24–29, and 30–35, the midpoints would be 2.5, 8.5, 14.5, 20.5, 26.5, and 32.5, even though all the data are whole numbers.

2. If we used different intervals in Example 5, the mean would come out to be a slightly different number. Verify that with the intervals 0–5, 6–11, 12–17, 18–23, 24–29, and 30–35, the mean in Example 5 is 16.1.

The mean of a random sample is a random variable, and for this reason it is sometimes called the **sample mean**. The sample mean is a random variable because it assigns a number to the experiment of taking a random sample. If a different random sample were taken, the mean would probably have a different value, with some values more probable than others. If another set of 30 business executives were selected in Example 1, the mean number of college credits in management recommended for a business major might be 13.22 or 17.69. It is unlikely that the mean would be as small as 1.21 or as large as 32.75, although these values are remotely possible.

We saw in Section 8.5 how to calculate the expected value of a random variable when we know its probability distribution. The expected value is sometimes called the **population mean**, denoted by the Greek letter μ. In other words,

$$E(x) = \mu.$$

Furthermore, it can be shown that the expected value of $\bar{x}$ is also equal to μ; that is,

$$E(\bar{x}) = \mu.$$

For instance, consider again the 30 business executives in Example 1. We found that $\bar{x} = 15.93$, but the value of μ, the average for all possible business executives, is unknown. If a good estimate of μ were needed, the best guess (based on this data) is 15.93.

Median

Asked by a reporter to give the average height of the players on his team, a Little League coach lined up his 15 players by increasing height. He picked the player in the middle and pronounced that player to be of average height. This kind of average, called the **median**, is defined as the middle entry in a set of data arranged in either increasing or decreasing order. If there is an even number of entries, the median is defined to be the mean of the two center entries.

Calculating the Median	
Odd Number of Entries	**Even Number of Entries**
1	2
3	3
Median = 4	4
7	7
8	9
	15

Median = $\dfrac{4 + 7}{2}$ = 5.5

EXAMPLE 6   Median

Find the median for each list of numbers.

(a) 11, 12, 17, **20**, 23, 28, 29

SOLUTION   The median is the middle number; in this case, 20. (Note that the numbers are already arranged in numerical order.) In this list, three numbers are smaller than 20 and three are larger.

(b) 15, 13, 7, 11, 19, 30, 39, 5, 10

SOLUTION   First arrange the numbers in numerical order, from smallest to largest.

$$5, 7, 10, 11, \mathbf{13}, 15, 19, 30, 39$$

The middle number, or median, can now be determined; it is 13.

(c) 47, 59, 32, 81, 74, 153

SOLUTION   Write the numbers in numerical order.

$$32, 47, \mathbf{59, 74}, 81, 153$$

There are six numbers here; the median is the mean of the two middle numbers.

YOUR TURN 3   Find the median of the data in Your Turn 1.

$$\text{Median} = \frac{59 + 74}{2} = \frac{133}{2} = 66\frac{1}{2}$$

TRY YOUR TURN 3

Both the mean and the median are examples of a **statistic**, which is simply a number that gives information about a sample. The next example shows a situation in which the median gives a truer representation than the mean.

EXAMPLE 7   Salaries

An office has 10 salespersons, 4 secretaries, the sales manager, and Jeff Weidenaar, who owns the business. Their annual salaries are as follows: secretaries, $35,000 each; salespersons, $45,000 each; manager, $55,000; and owner, $300,000. Calculate the mean and median salaries of those working in the office.

SOLUTION   The mean salary is

$$\bar{x} = \frac{(35,000)4 + (45,000)10 + 55,000 + 300,000}{16} = \$59,062.50$$

However, since 15 people earn less than this mean and only one earns more, the mean does not seem very representative. The median salary is found by ranking the salaries by size: $35,000, $35,000, $35,000, $35,000, $45,000, $45,000, . . . , $300,000. Because there are 16 salaries (an even number) in the list, the mean of the eighth and ninth entries will give the value of the median. The eighth and ninth entries are both $45,000, so the median is $45,000. In this example, the median gives a truer representation than the mean. One salary is much larger than the others and has a pronounced effect on the mean.

Mode   Katie's scores on ten class quizzes include one 7, two 8's, six 9's, and one 10 (out of 10 points possible). She claims that her average grade on quizzes is 9, because most of her scores are 9's. This kind of "average," found by selecting the most frequent entry, is called the **mode**.

EXAMPLE 8   Mode

Find the mode for each list of numbers.

(a) 57, 38, **55, 55**, 80, 87, 98, **55**, 57

SOLUTION   The number 55 occurs more often than any other, so it is the mode. It is not necessary to place the numbers in numerical order when looking for the mode.

(b) 182, *185*, 183, *185*, *187*, *187*, 189

    SOLUTION   Both 185 and 187 occur twice. This list has *two* modes.

(c) 10,708; 11,519; 10,972; 17,546; 13,905; 12,182

    SOLUTION   No number occurs more than once. This list has no mode. ▬

The mode has the advantages of being easily found and not being influenced by data that are very large or very small compared to the rest of the data. It is often used in samples where the data to be "averaged" are not numerical. A major disadvantage of the mode is that we cannot always locate exactly one mode for a set of values. There can be more than one mode, in the case of ties, or there can be no mode if all entries occur with the same frequency.

The mean is the most commonly used measure of central tendency. Its advantages are that it is easy to compute, it takes all the data into consideration, and it is reliable—that is, repeated samples are likely to give very similar means. A disadvantage of the mean is that it is influenced by extreme values, as illustrated in the previous salary example.

The median can be easy to compute and is influenced very little by extremes. Like the mode, the median can be found in situations where the data are not numerical. For example, in a taste test, people are asked to rank five soft drinks from the one they like best to the one they like least. The combined rankings then produce an ordered sample, from which the median can be identified. A disadvantage of the median is the need to rank the data in order; this can be difficult when the number of items is large.

EXAMPLE 9 **Seed Storage**

Seeds that are dried, placed in an airtight container, and stored in a cool, dry place remain ready to be planted for a long time. The table below gives the amount of time that each type of seed can be stored and still remain viable for planting. *Source: The Handy Science Answer Book.*

Storage Time			
Vegetable	**Years**	**Vegetable**	**Years**
Beans	3	Cucumbers	5
Cabbage	4	Melons	4
Carrots	1	Peppers	2
Cauliflower	4	Pumpkin	4
Corn	2	Tomatoes	3

Find the mean, median, and mode of the information in the table.

SOLUTION

Method 1
Calculating by Hand

The mean amount of time that the seeds can be stored is

$$\bar{x} = \frac{3 + 4 + 1 + 4 + 2 + 5 + 4 + 2 + 4 + 3}{10} = 3.2 \text{ years.}$$

After the numbers are arranged in order from smallest to largest, the middle number, or median, is found; it is 3.5.

The number 4 occurs more often than any other, so it is the mode.

Method 2
Graphing Calculator

Most scientific calculators have some statistical capability and can calculate the mean of a set of data; graphing calculators can often calculate the median as well. For example, Figure 3 shows the mean and the median for the data above calculated on a TI-84 Plus, where the data was stored in the list L_1. These commands are found under the LIST-MATH menu. This calculator does not include a command for finding the mode.

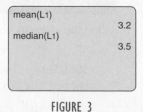

```
mean(L1)
                    3.2
median(L1)
                    3.5
```

FIGURE 3

Method 3 Spreadsheet	Using Microsoft Excel, place the data in cells A1 through A10. To find the mean of this data, type "=AVERAGE(A1:A10)" in cell A11, or any other unused cell, and then press Enter. The result of 3.2 will appear in cell A11. To find the median of this data, type "=MEDIAN(A1:A10)" in cell A12, or any other unused cell, and press Enter. The result of 3.5 will appear in cell A12. To find the mode of this data, type "=MODE(A1:A10)" in cell A13, or any other unused cell, and press Enter. The result of 4 will appear in cell A13.

9.1 EXERCISES

For Exercises 1–4, do the following:

  a. Group the data as indicated.

  b. Prepare a frequency distribution with a column for intervals and frequencies.

  c. Construct a histogram.

  d. Construct a frequency polygon.

1. Use six intervals, starting with 0–24.

7	105	116	73	129	26
29	44	126	82	56	137
43	73	65	141	79	74
121	12	46	37	85	82
2	99	85	95	90	38
86	147	32	84	13	100

2. Use seven intervals, starting with 30–39.

79	71	78	87	69	50	63	51
60	46	65	65	56	88	94	56
74	63	87	62	84	76	82	67
59	66	57	81	93	93	54	88
55	69	78	63	63	48	89	81
98	42	91	66	60	70	64	70
61	75	82	65	68	39	77	81
67	62	73	49	51	76	94	54
83	71	94	45	73	95	72	66
71	77	48	51	54	57	69	87

3. Repeat Exercise 1 using eight intervals, starting with 0–19.

4. Repeat Exercise 2 using six intervals, starting with 39–48.

5. How does a frequency polygon differ from a histogram?

6. Discuss the advantages and disadvantages of the mean as a measure of central tendency.

Find the mean for each list of numbers.

7. 8, 10, 16, 21, 25

8. 67, 89, 75, 86, 100, 93

9. 30,200; 23,700; 33,320; 29,410; 24,600; 27,750; 27,300; 32,680

10. 38,500; 39,720; 42,183; 21,982; 43,250

11. 9.4, 11.3, 10.5, 7.4, 9.1, 8.4, 9.7, 5.2, 1.1, 4.7

12. 15.3, 27.2, 14.8, 16.5, 31.8, 40.1, 18.9, 28.4, 26.3, 35.3

Find the mean for the following.

13.

Value	Frequency
4	6
6	1
9	3
15	2

14.

Value	Frequency
9	3
12	5
15	1
18	1

15. Find the mean for the data in Exercise 1 from the grouped frequency distribution found in each of the following exercises.

  a. Exercise 1

  b. Exercise 3

16. Find the mean for the data in Exercise 2 from the grouped frequency distribution found in each of the following exercises.

  a. Exercise 2

  b. Exercise 4

Find the median for each list of numbers.

17. 27, 35, 39, 42, 47, 51, 54

18. 596, 604, 612, 683, 719

19. 100, 114, 125, 135, 150, 172

20. 359, 831, 904, 615, 211, 279, 505

21. 28.4, 9.1, 3.4, 27.6, 59.8, 32.1, 47.6, 29.8

22. 0.2, 1.4, 0.6, 0.2, 2.5, 1.9, 0.8, 1.5

Use a graphing calculator or spreadsheet to calculate the mean and median for the data in the indicated exercises.

23. Exercise 1

24. Exercise 2

Find the mode or modes for each list of numbers.

25. 4, 9, 8, 6, 9, 2, 1, 3

26. 16, 15, 13, 15, 14, 13, 11, 15, 14

27. 55, 62, 62, 71, 62, 55, 73, 55, 71

28. 158, 162, 165, 162, 165, 157, 163

29. 6.8, 6.3, 6.3, 6.9, 6.7, 6.4, 6.1, 6.0

30. 22.35, 14.90, 17.85, 15.46, 14.91, 17.85, 21.35

31. When is the median the most appropriate measure of central tendency?

32. Under what circumstances would the mode be an appropriate measure of central tendency?

For grouped data, the *modal class* is the interval containing the most data values. Find the modal class for each collection of grouped data.

33. Use the distribution in Exercise 1.

34. Use the distribution in Exercise 2.

35. To predict the outcome of the next congressional election, you take a survey of your friends. Is this a random sample of the voters in your congressional district? Explain why or why not.

APPLICATIONS

Business and Economics

Wheat Production U.S. wheat prices and production figures for a recent decade are given in the following table. *Source: USDA.*

Year	Price ($ per bushel)	Production (millions of bushels)
2000	2.62	2228
2001	2.78	1947
2002	3.56	1606
2003	3.40	2344
2004	3.40	2157
2005	3.42	2103
2006	4.26	1808
2007	6.48	2051
2008	6.78	2499
2009	4.87	2216

Find the mean and median for the following.

36. Price per bushel of wheat

37. Wheat production

38. Salaries The total compensation (in millions of dollars) for the 10 highest paid CEOs in 2009 is given in the following table. *Source: The Huffington Post.*

Person, Company	Total Compensation
Carol Bartz, Yahoo	47.2
Leslie Moonves, CBS	42.9
Marc Casper, Thermo Fisher Scientific, Inc.	34.1
Phillipe Dauman, Viacom Inc.	33.9
J. Raymond Elliott, Boston Scientific Corp.	33.3
Ray Irani, Occidental Petroleum	31.4
Glen Senk, Urban Outfitters	29.9
Brian Roberts, Comcast	27.2
William Weldon, Johnson & Johnson	25.5
Louis Camilleri, Philip Morris International	24.4

a. Find the mean total compensation for this group of people.

b. Find the median total compensation for this group of people.

39. Household Income The total income for African American households making under $250,000 in 2008 is given in the following table. *Source: U.S. Census Bureau.*

Income Range	Midpoint Salary	Frequency (in thousands)
Under $20,000	$10,000	4378
$20,000–$39,999	$30,000	3891
$40,000–$59,999	$50,000	2500
$60,000–$79,999	$70,000	1497
$80,000–$99,999	$90,000	866
$100,000–$149,999	$125,000	998
$150,000–$199,999	$175,000	309
$200,000–$249,999	$225,000	79

Use the table to estimate the mean income for African American households in 2008.

40. Household Income The total income for white households making under $250,000 in 2008 is given in the following table. *Source: U.S. Census Bureau.*

Income Range	Midpoint Salary	Frequency (in thousands)
Under $20,000	$10,000	16,241
$20,000–$39,999	$30,000	20,487
$40,000–$59,999	$50,000	16,065
$60,000–$79,999	$70,000	12,746
$80,000–$99,999	$90,000	9080
$100,000–$149,999	$125,000	12,230
$150,000–$199,999	$175,000	4498
$200,000–$249,999	$225,000	1743

a. Use this table to estimate the mean income for white house-holds in 2008.

b. Compare this estimate with the estimate found in Exercise 39. Discuss whether this provides evidence that white American households have higher earnings than African American households.

41. Airlines The number of consumer complaints against the top U.S. airlines during the first six months of 2010 is given in the following table. *Source: U.S. Department of Transportation.*

Airline	Complaints	Complaints per 100,000 Passengers Boarding
Delta	1175	2.19
American	660	1.56
United	487	1.84
US Airways	428	1.69
Continental	350	1.64
Southwest	149	0.29
Skywest	77	0.65
American Eagle	68	0.87
Expressjet	56	0.70
Alaska	34	0.44

a. By considering the numbers in the column labeled "Complaints," calculate the mean and median number of complaints per airline.

b. Explain why the averages found in part a are not meaningful.

c. Find the mean and median of the numbers in the column labeled "Complaints per 100,000 Passengers Boarding." Discuss whether these averages are meaningful.

Life Sciences

42. Pandas The size of the home ranges (in square kilometers) of several pandas were surveyed over a year's time, with the following results.

Home Range	Frequency
0.1–0.5	11
0.6–1.0	12
1.1–1.5	7
1.6–2.0	6
2.1–2.5	2
2.6–3.0	1
3.1–3.5	1

a. Sketch a histogram and frequency polygon for the data.

b. Find the mean for the data.

43. Blood Types The number of recognized blood types varies by species, as indicated by the table below. Find the mean, median, and mode of this data. *Source: The Handy Science Answer Book.*

Animal	Number of Blood Types
Pig	16
Cow	12
Chicken	11
Horse	9
Human	8
Sheep	7
Dog	7
Rhesus monkey	6
Mink	5
Rabbit	5
Mouse	4
Rat	4
Cat	2

General Interest

44. Temperature The following table gives the number of days in June and July of recent years in which the temperature reached 90 degrees or higher in New York's Central Park. *Source: The New York Times and Accuweather.com.*

Year	Days	Year	Days	Year	Days
1972	11	1985	4	1998	5
1973	8	1986	8	1999	24
1974	11	1987	14	2000	3
1975	3	1988	21	2001	4
1976	8	1989	10	2002	13
1977	11	1990	6	2003	11
1978	5	1991	21	2004	1
1979	7	1992	4	2005	12
1980	12	1993	25	2006	5
1981	12	1994	16	2007	4
1982	11	1995	14	2008	10
1983	20	1996	0	2009	0
1984	7	1997	10	2010	20

a. Prepare a frequency distribution with a column for intervals and frequencies. Use six intervals, starting with 0–4.

b. Sketch a histogram and a frequency polygon, using the intervals in part a.

c. Find the mean for the original data.

d. Find the mean using the grouped data from part a.

e. Explain why your answers to parts c and d are different.

f. Find the median and the mode for the original data.

45. Temperature The table below gives the average monthly temperatures in degrees Fahrenheit for a certain area.

Month	Maximum	Minimum
January	39	16
February	39	18
March	44	21
April	50	26
May	60	32
June	69	37
July	79	43
August	78	42
September	70	37
October	51	31
November	47	24
December	40	20

Find the mean and median for the following.

a. The maximum temperature

b. The minimum temperature

46. Olympics The number of nations participating in the winter Olympic games, from 1968 to 2010, is given below. Find the following measures for the data. *Source: New York Times 2010 Almanac and International Olympic Committee.*

Year	Nations Participating	Year	Nations Participating
1968	37	1992	64
1972	35	1994	67
1976	37	1998	72
1980	37	2002	77
1984	49	2006	85
1988	57	2010	82

a. Mean   **b.** Median   **c.** Mode

47. Personal Wealth When Russian billionaire Roman Abramovich became governor of the Russian province Chukotka (in the Bering Straits, opposite Alaska), it instantly became the fourth most prosperous region in Russia, even though its 80,000 other residents are poor. Mr. Abramovich was then worth $5.7 billion. Suppose each of the 80,000 other residents of Chukotka was worth $100. *Source: National Public Radio.*

a. Calculate the average worth of a citizen of Chukotka.

b. What does this example tell you about the use of the mean to describe an average?

48. Personal Wealth *Washington Post* writer John Schwartz pointed out that if Microsoft Corp. cofounder Bill Gates, who, at the time, was reportedly worth $10 billion, lived in a town with 10,000 totally penniless people, the average personal wealth in the town would make it seem as if everyone were a millionaire. *Source: The Washington Post.*

a. Verify Schwartz's statement.

b. What would be the median personal wealth in this town?

c. What would be the mode for the personal wealth in this town?

d. In this example, which average is most representative: the mean, the median, or the mode?

49. Baseball Salaries The Major League Baseball team with the highest payroll in 2010 (and most other years) was the New York Yankees. The following table gives the salary of each Yankee in 2010. *Source: About.com.*

Name	Salary
Alex Rodriguez	$33,000,000
CC Sabathia	$24,285,714
Derek Jeter	$22,600,000
Mark Teixeira	$20,625,000
A.J. Burnett	$16,500,000
Mariano Rivera	$15,000,000
Jorge Posada	$13,100,000
Andy Pettitte	$11,750,000
Javier Vazquez	$11,500,000
Robinson Cano	$9,000,000
Nick Swisher	$6,850,000
Curtis Granderson	$5,500,000
Nick Johnson	$5,500,000
Damaso Marte	$4,000,000
Chan Ho Park	$1,200,000
Randy Winn	$1,100,000
Marcus Thames	$900,000
Sergio Mitre	$850,000
Joba Chamberlain	$487,975
Brett Gardner	$452,500
Phil Hughes	$447,000
Alfredo Aceves	$435,650
David Robertson	$426,650
Ramiro Pena	$412,100
Francisco Cervelli	$410,800

a. Find the mean, median, and mode of the salaries.

b. Which average best describes this data?

c. Why is there such a difference between the mean and the median?

50. SAT I: Reasoning Test Given the following sequence of numbers*

$$1, a, a^2, a^3, \ldots, a^n,$$

*Permission to reprint SAT materials does not constitute review or endorsement by Educational Testing Service or the College Board of this publication as a whole or of any other questions or testing information it may contain. This problem appeared, minus the *additional assumption*, on an SAT in 1996. Colin Rizzio, a high school student at the time, became an instant celebrity when he noticed that the additional assumption was needed to complete the problem. *Source: The New York Times.*

where n is a positive even integer, with the *additional assumption* that a is a positive number, the median is best described as

a. greater than $a^{n/2}$;

b. smaller than $a^{n/2}$;

c. equal to $a^{n/2}$.

d. The relationship cannot be determined from the information given.

YOUR TURN ANSWERS

1. 27

2. 19.85

3. 25

9.2 Measures of Variation

APPLY IT **How can we tell when a manufacturing process is out of control?**

To answer this question, which we will do in Example 6, we need to understand measures of variation, which tell us how much the numbers in a sample vary from the mean.

The mean gives a measure of central tendency of a list of numbers but tells nothing about the *spread* of the numbers in the list. For example, suppose three inspectors each inspect a sample of five restaurants in their region and record the number of health violations they find in each restaurant. Their results are recorded in the following table.

Three Sets of Data					
I	3	5	6	3	3
II	4	4	4	4	4
III	10	1	0	0	9

Each of these three samples has a mean of 4, and yet they are quite different; the amount of dispersion or variation within the samples is different. Therefore, in addition to a measure of central tendency, another kind of measure is needed that describes how much the numbers vary.

The largest number in sample I is 6, while the smallest is 3, a difference of 3. In sample II this difference is 0; in sample III, it is 10. The difference between the largest and smallest number in a sample is called the **range**, one example of a measure of variation. The range of sample I is 3, of sample II, 0, and of sample III, 10. The range has the advantage of being very easy to compute, and gives a rough estimate of the variation among the data in the sample. It depends only on the two extremes, however, and tells nothing about how the other data are distributed between the extremes.

EXAMPLE 1 **Range**

Find the range for each list of numbers.

(a) 12, 27, 6, 19, 38, 9, 42, 15

SOLUTION The highest number here is 42; the lowest is 6. The range is the difference between these numbers, or $42 - 6 = 36$.

(b) 74, 112, 59, 88, 200, 73, 92, 175

SOLUTION

$$\text{Range} = 200 - 59 = 141$$

The most useful measure of variation is the *standard deviation*. Before defining it, however, we must find the **deviations from the mean**, the differences found by subtracting the mean from each number in a sample.

EXAMPLE 2 **Deviations from the Mean**

Find the deviations from the mean for the numbers

$$32, 41, 47, 53, 57.$$

SOLUTION Adding these numbers and dividing by 5 gives a mean of 46. To find the deviations from the mean, subtract 46 from each number in the list. For example, as shown in the table to the left, the first deviation from the mean is $32 - 46 = -14$; the last is $57 - 46 = 11$.

To check your work, find the sum of these deviations. It should always equal 0. (The answer is always 0 because the positive and negative numbers cancel each other.)

Deviations from the Mean	
Number	Deviation From Mean
32	−14
41	−5
47	1
53	7
57	11
	0

To find a measure of variation, we might be tempted to use the mean of the deviations. As mentioned above, however, this number is always 0, no matter how widely the data are dispersed. One way to solve this problem is to use absolute value and find the mean of the absolute values of the deviations from the mean. Absolute value is awkward to work with algebraically, and there is an alternative approach that provides better theoretical results. In this method, the way to get a list of positive numbers is to square each deviation and then find the mean. When finding the mean of the squared deviations, most statisticians prefer to divide by $n - 1$, rather than n. We will give the reason later in this section. For the data above, this gives

$$\frac{(-14)^2 + (-5)^2 + 1^2 + 7^2 + 11^2}{5 - 1} = \frac{196 + 25 + 1 + 49 + 121}{4}$$

$$= 98.$$

This number, 98, is called the *variance* of the sample. Since it is found by averaging a list of squares, the variance of a sample is represented by s^2.

For a sample of n numbers $x_1, x_2, x_3, \ldots, x_n$, with mean $\overline{x}$, the variance is

$$s^2 = \frac{\Sigma(x - \overline{x})^2}{n - 1}.$$

The following shortcut formula for the variance can be derived algebraically from the formula above. You are asked for this derivation in Exercise 23.

Variance
The **variance** of a sample of n numbers $x_1, x_2, x_3, \ldots, x_n$, with mean $\overline{x}$, is

$$s^2 = \frac{\Sigma x^2 - n\overline{x}^2}{n - 1}.$$

To find the variance, we squared the deviations from the mean, so the variance is in squared units. To return to the same units as the data, we use the *square root* of the variance, called the *standard deviation*.

Standard Deviation
The **standard deviation** of a sample of n numbers $x_1, x_2, x_3, \ldots, x_n$, with mean $\overline{x}$, is

$$s = \sqrt{\frac{\Sigma x^2 - n\overline{x}^2}{n - 1}}.$$

As its name indicates, the standard deviation is the most commonly used measure of variation. The standard deviation is a measure of the variation from the mean. The size of the standard deviation tells us something about how spread out the data are from the mean.

EXAMPLE 3 Standard Deviation

Find the standard deviation of the numbers

$$7, 9, 18, 22, 27, 29, 32, 40.$$

SOLUTION

Method 1
Calculating by Hand

The mean of the numbers is

$$\frac{7 + 9 + 18 + 22 + 27 + 29 + 32 + 40}{8} = 23.$$

Arrange the work in columns, as shown in the table.

Standard Deviation Calculations	
Number, x	Square of the Number, x^2
7	49
9	81
18	324
22	484
27	729
29	841
32	1024
40	1600
	Total: 5132

The total of the second column gives $\sum x^2 = 5132$. Now, using the formula for variance with $n = 8$, the variance is

$$s^2 = \frac{\sum x^2 - n\bar{x}^2}{n - 1}$$

$$= \frac{5132 - 8(23)^2}{8 - 1}$$

$$\approx 128.6,$$

rounded, and the standard deviation is

$$\sqrt{128.57} \approx 11.3.$$

Method 2
Graphing Calculator

The data are entered into the L_5 list on a TI-84 Plus calculator. Figure 4 shows how the variance and standard deviation are then calculated through the LIST-MATH menu. Figure 5 shows an alternative method, going through the STAT menu, which calculates the mean, the standard deviation using both $n - 1$ (indicated by Sx) and n (indicated by σx) in the denominator, and other statistics.

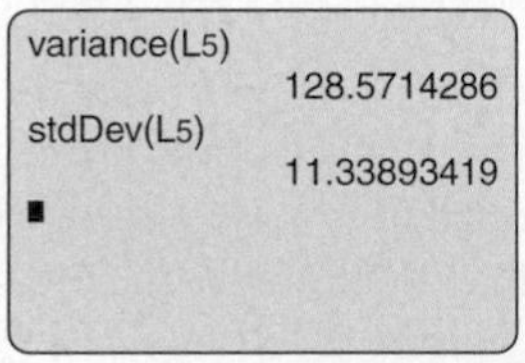

YOUR TURN 1 Find the range, variance, and standard deviation for the following list of numbers: 7, 11, 16, 17, 19, 35.

```
variance(L5)
                128.5714286
stdDev(L5)
                11.33893419
■
```

FIGURE 4

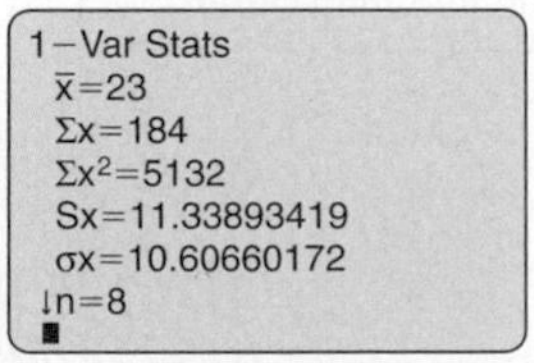

```
1-Var Stats
x̄=23
Σx=184
Σx²=5132
Sx=11.33893419
σx=10.60660172
↓n=8
■
```

FIGURE 5

Method 3
Spreadsheet

The data are entered in cells A1 through A8. Then, in cell A9, type "=VAR(A1:A8)" and press Enter. The standard deviation can be calculated by either taking the square root of cell A9 or by typing "=STDEV(A1:A8)" in cell A10 and pressing Enter.

TRY YOUR TURN 1

CAUTION Be careful to divide by $n - 1$, not n, when calculating the standard deviation of a sample. Many calculators are equipped with statistical keys that compute the variance and standard deviation. Some of these calculators use $n - 1$ and others use n for these computations; some may have keys for both. Check your calculator's instruction book before using a statistical calculator for the exercises.

One way to interpret the standard deviation uses the fact that, for many populations, most of the data are within three standard deviations of the mean. (See Section 9.3.) This implies that, in Example 3, most members of the population from which this sample is taken are between

$$\bar{x} - 3s = 23 - 3(11.3) = -10.9$$

and

$$\bar{x} + 3s = 23 + 3(11.3) = 56.9.$$

This has important implications for quality control. If the sample in Example 3 represents measurements of a product that the manufacturer wants to be between 5 and 45, the standard deviation is too large, even though all the numbers in the sample are within these bounds.

We saw in the previous section that the mean of a random sample is a random variable. It should not surprise you, then, to learn that the variance and standard deviation are also random variables. We will refer to the variance and standard deviation of a random sample as the **sample variance** and **sample standard deviation**.

Recall from the previous section that the sample mean $\bar{x}$ is not the same as the population mean μ, which is defined by $\mu = E(x)$, but that $\bar{x}$ gives a good approximation to μ because $E(\bar{x}) = \mu$. Similarly, there is a **population variance**, denoted σ^2, defined by $\sigma^2 = E(x - \mu)^2$, which measures the amount of variation in a population. The **population standard deviation** is simply σ, the square root of the population variance σ^2. (The Greek letter σ is the lowercase version of sigma. You have already seen Σ, the uppercase version.) In more advanced courses in statistics, it is shown that $E(s^2) = \sigma^2$. The reason many statisticians prefer $n - 1$ in the denominator of the standard deviation formula is that it makes $E(s^2) = \sigma^2$ true; this is not true if n is used in the denominator. It may surprise you, then, that $E(s) = \sigma$ is *false*, whether n or $n - 1$ is used. If n is large, the difference between $E(s)$ and σ is slight, so, in practice, the sample standard deviation s gives a good estimate of the population standard deviation σ.

For data in a grouped frequency distribution, a slightly different formula for the standard deviation is used.

FOR REVIEW

Recall from Section 8.5 that a random variable is a function that assigns a real number to each outcome of an experiment. When the experiment consists of drawing a random sample, the standard deviation and the variance are two real numbers assigned to each outcome. Every time the experiment is performed, the standard deviation and variance will most likely have different values, with some values more probable than others.

Standard Deviation for a Grouped Distribution
The standard deviation for a distribution with mean $\bar{x}$, where x is an interval midpoint with frequency f, and $n = \Sigma f$, is

$$s = \sqrt{\frac{\Sigma fx^2 - n\bar{x}^2}{n - 1}}.$$

The formula indicates that the product fx^2 is to be found for each interval. Then these products are summed, n times the square of the mean is subtracted, and the difference is divided by one less than the total frequency; that is, by $n - 1$. The square root of this result is s, the standard deviation.

CAUTION In calculating the standard deviation for either a grouped or ungrouped distribution, using a rounded value for the mean may produce an inaccurate value.

EXAMPLE 4 Standard Deviation for Grouped Data

Find the standard deviation for the grouped data of Example 5, Section 9.1.

SOLUTION Begin by forming columns for x (the midpoint of the interval), x^2, and fx^2. Then sum the f and fx^2 columns. Recall from Example 5 of Section 9.1 that $\bar{x} = 15.5$.

Standard Deviation for Grouped Data				
Interval	x	x^2	f	fx^2
0–4	2	4	3	12
5–9	7	49	4	196
10–14	12	144	6	864
15–19	17	289	8	2312
20–24	22	484	5	2420
25–29	27	729	3	2187
30–34	32	1024	1	1024
			Total: 30	Total: 9015

YOUR TURN 2 Find the standard deviation for the following grouped frequency distribution.

Interval	Frequency
0–6	2
7–13	4
14–20	7
21–27	10
28–34	3
35–41	1

The total of the fourth column gives $n = \Sigma f = 30$, and the total of the last column gives $\Sigma fx^2 = 9015$. Use the formula for standard deviation for a grouped distribution to find s.

$$s = \sqrt{\frac{\Sigma fx^2 - n\bar{x}^2}{n - 1}}$$

$$= \sqrt{\frac{9015 - 30(15.5)^2}{30 - 1}}$$

$$\approx 7.89$$

Verify that the standard deviation of the original, ungrouped data in Example 1 of Section 9.1 is 7.92. **TRY YOUR TURN 2**

EXAMPLE 5 Nathan's Hot Dog Eating Contest

Since 1916, Nathan's Famous Hot Dogs has held an annual hot dog eating contest, in which each contestant attempts to consume as many hot dogs with buns as possible in a 12-minute period. The following table contains a list of each year's winners since 1997, when the International Federation of Competitive Eating began officiating the contest. *Source: Wikipedia.* In what percent of the contests did the number of hot dogs eaten by the winner fall within one standard deviation of the mean number of hot dogs? Within two standard deviations?

Hot Dog Eating Contest Winners		
Year	Winner	Hot Dogs Eaten
1997	Hirofumi Nakajima	24.5
1998	Hirofumi Nakajima	19
1999	Steve Keiner	20.25
2000	Kazutoyo Arai	25.125
2001	Takeru Kobayashi	50
2002	Takeru Kobayashi	50.5
2003	Takeru Kobayashi	44.5
2004	Takeru Kobayashi	53.5
2005	Takeru Kobayashi	49
2006	Takeru Kobayashi	53.75
2007	Joey Chestnut	66
2008	Joey Chestnut	59
2009	Joey Chestnut	68
2010	Joey Chestnut	54

SOLUTION First, using the formulas for $\bar{x}$ and s, we calculate the mean and standard deviation:

$$\bar{x} \approx 45.51 \qquad \text{and} \qquad s \approx 16.57.$$

Subtracting one standard deviation from the mean and then adding one standard deviation to the mean, we find the lower and upper limits:

$$\bar{x} - s = 45.51 - 16.57 = 28.94 \quad \text{(lower)}$$

and

$$\bar{x} + s = 45.51 + 16.57 = 62.08. \quad \text{(upper)}$$

In 8 of the 14 contests, the number of hot dogs eaten by the winner is between 28.94 and 62.08. Therefore, in about 57% $(8/14 \approx 0.57)$ of the recent contests, the number of hot dogs consumed by the winner was within one standard deviation of the mean.

Likewise, subtracting 2 standard deviations from the mean and adding 2 standard deviations to the mean, we get a lower limit of 12.37, and an upper limit of 78.65. All 14 contests fall in this range, so in 100% of the recent contests, the number of hot dogs eaten by the winner was within two standard deviations of the mean. ▬▬

EXAMPLE 6 Quality Assurance

APPLY IT

Statistical process control is a method of determining when a manufacturing process is out of control, producing defective items. The procedure involves taking samples of a measurement on a product over a production run and calculating the mean and standard deviation of each sample. These results are used to determine when the manufacturing process is out of control. For example, three sample measurements from a manufacturing process on each of four days are given in the table below. The mean $\bar{x}$ and standard deviation s are calculated for each sample.

Samples from a Manufacturing Process											
Day	1			2			3			4		
Sample Number	1	2	3	1	2	3	1	2	3	1	2	3
Measurements	−3	0	4	5	−2	4	3	−1	0	4	−2	1
	0	5	3	4	0	3	−2	0	0	3	0	3
	2	2	2	3	1	4	0	1	−2	3	−1	0
$\bar{x}$	−1/3	7/3	3	4	−1/3	11/3	1/3	0	−2/3	10/3	−1	4/3
s	2.5	2.5	1	1	1.5	0.6	2.5	1	1.2	0.6	1	1.5

Next, the mean of the 12 sample means, $\overline{X}$, and the mean of the 12 sample standard deviations, $\bar{s}$, are found (using the formula for $\bar{x}$). Here, these measures are

$$\overline{X} = 1.3 \qquad \text{and} \qquad \bar{s} = 1.41.$$

The control limits for the sample means are given by

$$\overline{X} \pm k_1 \bar{s},$$

where k_1 is a constant that depends on the sample size, and can be found from a manual. *Source: Statistical Process Control.* For samples of size 3, $k_1 = 1.954$, so the control limits for the sample means are

$$1.3 \pm (1.954)(1.41).$$

The upper control limit is 4.06, and the lower control limit is −1.46.

Similarly, the control limits for the sample standard deviations are given by $k_2 \cdot \bar{s}$ and $k_3 \cdot \bar{s}$, where k_2 and k_3 also are values given in the same manual. Here, $k_2 = 2.568$ and $k_3 = 0$, with the upper and lower control limits for the sample standard deviations equal to

2.568(1.41) and 0(1.41), or 3.62 and 0. As long as the sample means are between -1.46 and 4.06 and the sample standard deviations are between 0 and 3.62, the process is in control.

9.2 EXERCISES

1. How are the variance and the standard deviation related?

2. Why can't we use the sum of the deviations from the mean as a measure of dispersion of a distribution?

Find the range and standard deviation for each set of numbers.

3. 72, 61, 57, 83, 52, 66, 85

4. 122, 132, 141, 158, 162, 169, 180

5. 241, 248, 251, 257, 252, 287

6. 51, 58, 62, 64, 67, 71, 74, 78, 82, 93

7. 3, 7, 4, 12, 15, 18, 19, 27, 24, 11

8. 17, 57, 48, 13, 26, 3, 36, 21, 9, 40

Use a graphing calculator or spreadsheet to calculate the standard deviation for the data in the indicated exercises.

9. Exercise 1 from Section 9.1

10. Exercise 44 from Section 9.1

Find the standard deviation for the following grouped data.

11. (From Exercise 1, Section 9.1)

Interval	Frequency
0–24	4
25–49	8
50–74	5
75–99	10
100–124	4
125–149	5

12. (From Exercise 2, Section 9.1)

Interval	Frequency
30–39	1
40–49	6
50–59	13
60–69	22
70–79	17
80–89	13
90–99	8

Chebyshev's theorem states that for any set of numbers, the fraction that will lie within k standard deviations of the mean (for $k > 1$) is at least

$$1 - \frac{1}{k^2}.$$

For example, at least $1 - 1/2^2 = 3/4$ of any set of numbers lie within 2 standard deviations of the mean. Similarly, for any probability distribution, the probability that a number will lie within k standard deviations of the mean is at least $1 - 1/k^2$. For example, if the mean is 100 and the standard deviation is 10, the probability that a number will lie within 2 standard deviations of 100, or between 80 and 120, is at least 3/4. Use Chebyshev's theorem to find the fraction of all the numbers of a data set that must lie within the following numbers of standard deviations from the mean.

13. 3

14. 4

15. 5

16. 6

In a certain distribution of numbers, the mean is 60 with a standard deviation of 8. Use Chebyshev's theorem to tell the probability that a number lies in each interval.

17. Between 36 and 84

18. Between 48 and 72

19. Less than 36 or more than 84

20. Less than 48 or more than 72

21. Discuss what the standard deviation tells us about a distribution.

22. Explain the difference between the sample mean and standard deviation, and the population mean and standard deviation.

23. Derive the shortcut formula for the variance

$$s^2 = \frac{\sum x^2 - n\bar{x}^2}{n - 1}$$

from the formula

$$s^2 = \frac{\sum (x - \bar{x})^2}{n - 1}$$

and the following summation formulas, in which c is a constant:

$$\sum cx = c\sum x, \quad \sum c = nc, \quad \text{and} \quad \sum(x \pm y) = \sum x \pm \sum y.$$

(*Hint:* Multiply out $(x - \bar{x})^2$.)

24. Consider the set of numbers 9,999,999, 10,000,000, and 10,000,001.

a. Calculate the variance by hand using the formula
$$\frac{\sum (x - \bar{x})^2}{n - 1}$$

b. Calculate the variance using your calculator and the shortcut formula.

c. There may be a discrepancy between your answers to parts a and b because of roundoff. Explain this discrepancy, and then discuss advantages and disadvantages of the shortcut formula.

APPLICATIONS

Business and Economics

25. Battery Life Forever Power Company analysts conducted tests on the life of its batteries and those of a competitor (Brand X). They found that their batteries had a mean life (in hours) of 26.2, with a standard deviation of 4.1. Their results for a sample of 10 Brand X batteries were as follows: 15, 18, 19, 23, 25, 25, 28, 30, 34, 38.

a. Find the mean and standard deviation for the sample of Brand X batteries.

b. Which batteries have a more uniform life in hours?

c. Which batteries have the highest average life in hours?

26. Sales Promotion The Quaker Oats Company conducted a survey to determine whether a proposed premium, to be included in boxes of cereal, was appealing enough to generate new sales. Four cities were used as test markets, where the cereal was distributed with the premium, and four cities as control markets, where the cereal was distributed without the premium. The eight cities were chosen on the basis of their similarity in terms of population, per capita income, and total cereal purchase volume. The results were as follows. *Source: Quaker Oats Company.*

	City	Percent Change in Average Market Share per Month
Test Cities	1	+18
	2	+15
	3	+7
	4	+10
Control Cities	1	+1
	2	−8
	3	−5
	4	0

a. Find the mean of the change in market share for the four test cities.

b. Find the mean of the change in market share for the four control cities.

c. Find the standard deviation of the change in market share for the test cities.

d. Find the standard deviation of the change in market share for the control cities.

e. Find the difference between the means of parts a and b. This difference represents the estimate of the percent change in sales due to the premium.

f. The two standard deviations from parts c and d were used to calculate an "error" of ± 7.95 for the estimate in part e. With this amount of error, what are the smallest and largest estimates of the increase in sales?

On the basis of the interval estimate of part f, the company decided to mass-produce the premium and distribute it nationally.

27. Process Control The following table gives 10 samples of three measurements, made during a production run.

	Sample Number									
1	**2**	**3**	**4**	**5**	**6**	**7**	**8**	**9**	**10**
2	3	−2	−3	−1	3	0	−1	2	0
−2	−1	0	1	2	2	1	2	3	0
1	4	1	2	4	2	2	3	2	2

Use the information in Example 6 to find the following.

a. Find the mean $\bar{x}$ for each sample of three measurements.

b. Find the standard deviation s for each sample of three measurements.

c. Find the mean $\bar{X}$ of the sample means.

d. Find the mean $\bar{s}$ of the sample standard deviations.

e. Using $k_1 = 1.954$, find the upper and lower control limits for the sample means.

f. Using $k_2 = 2.568$ and $k_3 = 0$, find the upper and lower control limits for the sample standard deviations.

28. Process Control Given the following measurements from later samples on the process in Exercise 27, decide whether the process is out of control. (*Hint:* Use the results of Exercise 27e and f.)

	Sample Number				
1	**2**	**3**	**4**	**5**	**6**
3	−4	2	5	4	0
−5	2	0	1	−1	1
2	1	1	−4	−2	−6

29. Washer Thickness An assembly-line machine turns out washers with the following thicknesses (in millimeters).

1.20	1.01	1.25	2.20	2.58	2.19	1.29	1.15
2.05	1.46	1.90	2.03	2.13	1.86	1.65	2.27
1.64	2.19	2.25	2.08	1.96	1.83	1.17	2.24

Find the mean and standard deviation of these thicknesses.

30. Unemployment The number of unemployed workers in the United States in recent years (in millions) is given below. *Source: Bureau of Labor Statistics.*

Year	Number Unemployed
2000	5.69
2001	6.80
2002	8.38
2003	8.77
2004	8.15
2005	7.59
2006	7.00
2007	7.08
2008	8.92
2009	14.27

a. Find the mean number unemployed (in millions) in this period. Which years had unemployment closest to the mean?

b. Find the standard deviation for the data.

c. In how many of these years is unemployment within 1 standard deviation of the mean?

d. In how many of these years is unemployment within 3 standard deviations of the mean?

Life Sciences

31. Blood pH   A medical laboratory tested 21 samples of human blood for acidity on the pH scale, with the following results.

7.1	7.5	7.3	7.4	7.6	7.2	7.3
7.4	7.5	7.3	7.2	7.4	7.3	7.5
7.5	7.4	7.4	7.1	7.3	7.4	7.4

a. Find the mean and standard deviation.

b. What percentage of the data is within 2 standard deviations of the mean?

32. Blood Types   The number of recognized blood types between species is given in the following table. In Exercise 43 of the previous section, the mean was found to be 7.38. *Source: The Handy Science Answer Book.*

Animal	Number of Blood Types
Pig	16
Cow	12
Chicken	11
Horse	9
Human	8
Sheep	7
Dog	7
Rhesus monkey	6
Mink	5
Rabbit	5
Mouse	4
Rat	4
Cat	2

a. Find the variance and the standard deviation of these data.

b. How many of these animals have blood types that are within 1 standard deviation of the mean?

33. Tumor Growth   The amount of time that it takes for various slow-growing tumors to double in size are listed in the following table. *Source: American Journal of Roentgen.*

Type of Cancer	Doubling Time (days)
Breast cancer	84
Rectal cancer	91
Synovioma	128
Skin cancer	131
Lip cancer	143
Testicular cancer	153
Esophageal cancer	164

a. Find the mean and standard deviation of these data.

b. How many of these cancers have doubling times that are within 2 standard deviations of the mean?

c. If a person had a nonspecified tumor that was doubling every 200 days, discuss whether this particular tumor is growing at a rate that would be expected.

General Interest

34. Box Office Receipts   The table below lists the 18 films in which actor Will Smith has starred through 2008, along with the gross domestic box office receipts and the year for each movie. *Source: The Movie Times.*

Movie	Domestic Box Office Receipts
Hancock, 2008	$227,946,274
Seven Pounds, 2008	$69,951,824
I Am Legend, 2007	$256,386,216
The Pursuit of Happyness, 2006	$162,586,036
Hitch, 2005	$177,575,142
Shark Tale, 2004	$161,412,000
I, Robot, 2004	$144,801,023
Bad Boys II, 2003	$138,540,870
Men in Black 2, 2002	$190,418,803
Ali, 2001	$58,200,000
The Legend of Bagger Vance, 2000	$30,695,000
Wild Wild West, 1999	$113,745,000
Enemy of the State, 1998	$111,544,000
Men in Black, 1997	$250,107,128
Independence Day, 1996	$306,124,000
Bad Boys, 1995	$65,807,000
Made in America, 1993	$44,942,000
Six Degrees of Separation, 1993	$6,410,000

a. Find the mean domestic box office receipts for Will Smith's movies. Which movie has box office receipts closest to the mean?

b. Find the standard deviation for the data.

c. What percent of the movies have box office receipts within 1 standard deviation of the mean? Within 2 standard deviations of the mean?

35. Baseball Salaries   The table in Exercise 49 in the previous section lists the salary for each player on the New York Yankees baseball team in 2010.

a. Calculate the standard deviation of these data.

b. What percent of the 2010 New York Yankee players have salaries that are beyond 2 standard deviations from the mean?

c. What does your answer to part b suggest?

36. Cookies   Marie Revak and Jihan Williams performed an experiment to determine whether Oreo Double Stuf cookies contain twice as much filling as traditional Oreo cookies. The table on the next page gives the results in grams of the amount of filling inside 49 traditional cookies and 52 Double Stuf cookies. *Source: The Mathematics Teacher.*

a. Find the mean, maximum, minimum, and standard deviation of the weights for traditional Oreo cookies.

b. Find the mean, maximum, minimum, and standard deviation of the weights for Oreo Double Stuf cookies.

c. What percent of the data of traditional Oreo cookies is within 2 standard deviations of the Double Stuf Oreo mean? (*Hint:* Use the mean and standard deviation for the Double Stuf data.)

d. What percent of the data of traditional Oreo cookies, when multiplied by 2, is within 2 standard deviations of the Double Stuf Oreo mean? (*Hint:* Use the mean and standard deviation for the Double Stuf data.)

e. Is there evidence that Double Stuf Oreos have twice as much filling as the traditional Oreo cookie? Explain.

Traditional	Traditional	Traditional	Double Stuf	Double Stuf	Double Stuf
2.9	2.4	2.7	4.7	6.5	5.8
2.8	2.8	2.8	6.5	6.3	5.9
2.6	3.8	2.6	5.5	4.8	6.2
3.5	3.1	2.6	5.6	3.3	5.9
3.0	2.9	3.0	5.1	6.4	6.5
2.4	3.0	2.8	5.3	5.0	6.5
2.7	2.1	3.5	5.4	5.3	6.1
2.4	3.8	3.3	5.4	5.5	5.8
2.5	3.0	3.3	3.5	5.0	6.0
2.2	3.0	2.8	5.5	6.0	6.2
2.6	2.8	3.1	6.5	5.7	6.2
2.6	2.9	2.6	5.9	6.3	6.0
2.9	2.7	3.5	5.4	6.0	6.8
2.6	3.2	3.5	4.9	6.3	6.2
2.6	2.8	3.1	5.6	6.1	5.4
3.1	3.1	3.1	5.7	6.0	6.6
2.9			5.3	5.8	6.2
			6.9		

YOUR TURN ANSWERS

1. 28, 92.7, 9.628   **2.** 8.52

9.3 The Normal Distribution

APPLY IT **What is the probability that a salesperson drives between 1200 miles and 1600 miles per month?**
This question will be answered in Example 4 by using the normal probability distribution introduced in this section.

Suppose a bank is interested in improving its services to customers. The manager decides to begin by finding the amount of time tellers spend on each transaction, rounded to the nearest minute. The times for 75 different transactions are recorded with the results shown in the following table. The frequencies listed in the second column are divided by 75 to find the empirical probabilities.

FOR REVIEW

Empirical probabilities, discussed in Section 7.3, are derived from grouped data by dividing the frequency or amount for each group by the total for all the groups. This gives one example of a probability distribution, discussed further in Sections 7.3 and 8.5.

Teller Transaction Times		
Time	Frequency	Probability
1	3	$3/75 = 0.04$
2	5	$5/75 \approx 0.07$
3	9	$9/75 = 0.12$
4	12	$12/75 = 0.16$
5	15	$15/75 = 0.20$
6	11	$11/75 \approx 0.15$
7	10	$10/75 \approx 0.13$
8	6	$6/75 = 0.08$
9	3	$3/75 = 0.04$
10	1	$1/75 \approx 0.01$

Figure 6(a) shows a histogram and frequency polygon for the data. The heights of the bars are the empirical probabilities rather than the frequencies. The transaction times are given to the nearest minute. Theoretically at least, they could have been timed to the nearest tenth of a minute, or hundredth of a minute, or even more precisely. In each case, a histogram and frequency polygon could be drawn. If the times are measured with smaller and smaller units, there are more bars in the histogram, and the frequency polygon begins to look more and more like the curve in Figure 6(b) instead of a polygon. Actually, it is possible for the transaction times to take on any real number value greater than 0. A distribution in which the outcomes can take any real number value within some interval is a **continuous distribution**. The graph of a continuous distribution is a curve.

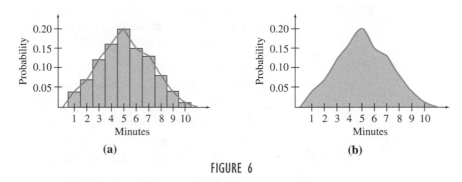

FIGURE 6

The distribution of heights (in inches) of college women is another example of a continuous distribution, since these heights include infinitely many possible measurements, such as 53, 58.5, 66.3, 72.666,..., and so on. Figure 7 shows the continuous distribution of heights of college women. Here the most frequent heights occur near the center of the interval shown.

Another continuous curve, which approximates the distribution of yearly incomes in the United States, is given in Figure 8. The graph shows that the most frequent incomes are grouped near the low end of the interval. This kind of distribution, where the peak is not at the center, is called **skewed**.

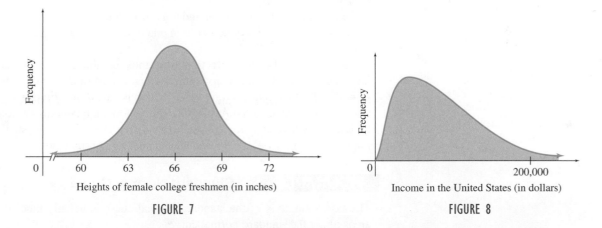

FIGURE 7   FIGURE 8

Many natural and social phenomena produce continuous probability distributions whose graphs can be approximated very well by bell-shaped curves, such as those shown in Figure 9 on the next page. Such distributions are called **normal distributions** and their graphs are called **normal curves**. Examples of distributions that are approximately normal are the heights of college women and the errors made in filling 1-lb cereal boxes. We use the Greek letters μ (mu) to denote the mean and σ (sigma) to denote the standard deviation of a normal distribution. The definitions of the mean and standard deviation of a continuous distribution require ideas from calculus, but the intuitive ideas are similar to those in the previous section.

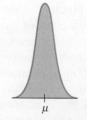

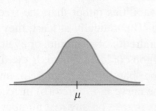

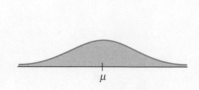

Three normal distributions

FIGURE 9

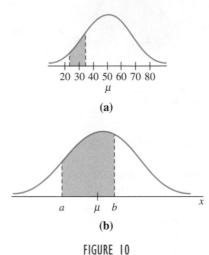

(a)

(b)

FIGURE 10

There are many normal distributions. Some of the corresponding normal curves are tall and thin and others are short and wide, as shown in Figure 9. But every normal curve has the following properties.

1. Its peak occurs directly above the mean μ.
2. The curve is symmetric about the vertical line through the mean (that is, if you fold the page along this line, the left half of the graph will fit exactly on the right half).
3. The curve never touches the x-axis—it extends indefinitely in both directions.
4. The area under the curve (and above the horizontal axis) is always 1. (This agrees with the fact that the sum of the probabilities in any distribution is 1.)

It can be shown that a normal distribution is completely determined by its mean μ and standard deviation σ.* A small standard deviation leads to a tall, narrow curve like the one in the center of Figure 9. A large standard deviation produces a flat, wide curve, like the one on the right in Figure 9.

Since the area under a normal curve is 1, parts of this area can be used to determine certain probabilities. For instance, Figure 10(a) is the probability distribution of the annual rainfall in a certain region. Calculus can be used to show that the probability that the annual rainfall will be between 25 in. and 35 in. is the area under the curve from 25 to 35. The general case, shown in Figure 10(b), can be stated as follows.

Normal Probability

The area of the shaded region under a normal curve from a to b is the probability that an observed data value will be between a and b.

To use normal curves effectively, we must be able to calculate areas under portions of these curves. These calculations have already been done for the normal curve with mean $\mu = 0$ and standard deviation $\sigma = 1$ (which is called the **standard normal curve**) and are available in a table in the Appendix. The following examples demonstrate how to use the table or a calculator or spreadsheet to find such areas. Later we shall see how the standard normal curve may be used to find areas under any normal curve.

EXAMPLE 1   Standard Normal Curve

The horizontal axis of the standard normal curve is usually labeled z. Find the following areas under the standard normal curve.

Method 1
Using a Table

(a) The area to the left of $z = 1.25$

    SOLUTION   Look up 1.25 in the normal curve table. (Find 1.2 in the left-hand column and 0.05 at the top, then locate the intersection of the corresponding row and column.)

*As is shown in more advanced courses, its graph is the graph of the function

$$f(x) = \frac{1}{\sigma\sqrt{2\pi}}\, e^{-(x-\mu)^2/(2\sigma^2)},$$

where $e \approx 2.71828$ is a real number.

The specified area is 0.8944, so the shaded area shown in Figure 11 is 0.8944. This area represents 89.44% of the total area under the normal curve, and so the probability that $z \leq 1.25$ is

$$P(z \leq 1.25) = 0.8944.$$

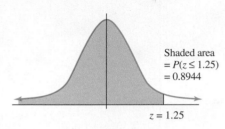

FIGURE 11

(b) The area to the right of $z = 1.25$

SOLUTION From part (a), the area to the left of $z = 1.25$ is 0.8944. The total area under the normal curve is 1, so the area to the right of $z = 1.25$ is

$$1 - 0.8944 = 0.1056.$$

See Figure 12, where the shaded area represents 10.56% of the total area under the normal curve, and the probability that $z \geq 1.25$ is $P(z \geq 1.25) = 0.1056$.

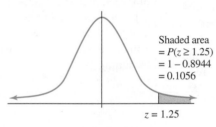

FIGURE 12

(c) The area between $z = -1.02$ and $z = 0.92$

SOLUTION To find this area, which is shaded in Figure 13, start with the area to the left of $z = 0.92$ and subtract the area to the left of $z = -1.02$. See the two shaded regions in Figure 14. The result is

$$P(-1.02 \leq z \leq 0.92) = 0.8212 - 0.1539 = 0.6673.$$

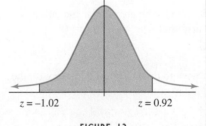

FIGURE 13

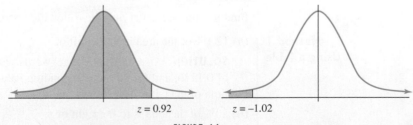

FIGURE 14

Method 2
Graphing Calculator

Because of convenience and accuracy, graphing calculators and computers have made normal curve tables less important. Figure 15 on the next page shows how parts (a) and (b) of this example can be done on a TI-84 Plus using the `normalcdf` command in the DISTR menu. In Figure 15, $-1E99$ stands for -1×10^{99}. The area between -1×10^{99} and 1.25 is essentially the same as the area to the left of 1.25. Similarly, the area between 1.25 and 1×10^{99} is essentially the same as the area to the right of 1.25. Verify the results of part (c) with a graphing calculator.

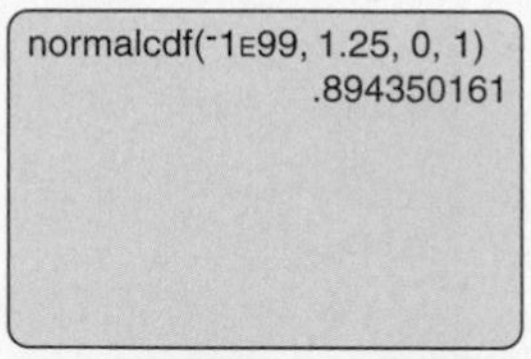

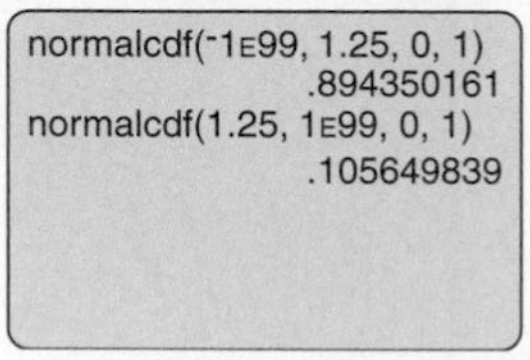

```
normalcdf(-1E99, 1.25, 0, 1)
                      .894350161
```

```
normalcdf(-1E99, 1.25, 0, 1)
                      .894350161
normalcdf(1.25, 1E99, 0, 1)
                      .105649839
```

FIGURE 15

**Method 3
Spreadsheet**

YOUR TURN 1 Find the following areas under the standard normal curve. **(a)** The area to the left of $z = -0.76$. **(b)** The area to the right of $z = -1.36$. **(c)** The area between $z = -1.22$ and $z = 1.33$.

Statistical software packages are widely used today. These packages are set up in a way that is similar to a spreadsheet, and they can be used to generate normal curve values. In addition, most spreadsheets can also perform a wide range of statistical calculations. For example, Microsoft Excel can be used to generate the answers to parts (a), (b), and (c) of this example. In any cell, type "=NORMDIST(1.25,0,1,1)" and press Enter. The value of 0.894350226 is returned. Notice that this value differs slightly from the value returned by a TI-84 Plus. The first three input values represent the z value, mean, and standard deviation. The fourth value is always either a 0 or 1. For applications in this text, we will always place a 1 in this position to indicate that we want the area to the left of the first input value. Similarly, by typing "=1-NORMDIST(1.25,0,1,1)" and pressing Enter, we find that the area to the right of $z = 1.25$ is 0.105649774.

TRY YOUR TURN 1

NOTE Notice in Example 1 that $P(z \leq 1.25) = P(z < 1.25)$. The area under the curve is the same, whether we include the endpoint or not. Notice also that $P(z = 1.25) = 0$, because no area is included.

CAUTION When calculating normal probabilities, it is wise to draw a normal curve with the mean and the z-scores every time. This will avoid confusion as to whether you should add or subtract probabilities.

EXAMPLE 2 Normal Probabilities

Find a value of z satisfying the following conditions.

**Method 1
Using a Table**

(a) 12.1% of the area is to the left of z.

SOLUTION Use the table backwards. Look in the body of the table for an area of 0.1210, and find the corresponding value of z using the left column and the top column of the table. You should find that $z = -1.17$ corresponds to an area of 0.1210.

(b) 20% of the area is to the right of z.

SOLUTION If 20% of the area is to the right, 80% is to the left. Find the value of z corresponding to an area of 0.8000. The closest value is $z = 0.84$.

**Method 2
Graphing Calculator**

Figure 16 illustrates how a TI-84 Plus can be used to find z values for the particular probabilities given in parts (a) and (b) of this example. The command invNorm is found in the DISTR menu.

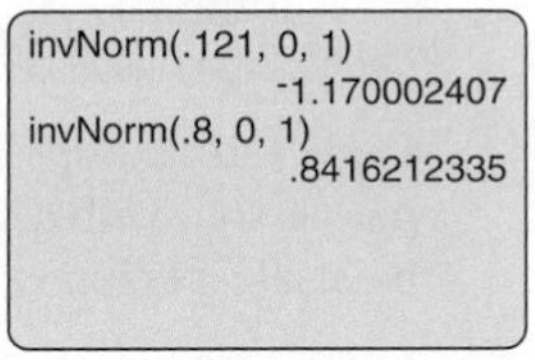

```
invNorm(.121, 0, 1)
              -1.170002407
invNorm(.8, 0, 1)
               .8416212335
```

FIGURE 16

Method 3
Spreadsheet

YOUR TURN 2 Find a value of z satisfying the following conditions. **(a)** 2.5% of the area is to the left of z. **(b)** 20.9% of the area is to the right of z

Microsoft Excel can also be used to generate the answers to parts (a) and (b) of this example. In any cell, type "=NORMINV(.121,0,1)" and press Enter. The value of -1.170002408 is returned. Similarly, by typing "=NORMINV(.8,0,1)" and pressing Enter, we find that the corresponding z value is 0.841621234.

TRY YOUR TURN 2

The key to finding areas under *any* normal curve is to express each number x on the horizontal axis in terms of standard deviation above or below the mean. The **z-score** for x is the number of standard deviations that x lies from the mean (positive if x is above the mean, negative if x is below the mean).

EXAMPLE 3 *z*-Scores

If a normal distribution has mean 50 and standard deviation 4, find the following z-scores.

(a) The z-score for $x = 46$

SOLUTION Since 46 is 4 units below 50 and the standard deviation is 4, 46 is 1 standard deviation below the mean. So, its z-score is -1.

(b) The z-score for $x = 60$

YOUR TURN 3 Find the z-score for $x = 20$ if a normal distribution has a mean 35 and standard deviation 20.

SOLUTION The z-score is 2.5 because 60 is 10 units above the mean (since $60 - 50 = 10$), and 10 units is 2.5 standard deviations (since $10/4 = 2.5$).

TRY YOUR TURN 3

In Example 3(b), we found the z-score by taking the difference between 60 and the mean and dividing this difference by the standard deviation. The same procedure works in the general case.

z-Score

If a normal distribution has mean μ and standard deviation σ, then the z-score for the number x is

$$z = \frac{x - \mu}{\sigma}.$$

The importance of z-scores lies in the following fact.

Area under a Normal Curve

The area under a normal curve between $x = a$ and $x = b$ is the same as the area under the standard normal curve between the z-score for a and the z-score for b.

Therefore, by converting to z-scores and using the table for the standard normal curve, we can find areas under any normal curve. Since these areas are probabilities, we can now handle a variety of applications.

EXAMPLE 4 **Sales**

Dixie Office Supplies finds that its sales force drives an average of 1200 miles per month per person, with a standard deviation of 150 miles. Assume that the number of miles driven by a salesperson is closely approximated by a normal distribution.

(a) Find the probability that a salesperson drives between 1200 miles and 1600 miles per month.

APPLY IT

SOLUTION Here $\mu = 1200$ and $\sigma = 150$, and we must find the area under the normal distribution curve between $x_1 = 1200$ and $x_2 = 1600$. We begin by finding the z-score for $x_1 = 1200$.

$$z_1 = \frac{x_1 - \mu}{\sigma} = \frac{1200 - 1200}{150} = \frac{0}{150} = 0$$

The z-score for $x_2 = 1600$ is

$$z_2 = \frac{x_2 - \mu}{\sigma} = \frac{1600 - 1200}{150} = \frac{400}{150} \approx 2.67.$$

From the table, the area to the left of $z_2 = 2.67$ is 0.9962, the area to the left of $z_1 = 0$ is 0.5000, and

$$0.9962 - 0.5000 = 0.4962.$$

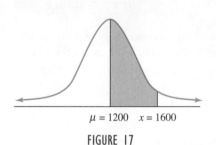

$\mu = 1200 \quad x = 1600$

FIGURE 17

Therefore, the probability that a salesperson drives between 1200 miles and 1600 miles per month is 0.4962. See Figure 17.

(b) Find the probability that a salesperson drives between 1000 miles and 1500 miles per month.

SOLUTION As shown in Figure 18, z-scores for both $x_1 = 1000$ and $x_2 = 1500$ are needed.

For $x_1 = 1000$,   For $x_2 = 1500$,

$$z_1 = \frac{1000 - 1200}{150} \qquad z_2 = \frac{1500 - 1200}{150}$$

$$= \frac{-200}{150} \qquad\qquad = \frac{300}{150}$$

$$z_1 \approx -1.33. \qquad\qquad z_2 = 2.00.$$

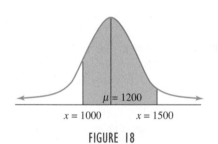

$\mu = 1200$

$x = 1000 \qquad x = 1500$

FIGURE 18

From the table, $z_1 = -1.33$ leads to an area of 0.0918, while $z_2 = 2.00$ corresponds to 0.9772. A total of $0.9772 - 0.0918 = 0.8854$, or 88.54%, of the drivers travel between 1000 and 1500 miles per month. The probability that a driver travels between 1000 miles and 1500 miles per month is 0.8854.

(c) Find the shortest and longest distances driven by the middle 95% of the sales force.

SOLUTION First, find the values of z that bound the middle 95% of the data. As Figure 19 illustrates, the lower z value, z_1, has 2.5% of the area to its left, and the higher z value, z_2, has 97.5% of the area to its left. Using the table backwards, we find that $z_1 = -1.96$ and $z_2 = 1.96$.

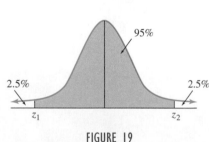

95%

2.5%   2.5%

$z_1 \qquad\qquad\qquad z_2$

FIGURE 19

The shortest distance is, therefore, 1.96 standard deviations *below* the mean, or

$$\text{Shortest} = \mu + z \cdot \sigma = 1200 + (-1.96) \cdot (150) = 906 \text{ miles.}$$

YOUR TURN 4 For Example 4, find the probability that a salesperson drives between 1275 and 1425 miles.

Likewise, the longest distance is 1.96 standard deviations *above* the mean, or

$$\text{Longest} = \mu + z \cdot \sigma = 1200 + (1.96) \cdot (150) = 1494 \text{ miles.}$$

Therefore, the distances driven by the middle 95% of the sales force are between 906 and 1494 miles.

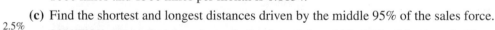

TRY YOUR TURN 4

✂ **TECHNOLOGY NOTE** Example 4 can also be done using a graphing calculator or computer, as described before, putting 1200 and 150 in place of 0 and 1 for the mean and standard deviation. The TI-84 Plus command `normalcdf(1000,1500,1200,150)` gives the answer 0.8860386561. This is more accurate than the value of 0.8854 found using the normal curve table, which required rounding the *z*-scores to two decimal places.

NOTE   The answers given to the exercises in this text are found using the normal curve table. If you use a graphing calculator or computer program, your answers may differ slightly.

As mentioned above, *z*-scores are the number of standard deviations from the mean, so $z = 1$ corresponds to 1 standard deviation above the mean, and so on. Looking up $z = 1.00$ and $z = -1.00$ in the table shows that

$$0.8413 - 0.1587 = 0.6826,$$

or 68.3% of the area under a normal curve lies within 1 standard deviation of the mean. Also, looking up $z = 2.00$ and $z = -2.00$ shows that

$$0.9772 - 0.0228 = 0.9544,$$

or 95.4% of the area lies within 2 standard deviations of the mean. These results, summarized in Figure 20, can be used to get a quick estimate of results when working with normal curves.

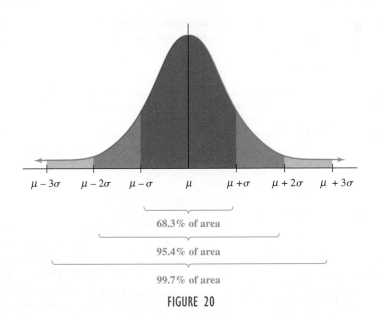

FIGURE 20

9.3   EXERCISES

1. The peak in a normal curve occurs directly above _____.

2. The total area under a normal curve (above the horizontal axis) is _____.

3. How are *z*-scores found for normal distributions where $\mu \neq 0$ or $\sigma \neq 1$?

4. How is the standard normal curve used to find probabilities for normal distributions?

Find the percent of the area under a normal curve between the mean and the given number of standard deviations from the mean.

5. 1.70

6. 0.93

7. −2.31

8. −1.45

Find the percent of the total area under the standard normal curve between each pair of z-scores.

9. $z = 0.32$ and $z = 3.18$

10. $z = 0.99$ and $z = 2.37$

11. $z = -1.83$ and $z = -0.91$

12. $z = -3.13$ and $z = -2.65$

13. $z = -2.95$ and $z = 2.03$

14. $z = -0.15$ and $z = 0.23$

Find a z-score satisfying the following conditions.

15. 5% of the total area is to the left of z.

16. 1% of the total area is to the left of z.

17. 10% of the total area is to the right of z.

18. 25% of the total area is to the right of z.

19. For any normal distribution, what is the value of $P(x \le \mu)$? $P(x \ge \mu)$?

20. Compare the probability that a number will lie within 2 standard deviations of the mean of a probability distribution using Chebyshev's theorem and using the normal distribution. (See Exercises 13–20, Section 9.2.) Explain what you observe.

21. Repeat Exercise 20 using 3 standard deviations.

APPLICATIONS

In all of the following applications, assume the distributions are normal. In each case, you should consider whether this is reasonable.

Business and Economics

Life of Light Bulbs   A certain type of light bulb has an average life of 500 hours, with a standard deviation of 100 hours. The length of life of the bulb can be closely approximated by a normal curve. An amusement park buys and installs 10,000 such bulbs. Find the total number that can be expected to last for each period of time.

22. At least 500 hours

23. Less than 500 hours

24. Between 680 and 780 hours

25. Between 350 and 550 hours

26. Less than 770 hours

27. More than 440 hours

28. Find the shortest and longest lengths of life for the middle 60% of the bulbs.

Quality Control   A box of oatmeal must contain 16 oz. The machine that fills the oatmeal boxes is set so that, on the average, a box contains 16.5 oz. The boxes filled by the machine have weights that can be closely approximated by a normal curve. What fraction of the boxes filled by the machine are underweight if the standard deviation is as follows?

29. 0.5 oz

30. 0.3 oz

31. 0.2 oz

32. 0.1 oz

Quality Control   The chickens at Colonel Thompson's Ranch have a mean weight of 1850 g, with a standard deviation of 150 g. The weights of the chickens are closely approximated by a normal curve. Find the percent of all chickens having weights in the following ranges.

33. More than 1700 g

34. Less than 1950 g

35. Between 1750 and 1900 g

36. Between 1600 and 2000 g

37. More than 2100 g or less than 1550 g

38. Find the smallest and largest weights for the middle 95% of the chickens.

39. University Expenditures   A 2000 report found that private colleges and universities in the United States had a mean expenditure of $169 million, with a standard deviation of $300 million. *Source: Higher Education.*

   a. If the data are normally distributed, what percent of a population is more than one standard deviation below the mean?

   b. What percent of private colleges and universities had an expenditure more than one standard deviation below the mean?

   c. Based on your answers to parts a and b, discuss whether expenditures at private colleges and universities in the United States are likely to be normally distributed. Explain why this might be expected.

40. Quality Control   A machine produces bolts with an average diameter of 0.25 in. and a standard deviation of 0.02 in. What is the probability that a bolt will be produced with a diameter greater than 0.3 in.?

41. Quality Control   A machine that fills quart milk cartons is set up to average 32.2 oz per carton, with a standard deviation of 1.2 oz. What is the probability that a filled carton will contain less than 32 oz of milk?

42. Grocery Bills   At the Discount Market, the average weekly grocery bill is $74.50, with a standard deviation of $24.30. What are the largest and smallest amounts spent by the middle 50% of this market's customers?

43. Grading Eggs   To be graded extra large, an egg must weigh at least 2.2 oz. If the average weight for an egg is 1.5 oz, with a standard deviation of 0.4 oz, how many eggs in a sample of five dozen would you expect to grade extra large?

Life Sciences

Vitamin Requirements   In nutrition, the Recommended Daily Allowance of vitamins is a number set by the government as a guide to an individual's daily vitamin intake. Actually, vitamin needs vary drastically from person to person, but the needs are very closely approximated by a normal curve. To calculate the Recommended Daily Allowance, the government first finds the average need for vitamins among people in the population and the standard deviation. The Recommended Daily Allowance is then defined as the mean plus 2.5 times the standard deviation.

44. What percent of the population will receive adequate amounts of vitamins under this plan?

Find the Recommended Daily Allowance for each vitamin in Exercises 45–47.

45. Mean = 1200 units; standard deviation = 60 units

46. Mean = 159 units; standard deviation = 12 units

47. Mean = 1200 units; standard deviation = 92 units

48. Blood Clotting The mean clotting time of blood is 7.45 seconds, with a standard deviation of 3.6 seconds. What is the probability that an individual's blood clotting time will be less than 7 seconds or greater than 8 seconds?

Social Sciences

Speed Limits New studies by Federal Highway Administration traffic engineers suggest that speed limits on many thoroughfares are set arbitrarily and often are artificially low. According to traffic engineers, the ideal limit should be the "85th percentile speed." This means the speed at or below which 85 percent of the traffic moves. Assuming speeds are normally distributed, find the 85th percentile speed for roads with the following conditions. *Source: Federal Highway Administration.*

49. The mean speed is 52 mph with a standard deviation of 8 mph.

50. The mean speed is 30 mph with a standard deviation of 5 mph.

Education The grading system known as "grading on the curve" is based on the assumption that grades are often distributed according to the normal curve and that a certain percent of a class should receive each grade, regardless of the performance of the class as a whole. The following is how one professor might grade on the curve.

Grade	Total Points
A	Greater than $\mu + (3/2)\sigma$
B	$\mu + (1/2)\sigma$ to $\mu + (3/2)\sigma$
C	$\mu - (1/2)\sigma$ to $\mu + (1/2)\sigma$
D	$\mu - (3/2)\sigma$ to $\mu - (1/2)\sigma$
F	Below $\mu - (3/2)\sigma$

What percent of the students receive the following grades?

51. A   **52.** B   **53.** C

54. Do you think this system would be more likely to be fair in a large freshman class in psychology or in a graduate seminar of five students? Why?

Education A teacher gives a test to a large group of students. The results are closely approximated by a normal curve. The mean is 76, with a standard deviation of 8. The teacher wishes to give A's to the top 8% of the students and F's to the bottom 8%. A grade of B is given to the next 20%, with D's given similarly. All other students get C's. Find the bottom cutoff (rounded to the nearest whole number) for the following grades.

55. A   **56.** B   **57.** C   **58.** D

59. Standardized Tests David Rogosa, a professor of educational statistics at Stanford University, has calculated the accuracy of tests used in California to abolish social promotion. Dr. Rogosa has claimed that a fourth grader whose true reading score is exactly at reading level (50th percentile—half of all the students read worse and half read better than this student) has a 58% chance of either scoring above the 55th percentile or below the 45th percentile on any one test. Assume that the results of a given test are normally distributed with mean 0.50 and standard deviation 0.09. *Source: The New York Times.*

a. Verify that Dr. Rogosa's claim is true.

b. Find the probability that this student will not score between the 40th and 60th percentile.

c. Using the results of parts a and b, discuss problems with the use of standardized testing to prevent social promotion.

General Interest

60. Christopher Columbus Before Christopher Columbus crossed the ocean, he measured the heights of the men on his three ships and found that they were normally distributed with a mean of 69.60 in. and a standard deviation of 3.20 in. What is the probability that a member of his crew had a height less than 66.27 in.? (The answer has another connection with Christopher Columbus!)

61. Lead Poisoning Historians and biographers have collected evidence that suggests that President Andrew Jackson suffered from lead poisoning. Recently, researchers measured the amount of lead in samples of Jackson's hair from 1815. The results of this experiment showed that Jackson had a mean lead level of 130.5 ppm. *Source: JAMA.*

a. If levels of lead in hair samples from that time period follow a normal distribution with mean 93 and standard deviation 16, find the probability that a randomly selected person from this time period would have a lead level of 130.5 ppm or higher. Does this provide evidence that Jackson suffered from lead poisoning during this time period?* *Source: Science.*

b. Today's typical lead levels follow a normal distribution with approximate mean 10 ppm and standard deviation 5 ppm. By these standards, calculate the probability that a randomly selected person from today would have a lead level of 130.5 or higher. (*Note:* These standards may not be valid for this experiment.) *Source: Clinical Chemistry*

c. Discuss whether we can conclude that Andrew Jackson suffered from lead poisoning.

62. Mercury Poisoning Historians and biographers have also collected evidence that suggests that President Andrew Jackson suffered from mercury poisoning. Recently, researchers measured the amount of mercury in samples of Jackson's hair from 1815. The results of this experiment showed that Jackson had a mean mercury level of 6.0 ppm. *Source: JAMA.*

a. If levels of mercury in hair samples from that time period follow a normal distribution with mean 6.9 and standard deviation 4.6, find the probability that a randomly selected person from that time period would have a mercury level of 6.0 ppm or higher. *Source: Science of the Total Environment.*

b. Discuss whether this provides evidence that Jackson suffered from mercury poisoning during this time period.

c. Today's accepted normal mercury levels follow a normal distribution with approximate mean 0.6 ppm and standard deviation 0.3 ppm. By present standards, is it likely that a

*Although this provides evidence that Andrew Jackson had elevated lead levels, the authors of the paper concluded that Andrew Jackson did not die from lead poisoning.

randomly selected person from today would have a mercury level of 6.0 ppm or higher?

d. Discuss whether we can conclude that Andrew Jackson suffered from mercury poisoning.

63. Barbie The popularity and voluptuous shape of Barbie dolls have generated much discussion about the influence these dolls may have on young children, particularly with regard to normal body shape. In fact, many people have speculated as to what Barbie's measurements would be if they were scaled to a common human height. Researchers have done this and have compared Barbie's measurements to the average 18- to 35-year-old woman, labeled Reference, and with the average model. The table below illustrates some of the results of their research, where each measurement is in centimeters. Assume that the distributions of measurements for the models and for the reference group follow a normal distribution with the given mean and standard deviation. *Source: Sex Roles.*

Measurement	Models Mean	Models s.d.	Reference Mean	Reference s.d.	Barbie
Head	50.0	2.4	55.3	2.0	55.0
Neck	31.0	1.0	32.7	1.4	23.9
Chest (bust)	87.4	3.0	90.3	5.5	82.3
Wrist	15.0	0.6	16.1	0.8	10.6
Waist	65.7	3.5	69.8	4.7	40.7

a. Find the probability of Barbie's head size or larger occurring for the reference group and for the models.

b. Find the probability of Barbie's neck size or smaller occurring for the reference group and for the models.

c. Find the probability of Barbie's bust size or larger occurring for the reference group and for the models.

d. Find the probability of Barbie's wrist size or smaller occurring for the reference group and for the models.

e. Find the probability of Barbie's waist size or smaller occurring for the reference group and for the models.

f. Compare the above values and discuss whether Barbie represents either the reference group or models. Any surprises?

64. Ken The same researchers from Exercise 63 wondered how the famous Ken doll measured up to average males and with Australian football players. The table below illustrates some of the results of their research, where each measurement is in centimeters. Assume that the distributions of measurements for the football players and for the reference group follow a normal distribution with the given mean and standard deviation. *Source: Sex Roles.*

Measurement	Football Mean	Football s.d.	Reference Mean	Reference s.d.	Ken
Head	52.1	2.3	53.7	2.9	53.0
Neck	34.6	1.8	34.2	1.9	32.1
Chest	92.3	3.5	91.2	4.8	75.0
Upper Arm	29.9	1.9	28.8	2.2	27.1
Waist	75.1	3.6	80.9	9.8	56.5

a. Find the probability of Ken's head size or larger occurring for the reference group and for the football players.

b. Find the probability of Ken's neck size or smaller occurring for the reference group and for the football players.

c. Find the probability of Ken's chest size or smaller occurring for the reference group and for the football players.

d. Find the probability of Ken's upper arm size or smaller occurring for the reference group and for the football players.

e. Find the probability of Ken's waist size or smaller occurring for the reference group and for the football players.

f. Compare the above values and discuss whether Ken's measurements are representative of either the reference group or football players. Then compare these results with the results of Exercise 63. Any surprises?

YOUR TURN ANSWERS

1. **(a)** 0.2236 **(b)** 0.9131 **(c)** 0.7970
2. **(a)** −1.96 **(b)** 0.81
3. −0.75 **4.** 0.2417

9.4 Normal Approximation to the Binomial Distribution

APPLY IT **What is the probability that at least 40 out of 100 drivers exceed the speed limit by at least 20 mph in Atlanta?**

This is a binomial probability problem with a large number of trials (100). In this section we will see how the normal curve can be used to approximate the binomial distribution, allowing us to answer this question in Example 2.

As we saw in Section 8.4 on Binomial Probability, many practical experiments have only two possible outcomes, sometimes referred to as success or failure. Such experiments are called Bernoulli trials or Bernoulli processes. Examples of Bernoulli trials include flipping a coin (with heads being a success, for instance, and tails a failure) or testing a computer chip coming off the assembly line to see whether or not it is defective. A binomial experiment consists of repeated independent Bernoulli trials, such as flipping a coin 10 times or taking a random sample of 20 computer chips from the assembly line. In Section 8.5 on Probability Distributions and Expected Value, we found the probability distribution for several binomial experiments, such as sampling five people with bachelor's degrees in education and counting how many are women. The probability distribution for a binomial experiment is known as a **binomial distribution**.

As another example, it was reported in a recent study that 40% of drivers in Atlanta exceed the speed limit by at least 20 mph. *Source: Laser Atlanta.* Suppose a state trooper wants to verify this statistic and records the speed of 10 randomly selected drivers. The trooper finds that 5 out of 10, or 50%, exceed the speed limit by at least 20 mph. How likely is this if the 40% figure is accurate? We can answer this question with the binomial probability formula from the previous chapter:

$$C(n, x) \cdot p^x \cdot (1 - p)^{n-x},$$

where n is the size of the sample (10 in this case), x is the number of speeders (5 in this example), and p is the probability that a driver is a speeder (0.40). This gives

$$P(x = 5) = C(10, 5) \cdot 0.40^5 \cdot (1 - 0.40)^5$$
$$= 252(0.01024)(0.07776) \approx 0.2007.$$

The probability is about 20%, so this result is not unusual.

Suppose that the state trooper takes a larger random sample of 100 drivers. What is the probability that 50 or more drivers speed if the 40% figure is accurate? Calculating $P(x = 50) + P(x = 51) + \cdots + P(x = 100)$ is a formidable task. One solution is provided by graphing calculators or computers. On the TI-84 Plus, for example, we can first calculate the probability that 49 or fewer drivers exceed the speed limit using the DISTR menu command `binomcdf(100,.40,49)`. Subtracting the answer from 1 gives a probability of 0.0271. But this high-tech method fails as n becomes larger; the command `binomcdf(1000000,.40,50000)` gives an error message. On the other hand, there is a low-tech method that works regardless of the size of n. It has further interest because it connects two different distributions: the normal and the binomial. The normal distribution is continuous, since the random variable can take on any real number. The binomial distribution is *discrete*, because the random variable can only take on integer values between 0 and n. Nevertheless, the normal distribution can be used to give a good approximation to binomial probability.

In order to use the normal approximation, we first need to know the mean and standard deviation of the binomial distribution. Recall from Section 8.5 that for the binomial distribution, $E(x) = np$. In Section 9.1, we referred to $E(x)$ as μ, and that notation will be used here. It is shown in more advanced courses in statistics that the standard deviation of the binomial distribution is given by $\sigma = \sqrt{np(1 - p)}$.

FOR REVIEW

Recall from Chapter 8 that the symbol $C(n, r)$ is defined as $\dfrac{n!}{r!(n - r)!}$. For example,

$$C(10, 5) = \frac{10!}{5! \, 5!}$$
$$= \frac{10 \cdot 9 \cdot 8 \cdot 7 \cdot 6}{5 \cdot 4 \cdot 3 \cdot 2 \cdot 1} = 252.$$

Mean and Standard Deviation for the Binomial Distribution

For the binomial distribution, the mean and standard deviation are given by

$$\boldsymbol{\mu = np} \quad \text{and} \quad \boldsymbol{\sigma = \sqrt{np(1 - p)}},$$

where n is the number of trials and p is the probability of success on a single trial.

EXAMPLE 1 Coin Flip

Suppose a fair coin is flipped 15 times.

(a) Find the mean and standard deviation for the number of heads.

SOLUTION Using $n = 15$ and $p = 1/2$, the mean is

$$\mu = np = 15\left(\frac{1}{2}\right) = 7.5.$$

The standard deviation is

$$\sigma = \sqrt{np(1 - p)} = \sqrt{15\left(\frac{1}{2}\right)\left(1 - \frac{1}{2}\right)}$$

$$= \sqrt{15\left(\frac{1}{2}\right)\left(\frac{1}{2}\right)} = \sqrt{3.75} \approx 1.94.$$

We expect, on average, to get 7.5 heads out of 15 tosses. Most of the time, the number of heads will be within 3 standard deviations of the mean, or between $7.5 - 3(1.94) = 1.68$ and $7.5 + 3(1.94) = 13.32$.

(b) Find the probability distribution for the number of heads, and draw a histogram of the probabilities.

SOLUTION The probability distribution is found by putting $n = 15$ and $p = 1/2$ into the formula for binomial probability. For example, the probability of 9 heads is given by

$$P(x = 9) = C(15, 9)\left(\frac{1}{2}\right)^9\left(1 - \frac{1}{2}\right)^6 \approx 0.15274.$$

Probabilities for the other values of x between 0 and 15, as well as a histogram of the probabilities, are shown in the table and in Figure 21. **TRY YOUR TURN 1**

In Figure 21, we have superimposed the normal curve with $\mu = 7.5$ and $\sigma = 1.94$ over the histogram of the distribution. Notice how well the normal distribution fits the binomial distribution. This approximation was first discovered in 1733 by Abraham De Moivre (1667–1754) for the case $p = 1/2$. The result was generalized by the French mathematician

YOUR TURN 1 Suppose a die is rolled 12 times. Find the mean and standard deviation of the number of sixes rolled.

Probability Distribution for Number of Heads	
x	$P(x)$
0	0.00003
1	0.00046
2	0.00320
3	0.01389
4	0.04166
5	0.09164
6	0.15274
7	0.19638
8	0.19638
9	0.15274
10	0.09164
11	0.04166
12	0.01389
13	0.00320
14	0.00046
15	0.00003

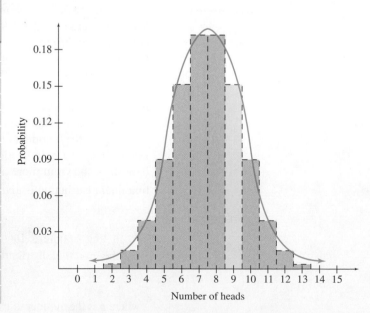

FIGURE 21

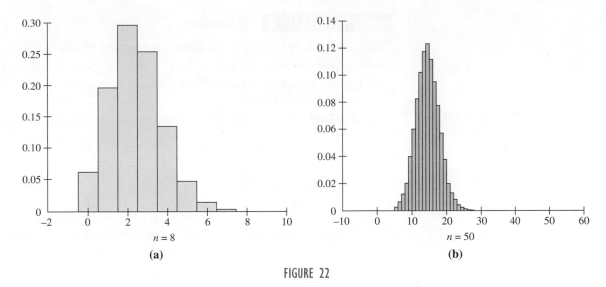

FIGURE 22

Pierre-Simon Laplace (1749–1827) in a book published in 1812.* As n becomes larger and larger, a histogram for the binomial distribution looks more and more like a normal curve. Histograms of the binomial distribution with $p = 0.3$, using $n = 8$ and $n = 50$ are shown in Figures 22(a) and (b), respectively.

The probability of getting exactly 9 heads in 15 tosses, or 0.15274, is the same as the area of the bar in blue in Figure 21. As the graph suggests, the area in blue is approximately equal to the area under the normal curve from $x = 8.5$ to $x = 9.5$. The normal curve is higher than the top of the bar in the left half but lower in the right half.

To find the area under the normal curve from $x = 8.5$ to $x = 9.5$, first find z-scores, as in the previous section. Use the mean and the standard deviation for the distribution, which we have already calculated, to get z-scores for $x_1 = 8.5$ and $x_2 = 9.5$.

$$
\begin{array}{ll}
\text{For } x_1 = 8.5, & \text{For } x_2 = 9.5, \\[4pt]
z_1 = \dfrac{8.5 - 7.5}{1.94} & z_2 = \dfrac{9.5 - 7.5}{1.94} \\[10pt]
= \dfrac{1.00}{1.94} & = \dfrac{2.00}{1.94} \\[10pt]
z_1 \approx 0.52. & z_2 \approx 1.03.
\end{array}
$$

From the table in the Appendix, $z_1 = 0.52$ gives an area of 0.6985, and $z_2 = 1.03$ gives 0.8485. The difference between these two numbers is the desired result.

$$P(z \le 1.03) - P(z \le 0.52) = 0.8485 - 0.6985 = 0.1500$$

This answer (0.1500) is not far from the more accurate answer of 0.15274 found in Example 1(b).

> **CAUTION** The normal curve approximation to a binomial distribution is quite accurate *provided that n is large and p is not close to 0 or 1.* As a rule of thumb, the normal curve approximation can be used as long as both np and $n(1 - p)$ are at least 5.

*Laplace's generalization, known as the Central Limit theorem, states that the distribution of the sample mean from *any* distribution approaches the normal distribution as the sample size increases. For more details, see any statistics book, such as *Elementary Statistics* (11th ed.) by Mario F. Triola, Pearson, 2009.

EXAMPLE 2   Speeding

Consider the random sample discussed earlier of 100 drivers in Atlanta, where 40% of the drivers exceed the speed limit by at least 20 mph.

(a) Use the normal distribution to approximate the probability that at least 50 drivers exceed the speed limit.

APPLY IT

SOLUTION   First find the mean and the standard deviation using $n = 100$ and $p = 0.40$.

$$\mu = 100(0.40) \qquad \sigma = \sqrt{100(0.40)(1 - 0.40)}$$
$$= 40 \qquad\qquad = \sqrt{100(0.40)(0.60)}$$
$$= \sqrt{24} \approx 4.899$$

As the graph in Figure 23 shows, we need to find the area to the right of $x = 49.5$ (since we want 50 or more speeders). The z-score corresponding to $x = 49.5$ is

$$z = \frac{49.5 - 40}{4.899} \approx 1.94.$$

From the table, $z = 1.94$ leads to an area of 0.9738, so

$$P(z > 1.94) = 1 - 0.9738 = 0.0262.$$

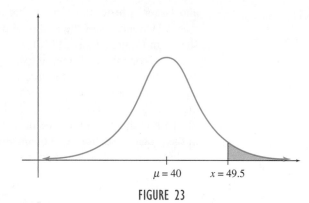

FIGURE 23

This value is close to the value of 0.0271 found earlier with the help of a graphing calculator. Either method tells us there is roughly a 3% chance of finding 50 or more speeders out of a random sample of 100. If the trooper found 50 or more speeders in his sample, he might suspect that either his sample is not truly random, or that the 40% figure for the percent of drivers who speed is too low.

(b) Find the probability of finding between 42 and 48 speeders in a random sample of 100.

SOLUTION   As Figure 24 shows, we need to find the area between $x_1 = 41.5$ and $x_2 = 48.5$.

$$\text{If } x_1 = 41.5, \text{ then } z_1 = \frac{41.5 - 40}{4.899} \approx 0.31.$$

$$\text{If } x_2 = 48.5, \text{ then } z_2 = \frac{48.5 - 40}{4.899} \approx 1.74.$$

Use the table to find that $z_1 = 0.31$ gives an area of 0.6217, and $z_2 = 1.74$ yields 0.9591. The final answer is the difference of these numbers, or

$$P(0.31 \le z \le 1.74) = P(z \le 1.74) - P(z \le 0.31)$$
$$= 0.9591 - 0.6217 = 0.3374.$$

The probability of finding between 42 and 48 speeders is about 0.3374. This is close to the value of 0.3352 found using the `binomcdf` command on a TI-84 Plus.

TRY YOUR TURN 2

YOUR TURN 2 Suppose that a test consists of 120 multiple choice questions, each with 5 answers from which to choose. Use the normal distribution to find the probability that if you guess at random, you get at least 32 correct.

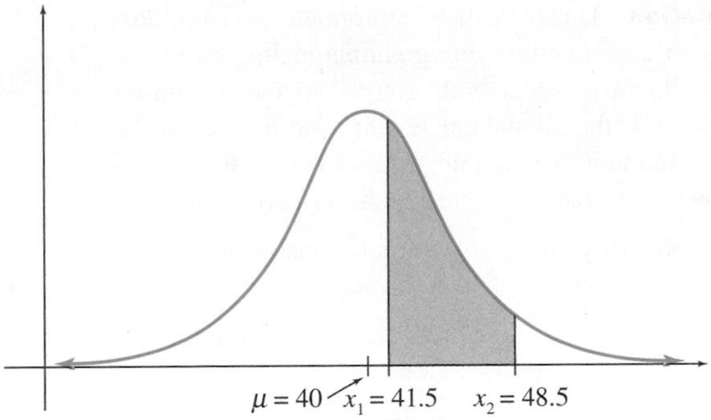

FIGURE 24

9.4   EXERCISES

1. What must be known to find the mean and standard deviation of a binomial distribution?

2. What is the rule of thumb for using the normal distribution to approximate a binomial distribution?

Suppose 16 coins are tossed. Find the probability of getting the following results (a) using the binomial probability formula and (b) using the normal curve approximation.

3. Exactly 4 heads   4. Exactly 10 heads

5. More than 12 tails   6. Fewer than 5 tails

For the remaining exercises in this section, use the normal curve approximation to the binomial distribution.

Suppose 1000 coins are tossed. Find the probability of getting the following results.

7. Exactly 500 heads   8. Exactly 510 heads

9. 475 heads or more   10. Fewer than 490 tails

A die is tossed 120 times. Find the probability of getting the following results.

11. Exactly twenty 5's   12. Exactly twenty-four 6's

13. More than fifteen 3's   14. Fewer than twenty-eight 6's

15. A reader asked Mr. Statistics (a feature in *Fortune* magazine) about the game of 26 once played in the bars of Chicago. The player chooses a number between 1 and 6 and then rolls a cup full of 10 dice 13 times. Out of the 130 numbers rolled, if the number chosen appears at least 26 times, the player wins. Calculate the probability of winning. *Source: Fortune.*

16. a. Try to use both the binomial probability formula and the normal approximation to the binomial to calculate the probability that exactly half of the coins come up heads if the following number of coins are flipped. You may run into problems using the binomial probability formula for part iii.

   i. 10   **ii.** 100   **iii.** 1000

   b. If you ran into problems using the binomial probability formula for part iii, tell what happened and explain why it happened.

c. Someone might speculate that with more coins, the probability that exactly half are heads goes up with the number of coin flips. Based on the results from part a, does this happen? Explain the error in the speculation.

APPLICATIONS

Business and Economics

17. **Quality Control**   Two percent of the quartz heaters produced in a certain plant are defective. Suppose the plant produced 10,000 such heaters last month. Find the probabilities that among these heaters, the following numbers were defective.

   a. Fewer than 170   **b.** More than 222

18. **Quality Control**   The probability that a certain machine turns out a defective item is 0.05. Find the probabilities that in a run of 75 items, the following results are obtained.

   a. Exactly 5 defectives

   b. No defectives

   c. At least 1 defective

19. **Survey Results**   A company is taking a survey to find out if people like its product. Their last survey indicated that 70% of the population like the product. Based on that, of a sample of 58 people, find the probabilities of the following.

   a. All 58 like the product.

   b. From 28 to 30 (inclusive) like the product.

20. **Minimum Wage**   A recent study of minimum wage earners found that 50.4% of them are 16 to 24 years old. Suppose a random sample of 600 minimum wage earners is selected. What is the probability that more than 330 of them are 16 to 24 years old? *Source: Bureau of Labor Statistics.*

Life Sciences

21. **Nest Predation**   For certain bird species, with appropriate assumptions, the number of nests escaping predation has a binomial distribution. Suppose the probability of success (that is, a nest escaping predation) is 0.3. Find the probability that at least half of 24 nests escape predation. *Source: The American Naturalist.*

22. Food Consumption Under certain appropriate assumptions, the probability of a competing young animal eating x units of food is binomially distributed, with n equal to the maximum number of food units the animal can acquire and p equal to the probability per time unit that an animal eats a unit of food. Suppose $n = 120$ and $p = 0.6$. *Source: The American Naturalist.*

a. Find the probability that an animal consumes exactly 80 units of food.

b. Suppose the animal must consume at least 70 units of food to survive. What is the probability that this happens?

23. Coconuts A 4-year review of trauma admissions to the Provincial Hospital, Alotau, Milne Bay Providence, reveals that 2.5% of such admissions were due to being struck by falling coconuts. *Source: The Journal of Trauma.*

a. Suppose 20 patients are admitted to the hospital during a certain time period. What is the probability that no more than 1 of these patients are there because they were struck by falling coconuts? Do not use the normal distribution here.

b. Suppose 2000 patients are admitted to the hospital during a longer time period. What is the approximate probability that no more than 70 of these patients are there because they were struck by falling coconuts?

24. Drug Effectiveness A new drug cures 80% of the patients to whom it is administered. It is given to 25 patients. Find the probabilities that among these patients, the following results occur.

a. Exactly 20 are cured.

b. All are cured.

c. No one is cured.

d. Twelve or fewer are cured.

25. Flu Inoculations A flu vaccine has a probability of 80% of preventing a person who is inoculated from getting the flu. A county health office inoculates 134 people. Find the probabilities of the following.

a. Exactly 10 of the people inoculated get the flu.

b. No more than 10 of the people inoculated get the flu.

c. None of the people inoculated get the flu.

26. Blood Types The blood types B− and AB− are the rarest of the eight human blood types, representing 1.5% and 0.6% of the population, respectively. *Source: The Handy Science Answer Book.*

a. If the blood types of a random sample of 1000 blood donors are recorded, what is the probability that 10 or more of the samples are AB−?

b. If the blood types of a random sample of 1000 blood donors are recorded, what is the probability that 20 to 40 inclusive of the samples are B−?

c. If a particular city had a blood drive in which 500 people gave blood and 3% of the donations were B−, would we have reason to believe that this town has a higher than normal number of donors who are B−? (*Hint:* Calculate the probability of 15 or more donors being B− for a random sample of 500 and then discuss the probability obtained.)

27. Motorcycles According to a recent report, 24.1% of non-fatal injuries suffered by motorcycle riders occur between 3 P.M. and 6 P.M. If 200 injured motorcyclists are surveyed, what is the probability that at most 50 were injured between 3 P.M. and 6 P.M.? *Source: Insurance Information Institute.*

Social Sciences

28. Straw Votes In one state, 55% of the voters expect to vote for Vale Leist. Suppose 1400 people are asked the name of the person for whom they expect to vote. Find the probability that at least 750 people will say that they expect to vote for Leist.

29. Smoking A recent study found that 46.3% of all ninth grade students in the United States have tried cigarette smoking, even if only one or two puffs. If 500 ninth grade students are surveyed, what is the probability that at most half have ever tried cigarettes? *Source: Centers for Disease Control and Prevention.*

30. Weapons and Youth A recent study found that 17.5% of all high school students in the United States have carried a weapon, including a gun, knife, or club. If 1200 high school students are surveyed, what is the probability that more than 200 students, but fewer than 250, have carried a weapon? *Source: Centers for Disease Control and Prevention.*

31. Election 2000 The Florida recount in the 2000 presidential election gave George W. Bush 2,912,790 votes and Al Gore 2,912,253 votes. What is the likelihood of the vote being so close, even if the electorate is evenly divided? Assume that the number of votes for Bush is binomially distributed with $n = 5,825,043$ (the sum of the votes for the two candidates) and $p = 0.5$. *Source: historycentral.com.*

a. Using the binomial probability feature on a graphing calculator, try to calculate $P(2,912,253 \le X \le 2,912,790)$. What happens?

b. Use the normal approximation to calculate the probability in part a.

General Interest

32. Homework Only 1 out of 12 American parents requires that children do their homework before watching TV. If your neighborhood is typical, what is the probability that out of 51 parents, 5 or fewer require their children to do homework before watching TV? *Source: Harper's.*

33. True-False Test A professor gives a test with 100 true-false questions. If 60 or more correct is necessary to pass, what is the probability that a student will pass by random guessing?

34. Hole in One In the 1989 U.S. Open, four golfers each made a hole in one on the same par-3 hole on the same day. *Sports Illustrated* writer R. Reilly stated the probability of a hole in one for a given golf pro on a given par-3 hole to be 1/3709. *Source: Sports Illustrated.*

a. For a specific par-3 hole, use the binomial distribution to find the probability that 4 or more of the 156 golf pros in the tournament field shoot a hole in one. *Source: School Science and Mathematics.*

b. For a specific par-3 hole, use the normal approximation to the binomial distribution to find the probability that 4 or more of the 156 golf pros in the tournament field shoot a hole in one. Why must we be very cautious when using this approximation for this application?

c. If the probability of a hole in one remains constant and is 1/3709 for any par-3 hole, find the probability that in 20,000

attempts by golf pros, there will be 4 or more hole in ones. Discuss whether this assumption is reasonable.

YOUR TURN ANSWERS

1. 2, 1.291   **2.** 0.0436

9 CHAPTER REVIEW

SUMMARY

In this chapter we introduced the field of statistics. Measures of central tendency, such as mean, median, and mode, were defined and illustrated by examples. We determined how much the numbers in a sample vary from the mean of a distribution by calculating the variance and standard deviation. The normal distribution,

perhaps the most important and widely used probability distribution, was defined and used to study a wide range of problems. The normal approximation to the binomial distribution was then developed, as were several important applications.

Mean   The mean of the n numbers $x_1, x_2, x_3, \ldots, x_n$ is

$$\bar{x} = \frac{\Sigma x}{n}.$$

Mean of a Grouped Distribution   The mean of a distribution, where x represents the midpoints, f the frequencies, and $n = \Sigma f$, is

$$\bar{x} = \frac{\Sigma xf}{n}.$$

Variance   The variance of a sample of n numbers $x_1, x_2, x_3, \ldots, x_n$, with mean $\bar{x}$, is

$$s^2 = \frac{\Sigma x^2 - n\bar{x}^2}{n - 1}.$$

Standard Deviation   The standard deviation of a sample of n numbers $x_1, x_2, x_3, \ldots, x_n$, with mean $\bar{x}$, is

$$s = \sqrt{\frac{\Sigma x^2 - n\bar{x}^2}{n - 1}}.$$

Standard Deviation for a Grouped Distribution   The standard deviation for a distribution with mean $\bar{x}$, where x is an interval midpoint with frequency f, and $n = \Sigma f$, is

$$s = \sqrt{\frac{\Sigma fx^2 - n\bar{x}^2}{n - 1}}.$$

Normal Distribution   The area of the shaded region under a normal curve from a to b is the probability that an observed data value will be between a and b.

z-scores   If a normal distribution has mean μ and standard deviation σ, then the z-score for the number x is

$$z = \frac{x - \mu}{\sigma}.$$

Area Under a Normal Curve   The area under a normal curve between $x = a$ and $x = b$ is the same as the area under the standard normal curve between the z-score for a and the z-score for b.

Mean and Standard Deviation for the Binomial Distribution   For the binomial distribution, the mean and standard deviation are given by

$$\mu = np \quad \text{and} \quad \sigma = \sqrt{np(1 - p)},$$

where n is the number of trials and p is the probability of success on a single trial.

KEY TERMS

9.1
random sample
grouped frequency distribution
frequency polygon
(arithmetic) mean
sample mean
population mean
median

statistic
mode

9.2
range
deviations from the mean
variance
standard deviation

sample variance
sample standard deviation
population variance
population standard deviation
Chebyshev's theorem

9.3
continuous distribution

skewed distribution
normal distribution
normal curve
standard normal curve
z-score

9.4
binomial distribution

REVIEW EXERCISES

CONCEPT CHECK

Determine whether each of the following statements is true or false, and explain why.

1. The mean, median, and mode of a normal distribution are all equal.

2. If the mean, median, and mode of a distribution are all equal, then the distribution must be a normal distribution.

3. If the means of two distributions are equal, then the variances must also be equal.

4. The sample mean $\overline{x}$ is not the same as the population mean μ.

5. A large variance indicates that the data are grouped closely together.

6. The mode of a distribution is the middle element of the distribution.

7. For a random variable X that is normally distributed, we know that $P(X \geq 2) = P(X \leq -2)$.

8. For a random variable X that is normally distributed with $\mu = 5$, we know that $P(X > 10) = P(X < 0)$.

9. The normal curve approximation to the binomial distribution should not be used on an experiment where $n = 30$ and $p = 0.1$.

10. The expected value of a sample mean is the population mean.

11. The expected value of a sample standard deviation is the population standard deviation.

12. For a standard normal random variable Z, $P(-1.5 < Z < 0) = 0.50 - P(Z > 1.5)$.

PRACTICE AND EXPLORATIONS

13. Discuss some reasons for organizing data into a grouped frequency distribution.

14. What is the rule of thumb for an appropriate interval in a grouped frequency distribution?

In Exercises 15 and 16, (a) write a frequency distribution, (b) draw a histogram, and (c) draw a frequency polygon.

15. The following numbers give the sales (in dollars) for the lunch hour at a local hamburger stand for the last 20 Fridays. Use intervals 450–474, 475–499, and so on.

480	451	501	478	512	473	509	515	458	566
516	535	492	558	488	547	461	475	492	471

16. The number of credits carried in one semester by students in a business mathematics class was as follows. Use intervals 9–10, 11–12, 13–14, 15–16.

10	9	16	12	13	15	13	16	15	11	13
12	12	15	12	14	10	12	14	15	15	13

Find the mean for the following.

17. 30, 24, 34, 30, 29, 28, 30, 29

18. 105, 108, 110, 115, 106, 110, 104, 113, 117

19.

Interval	Frequency
10–19	6
20–29	12
30–39	14
40–49	10
50–59	8

20.

Interval	Frequency
40–44	3
45–49	6
50–54	7
55–59	14
60–64	3
65–69	2

21. What do the mean, median, and mode of a distribution have in common? How do they differ? Describe each in a sentence or two.

Find the median and the mode (or modes) for each list of numbers.

22. 12, 17, 21, 23, 27, 27, 34

23. 38, 36, 42, 44, 38, 36, 48, 35

Find the modal class for the indicated distributions.

24. Exercise 20
25. Exercise 19

26. What is meant by the range of a distribution?

27. How are the variance and the standard deviation of a distribution related? What is measured by the standard deviation?

Find the range and standard deviation for each distribution.

28. 22, 27, 31, 35, 41

29. 26, 43, 51, 29, 37, 56, 29, 82, 74, 93

Find the standard deviation for the following.

30. Exercise 20   **31.** Exercise 19

32. Describe the characteristics of a normal distribution.

33. What is meant by a skewed distribution?

Find the following areas under the standard normal curve.

34. Between $z = 0$ and $z = 2.17$

35. To the left of $z = 0.84$

36. Between $z = -2.13$ and $z = 1.11$

37. Between $z = 1.53$ and $z = 2.82$

38. Find a z-score such that 7% of the area under the curve is to the right of z.

39. Why is the normal distribution not a good approximation of a binomial distribution that has a value of p close to 0 or 1?

40. Suppose a card is drawn at random from an ordinary deck 1,000,000 times with replacement.

  a. What is the probability that between 249,500 and 251,000 hearts (inclusive) are drawn?

  b. Why must the normal approximation to the binomial distribution be used to solve part a?

41. Suppose four coins are flipped and the number of heads counted. This experiment is repeated 20 times. The data might look something like the following. (You may wish to try this yourself and use your own results rather than these.)

Number of Heads	Frequency
0	1
1	5
2	7
3	5
4	2

  a. Calculate the sample mean $\bar{x}$ and sample standard deviation s.

  b. Calculate the population mean μ and population standard deviation σ for this binomial population.

  c. Compare your answer to parts a and b. What do you expect to happen?

42. Much of our work in Chapters 8 and 9 is interrelated. Note the similarities in the following parallel treatments of a frequency distribution and a probability distribution.

Frequency Distribution

Complete the table below for the following data. (Recall that x is the midpoint of the interval.)

14, 7, 1, 11, 2, 3, 11, 6, 10, 13, 11, 11, 16, 12, 9, 11, 9, 10, 7, 12, 9, 6, 4, 5, 9, 16, 12, 12, 11, 10, 14, 9, 13, 10, 15, 11, 11, 1, 12, 12, 6, 7, 8, 2, 9, 12, 10, 15, 9, 3

Interval	x	Tally	f	$x \cdot f$
1–3	2	HHT I	6	12
4–6				
7–9				
10–12				
13–15				
16–18				

Probability Distribution

A binomial distribution has $n = 10$ and $p = 0.5$. Complete the following table.

x	$P(x)$	$x \cdot P(x)$
0	0.001	
1	0.010	
2	0.044	
3	0.117	
4		
5		
6		
7		
8		
9		
10		

  a. Find the mean (or expected value) for each distribution.

  b. Find the standard deviation for each distribution.

  c. Use the normal approximation of the binomial probability distribution to find an interval centered on the mean that contains 95.44% of that distribution.

  d. Why can't we use the normal distribution to answer probability questions about the frequency distribution?

APPLICATIONS

Business and Economics

43. Stock Returns The annual returns of two stocks for 3 years are given below.

Stock	2009	2010	2011
Stock I	11%	-1%	14%
Stock II	9%	5%	10%

a. Find the mean and standard deviation for each stock over the 3-year period.

b. If you are looking for security (hence, less variability) with an average 8% return, which of these stocks should you choose?

44. Quality Control A machine that fills quart orange juice cartons is set to fill them with 32.1 oz. If the actual contents of the cartons vary normally, with a standard deviation of 0.1 oz, what percentage of the cartons contain less than a quart (32 oz)?

45. Quality Control About 4% of the frankfurters produced by a certain machine are overstuffed and thus defective. For a sample of 500 frankfurters, find the following probabilities—first by using the binomial probability formula, and then by using the normal approximation.

a. Twenty-five or fewer are overstuffed.

b. Exactly 25 are overstuffed.

c. At least 30 are overstuffed.

46. Bankruptcy The probability that a small business will go bankrupt in its first year is 0.21. For 50 such small businesses, find the following probabilities—first by using the binomial probability formula, and then by using the normal approximation.

a. Exactly 8 go bankrupt.

b. No more than 2 go bankrupt.

Life Sciences

47. Rat Diets The weight gains of 2 groups of 10 rats fed different experimental diets were as follows.

Diet	Weight Gains									
A	1	0	3	7	1	1	5	4	1	4
B	2	1	1	2	3	2	1	0	1	0

Compute the mean and standard deviation for each group.

a. Which diet produced the greatest mean gain?

b. Which diet produced the most consistent gain?

Chemical Effectiveness White flies are devastating California crops. An area infested with white flies is to be sprayed with a chemical that is known to be 98% effective for each application. Assume a sample of 1000 flies is checked.

48. Use the normal distribution to find the approximate probability that exactly 980 of the flies are killed in one application.

49. Use the normal distribution to find the approximate probability that no more than 986 of the flies are killed in one application.

50. Use the normal distribution to find the approximate probability that at least 975 of the flies are killed in one application.

51. Use the normal distribution to find the approximate probability that between 973 and 993 (inclusive) of the flies are killed in one application.

Social Sciences

Commuting Times The average resident of a certain East Coast suburb spends 42 minutes per day commuting, with a standard deviation of 12 minutes. Assume a normal distribution. Find the percent of all residents of this suburb who have the following commuting times.

52. At least 50 minutes per day

53. No more than 40 minutes per day

54. Between 32 and 40 minutes per day

55. Between 38 and 60 minutes per day

56. I.Q. Scores On standard IQ tests, the mean is 100, with a standard deviation of 15. The results are very close to fitting a normal curve. Suppose an IQ test is given to a very large group of people. Find the percent of those people whose IQ scores are as follows.

a. More than 130

b. Less than 85

c. Between 85 and 115

General Interest

57. Olympics The number of countries participating in the Summer Olympics since 1960 is given in the following table. *Source: The New York Times 2010 Almanac.*

Olympic City	Year	Number of Countries
Rome	1960	83
Tokyo	1964	93
Mexico City	1968	112
Munich	1972	121
Montreal	1976	92
Moscow	1980	80
Los Angeles	1984	140
Seoul	1988	159
Barcelona	1992	169
Atlanta	1996	197
Sydney	2000	199
Athens	2004	202
Beijing	2008	205

a. Find the mean, median, and mode of the data.

b. Find the standard deviation of the data.

c. What percent of the data is within 1 standard deviation of the mean?

d. What percent of the data is within 2 standard deviations of the mean?

58. Broadway A survey was given to 313 performers appearing in 23 Broadway companies. The percentage of performers injured during practice or a performance was 55.5%. If a random sample of 500 Broadway performers is taken, use the normal approximation to the binomial distribution to find the approximate probability that more than 300 performers have been injured. *Source: American Journal of Public Health.*

59. Broadway In the survey described in Exercise 58, the demographics of the Broadway performers were recorded as shown below. Assume that all of these demographics follow a normal distribution, an assumption that always must be verified prior to using it in real situations. *Source: American Journal of Public Health.*

	Mean	Standard Deviation
Dancer's Age (female)	28.0	5.5
Dancer's Age (male)	32.2	8.4
Height (in m) (female)	1.64	0.08
Duration as Professional in yr (female)	11.0	8.9
Total No. of Injuries as Performer (female)	3.0	2.2

a. Find the probability that a female dancer is 35 years old or older.

b. Find the probability that a male dancer is 35 years old or older.

c. Compare your answers to parts a and b.

d. Find the probability that a female performer is 1.4 m tall or taller.

e. Find the probability that a female performer has a career duration that is more than 1.5 standard deviations from the mean.

f. Would a female who has more than 6 injuries during her career be considered a rare event? Explain.

EXTENDED APPLICATION

STATISTICS IN THE LAW—THE *CASTANEDA* DECISION

Statistical evidence is now routinely presented in both criminal and civil cases. In this application we'll look at a famous case that established use of the binomial distribution and measurement by standard deviation as an accepted procedure.*

Defendants who are convicted in criminal cases sometimes appeal their conviction on the grounds that the jury that indicted or convicted them was drawn from a pool of jurors that does not represent the population of the district in which they live. These appeals almost always cite the Supreme Court's decision in *Castaneda v. Partida* [430 U.S. 482], a case that dealt with the selection of grand juries in the state of Texas. The decision summarizes the facts this way:

> After respondent, a Mexican-American, had been convicted of a crime in a Texas District Court and had exhausted his state remedies on his claim of discrimination in the selection of the grand jury that had indicted him, he filed a habeas corpus petition in the Federal District Court, alleging a denial of due process and equal protection under the Fourteenth Amendment, because of gross underrepresentation of Mexican-Americans on the county grand juries.

The case went to the Appeals Court, which noted that "the county population was 79% Mexican-American, but, over an 11-year period, only 39% of those summoned for grand jury service were Mexican-American," and concluded that together with other testimony about the selection process, "the proof offered by respondent was sufficient to demonstrate a prima facie case of intentional discrimination in grand jury selection. . . ."

The state appealed to the Supreme Court, and the Supreme Court needed to decide whether the underrepresentation of Mexican-Americans on grand juries was indeed too extreme to be an effect of chance. To do so, they invoked the binomial distribution. Here is the argument:

> Given that 79.1% of the population is Mexican-American, the expected number of Mexican-Americans among the 870 persons summoned to serve as grand jurors over the 11-year period is approximately 688. The observed number is 339. Of course, in any given drawing, some fluctuation from the expected number is predicted. The important point, however, is that the statistical model shows that the results of a random drawing are likely to fall in the vicinity of the expected value. . . .

*The *Castaneda* case and many other interesting applications of statistics in law are discussed in Finkelstein and Levin, *Statistics for Lawyers,* New York, Springer-Verlag, 1990. U.S. Supreme Court decisions are online at http://www.findlaw.com/casecode/supreme.html. In addition, most states now have important state court decisions online.

The measure of the predicted fluctuations from the expected value is the standard deviation, defined for the binomial distribution as the square root of the product of the total number in the sample (here 870) times the probability of selecting a Mexican-American (0.791) times the probability of selecting a non-Mexican-American (0.209) . . . . Thus, in this case the standard deviation is approximately 12. As a general rule for such large samples, if the difference between the expected value and the observed number is greater than two or three standard deviations, then the hypothesis that the jury drawing was random would be suspect to a social scientist. The 11-year data here reflect a difference between the expected and observed number of Mexican-Americans of approximately 29 standard deviations. A detailed calculation reveals that the likelihood that such a substantial departure from the expected value would occur by chance is less than 1 in 10^{140}.

The Court decided that the statistical evidence supported the conclusion that jurors were not randomly selected, and that it was up to the state to show that its selection process did not discriminate against Mexican-Americans. The Court concluded:

The proof offered by respondent was sufficient to demonstrate a prima facie case of discrimination in grand jury

selection. Since the State failed to rebut the presumption of purposeful discrimination by competent testimony, despite two opportunities to do so, we affirm the Court of Appeals' holding of a denial of equal protection of the law in the grand jury selection process in respondent's case.

EXERCISES

1. Check the Court's calculation of 29 standard deviations as the difference between the expected number of Mexican-Americans and the number actually chosen.

2. Where do you think the Court's figure of 1 in 10^{140} came from?

3. The *Castaneda* decision also presents data from a $2\frac{1}{2}$-year period during which the State District Judge supervised the selection process. During this period, 220 persons were called to serve as grand jurors, and only 100 of these were Mexican-American.

   a. Considering the 220 jurors as a random selection from a large population, what is the expected number of Mexican-Americans, using the 79.1% population figure?

   b. If we model the drawing of jurors as a sequence of 220 independent Bernoulli trials, what is the standard deviation of the number of Mexican-Americans?

   c. About how many standard deviations is the actual number of Mexican-Americans drawn (100) from the expected number that you calculated in part a?

   d. What does the normal distribution table at the back of the book tell you about this result?

4. The following information is from an appeal brought by Hy-Vee stores before the Iowa Supreme Court, appealing a ruling by the Iowa Civil Rights Commission in favor of a female employee of one of their grocery stores.

   In 1985, there were 112 managerial positions in the ten Hy-Vee stores located in Cedar Rapids. Only 6 of these managers were women. During that same year there were 294 employees; 206 were men and 88 were women.

   a. How far from the expected number of women in management was the actual number, assuming that gender had nothing to do with promotion? Measure the difference in standard deviations.

   b. Does this look like evidence of purposeful discrimination?

5. Go to the website WolframAlpha.com and enter "normal probability." Use the results to calculate the probability of being 12.3 standard deviations below the mean, as calculated in Exercise 3c. Compare this result with the result using a graphing calculator and using Excel.

DIRECTIONS FOR GROUP PROJECT

Suppose that you and three other students are serving as interns at a prestigious law firm. One of the partners is interested in the use of probability in court cases and would like the four of you to prepare a brief on the Castaneda *decision. She insists that you describe the case and highlight the mathematics used in your brief.*

Be sure to use the results from the case along with the results of Exercises 1–3 in preparing your brief. Also, make recommendations of other types of cases where probability may be used in law. Presentation software, such as Microsoft PowerPoint, should be used to present your brief to the partners of the firm.

10

Markov Chains

In a Markov process, the next state of the system you are analyzing depends only on the current state and a set of transition probabilities. For example, the length of a waiting line in a bank a minute from now depends on the current length and the probabilities of a customer arriving or completing a transaction. This *queuing chain* model is studied in the exercises in Section 1.

Stochastic processes are mathematical models that evolve over time in a probabilistic manner. In this chapter we study a special kind of stochastic process called a *Markov chain*, where the outcome of an experiment depends only on the outcome of the previous experiment. In other words, the next **state** of the system depends only on the present state, not on preceding states. Such experiments are common enough in applications to make their study worthwhile. Markov chains are named after the Russian mathematician A. A. Markov (1856–1922), who initiated the theory of stochastic processes.

10.1 Basic Properties of Markov Chains

APPLY IT

If we know the probability that the child of an upper-class parent becomes middle-class or lower-class, and we know similar information for the child of a middle-class or lower-class parent, what is the probability that the grandchild or great-grandchild of an upper-class parent is middle- or lower-class?

Using Markov chains, we will learn the answer to this question in Example 2.

Transition Matrix

In sociology, it is convenient to classify people by income as *lower-class*, *middle-class*, and *upper-class*. Sociologists have found that the strongest determinant of the income class of an individual is the income class of the individual's parents. For example, if an individual in the lower-income class is said to be in *state 1*, an individual in the middle-income class is in *state 2*, and an individual in the upper-income class is in *state 3*, then the following probabilities of change in income class from one generation to the next might apply.*

		Class Transition		
			Next Generation	
	State	1	2	3
Current Generation	1	0.65	0.28	0.07
	2	0.15	0.67	0.18
	3	0.12	0.36	0.52

FOR REVIEW

Recall from Chapter 7 that the probability of an event is a real number between 0 and 1. The closer the probability is to 1, the more likely the event will occur. Also, in a sample space consisting of n mutually exclusive events, the sum of the probabilities of these events equals 1.

This table shows that if an individual is in state 1 (lower-income class) then there is a probability of 0.65 that any offspring will be in the lower-income class, a probability of 0.28 that offspring will be in the middle-income class, and a probability of 0.07 that offspring will be in the upper-income class.

The symbol p_{ij} will be used for the probability of transition from state i to state j in one generation. For example, p_{23} represents the probability that a person in state 2 will have offspring in state 3; from the table above,

$$p_{23} = 0.18.$$

Also from the table, $p_{31} = 0.12$, $p_{22} = 0.67$, and so on.

*For an example with actual data, see Exercise 32 in Section 2 of this chapter. For another example, see the landmark study by Glass, D. V. and J. R. Hall, "Social Mobility in Great Britain: A Study of Intergenerational Changes in Status," in *Social Mobility in Great Britain*, D. V. Glass, ed., Routledge & Kegan Paul, 1954. This data is analyzed using Markov chains in *Finite Markov Chains* by John G. Kemeny and J. Laurie Snell, Springer-Verlag, 1976.

The information from the table can be written in other forms. Figure 1 is a **transition diagram** that shows the three states and the probabilities of going from one state to another.

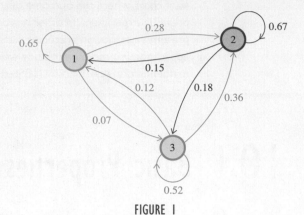

FIGURE 1

In a **transition matrix,** the states are indicated at the side and the top. If P represents the transition matrix for the table above, then

$$
\begin{array}{c c c c}
 & 1 & 2 & 3 \\
\begin{array}{c} 1 \\ 2 \\ 3 \end{array} &
\left[\begin{array}{c c c}
0.65 & 0.28 & 0.07 \\
0.15 & 0.67 & 0.18 \\
0.12 & 0.36 & 0.52
\end{array}\right] & = P.
\end{array}
$$

A transition matrix has several features:

1. It is square, since all possible states must be used both as rows and as columns.

2. All entries are between 0 and 1, inclusive, because all entries represent probabilities.

3. The sum of the entries in any row must be 1, since the numbers in the row give the probability of changing from the state at the left to one of the states indicated across the top.

Markov Chains
A transition matrix, such as matrix P above, also shows three key features of a Markov chain.

Markov Chain

A sequence of trials of an experiment is a **Markov chain** if

1. the outcome of each experiment is one of a set of discrete states;

2. the outcome of an experiment depends only on the present state and not on any past states;

3. the transition probabilities remain constant from one transition to the next.

For example, in transition matrix P with constant probabilities, a person is assumed to be in one of three discrete states (lower, middle, or upper income), with each offspring in one of these same three discrete states.

EXAMPLE 1 Dry Cleaners

A small town has only two dry cleaners, Johnson and NorthClean. Johnson's manager hopes to increase the firm's market share by conducting an extensive advertising campaign. After the campaign, a market research firm finds that there is a probability of 0.8 that a customer of Johnson's will bring his next batch of dirty clothes to Johnson and a 0.35 chance that a North-Clean customer will switch to Johnson for his next batch. Write a transition matrix showing this information.

YOUR TURN 1 In Example 1, suppose there is a probability of 0.68 that a customer of Johnson's will bring his next batch of dirty clothes to Johnson and a 0.21 chance that a NorthClean customer will switch to Johnson for his next batch. Write a transition matrix showing this information.

SOLUTION We must assume that the probability that a customer comes to a given dry cleaner depends only on where the last batch of clothes was taken. If there is a probability of 0.8 that a Johnson customer will return to Johnson, then there must be a $1 - 0.8 = 0.2$ chance that the customer will switch to NorthClean. In the same way, there is a $1 - 0.35 = 0.65$ chance that a NorthClean customer will return to NorthClean. These probabilities give the following transition matrix.

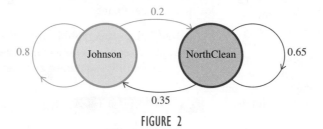

We shall come back to this transition matrix later in this section (Example 3).

Figure 2 shows a transition diagram with the probabilities of using each dry cleaner for the second batch of dirty clothes. **TRY YOUR TURN 1**

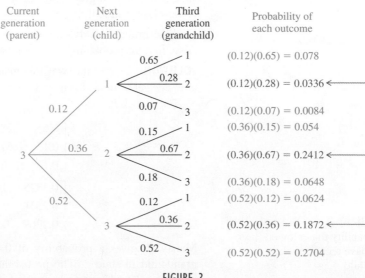

FIGURE 2

Look again at transition matrix P for income-class changes.

$$\begin{array}{c} 1 \\ 2 \\ 3 \end{array} \begin{bmatrix} 0.65 & 0.28 & 0.07 \\ 0.15 & 0.67 & 0.18 \\ 0.12 & 0.36 & 0.52 \end{bmatrix} = P$$

This matrix shows the probability of change in income class from one generation to the next. Now let us investigate the probabilities for changes in income class over *two* generations, assuming that the transition probabilities remain the same. For example, if a parent is in state 3 (the upper-income class), what is the probability that a grandchild will be in state 2?

To find out, start with a tree diagram, as shown in Figure 3. This diagram shows only the part of the tree that starts in state 3. The various probabilities come from transition matrix P.

FIGURE 3

Multiplication of matrices was covered in Chapter 2. To get the entry in row i, column j of a product, multiply row i of the first matrix times column j of the second matrix and add up the products. For example, to get the element in row 1, column 1 of P^2, where

$$P = \begin{bmatrix} 0.65 & 0.28 & 0.07 \\ 0.15 & 0.67 & 0.18 \\ 0.12 & 0.36 & 0.52 \end{bmatrix},$$

we calculate $(0.65)(0.65) + (0.28)(0.15) + (0.07)(0.12) = 0.4729 \approx 0.47$. To get row 3, column 2, the computation is $(0.12)(0.28) + (0.36)(0.67) + (0.52)(0.36) = 0.462$. You should review matrix multiplication by working out the rest of P^2 and verifying that it agrees with the result given in Example 2.

APPLY IT

The arrows point to the outcomes "grandchild in state 2"; the grandchild can get to state 2 after having had parents in either state 1, state 2, or state 3. The probability that a parent in state 3 will have a grandchild in state 2 is given by the sum of the probabilities indicated with arrows, or

$$0.0336 + 0.2412 + 0.1872 = 0.4620.$$

We used p_{ij} to represent the probability of changing from state i to state j in one generation. This notation can be used to write the probability that a parent in state 3 will have a grandchild in state 2:

$$p_{31} \cdot p_{12} + p_{32} \cdot p_{22} + p_{33} \cdot p_{32}.$$

This sum of products of probabilities should remind you of matrix multiplication—it is nothing more than one step in the process of multiplying matrix P by itself. In particular, it is row 3 of P times column 2 of P. If P^2 represents the matrix product $P \cdot P$, then P^2 gives the probabilities of a transition from one state to another in *two* repetitions of an experiment. Generalizing,

> P^n gives the probabilities of a transition from one state to another in n repetitions of an experiment, provided the transition probabilities remain constant from one repetition to the next.

EXAMPLE 2   Income-class Changes

Use the transition matrix P for income-class changes to find the following.

(a) Find the probability that the grandchild of an upper-class parent is middle-class. Also, find the probability that the grandchild is lower-class.

SOLUTION   To find the probability that the grandchild is middle- or lower-class, we calculate P^2 using matrix multiplication.

$$P^2 = \begin{bmatrix} 0.65 & 0.28 & 0.07 \\ 0.15 & 0.67 & 0.18 \\ 0.12 & 0.36 & 0.52 \end{bmatrix}\begin{bmatrix} 0.65 & 0.28 & 0.07 \\ 0.15 & 0.67 & 0.18 \\ 0.12 & 0.36 & 0.52 \end{bmatrix} = \begin{bmatrix} 0.4729 & 0.3948 & 0.1323 \\ 0.2196 & 0.5557 & 0.2247 \\ 0.1944 & 0.4620 & 0.3436 \end{bmatrix}.$$

The entry in row 3, column 2 of P^2 gives the probability that a person in state 3 will have a grandchild in state 2; that is, that an upper-class person will have a middle-class grandchild. This number, 0.4620, is the same result found through using the tree diagram.

Row 3, column 1 of P^2 gives the number 0.1944, the probability that a person in state 3 will have a grandchild in state 1; that is, that an upper-class person will have a lower-class grandchild.

(b) Find the probability that the great-grandchild of an upper-class parent is middle-class. Also, find the probability that the great-grandchild is lower-class.

SOLUTION   In the same way that matrix P^2 gives the probability of income-class changes after *two* generations, the matrix $P^3 = P \cdot P^2$ gives the probabilities after *three* generations.
For matrix P,

$$P^3 = P \cdot P^2 = \begin{bmatrix} 0.65 & 0.28 & 0.07 \\ 0.15 & 0.67 & 0.18 \\ 0.12 & 0.36 & 0.52 \end{bmatrix}\begin{bmatrix} 0.4729 & 0.3948 & 0.1323 \\ 0.2196 & 0.5557 & 0.2247 \\ 0.1944 & 0.4620 & 0.3436 \end{bmatrix}$$

$$\approx \begin{bmatrix} 0.3825 & 0.4446 & 0.1730 \\ 0.2531 & 0.5147 & 0.2322 \\ 0.2369 & 0.4877 & 0.2754 \end{bmatrix}.$$

YOUR TURN 2   In Example 2, find the probability that a lower-class person will have an upper-class great-grandchild.

Matrix P^3 gives a probability of 0.4877 that a person in state 3 will have a great-grandchild in state 2. The probability is 0.2369 that a person in state 3 will have a great-grandchild in state 1.   **TRY YOUR TURN 2**

A graphing calculator with matrix capability is useful for finding powers of a matrix. If you enter matrix A, then multiply by A, then multiply the product by A again, you get each new power in turn. You can also raise a matrix to a power just as you do with a number. On a TI-84 Plus, for example, P^3 to four decimal places is calculated as illustrated by Figure 4 where P was entered into the matrix [A] on the calculator.

```
round([A]³, 4)
[.3825   .4446    .1  ]
[.2531   .5147    .2▶ ]
[.2369   .4877    .27 ]
■
```

FIGURE 4

EXAMPLE 3 Dry Cleaners

Use the transition matrix for the dry cleaners, found in Example 1, to calculate the following probabilities.

$$\begin{array}{c} & \textit{Second Batch} \\ & \begin{array}{cc} \text{Johnson} & \text{NorthClean} \end{array} \\ \textit{First Batch} \begin{array}{c} \text{Johnson} \\ \text{NorthClean} \end{array} & \begin{bmatrix} 0.8 & 0.2 \\ 0.35 & 0.65 \end{bmatrix} \end{array}$$

(a) Find the probability that a person bringing his first batch of dry cleaning to Johnson will also bring his third batch to Johnson.

SOLUTION To find the probabilities for the third batch, the second stage of this Markov chain, find the square of the transition matrix. If C represents the transition matrix, then

$$C^2 = C \cdot C = \begin{bmatrix} 0.8 & 0.2 \\ 0.35 & 0.65 \end{bmatrix}\begin{bmatrix} 0.8 & 0.2 \\ 0.35 & 0.65 \end{bmatrix} = \begin{bmatrix} 0.71 & 0.29 \\ 0.5075 & 0.4925 \end{bmatrix}.$$

From C^2, the probability that a person bringing his first batch of clothes to Johnson will also bring his third batch to Johnson is 0.71. Likewise, the probability that a person bringing his first batch to NorthClean will bring his third batch to NorthClean is 0.4925.

(b) Find the probability that a person bringing his first batch of dry cleaning to NorthClean will bring his fourth batch to Johnson.

SOLUTION The cube of matrix C gives the probabilities for the fourth batch, the third step in our experiment.

$$C^3 = C \cdot C^2 \approx \begin{bmatrix} 0.6695 & 0.3305 \\ 0.5784 & 0.4216 \end{bmatrix}$$

YOUR TURN 3 In Example 3, find the probability that a person bringing his first batch to Johnson will bring his fifth batch to NorthClean.

Row 2, column 1 gives the number 0.5784, the probability that a person bringing his first batch to NorthClean will bring his fourth batch to Johnson. **TRY YOUR TURN 3**

Distribution of States

Look again at the transition matrix for income-class changes:

$$P = \begin{bmatrix} 0.65 & 0.28 & 0.07 \\ 0.15 & 0.67 & 0.18 \\ 0.12 & 0.36 & 0.52 \end{bmatrix}.$$

Suppose the table on the left gives the initial distribution of people in the three income classes.

To see how these proportions would change after one generation, use the tree diagram in Figure 5 on the next page. For example, to find the proportion of people in state 2 after one generation, add the numbers indicated with arrows.

$$0.0588 + 0.4556 + 0.0396 = 0.5540$$

In a similar way, the proportion of people in state 1 after one generation is

$$0.1365 + 0.1020 + 0.0132 = 0.2517,$$

Initial Distribution		
Class	**State**	**Proportion**
Lower	1	21%
Middle	2	68%
Upper	3	11%

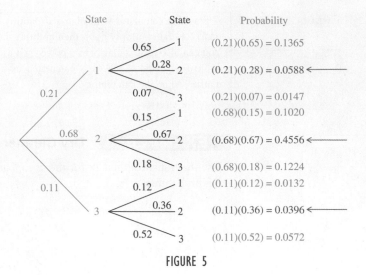

FIGURE 5

and the proportion of people in state 3 after one generation is

$$0.0147 + 0.1224 + 0.0572 = 0.1943.$$

The initial distribution of states, 21%, 68%, and 11%, becomes, after one generation, 25.17% in state 1, 55.4% in state 2, and 19.43% in state 3. These distributions can be written as *probability vectors* (where the percents have been changed to decimals rounded to the nearest hundredth)

$$[0.21 \quad 0.68 \quad 0.11] \quad \text{and} \quad [0.25 \quad 0.55 \quad 0.19],$$

respectively. A **probability vector** is a matrix of only one row, having nonnegative entries, with the sum of the entries equal to 1.

The work with the tree diagram to find the distribution of states after one generation is exactly the work required to multiply the initial probability vector, $X_0 = [0.21 \quad 0.68 \quad 0.11]$, and the transition matrix P:

$$X_0 \cdot P = [0.21 \quad 0.68 \quad 0.11] \begin{bmatrix} 0.65 & 0.28 & 0.07 \\ 0.15 & 0.67 & 0.18 \\ 0.12 & 0.36 & 0.52 \end{bmatrix} = [0.2517 \quad 0.5540 \quad 0.1943].$$

EXAMPLE 4 **Distribution of Income Classes**

Find the distribution of income classes after two generations.

SOLUTION To find the distribution of income classes after two generations, multiply the initial probability vector and the square of P, the matrix P^2. Using P^2 from Example 2,

$$X_0 \cdot P^2 = [0.21 \quad 0.68 \quad 0.11] \begin{bmatrix} 0.4729 & 0.3948 & 0.1323 \\ 0.2196 & 0.5557 & 0.2247 \\ 0.1944 & 0.4620 & 0.3436 \end{bmatrix}$$

YOUR TURN 4 In Example 4, find the distribution of income classes after three generations.

$$\approx [0.2700 \quad 0.5116 \quad 0.2184].$$

We could have also calculated this result by taking $X_0 \cdot P$ and multiplying it on the right by P, since $(X_0 \cdot P) \cdot P = X_0 \cdot P^2$. (See Exercise 27.) **TRY YOUR TURN 4**

In the next section we will develop a long-range prediction for the proportion of the population in each income class. The work in this section is summarized on the next page.

Suppose a Markov chain has initial probability vector

$$X_0 = [i_1 \ i_2 \ i_3 \cdots i_n]$$

and transition matrix P. The probability vector after n repetitions of the experiment is

$$X_0 \cdot P^n.$$

10.1 EXERCISES

Decide whether each matrix could be a probability vector.

1. $\left[\frac{2}{3} \ \ \frac{1}{2}\right]$

2. $\left[\frac{1}{2} \ \ 1\right]$

3. $[0 \ \ 1]$

4. $[0.1 \ \ 0.1]$

5. $[0.4 \ \ 0.2 \ \ 0]$

6. $\left[\frac{1}{4} \ \ \frac{1}{8} \ \ \frac{5}{8}\right]$

7. $[0.07 \ \ 0.04 \ \ 0.37 \ \ 0.52]$

8. $[0 \ \ -0.2 \ \ 0.6 \ \ 0.6]$

Decide whether each matrix could be a transition matrix, by definition. Sketch a transition diagram for any transition matrices.

9. $\begin{bmatrix} 0.6 & 0 \\ 0 & 0.6 \end{bmatrix}$

10. $\begin{bmatrix} \frac{3}{4} & \frac{1}{4} \\ 1 & 0 \end{bmatrix}$

11. $\begin{bmatrix} \frac{1}{3} & \frac{2}{3} \\ \frac{1}{2} & \frac{1}{2} \end{bmatrix}$

12. $\begin{bmatrix} \frac{1}{4} & \frac{3}{4} \\ 2 & 0 \end{bmatrix}$

13. $\begin{bmatrix} \frac{1}{3} & \frac{1}{3} & \frac{1}{3} \\ 0 & 1 & 0 \\ \frac{1}{2} & 0 & \frac{1}{2} \end{bmatrix}$

14. $\begin{bmatrix} \frac{1}{3} & \frac{1}{2} & 1 \\ 0 & 1 & 0 \\ \frac{1}{2} & \frac{1}{2} & 1 \end{bmatrix}$

In Exercises 15–18, decide whether each diagram is a transition diagram. Write any transition diagrams as transition matrices.

15.
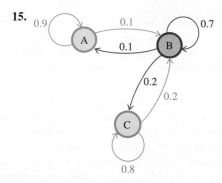

16.
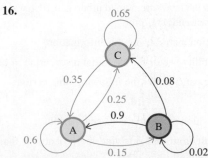

17.
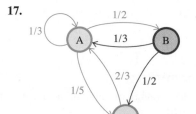

18.
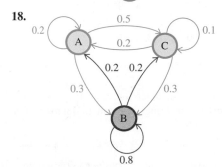

Find the first three powers of each transition matrix in Exercises 19–24 (for example, A, A^2, and A^3 in Exercise 19). For each transition matrix, find the probability that state 1 changes to state 2 after three repetitions of the experiment.

19. $A = \begin{bmatrix} 1 & 0 \\ 0.7 & 0.3 \end{bmatrix}$

20. $B = \begin{bmatrix} 0.8 & 0.2 \\ 0 & 1 \end{bmatrix}$

21. $C = \begin{bmatrix} 0 & 0 & 1 \\ 0.2 & 0.6 & 0.2 \\ 0.1 & 0.7 & 0.2 \end{bmatrix}$

22. $D = \begin{bmatrix} 0.3 & 0.2 & 0.5 \\ 0 & 0 & 1 \\ 0.6 & 0.1 & 0.3 \end{bmatrix}$

23. $E = \begin{bmatrix} 0.8 & 0.1 & 0.1 \\ 0.3 & 0.6 & 0.1 \\ 0 & 1 & 0 \end{bmatrix}$

24. $F = \begin{bmatrix} 0.01 & 0.9 & 0.09 \\ 0.72 & 0.1 & 0.18 \\ 0.34 & 0 & 0.66 \end{bmatrix}$

For each transition matrix, find the first five powers of the matrix. Then find the probability that state 2 changes to state 4 after 5 repetitions of the experiment.

25. $\begin{bmatrix} 0.1 & 0.2 & 0.2 & 0.3 & 0.2 \\ 0.2 & 0.1 & 0.1 & 0.2 & 0.4 \\ 0.2 & 0.1 & 0.4 & 0.2 & 0.1 \\ 0.3 & 0.1 & 0.1 & 0.2 & 0.3 \\ 0.1 & 0.3 & 0.1 & 0.1 & 0.4 \end{bmatrix}$

26. $\begin{bmatrix} 0.3 & 0.2 & 0.3 & 0.1 & 0.1 \\ 0.4 & 0.2 & 0.1 & 0.2 & 0.1 \\ 0.1 & 0.3 & 0.2 & 0.2 & 0.2 \\ 0.2 & 0.1 & 0.3 & 0.2 & 0.2 \\ 0.1 & 0.1 & 0.4 & 0.2 & 0.2 \end{bmatrix}$

27. a. Verify that $X_0 \cdot P^n$ can be computed in two ways: (1) by first multiplying P by itself n times, then multiplying X_0 times this result; and (2) by multiplying $X_0 \cdot P$, multiplying this result by P, and continuing to multiply by P a total of n times. (*Hint:* Use the fact that matrix multiplication is associative.)

b. Which of the two methods in part a is simpler? Explain your answer.

APPLICATIONS

Business and Economics

28. Dry Cleaning The dry cleaning example in the text used the following transition matrix:

$$\begin{array}{c} \\ \text{Johnson} \\ \text{NorthClean} \end{array} \begin{array}{cc} \text{Johnson} & \text{NorthClean} \\ \begin{bmatrix} 0.8 & 0.2 \\ 0.35 & 0.65 \end{bmatrix} \end{array}.$$

Suppose now that each customer brings in one batch of clothes per week. Use various powers of the transition matrix to find the probability that a customer initially bringing a batch of clothes to Johnson also brings a batch to Johnson after the following time periods.

a. 1 week   **b.** 2 weeks   **c.** 3 weeks   **d.** 4 weeks

e. What is the probability that a customer initially bringing a batch of clothes to NorthClean brings a batch to Johnson after 2 weeks?

29. Dry Cleaning Suppose Johnson has a 40% market share initially, with NorthClean having a 60% share. Use this information to write a probability vector; use this vector, along with the transition matrix from Exercise 28, to find the share of the market for each firm after each of the following time periods. (As in Exercise 28, assume that customers bring in one batch of dry cleaning per week.)

a. 1 week   **b.** 2 weeks   **c.** 3 weeks   **d.** 4 weeks

30. Insurance An insurance company classifies its drivers into three groups: G_0 (no accidents), G_1 (one accident), and G_2 (more than one accident). The probability that a G_0 driver will remain a G_0 after 1 year is 0.75, that the driver will become a G_1 is 0.20, and that the driver will become a G_2 is 0.05. A G_1 driver cannot become a G_0 (this company has a long memory). There is a 0.70 probability that a G_1 driver will remain a G_1. A G_2 driver must remain a G_2. Write a transition matrix using this information.

31. Insurance Suppose that the company in Exercise 30 accepts 50,000 new policyholders, all of whom are G_0 drivers. Find the number in each group after the following time periods.

a. 1 year   **b.** 2 years   **c.** 3 years   **d.** 4 years

32. Insurance The difficulty with the mathematical model in Exercises 30 and 31 is that no "grace period" is provided; there should be a certain positive probability of moving from G_1 or G_2 back to G_0. A new system with this feature might produce the following transition matrix.

$$\begin{bmatrix} 0.75 & 0.20 & 0.05 \\ 0.10 & 0.70 & 0.20 \\ 0.10 & 0.30 & 0.60 \end{bmatrix}$$

Suppose that when this new policy is adopted, the company has 50,000 policyholders, all in G_0. Find the number in each group after the following time periods.

a. 1 year   **b.** 2 years   **c.** 3 years

d. Write the transition matrix for a 2-year period.

e. Use your result from part d to find the probability that a driver in G_0 is still in G_0 two years later.

33. Solar Energy A community has set a goal of reducing their carbon footprint in 10 years. As part of their "Green in Ten" campaign, they have offered homeowners incentives to convert to solar energy for home heating. Community leaders can convince 5% of those who use electric heat and 10% of those who use fossil fuels to convert to solar heat annually. Leaders also realize that some homeowners will not cooperate and that 5% of those who use electric heat will instead convert to fossil fuels while 15% of those who use fossil fuels will convert to electric heat. Of course, those who use solar energy will continue using solar energy.

a. Write a transition matrix using this information, labeling the rows and columns e, f, and s for electric, fossil fuels, and solar, respectively.

At the beginning of the campaign, 35% of homeowners use electric heat, 60% use fossil fuels, and only 5% use solar heat. Find the number of homeowners using each type of heat after the following time periods.

b. 1 year   **c.** 2 years   **d.** 3 years

e. The goal of the community leaders is to have at least 50% of homeowners using solar energy for home heating in 10 years. If this trend continues, will they meet their goal? How many years will it take until they achieve this goal?

34. Land Use In one state, a Board of Realtors land use survey showed that 35% of all land was used for agricultural purposes, while 10% was urban. Ten years later, of the agricultural land, 15% had become urbanized and 80% had remained agricultural. (The remainder lay idle.) Of the idle land, 20% had become urbanized and 10% had been converted for agricultural use. Of the urban land, 90% remained urban and 10% was idle. Assume that these trends continue.

a. Write a transition matrix using this information.

b. Write a probability vector for the initial distribution of land.

Find the land use pattern after the following time periods.

c. 10 years   **d.** 20 years

e. Write the transition matrix for a 20-year period.

f. Use your result from part e to find the probability that an idle plot of land is still idle 20 years later.

35. Business The change in the size of businesses in a certain Canadian city from one year to the next can be described by a Markov chain. The businesses are classified into three categories based on size: small (2–10 employees), medium (11–43 employees), and large (44 or more employees). The transition matrix for a 1-year period is given below. *Source: Applied Statistics.*

	Small	Medium	Large
Small	0.9216	0.0780	0.0004
Medium	0.0460	0.8959	0.0581
Large	0.0003	0.0301	0.9696

Suppose that in 2011, there were 2094 small, 2363 medium, and 2378 large businesses. Based on this model, find the number of businesses of each type that would be expected in each of the following years.

a. 2012   **b.** 2013

c. Write the transition matrix for a 2-year period.

Based on your answer to part c, what percent of medium businesses are in each of the following categories after 2 years?

d. Small businesses   **e.** Large businesses

36. Queuing Chain In the queuing chain, we assume that people are queuing up to be served by, say, a bank teller. For simplicity, let us assume that once two people are in line, no one else can enter the line. Let us further assume that one person is served every minute, as long as someone is in line. Assume further that in any minute, there is a probability of $1/2$ that no one enters the line, a probability of $1/3$ that exactly one person enters the line, and a probability of $1/6$ that exactly two people enter the line, assuming there is room. If there is not enough room for two people, then the probability that one person enters the line is $1/2$. Let the state be given by the number of people in line.

a. Verify that the transition matrix is

$$
\begin{array}{c c}
 & \begin{array}{c c c} 0 & 1 & 2 \end{array} \\
\begin{array}{c} 0 \\ 1 \\ 2 \end{array} &
\left[\begin{array}{c c c}
\frac{1}{2} & \frac{1}{3} & \frac{1}{6} \\
\frac{1}{2} & \frac{1}{3} & \frac{1}{6} \\
0 & \frac{1}{2} & \frac{1}{2}
\end{array} \right].
\end{array}
$$

b. Find the transition matrix for a 2-minute period.

c. Use your result from part b to find the probability that a queue with no one in line has two people in line 2 minutes later.

Life Sciences

37. Immune Response A study of immune response in rabbits classified the rabbits into four groups, according to the strength of the response. From one week to the next, the rabbits changed classification from one group to another, according to the following transition matrix. *Source: Journal of Theoretical Biology.*

$$
\begin{array}{c c}
 & \begin{array}{c c c c} 1 & 2 & 3 & 4 \end{array} \\
\begin{array}{c} 1 \\ 2 \\ 3 \\ 4 \end{array} &
\left[\begin{array}{c c c c}
\frac{5}{7} & \frac{2}{7} & 0 & 0 \\
0 & \frac{1}{2} & \frac{1}{3} & \frac{1}{6} \\
0 & 0 & \frac{1}{2} & \frac{1}{2} \\
0 & 0 & \frac{1}{4} & \frac{3}{4}
\end{array} \right]
\end{array}
$$

a. Five weeks later, what proportion of the rabbits in group 1 were still in group 1?

b. In the first week, there were 9 rabbits in the first group, 4 in the second, and none in the third or fourth groups. How many rabbits would you expect in each group after 4 weeks?

c. By investigating the transition matrix raised to larger and larger powers, make a reasonable guess for the long-range probability that a rabbit in group 1 or 2 will still be in group 1 or 2 after an arbitrarily long time. Explain why this answer is reasonable.

Social Sciences

38. Housing Patterns In a survey investigating changes in housing patterns in one urban area, it was found that 75% of the population lived in single-family dwellings and 25% in multiple housing of some kind. Five years later, in a follow-up survey, of those who had been living in single-family dwellings, 90% still did so, but 10% had moved to multiple-family dwellings. Of those in multiple-family housing, 95% were still living in that type of housing, while 5% had moved to single-family dwellings. Assume that these trends continue.

a. Write a transition matrix for this information.

b. Write a probability vector for the initial distribution of housing.

What percent of the population can be expected in each category after the following time periods?

c. 5 years   **d.** 10 years

e. Write the transition matrix for a 10-year period.

f. Use your result from part e to find the probability that someone living in a single-family dwelling is still doing so 10 years later.

39. Migration A study found that the way people living in one type of neighborhood migrate to another could be described by a Markov chain. The study classified housing into five types:

I: Middle-class family households;

II: Upper-middle-class households;

III: Sound, rented, two-family dwelling units;

IV: Unsound, rented, multi-family dwelling units; and

V: Commercial.

The transition matrix for a 1-year period, based on data from Cedar Rapids, Iowa, is given below. *Source: Annals of the Association of American Geographers.*

	I	II	III	IV	V
I	0.71	0.19	0.03	0.07	0.00
II	0.52	0.31	0.04	0.13	0.00
III	0.40	0.16	0.22	0.22	0.00
IV	0.34	0.19	0.09	0.37	0.01
V	0.21	0.25	0.13	0.37	0.04

Suppose that in some year, 20% of the population lives in each type of housing. Determine the percent of the population that can be expected in each type of housing after the following times.

a. 1 year   **b.** 2 years

c. Write the transition matrix for a 2-year period.

Based on your answer to part c, what percent of those living in a commercial area can be expected to be living in each of the following types of housing after 2 years?

d. Middle-class family households

e. Unsound, rented, multi-family dwelling units.

40. Voting Trends At the end of June in a presidential election year, 40% of the voters were registered as liberal, 45% as conservative, and 15% as independent. Over a 1-month period, the liberals retained 80% of their constituency, while 15% switched to conservative and 5% to independent. The conservatives retained 70% and lost 20% to the liberals. The independents retained 60% and lost 20% each to the conservatives and liberals. Assume that these trends continue.

a. Write a transition matrix using this information.

b. Write a probability vector for the initial distribution.

Find the percent of each type of voter at the end of each month.

c. July **d.** August

e. September **f.** October

g. Write the transition matrix for a 2-month period.

General Interest

41. Cricket The results of cricket matches between England and Australia have been found to be modeled by a Markov chain. The probability that England wins, loses, or draws is based on the result of the previous game, with the following transition matrix: *Source: The Mathematical Gazette.*

	Wins	Loses	Draws
Wins	0.443	0.364	0.193
Loses	0.277	0.436	0.287
Draws	0.266	0.304	0.430

a. Compute the transition matrix for the game after the next one, based on the result of the last game.

b. Use your answer from part a to find the probability that, if England won the last game, England will win the game after the next one.

c. Use your answer from part a to find the probability that, if Australia won the last game, England will win the game after the next one.

42. Professional Football There has been an ongoing debate as to how NFL overtime games should be carried out. In the current system, a coin is tossed and the winner of the toss gets to decide whether to kickoff or to receive the ball. The first team to score wins. Based on data from the 2008 season, 36% of any given possession will result in the offense scoring, while 62% of the possessions result in a change of possession. The remaining 2% of the possessions result in the defense scoring (by way of a turnover returned for a touchdown or a safety). The following transition matrix gives the probability of going from each row state to each column state. The first row tells the probability of being in any given situation after one possession, assuming that team A starts with the ball. *Source: Mathematics in Sports.*

	Possession for A	Possession for B	A wins	B wins
Possession for A	0	0.62	0.36	0.02
Possession for B	0.62	0	0.02	0.36
A wins	0	0	1	0
B wins	0	0	0	1

a. Suppose that team A starts with the ball. Determine the probabilities that after one possession (i) team A wins, (ii) team B wins, and (iii) the game continues.

b. Determine the same probabilities after two possessions.

c. Determine the same probabilities after three possessions.

d. Which team has an advantage in this method of overtime?

YOUR TURN ANSWERS

1. $\begin{matrix} & J & N \\ J & 0.68 & 0.32 \\ N & 0.21 & 0.79 \end{matrix}$

2. 0.1730 3. 0.3487

4. [0.2785 0.4970 0.2245]

10.2 Regular Markov Chains

APPLY IT Given the transition probabilities for two dry cleaners, how can we predict the market share of each cleaner far into the future if we do not know the current proportions?
In Example 2 of this section, we will see how to answer this question.

If we start with a transition matrix P and an initial probability vector, we can use the nth power of P to find the probability vector for n repetitions of an experiment. In this section, we try to decide what happens to an initial probability vector in the long run—that is, as n gets larger and larger. Again, we assume that the transition probabilities remain constant from repetition to repetition.

For example, let us use the transition matrix associated with the dry cleaning example in the previous section.

$$P = \begin{bmatrix} 0.8 & 0.2 \\ 0.35 & 0.65 \end{bmatrix}$$

The initial probability vector, which gives the market share for each firm at the beginning of the experiment, is $v = \begin{bmatrix} 0.4 & 0.6 \end{bmatrix}$. The market shares shown in the following table were found by using powers of the transition matrix. (See Exercise 29 in the previous section.)

Weeks After Start	Market Share		
	Johnson	NorthClean	
0	0.4	0.6	v
1	0.53	0.47	vP^1
2	0.59	0.41	vP^2
3	0.61	0.39	vP^3
4	0.63	0.37	vP^4
5	0.63	0.37	vP^5
12	0.64	0.36	vP^{12}
13	0.64	0.36	vP^{13}

The results seem to approach the numbers in the probability vector $\begin{bmatrix} 0.64 & 0.36 \end{bmatrix}$.

What happens if the initial probability vector is different from $\begin{bmatrix} 0.4 & 0.6 \end{bmatrix}$? Suppose $v = \begin{bmatrix} 0.75 & 0.25 \end{bmatrix}$ is used; the same powers of the transition matrix as above give the following results.

Weeks After Start	Market Share		
	Johnson	NorthClean	
0	0.75	0.25	v
1	0.69	0.31	vP^1
2	0.66	0.34	vP^2
3	0.65	0.35	vP^3
4	0.64	0.36	vP^4
5	0.64	0.36	vP^5
6	0.64	0.36	vP^6

The results again seem to be approaching the numbers in the probability vector $\begin{bmatrix} 0.64 & 0.36 \end{bmatrix}$, the same numbers approached with the initial probability vector $\begin{bmatrix} 0.4 & 0.6 \end{bmatrix}$. In either case, the long-range trend is for a market share of about 64% for Johnson and 36% for NorthClean. The example above suggests that this long-range trend does not depend on the initial distribution of market shares. This means that if the initial market share for Johnson was less than 64%, the advertising campaign has paid off in terms of a greater long-range market share. If the initial share was more than 64%, the campaign did not pay off.

Regular Transition Matrices

One of the many applications of Markov chains is in finding long-range predictions. It is not possible to make long-range predictions with all transition matrices, but for a large set of transition matrices, long-range predictions *are* possible. Such predictions are always possible with **regular transition matrices**. A transition matrix is *regular* if some power of the matrix contains all positive entries. A Markov chain is a **regular Markov chain** if its transition matrix is regular.

EXAMPLE 1 Regular Transition Matrices

Decide whether the following transition matrices are regular.

(a) $A = \begin{bmatrix} 0.75 & 0.25 & 0 \\ 0 & 0.5 & 0.5 \\ 0.6 & 0.4 & 0 \end{bmatrix}$

SOLUTION Square A.

$$A^2 = \begin{bmatrix} 0.5625 & 0.3125 & 0.125 \\ 0.3 & 0.45 & 0.25 \\ 0.45 & 0.35 & 0.2 \end{bmatrix}$$

Since all entries in A^2 are positive, matrix A is regular.

(b) $B = \begin{bmatrix} 0.5 & 0 & 0.5 \\ 0 & 1 & 0 \\ 0 & 0 & 1 \end{bmatrix}$

YOUR TURN 1 Decide whether the following transition matrix is regular.

$$C = \begin{bmatrix} 0.2 & 0.1 & 0.7 \\ 0.3 & 0.7 & 0 \\ 0.2 & 0 & 0.8 \end{bmatrix}$$

SOLUTION Find various powers of B.

$$B^2 = \begin{bmatrix} 0.25 & 0 & 0.75 \\ 0 & 1 & 0 \\ 0 & 0 & 1 \end{bmatrix}; B^3 = \begin{bmatrix} 0.125 & 0 & 0.875 \\ 0 & 1 & 0 \\ 0 & 0 & 1 \end{bmatrix}; B^4 = \begin{bmatrix} 0.0625 & 0 & 0.9375 \\ 0 & 1 & 0 \\ 0 & 0 & 1 \end{bmatrix}$$

Further powers of B will still give the same zero entries, so no power of matrix B contains all positive entries. For this reason, B is not regular. **TRY YOUR TURN 1**

NOTE If a transition matrix P has some zero entries, and P^2 does as well, you may wonder how far you must compute P^n to be certain that the matrix is not regular. The answer is that if all zeros occur in the identical places in both P^n and P^{n+1} for any n, they will appear in those places for all higher powers of P, so P is not regular.

Suppose that v is any probability vector. It can be shown that for a regular Markov chain with a transition matrix P, there exists a single vector V that does not depend on v, such that $v \cdot P^n$ gets closer and closer to V as n gets larger and larger.

Equilibrium Vector of a Markov Chain

If a Markov chain with transition matrix P is regular, then there is a unique vector V such that, for any probability vector v and for large values of n,

$$v \cdot P^n \approx V.$$

Vector V is called the **equilibrium vector** or the **fixed vector** of the Markov chain.

In the example with Johnson Cleaners, the equilibrium vector V is approximately $\begin{bmatrix} 0.64 & 0.36 \end{bmatrix}$. Vector V can be determined by finding P^n for larger and larger values of n, and then looking for a vector that the product $v \cdot P^n$ approaches. Such an approach can be very tedious, however, and is prone to error. To find a better way, start with the fact that for a large value of n,

$$v \cdot P^n \approx V,$$

as mentioned above. From this result, $v \cdot P^n \cdot P \approx V \cdot P$, so that

$$v \cdot P^n \cdot P = v \cdot P^{n+1} \approx VP.$$

Since $v \cdot P^n \approx V$ for large values of n, it is also true that $v \cdot P^{n+1} \approx V$ for large values of n (the product $v \cdot P^n$ approaches V, so that $v \cdot P^{n+1}$ must also approach V). Thus, $v \cdot P^{n+1} \approx V$ and $v \cdot P^{n+1} \approx VP$, which suggests that

$$VP = V.$$

If a Markov chain with transition matrix P is regular, then there exists a probability vector V such that

$$VP = V.$$

This vector V gives the long-range trend of the Markov chain. Vector V is found by solving a system of linear equations, as shown in the next examples.

EXAMPLE 2 Dry Cleaners

Find the long-range trend for the Markov chain in the dry cleaning example with transition matrix

$$P = \begin{bmatrix} 0.8 & 0.2 \\ 0.35 & 0.65 \end{bmatrix}.$$

APPLY IT

SOLUTION This matrix is regular since all entries are positive. Let P represent this transition matrix, and let V be the probability vector $\begin{bmatrix} v_1 & v_2 \end{bmatrix}$. We want to find V such that

$$VP = V,$$

or

$$\begin{bmatrix} v_1 & v_2 \end{bmatrix} \begin{bmatrix} 0.8 & 0.2 \\ 0.35 & 0.65 \end{bmatrix} = \begin{bmatrix} v_1 & v_2 \end{bmatrix}.$$

Use matrix multiplication on the left.

$$\begin{bmatrix} 0.8v_1 + 0.35v_2 & 0.2v_1 + 0.65v_2 \end{bmatrix} = \begin{bmatrix} v_1 & v_2 \end{bmatrix}$$

Set corresponding entries from the two matrices equal to get

$$0.8v_1 + 0.35v_2 = v_1 \qquad \text{and} \qquad 0.2v_1 + 0.65v_2 = v_2.$$

Simplify each of these equations.

$$-0.2v_1 + 0.35v_2 = 0 \qquad \text{and} \qquad 0.2v_1 - 0.35v_2 = 0$$

These last two equations are really the same. In fact, the equations in the system obtained from $VP = V$ are always dependent, implying that there exists an infinite number of solutions to the system of equations. For this example, the solution set contains all v_1 and v_2 that satisfy $-0.2v_1 + 0.35v_2 = 0$, that is, all points on the line $-0.2v_1 + 0.35v_2 = 0$.

There is, however, one additional piece of information. The sum of the simple probabilities must be 1. For this example, $v_1 + v_2 = 1$. Combining this equation with the equation $-0.2v_1 + 0.35v_2 = 0$ will provide the long-range trend for the dry cleaners. See Figure 6.

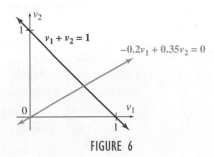

FIGURE 6

Notice in Figure 6 that the lines corresponding to the equations $-0.2v_1 + 0.35v_2 = 0$ and $v_1 + v_2 = 1$ cross at a point. To find the values of v_1 and v_2 where they cross, solve the system

$$v_1 + v_2 = 1$$
$$-0.2v_1 + 0.35v_2 = 0.$$

For a larger system, we would use the Gauss-Jordan method. For just two equations in two unknowns, however, it is simpler to use the echelon method. Multiply the first equation by 0.2 and add it to the second equation.

YOUR TURN 2 If the transition matrix is $\begin{bmatrix} 0.7 & 0.3 \\ 0.2 & 0.8 \end{bmatrix}$, find the long-range trend.

$$v_1 + v_2 = 1$$
$$0.55v_2 = 0.2 \qquad \textbf{0.2R}_1 + \textbf{R}_2 \rightarrow \textbf{R}_2$$
$$v_2 = \frac{0.2}{0.55} = \frac{4}{11} \approx 0.3636 \qquad \text{Solve the second equation for } v_2.$$

Since $v_1 = 1 - v_2$, we find that $v_1 = 7/11 \approx 0.6364$, and the equilibrium vector is $V = \begin{bmatrix} 7/11 & 4/11 \end{bmatrix} \approx \begin{bmatrix} 0.6364 & 0.3636 \end{bmatrix}$. **TRY YOUR TURN 2**

Some powers of the transition matrix P in Example 2 (with entries rounded to four decimal places) are shown here.

$$P^2 = \begin{bmatrix} 0.71 & 0.29 \\ 0.5075 & 0.4925 \end{bmatrix} \quad P^3 = \begin{bmatrix} 0.6695 & 0.3305 \\ 0.5784 & 0.4216 \end{bmatrix} \quad P^4 = \begin{bmatrix} 0.6513 & 0.3487 \\ 0.6103 & 0.3897 \end{bmatrix}$$

$$P^5 = \begin{bmatrix} 0.6431 & 0.3569 \\ 0.6246 & 0.3754 \end{bmatrix} \quad P^{10} = \begin{bmatrix} 0.6365 & 0.3635 \\ 0.6361 & 0.3639 \end{bmatrix} \quad P^{20} = \begin{bmatrix} 0.6364 & 0.3636 \\ 0.6364 & 0.3636 \end{bmatrix}$$

As these results suggest, higher and higher powers of the transition matrix P approach a matrix having all rows identical; these identical rows have as entries the entries of the equilibrium vector V. This agrees with the statement above: The initial state does not matter. Regardless of the initial probability vector, the system will approach a fixed vector V. This unexpected and remarkable fact is the basic property of regular Markov chains—*the limiting distribution is independent of the initial distribution*. This happens because some power of the transition matrix has all positive entries, so that all the initial probabilities are thoroughly mixed.

TECHNOLOGY NOTE

Figure 7 shows the result of taking P^{10} and P^{20} to four decimal places on a TI-84 Plus, where the transition matrix P from the dry cleaning example was entered into the matrix [A] on the calculator.

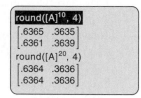

FIGURE 7

Let us summarize the results of this section.

Properties of a Regular Markov Chain

Suppose a regular Markov chain has a transition matrix P.

1. As n gets larger and larger, the product $v \cdot P^n$ approaches a unique vector V for any initial probability vector v. Vector V is called the *equilibrium vector* or *fixed vector*.

2. Vector V has the property that $VP = V$.

3. To find V, solve a system of equations obtained from the matrix equation $VP = V$ and from the fact that the sum of the entries of V is 1.

4. The powers P^n come closer and closer to a matrix whose rows are made up of the entries of the equilibrium vector V.

EXAMPLE 3 **Equilibrium Vector**

Find the equilibrium vector for the transition matrix

$$K = \begin{bmatrix} 0.2 & 0.6 & 0.2 \\ 0.1 & 0.1 & 0.8 \\ 0.3 & 0.3 & 0.4 \end{bmatrix}.$$

SOLUTION Matrix K has all positive entries and, thus, is regular. For this reason, an equilibrium vector V must exist such that $VK = V$. Let $V = \begin{bmatrix} v_1 & v_2 & v_3 \end{bmatrix}$.

Then

$$\begin{bmatrix} v_1 & v_2 & v_3 \end{bmatrix} \begin{bmatrix} 0.2 & 0.6 & 0.2 \\ 0.1 & 0.1 & 0.8 \\ 0.3 & 0.3 & 0.4 \end{bmatrix} = \begin{bmatrix} v_1 & v_2 & v_3 \end{bmatrix}.$$

Use matrix multiplication on the left.

$$\begin{bmatrix} 0.2v_1 + 0.1v_2 + 0.3v_3 & 0.6v_1 + 0.1v_2 + 0.3v_3 & 0.2v_1 + 0.8v_2 + 0.4v_3 \end{bmatrix}$$
$$= \begin{bmatrix} v_1 & v_2 & v_3 \end{bmatrix}$$

Setting corresponding entries equal gives the following equations:

$$0.2v_1 + 0.1v_2 + 0.3v_3 = v_1$$
$$0.6v_1 + 0.1v_2 + 0.3v_3 = v_2$$
$$0.2v_1 + 0.8v_2 + 0.4v_3 = v_3.$$

Simplifying these equations gives

$$-0.8v_1 + 0.1v_2 + 0.3v_3 = 0$$
$$0.6v_1 - 0.9v_2 + 0.3v_3 = 0$$
$$0.2v_1 + 0.8v_2 - 0.6v_3 = 0.*$$

Since V is a probability vector,

$$v_1 + v_2 + v_3 = 1.$$

This gives a system of four equations in three unknowns:

$$v_1 + v_2 + v_3 = 1$$
$$-0.8v_1 + 0.1v_2 + 0.3v_3 = 0$$
$$0.6v_1 - 0.9v_2 + 0.3v_3 = 0$$
$$0.2v_1 + 0.8v_2 - 0.6v_3 = 0.$$

This system can be solved with the Gauss-Jordan method presented earlier. Start with the augmented matrix

$$\begin{bmatrix} 1 & 1 & 1 & | & 1 \\ -0.8 & 0.1 & 0.3 & | & 0 \\ 0.6 & -0.9 & 0.3 & | & 0 \\ 0.2 & 0.8 & -0.6 & | & 0 \end{bmatrix}.$$

The Gauss-Jordan method will always yield a row of zeros at the bottom, since the system is dependent. (See the "For Review" in the margin; see also the footnote at the bottom of this page.) Nevertheless, there is one more equation than unknown, and the remaining

---

FOR REVIEW

The Gauss-Jordan method was discussed in Chapter 2. We will review it here by solving the system given in the text. Begin by ridding the matrix of decimals by multiplying by 10, and then dividing out any common factors.

$$\begin{matrix} \\ 10R_2 \to R_2 \\ 10R_3/3 \to R_3 \\ 10R_4/2 \to R_4 \end{matrix} \begin{bmatrix} 1 & 1 & 1 & | & 1 \\ -8 & 1 & 3 & | & 0 \\ 2 & -3 & 1 & | & 0 \\ 1 & 4 & -3 & | & 0 \end{bmatrix}$$

Next clear column 1 by combining multiples of row 1 with multiples of the other rows.

$$\begin{matrix} \\ 8R_1 + R_2 \to R_2 \\ -2R_1 + R_3 \to R_3 \\ -R_1 + R_4 \to R_4 \end{matrix} \begin{bmatrix} 1 & 1 & 1 & | & 1 \\ 0 & 9 & 11 & | & 8 \\ 0 & -5 & -1 & | & -2 \\ 0 & 3 & -4 & | & -1 \end{bmatrix}$$

Similarly, clear column 2 using multiples of row 2 as shown below.

$$\begin{matrix} -R_2 + 9R_1 \to R_1 \\ \\ 5R_2 + 9R_3 \to R_3 \\ -R_2 + 3R_4 \to R_4 \end{matrix} \begin{bmatrix} 9 & 0 & -2 & | & 1 \\ 0 & 9 & 11 & | & 8 \\ 0 & 0 & 46 & | & 22 \\ 0 & 0 & -23 & | & -11 \end{bmatrix}$$

Finally, clear column 3 and simplify with the row operations $R_3 + 23R_1 \to R_1$, $-11R_3 + 46R_2 \to R_2$, $R_3/2 \to R_3$, and $R_3 + 2R_4 \to R_4$. The result is

$$\begin{bmatrix} 207 & 0 & 0 & | & 45 \\ 0 & 414 & 0 & | & 126 \\ 0 & 0 & 23 & | & 11 \\ 0 & 0 & 0 & | & 0 \end{bmatrix}.$$

Thus, $v_1 = 45/207 = 5/23$; $v_2 = 126/414 = 7/23$; and $v_3 = 11/23$.

---

*As mentioned earlier, any system of equations obtained in this manner is dependent and will have an infinite number of solutions. To see the dependency, notice that $-R_1 = R_2 + R_3$. Since one row can be written as a combination of the other two, any one of the three equations can be dropped prior to moving on to the next step. Of course, the equation causes no harm and one may simply leave it in and proceed, as we shall do.

YOUR TURN 3 Find the equilibrium vector for the transition matrix

$$K = \begin{bmatrix} 0.2 & 0.1 & 0.7 \\ 0.3 & 0.7 & 0 \\ 0.2 & 0 & 0.8 \end{bmatrix}.$$

system, minus the row of zeros, will always have exactly one solution. The solution of this system is $v_1 = 5/23$, $v_2 = 7/23$, $v_3 = 11/23$, and

$$V = \begin{bmatrix} \dfrac{5}{23} & \dfrac{7}{23} & \dfrac{11}{23} \end{bmatrix} \approx \begin{bmatrix} 0.2174 & 0.3043 & 0.4783 \end{bmatrix}.$$

TRY YOUR TURN 3

This is a good place to use a graphing calculator that performs row operations. Refer to Section 2.2.

10.2 EXERCISES

Which of the following transition matrices are regular?

1. $\begin{bmatrix} 0.2 & 0.8 \\ 0.9 & 0.1 \end{bmatrix}$

2. $\begin{bmatrix} 0.28 & 0.72 \\ 0.47 & 0.53 \end{bmatrix}$

3. $\begin{bmatrix} 1 & 0 \\ 0.65 & 0.35 \end{bmatrix}$

4. $\begin{bmatrix} 0.55 & 0.45 \\ 0 & 1 \end{bmatrix}$

5. $\begin{bmatrix} 0 & 1 & 0 \\ 0.4 & 0.2 & 0.4 \\ 1 & 0 & 0 \end{bmatrix}$

6. $\begin{bmatrix} 0.3 & 0.5 & 0.2 \\ 1 & 0 & 0 \\ 0.5 & 0.1 & 0.4 \end{bmatrix}$

Find the equilibrium vector for each transition matrix.

7. $\begin{bmatrix} \frac{1}{4} & \frac{3}{4} \\ \frac{1}{2} & \frac{1}{2} \end{bmatrix}$

8. $\begin{bmatrix} \frac{2}{3} & \frac{1}{3} \\ \frac{1}{8} & \frac{7}{8} \end{bmatrix}$

9. $\begin{bmatrix} 0.4 & 0.6 \\ 0.3 & 0.7 \end{bmatrix}$

10. $\begin{bmatrix} 0.1 & 0.9 \\ 0.8 & 0.2 \end{bmatrix}$

11. $\begin{bmatrix} 0.1 & 0.1 & 0.8 \\ 0.4 & 0.3 & 0.3 \\ 0.1 & 0.2 & 0.7 \end{bmatrix}$

12. $\begin{bmatrix} 0.5 & 0.2 & 0.3 \\ 0.1 & 0.4 & 0.5 \\ 0.3 & 0.1 & 0.6 \end{bmatrix}$

13. $\begin{bmatrix} 0.25 & 0.35 & 0.4 \\ 0.1 & 0.3 & 0.6 \\ 0.55 & 0.4 & 0.05 \end{bmatrix}$

14. $\begin{bmatrix} 0.16 & 0.28 & 0.56 \\ 0.43 & 0.12 & 0.45 \\ 0.86 & 0.05 & 0.09 \end{bmatrix}$

Find the equilibrium vector for each transition matrix in Exercises 15–20. These matrices were first used in the exercises for Section 10.1. (Note: Not all of these transition matrices are regular, but equilibrium vectors still exist. Why doesn't this contradict the work of this section?)

15. Modified insurance categories (10.1 Exercise 32)

$$\begin{bmatrix} 0.75 & 0.20 & 0.05 \\ 0.10 & 0.70 & 0.20 \\ 0.10 & 0.30 & 0.60 \end{bmatrix}$$

16. Insurance categories (10.1 Exercise 30)

$$\begin{bmatrix} 0.75 & 0.20 & 0.05 \\ 0 & 0.70 & 0.30 \\ 0 & 0 & 1 \end{bmatrix}$$

17. Home heating systems (10.1 Exercise 33)

$$\begin{bmatrix} 0.9 & 0.05 & 0.05 \\ 0.15 & 0.75 & 0.1 \\ 0 & 0 & 1 \end{bmatrix}$$

18. Land use (10.1 Exercise 34)

$$\begin{bmatrix} 0.80 & 0.15 & 0.05 \\ 0 & 0.90 & 0.10 \\ 0.10 & 0.20 & 0.70 \end{bmatrix}$$

19. Voter registration (10.1 Exercise 40)

$$\begin{bmatrix} 0.80 & 0.15 & 0.05 \\ 0.20 & 0.70 & 0.10 \\ 0.20 & 0.20 & 0.60 \end{bmatrix}$$

20. Housing patterns (10.1 Exercise 38)

$$\begin{bmatrix} 0.90 & 0.10 \\ 0.05 & 0.95 \end{bmatrix}$$

21. Find the equilibrium vector for the transition matrix

$$\begin{bmatrix} p & 1 - p \\ 1 - q & q \end{bmatrix},$$

where $0 < p < 1$ and $0 < q < 1$. Under what conditions is this matrix regular?

22. Show that the transition matrix

$$K = \begin{bmatrix} \frac{1}{4} & 0 & \frac{3}{4} \\ 0 & 1 & 0 \\ 0 & 0 & 1 \end{bmatrix}$$

has more than one vector V such that $VK = V$. Why does this not violate the statements of this section?

23. Let

$$P = \begin{bmatrix} a_{11} & a_{12} \\ a_{21} & a_{22} \end{bmatrix}$$

be a regular matrix having *column* sums of 1. Show that the equilibrium vector for P is $\begin{bmatrix} 1/2 & 1/2 \end{bmatrix}$.

24. Notice in Example 3 that the system of equations $VK = V$, with the extra equation that the sum of the elements of V must equal 1, had exactly one solution. What can you say about the number of solutions to the system $VK = V$?

APPLICATIONS

Business and Economics

25. Quality Control The probability that a complex assembly line works correctly depends on whether the line worked correctly the last time it was used. There is a 0.9 chance that the line will work correctly if it worked correctly the time before, and a 0.8 chance that it will work correctly if it did *not* work correctly the time before. Set up a transition matrix with this information and find the long-range probability that the line will work correctly.

26. **Quality Control**   Suppose improvements are made in the assembly line of Exercise 25, so that the transition matrix becomes

$$\begin{array}{c} \\ \text{Works} \\ \text{Doesn't Work} \end{array} \begin{array}{cc} \text{Works} & \text{Doesn't Work} \\ \begin{bmatrix} 0.95 & 0.05 \\ 0.85 & 0.15 \end{bmatrix} \end{array}.$$

Find the new long-range probability that the line will work properly.

27. **a. Dry Cleaning**   Using the initial probability vector $[0.4 \quad 0.6]$, find the probability vector for the next 9 weeks in Example 2 (the dry cleaning example). Compute your answers to at least 6 decimal places.

   b. Using your results from part a, find the difference between the equilibrium proportion of customers going to Johnson and the proportion going there for each of the first 10 weeks. Be sure to compute the equilibrium proportions to at least 6 decimal places.

   c. Find the ratio between each difference calculated in part b and the difference for the previous week.

   d. Using your results from part c, explain how the probability vector approaches the equilibrium vector.

   e. Repeat parts a–d of this exercise using the initial probability vector $[0.75 \quad 0.25]$.

28. **Mortgage Refinancing**   In 2009, many homeowners refinanced their mortgages to take advantage of lower interest rates. Most of the mortgages could be classified into four groups: adjustable rate, 15-year fixed rate, 20-year fixed rate, and 30-year fixed rate. (For this exercise, the small number of loans that are not of those four types will be classified with the adjustable rate loans.) Sometimes when a homeowner refinanced, the new loan was the same type as the old loan, and sometimes it was different. The breakdown of the percent (in decimal form) in each category is shown in the following table. If these conversion rates were to persist, find the long-range trend for the percent of loans of each type. *Source: Freddie Mac.*

| | New Loan | | | |
Old Loan	Adjustable	15-Year Fixed	20-Year Fixed	30-Year Fixed
Adjustable	0.04	0.22	0.07	0.67
15-Year Fixed	0	0.70	0.02	0.28
20-Year Fixed	0.01	0.545	0.075	0.37
30-Year Fixed	0	0.145	0.055	0.80

29. **Business**   As we saw in the last section, the change in the size of businesses in a certain Canadian city from one year to the next could be described by a Markov chain. The businesses were classified into three categories based on size: small (2–10 employees), medium (11–43 employees) and large (44 or more employees). The transition period for a 1-year period is given in the next column. *Source: Applied Statistics.*

$$\begin{array}{c} \\ \text{Small} \\ \text{Medium} \\ \text{Large} \end{array} \begin{array}{ccc} \text{Small} & \text{Medium} & \text{Large} \\ \begin{bmatrix} 0.9216 & 0.0780 & 0.0004 \\ 0.0460 & 0.8959 & 0.0581 \\ 0.0003 & 0.0301 & 0.9696 \end{bmatrix} \end{array}$$

If these trends continue, what percent of the business will be small, medium, or large in the long run?

Life Sciences

30. **Research with Mice**   A large group of mice is kept in a cage having connected compartments A, B, and C. Mice in compartment A move to B with probability 0.3 and to C with probability 0.4. Mice in B move to A or C with probabilities of 0.15 and 0.55, respectively. Mice in C move to A or B with probabilities of 0.3 and 0.6, respectively. Find the long-range prediction for the fraction of mice in each of the compartments.

Social Sciences

31. **Family Structure**   Data from the National Longitudinal Survey of Youth (NLSY) was analyzed to derive the probabilities that children will repeat the child-raising choices of their parents. The living arrangements of a group of children at age 14 were recorded. The NLSY then tracked those same children and recorded the living arrangements for their children. Based on this data, the following transition matrix gives the probabilities of change in child-raising choices from one generation to the next. Find the long-term proportions of child-raising choices if this trend were to continue. *Source: Journal for Economic Educators.*

| | Second Generation Living Arrangements | | | | |
	Couple	Mother	Father	Relative	Other
Couple	0.566	0.309	0.083	0.023	0.019
Mother	0.392	0.453	0.066	0.061	0.028
Father	0.391	0.354	0.109	0.108	0.038
Relative	0.307	0.558	0.040	0.056	0.039
Other	0.337	0.320	0.025	0.252	0.066

32. **Class Mobility**   The following transition matrix shows the probabilities of a person attaining different income classes based on the parents' income class. The income classes, based on total household income, are defined as poor (lowest quintile), middle class (middle quintiles), and upper class (highest quintile). Find the long-range percent of poor, middle class, and upper class if these trends were to continue. *Source: Center for American Progress.*

$$\begin{array}{c} \\ \text{Poor} \\ \text{Middle} \\ \text{Upper} \end{array} \begin{array}{ccc} \text{Poor} & \text{Middle} & \text{Upper} \\ \begin{bmatrix} 0.415 & 0.526 & 0.059 \\ 0.175 & 0.655 & 0.170 \\ 0.061 & 0.520 & 0.419 \end{bmatrix} \end{array}$$

33. **Migration**   As we saw in the last section, a study found that the probability that people living in one type of neighborhood migrate to another could be described by a Markov chain. *Source: Annals of the Association of American Geographers.*

The study classified housing into five types:

I: Middle-class family households;

II: Upper-middle-class households;

III: Sound, rented, two-family dwelling units;

IV: Unsound, rented, multi-family dwelling units; and

V: Commercial.

The transition matrix for a 1-year period is given below.

$$
\begin{array}{c c c c c c}
 & \text{I} & \text{II} & \text{III} & \text{IV} & \text{V} \\
\text{I} & \begin{bmatrix} 0.71 & 0.19 & 0.03 & 0.07 & 0.00 \\
\text{II} & 0.52 & 0.31 & 0.04 & 0.13 & 0.00 \\
\text{III} & 0.40 & 0.16 & 0.22 & 0.22 & 0.00 \\
\text{IV} & 0.34 & 0.19 & 0.09 & 0.37 & 0.01 \\
\text{V} & 0.21 & 0.25 & 0.13 & 0.37 & 0.04 \end{bmatrix}
\end{array}
$$

If these trends continue, what are the long-term probabilities of people living in each type of housing?

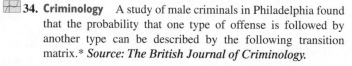 **34. Criminology** A study of male criminals in Philadelphia found that the probability that one type of offense is followed by another type can be described by the following transition matrix.* *Source: The British Journal of Criminology.*

	Nonindex	Injury	Theft	Damage	Combination
Nonindex	0.645	0.099	0.152	0.033	0.071
Injury	0.611	0.138	0.128	0.033	0.090
Theft	0.514	0.067	0.271	0.030	0.118
Damage	0.609	0.107	0.178	0.064	0.042
Combination	0.523	0.093	0.183	0.022	0.179

a. For a criminal who commits theft, what is the probability that his next crime is also a theft?

b. For a criminal who commits theft, what is the probability that his second crime after that is also a theft?

c. If these trends continue, what are the long-term probabilities for each type of crime?

35. Education At one liberal arts college, students are classified as humanities majors, science majors, or undecided. There is a 20% chance that a humanities major will change to a science major from one year to the next and a 45% chance that a humanities major will change to undecided. A science major will change to humanities with probability 0.15 and to undecided with probability 0.35. An undecided will switch to humanities or science with probabilities of 0.5 and 0.3, respectively.

a. Find the long-range prediction for the fraction of students in each of these three majors.

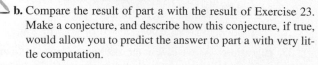 **b.** Compare the result of part a with the result of Exercise 23. Make a conjecture, and describe how this conjecture, if true, would allow you to predict the answer to part a with very little computation.

*The rounding was changed slightly so the rows of the transition matrix sum to 1.

36. Rumors The manager of the slot machines at a major casino makes a decision about whether or not to "loosen up" the slots so that the customers get a larger payback. The manager tells only one other person, a person whose word cannot be trusted. In fact, there is only a probability p, where $0 < p < 1$, that this person will tell the truth. Suppose this person tells several other people, each of whom tells several people, what the manager's decision is. Suppose there is always a probability p that the decision is passed on as heard. Find the long-range prediction for the fraction of the people who will hear the decision correctly. (*Hint:* Use a transition matrix; let the first row be $\begin{bmatrix} p & 1-p \end{bmatrix}$ and the second row be $\begin{bmatrix} 1-p & p \end{bmatrix}$.)

37. Education A study of students taking a 20-question chemistry exam tracked their progress from one testing period to the next. For simplicity, we have grouped students scoring from 0 to 5 in group 1, from 6 to 10 in group 2, from 11 to 15 in group 3, and from 15 to 20 in group 4. The result is the following transition matrix. *Source: National Forum of Teacher Education Journal.*

$$
\begin{array}{c c c c c}
 & 1 & 2 & 3 & 4 \\
1 & \begin{bmatrix} 0.065 & 0.585 & 0.34 & 0.01 \\
2 & 0.042 & 0.44 & 0.42 & 0.098 \\
3 & 0.018 & 0.276 & 0.452 & 0.254 \\
4 & 0 & 0.044 & 0.292 & 0.664 \end{bmatrix}
\end{array}
$$

a. Find the long-range prediction for the proportion of the students in each group.

b. The authors of this study were interested in the number of testing periods required before a certain proportion of the students had mastered the material. Suppose that once a student reaches group 4, the student is said to have mastered the material and is no longer tested, so the student stays in that group forever. Initially, all of the students in the study were in group 1. Find the number of testing periods you would expect for at least 70% of the students to have mastered the material. (*Hint:* Try increasing values of n in $x_0 \cdot P^n$.)

Physical Sciences

38. Weather The weather in a certain spot is classified as fair, cloudy without rain, or rainy. A fair day is followed by a fair day 60% of the time and by a cloudy day 25% of the time. A cloudy day is followed by a cloudy day 35% of the time and by a rainy day 25% of the time. A rainy day is followed by a cloudy day 40% of the time and by another rainy day 25% of the time. What proportion of days are expected to be fair, cloudy, and rainy over the long term?

General Interest

39. Ehrenfest Chain The model for the Ehrenfest chain consists of 2 boxes containing a total of n balls, where n is any integer greater than or equal to 2. In each turn, a ball is picked at random and moved from whatever box it is in to the other box. Let the state of the Markov process be the number of balls in the first box.

a. Verify that the probability of going from state i to state j is given by the following.

$$p_{ij} = \begin{cases} \frac{i}{n} & \text{if } i \geq 1 \text{ and } j = i - 1 \\ 1 - \frac{i}{n} & \text{if } i \leq n - 1 \text{ and } j = i + 1 \\ 1 & \text{if } i = 0 \text{ and } j = 1 \text{ or } i = n \text{ and } j = n - 1 \\ 0 & \text{otherwise.} \end{cases}$$

b. Verify that the transition matrix is given by

$$\begin{array}{c} \\ 0 \\ 1 \\ 2 \\ \vdots \\ n \end{array} \begin{array}{cccccc} 0 & 1 & 2 & 3 & \cdots & n \\ \begin{bmatrix} 0 & 1 & 0 & 0 & \cdots & 0 \\ \frac{1}{n} & 0 & 1-\frac{1}{n} & 0 & \cdots & 0 \\ 0 & \frac{2}{n} & 0 & 1-\frac{2}{n} & \cdots & 0 \\ \vdots & \vdots & \vdots & \vdots & \ddots & \vdots \\ 0 & 0 & 0 & 0 & \cdots & 0 \end{bmatrix} \end{array}$$

c. Write the transition matrix for the case $n = 2$.

d. Determine whether the transition matrix in part c is a regular transition matrix.

e. Determine an equilibrium vector for the matrix in part c. Explain what the result means.

40. Language One of Markov's own applications was a 1913 study of how often a vowel is followed by another vowel or a consonant by another consonant in Russian text. A similar study of a passage of English text revealed the following transition matrix.

$$\begin{array}{c} \\ \text{Vowel} \\ \text{Consonant} \end{array} \begin{array}{cc} \text{Vowel} & \text{Consonant} \\ \begin{bmatrix} 0.12 & 0.88 \\ 0.54 & 0.46 \end{bmatrix} \end{array}$$

Find the percent of letters in English text that are expected to be vowels.

41. Random Walk Many phenomena can be viewed as examples of a random walk. Consider the following simple example. A security guard can stand in front of any one of three doors 20 ft apart in front of a building, and every minute he decides whether to move to another door chosen at random. If he is at the middle door, he is equally likely to stay where he is, move to the door to the left, or move to the door to the right. If he is at the door on either end, he is equally likely to stay where he is or move to the middle door.

a. Verify that the transition matrix is given by

$$\begin{array}{c} \\ 1 \\ 2 \\ 3 \end{array} \begin{array}{ccc} 1 & 2 & 3 \\ \begin{bmatrix} \frac{1}{2} & \frac{1}{2} & 0 \\ \frac{1}{3} & \frac{1}{3} & \frac{1}{3} \\ 0 & \frac{1}{2} & \frac{1}{2} \end{bmatrix} \end{array}.$$

b. Find the long-range trend for the fraction of time the guard spends in front of each door.

YOUR TURN ANSWERS

1. C is regular.   **2.** [0.4 0.6]   **3.** [0.2069 0.0690 0.7241]

10.3 Absorbing Markov Chains

APPLY IT **If a gambler gambles until she either goes broke or wins some predetermined amount of money, what is the probability that she will eventually go broke?**

Using properties of absorbing Markov chains, we will answer this question in Example 2.

Suppose a Markov chain has transition matrix

$$\begin{array}{c} \\ 1 \\ 2 \\ 3 \end{array} \begin{array}{ccc} 1 & 2 & 3 \\ \begin{bmatrix} 0.3 & 0.6 & 0.1 \\ 0 & 1 & 0 \\ 0.6 & 0.2 & 0.2 \end{bmatrix} \end{array} = P.$$

The matrix shows that p_{12}, the probability of going from state 1 to state 2, is 0.6, and that p_{22}, the probability of staying in state 2, is 1. Thus, once state 2 is entered, it is impossible to leave. For this reason, state 2 is called an *absorbing state*. Figure 8 shows a transition diagram for this matrix. The diagram shows that it is not possible to leave state 2.

Generalizing from this example leads to the following definition.

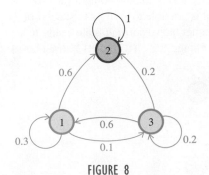

FIGURE 8

Absorbing State

State i of a Markov chain is an **absorbing state** if $p_{ii} = 1$.

Using the idea of an absorbing state, we can define an *absorbing Markov chain*.

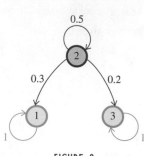

FIGURE 9

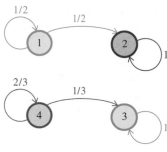

FIGURE 10

YOUR TURN 1 Identify all absorbing states in the following transition matrix. Decide whether the Markov chain is absorbing.

$$
\begin{array}{c}
 \\
1 \\
2 \\
3 \\
4
\end{array}
\begin{array}{cccc}
1 & 2 & 3 & 4 \\
\left[\begin{array}{cccc}
1 & 0 & 0 & 0 \\
\frac{1}{5} & \frac{2}{5} & 0 & \frac{2}{5} \\
0 & 0 & 1 & 0 \\
\frac{1}{4} & \frac{3}{4} & 0 & 0
\end{array}\right]
\end{array}
$$

Absorbing Markov Chain

A Markov chain is an **absorbing chain** if and only if the following two conditions are satisfied:

1. the chain has at least one absorbing state; and
2. it is possible to go from any nonabsorbing state to some absorbing state (perhaps in more than one step).

Note that the second condition does not mean that it is possible to go from any nonabsorbing state to *any* absorbing state, but it is possible to go to *some* absorbing state.

EXAMPLE 1 Absorbing Markov Chain

Identify all absorbing states in the Markov chains having the following matrices. Decide whether the Markov chain is absorbing.

$$
\textbf{(a)}\quad
\begin{array}{c}
 \\
1 \\
2 \\
3
\end{array}
\begin{array}{ccc}
1 & 2 & 3 \\
\left[\begin{array}{ccc}
1 & 0 & 0 \\
0.3 & 0.5 & 0.2 \\
0 & 0 & 1
\end{array}\right]
\end{array}
$$

SOLUTION Since $p_{11} = 1$ and $p_{33} = 1$, both state 1 and state 3 are absorbing states. (Once these states are reached, they cannot be left.) The only nonabsorbing state is state 2. There is a 0.3 probability of going from state 2 to the absorbing state 1, and a 0.2 probability of going from state 2 to state 3, so that it is possible to go from the nonabsorbing state to an absorbing state. This Markov chain is absorbing. The transition diagram is shown in Figure 9.

$$
\textbf{(b)}\quad
\begin{array}{c}
 \\
1 \\
2 \\
3 \\
4
\end{array}
\begin{array}{cccc}
1 & 2 & 3 & 4 \\
\left[\begin{array}{cccc}
\frac{1}{2} & \frac{1}{2} & 0 & 0 \\
0 & 1 & 0 & 0 \\
0 & 0 & 1 & 0 \\
0 & 0 & \frac{1}{3} & \frac{2}{3}
\end{array}\right]
\end{array}
$$

SOLUTION Since $p_{22} = 1$ and $p_{33} = 1$, both states 2 and 3 are absorbing. States 1 and 4 are nonabsorbing. As the transition diagram in Figure 10 shows, it is possible to go from state 1 to state 2 and from state 4 to state 3. Thus, it is possible to go from any nonabsorbing state to some absorbing state, so this Markov chain is absorbing.

$$
\textbf{(c)}\quad
\begin{array}{c}
 \\
1 \\
2 \\
3 \\
4
\end{array}
\begin{array}{cccc}
1 & 2 & 3 & 4 \\
\left[\begin{array}{cccc}
0.6 & 0 & 0.4 & 0 \\
0 & 1 & 0 & 0 \\
0.9 & 0 & 0.1 & 0 \\
0 & 0 & 0 & 1
\end{array}\right]
\end{array}
$$

SOLUTION States 2 and 4 are absorbing, with states 1 and 3 nonabsorbing. From state 1, it is possible to go only to states 1 or 3; from state 3 it is possible to go only to states 1 or 3. As the transition diagram in Figure 11 shows, neither nonabsorbing state leads to an absorbing state, so that this Markov chain is nonabsorbing. **TRY YOUR TURN 1**

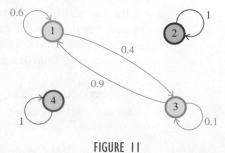

FIGURE 11

EXAMPLE 2 Gambler's Ruin

Suppose players A and B have a coin tossing game going on—a fair coin is tossed and the player predicting the toss correctly wins $1 from the other player. Suppose the players have a total of $6 between them and that the game goes on until one player has no money (is ruined).

(a) Write a transition matrix for this game.

SOLUTION Let us agree that the states of this system are the amounts of money held by player A. There are seven possible states: A can have 0, 1, 2, 3, 4, 5, or 6 dollars. When either state 0 or state 6 is reached, the game is over. In any other state, the amount of money held by player A will increase by $1, or decrease by $1, with each of these events having probability $1/2$ (since we assume a fair coin). For example, in state 3 (A has $3), there is a $1/2$ chance of changing to state 2 and a $1/2$ chance of changing to state 4. Thus, $p_{32} = 1/2$ and $p_{34} = 1/2$. The probability of changing from state 3 to any other state is 0. Using this information gives the following 7×7 transition matrix.

$$
\begin{array}{c c}
 & \begin{matrix} 0 & 1 & 2 & 3 & 4 & 5 & 6 \end{matrix} \\
\begin{matrix} 0 \\ 1 \\ 2 \\ 3 \\ 4 \\ 5 \\ 6 \end{matrix} &
\begin{bmatrix}
1 & 0 & 0 & 0 & 0 & 0 & 0 \\
\frac{1}{2} & 0 & \frac{1}{2} & 0 & 0 & 0 & 0 \\
0 & \frac{1}{2} & 0 & \frac{1}{2} & 0 & 0 & 0 \\
0 & 0 & \frac{1}{2} & 0 & \frac{1}{2} & 0 & 0 \\
0 & 0 & 0 & \frac{1}{2} & 0 & \frac{1}{2} & 0 \\
0 & 0 & 0 & 0 & \frac{1}{2} & 0 & \frac{1}{2} \\
0 & 0 & 0 & 0 & 0 & 0 & 1
\end{bmatrix} = P
\end{array}
$$

(b) Identify all absorbing states. Decide whether the Markov chain is absorbing.

SOLUTION Based on the rules of the game, states 0 and 6 are absorbing—once these states are reached, they can never be left, and the game is over. It is possible to get from one of the nonabsorbing states, 1, 2, 3, 4, or 5, to one of the absorbing states, so the Markov chain is absorbing.

(c) Estimate the long-term trend of the game.

APPLY IT

SOLUTION For the long-term trend of the game, find various powers of the transition matrix. A computer or a graphing calculator can be used to verify the following results.

$$
P^{20} =
\begin{bmatrix}
1.0000 & 0.0000 & 0.0000 & 0.0000 & 0.0000 & 0.0000 & 0.0000 \\
0.8146 & 0.0094 & 0.0000 & 0.0188 & 0.0000 & 0.0094 & 0.1479 \\
0.6385 & 0.0000 & 0.0282 & 0.0000 & 0.0282 & 0.0000 & 0.3052 \\
0.4625 & 0.0188 & 0.0000 & 0.0375 & 0.0000 & 0.0188 & 0.4625 \\
0.3052 & 0.0000 & 0.0282 & 0.0000 & 0.0282 & 0.0000 & 0.6385 \\
0.1479 & 0.0094 & 0.0000 & 0.0188 & 0.0000 & 0.0094 & 0.8146 \\
0.0000 & 0.0000 & 0.0000 & 0.0000 & 0.0000 & 0.0000 & 1.0000
\end{bmatrix}
$$

$$
P^{70} =
\begin{bmatrix}
1.0000 & 0.0000 & 0.0000 & 0.0000 & 0.0000 & 0.0000 & 0.0000 \\
0.8333 & 0.0000 & 0.0000 & 0.0000 & 0.0000 & 0.0000 & 0.1667 \\
0.6667 & 0.0000 & 0.0000 & 0.0000 & 0.0000 & 0.0000 & 0.3333 \\
0.5000 & 0.0000 & 0.0000 & 0.0000 & 0.0000 & 0.0000 & 0.5000 \\
0.3333 & 0.0000 & 0.0000 & 0.0000 & 0.0000 & 0.0000 & 0.6667 \\
0.1667 & 0.0000 & 0.0000 & 0.0000 & 0.0000 & 0.0000 & 0.8333 \\
0.0000 & 0.0000 & 0.0000 & 0.0000 & 0.0000 & 0.0000 & 1.0000
\end{bmatrix}
$$

As these results suggest, the system tends toward one of the absorbing states, so that the probability is 1 that one of the two gamblers will be wiped out. The probability that player

A will end in state 0 (player A is ruined) depends on player A's initial state. For example, if player A starts with $2 (state 2), there is a probability of 0.6667 of player A going broke (state 0), but if player A starts with $5, there is only a probability of 0.1667 of player A going broke.

EXAMPLE 3 Long-term Trend

Estimate the long-term trend for the following transition matrix.

$$P = \begin{bmatrix} 0.3 & 0.2 & 0.5 \\ 0 & 1 & 0 \\ 0 & 0 & 1 \end{bmatrix}$$

SOLUTION Both states 2 and 3 are absorbing, and since it is possible to go from non-absorbing state 1 to an absorbing state, the chain will eventually enter either state 2 or state 3. To find the long-term trend, let us find various powers of P.

$$P^2 = \begin{bmatrix} 0.09 & 0.26 & 0.65 \\ 0 & 1 & 0 \\ 0 & 0 & 1 \end{bmatrix} \qquad P^4 = \begin{bmatrix} 0.0081 & 0.2834 & 0.7085 \\ 0 & 1 & 0 \\ 0 & 0 & 1 \end{bmatrix}$$

$$P^8 = \begin{bmatrix} 0.0001 & 0.2857 & 0.7142 \\ 0 & 1 & 0 \\ 0 & 0 & 1 \end{bmatrix} \qquad P^{16} = \begin{bmatrix} 0.0000 & 0.2857 & 0.7143 \\ 0 & 1 & 0 \\ 0 & 0 & 1 \end{bmatrix}$$

Based on these powers, it appears that the transition matrix is getting closer and closer to the matrix

$$\begin{bmatrix} 0 & 0.2857 & 0.7143 \\ 0 & 1 & 0 \\ 0 & 0 & 1 \end{bmatrix}.$$

If the system is originally in state 1, there is no chance it will end up in state 1, but a 0.2857 chance that it will end up in state 2 and a 0.7143 chance it will end up in state 3. If the system was originally in state 2, it will end up in state 2; a similar statement can be made for state 3.

The examples suggest the following properties of absorbing Markov chains, which can be verified using more advanced methods.

1. Regardless of the original state of an absorbing Markov chain, in a finite number of steps the chain will enter an absorbing state and then stay in that state.

2. The powers of the transition matrix get closer and closer to some particular matrix.

3. The long-term trend depends on the initial state—changing the initial state can change the final result.

The third property distinguishes absorbing Markov chains from regular Markov chains, where the final result is independent of the initial state.

It would be preferable to have a method for finding the final probabilities of entering an absorbing state without finding all the powers of the transition matrix, as in Example 3. We do not really need to worry about the absorbing states (to enter an absorbing state is to stay there). Therefore, it is necessary only to work with the nonabsorbing states. To see how this is done, let us use as an example the transition matrix from the gambler's ruin problem in Example 2. Rewrite the matrix so that the rows and columns corresponding to the absorbing states come first.

$$
\begin{array}{c}
\\
\begin{array}{cc}
\text{Absorbing} & \text{Nonabsorbing}
\end{array}\\
\begin{array}{cc cccccc}
\ 0 & 6 & 1 & 2 & 3 & 4 & 5
\end{array}\\
\begin{array}{c}
0\\6\\1\\2\\3\\4\\5
\end{array}
\left[
\begin{array}{cc|ccccc}
1 & 0 & 0 & 0 & 0 & 0 & 0\\
0 & 1 & 0 & 0 & 0 & 0 & 0\\
\frac{1}{2} & 0 & 0 & \frac{1}{2} & 0 & 0 & 0\\
0 & 0 & \frac{1}{2} & 0 & \frac{1}{2} & 0 & 0\\
0 & 0 & 0 & \frac{1}{2} & 0 & \frac{1}{2} & 0\\
0 & 0 & 0 & 0 & \frac{1}{2} & 0 & \frac{1}{2}\\
0 & \frac{1}{2} & 0 & 0 & 0 & \frac{1}{2} & 0
\end{array}
\right] = P
\end{array}
$$

Let I_2 represent the 2×2 identity matrix in the upper left corner; let O represent the matrix of zeros in the upper right; let R represent the matrix in the lower left, and let Q represent the matrix in the lower right. Using these symbols, P can be written as

$$
P = \left[\begin{array}{c|c} I_2 & O \\ \hline R & Q \end{array}\right].
$$

The **fundamental matrix** for an absorbing Markov chain is defined as matrix F, where

$$
F = (I_n - Q)^{-1}.
$$

Here I_n is the $n \times n$ identity matrix corresponding in size to matrix Q, so that the difference $I_n - Q$ exists.

For the gambler's ruin problem, using I_5 gives

$$
F = \left(\begin{bmatrix}
1 & 0 & 0 & 0 & 0\\
0 & 1 & 0 & 0 & 0\\
0 & 0 & 1 & 0 & 0\\
0 & 0 & 0 & 1 & 0\\
0 & 0 & 0 & 0 & 1
\end{bmatrix}
-
\begin{bmatrix}
0 & \frac{1}{2} & 0 & 0 & 0\\
\frac{1}{2} & 0 & \frac{1}{2} & 0 & 0\\
0 & \frac{1}{2} & 0 & \frac{1}{2} & 0\\
0 & 0 & \frac{1}{2} & 0 & \frac{1}{2}\\
0 & 0 & 0 & \frac{1}{2} & 0
\end{bmatrix}\right)^{-1}
$$

$$
= \begin{bmatrix}
1 & -\frac{1}{2} & 0 & 0 & 0\\
-\frac{1}{2} & 1 & -\frac{1}{2} & 0 & 0\\
0 & -\frac{1}{2} & 1 & -\frac{1}{2} & 0\\
0 & 0 & -\frac{1}{2} & 1 & -\frac{1}{2}\\
0 & 0 & 0 & -\frac{1}{2} & 1
\end{bmatrix}^{-1}
$$

$$
\begin{array}{c}
\begin{array}{ccccc}
1 & 2 & 3 & 4 & 5
\end{array}\\
\begin{array}{c}
1\\2\\3\\4\\5
\end{array}
= \begin{array}{c}
\\
\begin{bmatrix}
\frac{5}{3} & \frac{4}{3} & 1 & \frac{2}{3} & \frac{1}{3}\\
\frac{4}{3} & \frac{8}{3} & 2 & \frac{4}{3} & \frac{2}{3}\\
1 & 2 & 3 & 2 & 1\\
\frac{2}{3} & \frac{4}{3} & 2 & \frac{8}{3} & \frac{4}{3}\\
\frac{1}{3} & \frac{2}{3} & 1 & \frac{4}{3} & \frac{5}{3}
\end{bmatrix}.
\end{array}
\end{array}
$$

The inverse was found using techniques from Chapter 2. Recall, we also discussed finding the inverse of a matrix with a graphing calculator there.

The fundamental matrix gives the expected number of visits to each state before absorption occurs. For example, if player A currently has $2, then the second row of the fundamental matrix just computed says that she expects to have $1 an average of $1\frac{1}{3}$ times, and to have $3 twice, before quitting the game because she either runs out of money or wins $6. The total number of times that player A expects to have various amounts of money before quitting the game is the sum of entries in row 2 of F: $(4/3) + (8/3) + 2 + (4/3) + (2/3) = 8$. In other words, if player A currently has $2, she can expect to stay in the game 8 more turns before either she or player B goes broke.

---FOR REVIEW---

To find the inverse of a matrix, we first form an augmented matrix by putting the original matrix on the left and the identity matrix on the right: $[A|I]$. The Gauss-Jordan method is used to turn the matrix on the left into the identity. The matrix on the right is then the inverse of the original matrix: $[I|A^{-1}]$.

To see why this is true, consider a Markov chain currently in state i. The expected number of times that the chain visits state j at this step is 1 for $j = i$ and 0 for all other states. The expected number of times that the chain visits state j at the next step is given by the element in row i, column j of the transition matrix Q. The expected number of times the chain visits state j two steps from now is given by the corresponding entry in the matrix Q^2. The expected number of visits in all steps is given by $I + Q + Q^2 + Q^3 + \cdots$. To find out whether this infinite sum is the same as $(I - Q)^{-1}$, multiply the sum by $(I - Q)$:

$$(I + Q + Q^2 + Q^3 + \cdots)(I - Q)$$
$$= I + Q + Q^2 + Q^3 + \cdots - Q - Q^2 - Q^3 + \cdots = I,$$

which verifies our result.

It can be shown that

$$P^k = \left[\begin{array}{c|c} I_m & O \\ \hline (I + Q + Q^2 + \cdots + Q^{k-1})R & Q^k \end{array}\right],$$

where I_m is the $m \times m$ identity matrix. As k gets larger and larger (denoted by $k \to \infty$), Q^k approaches O_n (denoted by $Q^k \to O_n$), the $n \times n$ zero matrix, and

$$P^k \to \left[\begin{array}{c|c} I_m & O \\ \hline FR & O_n \end{array}\right],$$

so we see that FR gives the probabilities that if the system was originally in a nonabsorbing state, it ends up in one of the absorbing states.*

Finally, use the fundamental matrix F along with matrix R found above to get the product FR.

$$FR = \begin{bmatrix} \frac{5}{3} & \frac{4}{3} & 1 & \frac{2}{3} & \frac{1}{3} \\ \frac{4}{3} & \frac{8}{3} & 2 & \frac{4}{3} & \frac{2}{3} \\ 1 & 2 & 3 & 2 & 1 \\ \frac{2}{3} & \frac{4}{3} & 2 & \frac{8}{3} & \frac{4}{3} \\ \frac{1}{3} & \frac{2}{3} & 1 & \frac{4}{3} & \frac{5}{3} \end{bmatrix} \begin{bmatrix} \frac{1}{2} & 0 \\ 0 & 0 \\ 0 & 0 \\ 0 & 0 \\ 0 & \frac{1}{2} \end{bmatrix} = \begin{array}{c} 1 \\ 2 \\ 3 \\ 4 \\ 5 \end{array} \begin{array}{cc} 0 & 6 \\ \begin{bmatrix} \frac{5}{6} & \frac{1}{6} \\ \frac{2}{3} & \frac{1}{3} \\ \frac{1}{2} & \frac{1}{2} \\ \frac{1}{3} & \frac{2}{3} \\ \frac{1}{6} & \frac{5}{6} \end{bmatrix} \end{array}$$

The product matrix FR gives the probability that if the system was originally in a nonabsorbing state, it ended up in either of the two absorbing states. For example, the probability is $2/3$ that if the system was originally in state 2, it ended up in state 0; the probability is $5/6$ that if the system was originally in state 5 it ended up in state 6, and so on. Based on the original statement of the gambler's ruin problem, if player A starts with \$2 (state 2), there is a $2/3$ chance of ending in state 0 (player A is ruined); if player A starts with \$5 (state 5) there is a $1/6$ chance of player A being ruined, and so on. (Note that these results agree with the estimates found in Example 2.)

In the fundamental matrix F, the sum of the elements in row i gives the expected number of steps for the matrix to enter an absorbing state from the ith nonabsorbing state. Thus, for $i = 2$, $(4/3) + (8/3) + 2 + (4/3) + (2/3) = 8$ steps will be needed on the average for the system to go from state 2 to an absorbing state (0 or 6).

Let us summarize what we have learned about absorbing Markov chains.

Properties of an Absorbing Markov Chain

1. Regardless of the initial state, in a finite number of steps the chain will enter an absorbing state and then stay in that state.

2. The powers of the transition matrix get closer and closer to some particular matrix.

3. The long-term trend depends on the initial state.

(continued)

*We have omitted details in these steps that can be justified using advanced techniques.

4. Let P be the transition matrix for an absorbing Markov chain. Rearrange the rows and columns of P so that the absorbing states come first. Matrix P will have the form

$$P = \left[\begin{array}{c|c} I_m & O \\ \hline R & Q \end{array}\right],$$

where I_m is an identity matrix, with m equal to the number of absorbing states, and O is a matrix of all zeros. The fundamental matrix is defined as

$$F = (I_n - Q)^{-1},$$

where I_n has the same size as Q. The element in row i, column j of the fundamental matrix gives the number of visits to state j that are expected to occur before absorption, given that the current state is state i.

5. The product FR gives the matrix of probabilities that a particular initial nonabsorbing state will lead to a particular absorbing state.

EXAMPLE 4 Long-term Trend

Find the long-term trend for the transition matrix

$$\begin{array}{c c c c} & 1 & 2 & 3 \\ \begin{array}{c} 1 \\ 2 \\ 3 \end{array} & \left[\begin{array}{ccc} 0.3 & 0.2 & 0.5 \\ 0 & 1 & 0 \\ 0 & 0 & 1 \end{array}\right] \end{array} = P$$

of Example 3.

SOLUTION Rewrite the matrix so that absorbing states 2 and 3 come first.

$$\begin{array}{c c c c} & 2 & 3 & 1 \\ \begin{array}{c} 2 \\ 3 \\ 1 \end{array} & \left[\begin{array}{cc|c} 1 & 0 & 0 \\ 0 & 1 & 0 \\ \hline 0.2 & 0.5 & 0.3 \end{array}\right] \end{array}$$

Here $R = [0.2 \quad 0.5]$ and $Q = [0.3]$. Find the fundamental matrix F.

$$F = (I_1 - Q)^{-1} = [1 - 0.3]^{-1} = [0.7]^{-1} = [1/0.7] = [10/7]$$

This tells us that the expected number of steps before ending up in an absorbing state is 10/7.

The product FR is

$$FR = [10/7][0.2 \quad 0.5] = [2/7 \quad 5/7].$$

If the system starts in the nonabsorbing state 1, there is a 2/7 chance of ending up in the absorbing state 2 and a 5/7 chance of ending up in the absorbing state 3.

TRY YOUR TURN 2

YOUR TURN 2 Find the long-term trend for the transition matrix

$$\left[\begin{array}{ccc} 1 & 0 & 0 \\ 0.1 & 0.6 & 0.3 \\ 0 & 0 & 1 \end{array}\right].$$

10.3 EXERCISES

Find all absorbing states for each transition matrix. Which are transition matrices for absorbing Markov chains?

1. $\left[\begin{array}{ccc} 0.25 & 0.05 & 0.7 \\ 0.35 & 0 & 0.65 \\ 0 & 0 & 1 \end{array}\right]$

2. $\left[\begin{array}{ccc} 0.15 & 0.35 & 0.5 \\ 0 & 1 & 0 \\ 0.3 & 0.3 & 0.4 \end{array}\right]$

3. $\left[\begin{array}{ccc} 1 & 0 & 0 \\ 0 & 0.25 & 0.75 \\ 0 & 0.85 & 0.15 \end{array}\right]$

4. $\left[\begin{array}{ccc} 0.4 & 0 & 0.6 \\ 0 & 1 & 0 \\ 0.5 & 0 & 0.5 \end{array}\right]$

5. $\begin{bmatrix} 0.2 & 0.5 & 0.1 & 0.2 \\ 0 & 1 & 0 & 0 \\ 0.9 & 0.02 & 0.04 & 0.04 \\ 0 & 0 & 0 & 1 \end{bmatrix}$ **6.** $\begin{bmatrix} 0.32 & 0.41 & 0.16 & 0.11 \\ 0.42 & 0.30 & 0 & 0.28 \\ 0 & 0 & 0 & 1 \\ 1 & 0 & 0 & 0 \end{bmatrix}$

Find the fundamental matrix F for the absorbing Markov chains with the matrices in Exercises 7–14. Also, find the product matrix FR.

7. $\begin{bmatrix} 1 & 0 & 0 \\ 0 & 1 & 0 \\ 0.15 & 0.35 & 0.5 \end{bmatrix}$ **8.** $\begin{bmatrix} 1 & 0 & 0 \\ 0.65 & 0.1 & 0.25 \\ 0 & 0 & 1 \end{bmatrix}$

9. $\begin{bmatrix} 1 & 0 & 0 \\ 0 & 1 & 0 \\ \frac{1}{2} & \frac{1}{6} & \frac{1}{3} \end{bmatrix}$ **10.** $\begin{bmatrix} 1 & 0 & 0 \\ \frac{5}{8} & \frac{1}{8} & \frac{1}{4} \\ 0 & 0 & 1 \end{bmatrix}$

11. $\begin{bmatrix} 1 & 0 & 0 & 0 \\ \frac{1}{3} & 0 & \frac{2}{3} & 0 \\ 0 & 0 & 1 & 0 \\ \frac{1}{4} & \frac{1}{4} & \frac{1}{4} & \frac{1}{4} \end{bmatrix}$ **12.** $\begin{bmatrix} \frac{1}{4} & \frac{1}{2} & 0 & \frac{1}{4} \\ 0 & 1 & 0 & 0 \\ 0 & 0 & 1 & 0 \\ \frac{1}{2} & 0 & 0 & \frac{1}{2} \end{bmatrix}$

13. $\begin{bmatrix} 1 & 0 & 0 & 0 & 0 \\ 0 & 1 & 0 & 0 & 0 \\ 0.1 & 0.2 & 0.3 & 0.2 & 0.2 \\ 0.3 & 0.5 & 0.1 & 0 & 0.1 \\ 0 & 0 & 0 & 0 & 1 \end{bmatrix}$

14. $\begin{bmatrix} 0.4 & 0.2 & 0.3 & 0 & 0.1 \\ 0 & 1 & 0 & 0 & 0 \\ 0 & 0 & 1 & 0 & 0 \\ 0.1 & 0.5 & 0.1 & 0.1 & 0.2 \\ 0 & 0 & 0 & 0 & 1 \end{bmatrix}$

15. a. Write a transition matrix for a gambler's ruin problem when player A and player B start with a total of $4. (See Example 2.)

b. Find matrix F for this transition matrix, and find the product matrix FR.

c. Suppose player A starts with $1. What is the probability of ruin for A?

d. Suppose player A starts with $3. What is the probability of ruin for A?

16. Suppose player B (Exercise 15) slips in a coin that is slightly "loaded"—such that the probability that B wins a particular toss changes from $1/2$ to $3/5$. Suppose that A and B start the game with a total of $5.

a. If B starts with $3, find the probability that A will be ruined.

b. If B starts with $1, find the probability that A will be ruined.

It can be shown that the probability of ruin for player A in a game such as the one described in this section is

$$x_a = \frac{b}{a+b} \text{ if } r = 1, \quad \text{and} \quad x_a = \frac{r^a - r^{a+b}}{1 - r^{a+b}} \text{ if } r \neq 1,$$

where a is the initial amount of money that player A has, b is the initial amount that player B has, $r = (1-p)/p$, and p is the probability that player A will win on a given play.

17. Find the probability that A will be ruined if $a = 10$, $b = 30$, and $p = 0.49$.

18. Find the probability in Exercise 17 if p changes to 0.50.

19. Complete the following chart, assuming $a = 10$ and $b = 10$.

p	0.1	0.2	0.3	0.4	0.5	0.6	0.7	0.8	0.9
x_a									

20. How can we calculate the expected total number of times a Markov chain will visit state j before absorption, regardless of the current state?

21. Suppose an absorbing Markov chain has only one absorbing state. What is the product FR?

APPLICATIONS

Business and Economics

22. Company Training Program A company with a new training program classified each worker into one of the following four categories: s_1, never in the program; s_2, currently in the program; s_3, discharged; s_4, completed the program. The transition matrix for this company is given below.

$$\begin{array}{c} \\ s_1 \\ s_2 \\ s_3 \\ s_4 \end{array} \begin{array}{cccc} s_1 & s_2 & s_3 & s_4 \\ \begin{bmatrix} 0.4 & 0.2 & 0.05 & 0.35 \\ 0 & 0.45 & 0.05 & 0.5 \\ 0 & 0 & 1 & 0 \\ 0 & 0 & 0 & 1 \end{bmatrix} \end{array}$$

a. Find F and FR.

b. Find the probability that a worker originally in the program is discharged.

c. Find the probability that a worker not originally in the program goes on to complete the program.

23. Solar Energy In Exercise 33 of the first section of this chapter, a community began a campaign to convince homeowners to convert their energy source for home heating to solar energy. The leaders estimated the probability of homeowners changing their home heating energy source each year, giving the following transition matrix.

$$\begin{array}{c} \\ \text{electric} \\ \text{fossil} \\ \text{solar} \end{array} \begin{array}{ccc} \text{electric} & \text{fossil} & \text{solar} \\ \begin{bmatrix} 0.9 & 0.05 & 0.05 \\ 0.15 & 0.75 & 0.1 \\ 0 & 0 & 1 \end{bmatrix} \end{array}$$

a. Find F and FR.

b. Find the probability that a homeowner in this community will eventually use solar energy for home heating.

c. Find the expected number of years until a homeowner who is presently using electric heat will convert to solar energy for a heating energy source.

Life Sciences

24. Medical Prognosis A study using Markov chains to estimate a patient's prognosis for improving under various treatment plans gives the following transition matrix as an example. *Source: Medical Decision Making.*

	Well	Ill	Dead
Well	0.3	0.5	0.2
Ill	0	0.5	0.5
Dead	0	0	1

a. Estimate the probability that a well person will eventually end up dead.

b. Verify your answer to part a using the matrix product *FR*.

c. Find the expected number of cycles that a well patient will continue to be well before dying and the expected number of cycles that a well patient will be ill before dying.

25. Contagion Under certain conditions, the probability that a person will get a particular contagious disease and die from it is 0.05, and the probability of getting the disease and surviving is 0.15. The probability that a survivor will infect another person who dies from it is also 0.05, that a survivor will infect another person who survives it is 0.15, and so on. A transition matrix using the following states is given below. A person in state 1 is one who gets the disease and dies, a person in state 2 gets the disease and survives, and a person in state 3 does not get the disease. Consider a chain of people, each of whom interacts with the previous person and may catch the disease from that individual, and then may infect the next person.

a. Verify that the transition matrix is as follows:

Second Person

		1	2	3
First Person	1	0.05	0.15	0.8
	2	0.05	0.15	0.8
	3	0	0	1

b. Find *F* and *FR*.

c. Find the probability that the disease eventually disappears.

d. Given a person who has the disease and survives, find the expected number of people in the chain who will get the disease until a person who does not get the disease is reached.

Social Sciences

26. Student Retention At a particular two-year college, a student has a probability of 0.25 of flunking out during a given year, a 0.15 probability of having to repeat the year, and a 0.6 probability of finishing the year. Use the states below.

State	Meaning
1	Freshman
2	Sophomore
3	Has flunked out
4	Has graduated

a. Write a transition matrix. Find *F* and *FR*.

b. Find the probability that a freshman will graduate.

c. Find the expected number of years that a freshman will be in college before graduating or flunking out.

27. Transportation The city of Sacramento completed a new light rail system to bring commuters and shoppers into the downtown area and relieve freeway congestion. City planners estimate that each year, 15% of those who drive or ride in an automobile will change to the light rail system; 80% will continue to use automobiles; and the rest will no longer go to the downtown area. Of those who use light rail, 5% will go back to using an automobile, 80% will continue to use light rail, and the rest will stay out of the downtown area. Assume those who do not go downtown will continue to stay out of the downtown area.

a. Write a transition matrix. Find *F* and *FR*.

b. Find the probability that a person who commuted by automobile ends up avoiding the downtown area.

c. Find the expected number of years until a person who commutes by automobile this year no longer enters the downtown area.

28. Education Careers Data has been collected on the likelihood that a teacher, or a student with a declared interest in teaching, will continue on that career path the following year. We have simplified the classification of the original data to four groups: high school and college students, new teachers, continuing teachers, and those who have quit the profession. The transition probabilities are given in the following matrix. *Source: National Forum of Educational Administration and Supervision Journal.*

	Student	New	Continuing	Quit
Student	0.70	0.11	0	0.19
New	0	0	0.86	0.14
Continuing	0	0	0.88	0.12
Quit	0	0	0	1

a. Find the expected number of years that a student with an interest in teaching will spend as a continuing teacher.

b. Find the expected number of years that a new teacher will spend as a continuing teacher.

c. Find the expected number of additional years that a continuing teacher will spend as a continuing teacher.

d. Notice that the answer to part b is larger than the answer to part a, and the answer to part c is even larger. Explain why this is to be expected.

e. What other states might be added to this model to make it more realistic? Discuss how this would affect the transition matrix. (See the Extended Application for this chapter.)

29. Rat Maze A rat is placed at random in one of the compartments of the maze pictured. The probability that a rat in compartment 1 will move to compartment 2 is 0.3, to compartment 3 is 0.2, and to compartment 4 is 0.1. A rat in compartment 2 will move to compartments 1, 4, or 5 with probabilities of 0.2, 0.6, and 0.1, respectively. A rat in compartment 3 cannot leave that compartment. A rat in compartment 4 will move to 1, 2, 3, or 5 with probabilities of 0.1, 0.1, 0.4, and 0.3, respectively. A rat in compartment 5 cannot leave that compartment.

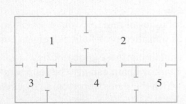

a. Set up a transition matrix using this information. Find matrices F and FR.

Find the probability that a rat ends up in compartment 5 if it was originally in the given compartment.

b. 1   **c.** 2   **d.** 3   **e.** 4

f. Find the expected number of times that a rat in compartment 1 will be in compartment 1 before ending up in compartment 3 or 5.

g. Find the expected number of times that a rat in compartment 4 will be in compartment 4 before ending up in compartment 3 or 5.

General Interest

30. Tennis Consider a game of tennis when each player has won at least two serves. After that, the first player to win two serves more than his opponent wins. There are five possibilities: Either the players are tied (deuce), the server is a point ahead (ad in), the other player is a point ahead (ad out), the server wins, or the server loses. If we assume that at each serve the server has a probability p of winning the next point, then the game can be modeled by a Markov chain. *Source: The Mathematics Teacher.*

a. Verify that the transition matrix from the server's point of view is given by

$$
\begin{array}{c c}
& \begin{array}{c c c c c} \text{Loss} & \text{Ad out} & \text{Deuce} & \text{Ad in} & \text{Win} \end{array} \\
\begin{array}{c} \text{Loss} \\ \text{Ad out} \\ \text{Deuce} \\ \text{Ad in} \\ \text{Win} \end{array} &
\left[\begin{array}{c c c c c}
1 & 0 & 0 & 0 & 0 \\
1-p & 0 & p & 0 & 0 \\
0 & 1-p & 0 & p & 0 \\
0 & 0 & 1-p & 0 & p \\
0 & 0 & 0 & 0 & 1
\end{array}\right]
\end{array}
$$

b. For the case in which $p = 0.6$, find the probability that the server will win when the score is ad out, deuce, and ad in.

31. Gambler's Ruin

a. Write a transition matrix for a gambler's ruin problem, where players A and B start with a total of \$7. (See Example 2.)

b. Find the probability of ruin for A if A starts with \$4.

c. Find the probability of ruin for A if A starts with \$5.

32. Professional Football In Exercise 42 of the first section of this chapter, the current method used by the NFL to determine the winner in an overtime game was analyzed. In this system, a coin is tossed and the winning team gets to decide whether to kickoff or to receive the ball. The first team to score wins. Based on data from the 2008 season, the following transition matrix was determined. The first row tells the probability of being in any given situation after one possession, assuming team A starts with the ball. *Source: Mathematics in Sports.*

	Possession for A	Possession for B	A wins	B wins
Possession for A	0	0.62	0.36	0.02
Possession for B	0.62	0	0.02	0.36
A wins	0	0	1	0
B wins	0	0	0	1

a. Find F and FR.

b. Find the probability that, if team A gets the ball first, team A wins.

c. Find the probability that, if team B gets the ball first, team B wins.

d. What is the expected number of possessions to determine a winner in overtime?

e. Determine the strategy that each team should take. What does this imply about the winner of the coin toss (the one who gets to decide who gets the ball first)?

YOUR TURN ANSWERS

1. States 1 and 3 are absorbing states. The Markov chain is absorbing.
2. If the system starts in state 2, there is a $1/4$ chance of ending up in state 1 and a $3/4$ chance of ending up in state 3.

10   CHAPTER REVIEW

SUMMARY

Markov chains are useful for modeling a sequence of trials of an experiment in which the following conditions hold:

- the outcome of each experiment is one of a finite number of possible states;
- the outcome of each experiment depends only on the present state; and
- the probabilities of moving from one state to another state, known as transition probabilities, remain constant over time.

The matrix containing these probabilities is called the transition matrix P. A Markov chain can also be represented by a transition diagram, in which the states, represented by circles, are connected by lines on which the transition probabilities are written. We saw in this chapter that to find the transition probabilities over n time periods, we raise the transition matrix to the nth power. Three other definitions presented in this chapter are the following:

- probability vector (a row matrix giving the probability of being in each state at a given time),
- regular Markov chain (some power of the transition matrix contains all positive entries), and
- absorbing Markov chain (one or more states, known as absorbing states, are impossible to leave and it is possible to go from any nonabsorbing state to some absorbing state).

A regular Markov chain has the following two properties:

- an equilibrium vector V can be found, which gives the long-range probability of being in each state;
- the powers of P^n approach a matrix whose rows make up the entries of V.

The probability that a particular nonabsorbing state will lead to a particular absorbing state can be calculated by methods summarized below. The applications in this chapter show the usefulness of Markov chains, whether regular or absorbing.

Probability Vector   If X_0 is the initial probability vector and P is the transition matrix, then the probability vector after n repetitions of the experiment is

$$X_0 \cdot P^n.$$

Equilibrium Vector   If a Markov chain with transition matrix P is regular, then the equilibrium vector V satisfies

$$VP = V.$$

Solve this system using the Gauss-Jordan method, with the added condition that the elements of V sum to 1.

Absorbing Markov Chain   Rearrange the rows and columns of P so it can be written as

$$\left[\begin{array}{c|c} I_m & O \\ \hline R & Q \end{array}\right],$$

where I_m is an identity matrix, m is the number of absorbing states, and O is a matrix of all zeros.

Fundamental Matrix   The row i, column j element of the fundamental matrix

$$F = (I_n - Q)^{-1},$$

where I_n has the same size as Q, gives the expected number of visits to state j before absorption, given the current state i. The probabilities that a particular initial nonabsorbing state will lead to a particular absorbing state is given by the product

$$FR.$$

KEY TERMS

state	Markov chain	regular Markov chain	absorbing chain
10.1	probability vector	equilibrium (or fixed) vector	fundamental matrix
transition diagram	**10.2**	**10.3**	
transition matrix	regular transition matrix	absorbing state	

REVIEW EXERCISES

CONCEPT CHECK

Determine whether each of the following statements is true or false, and explain why.

1. In a Markov chain, the outcome of an experiment might depend not only on the present state but also on the state before that.

2. In a Markov chain, the transition probabilities might vary from one transition to the next.

3. A transition diagram contains the same information as a transition matrix.

4. To find the transition probabilities in k repetitions of an experiment, one should multiply the transition matrix P by k.

5. A regular transition matrix has no 0 elements.

6. A transition matrix can have more than one equilibrium vector.

7. For a regular Markov chain, the matrix equation $VP = V$ has only one solution.

8. For any transition matrix, the matrix P^n has no 0 elements if n is made large enough.

9. For any transition matrix, the matrix P^n approaches a limiting value as n is made larger and larger.

10. A transition matrix that is not regular can have an equilibrium vector.

11. A Markov chain can have more than one absorbing state.

12. An absorbing Markov chain will eventually reach an absorbing state with a probability of 1.

PRACTICE AND EXPLORATIONS

13. How can you tell by looking at a matrix whether it represents the transition matrix for a Markov chain?

14. Under what conditions is the existence of an equilibrium vector guaranteed?

Decide whether each matrix could be a transition matrix. Sketch a transition diagram for any transition matrices.

15. $\begin{bmatrix} -0.2 & 1.2 \\ 0.8 & 0.2 \end{bmatrix}$

16. $\begin{bmatrix} 0.4 & 0.6 \\ 1 & 0 \end{bmatrix}$

17. $\begin{bmatrix} 0.8 & 0.2 & 0 \\ 0 & 1 & 0 \\ 0.1 & 0.4 & 0.5 \end{bmatrix}$

18. $\begin{bmatrix} 0.6 & 0.2 & 0.3 \\ 0.1 & 0.5 & 0.4 \\ 0.3 & 0.3 & 0.4 \end{bmatrix}$

For each transition matrix, (a) find the first three powers; and (b) find the probability that state 2 changes to state 1 after three repetitions of the experiment.

19. $C = \begin{bmatrix} 0.7 & 0.3 \\ 1 & 0 \end{bmatrix}$

20. $D = \begin{bmatrix} 0.4 & 0.6 \\ 0.5 & 0.5 \end{bmatrix}$

21. $E = \begin{bmatrix} 0.2 & 0.5 & 0.3 \\ 0.3 & 0.4 & 0.3 \\ 0 & 1 & 0 \end{bmatrix}$

22. $F = \begin{bmatrix} 0.14 & 0.18 & 0.68 \\ 0.35 & 0.28 & 0.37 \\ 0.71 & 0.22 & 0.07 \end{bmatrix}$

In Exercises 23–26, use the transition matrix P, along with the given initial distribution D, to find the distribution after two repetitions of the experiment. Also, predict the long-range distribution.

23. $D = \begin{bmatrix} 0.3 & 0.7 \end{bmatrix}$; $P = \begin{bmatrix} 0.2 & 0.8 \\ 0.5 & 0.5 \end{bmatrix}$

24. $D = \begin{bmatrix} 0.8 & 0.2 \end{bmatrix}$; $P = \begin{bmatrix} 0.9 & 0.1 \\ 0.2 & 0.8 \end{bmatrix}$

25. $D = \begin{bmatrix} 0.2 & 0.4 & 0.4 \end{bmatrix}$; $P = \begin{bmatrix} 0.7 & 0.1 & 0.2 \\ 0.3 & 0.3 & 0.4 \\ 0.4 & 0.5 & 0.1 \end{bmatrix}$

26. $D = \begin{bmatrix} 0.1 & 0.1 & 0.8 \end{bmatrix}$; $P = \begin{bmatrix} 0.2 & 0.1 & 0.7 \\ 0.1 & 0.1 & 0.8 \\ 0.5 & 0.1 & 0.4 \end{bmatrix}$

Decide whether each transition matrix is regular.

27. $\begin{bmatrix} 0 & 1 \\ 0.2 & 0.8 \end{bmatrix}$

28. $D = \begin{bmatrix} \frac{1}{3} & \frac{2}{3} \\ 1 & 0 \end{bmatrix}$

29. $\begin{bmatrix} 0.4 & 0.2 & 0.4 \\ 0 & 1 & 0 \\ 0.6 & 0.3 & 0.1 \end{bmatrix}$

30. $\begin{bmatrix} 1 & 0 & 0 \\ 0 & 1 & 0 \\ 0.3 & 0.5 & 0.2 \end{bmatrix}$

31. How can you tell from the transition matrix whether a Markov chain is regular or not?

32. How can you tell from the transition matrix whether a Markov chain is absorbing or not?

33. How can you tell from the transition matrix where the absorbing states are in an absorbing chain?

34. Can a Markov chain be both regular and absorbing? Explain.

Find all absorbing states for each matrix. Which are transition matrices for an absorbing Markov chain?

35. $\begin{bmatrix} 0 & 1 & 0 \\ 0.5 & 0.1 & 0.4 \\ 0 & 0 & 1 \end{bmatrix}$

36. $\begin{bmatrix} 0.2 & 0 & 0.8 \\ 0 & 1 & 0 \\ 0.7 & 0 & 0.3 \end{bmatrix}$

37. $\begin{bmatrix} 0.5 & 0.1 & 0.1 & 0.3 \\ 0 & 0 & 1 & 0 \\ 1 & 0 & 0 & 0 \\ 0.1 & 0.8 & 0.05 & 0.05 \end{bmatrix}$

38. $\begin{bmatrix} 0.2 & 0.3 & 0.4 & 0.1 \\ 0 & 1 & 0 & 0 \\ 0 & 0 & 0.2 & 0.8 \\ 0.3 & 0 & 0.6 & 0.1 \end{bmatrix}$

Find the fundamental matrix F for the absorbing Markov chains with matrices as follows. Also find the matrix FR.

39. $\begin{bmatrix} 0.2 & 0.45 & 0.35 \\ 0 & 1 & 0 \\ 0 & 0 & 1 \end{bmatrix}$

40. $\begin{bmatrix} 1 & 0 & 0 \\ 0 & 1 & 0 \\ 0.25 & 0.15 & 0.6 \end{bmatrix}$

41. $\begin{bmatrix} \frac{1}{5} & \frac{1}{5} & \frac{2}{5} & \frac{1}{5} \\ 0 & 1 & 0 & 0 \\ \frac{1}{2} & \frac{1}{4} & \frac{1}{8} & \frac{1}{8} \\ 0 & 0 & 0 & 1 \end{bmatrix}$

42. $\begin{bmatrix} 0.3 & 0.5 & 0.1 & 0.1 \\ 0.4 & 0.1 & 0.3 & 0.2 \\ 0 & 0 & 1 & 0 \\ 0 & 0 & 0 & 1 \end{bmatrix}$

APPLICATIONS

Business and Economics

Advertising Currently, 35% of all hot dogs sold in one area are made by Dogkins and 65% are made by Long Dog. Suppose that Dogkins starts a heavy advertising campaign, with the campaign producing the following transition matrix.

		After Campaign	
		Dogkins	Long Dog
Before	Dogkins	0.8	0.2
Campaign	Long Dog	0.45	0.55

43. Find the share of the market for each company

a. after one campaign;

b. after three such campaigns.

44. Predict the long-range market share for Dogkins.

Credit Cards A credit card company classifies its customers in a given month in three groups; nonusers, light users, and heavy users. The transition matrix for these states is

	Nonuser	Light	Heavy
Nonuser	0.8	0.15	0.05
Light	0.25	0.55	0.2
Heavy	0.04	0.21	0.75

Suppose the initial distribution for the three states is [0.4 0.4 0.2]. Find the distribution after each of the following periods.

45. 1 month **46.** 2 months **47.** 3 months

48. What is the long-range prediction for the distribution of users?

Life Sciences

49. Medical Prognosis A study of patients at the University of North Carolina Hospitals used a Markov chain model with three categories of patients: 0 (death), 1 (unfavorable status), and 2 (favorable status). The transition matrix for a cycle of 72 hours was as follows. *Source: Journal of Applied Statistics.*

$$\begin{array}{c} \\ 0 \\ 1 \\ 2 \end{array}\begin{array}{ccc} 0 & 1 & 2 \\ \begin{bmatrix} 1 & 0 & 0 \\ 0.085 & 0.779 & 0.136 \\ 0.017 & 0.017 & 0.966 \end{bmatrix} \end{array}$$

a. Find the fundamental matrix.

b. For a patient with a favorable status, find the expected number of cycles that the patient will continue to have that status before dying.

c. For a patient with an unfavorable status, find the expected number of cycles that the patient will have a favorable status before dying.

Medical Research A medical researcher is studying the risk of heart attack in men. She first divides men into three weight categories: thin, normal, and overweight. By studying the male ancestors, sons, and grandsons of these men, the researcher comes up with the following transition matrix.

	Thin	Normal	Overweight
Thin	0.3	0.5	0.2
Normal	0.2	0.6	0.2
Overweight	0.1	0.5	0.4

Find the probabilities of the following for a man of normal weight.

50. Thin son

51. Thin grandson

52. Thin great-grandson

Find the probabilities of the following for an overweight man.

53. Overweight son

54. Overweight grandson

55. Overweight great-grandson

Suppose that the distribution of men by weight is initially given by [0.2 0.55 0.25]. Find the following distributions.

56. After 1 generation

57. After 2 generations

58. After 3 generations

59. Find the long-range prediction for the distribution of weights.

Genetics Researchers sometimes study the problem of mating the offspring from the same two parents; two of these offspring are then mated and so on. Let A be a dominant gene for some trait, and a the recessive gene. The original offspring can carry genes AA, Aa, or aa. There are six possible ways that these offspring can mate.

State	Mating
1	AA and AA
2	AA and Aa
3	AA and aa
4	Aa and Aa
5	Aa and aa
6	aa and aa

60. Suppose that the offspring are randomly mated with each other. Verify that the transition matrix is given by the matrix below.

$$\begin{array}{c} \\ 1 \\ 2 \\ 3 \\ 4 \\ 5 \\ 6 \end{array}\begin{array}{cccccc} 1 & 2 & 3 & 4 & 5 & 6 \\ \begin{bmatrix} 1 & 0 & 0 & 0 & 0 & 0 \\ \frac{1}{4} & \frac{1}{2} & 0 & \frac{1}{4} & 0 & 0 \\ 0 & 0 & 0 & 1 & 0 & 0 \\ \frac{1}{16} & \frac{1}{4} & \frac{1}{8} & \frac{1}{4} & \frac{1}{4} & \frac{1}{16} \\ 0 & 0 & 0 & \frac{1}{4} & \frac{1}{2} & \frac{1}{4} \\ 0 & 0 & 0 & 0 & 0 & 1 \end{bmatrix} \end{array}$$

61. Identify the absorbing states.

62. Find matrix Q.

63. Find F and the product FR.

64. If two parents with the genes Aa are mated, find the number of pairs of offspring with these genes that can be expected before either the dominant or the recessive gene no longer appears.

65. If two parents with the genes Aa are mated, find the probability that the recessive gene will eventually disappear.

66. Genetics Suppose that a set of n genes includes m mutant genes. For the next generation, these genes are duplicated and a subset of n genes is selected from the $2n$ genes containing $2m$ mutant genes. Let the state be given by the number of mutant genes.

a. Verify that the transition probability from state i to state j is given by

$$p_{ij} = \frac{C(2i, j)\, C(2n - 2i, n - j)}{C(2n, n)}$$

where i and j are the number of mutant genes in this generation and the next, and where $C(n, r)$ represents the number of combinations of n objects taken r at a time, discussed in Chapter 8. (*Hint:* Let $C(n, r) = 0$ when $n < r$. If the current generation has i mutant genes, it has $n - i$ nonmutant genes. When the genes are duplicated, this results in $2i$ mutant genes and $2(n - i)$ nonmutant genes.)

b. What are the absorbing states in this chain?

c. Calculate the transition matrix for the case $n = 3$.

d. Find the fundamental matrix F and the product FR for $n = 3$.

e. If a set of 3 genes has 1 mutant gene, what is the probability that the mutant gene will eventually disappear?

f. If a set of 3 genes has 1 mutant gene, how many generations would be expected to have 1 mutant gene before either the mutant genes or the nonmutant genes disappear?

67. Class Mobility The following chart gives the probability that the child of a father in the bottom, second, middle, fourth, and top fifth of the U.S. population by income will end up in each of the five groups. *Source: The Institute for the Study of Labor.*

	Bottom	Second	Middle	Fourth	Top
Bottom	0.42	0.25	0.15	0.10	0.08
Second	0.19	0.28	0.21	0.18	0.14
Middle	0.19	0.19	0.26	0.20	0.16
Fourth	0.13	0.18	0.20	0.25	0.24
Top	0.10	0.12	0.19	0.23	0.36

a. Find the probability that the grandson of someone in the bottom income group is in the top income group.

b. Find the probability that the grandson of someone in the top income group is in the top income group.

c. Find the long-range probability that a male is in each income group if the trends given by the transition matrix continue.

d. Describe what the answers in part c tell you about the long-range distribution of income in the United States.

General Interest

68. Monopoly In an article on a Markov chain analysis of the game Monopoly, a simplified version of the game is presented. The board consists of the four squares shown. *Source: The College Mathematics Journal.*

POLICEMAN	COMMUNITY CHEST
JAIL	GO

Players move clockwise by flipping a coin and moving once for heads and twice for tails. If you land on the Policeman, you go directly to jail. Therefore, landing on the Policeman is equivalent to landing on Jail, so we need only consider three states in our Markov chain. If you land on the Community Chest, you pick a card that might give or take away money or might send you to Jail or to Go. The result is the following transition matrix.

$$\begin{array}{c c} & \begin{array}{ccc} \text{Jail} & \text{CC} & \text{Go} \end{array} \\ \begin{array}{c} \text{Jail} \\ \text{CC} \\ \text{Go} \end{array} & \begin{bmatrix} \frac{17}{32} & \frac{7}{16} & \frac{1}{32} \\ \frac{1}{2} & 0 & \frac{1}{2} \\ 1 & 0 & 0 \end{bmatrix} \end{array}$$

a. Explain rows 2 and 3 of the transition matrix.

b. There are 16 cards for the Community Chest. One sends you to Jail, one sends you to Go, and the others leave you on Community Chest. Use these facts to explain row 1 of the transition matrix.

c. Find the long-term probabilities for being on Jail, the Community Chest, and Go.

EXTENDED APPLICATION

A MARKOV CHAIN MODEL FOR TEACHER RETENTION

In an article published in the *Review of Public Personnel Administration*, Michael Reid and Raymond Taylor used a Markov chain model to describe the employment patterns for public school teachers in a New England school district. They identified 8 possible states for teachers in the system: newly employed, continuing from the previous year, on leave without pay, on sabbatical, ill for at least 30 days during the year, resigned, retired, and deceased. Each teacher could transition to 1 of these 8 states in the following year, and the researchers recorded the transition for each teacher from year 1 of the study to year 2. The researchers also noted how each teacher in the system in year 2 transitioned to year 3. This gave them two sets of transition frequencies, which they combined in order to estimate the transition probabilities from each of the 8 possible states to each other state. The resulting transition matrix, arranged with the absorbing states in the upper-left portion of the array, is given on the following page. *Source: Review of Public Personnel Administration.*

Note that there are 3 irreversible transitions: resigning, retiring, and dying, so the first 3 states listed are absorbing. Once you have entered the "resigned" state, you stay there forever (at least as far as the school system is concerned). The fourth column consists of

Teacher Transition								
	Resigned	Retired	Deceased	New	Continuing	On Leave	On Sabbatical	Ill
Resigned	1	0	0	0	0	0	0	0
Retired	0	1	0	0	0	0	0	0
Deceased	0	0	1	0	0	0	0	0
New	0.194	0	0	0	0.777	0	0	0.033
Continuing	0.040	0.016	0.002	0	0.896	0.022	0.007	0.017
On Leave	0.533	0	0	0	0.178	0.289	0	0
On Sabbatical	0	0	0	0	1	0	0	0
Ill	0	0.139	0	0	0.806	0	0	0.055

zeros, since no one already in the system can transition into the state of being a new teacher. In the exercises you will look at some of the other transition probabilities, including some of the forbidden transitions, such as from being on sabbatical to being on sabbatical again the next year.

Following the procedure outlined in Section 10.3, we can compute the fundamental matrix F, which looks like this:

$$F = \begin{bmatrix} 1 & 10.405 & 0.322 & 0.073 & 0.222 \\ 0 & 12.988 & 0.402 & 0.091 & 0.234 \\ 0 & 3.252 & 1.507 & 0.023 & 0.058 \\ 0 & 12.988 & 0.402 & 1.091 & 0.234 \\ 0 & 11.078 & 0.343 & 0.078 & 1.257 \end{bmatrix}$$

Recall that the rows and columns here represent the nonabsorbing states, in the same order in which they appear in the transition matrix. According to the properties of the fundamental matrix listed in Section 10.3, the entry in row A and column B represents the number of years during which a teacher currently in state A will be in state B before he or she exits the system into one of the absorbing states. For example, the 1 at the upper left indicates that a teacher who is new this year will spend exactly 1 year as a new teacher, which makes sense because you can only be new once. The entry 3.252 indicates that a teacher currently on leave without pay will, on the average, spend only about 3 years in the system, including the current year.

In the transition matrix, the matrix R sits under the identity matrix corresponding to the absorbing states. For the teacher transition matrix, R looks like this:

$$R = \begin{bmatrix} 0.194 & 0 & 0 \\ 0.04 & 0.016 & 0.002 \\ 0.533 & 0 & 0 \\ 0 & 0 & 0 \\ 0 & 0.139 & 0 \end{bmatrix}$$

In the exercises you will compute the product FR, which gives information about the distribution of absorbing states for each possible nonabsorbing initial state.

EXERCISES

1. If a teacher is currently ill, what is the probability that he or she will retire during the following year? What is the probability that he or she will be ill again in the following year?

2. The entry in row 2 and column 2 of F is 12.988. What information does this give about teachers who are currently actively teaching?

3. Row 2 of F is nearly the same as row 1. Is this an accident? Can you explain the one difference?

4. What does the entry 0.091 in row 2 of the fundamental matrix tell you?

5. Compute the matrix FR and answer the following questions:

   a. What is the probability that a teacher now on leave without pay will eventually resign? What is the probability that a teacher now on leave will resign the *following year*?

   b. Who is more likely to leave the system by retirement, a new teacher or a teacher who is currently teaching but has at least 1 year in the system?

   c. What is the probability that a new teacher will die on the job?

6. The study reported here collected two sets of transition data (year 1 to year 2 and year 2 to year 3). Why was this a good idea?

7. On the WolframAlpha.com website, the command $\{\{2,-1\},\{1,-3\}\}.\{\{1,2\},\{0,1\}\}$ gives the product of the matrices $\begin{bmatrix} 2 & -1 \\ 1 & -3 \end{bmatrix}$ and $\begin{bmatrix} 1 & 2 \\ 0 & 1 \end{bmatrix}$. Use this website and a similar command to determine FR in Exercise 5.

DIRECTIONS FOR GROUP PROJECT

Suppose your brother is president of a local school board and he has expressed concern about teacher retention at your former high school. Given that you are a college student and would like to make some money, you mention that you and your three friends are familiar with this problem and could lend some help to the school for a fee of $5000. His reply is that the School Board would have to agree to this. He invites you and your friends to the next School Board meeting to convince them that you can help them analyze their teacher retention issues. Prepare a presentation for the Board that describes the process of analyzing teacher retention, including the mathematics you will use for the analysis. Be sure to use Exercises 1–6 in making your presentation. Presentation software, such as Microsoft PowerPoint, should be used to present your case to the Board.

Game Theory

Game theory provides a framework for making decisions with incomplete information. For example, a hospital administrator must choose nursing staffing levels even though demand is not predictable. An exercise in Section 3 models this situation as a game by assigning payoffs to the possible degrees of underutilization or overcrowding and deriving an optimal strategy.

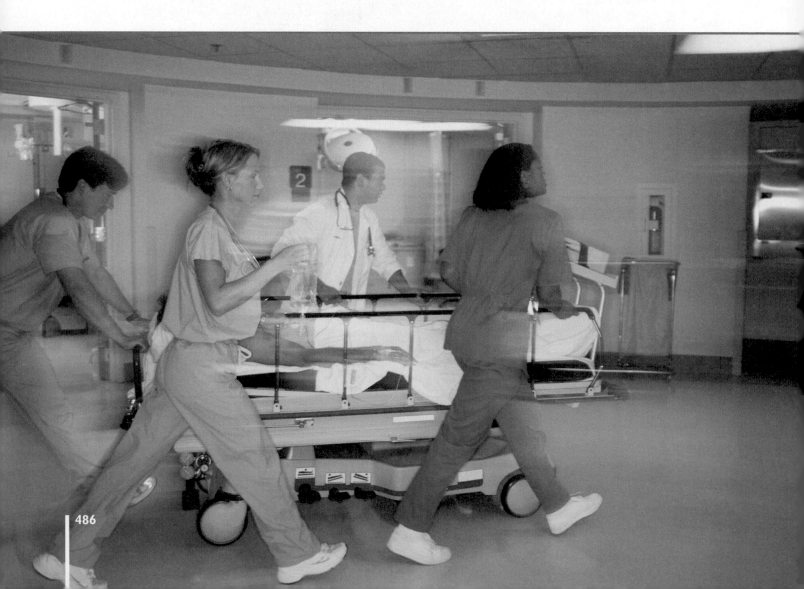

J ohn F. Kennedy once remarked that he had assumed that as president he would find it difficult to choose between distinct, opposite alternatives when a decision needed to be made. Actually, he found such decisions were easy to make; the hard decisions came when he was faced with choices that were not as clear-cut. Most decisions that we must make fall in the second category—decisions that must be made under conditions of uncertainty. *Game theory* is a branch of mathematics that provides a systematic way to attack problems of decision making when some alternatives are unclear or ambiguous.

11.1 Strictly Determined Games

APPLY IT

How can a football team decide which play to run when the play might have varying degrees of success, depending on what the opposing team does?
We will answer this question in Exercise 38.

A small manufacturer of Christmas cards must decide in February what type of cards to emphasize in her fall line. She has three possible strategies: emphasize modern cards, emphasize old-fashioned cards, or emphasize a mixture of the two. Her success is dependent on the state of the economy in December. If the economy is strong, she will do well with her modern cards, while in a weak economy people long for the old days and buy old-fashioned cards. In an in-between economy, her mixture of lines would do the best.

What should the manufacturer do? She should begin by carefully defining the problem. First, she must decide on the **states of nature**, the alternatives over which she has no control. Here, there are three: a weak economy, an in-between economy, or a strong economy. Next, she should list her **strategies**: emphasize modern cards, old-fashioned cards, or a mixture of the two. The consequences of each strategy under each state of nature are called **payoffs**. They can be summarized in a **payoff matrix**, as shown below. The numbers in the matrix represent her profits in thousands of dollars.

		States of Nature		
		Weak Economy	In-Between	Strong Economy
	Modern	40	85	120
Strategies	Old-Fashioned	90	45	85
	Mixture	75	105	65

(a) If the manufacturer is an optimist, she should aim for the biggest number on the matrix, 120 (representing $120,000 in profit). Her strategy in this case would be to produce modern cards.

(b) A pessimistic manufacturer wants to avoid the worst of all bad things that can happen. If she produces modern cards, the worst that can happen is a profit of $40,000. For old-fashioned cards, the worst is a profit of $45,000, while the worst that can happen from a mixture is a profit of $65,000. Her strategy here is to use a mixture.

(c) Suppose the manufacturer reads in a business magazine that leading experts think there is a 50% chance of a weak economy at Christmas, a 20% chance of an in-between

FOR REVIEW

Recall from Section 8.5 that if a random variable x can take on the n values $x_1, x_2, x_3, \ldots, x_n$, with probabilities $p_1, p_2, p_3, \ldots, p_n$, then the expected value of the random variable is

$$E(x) = x_1 p_1 + x_2 p_2 + x_3 p_3 + \cdots + x_n p_n.$$

In the modern Christmas card example, $n = 3$, $x_1 = 40$, $x_2 = 85$, $x_3 = 120$, $p_1 = 0.50$, $p_2 = 0.20$, and $p_3 = 0.30$.

economy, and a 30% chance of a strong economy. The manufacturer can now find her expected profit for each possible strategy.

Modern:	$40(0.50) + 85(0.20) + 120(0.30) = 73$	
Old-Fashioned:	$90(0.50) + 45(0.20) + 85(0.30) = 79.5$	
Mixture:	$75(0.50) + 105(0.20) + 65(0.30) = 78$	

Here the best strategy is old-fashioned cards; the expected profit is 79.5, or $79,500.

When a decision is made again and again, the meaning of the term *strategy* is more general: a strategy is then a rule for determining which choice is made each time the game is played. One possible strategy for the Christmas card manufacturer is to always choose modern cards; another is to rotate, using modern cards one year, old-fashioned the next, a mixture the third year, and so on; a third strategy is to make a random choice each year. If a decision is only made once, however, this meaning of the term *strategy* reduces to that of making a choice. In Section 11.2 we will see how more complex strategies can be selected.

The word *game* in the title of this chapter may have led you to think of chess, or perhaps some card game. While *game theory* does have some application to these recreational games, it was developed in the 1940s to analyze competitive situations in business, warfare, and social situations. Game theory was invented by John von Neumann (1903–1957), who, with Oskar Morgenstern, wrote the book entitled *Theory of Games and Economic Behavior* in 1944. **Game theory** is the study of how to make decisions when competing with an aggressive opponent.

Game theory is still an active area of research. In 1994, John F. Nash Jr. was a cowinner of the Nobel Prize in Economics for his work in game theory decades earlier, an event featured in the film *A Beautiful Mind*. In 2005, Robert J. Aumann and Thomas C. Schelling shared the same award for their more recent work in game theory.

A game can be set up with a payoff matrix, such as the one shown below. This game involves the two players A and B, and is called a **two-person game**. Player A can choose either row 1 or row 2, while player B can choose either column 1 or column 2. A player's choice is called a strategy, just as before. The payoff is at the intersection of the row and column selected. As a general agreement, a positive number represents a payoff from B to A; a negative number represents a payoff from A to B. For example, if A chooses row 2 and B chooses column 2, then B pays $4 to A.

$$\begin{array}{cc} & \begin{array}{cc} & B \end{array} \\ & \begin{array}{cc} 1 & 2 \end{array} \\ A \begin{array}{c} 1 \\ 2 \end{array} & \begin{bmatrix} 2 & -1 \\ -3 & 4 \end{bmatrix} \end{array}$$

EXAMPLE 1 Payoff Matrix

YOUR TURN 1 In Example 1, suppose A chooses row 1 and B chooses column 1. Who gets what?

In the payoff matrix just shown, suppose A chooses row 1 and B chooses column 2. Who gets what?

SOLUTION Row 1 and column 2 lead to the number -1. This number represents a payoff of $1 from A to B. **TRY YOUR TURN 1**

While the numbers in this payoff matrix represent money, they could just as easily represent goods or other property.

In the game above, no money enters the game from the outside; whenever one player wins, the other loses. Such a game is called a **zero-sum game**. The stock market is not a zero-sum game. Stocks can go up or down as a result of outside forces. Therefore, it is possible that all investors can make or lose money.

Only two-person zero-sum games are discussed in the rest of this chapter. Each player can have many different options. In particular, an $m \times n$ matrix game is one in which player A has m strategies (rows) and player B has n strategies (columns). We will always use rows for player A and columns for player B, so labels are not necessary on the matrix.

Dominated Strategies

In the rest of this section, the best possible strategy for each player is determined. Let us begin with the 3×3 game defined by the following matrix.

$$\begin{array}{c c c c} & 1 & 2 & 3 \\ 1 & \begin{bmatrix} -3 & -6 & 10 \\ 2 & 3 & 0 & -9 \\ 3 & 5 & -4 & -8 \end{bmatrix} \end{array}$$

From B's viewpoint, strategy 2 is better than strategy 1 no matter which strategy A selects. This can be seen by comparing the two columns. If A chooses row 1, receiving \$6 from A is better than receiving \$3; in row 2, breaking even is better than paying \$3; and in row 3, getting \$4 from A is better than paying \$5. Therefore, B should never select strategy 1. Strategy 2 is said to *dominate* strategy 1, and strategy 1—the **dominated strategy**—can be removed from consideration, producing the following reduced matrix.

$$\begin{array}{c c c} & 2 & 3 \\ 1 & \begin{bmatrix} -6 & 10 \\ 2 & 0 & -9 \\ 3 & -4 & -8 \end{bmatrix} \end{array}$$

Either player may have dominated strategies. In fact, after a dominated strategy for one player is removed, the other player may then have a dominated strategy where there was none before.

Dominated Strategies

A row for A **dominates** another row if every entry in the one row is *larger* than the corresponding entry in the other row. For a column of B to dominate another, each entry must be *smaller*.

In the 3×2 matrix above, neither player now has a dominated strategy. From A's viewpoint, strategy 1 is best if B chooses strategy 3, while strategy 2 is best if B chooses strategy 2. Verify that there are no dominated strategies for either player.

To find any dominated strategy, start with the first two rows. If the first row has an entry larger than the corresponding entry in the second row and another entry smaller than the corresponding entry in the second row, then neither row dominates. If this is not the case, then one row is dominated and may be removed. Continue this process for every pair of rows and every pair of columns.

EXAMPLE 2 Dominated Strategies

Find any dominated strategies in the games with the given payoff matrices.

(a)
$$\begin{array}{c c c c c} & 1 & 2 & 3 & 4 \\ 1 & \begin{bmatrix} -8 & -4 & -6 & -9 \\ 2 & -3 & 0 & -9 & 12 \end{bmatrix} \end{array}$$

SOLUTION Here every entry in column 3 is smaller than the corresponding entry in column 2. Thus, column 3 dominates column 2. (Notice that column 1 also dominates column 2.) By removing the dominated column 2, the final game is as follows.

$$\begin{array}{c c c c} & 1 & 3 & 4 \\ 1 & \begin{bmatrix} -8 & -6 & -9 \\ 2 & -3 & -9 & 12 \end{bmatrix} \end{array}$$

(b) $\begin{array}{c} \\ 1 \\ 2 \\ 3 \end{array} \begin{array}{cc} 1 & 2 \\ \left[\begin{array}{cc} 3 & -2 \\ 0 & 8 \\ 6 & 4 \end{array}\right] \end{array}$

YOUR TURN 2 Find any dominated strategies in the game with the given payoff matrix. Then write the payoff matrix with the dominated strategies removed.

$$\left[\begin{array}{ccc} 4 & -2 & -3 \\ 0 & -3 & 4 \\ -2 & -4 & -5 \end{array}\right]$$

SOLUTION Each entry in row 3 is greater than the corresponding entry in row 1, so that row 3 dominates row 1. Removing row 1 gives the following game.

$$\begin{array}{c} \\ 2 \\ 3 \end{array} \begin{array}{cc} 1 & 2 \\ \left[\begin{array}{cc} 0 & 8 \\ 6 & 4 \end{array}\right] \end{array}$$ **TRY YOUR TURN 2**

Strictly Determined Games
Which strategies should the players choose in the following game?

$$\begin{array}{c} \\ \\ A \begin{array}{c} 1 \\ 2 \\ 3 \end{array} \end{array} \begin{array}{ccc} & B & \\ 1 & 2 & 3 \\ \left[\begin{array}{ccc} -9 & -11 & 7 \\ 2 & 3 & 5 \\ -1 & 6 & -3 \end{array}\right] \end{array}$$

The goal of game theory is to find **optimum strategies**: those that are the most profitable to the respective players. The payoff that results from each player's choosing the optimum strategy is called the **value of the game**.

The simplest strategy for a player is to consistently choose a certain row (or column). Such a strategy is called a **pure strategy**, in contrast to strategies requiring the random choice of a row (or column); these alternative strategies are discussed in the next section.*

To choose a pure strategy in the game above, player A could choose row 1, in hopes of getting the payoff of $7. Player B would quickly discover this, however, and start playing column 2. By playing column 2, B would receive $11 from A. If A were to choose row 2 consistently, then B would again minimize outgo by choosing column 1 (a payoff of $2 by B to A is better than paying $3 or $5, respectively, to A). By choosing row 3 consistently, A would cause B to choose column 3. The table shows what B will do when A chooses a given row consistently.

Strategies		
If A Chooses Pure Strategy:	**Then B Will Choose:**	**With Payoff:**
Row 1	Column 2	$11 to B
Row 2	Column 1	$2 to A
Row 3	Column 3	$3 to B

Based on these results, A's optimum strategy is to choose row 2; in this way A will guarantee a minimum payoff of $2 per play of the game, no matter what B does.

The optimum pure strategy in this game for A (the *row* player) is found by identifying the *smallest* number in each row of the payoff matrix; the row giving the *largest* such number gives the optimum strategy.

By going through a similar analysis for player B, we find that B should choose the column that will minimize the amount A can win. In the game above, B will pay $2 to A if B consistently chooses column 1. By choosing column 2 consistently, B will pay $6 to A, and by choosing column 3, player B will pay $7 to A. The optimum strategy for B is thus to choose column 1—with each play of the game B will pay $2 to A.

*In this section we solve (find the optimum strategies for) only games that have optimum *pure* strategies.

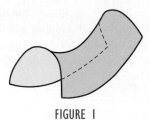

FIGURE 1

The optimum pure strategy in this game for B (the column player) is to identify the *largest* number in each column of the payoff matrix, and then choose the column producing the *smallest* such number.

In the game above, the entry 2 is both the *smallest* entry in its *row* and the *largest* entry in its *column.* Such an entry is called a *saddle point.* (The seat of a saddle is the maximum from one direction and the minimum from another direction. See Figure 1.) As Example 3(b) will show, there may be more than one such entry, but then the entries will have the same value.

> **Saddle Point**
>
> A **saddle point** is the smallest entry in its row and the largest entry in its column. In a game with a saddle point, the optimum pure strategy for player A is to choose the row containing the saddle point, while the optimum pure strategy for B is to choose the column containing the saddle point.

A game with a saddle point is called a **strictly determined game**. By using these optimum strategies, A and B will ensure that the same amount always changes hands with each play of the game; this amount, given by the saddle point, is the value of the game. The value of the game above is $2. A game having a value of 0 is a **fair game**; the game above is not fair. It is also not much fun for player B, who will lose $2 to player A every time. Finally, it is not very interesting to play the game, because players A and B will choose strategies 2 and 1, respectively, every time. Games in which there is no optimum pure strategy are more interesting to play again and again; we will study these in the next section.

To find the saddle point easily, underline the smallest number in each row, and circle the largest number in each column. If there is a number that is both underlined and circled, it is the saddle point. In case of a tie, underline or circle all numbers that equal the smallest number in the row or the largest number in the column.

EXAMPLE 3 **Saddle Points**

Find the saddle points in the following games. When a saddle point exists, find the strategy producing it and the value of the game, and determine whether the game is a fair game.

$$
\text{(a)} \quad
\begin{array}{c}
 & \begin{array}{cc} 1 & 2 \end{array} \\
\begin{array}{c} 1 \\ 2 \\ 3 \\ 4 \end{array} &
\left[\begin{array}{cc}
2 & 2 \\
0 & 4 \\
1 & 6 \\
3 & 7
\end{array}\right]
\end{array}
$$

SOLUTION

Method 1
Matrix

Underlining the smallest number in each row and circling the largest value in each column yields the following:

$$
\begin{array}{c}
 & \begin{array}{cc} 1 & 2 \end{array} \\
\begin{array}{c} 1 \\ 2 \\ 3 \\ 4 \end{array} &
\left[\begin{array}{cc}
\underline{2} & \underline{2} \\
\underline{0} & 4 \\
\underline{1} & 6 \\
\underline{③} & ⑦
\end{array}\right]
\end{array}.
$$

The number that is both circled and underlined is 3. Thus, 3 is the saddle point, and the game has a value of 3. Since the value is not 0, the game is not fair. The strategies producing the saddle point can be written $(4, 1)$. (Player A's strategy is written first.)

Method 2
Number Line

We've observed that player A looks at the minimum in each row and then chooses the maximum of those values. Similarly, player B looks at the maximum in each column and then chooses the minimum. We can illustrate this idea by drawing two number lines, one for player A, in which we put the minimum for each row, and one for player B, in which we put the maximum for each column, as we have done in Figure 2 for the matrix in this example.

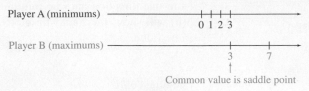

FIGURE 2

If the maximum in player A's number line equals the minimum in player B's number line, then that common value is the saddle point. So in this game, the saddle point is 3, which is in row 4 and column 1, so the strategy can be written as (4, 1).*

$$
\text{(b)} \quad
\begin{array}{c c}
 & \begin{array}{cccc} 1 & 2 & 3 & 4 \end{array} \\
\begin{array}{c} 1 \\ 2 \end{array} &
\begin{bmatrix} 4 & 6 & 4 & 12 \\ -8 & -9 & 3 & 2 \end{bmatrix}
\end{array}
$$

SOLUTION

Method 1
Matrix

Underlining the smallest number in each row and circling the largest value in each column yields the following:

$$
\begin{array}{c c}
 & \begin{array}{cccc} 1 & 2 & 3 & 4 \end{array} \\
\begin{array}{c} 1 \\ 2 \end{array} &
\begin{bmatrix} ④ & ⑥ & ④ & ⑫ \\ -8 & \underline{-9} & 3 & 2 \end{bmatrix}
\end{array}.
$$

The saddle point, 4, occurs with either of two strategies, $(1, 1)$ or $(1, 3)$. The value of the game is 4. The game is not a fair game because it does not have a value of 0.

Method 2
Number Line

In Figure 3, we have drawn a number line for player A containing the minimum in each row and a number line for player B showing the maximum in each row. Since 4 is the maximum in A's number line and the minimum in B's number line, 4 is a saddle point. The 4 occurs in row 1 and in columns 1 and 3, so either $(1, 1)$ or $(1, 3)$ is an optimum strategy.

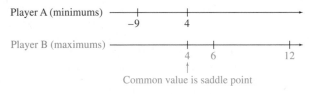

FIGURE 3

$$
\text{(c)} \quad
\begin{array}{c c}
 & \begin{array}{ccc} 1 & 2 & 3 \end{array} \\
\begin{array}{c} 1 \\ 2 \end{array} &
\begin{bmatrix} 3 & 6 & -2 \\ 8 & -3 & 5 \end{bmatrix}
\end{array}
$$

SOLUTION

Method 1
Matrix

Underlining the smallest number in each row and circling the largest value in each column yields the following:

$$
\begin{array}{c c}
 & \begin{array}{ccc} 1 & 2 & 3 \end{array} \\
\begin{array}{c} 1 \\ 2 \end{array} &
\begin{bmatrix} 3 & ⑥ & \underline{-2} \\ ⑧ & -3 & ⑤ \end{bmatrix}
\end{array}
$$

*This method is due to Professor Mary T. Treanor of Valparaiso University.

There is no number that is both the smallest number in its row and the largest number in its column, so the game has no saddle point. Since the game has no saddle point, it is not strictly determined. In the next section, methods are given for finding optimum strategies for such games.

Method 2
Number Line

In Figure 4, the largest number on player A's number line is −2, and the smallest on player B's is 5. Since these differ, there is no saddle point, and the game is not strictly determined.

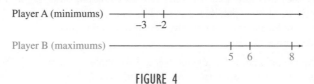

Player A (minimums)

Player B (maximums)

−3 −2

5 6 8

FIGURE 4

YOUR TURN 3 Find any saddle points in the following game. If a saddle point exists, find the strategy to produce it and the value of the game.

$$\begin{bmatrix} -2 & 1 & 0 \\ -3 & 2 & -5 \\ -1 & 0 & 2 \end{bmatrix}$$

TRY YOUR TURN 3

In Example 3(a), notice that row 4 dominates all other rows. Therefore, we may as well eliminate rows 1 through 3, resulting in the matrix $\begin{bmatrix} 3 & 7 \end{bmatrix}$. In this matrix, column 1 dominates column 2, so we may as well eliminate column 2, resulting in the 1×1 matrix $\begin{bmatrix} 3 \end{bmatrix}$, whose only entry is the saddle point. Successive elimination of dominated rows and columns in some cases gives another way to find a saddle point, if it exists.

11.1 EXERCISES

In the following game, decide on the payoff when the strategies of Exercises 1–6 are used.

B

		1	2	3
	1	7	−5	0
A	2	3	−3	8
	3	−1	5	11

1. $(1, 1)$ **2.** $(1, 2)$

3. $(2, 2)$ **4.** $(2, 3)$

5. $(3, 1)$ **6.** $(3, 2)$

7. Does the game have any dominated strategies?

8. Does it have a saddle point?

In Exercises 9–14, list which rows or columns dominate other rows or columns. (e.g., Row 2 dominates rows 1 and 3.) Then, remove any dominated strategies in the games. (From now on, we will save space by deleting the names of the strategies.)

9. $\begin{bmatrix} 0 & 9 & -3 \\ 3 & -8 & -2 \end{bmatrix}$ **10.** $\begin{bmatrix} 7 & 6 \\ -2 & -5 \\ 4 & 9 \end{bmatrix}$

11. $\begin{bmatrix} 3 & 6 \\ 1 & 4 \\ 4 & -2 \\ -4 & 0 \end{bmatrix}$ **12.** $\begin{bmatrix} 2 & -6 & 3 & 1 \\ -2 & 2 & 5 & 4 \\ 1 & -4 & 0 & 2 \end{bmatrix}$

13. $\begin{bmatrix} 8 & 12 & -7 \\ -2 & 1 & 4 \end{bmatrix}$ **14.** $\begin{bmatrix} 6 & 2 \\ -1 & 10 \\ 3 & 5 \end{bmatrix}$

For each game in Exercises 15–24, when the saddle point exists, find the strategies producing it and the value of the game. Identify any games that are strictly determined.

15. $\begin{bmatrix} -5 & 2 \\ 5 & 4 \end{bmatrix}$ **16.** $\begin{bmatrix} 8 & 6 \\ 12 & -4 \end{bmatrix}$

17. $\begin{bmatrix} 3 & -4 & 1 \\ 5 & 3 & -2 \end{bmatrix}$ **18.** $\begin{bmatrix} -4 & 2 & -3 & -7 \\ 4 & 3 & 5 & -9 \end{bmatrix}$

19. $\begin{bmatrix} -6 & 2 \\ -1 & -10 \\ 3 & 5 \end{bmatrix}$ **20.** $\begin{bmatrix} 1 & 4 & -3 & 1 & -1 \\ 2 & 5 & 0 & 4 & 10 \\ 1 & -3 & 2 & 5 & 2 \end{bmatrix}$

21. $\begin{bmatrix} 2 & 3 & 1 \\ -1 & 4 & -7 \\ 5 & 2 & 0 \\ 8 & -4 & -1 \end{bmatrix}$ **22.** $\begin{bmatrix} 3 & 8 & -4 & -9 \\ -1 & -2 & -3 & 0 \\ -2 & 6 & -4 & 5 \end{bmatrix}$

23. $\begin{bmatrix} -6 & 1 & 4 & 2 \\ 9 & 3 & -8 & -7 \end{bmatrix}$ **24.** $\begin{bmatrix} -3 & -2 & 6 \\ 2 & 0 & 2 \\ 5 & -2 & -4 \end{bmatrix}$

25. Consider the matrix of Exercise 21.

a. Remove any dominated columns, and write the resulting matrix.

b. Remove any dominated rows from the matrix in part a, and write the resulting matrix.

c. Remove any dominated columns from the matrix in part b, and write the resulting matrix.

d. Remove any dominated rows from the matrix in part c, and write the resulting matrix. Verify that the only entry left is the saddle point.

26. Repeat Exercise 25, starting with the matrix of Exercise 22.

27. Consider the matrix of Exercise 24.

    a. Remove any dominated columns, and write the resulting matrix.

    b. Remove any dominated rows, and write the resulting matrix.

    c. Compare the results of Exercises 25–27, and comment on whether it is always possible to use dominated rows or columns to reduce a payoff matrix to its optimum value in a strictly determined game.

28. Suppose the payoff matrix for a game has at least three rows. Also, suppose that row 1 dominates row 2 and that row 2 dominates row 3. Show that row 1 must dominate row 3.

APPLICATIONS

Business and Economics

29. Concert Preparations Hillsdale College has sold out all tickets for a jazz concert to be held in the stadium. If it rains, the show will have to be moved to the gym, which has a much smaller capacity. The dean must decide in advance whether to set up the seats and the stage in the gym or in the stadium, or both, just in case. The payoff matrix below shows the net profit in each case.

		States of Nature	
		Rain	No Rain
	Set Up in Stadium	$-\$1800$	$\$2400$
Strategies	Set Up in Gym	$\$1500$	$\$1500$
	Set Up in Both	$\$1200$	$\$2100$

What strategy should the dean choose if she is

a. an optimist?   **b.** a pessimist?

c. If the weather forecaster predicts rain with a probability of 0.6, what strategy should the dean choose to maximize expected profit? What is the maximum expected profit?

30. Machine Repairs An analyst must decide whether to recommend repairing a machine that is producing some defective items. He has already decided that there are three possibilities for the fraction of defective items: 0.01, 0.10, and 0.20. He may recommend two courses of action: repair the machine or make no repairs. The payoff matrix below represents the costs to the company in each case.

		States of Nature		
		0.01	0.10	0.20
	Repair	$-\$130$	$-\$130$	$-\$130$
Strategies	No Repair	$-\$25$	$-\$200$	$-\$500$

What strategy should the analyst recommend if he is

a. an optimist?   **b.** a pessimist?

c. Suppose the analyst is able to estimate probabilities for the three states of nature, as shown below.

Fraction of Defective Items	Probability
0.01	0.70
0.10	0.20
0.20	0.10

Which strategy should he recommend? Find the expected cost to the company if this strategy is chosen.

31. Marketing The research department of the Allied Manufacturing Company has developed a new process that it believes will result in an improved product. Management must decide whether to go ahead and market the new product. The new product may or may not be better than the old one. If the new product is better and the company decides to market it, sales should increase by $50,000. If it is not better and they replace the old product with the new product on the market, they will lose $25,000 to competitors. If they decide not to market the new product, they will lose a total of $40,000 if it is better and just research costs of $10,000 if it is not.

a. Prepare a payoff matrix.

b. If management believes there is a probability of 0.4 that the new product is better, find the expected profits under each strategy and determine the best action.

c. Find any dominated strategies. Is there a saddle point?

32. Machinery Overhaul Otis Taylor, a businessman, is planning to ship a used machine to his plant in Nigeria. He would like to use it there for the next 4 years. He must decide whether to overhaul the machine before sending it. The cost of overhaul is $2600. If the machine fails when in operation in Nigeria, it will cost him $6000 in lost production and repairs. He estimates the probability that it will fail at 0.55 if he does not overhaul it, and 0.1 if he does overhaul it. Neglect the possibility that the machine might fail more than once in the next 4 years.

a. Prepare a payoff matrix.

b. What should the businessman do to minimize his expected costs?

c. Find any dominated strategies. Is there a saddle point?

33. Competition Two merchants are planning to build competing stores to serve an area of three small cities. The fraction of the total population that live in each city is shown in the figure on the next page. If both merchants locate in the same city, merchant A will get 65% of the total business. If the merchants locate in different cities, each will get 80% of the business in the city it is in, and A will get 60% of the business from the city not containing B. Payoffs are measured by the number of percentage points above or below 50%. Write a payoff matrix for this game. Is this game strictly determined?

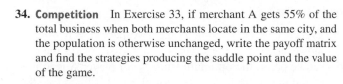

34. Competition In Exercise 33, if merchant A gets 55% of the total business when both merchants locate in the same city, and the population is otherwise unchanged, write the payoff matrix and find the strategies producing the saddle point and the value of the game.

Life Sciences

35. Global Warming The existence of global warming and appropriate governmental response to combat this phenomena are often debated. Greg Craven claims that it is not necessary to determine the existence of global warming, but instead it is pertinent to look at the possible costs for our action/inaction. The following payoff matrix represents the cost for taking action and for not taking any action, given global warming occurs or does not occur. Discuss how game theory might be used to approach the issue of global warming. Be sure to address the worst-case scenario and what solution would result from game theory. What are the strengths and problems with this approach? *Source: Greg Craven.*

		Global warming is . . .	
		True	**False**
Action taken to combat global warming?	**Yes**	High cost (but money well spent, since world is still livable)	High cost (money is wasted), may cause global economic recession
	No	Enormous cost (economic, political, social, environmental, and health catastrophes)	Everything's good

Social Sciences

36. Border Patrol A person attempting an illegal crossing of the border into the United States has two options. He can cross into unoccupied territory on foot at night, or he can cross at a regular entry point if he is hidden in a vehicle. His chances of detection depend on whether extra border guards are sent to the unoccupied territory or to the regular entry point. His chances of getting across the unoccupied territory are 50% if the extra guards are sent to the regular entry point and 40% if the extra guards are sent to the unoccupied territory. His chances of crossing at a regular entry point are 30% if the extra guards are sent to the unoccupied territory and 20% if the extra guards are sent to the regular entry point. Letting the payoffs be the probabilities of success, find the strategies producing the saddle point and the value of the game.

General Interest

37. War Games Two armies, A and B, are involved in a war game. Each army has available three different strategies, with payoffs as shown below. These payoffs represent square kilometers of land, with positive numbers representing gains by A.

$$\begin{bmatrix} -8 & -10 & 4 \\ 0 & -12 & 6 \\ 3 & -7 & 8 \end{bmatrix}$$

Find the strategies producing the saddle point and the value of the game.

38. APPLY IT Football When a football team has the ball and is planning its next play, it can choose one of several plays or strategies. The success of the chosen play depends largely on how well the other team "reads" the chosen play. Suppose a team with the ball (team A) can choose from three plays, while the opposition (team B) has four possible strategies. The numbers shown in the payoff matrix represent yards of gain to team A.

$$\begin{bmatrix} 9 & -3 & -4 & 16 \\ 12 & 9 & 6 & 8 \\ -5 & -2 & 3 & 18 \end{bmatrix}$$

Find the strategies producing the saddle point. Find the value of the game.

39. Children's Game In the children's game rock, paper, scissors, two players simultaneously extend one hand in the form of a fist (rock), a flat palm (paper), or two fingers extended (scissors). If they both make the same choice, the game is a tie. Otherwise, rock beats scissors (because rock can crush scissors), scissors beats paper (because scissors can cut paper), and paper beats rock (because paper can cover rock). Whoever wins gains one point; in the case of a tie, no one gains any points. Find the payoff matrix for this game. Is the game strictly determined?

40. Finger Game John and Joann play a finger matching game—each shows one or two fingers, simultaneously. If the sum of the number of fingers showing is even, Joann pays John that number of dollars; for an odd sum, John pays Joann. Find the payoff matrix for this game, making John the row player. Is the game strictly determined?

YOUR TURN ANSWERS

1. B pays $2 to A.
2. Rows 1 and 2 dominate row 3. Column 2 dominates column 1.
$$\begin{bmatrix} -2 & -3 \\ -3 & 4 \end{bmatrix}$$
3. Saddle point is -1 with strategy $(3,1)$. The value of the game is -1.

11.2 Mixed Strategies

How can a farmer decide whether or not to spray insecticide on his crops, given that neither strategy, when used consistently, is the best strategy? *We will answer this question in Example 3.*

As mentioned earlier, not every game has a saddle point. Two-person, zero-sum games still have optimum strategies, however, even if the strategy is not as simple as the ones we saw earlier. In a game with a saddle point, the optimum strategy for player A is to pick the row containing the saddle point and the optimum strategy for player B is to choose the column containing the saddle point. Such a strategy is called a *pure strategy,* since the same row and column are always chosen.

If there is no saddle point, then it will be necessary for both players to mix their strategies. For example, A will sometimes play row 1, sometimes row 2, and so on. If this were done in some specific pattern, the competitor would soon guess it and play accordingly.

For this reason, it is best to mix strategies according to previously determined probabilities. For example, if a player has only two strategies and has decided to play them with equal probability, the random choice could be made by tossing a fair coin, letting heads represent one strategy and tails the other. This would result in the two strategies being used about equally over the long run. On a particular play, however, it would not be possible to predetermine the strategy to be used. (Some other device, such as a spinner, is necessary if there are more than two strategies or if the probabilities are not $1/2$.)

EXAMPLE 1 **Expected Value**

Suppose a game has payoff matrix

$$\begin{bmatrix} -1 & 2 \\ 1 & 0 \end{bmatrix},$$

where the entries represent dollar winnings. Suppose player A chooses row 1 with probability $1/3$ and row 2 with probability $2/3$, and player B chooses each column with probability $1/2$. Find the expected value of the game.

SOLUTION Assume that rows and columns are chosen independently, so that

$$P(\text{row } 1, \text{column } 1) = P(\text{row } 1) \cdot P(\text{column } 1) = \frac{1}{3} \cdot \frac{1}{2} = \frac{1}{6}$$

$$P(\text{row } 1, \text{column } 2) = P(\text{row } 1) \cdot P(\text{column } 2) = \frac{1}{3} \cdot \frac{1}{2} = \frac{1}{6}$$

$$P(\text{row } 2, \text{column } 1) = P(\text{row } 2) \cdot P(\text{column } 1) = \frac{2}{3} \cdot \frac{1}{2} = \frac{1}{3}$$

$$P(\text{row } 2, \text{column } 2) = P(\text{row } 2) \cdot P(\text{column } 2) = \frac{2}{3} \cdot \frac{1}{2} = \frac{1}{3}.$$

The table below lists the probability of each possible outcome, along with the payoff to player A.

Probability of Payoffs for A		
Outcome	Probability of Outcome	Payoff for A
Row 1, column 1	1/6	−1
Row 1, column 2	1/6	2
Row 2, column 1	1/3	1
Row 2, column 2	1/3	0

The expected value of the game is given by the sum of the products of the probabilities and the payoffs, or

$$\text{Expected value} = \frac{1}{6}(-1) + \frac{1}{6}(2) + \frac{1}{3}(1) + \frac{1}{3}(0) = \frac{1}{2}.$$

In the long run, for a great many plays of the game, the payoff to A will average $1/2$ dollar per play of the game. It is important to note that as the mixed strategies used by A and B are changed, the expected value of the game may well change. (See Example 2.)

To generalize the work of Example 1, let the payoff matrix for a 2×2 game be

$$M = \begin{bmatrix} a_{11} & a_{12} \\ a_{21} & a_{22} \end{bmatrix}.$$

Let player A choose row 1 with probability p_1 and row 2 with probability p_2, where $p_1 + p_2 = 1$. Write these probabilities as the row matrix

$$A = \begin{bmatrix} p_1 & p_2 \end{bmatrix}.$$

Let player B choose column 1 with probability q_1 and column 2 with probability q_2, where $q_1 + q_2 = 1$. Write this as the column matrix

$$B = \begin{bmatrix} q_1 \\ q_2 \end{bmatrix}.$$

The probability of choosing row 1 and column 1 is

$$P(\text{row } 1, \text{column} 1) = P(\text{row } 1) \cdot P(\text{column } 1) = p_1 \cdot q_1.$$

In the same way, the probabilities of each possible outcome are shown in the table below, along with the payoff for each outcome.

Probability of Payoffs for A		
Outcome	Probability of Outcome	Payoff for A
Row 1, column 1	$p_1 \cdot q_1$	a_{11}
Row 1, column 2	$p_1 \cdot q_2$	a_{12}
Row 2, column 1	$p_2 \cdot q_1$	a_{21}
Row 2, column 2	$p_2 \cdot q_2$	a_{22}

The expected value for this game is

$$(p_1 \cdot q_1) \cdot a_{11} + (p_1 \cdot q_2) \cdot a_{12} + (p_2 \cdot q_1) \cdot a_{21} + (p_2 \cdot q_2) \cdot a_{22}.$$

This same result can be written as the matrix product

$$\text{Expected value} = \begin{bmatrix} p_1 & p_2 \end{bmatrix} \begin{bmatrix} a_{11} & a_{12} \\ a_{21} & a_{22} \end{bmatrix} \begin{bmatrix} q_1 \\ q_2 \end{bmatrix} = AMB.$$

The same method works for games larger than 2×2: Let the payoff matrix for a game have dimension $m \times n$; call this matrix $M = [a_{ij}]$. Let the mixed strategy for player A be given by the row matrix

$$A = \begin{bmatrix} p_1 & p_2 & p_3 & \cdots & p_m \end{bmatrix}$$

and the mixed strategy for player B be given by the column matrix

$$B = \begin{bmatrix} q_1 \\ q_2 \\ \vdots \\ q_n \end{bmatrix}.$$

The expected value for this game is the product

$$AMB = \begin{bmatrix} p_1 & p_2 & \cdots & p_m \end{bmatrix} \begin{bmatrix} a_{11} & a_{12} & \cdots & a_{1n} \\ a_{21} & a_{22} & \cdots & a_{2n} \\ \vdots & \vdots & & \vdots \\ a_{m1} & a_{m2} & \cdots & a_{mn} \end{bmatrix} \begin{bmatrix} q_1 \\ q_2 \\ \vdots \\ q_n \end{bmatrix}.$$

EXAMPLE 2 Expected Value

In the game in Example 1, having payoff matrix

$$M = \begin{bmatrix} -1 & 2 \\ 1 & 0 \end{bmatrix},$$

suppose player A chooses row 1 with probability 0.2, and player B chooses column 1 with probability 0.6. Find the expected value of the game.

SOLUTION If A chooses row 1 with probability 0.2, then row 2 is chosen with probability $1 - 0.2 = 0.8$, giving

$$A = \begin{bmatrix} 0.2 & 0.8 \end{bmatrix}.$$

In the same way,

$$B = \begin{bmatrix} 0.6 \\ 0.4 \end{bmatrix}.$$

The expected value of the game with these probability strategies is given by the product AMB, or

$$\begin{aligned} AMB &= \begin{bmatrix} 0.2 & 0.8 \end{bmatrix} \begin{bmatrix} -1 & 2 \\ 1 & 0 \end{bmatrix} \begin{bmatrix} 0.6 \\ 0.4 \end{bmatrix} \\ &= \begin{bmatrix} 0.6 & 0.4 \end{bmatrix} \begin{bmatrix} 0.6 \\ 0.4 \end{bmatrix} \\ &= \begin{bmatrix} 0.52 \end{bmatrix}. \end{aligned}$$

YOUR TURN 1 In Example 2, suppose player A chooses row 1 with probability 0.4, and player B chooses column 1 with probability 0.8. Find the expected value of the game.

On the average, these two strategies will produce a payoff of $0.52, or 52¢, for A for each play of the game. This payoff is slightly better than the 50¢ in Example 1.

TRY YOUR TURN 1

In Example 2, player B could reduce the payoff to A by changing strategy. (Check this by choosing different matrices for B.) For this reason, player A needs to develop an *optimum strategy*—a strategy that will produce the best possible payoff no matter what B does. Just as in the previous section, this is done by finding the largest of the smallest possible amounts that can be won.

To find values of p_1 and p_2 so that the probability vector $\begin{bmatrix} p_1 & p_2 \end{bmatrix}$ produces an optimum strategy, start with the payoff matrix

$$M = \begin{bmatrix} -1 & 2 \\ 1 & 0 \end{bmatrix}$$

and assume that A chooses row 1 with probability p_1. Since $p_1 + p_2 = 1$, we have $p_2 = 1 - p_1$, so player A's strategy can be written as $\begin{bmatrix} p_1 & 1 - p_1 \end{bmatrix}$. If player B chooses column 1, then player A's expectation is given by E_1, where

$$E_1 = -1 \cdot p_1 + 1 \cdot (1 - p_1) = 1 - 2p_1.$$

If B chooses column 2, then A's expectation is given by E_2, where

$$E_2 = 2 \cdot p_1 + 0 \cdot (1 - p_1) = 2p_1.$$

Draw graphs of $E_1 = 1 - 2p_1$ and $E_2 = 2p_1$; see Figure 5.

$E_1 = 1 - 2p_1$

$E_2 = 2p_1$

$\left(\dfrac{1}{4}, \dfrac{1}{2}\right)$

FIGURE 5

As mentioned previously, A needs to maximize the smallest amounts that can be won. On the graph, the smallest amounts that can be won are represented by the points of E_2 up to the intersection point. To the right of the intersection point, the smallest amounts that can be won are represented by the points of the line E_1. Player A can maximize the smallest amounts that can be won by choosing the point of intersection itself, the peak of the heavily shaded lines in Figure 5.

To find this point of intersection, find the simultaneous solution of the two equations. At the point of intersection, $E_1 = E_2$. Substitute $1 - 2p_1$ for E_1 and $2p_1$ for E_2.

$$E_1 = E_2$$
$$1 - 2p_1 = 2p_1$$
$$1 = 4p_1$$
$$\frac{1}{4} = p_1$$

By this result, player A should choose strategy 1 with probability 1/4 and strategy 2 with probability $1 - 1/4 = 3/4$. This will maximize A's expected winnings. To find the maximum winnings (which is also the value of the game), substitute 1/4 for p_1 in either E_1 or E_2. Choosing E_2 gives

$$E_2 = 2p_1 = 2\left(\frac{1}{4}\right) = \frac{1}{2};$$

that is, 1/2 dollar, or 50¢. Going through a similar argument for player B gives a different graph that includes the lines $E_1 = 2 - 3q_1$ and $E_2 = q_1$. (Verify this for yourself.) Yet the value also turns out to be 50¢, corresponding to a strategy in which player B chooses each column with probability 1/2. In Example 2, A's winnings were 52¢, but that was because B was not using his optimum strategy.

In the game above, player A can maximize expected winnings by playing row 1 with probability 1/4 and row 2 with probability 3/4. Such a strategy is called a **mixed strategy**. To actually decide which row to use on a given game, player A could use a spinner, such as the one in Figure 6, or use a random number generator on a calculator or a computer. A typical random number generator gives numbers between 0 and 1 with no apparent pattern or order. Player A could choose row 1 if the number generated were less than 0.25, and choose row 2 if the number is at least 0.25.

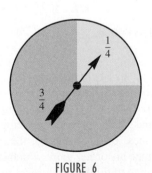

$\dfrac{1}{4}$

$\dfrac{3}{4}$

FIGURE 6

| CAUTION | The procedures developed in this section for finding optimum strategies do not apply if the game is strictly determined. In that case, use the method of Section 11.1 on Strictly Determined Games. |

EXAMPLE 3 Boll Weevils

Boll weevils threaten the cotton crop near Hattiesburg. Brian Williford owns a small farm; he can protect his crop by spraying with a potent (and expensive) insecticide. He can save money by not spraying, but he risks losing his crop to the boll weevils. What should his strategy be?

APPLY IT

SOLUTION Brian first sets up a payoff matrix. The numbers in the matrix represent his profits.

States of Nature

		Boll Weevil Attack	No Attack
Strategies	Spray	$14,000	$7000
	Don't Spray	−$3000	$8000

First, note that this game is not strictly determined. Thus, a mixed strategy will produce a better outcome for Brian. Let p_1 represent the probability that Brian chooses to spray, so that $1 - p_1$ is the probability that he chooses not to spray. If nature chooses a boll weevil attack, Brian's expected value is

$$E_1 = 14,000p_1 - 3000(1 - p_1)$$
$$= 14,000p_1 - 3000 + 3000p_1$$
$$E_1 = 17,000p_1 - 3000.$$

If nature chooses not to attack, Brian has an expected value of

$$E_2 = 7000p_1 + 8000(1 - p_1)$$
$$= 7000p_1 + 8000 - 8000p_1$$
$$E_2 = 8000 - 1000p_1.$$

As suggested by the work above, to maximize his expected profit, Brian should find the value of p_1 for which $E_1 = E_2$, as shown in Figure 7.

$$E_1 = E_2$$
$$17,000p_1 - 3000 = 8000 - 1000p_1$$
$$18,000p_1 = 11,000$$
$$p_1 = 11/18$$

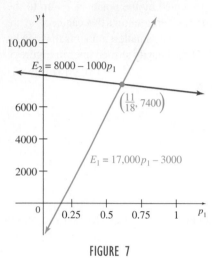

FIGURE 7

Thus, $p_2 = 1 - p_1 = 1 - 11/18 = 7/18$.

Brian will maximize his expected profit if he chooses to spray with probability $11/18$ and not spray with probability $7/18$. His expected profit from this mixed strategy, $[11/18 \quad 7/18]$, can be found by substituting $11/18$ for p_1 in either E_1 or E_2. If E_1 is chosen,

$$\text{Expected profit} = 17,000\left(\frac{11}{18}\right) - 3000 = \frac{133,000}{18} \approx \$7400.$$

YOUR TURN 2 In Example 3, suppose Brian's payoff matrix is as follows.

	Attack	No Attack
Spray	$8000	$2000
Don't spray	−$9000	$15,000

What should his strategy be?

If Brian makes this decision only once, some mathematicians (including these authors) believe it is meaningless to make a choice with probability $11/18$. This is because Brian either sprays or doesn't spray. Because the strategy to spray has a higher probability, he should spray and not use the spinner discussed earlier. To make a choice with a given probability only makes sense if the decision is made many times. In game theory, the other player (nature, in this case) is thought of as trying to maximize his or her gain, which is the same as minimizing the first player's gain. It may seem strange to think of nature as choosing a strategy to minimize Brian's profit, but that is a safe and rational strategy when no other information about the state of nature is known. **TRY YOUR TURN 2**

NOTE
Remember that the use of mixed strategies is meaningful only if a game is repeated many, many times.

We can continue to find optimal strategies for games that are not strictly determined (that is, the games that have no saddle points) by finding where the lines associated with the expected value of each player intersect. We could, however, develop a formula for the optimum strategy.

$$M = \begin{bmatrix} a_{11} & a_{12} \\ a_{21} & a_{22} \end{bmatrix},$$

the payoff matrix of the game. Assume that A chooses row 1 with probability p_1. The expected value for A, assuming that B plays column 1, is E_1, where

$$E_1 = a_{11} \cdot p_1 + a_{21} \cdot (1 - p_1).$$

The expected value for A if B chooses column 2 is E_2, where

$$E_2 = a_{12} \cdot p_1 + a_{22} \cdot (1 - p_1).$$

As above, the optimum strategy for player A is found by letting $E_1 = E_2$.

$$a_{11} \cdot p_1 + a_{21} \cdot (1 - p_1) = a_{12} \cdot p_1 + a_{22} \cdot (1 - p_1)$$

Solve this equation for p_1.

$$a_{11} \cdot p_1 + a_{21} - a_{21} \cdot p_1 = a_{12} \cdot p_1 + a_{22} - a_{22} \cdot p_1$$

$$a_{11} \cdot p_1 - a_{21} \cdot p_1 - a_{12} \cdot p_1 + a_{22} \cdot p_1 = a_{22} - a_{21}$$

$$p_1(a_{11} - a_{21} - a_{12} + a_{22}) = a_{22} - a_{21}$$

$$p_1 = \frac{a_{22} - a_{21}}{a_{11} - a_{21} - a_{12} + a_{22}}$$

Since $p_2 = 1 - p_1$,

$$p_2 = 1 - \frac{a_{22} - a_{21}}{a_{11} - a_{21} - a_{12} + a_{22}}$$

$$= \frac{a_{11} - a_{21} - a_{12} + a_{22} - (a_{22} - a_{21})}{a_{11} - a_{21} - a_{12} + a_{22}}$$

$$= \frac{a_{11} - a_{12}}{a_{11} - a_{21} - a_{12} + a_{22}}.$$

This result is valid only if $a_{11} - a_{21} - a_{12} + a_{22} \neq 0$; this condition is satisfied if the game is not strictly determined.

There is a similar result for player B, which is included in the following summary.

Optimum Strategies in a Non-Strictly Determined Game

Let a non–strictly determined game have payoff matrix

$$M = \begin{bmatrix} a_{11} & a_{12} \\ a_{21} & a_{22} \end{bmatrix}$$

and let $d = a_{11} - a_{21} - a_{12} + a_{22}$. The optimum strategy for player A is $[p_1 \quad p_2]$, where

$$p_1 = \frac{a_{22} - a_{21}}{d} \quad \text{and} \quad p_2 = \frac{a_{11} - a_{12}}{d}.$$

The optimum strategy for player B is $\begin{bmatrix} q_1 \\ q_2 \end{bmatrix}$, where

$$q_1 = \frac{a_{22} - a_{12}}{d} \quad \text{and} \quad q_2 = \frac{a_{11} - a_{21}}{d}.$$

The value of the game is

$$g = \frac{a_{11}a_{22} - a_{12}a_{21}}{d}.$$

NOTE **1.** It is not necessary to remember all of the given formulas. Once p_1 has been calculated, it is simple to calculate $p_2 = 1 - p_1$, rather than using the formula for p_2. Similarly, $q_2 = 1 - q_1$. Finally, g is the single element in the matrix product AMB, where

$$A = [p_1 \quad p_2] \quad \text{and} \quad B = \begin{bmatrix} q_1 \\ q_2 \end{bmatrix}.$$

2. Rather than memorize the formula above, you can calculate E_1 and E_2 and set them equal, as in Example 3.

CAUTION | The formulas above do not apply if the game is strictly determined. In that case, use the method of Section 11.1 on Strictly Determined Games.

EXAMPLE 4 Optimum Strategies

Suppose a game has the following payoff matrix.

$$\begin{bmatrix} 5 & -2 \\ -3 & -1 \end{bmatrix}$$

Find the optimum strategies and the value of the game.

SOLUTION

Method 1
Formulas | Here $a_{11} = 5$, $a_{12} = -2$, $a_{21} = -3$, and $a_{22} = -1$. To find the optimum strategy for player A, first find

$$\begin{aligned} d &= a_{11} - a_{21} - a_{12} + a_{22} \\ &= 5 - (-3) - (-2) + (-1) \\ &= 9. \end{aligned}$$

Next, calculate p_1.

$$p_1 = \frac{-1 - (-3)}{9} = \frac{2}{9}$$

Player A should play row 1 with probability 2/9 and row 2 with probability $1 - 2/9 = 7/9$.
   For player B,

$$q_1 = \frac{-1 - (-2)}{9} = \frac{1}{9}.$$

Player B should choose column 1 with probability 1/9 and column 2 with probability 8/9. The value of the game is

$$g = \frac{5(-1) - (-2)(-3)}{9} = -\frac{11}{9}.$$

On the average, B will receive 11/9 dollars from A per play of the game.

Method 2
Rows and Columns | There is an alternative to the formulas used in Method 1, in cases with a 2×2 matrix and no saddle point. Calculate the absolute value of the difference between elements in each row, and then swap the two numbers.

$$\begin{bmatrix} 5 & -2 \\ -3 & -1 \end{bmatrix} \begin{matrix} \rightarrow |5 - (-2)| = 7 \searrow 2 \\ \rightarrow |-3 - (-1)| = 2 \nearrow 7 \end{matrix}$$

The sum of the two numbers calculated is $7 + 2 = 9$. The values of p_1 and p_2 are then 2/9 and 7/9, respectively.
   Similarly, calculate the absolute value of the difference between elements in each column, and then swap the two numbers.

$$\begin{bmatrix} 5 & -2 \\ -3 & -1 \end{bmatrix}$$

$$\begin{matrix} \downarrow & \downarrow \\ |5 - (-3)| & |-2 - (-1)| \\ = 8 \searrow & = 1 \\ 1 \nearrow & 8 \end{matrix}$$

YOUR TURN 3 Find the optimum strategy and the value of the game with the following payoff matrix.

$$\begin{bmatrix} -4 & 2 \\ 1 & 0 \end{bmatrix}$$

The sum of the two numbers calculated is $1 + 8 = 9$, and the values of q_1 and q_2 are then $1/9$ and $8/9$, respectively. From these, the value of the game can be computed as

$$AMB = \begin{bmatrix} 2/9 & 7/9 \end{bmatrix} \begin{bmatrix} 5 & -2 \\ -3 & -1 \end{bmatrix} \begin{bmatrix} 1/9 \\ 8/9 \end{bmatrix} = [-11/9].$$

TRY YOUR TURN 3

To see why Method 2 in Example 4 works, consider the case in which $a_{11} > a_{12}$. Then the number computed in row 1 is $|a_{11} - a_{12}| = a_{11} - a_{12}$. Since there is no saddle point, a_{22} must be greater than a_{21}, because if both elements in column 2 were the smallest number in their respective rows, there would be a saddle point in column 2. So $a_{21} < a_{22}$, and $|a_{21} - a_{22}| = a_{22} - a_{21}$. The sum of the quantities in the two rows is then $(a_{11} - a_{12}) + (a_{22} - a_{21})$, which is equivalent to $d = a_{11} - a_{21} - a_{12} + a_{22}$. This method then gives $p_1 = (a_{22} - a_{21})/d$ and $p_2 = (a_{11} - a_{12})/d$, exactly as in Method 1. The analysis is similar for q_1 and q_2 and in the case in which $a_{11} < a_{12}$.

NOTE One limit of the formulas presented here is that they only work when the payoff matrix is 2×2. The method shown in Example 3, however, can be generalized to matrices with more than 2 columns, as long as there are still 2 rows. (See Exercises 27 and 28.)

11.2 EXERCISES

1. Suppose a game has payoff matrix $\begin{bmatrix} 3 & -4 \\ -5 & 2 \end{bmatrix}$. Suppose that player B uses the strategy $\begin{bmatrix} 0.4 \\ 0.6 \end{bmatrix}$. Find the expected value of the game if player A uses the following strategies.

a. $[0.5 \quad 0.5]$ **b.** $[0.1 \quad 0.9]$ **c.** $[0.8 \quad 0.2]$ **d.** $[0.2 \quad 0.8]$

2. Suppose a game has payoff matrix $\begin{bmatrix} 0 & -4 & 1 \\ 3 & 2 & -4 \\ 1 & -1 & 0 \end{bmatrix}$. Find the expected value of the game for the following strategies for players A and B.

a. $A = [0.1 \quad 0.4 \quad 0.5]; B = \begin{bmatrix} 0.2 \\ 0.4 \\ 0.4 \end{bmatrix}$

b. $A = [0.3 \quad 0.4 \quad 0.3]; B = \begin{bmatrix} 0.8 \\ 0.1 \\ 0.1 \end{bmatrix}$

Find the optimum strategies for player A and player B in the games in Exercises 3–14. Find the value of each game. (Be sure to look for a saddle point first.)

3. $\begin{bmatrix} 7 & 1 \\ 3 & 4 \end{bmatrix}$

4. $\begin{bmatrix} -4 & 6 \\ 3 & -4 \end{bmatrix}$

5. $\begin{bmatrix} -2 & 0 \\ 5 & -4 \end{bmatrix}$

6. $\begin{bmatrix} 6 & 3 \\ -1 & 10 \end{bmatrix}$

7. $\begin{bmatrix} 4 & -3 \\ -1 & 9 \end{bmatrix}$

8. $\begin{bmatrix} 0 & 9 \\ 4 & 0 \end{bmatrix}$

9. $\begin{bmatrix} -1 & 2 \\ 3 & 5 \end{bmatrix}$

10. $\begin{bmatrix} 4 & 1/5 \\ 2/3 & -1 \end{bmatrix}$

11. $\begin{bmatrix} 8/3 & -1/2 \\ 3/4 & -5/12 \end{bmatrix}$

12. $\begin{bmatrix} -1/2 & 2/3 \\ 7/8 & -3/4 \end{bmatrix}$

13. $\begin{bmatrix} -2 & 1/2 \\ 0 & -3 \end{bmatrix}$

14. $\begin{bmatrix} 8 & 18 \\ -4 & 2 \end{bmatrix}$

Remove any dominated strategies, and then find the optimum strategy for each player and the value of the game.

15. $\begin{bmatrix} -4 & 9 \\ 3 & -5 \\ 8 & 7 \end{bmatrix}$

16. $\begin{bmatrix} 3 & 4 & -1 \\ -2 & 1 & 0 \end{bmatrix}$

17. $\begin{bmatrix} 8 & 6 & 3 \\ -1 & -2 & 4 \end{bmatrix}$

18. $\begin{bmatrix} -1 & 6 \\ 8 & 3 \\ -2 & 5 \end{bmatrix}$

19. $\begin{bmatrix} 9 & -1 & 6 \\ 13 & 11 & 8 \\ 6 & 0 & 9 \end{bmatrix}$ **20.** $\begin{bmatrix} 4 & 8 & -3 \\ 2 & -1 & 1 \\ 7 & 9 & 0 \end{bmatrix}$

21. Verify that the optimum strategy for player B in a non–strictly determined game is as given in the text.

22. Verify that the value of a non–strictly determined game is as given in the text when players A and B play their optimum strategies.

23. Some people claim that even if a mixed strategy is used only once, the player should use a spinner to determine what choice to make. Write a few sentences giving your reaction to this claim.

24. For a game with optimum strategy $\begin{bmatrix} \frac{1}{2} & \frac{1}{2} \end{bmatrix}$, a player could flip a coin to determine which choice to make. A nonrandom method would be to alternate between the two choices. Discuss any disadvantages of this nonrandom method. Would it make a difference whether or not the opposing player is intelligent? Explain.

25. Why doesn't the reasoning in this section apply to strictly determined games?

26. a. Prove that if player A has the optimum strategy $\begin{bmatrix} p_1 & p_2 \end{bmatrix}$ as given in the text, then $AM = \begin{bmatrix} g & g \end{bmatrix}$, where g is the value of the game.

  b. Prove that if player B has the optimum strategy $\begin{bmatrix} q_1 \\ q_2 \end{bmatrix}$ given in the text, then $MB = \begin{bmatrix} g \\ g \end{bmatrix}$, where g is the value of the game.

27. Consider the matrix

$$\begin{bmatrix} 1 & 2 & 3 \\ 4 & 3 & 1 \end{bmatrix}.$$

  a. Letting p_1 be the probability of choosing row 1 and $1 - p_1$ be the probability of choosing row 2 (as in Example 3), find the expected values E_1, E_2, and E_3, respectively.

  b. Sketch graphs of E_1, E_2, and E_3, as in Example 3.

  c. Unlike Example 3, there are now three intersection points for $0 \le p_1 \le 1$. From your answer to part b, find the value of p_1 that maximizes the minimum expected value that the row player receives.

28. Repeat the instructions for Exercise 27, using the matrix

$$\begin{bmatrix} -1 & 5 & 1 \\ 3 & 1 & 2 \end{bmatrix}.$$

29. A reader wrote to the "Ask Marilyn" column in *Parade* magazine, "Say you're in a public library, and a beautiful stranger strikes up a conversation with you. She says: 'Let's show pennies to each other, either heads or tails. If we both show heads, I pay you $3. If we both show tails, I pay you $1. If they don't match, you pay me $2.' . . . As the game is quiet, you can play in the library. But should you?" *Source: Parade Magazine.*

  a. Find the optimum strategy for you and for the stranger, as well as the value of the game. Should you play?

  b. Marilyn replied, "She can win easily. One way: If she shows you twice as many tails as heads, she wins an average of $1 for every six plays." What strategy is Marilyn assuming that you use?

  c. If the stranger did use the strategy in part b, what strategy should you use to maximize the value of the game for you? What is the value in that case?

APPLICATIONS

Business and Economics

30. Advertising Suppose Allied Manufacturing Company decides to put its new product on the market with a big television and radio advertising campaign. At the same time, the company finds out that its major competitor, Bates Manufacturing, also has decided to launch a big advertising campaign for a similar product. The payoff matrix shows the increased sales (in millions) for Allied, as well as the decreased sales for Bates.

Bates

		T.V.	Radio
Allied	T.V.	1.0	−0.7
	Radio	−0.5	0.6

Find the optimum strategy for Allied Manufacturing and the value of the game.

31. Pricing The payoffs in the matrix below represent the differences between Boeing Aircraft Company's profit and its competitor's profit for two prices (in millions) on commercial jet transports, with positive payoffs being in Boeing's favor. What should Boeing's price strategy be? *Source: Management Sciences Models and Techniques.*

Competitor's Price Strategy

		4.9	4.75
Boeing's Strategy	4.9	2	−4
	4.75	0	2

32. Sales *The Huckster* Merrill has a concession at Yankee Stadium for the sale of sunglasses and umbrellas. The business places quite a strain on him, the weather being what it is. He has observed that he can sell about 500 umbrellas when it rains and about 100 when it is sunny; in the latter case he can also sell 1000 sunglasses. Umbrellas cost him $5 and sell for $10; sunglasses cost $2 and sell for $5. He is willing to invest $2500 in the project. Everything that is not sold is considered a total loss.

He assembles the facts regarding profit in a matrix. He immediately notices that this is a mixed-strategy game, and he should be able to find a stabilizing strategy that will save him from the vagaries of the weather. *Source: The Compleat Strategyst.*

  a. Determine the payoff matrix for this game.

  b. Find the best mixed strategy for Merrill.

Life Sciences

33. Choosing Medication The number of cases of African flu has reached epidemic levels. The disease is known to have two strains with similar symptoms. Dr. Goedeker has two medicines available; the first is 75% effective against the first strain and 40% effective against the second. The second medicine is completely effective against the second strain but ineffective against the first.

  a. Determine the payoff matrix giving the effectiveness for the two medicines.

b. Decide which medicine she should use and the results she can expect.

General Interest

34. Cats When two cats, Euclid and Jamie, play together, their game involves facing each other while several feet apart; each cat must then decide whether to pounce or to "freeze"—to stay motionless until the other cat pounces. Euclid weighs 2 lb more than Jamie, so if they both pounce, Jamie is squashed and Euclid gains 3 points. If they both freeze, Euclid remains in control of the area and gains 2 points. If Euclid pounces while Jamie freezes, Jamie can put up a good defense, so Euclid only gains 1 point. Euclid is poor at defense, so if he freezes while Jamie pounces, he loses 2 points. Find the optimum strategy for each cat, and find the value of the game.

35. Coin Game In the game of matching coins, each of two players flips a coin. If both coins match (both show heads or both show tails), player A wins $1. If there is no match, player B wins $1.

a. Determine the payoff matrix.

b. Find the optimum strategies for the two players and the value of the game.

36. Finger Game Players A and B play a game in which they show either 1 or 2 fingers at the same time. If there is a match, A wins the amount of dollars equal to the total number of fingers shown. If there is no match, B wins the amount of dollars equal to the number of fingers shown.

a. Write the payoff matrix.

b. Find optimum strategies for A and B and the value of the game.

37. Finger Game Repeat Exercise 36 if each player may show either 0 or 2 fingers with the same payoffs.

▮ YOUR TURN ANSWERS

1. 32 cents
2. Brian should spray with probability 4/5 and not spray with probability 1/5.
3. Player A should choose row 1 with probability 1/7 and row 2 with probability 6/7. Player B should choose column 1 with probability 2/7 and column 2 with probability 5/7. The value of the game is 2/7.

11.3 Game Theory and Linear Programming

APPLY IT   How can a company decide in which cities it should advertise?
Using game theory and linear programming, we will answer this question in Exercise 15.

Until now, we have not learned how to solve games in which one player has more than two choices, but for which neither player has an optimum pure strategy. In this section, we shall see how the techniques of linear programming can be used to solve games in which each player has an arbitrary number of choices.

If there is no saddle point, we must find the probabilities for mixing strategies to obtain an optimum solution and the value of the game. Linear programming can be applied to find the optimum solution and the value of the game, as shown in the following example.

EXAMPLE 1   Optimum Strategies (Using the Graphing Method)

Use linear programming to find the optimum strategy for the payoff matrix given in Example 1 in the previous section.

SOLUTION The payoff matrix is

$$M = \begin{bmatrix} -1 & 2 \\ 1 & 0 \end{bmatrix}.$$

When player A chose row 1 with probability p_1 and row 2 with probability p_2, we found that his expectations when B chose columns 1 and 2 were as follows.

$$E_1 = -1 \cdot p_1 + 1 \cdot p_2 = -p_1 + p_2$$
$$E_2 = 2 \cdot p_1 + 0 \cdot p_2 = 2p_1$$

Player A wishes his expected gain to be as large as possible, so he should maximize the minimum of the expected gains.

Let g represent the minimum of the expected gains, so that

$$E_1 = -p_1 + p_2 \geq g$$
$$E_2 = 2p_1 \geq g.$$

We can simplify this system of linear inequalities by dividing both inequalities by g. As long as $g > 0$, which we shall show momentarily, this yields

$$\frac{E_1}{g} = -\left(\frac{p_1}{g}\right) + \left(\frac{p_2}{g}\right) \geq 1$$
$$\frac{E_2}{g} = 2\left(\frac{p_1}{g}\right) \phantom{+ \left(\frac{p_2}{g}\right)} \geq 1.$$

Then denote p_1/g by x and p_2/g by y.

$$E_1/g = -x + y \geq 1$$
$$E_2/g = 2x \geq 1$$

To determine the objective function, note that $x + y = p_1/g + p_2/g = (p_1 + p_2)/g = 1/g$. Since the goal is to maximize g, which is the same as minimizing $1/g$, we can rewrite the problem as the following linear programming problem:

$$\begin{aligned}
\text{Minimize} \quad & w = x + y \\
\text{subject to:} \quad & -x + y \geq 1 \\
& 2x \geq 1 \\
\text{with} \quad & x \geq 0, y \geq 0,
\end{aligned}$$

where $w = 1/g$. This linear programming problem can be solved by the graphical method described in Chapter 3. The graph is shown in Figure 8. There is only one corner point, $(1/2, 3/2)$, and the value of w there is 2. Thus the value of the game is $g = 1/w = 1/2$, and the optimum strategy for player A is $p_1 = gx = (1/2)(1/2) = 1/4$, $p_2 = gy = (1/2)(3/2) = 3/4$. Notice that this agrees with the result found in the previous section.

A similar analysis can be done for player B. When B chooses columns 1 and 2 with probabilities q_1 and q_2, respectively, then her expectations when A chooses rows 1 and 2 are

$$E_1 = -1 \cdot q_1 + 2 \cdot q_2 = -q_1 + 2q_2$$
$$E_2 = 1 \cdot q_1 + 0 \cdot q_2 = q_1.$$

Player B wishes the payoff to be as small as possible, so she should minimize the maximum of the expected gains:

$$E_1 = -q_1 + 2q_2 \leq g$$
$$E_2 = q_1 \leq g.$$

As before, we divide both inequalities by g and denote q_1/g by x and q_2/g by y, yielding:

$$E_1/g = -x + 2y \leq 1$$
$$E_2/g = x \leq 1.$$

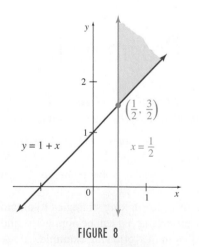

FIGURE 8

As before, $x + y = q_1/g + q_2/g = (q_1 + q_2)/g = 1/g$. Player B's goal is to minimize g, which is the same as maximizing $1/g$. (Recall from the previous page that $g > 0$.) We, therefore, have the following linear programming problem:

$$\text{Maximize} \quad z = x + y$$
$$\text{subject to:} \quad -x + 2y \leq 1$$
$$\qquad\qquad\quad x \qquad \leq 1$$
$$\text{with} \qquad x \geq 0, y \geq 0,$$

where $z = 1/g$. This problem can also be solved by the graphical method; do this, and verify that $x = y = 1$ and $z = 2$. As before, the value of the game is $g = 1/z = 1/2$, and the optimum strategy for B is $q_1 = gx = (1/2)(1) = 1/2$, $q_2 = gy = (1/2)(1) = 1/2$. The value of the game is the same for B as for A, as must be the case, and B's strategy is the same as in the previous section. ◼

Most significantly, notice in Example 1 that the linear programming problem for player A is the dual of the linear programming problem for player B. It can be shown that, under circumstances to be discussed shortly, this remarkable fact is always true. Notice further that A wants to maximize the minimum value of the game, while B wants to minimize the maximum of the game. That these two values are equal is an illustration of the **minimax principle**, which follows from the theorem of duality discussed in Chapter 4.

If the value of a zero-sum game is positive, the game can always be solved by linear programming techniques, in which case the solution of the row player is the dual of the solution of the column player. To guarantee that the value of the game is positive, find the most negative number in the payoff matrix and add its absolute value to all entries in the matrix. For example, in the previous game, -1 was the most negative entry. Adding 1 to all entries yields the payoff matrix

$$N = \begin{bmatrix} 0 & 3 \\ 2 & 1 \end{bmatrix}.$$

Solving this game as before would result in the same optimum strategies, but the value of the game would be $3/2$, that is, 1 greater than it was before. We would then subtract 1 from $3/2$ to get $1/2$, the value of the original game.

Just as linear programming problems with more than two variables could be solved by the simplex method, we can now use the simplex method to solve games involving more than two choices for each player. Before giving such an example, let us review the steps in solving a matrix game.

CAUTION | Remember that if there are any negative entries in the payoff matrix, you need to add a positive number to all entries so that all entries are nonnegative. If you fail to do so, it is possible that the value of the game is 0 or negative, in which case the linear programming approach will fail. Also remember at the end, when you find the value of the game, to subtract the positive number that you previously added.

Solving a Zero-Sum Game

1. Set up the payoff matrix M.

2. Remove any dominated rows or columns.

3. Check to see if there is a saddle point. If so, the optimum strategies are determined by the row and column in which the saddle point is located.

4. If there is no saddle point and negative numbers are present, add a positive number to all entries in the payoff matrix so that all entries are nonnegative. Denote the elements of the resulting $m \times n$ matrix by a_{ij}.

(continued)

5. Set up the linear programming problem:

$$\text{Maximize} \quad z = x_1 + \cdots + x_n$$

$$\text{subject to:} \quad a_{11}x_1 + \cdots + a_{1n}x_n \leq 1$$

$$\vdots \qquad \qquad \vdots$$

$$a_{m1}x_1 + \cdots + a_{mn}x_n \leq 1$$

$$\text{with} \quad x_1 \geq 0, \ldots, x_n \geq 0.$$

6. After finding the solution z to the linear programming problem, the value of the modified game is given by $g = 1/z$. The optimum strategy for player B is found by multiplying the solution to the linear programming problem by g: $q_i = gx_i$, $i = 1, \ldots, n$. The optimum strategy for player A is found by multiplying the solution to the dual problem by g: $p_j = gy_j$, $j = 1, \ldots, m$.

7. Subtract from g the positive number added in step 4 to find the value of the original game.

┌ FOR REVIEW ─────

The steps of the simplex algorithm can be stated briefly as follows.

1. Convert each constraint in the maximization problem into an equation by adding a slack variable.

2. Set up the initial tableau, with the negative of the payoffs in the bottom row.

3. Choose the column with the most negative indicator. Divide each element from that column into the corresponding element from the last column. The pivot is the element with the smallest quotient. Use row operations to change all elements in the pivot column other than the pivot to 0 by adding a suitable multiple of the pivot row to a positive multiple of each row.

4. When there are no more negative numbers in the bottom row, the values of $x_1, \ldots, x_n$ can be read using the appropriate columns in combination with the last column.

5. If the z column contains a number other than 1, divide the last row by that number. The value of z, which is the same as the value of w, is given in the lower right-hand corner. The solution to $y_1, \ldots, y_n$ can be read from the bottom row of the columns corresponding to the slack variables.

EXAMPLE 2 Optimum Strategies (Using the Simplex Method)

Find the optimum strategies and the value of the game that was introduced in Section 11.1 on Strictly Determined Games but was never solved.

SOLUTION The payoff matrix for this game, after deleting the column that was dominated by another column, was

$$\begin{bmatrix} -6 & 10 \\ 0 & -9 \\ -4 & -8 \end{bmatrix}.$$

Verify that this game has no saddle point. Because of the negative entries, we will add 9 to all entries to make them all nonnegative. The resulting payoff matrix is

$$\begin{bmatrix} 3 & 19 \\ 9 & 0 \\ 5 & 1 \end{bmatrix}.$$

The linear programming problem to be solved is

$$\text{Maximize} \quad z = x_1 + x_2$$

$$\text{subject to:} \quad 3x_1 + 19x_2 \leq 1$$

$$9x_1 \qquad \leq 1$$

$$5x_1 + \quad x_2 \leq 1$$

$$\text{with} \quad x_1 \geq 0, x_2 \geq 0.$$

The initial tableau is

$$\begin{array}{cccccc} x_1 & x_2 & s_1 & s_2 & s_3 & z \\ \begin{bmatrix} 3 & 19 & 1 & 0 & 0 & 0 & | & 1 \\ 9 & 0 & 0 & 1 & 0 & 0 & | & 1 \\ 5 & 1 & 0 & 0 & 1 & 0 & | & 1 \\ \hline -1 & -1 & 0 & 0 & 0 & 1 & | & 0 \end{bmatrix} \end{array}.$$

The first and second columns both contain -1 at the bottom; since these are equally negative, we arbitrarily choose the first column. The smallest ratio is formed by the 9 in row 2. Making this the pivot yields the following matrix.

$$
\begin{array}{c}
-R_2 + 3R_1 \to R_1 \\
\\
-5R_2 + 9R_3 \to R_3 \\
R_2 + 9R_4 \to R_4
\end{array}
\begin{array}{cccccccc}
& x_1 & x_2 & s_1 & s_2 & s_3 & z & \\
\left[\begin{array}{cccccc|c}
0 & 57 & 3 & -1 & 0 & 0 & 2 \\
9 & 0 & 0 & 1 & 0 & 0 & 1 \\
0 & 9 & 0 & -5 & 9 & 0 & 4 \\
0 & -9 & 0 & 1 & 0 & 9 & 1
\end{array}\right]
\end{array}
$$

The next pivot is the 57 in row 1, column 2.

$$
\begin{array}{c}
\\
\\
-3R_1 + 19R_3 \to R_3 \\
3R_1 + 19R_4 \to R_4
\end{array}
\begin{array}{cccccccc}
& x_1 & x_2 & s_1 & s_2 & s_3 & z & \\
\left[\begin{array}{cccccc|c}
0 & 57 & 3 & -1 & 0 & 0 & 2 \\
9 & 0 & 0 & 1 & 0 & 0 & 1 \\
0 & 0 & -9 & -92 & 171 & 0 & 70 \\
0 & 0 & 9 & 16 & 0 & 171 & 25
\end{array}\right]
\end{array}
$$

Since there are no more negative numbers in the bottom row, we are done pivoting. In each column with only one nonzero element, convert that element to a 1 by multiplying the corresponding row by the appropriate constant.

$$
\begin{array}{c}
\frac{1}{57}R_1 \to R_1 \\
\frac{1}{9}R_2 \to R_2 \\
\frac{1}{171}R_3 \to R_3 \\
\frac{1}{171}R_4 \to R_4
\end{array}
\begin{array}{cccccccc}
& x_1 & x_2 & s_1 & s_2 & s_3 & z & \\
\left[\begin{array}{cccccc|c}
0 & 1 & \frac{1}{19} & -\frac{1}{57} & 0 & 0 & \frac{2}{57} \\
1 & 0 & 0 & \frac{1}{9} & 0 & 0 & \frac{1}{9} \\
0 & 0 & -\frac{1}{19} & -\frac{92}{171} & 1 & 0 & \frac{70}{171} \\
0 & 0 & \frac{1}{19} & \frac{16}{171} & 0 & 1 & \frac{25}{171}
\end{array}\right]
\end{array}
$$

From the bottom row, $z = 25/171$, so

$$g = \frac{1}{z} = \frac{171}{25}.$$

The values of y_1, y_2, and y_3 are read from the bottom of the columns for the three slack variables.

$$y_1 = \frac{1}{19}, \qquad y_2 = \frac{16}{171}, \qquad y_3 = 0$$

We find the values of p_1, p_2, and p_3 by multiplying the values of y_1, y_2, and y_3 by g.

$$p_1 = \left(\frac{1}{19}\right)\left(\frac{171}{25}\right) = \frac{9}{25}, \quad p_2 = \left(\frac{16}{171}\right)\left(\frac{171}{25}\right) = \frac{16}{25}, \quad p_3 = (0)\left(\frac{171}{25}\right) = 0$$

Next find x_1 and x_2 by using the first and second columns combined with the last column.

$$x_1 = \frac{1}{9}, \quad x_2 = \frac{2}{57}$$

Find the values of q_1 and q_2 by multiplying the values of x_1 and x_2 by g.

$$q_1 = \left(\frac{1}{9}\right)\left(\frac{171}{25}\right) = \frac{19}{25}, \quad q_2 = \left(\frac{2}{57}\right)\left(\frac{171}{25}\right) = \frac{6}{25}$$

YOUR TURN 1 Find the optimum strategies and the value of the game with the following payoff matrix.
$$\begin{bmatrix} -2 & 1 & 0 \\ 1 & -4 & -2 \end{bmatrix}$$

Finally, the value of the game is found by subtracting from g the 9 that was added at the beginning, yielding $(171/25) - 9 = -54/25$.

To summarize, the optimum strategy for player A is $(9/25, 16/25, 0)$, and the optimum strategy for player B is $(19/25, 6/25)$. When these strategies are used, the value of the game is $-54/25$. **TRY YOUR TURN 1**

Some applications of these techniques are given in the exercises. For many realistic games, the details of using the simplex method become very tedious, and so a graphing calculator or a computer is of great help. With the aid of a program to perform the simplex algorithm, the optimum strategies for games of virtually any size can be found.

11.3 EXERCISES

Use the graphical method to find the optimum strategy for players A and B and the value of the game for each payoff matrix.

1. $\begin{bmatrix} 1 & 2 \\ 4 & 1 \end{bmatrix}$

2. $\begin{bmatrix} 6 & 2 \\ 0 & 3 \end{bmatrix}$

3. $\begin{bmatrix} 2 & -2 \\ -1 & 6 \end{bmatrix}$

4. $\begin{bmatrix} -1 & 5 \\ 1 & -8 \end{bmatrix}$

5. $\begin{bmatrix} 7 & -8 \\ -3 & 3 \end{bmatrix}$

6. $\begin{bmatrix} -4 & 1 \\ 5 & 0 \end{bmatrix}$

Use the simplex method to find the optimum strategy for players A and B and the value of the game for each payoff matrix.

7. $\begin{bmatrix} 3 & -4 & 1 \\ 5 & 3 & -2 \end{bmatrix}$

8. $\begin{bmatrix} -5 & 1 & 4 & 2 \\ 9 & 3 & -8 & -7 \end{bmatrix}$

9. $\begin{bmatrix} -1 & 1 & 4 \\ 3 & -2 & -3 \end{bmatrix}$

10. $\begin{bmatrix} 1 & 0 \\ -2 & 4 \\ -1 & -1 \end{bmatrix}$

11. $\begin{bmatrix} 1 & 0 & -1 \\ -1 & 0 & 1 \\ 2 & -1 & 2 \end{bmatrix}$

12. $\begin{bmatrix} 2 & -1 & 1 \\ 0 & 2 & 3 \\ 4 & 1 & 0 \end{bmatrix}$

APPLICATIONS

In Exercises 13–26, use the graphical method when the payoff matrix is a 2×2 matrix or can be reduced to one after removing rows or columns that are dominated. Otherwise, use the simplex method.

Business and Economics

13. Contractor Bidding Darien Estes, a contractor, prepares to bid on a job. If all goes well, his bid should be $30,000, which will cover his costs plus his usual profit margin of $4500. If a threatened labor strike actually occurs, however, his bid should be $40,000 to give him the same profit. If there is a strike and he bids $30,000, he will lose $5500. If he bids $40,000 and there is no strike, his bid will be too high, and he will lose the job entirely. Find the optimum strategy for Darien, and find the value of the game.

14. Negotiation In negotiating a labor contract, a labor union has considered four different approaches, based on different positions the union can take on various issues. Management can take three different approaches in the negotiations. The final contract will be regarded as either favorable to the union, favorable to the management, or neither, leading to the following payoff matrix.

Management Strategies

$$\begin{array}{c} \\ Labor\ Strategies \end{array} \begin{array}{c} \\ 1 \\ 2 \\ 3 \\ 4 \end{array} \begin{array}{ccc} 1 & 2 & 3 \\ \begin{bmatrix} -1 & -1 & 1 \\ -1 & 0 & 0 \\ 1 & -1 & -1 \\ 1 & 1 & -1 \end{bmatrix} \end{array}$$

Find the optimum strategies for labor and management, and find the value of the game.

15. Advertising Marketing executives for two competing companies are trying to decide whether to advertise in Atlanta, Boston, or Cleveland, given that each company can afford to target only one city at a time. General Items Company has a leading market share and has found that it will earn an additional profit of $10,000, $8000, or $6000 per week if they advertise in Atlanta, Boston, or Cleveland, respectively. Original Imitators, Inc., has a smaller portion of the market, but its executives can cut General Items' additional profit in half and gain that profit themselves if they run competing ads in the same city as General Items. If they run ads in a different city, their ads seem to have no impact.

a. Set up the profit matrix for this game.

b. APPLY IT Find the optimum strategy for each company and the value of the game.

16. Marketing Solve the Christmas card problem discussed in Section 11.1, for which we had the following payoff matrix.

States of Nature

		Weak Economy	In-Between	Strong Economy
Strategies	Modern	40	85	120
	Old-Fashioned	90	45	85
	Mixture	75	105	65

17. Competition Two merchants are planning to build competing stores to serve an area of three small cities. The fraction of the total population that live in each city is shown in the figure on the next page. If both merchants locate in the same city, merchant A will get 65% of the total business. If the merchants locate in different cities, each will get 80% of the business in the city where it is located, and 20% of the business in the city where the other merchant is located. In the city where neither is located, A will get 60% of the business and B will get 40%. Payoffs are measured by the number of percentage points above or below 50%. Find an optimum strategy for each merchant, and find the value of this game. (See Exercise 33 in Section 11.1.)

Life Sciences

18. Nurse Staffing Hospitals need to decide how many nurses to assign on a given day, even though they don't know the demand in advance. An article assigned the following points to various scenarios. *Source: Journal of Nursing Administration.*

Adequate staff, fully utilized	10
Moderate underutilization	8
Moderate overcrowding	6
Extreme underutilization	5
Extreme overcrowding	2
Dangerous overcrowding	1

The article gave a payoff matrix as follows, where the demand can be low, medium, or high, and anywhere from 1 to 4 nurses can be assigned.

		Demand		
		Low	Medium	High
	4	5	8	10
Number of Nurses	3	8	10	6
	2	10	6	2
	1	6	2	1

a. Remove any dominated strategies.

b. Find the optimum strategy for the hospital.

Social Sciences

19. Jamaican Fishing A study of fishing in a small Jamaican village found that the fishermen used three strategies for locating their fishing pots. Some fished in the inside banks with their safe currents and poorer fishing. Others fished in the outside banks, which yield more and better fish but have dangerous strong currents. Still others put one-third of their pots in the inner banks and the rest in the outer banks. The payoffs per month (in pounds) depended on whether or not there was a strong current, as indicated by the following payoff matrix: *Source: Papers in Caribbean Anthropology.*

		Environment	
		Current	No Current
	Inside	17.3	11.5
Fisherman	Outside	−4.4	20.6
	Mixture	5.2	17.0

a. Find the optimum strategy for the fishermen and for the environment, as well as the expected payoff to the fishermen.

b. The environment actually had current 25% of the time and no current 75% of the time. What should the fishermen's strategy be in this case, and what is the payoff?

c. Based on your results from parts a and b, discuss the wisdom of considering the environment as a player in the game.

20. Conflict In an article on the sociology of conflict, the author proposes a game in which a member of a race (she calls them the green race) visits a country in which some natives are pro-green, some indifferent, and some antigreen. The green visitor, unable to determine which group a native is from, has three options: act appeasing, expect civility but not fight for it, and expect civility and fight for it if he doesn't get it. The author suggests the following payoff matrix.

	Pro-green	Indifferent	Anti-green
Appease	0	0	−4
No Fight	0	−3	−1
Fight	−2	−1	−1

What is the visitor's optimum strategy, and what is the value of the game? *Source: American Journal of Sociology.*

21. Education In an accounting class, the instructor, Eric Olson, permits the students to bring a calculator or a reference book (but not both) to an examination. The examination will emphasize either numerical problems or definitions. In trying to decide which aid to take to an examination, a student first decides on the utilities shown in the following payoff matrix.

		Exam's Emphasis	
		Numbers	Definitions
Student's Choice	Calculator	50	0
	Book	10	40

What is the student's optimum strategy, and what is the value of the game?

General Interest

22. Military Science The Colonel Blotto game is a type of military strategy game. Two opposing armies are approaching two posts. Colonel Blotto has 4 regiments under his command, while his opponent, Captain Kije, has 3 regiments. Each commander must decide how many regiments to send to each post. The army that sends more regiments to a post not only captures that post but also captures the losing army's regiments. If both armies send the same number of regiments to a post, there is a stand-off, and neither army wins. The payoff is one point for capturing the post and one point for each regiment captured. *Source: Mathematical Methods and Theory in Games, Programming, and Economics.*

a. Set up the payoff matrix for this game. (*Hint:* Colonel Blotto has five choices, and Captain Kije has four.)

b. Find the optimum strategy for each commander and the value of the game.

c. Show that if Colonel Blotto uses the strategy found in part b, then any strategy used by Captain Kije results in the same payoff. (*Hint:* Show that $AM = (14/9)R$, where R is a row matrix consisting of all 1's, and then use the fact that $RB = [1]$.)

d. Based on the result of part c, what can you conclude about the uniqueness of the optimum strategy found by linear programming?

23. Soccer A study of penalty kicks in soccer found the following percentages of shots that a goal was scored, based on what direction (left, middle, or right) the kicker kicks and the goalie moves. The data are based on French and Italian soccer players. *Source: The American Economic Review.*

Goalie

		Left	Middle	Right
	Left	63.2	100	94.1
Kicker	Middle	81.2	0	89.3
	Right	89.5	100	44.0

Find the optimum strategy for the kicker and the goalie, as well as the expected payoff to the kicker.

24. **Card Games** Player A deals Player B one of three cards—ace, king, or queen—at random, face down. B looks at the card. If it is an ace, B must say "ace"; if it is a king, B can say either "king" or "ace"; if it is a queen, B can say either "queen" or "ace." If B says "ace," A can either believe him and give him $1, or ask him to show his card. If it is an ace, A must pay B $2; but if it is not, B pays A $2. If B says "king," neither side loses anything; but if he says "queen," B must pay A $1. *Source: Games, Theory, and Applications.*

   a. Set up the payoff matrix for this game. (*Hint:* Consider these two choices for A: (1) believe B when B says "ace"; (2) ask B to show his card when B says "ace." Consider the following four choices for B: (1) always tell the truth; (2) lie only if the card is a queen; (3) lie only if the card is a king; (4) lie if the card is a queen or a king. For each entry in the payoff matrix, find the average payoff over the three possible outcomes.)

   b. Find the optimum strategy for each player and the value of the game.

25. **Tennis** In the game of tennis, a skilled player strategically chooses to serve the ball to the left or to the right of the opponent. Accordingly, the person receiving the serve needs to decide whether the ball will go left or right. If the serve goes left and the receiver moves right, then the advantage goes to the server. On the other hand, if the ball goes left and the receiver is also moving

in that direction, the advantage goes to the receiver. This situation has been modeled for professional players at Wimbledon using game theory. Suppose that for two hypothetical players, the fraction of the time that a server wins a point is given in the following payoff matrix. *Source: The American Economic Review.*

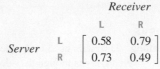

Receiver

		L	R
Server	L	0.58	0.79
	R	0.73	0.49

   a. Find the optimum strategy for each player and the value of the game.

   b. Discuss the possible wisdom of using these mixed strategies, in light of the fraction of the time that a particular strategy pays off for a particular player.

26. **Children's Game** In the children's game rock, paper, scissors, two players simultaneously extend one hand in the form of a fist (rock), a flat palm (paper), or two fingers extended (scissors). If they both make the same choice, the game is a tie. Otherwise, rock beats scissors (because rock can crush scissors), scissors beats paper (because scissors can cut paper), and paper beats rock (because paper can cover rock). Whoever wins gains one point; in the case of a tie, no one gains any points. (See Exercise 39 in Section 11.1, Strictly Determined Games.)

   a. Find the optimum strategy for each player and the value of the game.

   b. In hindsight, how could you have guessed the answer to part a without doing any work?

■ YOUR TURN ANSWERS

1. The strategy for player A is $(5/8, 3/8)$, and the strategy for player B is $(5/8, 3/8, 0)$. When these strategies are used, the value of the game is $-7/8$.

CHAPTER REVIEW

SUMMARY

Game theory is a tool for analyzing situations in which a person, known as a player, must choose from a finite set of choices. The payoff depends not only on what the player chooses but also on the choice made by a second player, usually considered the first player's opponent. The payoff matrix summarizes these payoffs. In many applications, the player is not a person but a company, an army, or some other organization. We introduced the following terms:

• zero-sum game (whatever one players wins is what the other player loses),
• dominated strategy (a row or column that never gives a better payoff than some other row or column, which is said to dominate it),

• saddle point (an entry that is the smallest in its row and the largest in its column), and
• strictly determined game (a game with a saddle point).

We saw that in a strictly determined game, each player has a single optimum choice determined by the saddle point. Otherwise, the best choice is a mixed strategy: The players randomly make choices according to certain probabilities. For a game in which each player has only two choices, we have developed formulas giving those probabilities. When at least one player has more than two choices, the optimum mixed strategies can be found using the simplex method.

Optimum Strategy in a 2 × 2 Non–Strictly Determined Game

With payoff matrix $M = \begin{bmatrix} a_{11} & a_{12} \\ a_{21} & a_{22} \end{bmatrix}$, let $d = a_{11} - a_{21} - a_{12} + a_{22}$.

The optimum strategy for player A is $[p_1 \quad p_2]$, where

$$p_1 = \frac{a_{22} - a_{21}}{d} \quad \text{and} \quad p_2 = \frac{a_{11} - a_{12}}{d}.$$

The optimum strategy for player B is $\begin{bmatrix} q_1 \\ q_2 \end{bmatrix}$, where

$$q_1 = \frac{a_{22} - a_{12}}{d} \quad \text{and} \quad q_2 = \frac{a_{11} - a_{21}}{d}.$$

The value of the game is

$$g = \frac{a_{11}a_{22} - a_{12}a_{21}}{d}.$$

Solving a Zero-Sum Game

1. Set up the payoff matrix M.

2. Remove any dominated rows or columns.

3. Check to see if there is a saddle point.

4. If there is no saddle point, add a positive number to all entries in the payoff matrix so that all entries are positive.

5. Set up the linear programming problem:

$$
\begin{aligned}
\text{Maximize} \quad & z = x_1 + \cdots + x_n \\
\text{subject to:} \quad & a_{11}x_1 + \cdots + a_{1n}x_n \leq 1 \\
& \qquad \vdots \qquad\qquad\quad \vdots \\
& a_{m1}x_1 + \cdots + a_{mn}x_n \leq 1 \\
\text{with} \quad & x_1 \geq 0, \cdots, x_n \geq 0.
\end{aligned}
$$

6. After finding the solution z to the linear programming problem, the value of the modified game is given by

$$g = 1/z.$$

The optimum strategy for player B is found by multiplying the solution to the linear programming problem by g:

$$q_i = g x_i, \quad i = 1, \ldots, n.$$

The optimum strategy for player A is found by multiplying the solution to the dual program by g:

$$p_j = g y_j, \quad j = 1, \ldots, m.$$

7. Subtract from g the positive number added in step 4 to find the value of the original game.

KEY TERMS

11.1
states of nature
strategies
payoff
payoff matrix
game theory

two-person game
zero-sum game
dominated strategy
dominates
optimum strategy
value of the game

pure strategy
saddle point
strictly determined
  game
fair game

11.2
mixed strategy

11.3
minimax principle

REVIEW EXERCISES

CONCEPT CHECK

Determine whether each of the statements is true or false, and explain why.

1. Game theory, as described in this chapter, can be applied to any number of players.

2. Some payoff matrices do not have a dominated strategy.

3. In a strictly determined game, each player has an optimum pure strategy.

4. A strictly determined game always has a saddle point.

5. In a fair game, the elements of the payoff matrix must sum to 0.

6. It is possible that the optimum strategy for one player in a game is pure while the optimum strategy for the other player is mixed.

7. A payoff matrix always has at least one dominated row or column.

8. When finding the optimum strategies with linear programming, the solution for the column player is found from the maximization problem, and the solution for the row player is found from the solution of the dual problem.

9. Linear programming can be used to find optimum mixed strategies when the players have more than two choices.

10. Linear programming can only be used to find optimum strategies when the value of the game is positive or has been made positive by adding a positive number to all entries in the payoff matrix.

PRACTICE AND EXPLORATIONS

11. How can you determine from the payoff matrix whether a game is strictly determined?

12. Briefly explain any advantages of removing from the payoff matrix any dominated strategies.

Use the following payoff matrix to determine the payoff if each of the strategies in Exercises 13–16 is used.

$$\begin{bmatrix} -3 & 5 & -6 & 2 \\ 0 & -1 & 9 & 5 \\ 2 & 6 & -4 & 3 \end{bmatrix}$$

13. $(1, 1)$

14. $(1, 4)$

15. $(2, 3)$

16. $(3, 4)$

17. Are there any dominated strategies in this game?

18. Is there a saddle point?

Remove any dominated strategies in the following games.

19. $\begin{bmatrix} -11 & 6 & 8 & 9 \\ -10 & -12 & 3 & 2 \end{bmatrix}$

20. $\begin{bmatrix} -1 & 9 & 0 \\ 4 & -10 & 6 \\ 8 & -6 & 7 \end{bmatrix}$

21. $\begin{bmatrix} -2 & 4 & 1 \\ 3 & 2 & 7 \\ -8 & 1 & 6 \\ 0 & 3 & 9 \end{bmatrix}$

22. $\begin{bmatrix} 3 & -1 & 4 \\ 0 & 4 & -1 \\ 1 & 2 & -3 \\ 0 & 0 & 2 \end{bmatrix}$

For the following games, find the strategies producing any saddle points. Give the value of the game. Identify any fair games.

23. $\begin{bmatrix} 5 & -4 \\ 3 & -3 \end{bmatrix}$

24. $\begin{bmatrix} -4 & 0 & -5 & 2 \\ 6 & 9 & 8 & 4 \end{bmatrix}$

25. $\begin{bmatrix} -4 & -1 \\ 6 & 0 \\ 8 & -3 \end{bmatrix}$

26. $\begin{bmatrix} -2 & 0 & -3 \\ -4 & 3 & -1 \\ 2 & 6 & 4 \end{bmatrix}$

27. $\begin{bmatrix} -1 & 4 & 3 & -4 \\ 8 & 1 & 2 & -7 \end{bmatrix}$

28. $\begin{bmatrix} 2 & -9 \\ 7 & 1 \\ 4 & 2 \end{bmatrix}$

Find the optimum strategies for each game using the method of Section 11.2. Find the value of the game.

29. $\begin{bmatrix} 1 & 0 \\ -2 & 3 \end{bmatrix}$

30. $\begin{bmatrix} 2 & -3 \\ -3 & 5 \end{bmatrix}$

31. $\begin{bmatrix} -3 & 5 \\ 1 & 0 \end{bmatrix}$

32. $\begin{bmatrix} 8 & -3 \\ -6 & 2 \end{bmatrix}$

For each game, remove any dominated strategies, then solve the game using the method of Section 11.2. Find the value of the game.

33. $\begin{bmatrix} -4 & 8 & 0 \\ -2 & 9 & -3 \end{bmatrix}$

34. $\begin{bmatrix} 1 & 0 & 3 & -3 \\ 4 & -2 & 4 & -1 \end{bmatrix}$

35. $\begin{bmatrix} 2 & -1 \\ -4 & 5 \\ -1 & -2 \end{bmatrix}$

36. $\begin{bmatrix} 8 & -6 \\ 4 & -8 \\ -9 & 9 \end{bmatrix}$

Find the optimum strategies for each game using the graphical method of linear programming. Find the value of the game.

37. $\begin{bmatrix} -4 & 2 \\ 3 & -5 \end{bmatrix}$

38. $\begin{bmatrix} -2 & 2 \\ 3 & 1 \end{bmatrix}$

39. $\begin{bmatrix} 1 & 0 \\ -3 & 4 \end{bmatrix}$

40. $\begin{bmatrix} 0 & -2 \\ -1 & 3 \end{bmatrix}$

Find the optimum strategies for each game using the simplex method. Find the value of the game.

41. $\begin{bmatrix} 2 & 1 & -1 \\ -3 & -2 & 0 \end{bmatrix}$

42. $\begin{bmatrix} 1 & -3 \\ -4 & 2 \\ -2 & 1 \end{bmatrix}$

43. $\begin{bmatrix} -2 & 1 & 0 \\ 2 & 0 & -2 \\ 0 & -1 & 3 \end{bmatrix}$

44. $\begin{bmatrix} 2 & 1 & -1 \\ 0 & 1 & 2 \\ -1 & 2 & 0 \end{bmatrix}$

45. Under what conditions is it necessary to use the simplex algorithm to solve a game?

46. Suppose you live in a country where the chance of rain is 90% every day. Should you carry an umbrella every day? Now suppose you lose 100 points if it rains and you don't carry an umbrella. Suppose further that if you carry an umbrella and it doesn't rain, you lose 1 point due to the inconvenience. Otherwise you neither gain nor lose any points. According to game theory, what should you do? Discuss any discrepancies with your first answer.

APPLICATIONS

Business and Economics

Labor Relations In labor-management relations, both labor and management can adopt either a friendly or a hostile attitude. The results are shown in the following payoff matrix. The numbers give the wage gains made by an average worker.

		Management	
		Friendly	Hostile
Labor	Friendly	$700	$900
	Hostile	$500	$1100

47. Suppose the chief negotiator for labor is an optimist. What strategy should he choose?

48. What strategy should he choose if he is a pessimist?

49. The chief negotiator for labor feels that there is a 40% chance that the company will be hostile. What strategy should he adopt? What is the expected payoff?

50. Just before negotiations begin, a new management is installed in the company. There is now a 60% chance that the new management will be hostile. What strategy should be adopted by labor? What is the expected payoff?

51. Find the optimum strategy for labor and management, and find the value of the game.

Investment Victor de Bouchel, who has inherited $10,000 from his rich grandmother, consults the firm of J. K. Knowitall & Company for advice on how to invest the sum. The company provides the following estimates of the gains he might make in five years by investing the $10,000 in two different types of stocks, each dependent on the state of the economy.

$$
\begin{array}{cc}
 & \textit{Economy} \\
 & \begin{array}{cc} \text{Inflationary} & \text{Stable} \end{array} \\
\textit{Stocks} \begin{array}{c} \text{Blue-Chip} \\ \text{Growth} \end{array} & \begin{bmatrix} 2800 & 3200 \\ 5000 & -2000 \end{bmatrix}
\end{array}
$$

52. Find the optimum strategy for Victor and the value of the game using the method found in Section 11.2.

53. Find the optimum strategy for Victor and the value of the game using the graphical method.

54. Find the optimum strategy for Victor and the value of the game using the simplex algorithm.

Social Sciences

Politics Martha McDonald, a candidate for city council, must decide whether to come out in favor of a new factory, be opposed to it, or waffle on the issue. The change in votes for Martha depends on what her opponent, Kevin Wallace, does, with payoffs as shown.

$$
\begin{array}{cc}
 & \textit{Kevin} \\
 & \begin{array}{ccc} \text{Favors} & \text{Waffles} & \text{Opposes} \end{array} \\
\textit{Martha} \begin{array}{c} \text{Favors} \\ \text{Waffles} \\ \text{Opposes} \end{array} & \begin{bmatrix} 0 & 1200 & 4200 \\ -1000 & 0 & 600 \\ -5000 & -2000 & 0 \end{bmatrix}
\end{array}
$$

55. What should Martha do if she is an optimist?

56. What should she do if she is a pessimist?

57. Suppose Martha's campaign manager feels there is a 40% chance that Kevin will favor the plant and a 35% chance that he will waffle. What strategy should Martha adopt? What is the expected change in the number of votes?

58. Kevin conducts a new poll that shows strong support for the new factory. This changes the probability that he will oppose the factory to 0 and the probability that he will waffle to 0.3. What strategy should Martha adopt? What is the expected change in the number of votes now?

59. Find the optimum strategy for Martha and her opponent, and find the value of the game.

General Interest

Military The little kingdom of Ravogna has two military installations, one three times as valuable as the other. The army has the capability to successfully defend either one of the installations, but not both at the same time. Rontovia, a country which has historically been antagonistic toward Ravogna, is capable of attacking either installation, but not both. The payoff matrix below indicates the respective values of the installations to Ravogna. Installation No. 2, with a value of 1, is the lesser installation.

$$
\begin{array}{cc}
 & \begin{array}{c} \textit{Rontovia:} \\ \textit{Attack Installation} \end{array} \\
 & \begin{array}{cc} 1 & 2 \end{array} \\
\begin{array}{c} \textit{Ravogna:} \\ \textit{Defend Installation} \end{array} \begin{array}{c} 1 \\ 2 \end{array} & \begin{bmatrix} 4 & 1 \\ 3 & 4 \end{bmatrix}
\end{array}
$$

60. Find the optimum strategy for each country and the value of the game using the method described in Section 11.2, Mixed Strategies.

61. Find the optimum strategy for each country and the value of the game using the graphical method.

62. Find the optimum strategy for each country and the value of the game using the simplex method.

63. **Theology** In his book entitled *Pensées*, the French mathematician Blaise Pascal (1623–1662) described what has since been referred to as Pascal's Wager. Pascal observed that everyone must wager on whether God exists. If you decide that God exists and in fact God does exist, you gain "an infinity of an infinitely happy life"; if you are wrong, you lose nothing of great value, having lived a good life without receiving any reward for it. If you choose not to believe in God and God actually does exist, you will have lost something of infinite value. Pascal concludes that believing in God is by far the most rational action. Write a few sentences discussing Pascal's Wager, including how the payoff matrix might be set up and what the optimum strategy would be. What is your reaction to Pascal's conclusion?

EXTENDED APPLICATION

THE PRISONER'S DILEMMA—NON-ZERO-SUM GAMES IN ECONOMICS

In this chapter we have looked at two-person, zero-sum games, games in which the total payoff is zero, so that what one player wins the other loses. When economists use game theory to model the marketplace, they encounter many non-zero-sum games, games in which the total payoff to all players might be positive or negative. For example, the total value of the stock market increases in periods of optimistic buying and plunges when everyone decides to sell, so the total payoff from investing is not fixed at 0. These games are more complicated to analyze, but the idea of dominated strategies introduced in Section 11.1 turns out to be useful. In this application we look at one of the simplest and most famous non-zero-sum games, the Prisoner's Dilemma.

The classic scenario goes like this: Mike and Ike are arrested and charged with committing a murder. Taken to separate cells, each has to decide whether to deny any involvement or confess. If they both deny any involvement, there is still enough evidence to convict them both of manslaughter, and they'll both get 5-year sentences. If Mike confesses and implicates Ike as the actual murderer, while Ike claims innocence, Mike will get 1 year and Ike will get 30 years. If instead Ike cuts a deal with the prosecutors, he'll get the light sentence and Mike will get the long one. If they both confess, each implicating the other, they'll both get 10-year sentences. Now *each* player has a payoff matrix in which all the payoffs are negative. Letting D stand for "deny" and C for "confess," Mike's matrix looks like this:

Mike's Payoffs

Ike's Strategy

$$\text{Mike's Strategy} \begin{array}{c} \\ C \\ D \end{array} \begin{array}{cc} C & D \\ \begin{bmatrix} -10 & -1 \\ -30 & -5 \end{bmatrix} \end{array}$$

What should Mike do? Using the idea of a dominated strategy, Mike notices that each entry in his first row is larger than each entry in the second row, so D is a dominated strategy: If Ike plays C, Mike will do better with C, and if Ike plays D, Mike will also do better with C, so it looks like C is Mike's best choice. Ike's matrix is similar:

Ike's Payoffs

Ike's Strategy

$$\text{Mike's Strategy} \begin{array}{c} \\ C \\ D \end{array} \begin{array}{cc} C & D \\ \begin{bmatrix} -10 & -1 \\ -30 & -5 \end{bmatrix} \end{array}$$

For Ike, the first column (strategy C) dominates the second column (D), so he also decides to play C. The result is that both prisoners get 10-year sentences; they both lose, but neither gets the worst possible outcome, the 30-year sentence. If neither player can predict what the other will do, confessing is a sensible strategy for both players. But notice that if they could agree beforehand to deny any involvement, they would *both* do better, getting only 5-year sentences!

Game theorists call a game in which players can make binding agreements a *cooperative game*. If Mike and Ike play noncoopera-

tively, they'll both end up with 10-year sentences, but if they play cooperatively, they can engineer a better outcome for each of them. There's just one catch: Once they have agreed to deny involvement, each player has a strong incentive to *defect* from the agreement and plead guilty, getting off with a 1-year sentence and leaving the other player with 30 years in jail. And, of course, if they *both* defect they'll end up with 10-year sentences after all.

The Prisoner's Dilemma models economic transactions in which cooperating may lead to greater gains for both players than competing, but in which the success of the cooperative strategy depends on trust. For example, suppose two art dealers have private information about a painting up for auction. They know that they'll be able to resell it for $1 million. If they bid against each other, the winning bidder will probably end up paying close to $1 million, so the profit from resale will be small. But suppose the dealers agree beforehand that only one will bid and that they will split the profits from resale. Without competition, the winning bidder may pay much less than $1 million, and both dealers will make out well. (This arrangement is called a *ring* in the auction business, and, of course, the auction houses don't like rings!) As with the Prisoner's Dilemma, trust is essential to the success of the ring. After all, the winning bidder could just deny that there was any agreement and pocket the entire profit.

In the Exercises we'll look at some variations of a two-person, non-zero-sum game called the Restaurant Game. Again, though the game is simple, the issues it raises appear in many settings involving competition and negotiation, including an arms race, labor–management bargaining, and sharing of resources between species occupying the same territory. Game theory has become an important tool in economics, sociology, and evolutionary biology.*

EXERCISES

The Restaurant Game   Linda and Mel both like to eat out; Linda likes the neighborhood Chinese restaurant and Mel likes the French one. At 5 o'clock they each announce their pick for the evening's restaurant. If they pick the same restaurant, they go out, otherwise they stay home and microwave some frozen dinners. Each player has two strategies, C for Chinese restaurant and F for French restaurant.

1. Since they like to eat out, each prefers a restaurant meal to the frozen dinner, but they enjoy their favorite food much more than the other type. Suppose that Linda and Mel have the following payoff matrices, where the numbers represent degree of enjoyment:

Linda's Payoffs

Mel's Strategy

$$\text{Linda's Strategy} \begin{array}{c} \\ C \\ F \end{array} \begin{array}{cc} C & F \\ \begin{bmatrix} 5 & 0 \\ 0 & 2 \end{bmatrix} \end{array}$$

*For more on the basics of Game Theory, see Morris, Peter, *Introduction to Game Theory*, New York: Springer, 1994. Morris has a good discussion of both zero-sum and non-zero-sum games. The Restaurant Game presented in the Exercises is a version of Morris's Battle of the Buddies.

Mel's Payoffs

Mel's Strategy

		C	F
Linda's Strategy	C	2	0
	F	0	5

Does either player have a dominated strategy? How should they resolve their dilemma?

2. If Linda likes French food more than Mel likes Chinese food, their matrices might look like this:

Linda's Payoffs

Mel's Strategy

		C	F
Linda's Strategy	C	5	0
	F	0	3

Mel's Payoffs

Mel's Strategy

		C	F
Linda's Strategy	C	1	0
	F	0	5

Does either player have a dominated strategy? If they decide to cooperate, how would they pick their restaurants for maximum combined enjoyment?

3. The suspects in the Prisoner's Dilemma play their game only once, but Linda and Mel repeat the Restaurant Game each night. As we noted in Section 11.2, this gives them the option of using mixed strategies. Suppose Linda chooses Chinese with probability 0.8 and French with probability 0.2, while Mel chooses Chinese with probability 0.1 and French with probability 0.9. What is Linda's expected payoff in "enjoyment units"? What is Mel's expected payoff? Who does better?

4. Suppose Linda knows that Mel is going to stick to his strategy (Chinese with probability 0.1 and French with probability 0.9). What strategy maximizes her enjoyment? What is her expected payoff? If she plays this way, what might Mel do?

5. In repeated games, a player can use an *adaptive* strategy that changes based on past history. For example, if Mel seems to be following an "always French" strategy, Linda could decide to pick Chinese a few times in a row. What are the possible hazards and benefits of this change in strategy?

6. Suppose Linda and Mel decide to cooperate. They like eating out, so they'll eat out every night, using a spinner to determine the type of restaurant. Given that Linda likes French more than Mel likes Chinese, what do you think would be a fair way of making the choice? Consider a picture like Figure 6 in Section 11.2.

DIRECTIONS FOR GROUP PROJECT

Suppose that you work for the prosecutor's office and after talking with two detectives assigned to the case, you believe that Ike and Mike have developed a plan of cooperation. Prepare a role-playing activity between you, the detectives, and Mike to try to convince him to play the "game" uncooperatively. That is, convince Mike to confess and implicate Ike.

Appendix A

Solutions to Prerequisite Skills Diagnostic Test *(with references to Ch. R)*

For more practice on the material in questions 1–4, see *Beginning and Intermediate Algebra* (5th ed.) by Margaret L. Lial, John Hornsby, and Terry McGinnis, Pearson, 2012.

1. $10/50 = 0.20 = 20\%$

2.
$$\frac{13}{7} - \frac{2}{5} = \frac{13}{7} \cdot \frac{5}{5} - \frac{2}{5} \cdot \frac{7}{7} \qquad \text{Get a common denominator.}$$
$$= \frac{65}{35} - \frac{14}{35}$$
$$= \frac{51}{35}$$

3. The total number of apples and oranges is $x + y$, so $x + y = 75$.

4. The sentence can be rephrased as "The number of students is at least four times the number of professors," or $s \geq 4p$.

5.
$$7k + 8 = -4(3 - k)$$
$$7k + 8 = -12 + 4k \qquad\qquad\qquad\qquad\quad \text{Multiply out.}$$
$$7k - 4k + 8 - 8 = -12 - 8 + 4k - 4k \qquad \text{Subtract 8 and } 4k \text{ from both sides.}$$
$$3k = -20 \qquad\qquad\qquad\qquad\qquad\quad\ \text{Simplify.}$$
$$k = -20/3 \qquad\qquad\qquad\qquad\qquad \text{Divide both sides by 3.}$$

For more practice, see Sec. R.4.

6.
$$\frac{5}{8}x + \frac{1}{16}x = \frac{11}{16} + x$$

$$\frac{5}{8}x + \frac{1}{16}x - x = \frac{11}{16} \qquad\qquad\qquad \text{Subtract } x \text{ from both sides.}$$

$$\frac{5x}{8} \cdot \frac{2}{2} + \frac{x}{16} - x \cdot \frac{16}{16} = \frac{11}{16} \qquad\quad \text{Get a common denominator.}$$

$$\frac{10x}{16} + \frac{x}{16} - \frac{16x}{16} = \frac{11}{16} \qquad\qquad \text{Simplify.}$$

$$\frac{-5x}{16} = \frac{11}{16} \qquad\qquad\qquad\qquad \text{Simplify.}$$

$$x = \frac{11}{16} \cdot \frac{16}{-5} = -\frac{11}{5} \qquad\qquad \text{Multiply both sides by the reciprocal of } -5/16.$$

For more practice, see Sec. R.4.

7. The interval $-2 < x \leq 5$ is written as $(-2, 5]$. For more practice, see Sec. R.5.

8. The interval $(-\infty, -3]$ is written as $x \leq -3$. For more practice, see Sec. R.5.

9.
$$5(y - 2) + 1 \leq 7y + 8$$
$$5y - 9 \leq 7y + 8 \qquad\qquad\qquad\qquad \text{Multiply out and simplify.}$$
$$5y - 7y - 9 + 9 \leq 7y - 7y + 8 + 9 \qquad \text{Subtract } 7y \text{ from both sides and add 9.}$$
$$-2y \leq 17 \qquad\qquad\qquad\qquad\qquad\ \text{Simplify.}$$
$$y \geq -17/2 \qquad\qquad\qquad\qquad\qquad \text{Divide both sides by 3.}$$

10.
$$\frac{2}{3}(5p - 3) > \frac{3}{4}(2p + 1)$$

$$\frac{10p}{3} - 2 > \frac{3p}{2} + \frac{3}{4} \qquad\qquad\qquad\qquad \text{Multiply out and simplify.}$$

$$\frac{10p}{3} - \frac{3p}{2} - 2 + 2 > \frac{3p}{2} - \frac{3p}{2} + \frac{3}{4} + 2 \qquad \text{Subtract } 3p/2 \text{ from both sides, and add 2.}$$

$$\frac{10p}{3} \cdot \frac{2}{2} - \frac{3p}{2} \cdot \frac{3}{3} > \frac{3}{4} + 2 \cdot \frac{4}{4} \qquad\qquad \text{Simplify and get a common denominator.}$$

$$\frac{11p}{6} > \frac{11}{4} \qquad\qquad\qquad\qquad\qquad\qquad \text{Simplify.}$$

$$p > \frac{11}{4} \cdot \frac{6}{11} \qquad\qquad\qquad\qquad\qquad\ \text{Multiply both sides by the reciprocal of } 11/6.$$

$$p > \frac{3}{2} \qquad\qquad\qquad\qquad\qquad\qquad\quad \text{Simplify.}$$

For more practice, see Sec. R.5.

Appendix B
Learning Objectives

CHAPTER R: Algebra Reference

R.1: Polynomials

1. Simplify polynomials

R.2: Factoring

1. Factor polynomials

R.3: Rational Expressions

1. Simplify rational expression using properties of rational expressions and order of operations

R.4: Equations

1. Solve linear, quadratic, and rational equations

R.5: Inequalities

1. Solve linear, quadratic, and rational inequalities

2. Graph the solution of linear, quadratic, and rational inequalities

3. Write the solutions of linear, quadratic, and rational inequalities using interval notation

R.6: Exponents

1. Evaluate exponential expressions

2. Simplify exponential expressions

R.7: Radicals

1. Simplify radical expressions

2. Rationalize the denominator in radical expressions

3. Rationalize the numerator in radical expressions

CHAPTER 1:   Linear Functions

1.1: Slopes and Equations of Lines

1. Find the slope of a line

2. Find the equation of a line using a point and the slope

3. Find the equation of parallel and perpendicular lines

4. Graph the equation of a line

5. Solve application problems using linear functions

1.2: Linear Functions and Applications

1. Evaluate linear functions

2. Write equations for linear models

3. Solve application problems

1.3: The Least Squares Line

1. Interpret the different value meanings for linear correlation

2. Draw a scatterplot

3. Calculate the correlation coefficient

4. Calculate the least squares line

5. Compute the response variable in a linear model

CHAPTER 2:   Systems of Linear Equations and Matrices

2.1: Solution of Linear Systems by the Echelon Method

1. Apply system transformation operations on matrices

2. Solve linear systems using the echelon method

3. Solve application problems

2.2: Solution of Linear Systems by the Gauss-Jordan Method

1. Perform row operations on matrices

2. Solve linear systems by the Gauss-Jordan method

3. Solve application problems

2.3: Addition and Subtraction of Matrices

1. Identify the size of a matrix

2. Add and subtract matrices

3. Solve application problems

2.4: Multiplication of Matrices

1. Multiply a matrix by a constant value

2. Find the product two matrices

3. Simplify matrix expressions

4. Solve application problems

2.5: Matrix Inverses

1. Determine if two matrices are inverses of each other

2. Find the inverse of a matrix (if it exists)

3. Solve a system by using the inverse

4. Solve application problems

2.6: Input-Output Models

1. Find the production matrix for an input-output model

2. Find the demand matrix for an input-output model

3. Solve application problems

CHAPTER 3:   Linear Programming: The Graphical Method

3.1: Graphing Linear Inequalities

1. Graph linear inequalities

2. Graph a system of linear inequalities

3. Determine the feasible region for a system of linear inequalities

4. Solve application problems

3.2: Solving Linear Programming Problems Graphically

1. Determine corner points

2. Solve linear programming problems graphically

3.3: Applications of Linear Programming

1. Solve application problems

CHAPTER 4: Linear Programming: The Simplex Method

4.1: Slack Variables and the Pivot

1. Determine the number of slack variables needed for a linear programming problem
2. Add slack variables to a linear programming problem
3. Generate the initial simplex tableau
4. Identify the pivot and find the resulting matrix
5. Solve linear programming problems using the simplex tableau
6. Solve application problems

4.2: Maximization Problems

1. Solve maximization problems using the simplex tableau and simplex method
2. Solve maximization application problems

4.3: Minimization Problems; Duality

1. Find the transpose of a matrix
2. Generate the dual problem
3. Solve minimization problems using the simplex method
4. Solve minimization application problems

4.4: Nonstandard Problems

1. Solve nonstandard linear programming problems

CHAPTER 5: Mathematics of Finance

5.1: Simple and Compound Interest

1. Compute the simple and compound interest
2. Compute the effective rate
3. Solve application problems

5.2: Future Value of an Annuity

1. Find terms of a geometric sequence
2. Find the sum of a geometric sequence
3. Find the future value of an annuity
4. Find the future value of an annuity due
5. Find the payment for a sinking fund
6. Solve application problems

5.3: Present Value of an Annuity; Amortization

1. Find the present value of an annuity
2. Find the payment needed for an amortized load
3. Generate an amortization table
4. Solve application problems

CHAPTER 6: Logic

6.1: Statements

1. Identify statements and compound statements
2. Generate a negation for a statement
3. Convert between statements and symbols

4. Determine the truth value of a compound statement

5. Solve application problems

6.2: Truth Tables and Equivalent Statements

1. Generate a truth table for a compound statement

2. Write the negation for a compound statement

3. Simplify compound statements using De Morgan's law

4. Determine if compound statements are true or false

5. Solve application problems

6.3: The Conditional and Circuits

1. Identify tautologies

2. Identify contradictions

3. Write equivalent statements

4. Simplify circuit problems

5. Draw circuits for statements

6. Solve application problems

6.4: More on the Conditional

1. Generate inverse, converse, and contrapositive statement

2. Solve application problems

6.5: Analyzing Arguments and Proofs

1. Identify valid and invalid arguments

2. Use truth tables to prove rules of logic

3. Solve application problems

6.6: Analyzing Arguments with Quantifiers

1. Use Euler diagrams to test the validity of arguments

2. Construct valid arguments based on Euler diagrams

3. Solve application problems

CHAPTER 7:   Sets and Probability

7.1: Sets

1. Understand set notation and terminology

2. Find the union and intersection of sets

3. Solve application problems

7.2: Applications of Venn Diagrams

1. Draw and interpret the Venn diagram for a set

2. Find the number of elements in a set

3. Solve application problems

7.3: Introduction to Probability

1. Generate the sample space for a given experiment

2. Find the probability of an event

3. Identify empirical probabilities

4. Solve application problems

7.4: Basic Concepts of Probability

1. Identify mutually exclusive events

2. Find the probability of an event using probability rules

3. Find the odds of an event

4. Solve application problems

7.5: Conditional Probability; Independent Events

1. Identify independent events

2. Use the product rule to find the probability of an event

3. Find the conditional probability of an event

4. Solve application problems

7.6: Bayes' Theorem

1. Use Bayes' theorem

2. Solve application problems

CHAPTER 8:   Counting Principles, Further Probability Topics

8.1: The Multiplication Principle; Permutations

1. Compute the factorial of a number

2. Use the multiplication principle to count outcomes

3. Find the permutation of an outcome

4. Solve application problems

8.2: Combinations

1. Find the combinations of an outcome

2. Solve problems using combinations or permutations

3. Solve application problems

8.3: Probability Applications of Counting Principles

1. Solve probability applications using counting principles

8.4: Binomial Probability

1. Find the probability of binomial experiments

2. Use technology to find binomial probabilities

3. Solve application problems

8.5: Probability Distribution; Expected Value

1. Determine the random variable for a probability distribution

2. Draw a histogram for a probability distribution

3. Generate probability distributions

4. Find the expected value of a probability distribution

5. Solve application problems

CHAPTER 9:   Statistics

9.1: Frequency Distributions; Measures of Central Tendency

1. Generate a frequency distribution table

2. Find the mean, median, and mode of a data set

3. Solve application problems

9.2: Measures of Variation

1. Find the range of a data set

2. Find the standard deviation of a data set

3. Solve application problems

9.3: The Normal Distribution

1. Find the area (probability) under the normal curve

2. Understand the standard normal curve

3. Find the z-score associated with a given area (probability) under the normal curve

4. Solve application problems

9.4: Normal Approximation to the Binomial Distribution

1. Use the normal curve distribution to find binomial distribution probabilities

2. Solve application problems

CHAPTER 10:   Markov Chains

10.1: Basic Properties of Markov Chains

1. Understand probability vectors and transition matrices

2. Solve application problems

10.2: Regular Markov Chains

1. Identify regular transition matrices

2. Find the equilibrium vector for transition matrices

3. Solve application problems

10.3: Absorbing Markov Chains

1. Find the absorbing states for transition matrices

2. Find the fundamental matrix for a Markov chain

3. Solve application problems

CHAPTER 11:   Game Theory

11.1: Strictly Determined Games

1. Find the payoff for a given strategy

2. Understand saddle points and their relation to strategies

3. Solve application problems

11.2: Mixed Strategies

1. Find the expected value of a game

2. Find optimum strategies for players

3. Solve application problems

11.3: Game Theory and Linear Programming

1. Find optimum strategies graphically

2. Find optimum strategies using the simplex method

3. Solve application problems

Appendix C
Logarithms

In Sections 5.1 and 5.3, we encountered problems in the mathematics of finance that required solving an exponential equation. Such equations may be solved using **logarithms**, whose properties are briefly reviewed in this appendix.

> **Logarithm**
> For $a > 0$, $a \neq 1$, and $x > 0$,
> $$y = \log_a x \quad \text{means} \quad a^y = x.$$

(Read $y = \log_a x$ as "y is the logarithm of x to the base a.") For example, the exponential statement $2^4 = 16$ can be translated into the logarithmic statement $4 = \log_2 16$. A logarithm is an exponent: **$\log_a x$ is the exponent used with the base a to get x.**

EXAMPLE 1 Equivalent Expressions

This example shows the same statements written in both exponential and logarithmic forms.

Exponential Form	Logarithmic Form
(a) $3^2 = 9$	$\log_3 9 = 2$
(b) $(1/5)^{-2} = 25$	$\log_{1/5} 25 = -2$
(c) $4^{-3} = 1/64$	$\log_4(1/64) = -3$
(d) $5^0 = 1$	$\log_5 1 = 0$

EXAMPLE 2 Evaluating Logarithms

Evaluate each of the following logarithms.

(a) $\log_4 64$

SOLUTION We seek a number x such that $4^x = 64$. Since $4^3 = 64$, we conclude that $\log_4 64 = 3$.

(b) $\log_2(-8)$

SOLUTION We seek a number x such that $2^x = -8$. Since 2^x is positive for all real numbers x, we conclude that $\log_2(-8)$ is undefined. (Actually, $\log_2(-8)$ can be defined if we use complex numbers, but in this textbook, we restrict ourselves to real numbers.)

Properties of Logarithms

The usefulness of logarithmic functions depends in large part on the following **properties of logarithms**.

> **Properties of Logarithms**
> Let x and y be any positive real numbers and r be any real number. Let a be a positive real number, $a \neq 1$. Then
>
> **a.** $\log_a xy = \log_a x + \log_a y$
>
> **b.** $\log_a \dfrac{x}{y} = \log_a x - \log_a y$
>
> **c.** $\log_a x^r = r \log_a x$
>
> **d.** $\log_a a = 1$
>
> **e.** $\log_a 1 = 0$
>
> **f.** $\log_a a^r = r$.

EXAMPLE 3   **Properties of Logarithms**

If all the following variable expressions represent positive numbers, then for $a > 0$, $a \neq 1$, the statements in (a)–(c) are true.

(a) $\log_a x + \log_a(x - 1) = \log_a x(x - 1)$

(b) $\log_a \dfrac{x^2 - 4x}{x + 6} = \log_a(x^2 - 4x) - \log_a(x + 6)$

(c) $\log_a(9x^5) = \log_a 9 + \log_a(x^5) = \log_a 9 + 5 \cdot \log_a x$

Evaluating Logarithms

Since our number system has base 10, logarithms to base 10 were most convenient for numerical calculations and so base 10 logarithms were called **common logarithms**. Common logarithms are still useful in other applications. For simplicity,

$$\log_{10} x \text{ is abbreviated } \log x.$$

Most practical applications of logarithms use the number e as base. (Recall that to 7 decimal places, $e = 2.7182818$.) Logarithms to base e are called **natural logarithms**, and

$$\log_e x \text{ is abbreviated } \ln x$$

(read "el-en x").

NOTE Keep in mind that $\ln x$ is a logarithmic function. Therefore, all of the properties of logarithms given previously are valid when a is replaced with e and $\log_e$ is replaced with $\ln$.

A calculator can be used to find both common and natural logarithms. For example, using a calculator and 4 decimal places, we get the following values.

$$\log 2.34 = 0.3692, \qquad \log 594 = 2.7738, \qquad \text{and} \qquad \log 0.0028 = -2.5528.$$

$$\ln 2.34 = 0.8502, \qquad \ln 594 = 6.3869, \qquad \text{and} \qquad \ln 0.0028 = -5.8781.$$

Notice that logarithms of numbers less than 1 are negative when the base is greater than 1.

Exponential Equations

Equations such as $3^x = 5$ can be solved approximately with a graphing calculator, but an algebraic method is also useful, particularly when the equation involves variables such as a and b rather than just numbers such as 3 and 5. A general method for solving these equations is shown in the following example.

EXAMPLE 4   **Solving Exponential Equations**

Solve each equation.

(a) $3^x = 5$

SOLUTION   Taking natural logarithms (logarithms to any base could be used) on both sides gives

$$\ln 3^x = \ln 5$$

$$x \ln 3 = \ln 5 \qquad\qquad \ln u^r = r \ln u$$

$$x = \frac{\ln 5}{\ln 3} \approx 1.465$$

(b) $3^{2x} = 4^{x+1}$

SOLUTION Taking natural logarithms on both sides gives

$$\ln 3^{2x} = \ln 4^{x+1}$$

$$2x \ln 3 = (x + 1) \ln 4 \qquad \text{In } u^r = r \ln u$$

$$(2 \ln 3)x = (\ln 4)x + \ln 4$$

$$(2 \ln 3)x - (\ln 4)x = \ln 4 \qquad \text{Subtract } (\ln 4)x \text{ from both sides.}$$

$$(2 \ln 3 - \ln 4)x = \ln 4 \qquad \text{Factor } x.$$

$$x = \frac{\ln 4}{2 \ln 3 - \ln 4}. \qquad \text{Divide both sides by 2 In 3 − In 4.}$$

Use a calculator to evaluate the logarithms, then divide, to get

$$x \approx \frac{1.3863}{2(1.0986) - 1.3863} \approx 1.710.$$

(c) $5e^{0.01x} = 9$

SOLUTION

$$e^{0.01x} = \frac{9}{5} = 1.8 \qquad \text{Divide both sides by 5.}$$

$$\ln e^{0.01x} = \ln 1.8 \qquad \text{Take natural logarithms on both sides.}$$

$$0.01x = \ln 1.8 \qquad \text{In } e^u = u$$

$$x = \frac{\ln 1.8}{0.01} \approx 58.779$$

EXERCISES

Write each exponential equation in logarithmic form.

1. $5^3 = 125$ **2.** $7^2 = 49$

3. $3^4 = 81$ **4.** $2^7 = 128$

5. $3^{-2} = \frac{1}{9}$ **6.** $\left(\frac{5}{4}\right)^{-2} = \frac{16}{25}$

Write each logarithmic equation in exponential form.

7. $\log_2 32 = 5$ **8.** $\log_3 81 = 4$

9. $\ln \frac{1}{e} = -1$ **10.** $\log_2 \frac{1}{8} = -3$

11. $\log 100,000 = 5$ **12.** $\log 0.001 = -3$

Evaluate each logarithm without using a calculator.

13. $\log_8 64$ **14.** $\log_9 81$

15. $\log_4 64$ **16.** $\log_3 27$

17. $\log_2 \frac{1}{16}$ **18.** $\log_3 \frac{1}{81}$

19. $\log_2 \sqrt[3]{\frac{1}{4}}$ **20.** $\log_8 \sqrt[4]{\frac{1}{2}}$

21. $\ln e$ **22.** $\ln e^3$

23. $\ln e^{5/3}$ **24.** $\ln 1$

25. Is the "logarithm to the base 3 of 4" written as $\log_4 3$ or $\log_3 4$?

26. Write a few sentences describing the relationship between e^x and $\ln x$.

Use the properties of logarithms to write each expression as a sum, difference, or product of simpler logarithms. For example, $\log_2(\sqrt{3}x) = \frac{1}{2} \log_2 3 + \log_2 x$.

27. $\log_5(3k)$ **28.** $\log_9(4m)$

29. $\log_3 \frac{3p}{5k}$ **30.** $\log_7 \frac{15p}{7y}$

31. $\ln \frac{3\sqrt{5}}{\sqrt[3]{6}}$ **32.** $\ln \frac{9\sqrt[3]{5}}{\sqrt[4]{3}}$

Solve each equation in Exercises 33–40. Round decimal answers to four decimal places.

33. $2^x = 6$ **34.** $5^x = 12$

35. $e^{k-1} = 6$ **36.** $e^{2y} = 15$

37. $3^{x+1} = 5^x$ **38.** $2^{x+1} = 6^{x-1}$

39. $5(0.10)^x = 4(0.12)^x$ **40.** $1.5(1.05)^x = 2(1.01)^x$

Appendix D

MathPrint Operating System for TI-84 and TI-84 Plus Silver Edition

The graphing calculator screens in this text display math in the format of the TI MathPrint operating system. With MathPrint, the math looks more like that seen in a printed book. You can obtain MathPrint and install it by following the instructions in the *Graphing Calculator and Spreadsheet Manual*. Only the TI-84 family of graphing calculators can be updated with the MathPrint operating system. If you own a TI-83 graphing calculator, you can use this brief appendix to help you "translate" what you see in the Classic mode shown on your calculator.

Translating between MathPrint Mode and Classic Mode
The following table compares displays of several types in MathPrint mode and Classic mode (on a calculator without MathPrint installed).

Feature	MathPrint	Classic (MathPrint not installed)
Improper fractions	$\dfrac{2}{5} - \dfrac{1}{3}$ $\dfrac{1}{15}$ Press **ALPHA** and F1 and select 1: n/d.	2/5 − 1/3 ▶ Frac 1/15 Enter an expression and press **MATH** and select 1:▶Frac.
Mixed fractions	$2\frac{1}{5} * (3\frac{2}{3})$ $\dfrac{121}{15}$ Press **ALPHA** and F1 and select 2: Un/d.	Not Supported
Absolute values	|10 − 15| 5 Press **ALPHA** and F2 and select 1:abs(.	abs(10 − 15) 5 Press **MATH** and ▶ and select 1:abs(.
Summation	$\sum\limits_{I=1}^{10} (I^2)$ 385 Press **ALPHA** and F2 and select 2:Σ(.	sum(seq(I²,I,1,10) 385 Press **2ND** and LIST and then ▶ twice and select 5:sum(for sum, and press **2ND** and LIST and then ▶ and select 5:seq(for sequence.
Numerical derivatives	$\left.\dfrac{d}{dX}(X^2)\right\|_{X=3}$ 6 Press **ALPHA** and F2 and select 3:nDeriv(.	nDeriv(X²,X,3) 6 Press **MATH** and select 8:nDeriv(.
Numerical values of integrals	$\int_{1}^{5}(X^2)dX$ 41.33333333 Press **ALPHA** and F2 and select 4:fnInt(.	fnInt(X²,X,1,5) 41.33333333 Press **MATH** and select 9:fnInt(.

(continued)

Feature	MathPrint	Classic (MathPrint not installed)
Logarithms	$\log_2(32)$ 5 Press **ALPHA** and F2 and select 5:logBASE(.	Evaluating logs with bases other than 10 or e cannot be done on a graphing calculator if the MathPrint operating system is not installed. To evaluate $\log_2 32$, use the change-of-base formula: $\log(32)/\log(2)$ 5

The Y= Editor

MathPrint features can be accessed from the Y= editor as well as from the home screen. The following table shows examples that illustrate differences between MathPrint in the Y= editor and Classic mode.

Feature	MathPrint	Classic (MathPrint not installed)
Graphing the derivative of $y = x^2$	Plot 1 Plot 2 Plot 3 $\backslash Y_1 = \frac{d}{dX}(X^2)\big|_{X=X}$ $\backslash Y_2 =$	Plot 1 Plot 2 Plot 3 $\backslash Y_1 =$ nDeriv(X^2,X,X) $\backslash Y_2 =$
Graphing an antiderivative of $y = x^2$	Plot 1 Plot 2 Plot 3 $\backslash Y_1 = \int_0^X (X^2)dX$ $\backslash Y_2 =$	Plot 1 Plot 2 Plot 3 $\backslash Y_1 =$ fnInt(X^2,X,0,X) $\backslash Y_2 =$

Appendix E

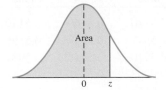

Area Under a Normal Curve to the Left of z, where $z = \dfrac{x - \mu}{\sigma}$

z	0.00	0.01	0.02	0.03	0.04	0.05	0.06	0.07	0.08	0.09
−3.4	0.0003	0.0003	0.0003	0.0003	0.0003	0.0003	0.0003	0.0003	0.0003	0.0002
−3.3	0.0005	0.0005	0.0005	0.0004	0.0004	0.0004	0.0004	0.0004	0.0004	0.0003
−3.2	0.0007	0.0007	0.0006	0.0006	0.0006	0.0006	0.0006	0.0005	0.0005	0.0005
−3.1	0.0010	0.0009	0.0009	0.0009	0.0008	0.0008	0.0008	0.0008	0.0007	0.0007
−3.0	0.0013	0.0013	0.0013	0.0012	0.0012	0.0011	0.0011	0.0011	0.0010	0.0010
−2.9	0.0019	0.0018	0.0017	0.0017	0.0016	0.0016	0.0015	0.0015	0.0014	0.0014
−2.8	0.0026	0.0025	0.0024	0.0023	0.0023	0.0022	0.0021	0.0021	0.0020	0.0019
−2.7	0.0035	0.0034	0.0033	0.0032	0.0031	0.0030	0.0029	0.0028	0.0027	0.0026
−2.6	0.0047	0.0045	0.0044	0.0043	0.0041	0.0040	0.0039	0.0038	0.0037	0.0036
−2.5	0.0062	0.0060	0.0059	0.0057	0.0055	0.0054	0.0052	0.0051	0.0049	0.0048
−2.4	0.0082	0.0080	0.0078	0.0075	0.0073	0.0071	0.0069	0.0068	0.0066	0.0064
−2.3	0.0107	0.0104	0.0102	0.0099	0.0096	0.0094	0.0091	0.0089	0.0087	0.0084
−2.2	0.0139	0.0136	0.0132	0.0129	0.0125	0.0122	0.0119	0.0116	0.0113	0.0110
−2.1	0.0179	0.0174	0.0170	0.0166	0.0162	0.0158	0.0154	0.0150	0.0146	0.0143
−2.0	0.0228	0.0222	0.0217	0.0212	0.0207	0.0202	0.0197	0.0192	0.0188	0.0183
−1.9	0.0287	0.0281	0.0274	0.0268	0.0262	0.0256	0.0250	0.0244	0.0239	0.0233
−1.8	0.0359	0.0352	0.0344	0.0336	0.0329	0.0322	0.0314	0.0307	0.0301	0.0294
−1.7	0.0446	0.0436	0.0427	0.0418	0.0409	0.0401	0.0392	0.0384	0.0375	0.0367
−1.6	0.0548	0.0537	0.0526	0.0516	0.0505	0.0495	0.0485	0.0475	0.0465	0.0455
−1.5	0.0668	0.0655	0.0643	0.0630	0.0618	0.0606	0.0594	0.0582	0.0571	0.0559
−1.4	0.0808	0.0793	0.0778	0.0764	0.0749	0.0735	0.0722	0.0708	0.0694	0.0681
−1.3	0.0968	0.0951	0.0934	0.0918	0.0901	0.0885	0.0869	0.0853	0.0838	0.0823
−1.2	0.1151	0.1131	0.1112	0.1093	0.1075	0.1056	0.1038	0.1020	0.1003	0.0985
−1.1	0.1357	0.1335	0.1314	0.1292	0.1271	0.1251	0.1230	0.1210	0.1190	0.1170
−1.0	0.1587	0.1562	0.1539	0.1515	0.1492	0.1469	0.1446	0.1423	0.1401	0.1379
−0.9	0.1841	0.1814	0.1788	0.1762	0.1736	0.1711	0.1685	0.1660	0.1635	0.1611
−0.8	0.2119	0.2090	0.2061	0.2033	0.2005	0.1977	0.1949	0.1922	0.1894	0.1867
−0.7	0.2420	0.2389	0.2358	0.2327	0.2296	0.2266	0.2236	0.2206	0.2177	0.2148
−0.6	0.2743	0.2709	0.2676	0.2643	0.2611	0.2578	0.2546	0.2514	0.2483	0.2451
−0.5	0.3085	0.3050	0.3015	0.2981	0.2946	0.2912	0.2877	0.2843	0.2810	0.2776
−0.4	0.3446	0.3409	0.3372	0.3336	0.3300	0.3264	0.3228	0.3192	0.3156	0.3121
−0.3	0.3821	0.3783	0.3745	0.3707	0.3669	0.3632	0.3594	0.3557	0.3520	0.3483
−0.2	0.4207	0.4168	0.4129	0.4090	0.4052	0.4013	0.3974	0.3936	0.3897	0.3859
−0.1	0.4602	0.4562	0.4522	0.4483	0.4443	0.4404	0.4364	0.4325	0.4286	0.4247
−0.0	0.5000	0.4960	0.4920	0.4880	0.4840	0.4801	0.4761	0.4721	0.4681	0.4641

Area Under a Normal Curve *(continued)*

z	0.00	0.01	0.02	0.03	0.04	0.05	0.06	0.07	0.08	0.09
0.0	0.5000	0.5040	0.5080	0.5120	0.5160	0.5199	0.5239	0.5279	0.5319	0.5359
0.1	0.5398	0.5438	0.5478	0.5517	0.5557	0.5596	0.5636	0.5675	0.5714	0.5753
0.2	0.5793	0.5832	0.5871	0.5910	0.5948	0.5987	0.6026	0.6064	0.6103	0.6141
0.3	0.6179	0.6217	0.6255	0.6293	0.6331	0.6368	0.6406	0.6443	0.6480	0.6517
0.4	0.6554	0.6591	0.6628	0.6664	0.6700	0.6736	0.6772	0.6808	0.6844	0.6879
0.5	0.6915	0.6950	0.6985	0.7019	0.7054	0.7088	0.7123	0.7157	0.7190	0.7224
0.6	0.7257	0.7291	0.7324	0.7357	0.7389	0.7422	0.7454	0.7486	0.7517	0.7549
0.7	0.7580	0.7611	0.7642	0.7673	0.7704	0.7734	0.7764	0.7794	0.7823	0.7852
0.8	0.7881	0.7910	0.7939	0.7967	0.7995	0.8023	0.8051	0.8078	0.8106	0.8133
0.9	0.8159	0.8186	0.8212	0.8238	0.8264	0.8289	0.8315	0.8340	0.8365	0.8389
1.0	0.8413	0.8438	0.8461	0.8485	0.8508	0.8531	0.8554	0.8577	0.8599	0.8621
1.1	0.8643	0.8665	0.8686	0.8708	0.8729	0.8749	0.8770	0.8790	0.8810	0.8830
1.2	0.8849	0.8869	0.8888	0.8907	0.8925	0.8944	0.8962	0.8980	0.8997	0.9015
1.3	0.9032	0.9049	0.9066	0.9082	0.9099	0.9115	0.9131	0.9147	0.9162	0.9177
1.4	0.9192	0.9207	0.9222	0.9236	0.9251	0.9265	0.9278	0.9292	0.9306	0.9319
1.5	0.9332	0.9345	0.9357	0.9370	0.9382	0.9394	0.9406	0.9418	0.9429	0.9441
1.6	0.9452	0.9463	0.9474	0.9484	0.9495	0.9505	0.9515	0.9525	0.9535	0.9545
1.7	0.9554	0.9564	0.9573	0.9582	0.9591	0.9599	0.9608	0.9616	0.9625	0.9633
1.8	0.9641	0.9649	0.9656	0.9664	0.9671	0.9678	0.9686	0.9693	0.9699	0.9706
1.9	0.9713	0.9719	0.9726	0.9732	0.9738	0.9744	0.9750	0.9756	0.9761	0.9767
2.0	0.9772	0.9778	0.9783	0.9788	0.9793	0.9798	0.9803	0.9808	0.9812	0.9817
2.1	0.9821	0.9826	0.9830	0.9834	0.9838	0.9842	0.9846	0.9850	0.9854	0.9857
2.2	0.9861	0.9864	0.9868	0.9871	0.9875	0.9878	0.9881	0.9884	0.9887	0.9890
2.3	0.9893	0.9896	0.9898	0.9901	0.9904	0.9906	0.9909	0.9911	0.9913	0.9916
2.4	0.9918	0.9920	0.9922	0.9925	0.9927	0.9929	0.9931	0.9932	0.9934	0.9936
2.5	0.9938	0.9940	0.9941	0.9943	0.9945	0.9946	0.9948	0.9949	0.9951	0.9952
2.6	0.9953	0.9955	0.9956	0.9957	0.9959	0.9960	0.9961	0.9962	0.9963	0.9964
2.7	0.9965	0.9966	0.9967	0.9968	0.9969	0.9970	0.9971	0.9972	0.9973	0.9974
2.8	0.9974	0.9975	0.9976	0.9977	0.9977	0.9978	0.9979	0.9979	0.9980	0.9981
2.9	0.9981	0.9982	0.9982	0.9983	0.9984	0.9984	0.9985	0.9985	0.9986	0.9986
3.0	0.9987	0.9987	0.9987	0.9988	0.9988	0.9989	0.9989	0.9989	0.9990	0.9990
3.1	0.9990	0.9991	0.9991	0.9991	0.9992	0.9992	0.9992	0.9992	0.9993	0.9993
3.2	0.9993	0.9993	0.9994	0.9994	0.9994	0.9994	0.9994	0.9995	0.9995	0.9995
3.3	0.9995	0.9995	0.9995	0.9996	0.9996	0.9996	0.9996	0.9996	0.9996	0.9997
3.4	0.9997	0.9997	0.9997	0.9997	0.9997	0.9997	0.9997	0.9997	0.9997	0.9998

Answers to Selected Exercises

Answers to selected writing exercises are provided.

Answers to Prerequisite Skills Diagnostic Test

1. 20% **2.** 51/35 **3.** $x + y = 75$ **4.** $s \geq 4p$ **5.** $-20/3$ (Sec. R.4) **6.** $-11/5$ (Sec. R.4) **7.** $(-2, 5]$ (Sec. R.5) **8.** $x \leq -3$
(Sec. R.5) **9.** $y \geq -17/2$ (Sec. R.5) **10.** $p > 3/2$ (Sec. R.5)

Chapter R Algebra Reference

Exercises R.1 (page R-5)

For exercises . . .	1–6	7,8,15–22	9–14	23–26
Refer to example . . .	2	3	4	5

1. $-x^2 + x + 9$ **2.** $-6y^2 + 3y + 10$ **3.** $-16q^2 + 4q + 6$
4. $9r^2 - 4r + 19$ **5.** $-0.327x^2 - 2.805x - 1.458$ **6.** $0.8r^2 + 3.6r - 1.5$ **7.** $-18m^3 - 27m^2 + 9m$ **8.** $-12x^4 + 30x^2 + 36x$
9. $9t^2 + 9ty - 10y^2$ **10.** $18k^2 - 7kq - q^2$ **11.** $4 - 9x^2$ **12.** $36m^2 - 25$ **13.** $(6/25)y^2 + (11/40)yz + (1/16)z^2$
14. $(15/16)r^2 - (7/12)rs - (2/9)s^2$ **15.** $27p^3 - 1$ **16.** $15p^3 + 13p^2 - 10p - 8$ **17.** $8m^3 + 1$
18. $12k^4 + 21k^3 - 5k^2 + 3k + 2$ **19.** $3x^2 + xy + 2xz - 2y^2 - 3yz - z^2$ **20.** $2r^2 + 2rs - 5rt - 4s^2 + 8st - 3t^2$
21. $x^3 + 6x^2 + 11x + 6$ **22.** $x^3 - 2x^2 - 5x + 6$ **23.** $x^2 + 4x + 4$ **24.** $4a^2 - 16ab + 16b^2$ **25.** $x^3 - 6x^2y + 12xy^2 - 8y^3$
26. $27x^3 + 27x^2y + 9xy^2 + y^3$

Exercises R.2 (page R-7)

For exercises . . .	1–4	5–15	16–20	21–32
Refer to example . . .	1	3	2nd CAUTION	4

1. $7a^2(a + 2)$ **2.** $3y(y^2 + 8y + 3)$ **3.** $13p^2q(p^2q - 3p + 2q)$
4. $10m^2(6m^2 - 12mn + 5n^2)$ **5.** $(m + 2)(m - 7)$ **6.** $(x + 5)(x - 1)$ **7.** $(z + 4)(z + 5)$ **8.** $(b - 7)(b - 1)$
9. $(a - 5b)(a - b)$ **10.** $(s - 5t)(s + 7t)$ **11.** $(y - 7z)(y + 3z)$ **12.** $(3x + 7)(x - 1)$ **13.** $(3a + 7)(a + 1)$
14. $(5y + 2)(3y - 1)$ **15.** $(7m + 2n)(3m + n)$ **16.** $6(a - 10)(a + 2)$ **17.** $3m(m + 3)(m + 1)$ **18.** $2(2a + 3)(a + 1)$
19. $2a^2(4a - b)(3a + 2b)$ **20.** $12x^2(x - y)(2x + 5y)$ **21.** $(x + 8)(x - 8)$ **22.** $(3m + 5)(3m - 5)$
23. $10(x + 4)(x - 4)$ **24.** Prime **25.** $(z + 7y)^2$ **26.** $(s - 5t)^2$ **27.** $(3p - 4)^2$ **28.** $(a - 6)(a^2 + 6a + 36)$
29. $(3r - 4s)(9r^2 + 12rs + 16s^2)$ **30.** $3(m + 5)(m^2 - 5m + 25)$ **31.** $(x - y)(x + y)(x^2 + y^2)$
32. $(2a - 3b)(2a + 3b)(4a^2 + 9b^2)$

Exercises R.3 (page R-10)

For exercises . . .	1–12	13–38
Refer to example . . .	1	2

1. $v/7$ **2.** $5p/2$ **3.** $8/9$ **4.** $2/(t + 2)$ **5.** $x - 2$ **6.** $4(y + 2)$ **7.** $(m - 2)/(m + 3)$
8. $(r + 2)/(r + 4)$ **9.** $3(x - 1)/(x - 2)$ **10.** $(z - 3)/(z + 2)$ **11.** $(m^2 + 4)/4$ **12.** $(2y + 1)/(y + 1)$ **13.** $3k/5$
14. $25p^2/9$ **15.** $9/(5c)$ **16.** 2 **17.** $1/4$ **18.** $3/10$ **19.** $2(a + 4)/(a - 3)$ **20.** $2/(r + 2)$ **21.** $(k - 2)/(k + 3)$
22. $(m + 6)/(m + 3)$ **23.** $(m - 3)/(2m - 3)$ **24.** $2(2n - 1)/(3n - 5)$ **25.** 1 **26.** $(6 + p)/(2p)$
27. $(12 - 15y)/(10y)$ **28.** $137/(30m)$ **29.** $(3m - 2)/[m(m - 1)]$ **30.** $(r - 6)/[r(2r + 3)]$ **31.** $14/[3(a - 1)]$
32. $23/[20(k - 2)]$ **33.** $(7x + 1)/[(x - 2)(x + 3)(x + 1)]$ **34.** $(y^2 + 1)/[(y + 3)(y + 1)(y - 1)]$
35. $k(k - 13)/[(2k - 1)(k + 2)(k - 3)]$ **36.** $m(3m - 19)/[(3m - 2)(m + 3)(m - 4)]$ **37.** $(4a + 1)/[a(a + 2)]$
38. $(5x^2 + 4x - 4)/[x(x - 1)(x + 1)]$

Exercises R.4 (page R-16)

For exercises . . .	1–8	9–26	27–37
Refer to example . . .	2	3–5	6,7

1. -12 **2.** $3/4$ **3.** 12 **4.** $-3/8$ **5.** $-7/8$ **6.** $-6/11$ **7.** 4 **8.** $-10/19$ **9.** $-3, -2$
10. $-1, 3$ **11.** 7 **12.** $-2, 5/2$ **13.** $-1/4, 2/3$ **14.** $2, 5$ **15.** $-3, 3$ **16.** $-4, 1/2$ **17.** $0, 4$ **18.** $(5 + \sqrt{13})/6 \approx 1.434$,
$(5 - \sqrt{13})/6 \approx 0.232$ **19.** $(2 + \sqrt{10})/2 \approx 2.581, (2 - \sqrt{10})/2 \approx -0.581$ **20.** $(-1 + \sqrt{5})/2 \approx 0.618, (-1 - \sqrt{5})/2 \approx$
-1.618 **21.** $5 + \sqrt{5} \approx 7.236, 5 - \sqrt{5} \approx 2.764$ **22.** $(4 + \sqrt{6})/5 \approx 1.290, (4 - \sqrt{6})/5 \approx 0.310$ **23.** $1, 5/2$ **24.** No real
number solutions **25.** $(-1 + \sqrt{73})/6 \approx 1.257, (-1 - \sqrt{73})/6 \approx -1.591$ **26.** $-1, 0$ **27.** 3 **28.** 12 **29.** $-59/6$ **30.** 6
31. 3 **32.** $-5/2$ **33.** $2/3$ **34.** 1 **35.** 2 **36.** No solution **37.** No solution

Exercises R.5 (page R-21)

For exercises . . .	1–14	15–26	27–38	39–42	43–54
Refer to example . . .	Figure 1, Example 2	2	3	4	5–7

1. $(-\infty, 4)$

2. $[-3, \infty)$ **3.** $[1, 2)$ **4.** $[-2, 3]$

5. $(-\infty, -9)$ **6.** $[6, \infty)$ **7.** $-7 \leq x \leq -3$ **8.** $4 \leq x < 10$

9. $x \leq -1$ **10.** $x > 3$ **11.** $-2 \leq x < 6$ **12.** $0 < x < 8$ **13.** $x \leq -4$ or $x \geq 4$ **14.** $x < 0$ or $x \geq 3$

15. $(-\infty, 2]$ **16.** $(-\infty, 1)$ **17.** $(3, \infty)$

18. $(-\infty, 1]$

19. $(1/5, \infty)$

20. $(1/3, \infty)$

21. $(-4, 6)$

22. $[7/3, 4]$

23. $[-5, 3)$

24. $[-1, 2]$

25. $[-17/7, \infty)$

26. $(-\infty, 50/9]$

27. $(-5, 3)$

28. $(-\infty, -6] \cup [1, \infty)$

29. $(1, 2)$

30. $(-\infty, -4) \cup (1/2, \infty)$

31. $(-\infty, -4) \cup (4, \infty)$

32. $[-3/2, 5]$

33. $(-\infty, -1] \cup [5, \infty)$

34. $[-1/2, 2/5]$

35. $(-\infty, -1) \cup (1/3, \infty)$

36. $(-\infty, -2) \cup (5/3, \infty)$

37. $(-\infty, -3] \cup [3, \infty)$

38. $(-\infty, 0) \cup (16, \infty)$

39. $[-2, 0] \cup [2, \infty)$

40. $(-\infty, -4] \cup [-3, 0]$

41. $(-\infty, 0) \cup (1, 6)$

42. $(-1, 0) \cup (4, \infty)$

43. $(-5, 3]$ **44.** $(-\infty, -1) \cup (1, \infty)$ **45.** $(-\infty, -2)$

46. $(-2, 3/2)$ **47.** $[-8, 5)$ **48.** $(-\infty, -3/2) \cup [-13/9, \infty)$ **49.** $[2, 3)$ **50.** $(-\infty, -1)$ **51.** $(-2, 0] \cup (3, \infty)$

52. $(-4, -2) \cup (0, 2)$ **53.** $(1, 3/2]$ **54.** $(-\infty, -2) \cup (-2, 2) \cup [4, \infty)$

Exercises R.6 (page R-25)

For exercises . . .	1–8	9–26	27–36	37–50	51–56
Refer to example . . .	1	2	3,4	5	6

1. $1/64$ **2.** $1/81$ **3.** 1   **4.** 1   **5.** $-1/9$ **6.** $1/9$ **7.** 36   **8.** $27/64$
9. $1/64$ **10.** 8^5 **11.** $1/10^8$ **12.** 7   **13.** x^2 **14.** 1   **15.** $2^3 k^3$ **16.** $1/(3z^7)$ **17.** $x^5/(3y^3)$ **18.** $m^3/5^4$ **19.** $a^3 b^6$
20. $49/(c^6 d^4)$ **21.** $(a+b)/(ab)$ **22.** $(1-ab^2)/b^2$ **23.** $2(m-n)/[mn(m+n^2)]$ **24.** $(3n^2+4m)/(mn^2)$
25. $xy/(y-x)$ **26.** $y^4/(xy-1)^2$ **27.** 11   **28.** 3   **29.** 4   **30.** -25 **31.** $1/2$ **32.** $4/3$ **33.** $1/16$ **34.** $1/5$ **35.** $4/3$
36. $1000/1331$ **37.** 9   **38.** 3   **39.** 64   **40.** 1   **41.** x^4/y^4 **42.** b/a^3 **43.** r **44.** $12^3/y^8$ **45.** $3k^{3/2}/8$ **46.** $1/(2p^2)$
47. $a^{2/3}b^2$ **48.** $y^2/(x^{1/6}z^{5/4})$ **49.** $h^{1/3}t^{1/5}/k^{2/5}$ **50.** $m^3 p/n$ **51.** $3x(x^2+3x)^2(x^2-5)$ **52.** $6x(x^3+7)(-2x^3-5x+7)$
53. $5x(x^2-1)^{-1/2}(x^2+1)$ **54.** $3(6x+2)^{-1/2}(27x+5)$ **55.** $(2x+5)(x^2-4)^{-1/2}(4x^2+5x-8)$
56. $(4x^2+1)(2x-1)^{-1/2}(36x^2-16x+1)$

Exercises R.7 (page R-28)

For exercises . . .	1–22	23–26	27–40	41–44
Refer to example . . .	1,2	3	4	5

1. 5   **2.** 6   **3.** -5 **4.** $5\sqrt{2}$ **5.** $20\sqrt{5}$ **6.** $4y^2\sqrt{2y}$ **7.** 9   **8.** 8
9. $7\sqrt{2}$ **10.** $9\sqrt{3}$ **11.** $9\sqrt{7}$ **12.** $-2\sqrt{7}$ **13.** $5\sqrt[3]{2}$ **14.** $3\sqrt[3]{5}$ **15.** $xyz^2\sqrt{2x}$ **16.** $4r^3s^4t^6\sqrt{10rs}$ **17.** $4xy^2z^3\sqrt[3]{2y^2}$
18. $x^2yz^2\sqrt[4]{y^3z^3}$ **19.** $ab\sqrt{ab}(b-2a^2+b^3)$ **20.** $p^2\sqrt{pq}(pq-q^4+p^2)$ **21.** $\sqrt[6]{a^5}$ **22.** $b^2\sqrt[4]{b}$ **23.** $|4-x|$
24. $|3y+5|$ **25.** Cannot be simplified   **26.** Cannot be simplified   **27.** $5\sqrt{7}/7$ **28.** $\sqrt{10}/2$ **29.** $-\sqrt{3}/2$ **30.** $\sqrt{2}$
31. $-3(1+\sqrt{2})$ **32.** $-5(2+\sqrt{6})/2$ **33.** $3(2-\sqrt{2})$ **34.** $(5-\sqrt{10})/3$ **35.** $(\sqrt{r}+\sqrt{3})/(r-3)$
36. $5(\sqrt{m}+\sqrt{5})/(m-5)$ **37.** $\sqrt{y}+\sqrt{5}$ **38.** $(z+\sqrt{5z}-\sqrt{z}-\sqrt{5})/(z-5)$ **39.** $-2x-2\sqrt{x(x+1)}-1$
40. $[p^2+p+2\sqrt{p(p^2-1)}-1]/(-p^2+p+1)$ **41.** $-1/[2(1-\sqrt{2})]$ **42.** $1/(3+\sqrt{3})$
43. $-1/[2x-2\sqrt{x(x+1)}+1]$ **44.** $2/[p+\sqrt{p(p-2)}]$

Chapter 1 Linear Functions

Exercises 1.1 (page 13)

For exercises . . .	1–4	5–8	13,29,30	14,31–34	15–17	18,27	19–24	25,26	28	45–60	61–75
Refer to example . . .	1	3	8	9	4	7	5	2	6	11–13	10,14

1. $3/5$ **3.** Not defined   **5.** 1
7. $5/9$ **9.** Not defined   **11.** 0   **13.** 2   **15.** $y=-2x+5$ **17.** $y=-7$ **19.** $y=-(1/3)x+10/3$ **21.** $y=6x-7/2$

23. $x = -8$ **25.** $x + 2y = -6$ **27.** $x = -6$ **29.** $3x + 2y = 0$ **31.** $x - y = 7$ **33.** $5x - y = -4$ **35.** No   **39.** a   **41.** -4

45.
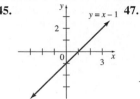

47.
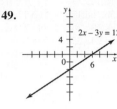

49.
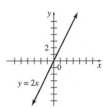

51.
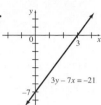

53.
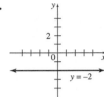

55.
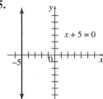

57.
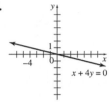

59.

61. a. 12,000 $y = 12,000x + 3000$ **b.** 8 years 1 month   **63. a.** $y = 4.612t + 86.164$ **b.** 178.4, which is slightly more than the actual CPI.   **c.** It is increasing at a rate of approximately 4.6 per year.   **65. a.** $u = 0.85(220 - x) = 187 - 0.85x$, $l = 0.7(220 - x) = 154 - 0.7x$ **b.** 140 to 170 beats per minute.   **c.** 126 to 153 beats per minute.   **d.** The women are 16 and 52. Their pulse is 143 beats per minute.   **67.** Approximately 86 yr   **69. a.** $y = 0.3444t + 14.1$ **b.** 2022 **71. a.** $y = 14,792.05t - 490,416$ **b.** 1,210,670   **73. a.** There appears to be a linear relationship.   **b.** $y = 76.9x$ **c.** About 780 megaparsecs (about 1.5×10^{22} mi)   **d.** About 12.4 billion yr

75. a.
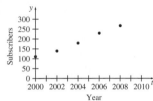

Yes, the data are approximately linear.

b. $y = 1133.4t + 16,072$; the slope 1133.4 indicates that tuition and fees have increased approximately $1133 per year.   **c.** The year 2025 is too far in the future to rely on this equation to predict costs; too many other factors may influence these costs by then.

Exercises 1.2 (page 23)

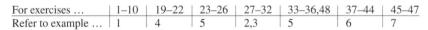

For exercises …	1–10	19–22	23–26	27–32	33–36,48	37–44	45–47
Refer to example …	1	4	5	2,3	5	6	7

1. -3 **3.** 22   **5.** 0   **7.** -4 **9.** $7 - 5t$ **11.** True
13. True   **19.** If $R(x)$ is the cost of renting a snowboard for x hours, then $R(x) = 2.25x + 10$.   **21.** If $C(x)$ is the cost of parking a car for x half-hours, then $C(x) = 0.75x + 2$.   **23.** $C(x) = 30x + 100$ **25.** $C(x) = 75x + 550$ **27. a.** $16   **b.** $11   **c.** $6 **d.** 640 watches   **e.** 480 watches   **f.** 320 watches

g.
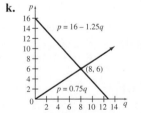

h. 0 watches
i. About 1333 watches
j. About 2667 watches

k.
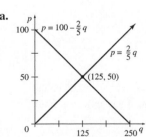

l. 800 watches, $6

29. a.

b. 125 tubs, $50

31. $D(q) = 6.9 - 0.4q$ **33. a.** $C(x) = 3.50x + 90$ **b.** 17 shirts **c.** 108 shirts   **35. a.** $C(x) = 0.097x + 1.32$ **b.** $1.32   **c.** $98.32 **d.** $98.42   **e.** 9.7¢   **f.** 9.7¢, the cost of producing one additional cup of coffee would be 9.7¢.   **37. a.** 2 units   **b.** $980   **c.** 52 units **39.** Break-even quantity is 45 units; don't produce; $P(x) = 20x - 900$ **41.** Break-even quantity is -50 units; impossible to make a profit when $C(x) > R(x)$ for all positive x; $P(x) = -10x - 500$ (always a loss) **43.** 5   **45. a.** 14.4°C   **b.** -28.9°C   **c.** 122°C   **47.** $-40°$

Exercises 1.3 (page 32)

For exercises . . .	3b,4,10d, 11d,12e,13e, 15b,16d,17b, 18d,19c,20b, 22d,24c,25d, 26a	3c,10a,11a, 12a,13a,15c, 16b,17c,18a, 19a,21c,22a, 23a,24b, 25a,26b	3d,10b,11b, 12c,13c,18b, 21d,22bc,23b, 26c	5ab,6ab, 7a,8a, 14ab,21ab	5c,6c, 15e	10c,11c, 12d,13d, 18c,23c
Refer to example . . .	4	1	2	1,4	5	3

3. a. 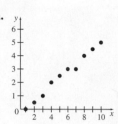 **b.** 0.993 **c.** $Y = 0.555x - 0.5$ 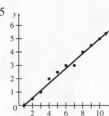 **d.** 5.6

5. a. $Y = 0.9783x + 0.0652, 0.9783$ **b.** $Y = 1.5, 0$ **c.** The point $(9,9)$ is an outlier that has a strong effect on the least squares line and the correlation coefficient.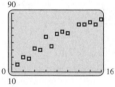

7. a. 0.7746 **b.**

11. a. $Y = -0.1534x + 11.36$ **b.** About 6760 **c.** 2025 **d.** -0.9890 **13. a.** $Y = 97.73x + 1833.3$ **b.** About \$97.73 billion per year **c.** \$3299 billion **d.** 2023 **e.** 0.9909

15. a. 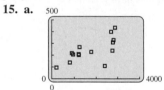 They lie in a linear pattern. **b.** $r = 0.693$; there is a positive correlation between the price and the distance. **c.** $Y = 0.0738x + 111.83$; the marginal cost is 7.38 cents per mile. **d.** In 2000 marginal cost was 2.43 cents per mile; it has increased to 7.38 cents per mile. **e.** Phoenix

17. a. 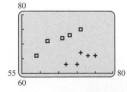 **b.** 0.959, yes **c.** $Y = 3.98x + 22.7$ 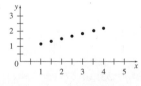 **19. a.** $Y = -0.1175x + 17.74$ **b.** 14.2 **c.** -0.9326

21. a. $Y = -0.08915x + 74.28, r = -0.1035$. The taller the student, the shorter is the ideal partner's height.
b. Females: $Y = 0.6674x + 27.89, r = 0.9459$; males: $Y = 0.4348x + 34.04, r = 0.7049$

23. a. 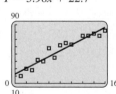 **b.** $Y = 0.366x + 0.803$; the line seems to fit the data.

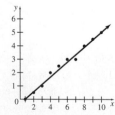

c. $r = 0.995$ indicates a good fit, which confirms the conclusion in part b. **25. a.** -0.995; yes **b.** $Y = -0.0769x + 5.91$ **c.** 2.07 points
27. a. 4.298 miles per hour **b.** 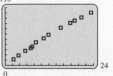 ; yes **c.** $Y = 4.317x + 3.419$ **d.** 0.9971; yes **e.** 4.317 miles per hour

Chapter 1 Review Exercises (page 39)

For exercises . . .	1–6,15–44	7–10,13,45–59b,62,63	11,12,14,59c–61,64,65
Refer to section . . .	1	2	3

1. False **2.** False **3.** True **4.** False **5.** True **6.** False **7.** True **8.** False **9.** False **10.** False **11.** False **12.** True
15. 1 **17.** $-2/11$ **19.** $-4/3$ **21.** 0 **23.** 5 **25.** $y = (2/3)x - 13/3$ **27.** $y = -x - 3$ **29.** $y = -10$ **31.** $2x - y = 10$
33. $x = -1$ **35.** $y = -5$

37. 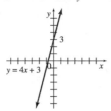 **39.** 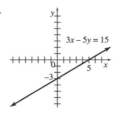 **41.** 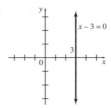 **43.**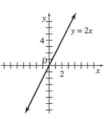

45. a. $E = 352 + 42x$ (where x is in thousands) **b.** $R = 130x$ (where x is in thousands) **c.** More than 4000 chips
47. $S(q) = 0.5q + 10$ **49.** \$41.25, 62.5 diet pills **51.** $C(x) = 180x + 2000$ **53.** $C(x) = 46x + 120$ **55. a.** 40 pounds
b. \$280 **57.** $y = 7.23t + 11.9$ **59. a.** $y = 836x + 7500$ **b.** $y = 795x + 8525$ **c.** $Y = 843.7x + 7662$
d.

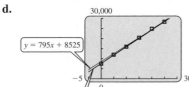

f. 0.9995 **61. a.** $Y = 0.9724x + 31.43$ **b.** About 216 **c.** 0.9338 **63.** $y = 1.22t + 48.9$
65. a. 0.6998; yes, but the fit is not very good. **b.**
c. $Y = 3.396x + 117.2$ **d.** \$3396

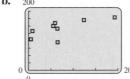

Chapter 2 Systems of Linear Equations and Matrices

Exercises 2.1 (page 52)

For exercises . . .	1–20	23–26,29–33,40–43	35–39,44–48
Refer to example . . .	1–3	5	4

1. $(3, 2)$ **3.** $(1, 3)$ **5.** $(-2, 0)$ **7.** $(0, 2)$ **9.** $(3, -2)$ **11.** $(2, -2)$
13. No solution **15.** $((2y - 4)/3, y)$ **17.** $(4, 1)$ **19.** $(7, -2)$ **21.** No **23.** 4 **25.** 3 **29.** $(8z - 4, 3 - 5z, z)$
31. $(3 - z, 4 - z, z)$ **35.** \$27 **37.** 400 main floor, 200 balcony **39.** Not possible; inconsistent system **41.** Either 10 buffets,
5 chairs, and no tables, or 11 buffets, 1 chair, and 1 table **43.** $z + 80$ long-sleeve blouses, $260 - 2z$ short-sleeve blouses, and
z sleeveless blouses with $0 \le z \le 130$. **45. a.** March 23, March 19 **b.** 1991 **47.** 36 field goals, 28 foul shots

Exercises 2.2 (page 65)

For exercises . . .	1–8,11–42	44–70
Refer to example . . .	1–5	6

1. $\begin{bmatrix} 3 & 1 & | & 6 \\ 2 & 5 & | & 15 \end{bmatrix}$ **3.** $\begin{bmatrix} 2 & 1 & 1 & | & 3 \\ 3 & -4 & 2 & | & -7 \\ 1 & 1 & 1 & | & 2 \end{bmatrix}$ **5.** $x = 2, y = 3$ **7.** $x = 4, y = -5, z = 1$ **9.** Row operations

11. $\begin{bmatrix} 3 & 7 & 4 & | & 10 \\ 0 & 1 & -5 & | & -8 \\ 0 & 4 & 5 & | & 11 \end{bmatrix}$ **13.** $\begin{bmatrix} 1 & 0 & 0 & | & -3 \\ 0 & 3 & 2 & | & 5 \\ 0 & 5 & 3 & | & 7 \end{bmatrix}$ **15.** $\begin{bmatrix} 1 & 0 & 0 & | & 6 \\ 0 & 5 & 0 & | & 9 \\ 0 & 0 & 4 & | & 8 \end{bmatrix}$ **17.** $(2, 3)$ **19.** $(1, 6)$ **21.** No solution

23. $((3y + 1)/6, y)$ **25.** $(4, 1, 0)$ **27.** No solution **29.** $(-1, 23, 16)$ **31.** $((-9z + 5)/23, (10z - 3)/23, z)$
33. $((-2z + 62)/35, (3z + 5)/7, z)$ **35.** $((9 - 3y - z)/2, y, z)$ **37.** $(0, 2, -2, 1)$; the answers are given in the order x, y, z, w.
39. $(-w - 3, -4w - 19, -3w - 2, w)$ **41.** $(28.9436, 36.6326, 9.6390, 37.1036)$ **43.** row 1: 3/8, 1/6, 11/24; row 2: 5/12, 1/3,
1/4; row 3: 5/24, 1/2, 7/24 **45.** \$2000 in U.S. Savings bonds, \$4000 in mutual funds, \$4000 in money market **47.** 2000 chairs,
1600 cabinets, and 2500 buffets. **49. a.** 5 trucks, 2 vans, and 3 SUVs. **b.** Use 2 trucks, 6 vans, and 1 SUV, or use 5 trucks, 2 vans,
and 3 SUVs. **51. a.** \$12,000 at 8%, \$7000 at 9%, and \$6000 at 10% **b.** The amount borrowed at 10% must be less than or equal
to \$9500. If $z = 5000, they borrowed \$11,000 at 8% and \$9000 at 9%. **c.** No solution. **d.** The total annual interest would be
\$2220, not \$2190, as specified as one of the conditions. **53.** The first supplier should send 40 units to Roseville and 35 units to
Akron. The second supplier should send 0 units to Roseville and 40 units to Akron **55. a.** 26 **b.** 0 two-person, 50 four-person,
and 0 six-person tents **c.** 25 two-person, 0 four-person, and 25 six-person tents **57. a.** 400/9 g of group A, 400/3 g of group B,
and 2000/9 g of group C. **b.** For any positive number z of grams of group C, there should be z grams less than 800/3 g of group A

and 400/3 g of group B.   **c.** No   **59.** About 244 fish of species A, 39 fish of species B, and 101 fish of species C.   **61. b.** 7,206,360 white cows, 4,893,246 black cows, 3,515,820 spotted cows, and 5,439,213 brown cows   **63. a.** $r = 175,000, b = 375,000$
65. 8.15, 4.23, 3.15   **67. a.** 24 balls, 57 dolls, and 19 cars   **b.** None   **c.** 48   **d.** 5 balls, 95 dolls, and 0 cars   **e.** 52 balls, 1 doll, and 47 cars   **69. a.** (1, 1, 1, 1); the strategy required to turn all the lights out is to push every button one time.   **b.** (0, 1, 1, 0); the strategy required to turn all the lights out is to push the button in the first row, second column, and push the button in the second row, first column.

Exercises 2.3 (page 74)

For exercises . . .	1–6,15–20	7–14	21–32	39,40,43	41,48	42	44–46
Refer to example . . .	3	2	4,6	5	1	5,7	7

1. False; not all corresponding elements are equal.
3. True   **5.** True   **7.** 2×2; square   **9.** 3×4 **11.** 2×1; column   **13.** Undefined   **15.** $x = 4, y = -8, z = 1$
17. $s = 10, t = 0, r = 7$ **19.** $a = 20, b = 5, c = 0, d = 4, f = 1$

21. $\begin{bmatrix} 10 & 4 & -5 & -6 \\ 4 & 5 & 3 & 11 \end{bmatrix}$ **23.** Not possible   **25.** $\begin{bmatrix} 1 & 5 & 6 & -9 \\ 5 & 7 & 2 & 1 \\ -7 & 2 & 2 & -7 \end{bmatrix}$ **27.** $\begin{bmatrix} 3 & 4 \\ 4 & 8 \end{bmatrix}$ **29.** $\begin{bmatrix} 10 & -2 \\ 10 & 9 \end{bmatrix}$

31. $\begin{bmatrix} -12x + 8y & -x + y \\ x & 8x - y \end{bmatrix}$ **33.** $\begin{bmatrix} -x & -y \\ -z & -w \end{bmatrix}$ **39. a.** Chicago: $\begin{bmatrix} 4.05 & 7.01 \\ 3.27 & 3.51 \end{bmatrix}$, Seattle: $\begin{bmatrix} 4.40 & 6.90 \\ 3.54 & 3.76 \end{bmatrix}$ **b.** $\begin{bmatrix} 4.42 & 7.43 \\ 3.38 & 3.62 \end{bmatrix}$

41. a. $\begin{bmatrix} 2 & 1 & 2 & 1 \\ 3 & 2 & 2 & 1 \\ 4 & 3 & 2 & 1 \end{bmatrix}$ **b.** $\begin{bmatrix} 5 & 0 & 7 \\ 0 & 10 & 1 \\ 0 & 15 & 2 \\ 10 & 12 & 8 \end{bmatrix}$ **c.** $\begin{bmatrix} 8 \\ 4 \\ 5 \end{bmatrix}$ **43. a.** 8   **b.** 3   **c.** $\begin{bmatrix} 85 & 15 \\ 27 & 73 \end{bmatrix}$ **d.** Yes

45. a. $\begin{bmatrix} 60.0 & 68.3 \\ 63.8 & 72.5 \\ 64.5 & 73.6 \\ 68.3 & 75.2 \end{bmatrix}$ **b.** $\begin{bmatrix} 68.0 & 75.6 \\ 70.7 & 78.1 \\ 72.7 & 79.4 \\ 74.9 & 80.1 \end{bmatrix}$ **c.** $\begin{bmatrix} -8.0 & -7.3 \\ -6.9 & -5.6 \\ -8.2 & -5.8 \\ -6.6 & -4.9 \end{bmatrix}$ **47. a.** $\begin{bmatrix} 51.2 & 7.9 \\ 59.8 & 11.1 \\ 66.2 & 11.3 \\ 73.8 & 13.2 \\ 78.5 & 16.5 \\ 83.0 & 19.6 \end{bmatrix}$ **b.** $\begin{bmatrix} 45.3 & 7.9 \\ 47.9 & 8.5 \\ 50.8 & 9.2 \\ 53.4 & 9.3 \\ 57.0 & 10.6 \\ 62.3 & 13.3 \end{bmatrix}$ **c.** $\begin{bmatrix} 5.9 & 0 \\ 11.9 & 2.6 \\ 15.4 & 2.1 \\ 20.4 & 3.9 \\ 21.5 & 5.9 \\ 20.7 & 6.3 \end{bmatrix}$

Exercises 2.4 (page 83)

For exercises . . .	1–6,49,50	7–12	15–31,51	43–48
Refer to example . . .	1	4	2,3,5	6

1. $\begin{bmatrix} -4 & 8 \\ 0 & 6 \end{bmatrix}$ **3.** $\begin{bmatrix} 12 & -24 \\ 0 & -18 \end{bmatrix}$ **5.** $\begin{bmatrix} -22 & -6 \\ 20 & -12 \end{bmatrix}$ **7.** 2×2; 2×2

9. 3×4; BA does not exist.   **11.** AB does not exist; 3×2 **13.** columns; rows   **15.** $\begin{bmatrix} 8 \\ -1 \end{bmatrix}$ **17.** $\begin{bmatrix} 14 \\ -23 \end{bmatrix}$ **19.** $\begin{bmatrix} -7 & 2 & 8 \\ 27 & -12 & 12 \end{bmatrix}$

21. $\begin{bmatrix} -2 & 10 \\ 0 & 8 \end{bmatrix}$ **23.** $\begin{bmatrix} 13 & 5 \\ 25 & 15 \end{bmatrix}$ **25.** $\begin{bmatrix} 13 \\ 29 \end{bmatrix}$ **27.** $\begin{bmatrix} 7 \\ -33 \\ 4 \end{bmatrix}$ **29.** $\begin{bmatrix} 22 & -8 \\ 11 & -4 \end{bmatrix}$ **31. a.** $\begin{bmatrix} 16 & 22 \\ 7 & 19 \end{bmatrix}$ **b.** $\begin{bmatrix} 5 & -5 \\ 0 & 30 \end{bmatrix}$ **c.** No   **d.** No

39. a. $\begin{bmatrix} 6 & 106 & 158 & 222 & 28 \\ 120 & 139 & 64 & 75 & 115 \\ -146 & -2 & 184 & 144 & -129 \\ 106 & 94 & 24 & 116 & 110 \end{bmatrix}$ **b.** Does not exist   **c.** No   **41. a.** $\begin{bmatrix} -1 & 5 & 9 & 13 & -1 \\ 7 & 17 & 2 & -10 & 6 \\ 18 & 9 & -12 & 12 & 22 \\ 9 & 4 & 18 & 10 & -3 \\ 1 & 6 & 10 & 28 & 5 \end{bmatrix}$

b. $\begin{bmatrix} -2 & -9 & 90 & 77 \\ -42 & -63 & 127 & 62 \\ 413 & 76 & 180 & -56 \\ -29 & -44 & 198 & 85 \\ 137 & 20 & 162 & 103 \end{bmatrix}$ **c.** $\begin{bmatrix} -56 & -1 & 1 & 45 \\ -156 & -119 & 76 & 122 \\ 315 & 86 & 118 & -91 \\ -17 & -17 & 116 & 51 \\ 118 & 19 & 125 & 77 \end{bmatrix}$ **d.** $\begin{bmatrix} 54 & -8 & 89 & 32 \\ 114 & 56 & 51 & -60 \\ 98 & -10 & 62 & 35 \\ -12 & -27 & 82 & 34 \\ 19 & 1 & 37 & 26 \end{bmatrix}$ **e.** $\begin{bmatrix} -2 & -9 & 90 & 77 \\ -42 & -63 & 127 & 62 \\ 413 & 76 & 180 & -56 \\ -29 & -44 & 198 & 85 \\ 137 & 20 & 162 & 103 \end{bmatrix}$

f. Yes **43. a.**

$$\begin{array}{c} \\ \text{Dept. 1} \\ \text{Dept. 2} \\ \text{Dept. 3} \\ \text{Dept. 4} \end{array} \begin{array}{cc} A & B \\ \begin{bmatrix} 57 & 70 \\ 41 & 54 \\ 27 & 40 \\ 39 & 40 \end{bmatrix} \end{array}$$

b. Supplier A: $164; Supplier B: $204; Supplier A **45. a.** $\begin{bmatrix} 4.24 & 6.95 \\ 3.42 & 3.64 \end{bmatrix}$ **b.** $\begin{bmatrix} 4.41 & 7.17 \\ 3.46 & 3.69 \end{bmatrix}$

47. a. $\begin{bmatrix} 80 & 40 & 120 \\ 60 & 30 & 150 \end{bmatrix}$ **b.** $\begin{bmatrix} 1/2 & 1/5 \\ 1/4 & 1/5 \\ 1/4 & 3/5 \end{bmatrix}$ **c.** $PF = \begin{bmatrix} 80 & 96 \\ 75 & 108 \end{bmatrix}$; the rows give the average price per pair of footwear sold by each store, and the columns give the state. **d.** $108

49. a. $\begin{bmatrix} 20 & 52 & 27 \\ 25 & 62 & 35 \\ 30 & 72 & 43 \end{bmatrix}$; the rows give the amounts of fat, carbohydrates, and protein, respectively, in each of the daily meals.

b. $\begin{bmatrix} 75 \\ 45 \\ 70 \\ 168 \end{bmatrix}$; the rows give the number of calories in one exchange of each of the food groups. **c.** The rows give the number of calories in each meal.

51. $\begin{bmatrix} 66.7 & 74.4 \\ 69.6 & 77.2 \\ 71.3 & 78.4 \\ 73.8 & 79.3 \end{bmatrix}$ **53. a.** $\begin{bmatrix} 0.0346 & 0.0118 \\ 0.0174 & 0.0073 \\ 0.0189 & 0.0059 \\ 0.0135 & 0.0083 \\ 0.0099 & 0.0103 \end{bmatrix}$; $\begin{bmatrix} 361 & 2038 & 286 & 227 & 460 \\ 473 & 2494 & 362 & 252 & 484 \\ 627 & 2978 & 443 & 278 & 499 \\ 803 & 3435 & 524 & 314 & 511 \\ 1013 & 3824 & 591 & 344 & 522 \end{bmatrix}$ **b.**

	Births	Deaths
1970	60.98	27.45
1980	74.80	33.00
1990	90.58	39.20
2000	106.75	45.51
2010	122.57	51.59

Exercises 2.5 (page 94)

For exercises...	1–8	11–26,51–55	27–42,56–58	59–63	64,65
Refer to example...	1	2,3	4,6	5	7

1. Yes **3.** No **5.** No **7.** Yes **9.** No; the row of all zeros makes it impossible to get all the 1's in the diagonal of the identity matrix, no matter what matrix is used as an inverse.

11. $\begin{bmatrix} 0 & 1/2 \\ -1 & 1/2 \end{bmatrix}$ **13.** $\begin{bmatrix} 2 & 1 \\ 5 & 3 \end{bmatrix}$ **15.** No inverse **17.** $\begin{bmatrix} 1 & 0 & 0 \\ 0 & -1 & 0 \\ -1 & 0 & 1 \end{bmatrix}$ **19.** $\begin{bmatrix} 15 & 4 & -5 \\ -12 & -3 & 4 \\ -4 & -1 & 1 \end{bmatrix}$ **21.** No inverse

23. $\begin{bmatrix} -11/2 & -1/2 & 5/2 \\ 1/2 & 1/2 & -1/2 \\ -5/2 & 1/2 & 1/2 \end{bmatrix}$ **25.** $\begin{bmatrix} 1/2 & 1/2 & -1/4 & 1/2 \\ -1 & 4 & -1/2 & -2 \\ -1/2 & 5/2 & -1/4 & -3/2 \\ 1/2 & -1/2 & 1/4 & 1/2 \end{bmatrix}$ **27.** $(5, 1)$ **29.** $(2, 1)$ **31.** $(15, 21)$

33. No inverse, $(-8y - 12, y)$ **35.** $(-8, 6, 1)$ **37.** $(-36, 8, -8)$ **39.** No inverse, no solution for system

41. $(-7, -34, -19, 7)$

51. $\begin{bmatrix} -0.0447 & -0.0230 & 0.0292 & 0.0895 & -0.0402 \\ 0.0921 & 0.0150 & 0.0321 & 0.0209 & -0.0276 \\ -0.0678 & 0.0315 & -0.0404 & 0.0326 & 0.0373 \\ 0.0171 & -0.0248 & 0.0069 & -0.0003 & 0.0246 \\ -0.0208 & 0.0740 & 0.0096 & -0.1018 & 0.0646 \end{bmatrix}$

53. $\begin{bmatrix} 0.0394 & 0.0880 & 0.0033 & 0.0530 & -0.1499 \\ -0.1492 & 0.0289 & 0.0187 & 0.1033 & 0.1668 \\ -0.1330 & -0.0543 & 0.0356 & 0.1768 & 0.1055 \\ 0.1407 & 0.0175 & -0.0453 & -0.1344 & 0.0655 \\ 0.0102 & -0.0653 & 0.0993 & 0.0085 & -0.0388 \end{bmatrix}$ **55.** Yes

57. $\begin{bmatrix} 1.51482 \\ 0.053479 \\ -0.637242 \\ 0.462629 \end{bmatrix}$ **59. a.** $\begin{bmatrix} 72 \\ 48 \\ 60 \end{bmatrix}$ **b.** $\begin{bmatrix} 2 & 4 & 2 \\ 2 & 1 & 2 \\ 2 & 1 & 3 \end{bmatrix} \begin{bmatrix} x_1 \\ x_2 \\ x_3 \end{bmatrix} = \begin{bmatrix} 72 \\ 48 \\ 60 \end{bmatrix}$ **c.** 8 type I, 8 type II, and 12 type III

61. a. $10,000 at 6%, $10,000 at 6.5%, and $5000 at 8%   **b.** $14,000 at 6%, $9000 at 6.5%, and $7000 at 8%   **c.** $24,000 at 6%, $4000 at 6.5%, and $12,000 at 8%   **63. a.** 50 Super Vim, 75 Multitab, and 100 Mighty Mix   **b.** 75 Super Vim, 50 Multitab, and 60 Mighty Mix   **c.** 80 Super Vim, 100 Multitab, and 50 Mighty Mix   **65. a.** 262, −161, −12, 186, −103, −22, 264, −168, −9, 208, −134, −5, 224, −152, 5, 92, −50, −3   **b.** $\begin{bmatrix} 1.75 & 2.5 & 3 \\ -0.25 & -0.5 & 0 \\ -0.25 & -0.5 & -1 \end{bmatrix}$ **c.** happy birthday

Exercises 2.6 (page 102)

For exercises . . .	1–6,9–12,17–25	7,8,26–29	13–16
Refer to example . . .	4	5	3,4

1. $\begin{bmatrix} 60 \\ 50 \end{bmatrix}$ **3.** $\begin{bmatrix} 6.43 \\ 26.12 \end{bmatrix}$ **5.** $\begin{bmatrix} 10 \\ 18 \\ 10 \end{bmatrix}$ **7.** 33:47:23   **9.** $\begin{bmatrix} 7697 \\ 4205 \\ 6345 \\ 4106 \end{bmatrix}$ (rounded)   **11.** About 1440 metric tons of wheat and 1938 metric tons of oil.   **13.** About 1506 units of agriculture, 1713 units of manufacturing, and 1412 units of transportation.   **15.** About 3077 units of agriculture, about 2564 units of manufacturing, and about 3179 units of transportation   **17. a.** 7/4 bushels of yams and 15/8 ≈ 2 pigs   **b.** 167.5 bushels of yams and 153.75 ≈ 154 pigs   **19.** About 848 units of agriculture, about 516 units of manufacturing, and about 2970 units of households   **21.** About 195 million pounds of agriculture, about 26 million pounds of manufacturing, and about 13.6 million pounds of energy   **23.** In millions of dollars, the amounts are about 532 for natural resources, about 481 for manufacturing, about 805 for trade and services, and about 1185 for personal consumption.

25. a. $\begin{bmatrix} 1.67 & 0.56 & 0.56 \\ 0.19 & 1.17 & 0.06 \\ 3.15 & 3.27 & 4.38 \end{bmatrix}$ **b.** These multipliers imply that if the demand for one community's output increases by $1, then the output in the other community will increase by the amount in the row and column of that matrix. For example, if the demand for Hermitage's output increases by $1, then output from Sharon will increase by $0.56, Farrell by $0.06, and Hermitage by $4.38.   **27.** 3 units of coal to every 4 units of steel   **29.** 6 units of mining to every 8 units of manufacturing and 5 units of communication

Chapter 2 Review Exercises (page 106)

For exercises . . .	1,3,10,13, 17,23–28, 67,68,74, 75,79,80,82	2,5,29–33, 36,69,73	4,19–22, 66,77,78, 81,83	6,7,18,34, 35,37–43, 76	8,9,11,12 14,15, 44–62,70, 74	16,63,64, 71,72
Refer to section . . .	2	3	1	4	5	6

1. False   **2.** False   **3.** True   **4.** True   **5.** False   **6.** True   **7.** False   **8.** False   **9.** True   **10.** False   **11.** False   **12.** False   **13.** True   **14.** True   **15.** False   **16.** True   **19.** $(1, -4)$ **21.** $((34 - 28z)/50, (53 - z)/25, z)$ **23.** $(-9, 3)$ **25.** $(7, -9, -1)$ **27.** $(6 - 7z/3, 1 + z/3, z)$ **29.** 2×2 (square); $a = 2, b = 3, c = 5, q = 9$ **31.** 1×4 (row); $m = 6, k = 3, z = -3, r = -9$

33. $\begin{bmatrix} 9 & 10 \\ -3 & 0 \\ 10 & 16 \end{bmatrix}$ **35.** $\begin{bmatrix} 23 & 20 \\ -7 & 3 \\ 24 & 39 \end{bmatrix}$ **37.** $\begin{bmatrix} -17 & 20 \\ 1 & -21 \\ -8 & -17 \end{bmatrix}$ **39.** Not possible   **41.** $[9]$ **43.** $[-14 \ -19]$ **45.** No inverse

47. $\begin{bmatrix} 7 & -3 \\ -2 & 1 \end{bmatrix}$ **49.** No inverse   **51.** $\begin{bmatrix} 2/3 & 0 & -1/3 \\ 1/3 & 0 & -2/3 \\ -2/3 & 1 & 1/3 \end{bmatrix}$ **53.** No inverse   **55.** $X = \begin{bmatrix} 1 \\ -13 \end{bmatrix}$ **57.** $X = \begin{bmatrix} -22 \\ -18 \\ 15 \end{bmatrix}$

59. $(2, 1)$ **61.** $(-1, 0, 2)$ **63.** $\begin{bmatrix} 218.1 \\ 318.3 \end{bmatrix}$ **65. a.** $(2, 3, -1)$ **b.** $(2, 3, -1)$ **c.** $\begin{bmatrix} 1 & 2 & 1 \\ 2 & -1 & -1 \\ 3 & -3 & 2 \end{bmatrix} \begin{bmatrix} x \\ y \\ z \end{bmatrix} = \begin{bmatrix} 7 \\ 2 \\ -5 \end{bmatrix}$

d. $\begin{bmatrix} 5/22 & 7/22 & 1/22 \\ 7/22 & 1/22 & -3/22 \\ 3/22 & -9/22 & 5/22 \end{bmatrix}$ **e.** $(2, 3, -1)$ **67.** 5 blankets, 3 rugs, 8 skirts   **69.** $\begin{bmatrix} 1.33 & 17.6 & 152,000 & 26.75 & +1.88 \\ 1.00 & 20.0 & 238,200 & 32.36 & -1.50 \\ 0.79 & 25.4 & 39,110 & 16.51 & -0.89 \\ 0.27 & 21.2 & 122,500 & 28.60 & +0.75 \end{bmatrix}$

71. a. $\begin{array}{c} \\ c \\ g \end{array}\begin{array}{cc} c & g \\ \end{array}$ $\begin{bmatrix} 0 & 1/2 \\ 2/3 & 0 \end{bmatrix}$ **b.** 1200 units of cheese; 1600 units of goats   **73.** $\begin{bmatrix} 8 & 8 & 8 \\ 10 & 5 & 9 \\ 7 & 10 & 7 \\ 8 & 9 & 7 \end{bmatrix}$ **75. a.** No   **b. (i)** 0.23, 0.37, 0.42; A is healthy; B and D are tumorous; C is bone.   **(ii)** 0.33, 0.27, 0.32; A and C are tumorous, B could be healthy or tumorous, D is bone.   **c.** 0.2, 0.4, 0.45, 0.3; A is healthy; B and C are bone; D is tumorous.   **d.** One example is to choose beams 1, 2, 3, and 6.   **77.** $W_1 = W_2 = 100\sqrt{3}/3 \approx 58$ lb   **79. a.** $C = 0.010985t^2 + 0.8803t + 317$ **b.** 2095   **81.** There are y girls and $2500 - 1.5y$ boys, where y is any even integer between 0 and 1666.   **83.** A chocolate wafer weighs 4.08 g, and a single layer of vanilla cream weighs 3.17 g.

Chapter 3 Linear Programming: The Graphical Method

Exercises 3.1 (page 118)

For exercises . . .	1–20	21–32	33–36	38–45
Refer to example . . .	1,2	3–5	3	6

1. 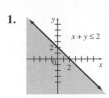 $x + y \le 2$

3. 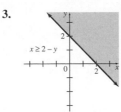 $x \ge 2 - y$

5. 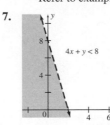 $4x - y < 6$

7. 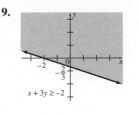 $4x + y < 8$

9. 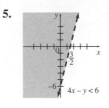 $x + 3y \ge -2$

11. 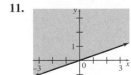 $x \le 3y$

13. 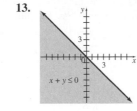 $x + y \le 0$

15. 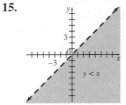 $y < x$

17. 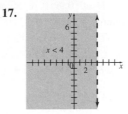 $x < 4$

19. 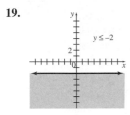 $y \le -2$

21. 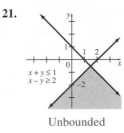 $x + y \le 1$ $x - y \ge 2$
Unbounded

23. 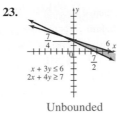 $x + 3y \le 6$ $2x + 4y \ge 7$
Unbounded

25. 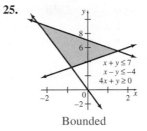 $x + y \le 7$ $x - y \le -4$ $4x + y \ge 0$
Bounded

27. 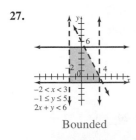 $-2 < x < 3$ $-1 \le y \le 5$ $2x + y < 6$
Bounded

29. 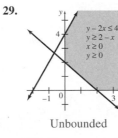 $y - 2x \le 4$ $y \ge 2 - x$ $x \ge 0$ $y \ge 0$
Unbounded

31. 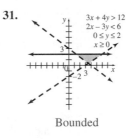 $3x + 4y > 12$ $2x - 3y < 6$ $0 \le y \le 2$ $x \ge 0$
Bounded

33.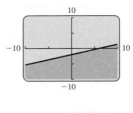

35.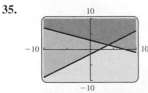

37. **B:** $\le, \le, \le$; **C:** $\ge, \ge, \le$; **D:** $\le, \ge, \le$; **E:** $\le, \le, \ge$; **F:** $\le, \ge, \ge$; **G:** $\ge, \ge, \ge$

39. a.

	Shawls	Afghans		Total
Number Made	x	y		
Spinning Time	1	2	$\le$	8
Dyeing Time	1	1	$\le$	6
Weaving Time	1	4	$\le$	14

b. 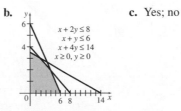 $x + 2y \le 8$ $x + y \le 6$ $x + 4y \le 14$ $x \ge 0, y \ge 0$

c. Yes; no

41. a. $x \ge 4y$; $0.06x + 0.08y \ge 1.6$; $x + y \le 30$; $x \ge 0$; $y \ge 0$

b. 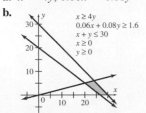 $x \ge 4y$ $0.06x + 0.08y \ge 1.6$ $x + y \le 30$ $x \ge 0$ $y \ge 0$

43. a. $x \le (1/2)y$; $x + y \le 800$; $x \ge 0$; $y \ge 0$

b. 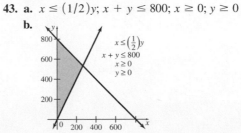 $x \le \left(\frac{1}{2}\right)y$ $x + y \le 800$ $x \ge 0$ $y \ge 0$

45. a. $x + y \geq 7$; $2x + y \geq 10$; $x + y \leq 9$; $x \geq 0$; $y \geq 0$ **b.**

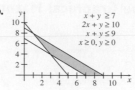

Exercises 3.2 (page 124)

For exercises . . .	1–16
Refer to example . . .	3

1. a. Maximum of 29 at $(7, 4)$; minimum of 10 at $(0, 5)$ **b.** Maximum of 35 at $(3, 8)$; minimum of 8 at $(4, 1)$ **3. a.** Maximum of 9 at $(0, 12)$; minimum of 0 at $(0, 0)$ **b.** Maximum of 12 at $(8, 0)$; minimum of 0 at $(0, 0)$
5. a. No maximum; minimum of 16 at $(0, 8)$ **b.** No maximum; minimum of 18 at $(3, 4)$ **c.** No maximum; minimum of 21 at $(13/2, 2)$ **d.** No maximum; minimum of 12 at $(12, 0)$ **7.** Minimum of 24 when $x = 6$ and $y = 0$ **9.** Maximum of 46 when $x = 6$ and $y = 8$ **11.** Maximum of 1500 when $x = 150$ and $y = 0$, as well as when $x = 50$ and $y = 100$ and all points on the line between   **13.** No solution   **15. a.** Maximum of 204 when $x = 18$ and $y = 2$ **b.** Maximum of 588/5 when $x = 12/5$ and $y = 39/5$ **c.** Maximum of 102 when $x = 0$ and $y = 17/2$ **17.** b

Exercises 3.3 (page 131)

For exercises . . .	1–6	7–9,16,21–25	10–15,17–20, 26	27
Refer to example . . .	1–3	3	1,2	2

1. Let x be the number of product A produced and y be the number of product B. Then $3x + 5y \leq 60$.   **3.** Let x be the number of calcium carbonate supplements and y be the number of calcium citrate supplements. Then $600x + 250y \geq 1500$.   **5.** Let x be the number of pounds of \$8 coffee and y be the number of \$10 coffee. Then $x + y \geq 40$.   **7.** 51 to plant I and 32 to plant II, for a minimum cost of \$2810   **9. a.** 6 units of policy A and 16 units of policy B, for a minimum premium cost of \$940   **b.** 30 units of policy A and 0 units of policy B, for a minimum premium cost of \$750   **11. a.** 500 type I and 1000 type II   **b.** Maximum revenue is \$275.   **c.** If the price of the type I bolt exceeds 20¢, then it is more profitable to produce 1050 type I bolts and 450 type II bolts.   **13. a.** 120 kg of the half-and-half mix and 120 kg of the other mix, for a maximum revenue of \$1980   **b.** 0 kg of the half-and-half mix and 200 kg of the other mix, for a maximum revenue of \$2200   **15. a.** 40 gal from dairy I and 60 gal from dairy II, for a maximum butterfat of 3.4 gal   **b.** 10 gal from dairy I and 20 gal from dairy 2. No.   **17.** \$10 million in bonds and \$20 million in mutual funds, or \$5 million in bonds and \$22.5 million in mutual funds (or any solution on the line in between those two points), for a maximum annual interest of \$2 million   **19.** a   **21. a.** Three of pill 1 and two of pill 2, for a minimum cost of \$1.05 per day   **b.** 360 surplus IU of vitamin A. No.   **23.** 4 ounces of fruit and 2 ounces of nuts, for a minimum of 140 calories   **25.** 0 plants and 18 animals, for a minimum of 270 hours

Chapter 3 Review Exercises (page 135)

For exercises . . .	1,6,8,10,12–14,21–36	2,7,9,15–20	3,4,5,37–46	11
Refer to Section . . .	2	1	3	2,3

1. False   **2.** True   **3.** False   **4.** False   **5.** False
6. False   **7.** True   **8.** False   **9.** False   **10.** True   **11.** True   **12.** True   **13.** True

15.

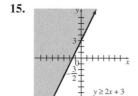

$y \geq 2x + 3$

17.

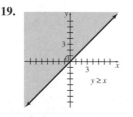

$2x + 6y \leq 8$

19.

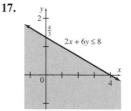

$y \geq x$

21.

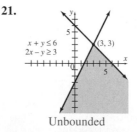

$x + y \leq 6$
$2x - y \geq 3$

$(3, 3)$

Unbounded

23.

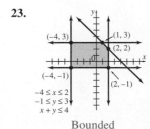

$(-4, 3)$ $(1, 3)$ $(2, 2)$
$(-4, -1)$ $(2, -1)$
$-4 \leq x \leq 2$
$-1 \leq y \leq 3$
$x + y \leq 4$

Bounded

25.

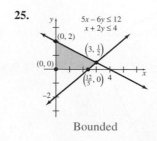

$5x - 6y \leq 12$
$x + 2y \leq 4$
$(0, 2)$ $\left(3, \frac{1}{2}\right)$
$(0, 0)$ $\left(\frac{12}{5}, 0\right)$

Bounded

27. Maximum of 22 at $(3, 4)$; minimum of 0 at $(0, 0)$
29. Maximum of 24 at $(0, 6)$ **31.** Minimum of 40 at any point on the segment connecting $(0, 20)$ and $(10/3, 40/3)$

35. a.

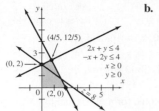

b.

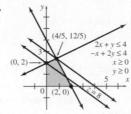

37. Let x = number of batches of cakes and y = number of batches of cookies. Then $x \geq 0$, $y \geq 0$, $2x + (3/2)y \leq 15$, and $3x + (2/3)y \leq 13$.

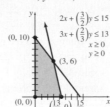

39. a. 3 batches of cakes and 6 batches of cookies, for a maximum profit of $210 **b.** If the profit per batch of cookies increases by more than $2.50 (to above $22.50), then it will be more profitable to make 10 batches of cookies and no batches of cake. **41.** 7 packages of gardening mixture and 2 packages of potting mixture, for a maximum income of $31 **43.** Produce no runs of type I and 7 runs of type II, for a minimum cost of $42,000. **45.** 0 acres for millet and 2 acres for wheat, for a maximum harvest of 1600 lb

Chapter 4 Linear Programming: The Simplex Method

Exercises 4.1 (page 148)

For exercises . . .	1–14	15–18	19–24	27–31
Refer to example . . .	1	3	4	2

1. $x_1 + 2x_2 + s_1 = 6$ **3.** $2.3x_1 + 5.7x_2 + 1.8x_3 + s_1 = 17$ **5. a.** 3
b. s_1, s_2, s_3 **c.** $2x_1 + 3x_2 + s_1 = 15$; $4x_1 + 5x_2 + s_2 = 35$; $x_1 + 6x_2 + s_3 = 20$ **7. a.** 2 **b.** s_1, s_2
c. $7x_1 + 6x_2 + 8x_3 + s_1 = 118$; $4x_1 + 5x_2 + 10x_3 + s_2 = 220$

9.
$$\begin{array}{ccccc|c} x_1 & x_2 & s_1 & s_2 & z & \\ 4 & 2 & 1 & 0 & 0 & 5 \\ 1 & 2 & 0 & 1 & 0 & 4 \\ \hline -7 & -1 & 0 & 0 & 1 & 0 \end{array}$$

11.
$$\begin{array}{cccccc|c} x_1 & x_2 & s_1 & s_2 & s_3 & z & \\ 1 & 1 & 1 & 0 & 0 & 0 & 10 \\ 5 & 2 & 0 & 1 & 0 & 0 & 20 \\ 1 & 2 & 0 & 0 & 1 & 0 & 36 \\ \hline -1 & -3 & 0 & 0 & 0 & 1 & 0 \end{array}$$

13.
$$\begin{array}{ccccc|c} x_1 & x_2 & s_1 & s_2 & z & \\ 3 & 1 & 1 & 0 & 0 & 12 \\ 1 & 1 & 0 & 1 & 0 & 15 \\ \hline -2 & -1 & 0 & 0 & 1 & 0 \end{array}$$

15. $x_1 = 0$, $x_2 = 4$, $x_3 = 0$, $s_1 = 0$, $s_2 = 8$, $z = 28$ **17.** $x_1 = 0$, $x_2 = 0$, $x_3 = 8$, $s_1 = 0$, $s_2 = 6$, $s_3 = 7$, $z = 12$
19. $x_1 = 0$, $x_2 = 20$, $x_3 = 0$, $s_1 = 16$, $s_2 = 0$, $z = 60$ **21.** $x_1 = 0$, $x_2 = 0$, $x_3 = 12$, $s_1 = 0$, $s_2 = 9$, $s_3 = 8$, $z = 36$
23. $x_1 = 0$, $x_2 = 250$, $x_3 = 0$, $s_1 = 0$, $s_2 = 50$, $s_3 = 200$, $z = 1000$

27. If x_1 is the number of simple figures, x_2 is the number of figures with additions, and x_3 is the number of computer drawn sketches, find $x_1 \geq 0$, $x_2 \geq 0$, and $x_3 \geq 0$ such that $20x_1 + 35x_2 + 60x_3 \leq 2200$, $x_1 + x_2 + x_3 \leq 400$, $x_3 \leq x_1 + x_2$, $x_1 \geq 2x_2$, and $z = 95x_1 + 200x_2 + 325x_3$ is maximized; $20x_1 + 35x_2 + 60x_3 + s_1 = 2200$, $x_1 + x_2 + x_3 + s_2 = 400$, $-x_1 - x_2 + x_3 + s_3 = 0$, $-x_1 + 2x_2 + s_4 = 0$.

$$\begin{array}{cccccccc|c} x_1 & x_2 & x_3 & s_1 & s_2 & s_3 & s_4 & z & \\ 20 & 35 & 60 & 1 & 0 & 0 & 0 & 0 & 2200 \\ 1 & 1 & 1 & 0 & 1 & 0 & 0 & 0 & 400 \\ -1 & -1 & 1 & 0 & 0 & 1 & 0 & 0 & 0 \\ -1 & 2 & 0 & 0 & 0 & 0 & 1 & 0 & 0 \\ \hline -95 & -200 & -325 & 0 & 0 & 0 & 0 & 1 & 0 \end{array}$$

29. If x_1 is the number of redwood tables made, x_2 is the number of stained Douglas fir tables made, and x_3 is the number of stained white spruce tables made, find $x_1 \geq 0$, $x_2 \geq 0$, and $x_3 \geq 0$ such that $8x_1 + 7x_2 + 8x_3 \leq 720$, $2x_2 + 2x_3 \leq 480$, $159x_1 + 138.85x_2 + 129.35x_3 \leq 15,000$, and $z = x_1 + x_2 + x_3$ is maximized; $8x_1 + 7x_2 + 8x_3 + s_1 = 720$, $2x_2 + 2x_3 + s_2 = 480$, $159x_1 + 138.85x_2 + 129.35x_3 + s_3 = 15,000$.

$$\begin{array}{ccccccc|c} x_1 & x_2 & x_3 & s_1 & s_2 & s_3 & z & \\ 8 & 7 & 8 & 1 & 0 & 0 & 0 & 720 \\ 0 & 2 & 2 & 0 & 1 & 0 & 0 & 480 \\ 159 & 138.85 & 129.35 & 0 & 0 & 1 & 0 & 15,000 \\ \hline -1 & -1 & -1 & 0 & 0 & 0 & 1 & 0 \end{array}$$

31. If x_1 is the number of newspaper ads run, x_2 is the number of Internet banner ads run, and x_3 is the number of TV ads run, find $x_1 \geq 0$, $x_2 \geq 0$, and $x_3 \geq 0$, such that $400x_1 + 20x_2 + 2000x_3 \leq 8000$, $x_1 \leq 30$, $x_2 \leq 60$, $x_3 \leq 10$, and $z = 4000x_1 + 3000x_2 + 10,000x_3$ is maximized; $400x_1 + 20x_2 + 2000x_3 + s_1 = 8000$, $x_1 + s_2 = 30$, $x_2 + s_3 = 60$, $x_3 + s_4 = 10$.

$$\begin{array}{cccccccc|c} x_1 & x_2 & x_3 & s_1 & s_2 & s_3 & s_4 & z & \\ 400 & 20 & 2000 & 1 & 0 & 0 & 0 & 0 & 8000 \\ 1 & 0 & 0 & 0 & 1 & 0 & 0 & 0 & 30 \\ 0 & 1 & 0 & 0 & 0 & 1 & 0 & 0 & 60 \\ 0 & 0 & 1 & 0 & 0 & 0 & 1 & 0 & 10 \\ \hline -4000 & -3000 & -10,000 & 0 & 0 & 0 & 0 & 1 & 0 \end{array}$$

Exercises 4.2 (page 157)

For exercises . . .	1–40
Refer to example . . .	1

1. Maximum is 30 when $x_1 = 10$, $x_2 = 0$, $x_3 = 0$, $s_1 = 6$, and $s_2 = 0$.   **3.** Maximum is 8 when $x_1 = 4$, $x_2 = 0$, $s_1 = 8$, $s_2 = 2$, and $s_3 = 0$.   **5.** Maximum is 264 when $x_1 = 16$, $x_2 = 4$, $x_3 = 0$, $s_1 = 0$, $s_2 = 16$, and $s_3 = 0$.
7. Maximum is 25 when $x_1 = 0$, $x_2 = 5$, $s_1 = 20$, and $s_2 = 0$.   **9.** Maximum is 120 when $x_1 = 0$, $x_2 = 10$, $s_1 = 0$, $s_2 = 40$, and $s_3 = 4$.   **11.** Maximum is 944 when $x_1 = 118$, $x_2 = 0$, $x_3 = 0$, $s_1 = 0$, and $s_2 = 102$.   **13.** Maximum is 3300 when $x_1 = 240$, $x_2 = 60$, $x_3 = 0$, $x_4 = 0$, $s_1 = 0$, and $s_2 = 0$.   **15.** No maximum   **17.** Maximum is 70,818.18 when $x_1 = 181.82$, $x_2 = 0$, $x_3 = 454.55$, $x_4 = 0$, $x_5 = 1363.64$, $s_1 = 0$, $s_2 = 0$, $s_3 = 0$, and $s_4 = 0$.   **23.** 6 churches and 2 labor unions, for a maximum of $1000 per month   **25. a.** Assemble 1000 Royal Flush poker sets, 3000 Deluxe Diamond poker sets, and 0 Full House poker sets, for a maximum profit of $104,000.   **b.** $s_4 = 1000$; there are 1000 unused dealer buttons.   **27. a.** No racing or touring bicycles and 2700 mountain bicycles   **b.** Maximum profit is $59,400   **c.** No; there are 1500 units of aluminum left; $s_2 = 1500$.   **29. a.** 17 newspaper ads, 60 Internet banner ads, and no TV ads, for a maximum exposure of 248,000   **31. a.** 3   **b.** 4   **c.** 3   **33.** $200, $66.67, $300, $100   **35.** Rachel should run 3 hours, bike 4 hours, and walk 8 hours, for a maximum calorie expenditure of 6313 calories.   **37. a.** None of species A, 114 of species B, and 291 of species C, for a maximum combined weight of 1119.72 kg   **b.** No; there are 346 units of Food II available.   **c.** Many answers are possible.   **d.** Many answers are possible.   **39.** 12 minutes to the senator, 9 minutes to the congresswoman and 6 minutes to the governor, for a maximum of 1,050,000 viewers.

Exercises 4.3 (page 168)

For exercises . . .	1–4	5–8	9–16	19–29
Refer to example . . .	2	3	4	5

1. $\begin{bmatrix} 1 & 3 & 1 \\ 2 & 2 & 10 \\ 3 & 1 & 0 \end{bmatrix}$ **3.** $\begin{bmatrix} 4 & 7 & 5 \\ 5 & 14 & 0 \\ -3 & 20 & -2 \\ 15 & -8 & 23 \end{bmatrix}$ **5.** Minimize $w = 5y_1 + 4y_2 + 15y_3$ subject to $y_1 + y_2 + 2y_3 \geq 4$, $y_1 + y_2 + y_3 \geq 3$, $y_1 + 3y_3 \geq 2$, with $y_1 \geq 0$, $y_2 \geq 0$, and $y_3 \geq 0$.   **7.** Maximize $z = 150x_1 + 275x_2$ subject to $x_1 + 2x_2 \leq 3$, $x_1 + 2x_2 \leq 6$, $x_1 + 3x_2 \leq 4$, $x_1 + 4x_2 \leq 1$, with $x_1 \geq 0$ and $x_2 \geq 0$.
9. Minimum is 14 when $y_1 = 0$ and $y_2 = 7$.   **11.** Minimum is 40 when $y_1 = 10$ and $y_2 = 0$.   **13.** Minimum is 30 when $y_1 = 5$ and $y_2 = 0$ or when $y_1 = 0$ and $y_2 = 3$, or any point on the line segment between $(5, 0)$ and $(0,3)$.   **15.** Minimum is 100 when $y_1 = 0$, $y_2 = 100$, and $y_3 = 0$.   **17.** a   **19. a.** 25 units of regular beer and 20 units of light beer, for a minimum cost of $1,800,000   **b.** The shadow cost is $0.10; total production cost is $1,850,000.   **21. a.** Minimize $w = 100y_1 + 20,000y_2$ subject to $y_1 + 400y_2 \geq 120$, $y_1 + 160y_2 \geq 40$, $y_1 + 280y_2 \geq 60$, with $y_1 \geq 0$, $y_2 \geq 0$.   **b.** 52.5 acres of potatoes and no corn or cabbage, for a profit of $6300.   **c.** 47.5 acres of potatoes and no corn or cabbage, for a profit of $5700.   **23.** 8 political interviews and no market interviews are done, for a minimum of 360 minutes.   **25. a.** 1 bag of feed 1 and 2 bags of feed 2   **b.** 1.4 (or 7/5) bags of feed 1 and 1.2 (or 6/5) bags of feed 2 should be used, for a minimum cost of $6.60   **27.** She should spend 30 minutes walking, 197.25 minutes cycling, and 75.75 minutes swimming, for a minimum time of 303 minutes per week.   **29.** 8/3 units of ingredient I and 4 units of ingredient III, for a minimum cost of $30.67.

Exercises 4.4 (page 176)

For exercises . . .	1–16	17–20,24,27,31,32	23,25,26,28,29,30,33
Refer to example . . .	1	3	2

1. $2x_1 + 3x_2 + s_1 = 8$; $x_1 + 4x_2 - s_2 = 7$
3. $2x_1 + x_2 + 2x_3 + s_1 = 50$; $x_1 + 3x_2 + x_3 - s_2 = 35$; $x_1 + 2x_2 - s_3 = 15$ **5.** Change the objective function to maximize $z = -3y_1 - 4y_2 - 5y_3$. The constraints are not changed.   **7.** Change the objective function to maximize $z = -y_1 - 2y_2 - y_3 - 5y_4$. The constraints are not changed.   **9.** Maximum is 480 when $x_1 = 40$ and $x_2 = 0$.   **11.** Maximum is 750 when $x_1 = 0$, $x_2 = 150$, and $x_3 = 0$.   **13.** Maximum is 135 when $x_1 = 30$ and $x_2 = 5$.   **15.** Minimum is 108 when $y_1 = 0$, $y_2 = 9$, and $y_3 = 0$.   **17.** Maximum is 400/3 when $x_1 = 100/3$ and $x_2 = 50/3$.   **19.** Minimum is 512 when $y_1 = 6$, $y_2 = 8$, and $y_3 = 0$.   **23. a.** Ship 200 barrels of oil from supplier S_1 to distributor D_1; ship 2800 barrels of oil from supplier S_2 to distributor D_1; ship 2800 barrels of oil from supplier S_1 to distributor D_2; ship 2200 barrels of oil from supplier S_2 to distributor D_2. Minimum cost is $180,400.   **b.** $s_3 = 2000$; S_1 could furnish 2000 more barrels of oil.   **25.** Make $3,000,000 in commercial loans and $22,000,000 in home loans, for a maximum return of $2,940,000.   **27.** Use 1500 lb of bluegrass, 2700 lb of rye, and 1800 lb of Bermuda, for a mininum cost of $834.   **29. a.** Ship 2 computers from W_1 to D_1, 20 computers from W_1 to D_2, 30 computers from W_2 to D_1, and 0 computers from W_2 to D_2, for a minimum cost of $628.   **b.** $s_3 = 3$; warehouse W_1 has 3 more computers that it could ship.   **31.** Use 59.21 kg of chemical I, 394.74 kg of chemical II, and 296.05 kg of chemical III, for a minimum cost of $600.39   **33.** 5/3 oz of I, 20/3 oz of II, and 5/3 oz of III, for a minimum cost of $1.55 per gal; 10 oz of the additive should be used per gal of gasoline.

Chapter 4 Review Exercises (page 180)

For exercises . . .	1,15,17,18,21–24	2,3,5–11,13,37–39,41–43,47	4,12	14,16,19,20,29–32,40,44	25–28,33–36,45,46
Refer to section . . .	1	2	3	4	3, 4

1. True   **2.** False   **3.** True   **4.** False   **5.** False   **6.** True   **7.** True   **8.** False   **9.** False   **10.** True   **11.** False   **12.** True
13. False   **14.** True   **15.** When the problem has more than two variables   **17. a.** $4x_1 + 6x_2 + s_1 = 60$; $3x_1 + x_2 + s_2 = 18$;

$2x_1 + 5x_2 + s_3 = 20; x_1 + x_2 + s_4 = 15$ **b.**

	x_1	x_2	s_1	s_2	s_3	s_4	z	
	4	6	1	0	0	0	0	60
	3	1	0	1	0	0	0	18
	2	5	0	0	1	0	0	20
	1	1	0	0	0	1	0	15
	−2	−7	0	0	0	0	1	0

19. a. $x_1 + x_2 + x_3 + s_1 = 90; 2x_1 + 5x_2 + x_3 + s_2 = 120; x_1 + 3x_2 - s_3 = 80$ **b.**

	x_1	x_2	x_3	s_1	s_2	s_3	z	
	1	1	1	1	0	0	0	90
	2	5	1	0	1	0	0	120
	1	3	0	0	0	−1	0	80
	−5	−8	−6	0	0	0	1	0

21. Maximum is 33 when $x_1 = 3$, $x_2 = 0$, $x_3 = 3$, $s_1 = 0$, and $s_2 = 0$.   **23.** Maximum is 76.67 when $x_1 = 6.67$, $x_2 = 0$, $x_3 = 21.67$, $s_1 = 0$, $s_2 = 0$, and $s_3 = 35$.   **25. Dual Method** Solve the dual problem: Maximize $17x_1 + 42x_2$ subject to $x_1 + 5x_2 \le 10$, $x_1 + 8x_2 \le 15$.   **Method of Section 4.4** Change the objective function to maximize $z = -10y_1 - 15y_2$. The constraints are not changed. Minimum is 170 when $y_1 = 17$ and $y_2 = 0$.   **27. Dual Method** Solve the dual problem: Maximize $48x_1 + 12x_2 + 10x_3 + 30x_4$ subject to $x_1 + x_2 + 3x_4 \le 7$, $x_1 + x_2 \le 2$, $2x_1 + x_3 + x_4 \le 3$.   **Method of Section 4.4** Change the objective function to maximize $z = -7y_1 - 2y_2 - 3y_3$. The constraints are not changed. Minimum is 98 when $y_1 = 4$, $y_2 = 8$, and $y_3 = 18$.   **29.** Maximum of 480 when $x_1 = 24$ and $x_2 = 0$ **31.** Maximum of 102 when $x_1 = 0$ and $x_2 = 8.5$ **33.** Problems with constraints involving "$\le$" can be solved using slack variables, while those involving "$\ge$" or "$=$" can be solved using surplus and artificial variables, respectively.   **35. a.** Maximize $z = 6x_1 + 7x_2 + 5x_3$, subject to $4x_1 + 2x_2 + 3x_3 \le 9$, $5x_1 + 4x_2 + x_3 \le 10$, with $x_1 \ge 0$, $x_2 \ge 0$, $x_3 \ge 0$.   **b.** The first constraint would be $4x_1 + 2x_2 + 3x_3 \ge 9$.   **c.** $x_1 = 0$, $x_2 = 2.1$, $x_3 = 1.6$, and $z = 22.7$ **d.** Minimize $w = 9y_1 + 10y_2$, subject to $4y_1 + 5y_2 \ge 6$, $2y_1 + 4y_2 \ge 7$, $3y_1 + y_2 \ge 5$, with $y_1 \ge 0$, $y_2 \ge 0$.   **e.** $y_1 = 1.3$, $y_2 = 1.1$, and $w = 22.7$ **37. a.** Let $x_1 = $ number of cake plates, $x_2 = $ number of bread plates, and $x_3 = $ number of dinner plates.   **b.** $z = 15x_1 + 12x_2 + 5x_3$ **c.** $15x_1 + 10x_2 + 8x_3 \le 1500$; $5x_1 + 4x_2 + 4x_3 \le 2700$; $6x_1 + 5x_2 + 5x_3 \le 1200$ **39. a.** Let $x_1 = $ number of gallons of Fruity wine and $x_2 = $ number of gallons of Crystal wine to be made.   **b.** $z = 12x_1 + 15x_2$ **c.** $2x_1 + x_2 \le 110$; $2x_1 + 3x_2 \le 125$; $2x_1 + x_2 \le 90$ **41.** Produce no cake plates, 150 bread plates, and no dinner plates, for a maximum profit of $1800.   **43.** 36.25 gal of Fruity and 17.5 gal of Crystal, for a maximum profit of $697.50

45. a and b Produce 660 cases of corn, 0 cases of beans, and 340 cases of carrots, for a minimum cost of $15,100.   **c.** $16,100

47. Ginger should do $5\frac{1}{3}$ hours of tai chi, $2\frac{2}{3}$ hours of riding a unicycle, and 2 hours of fencing, for a maximum calorie expenditure of $2753\frac{1}{3}$ calories.

Chapter 5 Mathematics of Finance

Exercises 5.1 (page 197)

For exercises...	5–10,49,50	11,12	13,14,51–53	19–24,54a,b,55–65	25–28,54c,d,66,77	29–32,67–69	33–38,70,71	41–44	45,46	47,48,63e
Refer to example...	1	2	3	4,5	6	7,8	9,10	11	12	13

1. The interest rate and number of compounding periods   **5.** $562.50   **7.** $59.79   **9.** $50.79   **11.** $3176.95; $51.95   **13.** 7.5%
17. t is the number of years, while n is the number of compounding periods.   **19.** $1593.85; $593.85   **21.** $890.82; $420.82
23. $12,630.55; $4130.55   **25.** $4.75%   **27.** 5.75%   **29.** 4.06%   **31.** 7.38%   **33.** $9677.13   **35.** $1246.33   **37.** $6864.08
41. 15 years   **43.** 24 years, 11 months   **45. a.** 21.35 years   **b.** 21.21 years   **47. a.** $7269.94   **b.** 3.15%   **c.** 19.29 years
49. $7534.80; $334.80   **51.** 6.8%   **53.** 11.4%   **55.** $41,325.95   **57.** $1000 now   **59.** $136,110.16   **61.** 9.31×10^{31}
63. a. $16,288.95   **b.** $16,436.19   **c.** $16,470.09   **d.** $16,486.65   **e.** $16,487.21   **65.** 5/4   **67.** 2.48, 5.01, 4.18, 4.43, 5.15
69. 5.64%, 5.63%, Centennial Bank   **71.** $22,829.89   **73.** About 18 years   **75.** About 12 years   **77. a.** 20.7%   **b.** $39.6 billion

Exercises 5.2 (page 207)

For exercises...	1–8	9–14	17–26,47–55,65,68	29,30,63,64	31–38,56–58,66,67	39–46,59–62
Refer to example...	1	2	3	4	5	6

1. 48   **3.** −648   **5.** 81   **7.** 1
9. 15   **11.** 156/25   **13.** −208   **17.** $437.46   **19.** $2,154,099.15   **21.** $180,307.41; $128,800; $51,507.41   **23.** $28,438.21;
$19,200; $9238.21   **25.** $1,145,619.96; $768,000; $377,619.96   **29.** 4.19%   **31.** $628.25   **33.** $952.62   **35.** $7382.54
37. $5970.23   **39.** $6294.79   **41.** $136,785.74   **43.** $26,874.97; $18,000; $8874.97   **45.** $15,662.40; $12,000; $3662.40
47. a. $149,850.69   **b.** $137,895.79   **c.** $11,954.90   **49.** $197,750.47   **51.** $239,315.17   **53.** $312,232.31; $212,232.31
55. $432,548.65; $332,548.65   **57.** $1349.48   **59.** $3777.89   **61.** $67,940.98   **63.** 6.5%   **65. a.** 7 yr **b.** 9 yr   **67. a.** $120
b. $681.83 except the last payment, which is $681.80

Exercises 5.3 (page 215)

3. $8994.25 **5.** $209,302.93
7. $170,275.47 **9.** $111,183.87

For exercises . . .	3–10,41	11–16a,b,27–30,35–40, 42,43,47,48a,49,50a–c, 51a,c,d,52,53	11–16c,21–24,44–46, 54,55	17–20	31–34	48b,49,50d,51b
Refer to example . . .	1	2,3	5	4	6	after Example 3

11. a. $438.81 **b.** $2632.86;
$132.86 **c.** $2632.88; $132.88 **13. a.** $10,734.93 **b.** $128,819.16; $38,819.16 **c.** $128,819.20; $38,819.20 **15. a.** $542.60
b. $9766.80; $2366.80 **c.** $9766.88; $2366.88 **17.** $73,015.71 **19.** $5368.98 **21.** $7.61 **23.** $35.24 **25.** $6699
27. $1407.76; $422,328; $223,328 **29.** $1590.82; $572,695.20; $319,659.20 **31. a.** 8 years **b.** $113,086.84 **c.** $15,732.36
33. a. 11 semiannual periods **b.** $8760.50 **c.** $1006.38 **35.** $1856.49; $114,168.20. The payments are $537.48 more than for the
30-yr loan, but the total interest paid is $140,675.40 less. **37. a.** $335.25 **b.** $2092 **39. a.** $844.95, $30,418.20 **b.** $655.92,
$31,484.16 **41. a.** $623,110.52 **b.** $456,427.28 **c.** $563,757.78 **d.** $392,903.18 **43.** $381.74, $59,522

45.

Payment Number	Amount of Payment	Interest for Period	Portion to Principal	Principal at End of Period
0	—	—	—	$110,000.00
1	$14,794.23	$4400.00	$10,394.23	$99,605.77
2	$14,794.23	$3984.23	$10,810.00	$88,795.77
3	$14,794.23	$3551.83	$11,242.40	$77,553.37
4	$14,794.23	$3102.13	$11,692.10	$65,861.27

47. a. $32.49 **b.** $195.52; $10.97
49. a. $1959.99; $127,798.20
b. $1677.54; $177,609.60 **c.** $1519.22;
$230,766.00 **d.** After 157 payments
51. a. $1121.63; $403,786.80;
$253,786.80 **b.** $115,962.66;
$201,893.40 **c.** $732.96; $267,265.60
d. $1010.16; $186,328.80
53. a. $17,584.58 **b.** $15,069.31
57. a. $25,000 **b.** $40,000

55.

Payment Number	Amount of Payment	Interest for Period	Portion to Principal	Principal at End of Period
0	—	—	—	$4836.00
1	$585.16	$175.31	$409.85	$4426.15
2	$585.16	$160.45	$424.71	$4001.43
3	$585.16	$145.05	$440.11	$3561.32
4	$585.16	$129.10	$456.06	$3105.26
5	$585.16	$112.57	$472.59	$2632.67
6	$585.16	$95.43	$489.73	$2142.94
7	$585.16	$77.68	$507.48	$1635.46
8	$585.16	$59.29	$525.87	$1109.59
9	$585.16	$40.22	$544.94	$564.65
10	$585.12	$20.47	$564.65	$0.00

Chapter 5 Review Exercises (page 219)

1. True **2.** False **3.** True
4. False **5.** True **6.** True

For exercises . . .	1,11–28,64–69,74,75,81a	2–4,29–48,70,71,80,81c,e	5–10,49–63,72,73,76–79,81b,d
Refer to example . . .	1	2	3

7. True **8.** False **9.** False **10.** True **11.** $636.12 **13.** $1290.11 **15.** Compound interest **17.** $33,691.69 **19.** $77,860.80
21. $5244.50 **23.** $4725.22 **25.** $27,624.86 **27.** $1067.71 **29.** 2, 6, 18, 54, 162 **31.** −96 **33.** −120 **35.** 40.56808
39. $23,559.98; $5527.98 **41.** $12,302.78; $1118.78 **43.** $160,224.29; $5524.29 **45.** $955.61 **47.** $6156.14 **49.** $2945.34
51. $56,711.93 **53.** A home loan and an auto loan **55.** $302.59; $431.08 **57.** $1796.20; $5871.40 **59.** $1140.50; $410,580;
$233,470 **61.** $132.99 **63.** $1535.61 **65.** $10,203.80; $383.80 **67.** 8.21% **69.** $2298.58 **71.** $107,892.82; $32,892.82
73. $8751.91; $13,263.37 **75.** 1.50% and 1.46%; Ascencia **77. a.** $266.67, $16,000 (or $16,000.20 using the rounded payments)
b. $249.59, $17,970.48 **c.** $283.64, $13,614.72 **79. a.** $954.42 **b.** $817.92 **c.** Method 1: $109,563.99; Method 2: $109,565.13
d. $9650 **e.** Method 1: $118,786.01; Method 2: $118,784.87 **81. a.** 9.569% **b.** $896.44 **c.** $626,200.88 **d.** $1200.39
e. $478,134.14 **f.** Sue is ahead by $148,066.74.

Chapter 6 Logic

Exercises 6.1 (page 231)

1. Statement, not compound **3.** Not a statement **5.** Statement,
compound **7.** Not a statement **9.** Statement, compound
11. Statement, not compound **13.** Statement, compound

For exercises . . .	1,14, 65,66, 75,76, 78,79	15,16, 67,77, 80	17–20	23–28, 68–72, 83–88	35–46, 49–56, 73,74, 89,90	57–64
Refer to example . . .	1	2	3	4	6	7

15. My favorite flavor is not chocolate. **17.** $y \leq 12$ **19.** $q < 5$
23. I'm not getting better. **25.** I'm not getting better or my parrot is dead. **27.** It is not the case that both I'm getting better and my
parrot is not dead. **29.** False **31.** True **33.** Both components are false. **35.** True **37.** True **39.** False **41.** True **43.** True
45. True **47.** Disjunction **49.** True **51.** False **53.** True **55.** True **57.** False **59.** True **61.** True **63.** True **65.** b, c, d

67. An individual has to be your biological child to be a "qualifying" child.   **69.** $a \wedge j$ **71.** $\sim a \vee j$ **73.** 69, 71, and 72
75. b, c, d, e   **77.** You may not find that exercise helps you cope with stress.   **79.** c, d   **83.** $n \wedge \sim m$ **85.** $\sim n \vee m$
87. $\sim n \wedge \sim m$ or $\sim(n \vee m)$ **89.** 85

Exercises 6.2 (page 238)

1. 4   **3.** 16   **5.** 128   **7.** 6

For exercises . . .	9–18	19–24,42,43,48	25–34,40,41, 44,46,47,49
Refer to example . . .	1,2	3,4	6

9.

p	q	$\sim p$	$\sim p \wedge q$
T	T	F	F
T	F	F	F
F	T	T	T
F	F	T	F

11.

p	q	$p \wedge q$	$\sim(p \wedge q)$
T	T	T	F
T	F	F	T
F	T	F	T
F	F	F	T

13.

p	q	$\sim p$	$\sim q$	$q \vee \sim p$	$(q \vee \sim p) \vee \sim q$
T	T	F	F	T	T
T	F	F	T	F	T
F	T	T	F	T	T
F	F	T	T	T	T

In Exercises 15–23, we are using the alternate method to save space.

15.

p	q	$\sim q$	$\wedge$	$(\sim p \vee q)$
T	T	F	F	F T T
T	F	T	F	F F F
F	T	F	F	T T T
F	F	T	T	T T F
		①	④	② ③ ②

17.

p	q	$(p \vee \sim q)$	$\wedge$	$(p \wedge q)$
T	T	T T F	T	T T T
T	F	T T T	F	T F F
F	T	F F F	F	F F T
F	F	F T T	F	F F F
		① ② ①	⑤	③ ④ ③

19.

p	q	r	$(\sim p \wedge q)$	$\wedge$	r
T	T	T	F F T	F	T
T	T	F	F F T	F	F
T	F	T	F F F	F	T
T	F	F	F F F	F	F
F	T	T	T T T	T	T
F	T	F	T T T	F	F
F	F	T	T F F	F	T
F	F	F	T F F	F	F
			① ② ①	④	③

21.

p	q	r	$(\sim p \wedge \sim q)$	$\vee$	$(\sim r \vee \sim p)$
T	T	T	F F F	F	F F F
T	T	F	F F F	T	T T F
T	F	T	F F T	F	F F F
T	F	F	F F T	T	T T F
F	T	T	T F F	T	F T T
F	T	F	T F F	T	T T T
F	F	T	T T T	T	F T T
F	F	F	T T T	T	T T T
			① ② ①	⑤	③ ④ ③

23.

p	q	r	s	$\sim(\sim p \wedge \sim q)$	$\vee$	$(\sim r \vee \sim s)$
T	T	T	T	T F F F	T	F F F
T	T	T	F	T F F F	T	F T T
T	T	F	T	T F F F	T	T T F
T	T	F	F	T F F F	T	T T T
T	F	T	T	T F F T	T	F F F
T	F	T	F	T F F T	T	F T T
T	F	F	T	T F F T	T	T T F
T	F	F	F	T F F T	T	T T T
F	T	T	T	T T F F	T	F F F
F	T	T	F	T T F F	T	F T T
F	T	F	T	T T F F	T	T T F
F	T	F	F	T T F F	T	T T T
F	F	T	T	F T T T	F	F F F
F	F	T	F	F T T T	T	F T T
F	F	F	T	F T T T	T	T T F
F	F	F	F	F T T T	T	T T T
				③ ① ② ①	⑥	④ ⑤ ④

25. It's not a vacation, or I am not having fun.
27. The door was locked and the thief didn't break a window.   **29.** I'm not ready to go, or Naomi Bahary is
31. $12 \le 4$ and $8 \ne 9$ **33.** Neither Larry nor Moe is
out sick today   **35.**

p	q	$p \veebar q$
T	T	F
T	F	T
F	T	T
F	F	F

37. True   **39. a.** False   **b.** True   **c.** False   **d.** True
41. Service will not be performed at the location, and the store may not send the Covered Equipment to an Apple repair service location to be repaired.

43. $s \vee (r \wedge \sim q)$

s	r	q	s	∨	(r ∧ ~q)
T	T	T	T	T	T F F
T	T	F	T	T	T T T
T	F	T	T	T	F F F
T	F	F	T	T	F F T
F	T	T	F	F	T F F
F	T	F	F	T	T T T
F	F	T	F	F	F F F
F	F	F	F	F	F F T
			①	④	② ③ ②

The guarantee would be false if you are not completely satisfied, and they either don't refund your money or ask you questions.
45. Inclusive **47.** Liberty without learning is not always in peril, or learning without liberty is not always in vain.
49. You cannot reroll the die again for your Large Straight and you cannot set aside the 2 Twos and roll for your Twos or for 3 of a Kind.

Exercises 6.3 (page 247)

1. True **3.** True **5.** True **9.** True
11. True **13.** If she dances tonight, then I'm leaving early and he sings loudly.
15. If he doesn't sing loudly, then she dances tonight or I'm not leaving early. **17.** $d \vee (f \rightarrow g)$ **19.** $\sim f \rightarrow g$ **21.** False **23.** False **25.** True **27.** True

For exercises . . .	9–12	21–30, 91	33–42, 51–68	43–46, 89,90, 92,93	47–50	69–76, 87,88	77–86
Refer to example . . .	3	4	5	6	7	8,9	10

33.

p	q	~q	→	p
T	T	F	T	T
T	F	T	T	T
F	T	F	T	F
F	F	T	F	F
		①	②	①

35.

p	(p ∨ ~p)	→	(p ∧ ~p)
T	T T F	F	T F F
F	F T T	F	F F T
	① ② ①	⑤	③ ④ ③

It is a contradiction.

37.

p	q	(p ∨ q)	→	(q ∨ p)
T	T	T T T	T	T T T
T	F	T T F	T	F T T
F	T	F T T	T	T T F
F	F	F F F	T	F F F
		① ② ①	⑤	③ ④ ③

It is a tautology.

39.

p	q	r	r	→	(p ∧ ~q)
T	T	T	T	F	T F F
T	T	F	F	T	T F F
T	F	T	T	T	T T T
T	F	F	F	T	T T T
F	T	T	T	F	F F F
F	T	F	F	T	F F F
F	F	T	T	F	F F T
F	F	F	F	T	F F T
			①	④	② ③ ②

41.

p	q	r	s	(~r → s)	∨	(p → ~q)
T	T	T	T	F T T	T	T F F
T	T	T	F	F T F	T	T F F
T	T	F	T	T T T	T	T F F
T	T	F	F	T F F	F	T F F
T	F	T	T	F T T	T	T T T
T	F	T	F	F T F	T	T T T
T	F	F	T	T T T	T	T T T
T	F	F	F	T F F	T	T T T
F	T	T	T	F T T	T	F T F
F	T	T	F	F T F	T	F T F
F	T	F	T	T T T	T	F T F
F	T	F	F	T F F	T	F T F
F	F	T	T	F T T	T	F T T
F	F	T	F	F T F	T	F T T
F	F	F	T	T T T	T	F T T
F	F	F	F	T F F	T	F T T
				① ② ①	⑤	③ ④ ③

43. Your eyes are not bad or your whole body will be full of darkness. **45.** I don't have the money or I'd buy that car.
47. You ask me and I do not do it. **49.** You don't love me and I will be happy. **51.** Equivalent **53.** Not equivalent
55. Equivalent **57.** Not equivalent

59.

p	q	p ∧ q	~(p → ~q)
T	T	T T T	T T F F
T	F	T F F	F T T T
F	T	F F T	F F T F
F	F	F F F	F F T T
		① ② ①	⑤ ③ ④ ③

The columns labeled 2 and 5 are identical.

61.

p	q	p ∨ q	q ∨ p
T	T	T T T	T T T
T	F	T T F	F T T
F	T	F T T	T T F
F	F	F F F	F F F
		① ② ①	③ ④ ③

The columns labeled 2 and 4 are identical.

63.

p	q	r	(p ∨ q) ∨ r	p ∨ (q ∨ r)
T	T	T	T T T T T	T T T T T
T	T	F	T T T T F	T T T T F
T	F	T	T T F T T	T T F T T
T	F	F	T T F T F	T T F F F
F	T	T	F T T T T	F T T T T
F	T	F	F T T T F	F T T T F
F	F	T	F F F T T	F T F T T
F	F	F	F F F F F	F F F F F
			① ② ① ④ ③	⑤ ⑧ ⑥ ⑦ ⑥

The columns labeled 4 and 8 are identical.

65.

p	q	r	p ∨ (q ∧ r)	(p ∨ q) ∧ (p ∨ r)
T	T	T	T T T T T	T T T T T T T
T	T	F	T T T F F	T T T T T T F
T	F	T	T T F F T	T T F T T T T
T	F	F	T T F F F	T T F T T T F
F	T	T	F T T T T	F T T T F T T
F	T	F	F F T F F	F T T F F F F
F	F	T	F F F F T	F F F F F T T
F	F	F	F F F F F	F F F F F F F
			① ④ ② ③ ②	⑤ ⑥ ⑤ ⑨ ⑦ ⑧ ⑦

The columns labeled 4 and 9 are identical.

67.

p	q	(p ∧ q) ∨ p
T	T	T T T T T
T	F	T F F T T
F	T	F F T F F
F	F	F F F F F
		① ② ① ④ ③

The *p* column and the column labeled 4 are identical.

69. $(p \wedge q) \vee (p \wedge {\sim}q) \equiv p$ **71.** $p \vee ({\sim}q \wedge r)$ **73.** $(p \vee q) \vee {\sim}p \equiv T$ **75.** $[(p \wedge q) \vee (p \wedge p)] \vee (r \wedge {\sim}r) \equiv p$
77. $p \wedge (q \vee {\sim}p) \equiv p \wedge p$ **79.** $(p \vee q) \wedge ({\sim}p \wedge {\sim}q) \equiv F$ **81.** $\{(p \vee q) \wedge r\} \wedge {\sim}p \equiv r \wedge ({\sim}p \wedge q)$

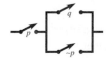 

83. ${\sim}q \rightarrow ({\sim}p \rightarrow q) \equiv p \vee q$ **85.** $[(p \wedge q) \vee p] \wedge [(p \vee q) \wedge q] \equiv p \wedge q$ **87.** False

  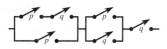

89. a. You are not married at the end of the year, or you may file a joint return with your spouse. **b.** A bequest received by an executor from an estate is compensation for services, or it is tax free. **c.** A course does not improve your current job skills or does not lead to qualification for a new profession, or the course is not deductible. **91. a.** $(\nu \vee p) \rightarrow (s \wedge g)$ **b.** True **d.** The value of my portfolio exceeds \$100,000 or the price of my stock in Ford Motor Company falls below \$50 per share, and I will not sell my shares of Ford stock or I will not give the proceeds to the United Way. **93. a.** If you cannot file a civil lawsuit yourself, then your attorney can do it for you. **b.** If your driver's license does not come with restrictions, then restrictions may sometimes be added on later. **c.** If you can marry when you're not at least 18 years old, then you have the permission of your parents or guardian.

Exercises 6.4 (page 254)

For exercises . . .	1–10,42,43, 45,47,48	13–29,41,44,46, 55–57	32–38
Refer to example . . .	4	1,2	5

1. a. *Converse:* If I don't see it, then the exit is ahead. **b.** *Inverse:* If the exit is not ahead, then I see it. **c.** *Contrapositive:* If I see it, then the exit is not ahead. **3. a.** *Converse:* If I cleaned the house, then I knew you were coming. **b.** *Inverse:* If I didn't know you were coming, I wouldn't have cleaned the house. **c.** *Contrapositive:* If I didn't clean the house, then I didn't know you were coming. **5. a.** *Converse:* If you wear a pocket protector, then you are a mathematician.
b. *Inverse:* If you are not a mathematician, then you do not wear a pocket protector. **c.** *Contraposition:* If you do not wear a pocket protector, then you are not a mathematician. **7. a.** *Converse:* ${\sim}q \rightarrow p$. **b.** *Inverse:* ${\sim}p \rightarrow q$. **c.** *Contrapositive:* $q \rightarrow {\sim}p$.
9. a. *Converse:* $(q \vee r) \rightarrow p$. **b.** *Inverse:* ${\sim}p \rightarrow {\sim}(q \vee r)$ or ${\sim}p \rightarrow ({\sim}q \wedge {\sim}r)$. **c.** *Contrapositive:* $({\sim}q \wedge {\sim}r) \rightarrow {\sim}p$.
13. If you sign, then you accept the conditions. **15.** If you can take this course pass/fail, then you have prior permission. **17.** If the temperature is below 10°, then you can skate on the pond. **19.** If someone eats 10 hot dogs, then he or she will get sick. **21.** If you

travel to France, then you have a valid passport. **23.** If a number has a real square root, then it is nonnegative. **25.** If someone is a bride, then she is beautiful. **27.** If the sum of a number's digits is divisible by 3, then it is divisible by 3. **29.** d **33.** True **35.** False **37.** False

39.

p	q	$(\sim p \wedge q)$	$\leftrightarrow$	$(p \rightarrow q)$
T	T	F F T	F	T T T
T	F	F F F	T	T F F
F	T	T T T	T	F T T
F	F	T F F	F	F T F
		① ② ①	⑤	③ ④ ③

41. a. If it is an employee contribution, then it must be reported on From 8889. **b.** If certain tax benefits may be claimed by married persons, then they file jointly. **c.** If he or she provides over half of his or her own support, then the child is not a qualifying child. **43.** *Converse:* If we close your account without notice, then your account is in default. *Inverse:* If your account is not in default, then we may not close your account without notice. *Contrapositive:* If we do not close your account without notice, then your account is not in default. The original statement and the contrapositive are equivalent, and the converse and inverse are equivalent. **45. a.** $p \rightarrow (q \wedge r)$ **b.** If the most persistent does not stand to gain an extra meal or it does not eat at the expense of another, then there are not triplets. **47. a.** *Converse:* If you can't get married again, then you are married. *Inverse:* If you aren't married, then you can get married again. *Contrapositive:* If you can get married again, then you are not married. **b.** *Converse:* If you are protected by the Fair Credit Billing Act, then you pay for your purchase with a credit card. *Inverse:* If you do not pay for your purchase with a credit card, then you are not protected by the Fair Credit Billing Act. *Contrapositive:* If you are not protected by the Fair Credit Billing Act, then you do not pay for your purchase with a credit card. **c.** *Converse:* If you're expected to make a reasonable effort to locate the owner, then you hit a parked car. *Inverse:* If you did not hit a parked car, then you are not expected to make a reasonable effort to locate the owner. *Contrapositive:* If you are not expected to make a reasonable effort to locate the owner, then you did not hit a parked car. The original statement and the contrapositive are equivalent, and the converse and inverse are equivalent.

49. a. $d \leftrightarrow a$,

d	a	$d \leftrightarrow a$
T	T	T
T	F	F
F	T	F
F	F	T

b. It is false **51.** If a country has democracy, then it has a high level of education. *Converse:* If a country has a high level of education, then it has democracy. *Inverse:* If a country does not have democracy, then it does not have a high level of education. *Contrapositive:* If a country does not have a high level of education, then it does not have a democracy. The contrapositive is equivalent to the original. **53.** D, 7 **55. a.** If nothing is ventured, then nothing is gained. If something is gained, then something is ventured. Something is ventured or nothing is gained. **b.** If something is one of the best things in life, then it is free. If something is not free, then it's not one of the best things in life. Something is not one of the best things in life or it is free. **c.** If something is a cloud, then it has a silver lining. If something doesn't have a silver lining, then it's not a cloud. Something is not a cloud or it doesn't have a silver lining. **57. a.** If you can score in this box, then the dice show any sequence of four numbers. You cannot score in this box, or the dice show any sequence of four numbers. **b.** If two or more words are formed in the same play, then each is scored. Two or more words are not formed in the same play, or each is scored. **c.** If words are labeled as a part of speech, then they are permitted. Words are not labeled as parts of speech, or they are permitted.

Exercises 6.5 (page 264)

1. Valid; Reasoning by Transitivity
3. Valid; Modus Ponens
5. Invalid; Fallacy of the Converse
7. Valid; Modus Tollens

For exercises . . .	1,2	3,4	5,6	7,8	9,10	11,12	13–16, 31,32, 35	17–24,29,30, 33,34, 36–44
Refer to example . . .	4	Modus Ponens	1	2	Fallacy of the Inverse	3	5,8	6,7

9. Invalid; Fallacy of the Inverse   **11.** Valid; Disjunctive Syllogism   **13.** Invalid; $p = $ T, $q = $ T   **15.** Invalid; $p = $ F, $q = $ F

17. Valid.
1. $\sim p \rightarrow \sim q$	Premise
2. q	Premise
3. p	1, 2, Modus Tollens

19. Valid.
1. $p \rightarrow q$	Premise
2. $\sim q$	Premise
3. $\sim p \rightarrow r$	Premise
4. $\sim p$	1, 2, Modus Tollens
5. r	3, 4, Modus Ponens

21. Valid.
1. $p \rightarrow q$	Premise
2. $q \rightarrow r$	Premise
3. $\sim r$	Premise
4. $p \rightarrow r$	1, 2, Transitivity
5. $\sim p$	3, 4, Modus Tollens

23. Valid.
1. $p \rightarrow q$	Premise
2. $q \rightarrow \sim r$	Premise
3. p	Premise
4. $r \vee s$	Premise
5. q	1, 3, Modus Ponens
6. $\sim r$	2, 5, Modus Ponens
7. s	4, 6, Disjunctive Syllogism

25.

p	q	(p ∧ q) → p
T	T	T T T T T
T	F	T F F T T
F	T	F F T T F
F	F	F F F T F
		① ② ① ③ ②

27.

p	q	(p ∧ q) → (p ∧ q)
T	T	T T T T T T T
T	F	T F F T T F F
F	T	F F T T F F T
F	F	F F F T F F F
		① ② ① ⑤ ③ ④ ③

29. Valid.
1. $a \rightarrow s$ Premise
2. $v \lor a$ Premise
3. $\sim v$ Premise
4. a 2, 3, Disjunctive Syllogism
5. s 1, 4, Modus Ponens

31. Invalid; $b =$ "it is a bear market" $=$ T, $p =$ "prices are rising" $=$ F, $i =$ "investor will sell stocks" $=$ T

33. Valid.
1. $s \lor i$ Premise
2. $s \rightarrow (l \land b)$ Premise
3. $\sim l \lor \sim b$ Premise
4. $\sim (l \land b)$ 3, DeMorgan's Law
5. $\sim s$ 2, 4, Modus Tollens
6. i 1, 5, Disjunctive Syllogism

35. Invalid; $M =$ "I am married to you" $=$ T, $o =$ "we are one" $=$ T, $r =$ "you are really a part of me" $=$ F

37. Valid.
1. $y \lor \sim p$ Premise
2. $\sim p \rightarrow \sim n$ Premise
3. n Premise
4. p 2, 3, Modus Tollens
5. y 1, 4, Disjunctive Syllogism

39. a. $d \rightarrow \sim w$ **b.** $o \rightarrow w$ or $\sim w \rightarrow \sim o$ **c.** $p \rightarrow d$ **d.** $p \rightarrow \sim o$, *Conclusion:* If it is my poultry, then it is not an officer. In Lewis Carroll's words, "My poultry are not officers." **41. a.** $b \rightarrow \sim t$ or $t \rightarrow \sim b$ **b.** $w \rightarrow c$ **c.** $\sim b \rightarrow h$ **d.** $\sim w \rightarrow \sim p$, or $p \rightarrow w$ **e.** $c \rightarrow t$ **f.** $p \rightarrow h$, *Conclusion:* If one is a pawnbroker, then one is honest. In Lewis Carroll's words, "No pawnbroker is dishonest." **43. a.** $d \rightarrow p$ **b.** $\sim t \rightarrow \sim i$ **c.** $r \rightarrow \sim f$ or $f \rightarrow \sim r$ **d.** $o \rightarrow d$ or $\sim d \rightarrow \sim o$ **e.** $\sim c \rightarrow i$ **f.** $b \rightarrow s$ **g.** $p \rightarrow f$ **h.** $\sim o \rightarrow \sim c$ or $c \rightarrow o$ **i.** $s \rightarrow \sim t$ or $t \rightarrow \sim s$ **j.** $b \rightarrow \sim r$, *Conclusion:* If it is written by Brown, then I can't read it. In Lewis Carroll's words, "I cannot read any of Brown's letters."

Exercises 6.6 (page 272)

1. a. $\exists x \, [b(x) \land s(x)]$ **b.** $\forall x \, [b(x) \rightarrow \sim s(x)]$
c. No books are bestsellers. **3. a.** $\forall x \, [c(x) \rightarrow \sim s(x)]$
b. $\exists x \, [c(x) \land s(x)]$ **c.** There is a CEO who sleeps well at night. **5. a.** $\forall x \, [l(x) \rightarrow b(x)]$
b. $\exists x \, [l(x) \land \sim b(x)]$ **c.** There is a leaf that's not brown.

For exercises . . .	1–6,34,42	7,8,23	9,10, 21	11,12, 24	13–20 37–41 43–48	22,25–32
Refer to example . . .	1	2	4	3	6	5

7. a. $\forall x[g(x) \rightarrow f(x)]$
$$\frac{g(t)}{f(t)}$$
b.   Valid

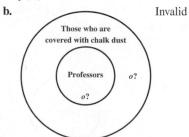

9. a. $\forall x[p(x) \rightarrow c(x)]$
$$\frac{c(o)}{p(o)}$$
b.   Invalid

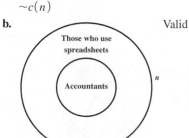

11. a. $\forall x \, [c(x) \rightarrow p(x)]$
$$\frac{\sim p(n)}{\sim c(n)}$$
b.   Valid

13. a. $\exists x\, [t(x) \wedge s(x)]$

$\dfrac{\forall x\, [t(x) \rightarrow b(x)]}{\exists x\, [s(x) \wedge b(x)]}$

15. a. $\exists x\, [w(x) \wedge l(x)]$

$\dfrac{w(m)}{l(m)}$

b.   Valid

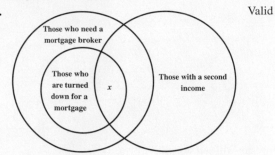

b.   Invalid

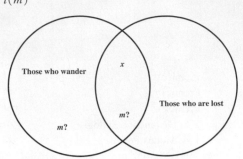

17. a. $\exists x\, [p(x) \wedge u(x)]$ **b.**

$\dfrac{\exists x\, [p(x) \wedge r(x)]}{\exists x\, [u(x) \wedge r(x)]}$

Invalid

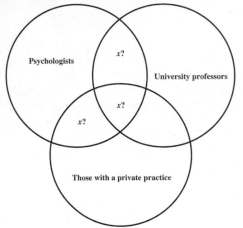

19. a. $\forall x\, [a(x) \vee i(x)]$ **b.**

$\dfrac{\exists x\, [\sim a(x)]}{\exists x\, [i(x)]}$

Valid   **21.** Yes

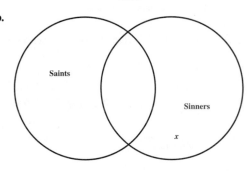

23. All major league baseball players earn at least $300,000 a year.

Ryan Howard is a major league baseball player.

Ryan Howard earns at least $300,000 a year.

a citizen of the United States, and was, when elected, an inhabitant of that State in which he was chosen.

25. Valid   **27.** Invalid   **29.** Invalid   **31.** Invalid   **37.** a, c, d

39. a. $\forall x\{r(x) \rightarrow [a(x) \wedge c(x) \wedge i(x)]\}$ **b.** John Boehner has attained to the age of twenty-five years, and been seven years

41. a. $\forall x\, \{s(x) \rightarrow \sim[t(x) \vee a(x) \vee c(x)]\}$ **b.** Texas shall not enter into any treaty, alliance, or confederation.   **c.**

c.

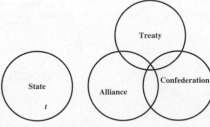

43. Invalid   **45.** Invalid   **47.** Valid

Chapter 6 Review Exercises (Page 276)

1. True   **3.** False   **5.** False   **7.** False
9. False   **11.** True   **13.** She doesn't pay
me and I have enough cash.   **15.** $l \wedge w$
17. $l \rightarrow \sim w$ **19.** He doesn't lose the
election and he wins the hearts of the voters.   **21.** True   **23.** True

For exercises . . .	15,19–22, 26,60–62, 69,70	14,27 39,40, 66	13,16,17,28, 35–38,41,42, 57–59,63,67	18,23–25, 29–34, 64,65	43–52	53–56,68 71–74
Refer to section . . .	1	2	3	4	5	6

27.

p	q	p	$\wedge$	$(\sim p \vee q)$
T	T	T	T	F T T
T	F	T	F	F F F
F	T	F	F	T T T
F	F	F	F	T T F
		①	④	② ③ ②

The statement is not a tautology.
29. If someone is a mathematician, then that person is loveable.   **31.** If a system has a
unique solution, then it has at least as many equations as unknowns.   **33. a.** If we need to
change the way we do business, then the proposed regulations have been approved.
b. If the proposed regulations have not been approved, then we do not need to change the
way we do business.   **c.** If we do not need to change the way we do business, then the proposed regulations have not been approved.

35. $(p \wedge p) \wedge (\sim p \vee q) \equiv p \wedge q$ **37.**

$(p \wedge q) \vee (p \wedge p) \equiv p$

39.

p	q	$p \veebar q$	$(p \vee q) \wedge \sim(p \wedge q)$
T	T	T F T	T T T F F T T T
T	F	T T F	T T F T T T F F
F	T	F T T	F T T T T F F T
F	F	F F F	F F F F T F F F
		① ② ①	③ ④ ③ ⑧ ⑦ ⑤ ⑥ ⑤

The columns labeled 2 and 8 are identical.
41. a. Yes **b.** No   **43.** Valid; Modus Ponens   **45.** Valid; Disjunctive
Syllogism   **47.** Invalid; Fallacy of the Converse

49. Valid.

1. $h \rightarrow t$ — Premise
2. $r \rightarrow \sim t$ — Premise
3. r — Premise
4. $\sim t$ — 2, 3, Modus Ponens
5. $\sim h$ — 1, 4, Modus Tollens

51. Invalid; $p = $ F, $q = $ F

53. a. $\forall x [d(x) \rightarrow l(x)]$ **b.** $\exists x [d(x) \wedge \sim l(x)]$ **c.** There is a dog that doesn't have a license.
55. a. $\forall x [f(x) \rightarrow w(x)]$ **b.**

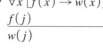

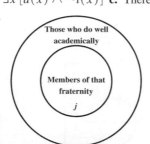

Valid

57.

p	q	r	$p \rightarrow$	$(q \rightarrow r)$	$(p \rightarrow q) \rightarrow r$
T	T	T	T T	T T T	T T T T T
T	T	F	T F	T F F	T T T F F
T	F	T	T T	F T T	T F F T T
T	F	F	T T	F T F	T F F T F
F	T	T	F T	T T T	F T T T T
F	T	F	F T	T F F	F T T F F
F	F	T	F T	F T T	F T F T T
F	F	F	F T	F T F	F T F F F
			① ④	② ③ ②	⑤ ⑥ ⑤ ⑧ ⑦

No   **59. a.**

p	q	$(p \wedge \sim p) \rightarrow q$
T	T	T F F	T T
T	F	T F F	T F
F	T	F F T	T T
F	F	F F T	T F
		① ② ①	④ ③

61. b, c

63. You do not use the Tax Table or you do not have to compute your
tax mathematically.   **65. a.** If you exercise regularly, then your heart
becomes stronger and more efficient.   **b.** If you are a teenager, then
you need to be aware of the risks of drinking and driving.   **c.** If you
are visiting a country that has a high incidence of infectious diseases, then you may need extra immunizations.   **d.** If you have good
health, then you have food.   **67.** $(w \rightarrow d) \rightarrow v$ **69.** b,c,d   **71. a.** $\sim s \rightarrow g$ **b.** $l \rightarrow \sim g$ **c.** $w \rightarrow l \equiv \sim l \rightarrow \sim w$
d. $\sim s \rightarrow \sim w$, *Conclusion:* If the puppy does not lie still, it does not care to do worsted work. In Lewis Carroll's words, "Puppies that
will not lie still never care to do worsted work."   **73. a.** $f \rightarrow t \equiv \sim t \rightarrow \sim f$ **b.** $\sim a \rightarrow \sim g \equiv g \rightarrow a$ **c.** $w \rightarrow f$
d. $t \rightarrow \sim g \equiv g \rightarrow \sim t$ **e.** $a \rightarrow w \equiv \sim w \rightarrow \sim a$ **f.** $g \rightarrow \sim e$, *Conclusion:* If the kitten will play with a gorilla, it does not have
green eyes. In Lewis Carroll's words, "No kitten with green eyes will play with a gorilla."

Chapter 7 Sets and Probability

Exercises 7.1 (page 290)

For exercises . . .	5–19	21–24,63,69	25–44	47–50,74–76,78,79	53–56,59–62,64–68,71–76,78,79	70
Refer to example . . .	2	4	5,6,7	9	8	3

1. False **3.** True **5.** True
7. True **9.** False **11.** $\subseteq$ **13.** $\not\subseteq$ **15.** $\subseteq$ **17.** $\subseteq$ **19.** $\subset;\subset;\not\subseteq;\not\subseteq;\subset;\not\subseteq;\subset;\not\subseteq$ **21.** 64 **23.** 8 **25.** $\cap$ **27.** $\cup$ **29.** $\cap$
31. $\cup$ or $\cap$ **35.** $\{2,4,6\}$ **37.** $\{1,3,5,7,9\}$ **39.** $\{1,7,9\}$ **41.** $\{2,3,4,6\}$ **43.** $\{7,8\}$ **45.** $\{3,6,9\} = A$ **47.** All students
in this school not taking this course **49.** All students in this school taking accounting and zoology **51.** C and D, B and E, C and
E, D and E **53.** B' is the set of all stocks on the list with a closing price below \$60 or above \$70; $B' = \{$AT&T, Coca-Cola,
FedEx, Disney$\}$. **55.** $(A \cap B)'$ is the set of all stocks on the list that do not have both a high price greater than \$50 and a closing
price between \$60 and \$70; $A \cap B = \{$AT&T, Coca-Cola, FedEx, Disney$\}$. **57. a.** True **b.** True **c.** False **d.** False **e.** True
f. True **g.** False **59.** $\{$Exxon Mobil, General Electric, JPMorgan Chase & Co.$\}$ **61.** $\{$IBM, Hewlett-Packard, Home Depot,
Aflac$\}$ **63.** $2^9 = 512$ **65.** $\{i, m, h\}$ **67.** U **69.** $2^{51} \approx 2.252 \times 10^{15}$ **71.** $\{$USA, TLC, TBS$\}$ **73.** $\{$Discovery, TNT$\}$
75. $\{$TNT, USA, TBS, Discovery$\}$; the set of networks that features sports or that have more than 97.6 million viewers.
79. a. The set of states who are not among those whose name contains the letter "e" or who are more than 4 million in population,
and who also have an area of more than 40,000 square miles. **b.** $\{$Alaska$\}$

Exercises 7.2 (page 298)

For exercises . . .	1–8,25–28	11–20,29–32	21–24	39,44,45,59	38	40,41,46–58	42,43,60,61
Refer to example . . .	1	2	6	4	5	7	8

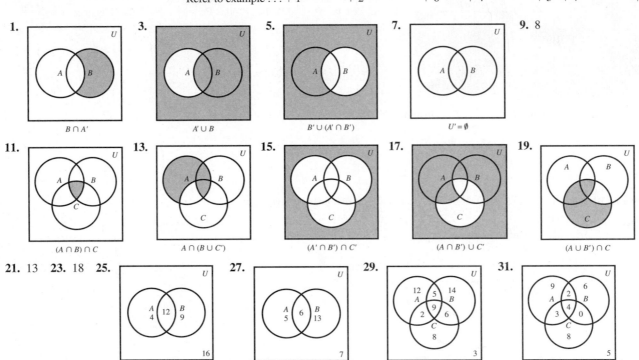

1. $B \cap A'$ **3.** $A' \cup B$ **5.** $B' \cup (A' \cap B')$ **7.** $U' = \emptyset$ **9.** 8
11. $(A \cap B) \cap C$ **13.** $A \cap (B \cup C')$ **15.** $(A' \cap B') \cap C'$ **17.** $(A \cap B') \cup C'$ **19.** $(A \cup B') \cap C$

21. 13 **23.** 18 **25.** [U: A 4, 12, B 9, 16] **27.** [U: A 5, 6, B 13, 7] **29.** [U: A 12, 5, 14, 2, 9, 6, C 8, 3] **31.** [U: A 9, 2, 6, 3, 4, 0, C 8, 5]

39. a. 12 **b.** 18 **c.** 37 **d.** 97 **41. a.** 2 **b.** 60 **c.** 8 **d.** 100 **e.** 27 **f.** All those who invest in stocks or bonds and are
age 18–29 **43. a.** 17 **b.** 2 **c.** 14 **45. a.** 54 **b.** 17 **c.** 10 **d.** 7 **e.** 15 **f.** 3 **g.** 12 **h.** 1 **47. a.** 34 **b.** 16 **c.** 431
d. 481 **49.** 110.6 million **51.** 85.4 million **53.** 71.0 million **55.** 25,568,000; Blacks or Hispanics who never married
57. 203,803,000; Whites or divorced/separated people who are not Asian/Pacific Islanders **59. a.** 342 **b.** 192
c. 72 **d.** 86 **61. a.** 89 **b.** 32 **c.** 26 **d.** 30 **e.** 22 **f.** 21

Exercises 7.3 (page 307)

For exercises . . .	3–10,13–18	13–18	19–24	25–34	35–40,53,62,63	41–48,52,56,57–61	51,55,56
Refer to example . . .	1	2,3	6	7	6,7	8	4

3. $\{$January, February, March, . . . , December$\}$ **5.** $\{0, 1, 2, 3, \ldots, 79, 80\}$ **7.** $\{$go ahead, cancel$\}$ **9.** $\{(h,1), (h,2), (h,3),$
$(h,4), (h,5), (h,6), (t,1), (t,2), (t,3), (t,4), (t,5), (t,6)\}$ **13.** $\{$AB, AC, AD, AE, BC, BD, BE, CD, CE, DE$\}$, 10, yes
a. $\{$AC, BC, CD, CE$\}$ **b.** $\{$AB, AC, AD, AE, BC, BD, BE, CD, CE$\}$ **c.** $\{$AC$\}$ **15.** $\{(1,2), (1,3), (1,4),$
$(1,5), (2,3), (2,4), (2,5), (3,4), (3,5), (4,5)\}$, 10, yes **a.** $\{(2,4)\}$ **b.** $\{(1,2), (1,4), (2,3), (2,5), (3,4), (4,5)\}$ **c.** $\emptyset$
17. $\{hh, thh, hth, tthh, thth, htth, ttth, ttht, thtt, httt, tttt\}$, 11, no **a.** $\{tthh, thth, htth, ttth, ttht, thtt, httt, tttt\}$ **b.** $\{hh, thh, hth,$
$tthh, thth, htth\}$ **c.** $\{tttt\}$ **19.** 1/6 **21.** 2/3 **23.** 1/3 **25.** 1/13 **27.** 1/26 **29.** 1/52 **31.** 2/13 **33.** 7/13 **35.** 3/20

37. 1/4 **39.** 3/5 **41.** Not empirical. **43.** Empirical **45.** Empirical **47.** Not empirical. **49.** The outcomes are not equally likely. **51. a.** Worker is male. **b.** Worker is female and has worked less than 5 years. **c.** Worker is female or does not contribute to a voluntary retirement plan. **d.** Worker has worked 5 years or more. **e.** Worker has worked less than 5 years or has contributed to a voluntary retirement plan. **f.** Worker has worked 5 years or more and does not contribute to a voluntary retirement plan. **53. a.** 8/15 **b.** 1/10 **55. a.** Person is not overweight. **b.** Person has a family history of heart disease and is overweight. **c.** Person smokes or is not overweight. **57. a.** 0.2540 **b.** 0.4851 **c.** 0.9225 **59.** 0.17 **61. a.** 0.3151 **b.** 0.3719 **c.** 0.3700 **d.** Calvary **e.** I Corps **63. a.** 25/57 **b.** 32/57 **c.** 4/19

Exercises 7.4 (page 316)

For exercises . . .	9–20,23,24,54,55,	21,22,47,52,53,57–59,61–64	27–32,48,65,70	35–40,49–51,56,66–69	60,71
Refer to example . . .	1,2,3,4	9	5	8	6,7

3. No **5.** No **7.** Yes **9. a.** 1/36 **b.** 1/12 **c.** 1/9 **d.** 5/36 **11. a.** 5/18 **b.** 5/12 **c.** 11/36 **13.** 5/18 **15. a.** 2/13 **b.** 7/13 **c.** 3/26 **d.** 3/4 **e.** 11/26 **17. a.** 5/13 **b.** 7/13 **c.** 3/13 **19. a.** 1/10 **b.** 2/5 **c.** 7/20 **21. a.** 0.51 **b.** 0.25 **c.** 0.10 **d.** 0.84 **23. a.** 5/9 **b.** 5/9 **c.** 5/9 **27.** 1 to 5 **29.** 2 to 1 **31. a.** 1 to 5 **b.** 11 to 7 **c.** 2 to 7 **d.** 7 to 2 **35.** Possible **37.** Not possible; the sum of the probabilities is less than 1. **39.** Not possible; a probability cannot be negative. **41. a.** 0.2778 **b.** 0.4167 **43. a.** 0.0463 **b.** 0.2963 **47.** 0.84 **49. a.** 0.62 **b.** 0.54 **c.** 0.43 **d.** 0.19 **51. a.** 0.061 **b.** 0.761 **c.** 0.822 **d.** 0.535 **53. a.** 0.961 **b.** 0.487 **c.** 0.513 **d.** 0.509 **e.** 0.004 **f.** 0.548 **55. a.** 1/4 **b.** 1/2 **c.** 1/4 **57.** a **59.** c **61. a.** 0.4 **b.** 0.1 **c.** 0.6 **d.** 0.9 **63.** 0 **65.** 2/5 **67. a.** 0.866 **b.** 0.478 **69. a.** 23/55 **b.** 67/220 **c.** 159/220 **71.** 0.0000000051; 0.0000012; 0.0063; 0.0166

Exercises 7.5 (page 330)

For exercises . . .	1–12	13–16	23,24,44–46, 49–53,63,71,72, 79–82,84,87–89	43,69,86	36–40,47, 54–60,62,64–67, 69,73,74,77	29,30,41,42, 75,76	48,61,68,78, 83,85,86
Refer to example . . .	3,4	8	5,6,7	2	1	9	10

1. 0 **3.** 1 **5.** 1/3 **7.** 0 **9.** 4/17 **11.** 11/51 **13.** 8/663 **15.** 25/102 **19.** Independent **21.** Dependent **23. a.** 1/4 **b.** 1/2 **25. a.** Many answers are possible **b.** Many answers are possible **29.** 1/20, 2/5 **31.** Second booth **33.** No, these events are not independent **35.** Yes **37.** The probability that a customer cashing a check will fail to make a deposit is 1/3. **39.** The probability that a customer making a deposit will not cash a check is 1/4. **41. a.** 0.3570 **43.** 0.875 **45.** 0.06 **47.** 2/3 **49.** 1/4 **51.** 1/4 **53.** 1/7 **55.** 0.039 **57.** 0.491 **59.** 0.072 **61.** Yes **63. a.** 0.3367 **b.** 0.6617 **c.** No **65.** 7/229 **67.** 191/229 **69.** e **71. a.** 0.5065 **c.** 0.2872 **d.** $p(1 - P(B)) + (1 - p)(1 - P(B))^2$ **e.** 0.2872 **f.** $2(1 - p)P(B)(1 - P(B))$ **g.** 0.2872 **73. a.** 0.2059 **b.** 0.1489 **c.** 0.0410 **d.** 0.2755 **e.** No **75. a.** 0.58 **77. a.** 0.2713 **b.** 0.3792 **c.** 0.6246 **d.** 0.3418 **e.** 0.6897 **f.** 0.5137 **g.** Not independent **79. a.** 0.052 **b.** 0.476 **81. a.** 7/10 **b.** 2/15 **83.** 10^{-12} **85.** No **87.** 0 points: 0.4; 1 point: 0.24; 2 points: 0.36. **89. c.** They are the same. **d.** The 2-points first strategy has a smaller probability of losing.

Exercises 7.6 (page 340)

For exercises . . .	1,2,9,10–13,23–26, 29–32,36,39	3–8,14–17,19 27,28	18,20–22,33–35, 37,38
Refer to example . . .	1	2	3

1. 1/3 **3.** 3/19 **5.** 21/38 **7.** 8/17 **9.** 85% **11.** 0.0727 **13.** 0.1765 **15.** 0.3636 **17.** 2/7 **19.** d **21.** c **23.** 0.0478 **25. a.** 0.039 **b.** 0.999 **c.** 0.001 **d.** 0.110 **27.** d **29.** b **31. a.** 0.1870 **b.** 0.9480 **33.** 0.1285 **35.** 0.1392 **37.** 0.5549 **39.** 9.9×10^{-5}

Chapter 7 Review Exercises (page 345)

For exercises . . .	1–5,13–28, 59,87,90	6,43–52, 60,61, 69,70,88, 90,103,104	7,8,53,54, 65–68, 71–73, 77,78 91,93, 95,96	9–11,55–58, 62,63,74–76, 86,90,94, 101,102, 107,108	12,79,80, 81–85, 89,97,98	39–42, 92, 99, 105,106, 109
Refer to example . . .	1	3	4	5	6	2

1. True **2.** True **3.** False **4.** False **5.** False **6.** True **7.** False **8.** False **9.** False **10.** True **11.** False **12.** True **13.** False **15.** False **17.** True **19.** False **21.** False **23.** 32 **25.** {a, b, g} **27.** {c, d} **29.** {a, b, e, f, g, h} **31.** U **33.** All female employees in the accounting department **35.** All employees who are in the accounting department or who have MBA degrees **37.** All male employees who are not in the sales department

39.

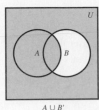

$A \cup B'$

41.

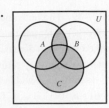

$(A \cap B) \cup C$

43. $\{1, 2, 3, 4, 5, 6\}$ **45.** $\{0, 0.5, 1, 1.5, 2, \ldots, 299.5, 300\}$
47. $\{(3, r), (3, g), (5, r), (5, g), (7, r), (7, g), (9, r), (9, g), (11, r), (11, g)\}$
49. $\{(3, g), (5, g), (7, g), (9, g), (11, g)\}$
51. 1/4 **53.** 11/26 **55.** 1/2 **57.** 1 **63.** No; yes **65.** 1 to 3 **67.** 2 to 11
69. 5/36 **71.** 1/6 **73.** 1/6 **75.** 2/11 **77. a.** 0.66 **b.** 0.29 **c.** 0.71
d. 0.34 **79.** 1/7 **81.** 0.8736 **83.** 0.3077 **85.** 0.87 **87. a.** $(E \cup F)'$ or
$E' \cap F'$ **b.** $E \cup F$ **89. a.** 0.0297 **b.** 0.0909 **c.** 0.2626 **d.** No **91.** b

93. b **95. a.**

	N_2	T_2
N_1	N_1N_2	N_1T_2
T_1	T_1N_2	T_1T_2

b. 1/4 **c.** 1/2 **d.** 1/4 **97.** c **99. a.** 53 **b.** 7 **c.** 12 **d.** 26 **101.** 0.6279
103. 0.90 **105. a.** 4 **b.** 18 **107.** No; 2/3 **109.** d

Chapter 8 Counting Principles; Further Probability Topics

Exercises 8.1 (page 359)

For exercises . . .	1–4,37,39,54,55	5–12,19,20,32,34,40	13–16,30,31,35,43–53	19–24	23,24,33,36,38,41,42
Refer to example . . .	4,5	6	1	9,10	8

1. 720 **3.** 1.308×10^{12}
5. 156 **7.** 1.024×10^{25} **9.** 1 **11.** n **13.** 36 **15.** 20 **19.** one **21. a.** 840 **b.** 180 **c.** 420 **23. a.** 362,880 **b.** 1728
c. 1260 **d.** 24 **e.** 144 **25.** Multiply by 10 **27. a.** 2 **b.** 6 **c.** 18 **29.** Undefined **31. a.** 42 **b.** 28 **33. a.** 39,916,800
b. 172,800 **c.** 86,400 **35.** No; use at least 4 initials **37.** 720 **39.** 3.352×10^{10} **41. a.** 120 **b.** 48 **43. a.** 27,600 **b.** 35,152
c. 1104 **45. a.** 160; 8,000,000 **b.** Some numbers, such as 911, 800, and 900, are reserved for special purposes. **47. a.** 17,576,000
b. 17,576,000 **c.** 456,976,000 **49.** 100,000; 90,000 **51.** 81 **53.** 1,572,864; no **55. a.** 1.216×10^{17} **b.** 43,589,145,600

Exercises 8.2 (page 367)

For exercises . . .	3–10,39,48,58	11,12,33,35,36, 40–42,45,54–56	14,32,37,43,44, 49,53	15,16	17–24	17–24,29–31,34, 38,46,47,50–52,57
Refer to example . . .	1	6	2,3	7	4	5

3. 56 **5.** 1.761×10^{12} **7.** 1
9. n **11.** 1716 **13. a.** 10
b. 7 **15. a.** 9 **b.** 6 **c.** 3; yes, from both **17.** Combinations; **a.** 126 **b.** 462 **c.** 4620 **19.** Permutations; 479,001,600
21. Combinations; **a.** 120 **b.** 1820 **c.** 36 **23.** Combinations; **a.** 10 **b.** 0 **c.** 1 **d.** 10 **e.** 30 **f.** 15 **g.** 0 **27. a.** 30
b. $n(n - 1)$ **29.** 336 **31. a.** 720 **b.** 360 **33. a.** 40 **b.** 20 **c.** 7 **35.** 4,115,439,900 **37. a.** 84 **b.** 10 **c.** 40 **d.** 74
39. No; 14,190 **41. a.** 48 **b.** 658,008 **c.** 652,080 **d.** 844,272 **e.** 79,092 **43.** 50 **45. a.** 15,504 **b.** 816 **47.** 558 **49.** a 26
b. 26 **51. a.** 3,089,157,760 **b.** 1,657,656,000 **53. a.** 838 **b.** 20,695,218,670 **55. a.** 84 **b.** 729 **d.** 680 **e.** 120
57. a. 6.402×10^{15} **b.** 3.135×10^{10}

Exercises 8.3 (page 378)

For exercises . . .	1–10,36–41,45,47,65,66,68–71,73	11–18,46–64	19–22,42–44,46,72	25–28,33	29–32
Refer to example . . .	1,2,3	4	5	6	7

1. 7/33 **3.** 14/55
5. 0.008127 **7.** 0.2980 **9.** 0.2214 **11.** 1326 **13.** 33/221 **15.** 52/221 **17.** 130/221 **19.** 8.417×10^{-8} **21.** 18,975/28,561
25. $1 - P(365, 43)/365^{43}$ **27.** 1 **29.** 13/34 **31.** 1/180 **33.** 0.6501 **35.** 3/5 and 2/5 **37.** 36/55 **39.** 21/55 **41.** 7/22
43. 1/3 **45. a.** 225/646 **b.** 15/323 **c.** 225/2584 **d.** 0 **e.** 1/2 **f.** 175/2584 **g.** 503/646 **47. a.** 0.0542 **b.** 0.0111 **c.** 0.0464
d. 0.1827 **e.** 0.3874 **f.** 0.8854 **49.** 1.385×10^{-5} **51.** 0.0039 **53.** 0.0475 **55.** 1.575×10^{-12} **57.** 0.0402 **59.** 0.4728
61. 0.0493 **63.** 0.0306 **65.** 0.0640 **67.** 4.980×10^{-7} **69.** The probability of picking 5 out of 52 is higher, 1/2,598,960 compared
with 1/13,983,816. **71. a.** 0.01642 **b.** 0.01231 **73. a.** 28 **b.** 268,435,456 **c.** 40,320 **d.** 1.502×10^{-4} **e.** $n!/2^{n(n - 1)/2}$
75. a. 3.291×10^{-6} **b.** 7.962×10^{-12} **c.** 5.927×10^{-11} **d.** 5.524×10^{26}

Exercises 8.4 (page 386)

For exercises . . .	1–4,9–12,15,16,24,25,29–39,33,34, 37,40,41,43,44,47,59,60	5–8,13,14,17,18,26–28,31,32, 35,36,38,39,42,45,46,48–51, 54,55,57,58,61–63	52,53,64,65
Refer to example . . .	2,3	4,5	7

1. 5/16 **3.** 1/32 **5.** 3/16
7. 13/16 **9.** 4.594×10^{-10}
11. 0.2692 **13.** 0.8748
15. 0.0002441 **17.** 0.9624 **23.** The potential callers are not likely to have birthdates that are evenly distributed throughout the
twentieth century. **25.** 0.6014 **27.** 0.0876 **29.** 0.1721 **31.** 0.7868 **33.** 0.2458 **35.** 0.0170 **37.** 0.3585 **39. a.** 0.1488
b. 0.0213 **c.** 0.9787 **41. a.** 0.0222 **b.** 0.1797 **c.** 0.7766 **43.** 0.0874 **45.** 0.9126 **47.** 0.2183 **49.** 0.0025 **51. a.** 0.0478
b. 0.9767 **c.** 0.8971 **53. a.** 1 chance in 1024 **b.** About 1 chance in 1.1×10^{12} **c.** About 1 chance in 2.587×10^6 **55.** 0.9523
57. e **59.** 0.0305 **61.** 0.8676 **63. a.** 0.2083 **b.** 0.5902 **c.** 0.1250 **d.** 0.8095 **65. a.** 0.5260 **b.** 0.9343 **67. a.** 0.125, 0.25,
0.3125, 0.3125 **b.** 0.2893, 0.3222, 0.2353, 0.1531

Exercises 8.5 (page 395)

For exercises . . .	1–4	5–8,29	9–16,30,32,36,37,42	17,18,26,45,56	19–25,29,31,38,43,44,47–54,57	33,39,40,55,56
Refer to example . . .	1	2	3	5	4,7	8

1.

Number of Heads	0	1	2	3	4
Probability	1/16	1/4	3/8	1/4	1/16

3.

Number of Aces	0	1	2	3
Probability	0.7826	0.2042	0.0130	0.0002

5. 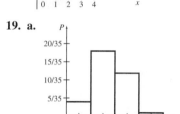 **7.** 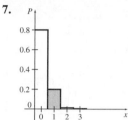 **9.** 3.6 **11.** 14.49 **13.** 2.7 **15.** 18 **17.** 0; yes

19. a.   **b.** 9/7 **21. a.** **b.** 2/3
23. 3/4
25. 1/2

x	0	1	2	3	4
$P(x)$	625/1296	125/324	25/216	5/324	1/1296

29. a.

Sum	5	6	7	8	9
Probability	1/6	1/6	1/3	1/6	1/6

b. 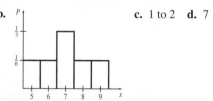 **c.** 1 to 2 **d.** 7

31. $54,000 **33. a.** 30 **b.** 10 **35.** e **37. a.** $68.51; $72.84 **b.** Amoxicillin **39.** 185 **41. a.** 0.007230 **b.** 5.094×10^{-4}
c. 5.5×10^{-5} **d.** 0.1143 **43. a.** **b.** 1.2 **c.** 1.2

Number of Cats	0	1	2	3	4
Probability	0.2401	0.4116	0.2646	0.0756	0.0081

45. −$0.72; no **47.** −$0.82 **49.** −$0.053 **51.** −$0.50 **53.** −$0.90; no **55. a.** 10/3 **b.** 50/9
57. a. **b.** 1.37

x	0	1	2	3	4
$P(x)$	0.1875	0.3897	0.3038	0.1053	0.0137

Chapter 8 Review Exercises (page 401)

For exercises . . .	1,3,13,14,19,20,53	2,15–18,21,22,54	4–6,25–30,35–40,69 76,78,79,82	7,8,31–34,49,50, 55–60,67,68,70–73	9–12,41–48,61,62, 67,70–75,77,78, 80,81
Refer to section . . .	1	2	3	4	5

1. True **2.** True **3.** True **4.** True **5.** False **6.** True **7.** True **8.** False **9.** True **10.** False **11.** True **12.** False
13. 720 **15.** 220 **17. a.** 90 **b.** 10 **c.** 120 **d.** 220 **19. a.** 120 **b.** 24 **21. a.** 840 **b.** 2045 **25.** 2/143 **27.** 21/143
29. 5/13 **31.** 5/16 **33.** 11/32 **35.** 25/102 **37.** 15/34 **39.** 546/1326

41. a.

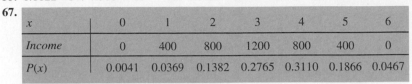

Number of Heads	0	1	2	3
Probability	0.125	0.375	0.375	0.125

b. 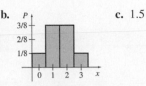   **c.** 1.5

43. 0.6   **45.** −$0.833; no   **47. a.** 0.231   **b.** 0.75   **49.** 31/32   **51. a.** $C(n, 0)$, or 1; $C(n, 1)$, or n; $C(n, 2)$; $C(n, n)$, or 1
b. $C(n, 0) + C(n, 1) + C(n, 2) + \cdots + C(n, n)$ **e.** The sum of the elements in row n of Pascal's triangle is 2^n.   **53.** 48
55. 0.1122   **57.** 7.580×10^{-7} **59.** 0.6187   **61.** 2   **63.** d   **65.** e

67.

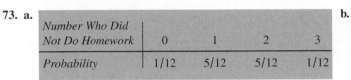

x	0	1	2	3	4	5	6
Income	0	400	800	1200	800	400	0
$P(x)$	0.0041	0.0369	0.1382	0.2765	0.3110	0.1866	0.0467

a. $780.60   **b.** $720; $856.32; $868.22; 5
69. 0.1875

71. a.

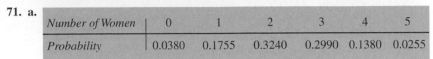

Number of Women	0	1	2	3	4	5
Probability	0.0380	0.1755	0.3240	0.2990	0.1380	0.0255

b.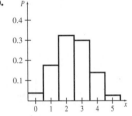

c. 2.4

73. a.

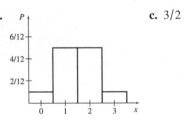

Number Who Did Not Do Homework	0	1	2	3
Probability	1/12	5/12	5/12	1/12

b. 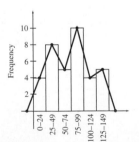   **c.** 3/2

75. −$0.71   **77. a.** 10.95   **b.** 2.56   **79. a.** 0.4799; 0.6533   **c.** 0.3467; 0.003096

Chapter 9 Statistics

Exercises 9.1 (page 415)

For exercises . . .	1a,2a,44a	1b,2b,42a, 44b	7–12,39,40, 47	13–16,42b, 44d	17–22,48b, 50	23,24,45,46, 49	25–30,44c, 44f,48c	36–38,41,43 45,46,48
Refer to example . . .	1	2	3,7	4,5	6,7	9	8	3,6

1. a.–b.

Interval	Frequency
0–24	4
25–49	8
50–74	5
75–99	10
100–124	4
125–149	5

c.–d.

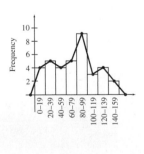

3. a.–b.

Interval	Frequency
0–19	4
20–39	5
40–59	4
60–79	5
80–99	9
100–119	3
120–139	4
140–159	2

c.–d.

7. 16   **9.** 28,620   **11.** 7.68   **13.** 7.25   **15. a.** 73.81   **b.** 74.5   **17.** 42   **19.** 130   **21.** 29.1   **23.** 73.86, 80.5   **25.** 9   **27.** 55
and 62   **29.** 6.3   **33.** 75–99   **37.** 2095.9 million bushels, 2130 million bushels   **39.** $45,794   **41. a.** 348.4, 249.5   **c.** 1.187, 1.215
43. 7.38; 7; 7, 5, 4   **45. a.** 55.5°F; 50.5°F   **b.** 28.9°F; 28.5°F   **47. a.** $71,349   **49. a.** $8,253,336, $5,500,000, $5,500,000

Exercises 9.2 (page 425)

For exercises . . .	3–8	9,10,25,26,29	11,12	27,28	30–36
Refer to example . . .	1,3	3	4	6	5

1. The standard deviation is the square root of the variance.
3. 33; 12.6   **5.** 46; 16.1   **7.** 24; 8.1   **9.** 40.05   **11.** 39.4   **13.** 8/9   **15.** 24/25   **17.** At least 8/9   **19.** No more than 1/9

25. a. Mean = 25.5 hr; standard deviation = 7.2 hr **b.** Forever Power **c.** Forever Power **27. a.** 1/3; 2; −1/3; 0; 5/3; 7/3; 1; 4/3; 7/3; 2/3 **b.** 2.1; 2.6; 1.5; 2.6; 2.5; 0.6; 1.0; 2.1; 0.6; 1.2 **c.** 1.13 **d.** 1.68 **e.** 4.41; −2.15 **f.** 4.31; 0 **29.** Mean = 1.816 mm; standard deviation = 0.4451 mm. **31. a.** Mean = 7.3571; standard deviation = 0.1326 **b.** 100% **33. a.** 127.7 days; 30.16 days **b.** Seven **35. a.** $9,267,188 **b.** 4%

Exercises 9.3 (page 435)

For exercises . . .	5–14	15–18	22–64
Refer to example . . .	1	2	4

1. the mean **3.** z-scores are found with the formula $z = (x − \mu)/\sigma$. **5.** 45.54%
7. 48.96% **9.** 37.38% **11.** 14.78% **13.** 97.72% **15.** −1.64 or −1.65 **17.** 1.28 **19.** 0.5; 0.5 **21.** 0.8889; 0.9974 **23.** 5000
25. 6247 **27.** 7257 **29.** 0.1587 **31.** 0.0062 **33.** 84.13% **35.** 37.79% **37.** 7.03% **39. a.** 15.87% **b.** 0% **41.** 0.4325
43. About 2 **45.** 1350 units **47.** 1430 units **49.** 60.32 mph **51.** 6.68% **53.** 38.3% **55.** 87 **57.** 71 **59. b.** 55%
61. a. About 0.01; yes **b.** Essentially 0 **63. a.** 0.5596; 0.0188 **b.** Essentially 0; essentially 0 **c.** 0.9265; 0.9554
d. Essentially 0; essentially 0 **e.** Essentially 0; essentially 0

Exercises 9.4 (page 443)

For exercises . . .	3–34
Refer to example . . .	2

1. The number of trials and the probability of success on each trial **3. a.** 0.0278 **b.** 0.0279 **5. a.** 0.0106
b. 0.0122 **7.** 0.0240 **9.** 0.9463 **11.** 0.0956 **13.** 0.8643 **15.** 0.1841 **17. a.** 0.0146 **b.** 0.0537 **19. a.** Essentially 0
b. 0.0018 **21.** 0.0274 **23. a.** 0.9118 **b.** 0.9984 **25. a.** 0.0001 **b.** 0.0002 **c.** 0.0000 **27.** 0.6480 **29.** 0.9554
31. a. The numbers are too large for the calculator to handle. **b.** 0.1742 **33.** 0.0287

Chapter 9 Review Exercises (page 446)

For exercises . . .	4,6,10,13–25	3,5,11,26–33	7,8,12,34–38,44,52–56,58,59	1,2,9,39–41,45,46,48–51	43,47,57
Refer to section . . .	1	2	3	4	1,2

1. True **2.** False **3.** False **4.** True **5.** False **6.** False **7.** False **8.** True **9.** True **10.** True **11.** False **12.** True

15. a.

Sales	Frequency
450–474	5
475–499	6
500–524	5
525–549	2
550–574	2

b.–c.
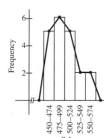

17. 29.25 **19.** 34.9 **23.** 38; 36 and 38 **25.** 30–39
29. 67; 23.9 **31.** 12.6 **33.** A skewed distribution has the largest frequency at one end. **35.** 0.7995 **37.** 0.0606
39. Because the histogram is skewed, not close to the shape of a normal distribution **41. a.** 2.1; 1.07 **b.** 2; 1
c. Answers to parts a and b should be close to each other.
43. a. Stock I: 8%, 7.9%; Stock II: 8%, 2.6% **b.** Stock II
45. a. 0.8924; 0.8962 **b.** 0.0446; 0.0477 **c.** 0.0196; 0.0150
47. Diet A: $\bar{x} = 2.7$, $s = 2.26$; Diet B: $\bar{x} = 1.3$, $s = 0.95$
a. Diet A **b.** Diet B **49.** 0.9292 **51.** 0.9534 **53.** 43.25% **55.** 56.25% **57. a.** 142.46; 140; no mode **b.** 48.67
c. 38.5% **d.** 100% **59. a.** 0.1020 **b.** 0.3707 **d.** 0.9987 **e.** 0.1336

Chapter 10 Markov Chains

Exercises 10.1 (page 459)

For exercises . . .	9–18,30,33a,34a,36a,38a,40a	19–26,28,32d,e,34e,f,35c,d,e,36b,c, 37a,38e,f,39c,d,e,40g,41,42	29,31,32a,b,c,33b,c,d,e, 34b,c,d,35a,b,37b,c, 38b,c,d,39a,b,40b,c,d,e,f
Refer to example . . .	1	2,3	4

1. No **3.** Yes **5.** No
7. Yes **9.** No

11. Yes **13.** Yes **15.** Yes; $\begin{bmatrix} 0.9 & 0.1 & 0 \\ 0.1 & 0.7 & 0.2 \\ 0 & 0.2 & 0.8 \end{bmatrix}$ **17.** No

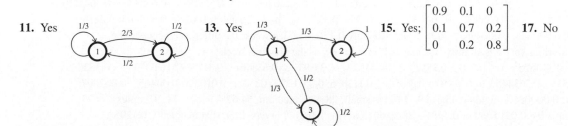

19. $A = \begin{bmatrix} 1 & 0 \\ 0.7 & 0.3 \end{bmatrix}$; $A^2 = \begin{bmatrix} 1 & 0 \\ 0.91 & 0.09 \end{bmatrix}$; $A^3 = \begin{bmatrix} 1 & 0 \\ 0.973 & 0.027 \end{bmatrix}$; 0

21. $C = \begin{bmatrix} 0 & 0 & 1 \\ 0.2 & 0.6 & 0.2 \\ 0.1 & 0.7 & 0.2 \end{bmatrix}$; $C^2 = \begin{bmatrix} 0.1 & 0.7 & 0.2 \\ 0.14 & 0.5 & 0.36 \\ 0.16 & 0.56 & 0.28 \end{bmatrix}$; $C^3 = \begin{bmatrix} 0.16 & 0.56 & 0.28 \\ 0.136 & 0.552 & 0.312 \\ 0.14 & 0.532 & 0.328 \end{bmatrix}$; 0.56

23. $E = \begin{bmatrix} 0.8 & 0.1 & 0.1 \\ 0.3 & 0.6 & 0.1 \\ 0 & 1 & 0 \end{bmatrix}$; $E^2 = \begin{bmatrix} 0.67 & 0.24 & 0.09 \\ 0.42 & 0.49 & 0.09 \\ 0.3 & 0.6 & 0.1 \end{bmatrix}$; $E^3 = \begin{bmatrix} 0.608 & 0.301 & 0.091 \\ 0.483 & 0.426 & 0.091 \\ 0.42 & 0.49 & 0.09 \end{bmatrix}$; 0.301

25. The first power is the given transition matrix; $\begin{bmatrix} 0.2 & 0.15 & 0.17 & 0.19 & 0.29 \\ 0.16 & 0.2 & 0.15 & 0.18 & 0.31 \\ 0.19 & 0.14 & 0.24 & 0.21 & 0.22 \\ 0.16 & 0.19 & 0.16 & 0.2 & 0.29 \\ 0.16 & 0.19 & 0.14 & 0.17 & 0.34 \end{bmatrix}$; $\begin{bmatrix} 0.17 & 0.178 & 0.171 & 0.191 & 0.29 \\ 0.171 & 0.178 & 0.161 & 0.185 & 0.305 \\ 0.18 & 0.163 & 0.191 & 0.197 & 0.269 \\ 0.175 & 0.174 & 0.164 & 0.187 & 0.3 \\ 0.167 & 0.184 & 0.158 & 0.182 & 0.309 \end{bmatrix}$;

$\begin{bmatrix} 0.1731 & 0.175 & 0.1683 & 0.188 & 0.2956 \\ 0.1709 & 0.1781 & 0.1654 & 0.1866 & 0.299 \\ 0.1748 & 0.1718 & 0.1753 & 0.1911 & 0.287 \\ 0.1712 & 0.1775 & 0.1667 & 0.1875 & 0.2971 \\ 0.1706 & 0.1785 & 0.1641 & 0.1858 & 0.301 \end{bmatrix}$; $\begin{bmatrix} 0.1719 & 0.1764 & 0.1678 & 0.1878 & 0.2961 \\ 0.1717 & 0.1769 & 0.1667 & 0.1872 & 0.2975 \\ 0.1729 & 0.1749 & 0.1701 & 0.1888 & 0.2933 \\ 0.1719 & 0.1765 & 0.1671 & 0.1874 & 0.2970 \\ 0.1714 & 0.1773 & 0.1663 & 0.1870 & 0.2981 \end{bmatrix}$; 0.1872

29. a. 53% for Johnson and 47% for NorthClean   **b.** 58.85% for Johnson and 41.15% for NorthClean   **c.** 61.48% for Johnson and 38.52% for NorthClean   **d.** 62.67% for Johnson and 37.33% for NorthClean   **31. a.** 37,500; 10,000; 2500   **b.** 28,125; 14,500; 7375   **c.** 21,094; 15,775; 13,131   **d.** 15,820; 15,261; 18,918 (rounded numbers do not sum to 50,000)

33. a. $\begin{matrix} & \begin{matrix} e & f & s \end{matrix} \\ \begin{matrix} e \\ f \\ s \end{matrix} & \begin{bmatrix} 0.9 & 0.05 & 0.05 \\ 0.15 & 0.75 & 0.1 \\ 0 & 0 & 1 \end{bmatrix} \end{matrix}$.   **b.** $\begin{bmatrix} 0.405 & 0.4675 & 0.1275 \end{bmatrix}$ **c.** $\begin{bmatrix} 0.4346 & 0.3709 & 0.1945 \end{bmatrix}$ **d.** $\begin{bmatrix} 0.4468 & 0.2999 & 0.2533 \end{bmatrix}$ **e.** Yes; 9 years

35. a. 2039 small, 2352 medium, 2444 large   **b.** 1988 small, 2340 medium, 2507 large   **c.** $\begin{bmatrix} 0.8529 & 0.1418 & 0.0053 \\ 0.0836 & 0.8080 & 0.1084 \\ 0.0020 & 0.0562 & 0.9419 \end{bmatrix}$ **d.** 8.36%   **e.** 10.84%   **37. a.** 0.1859   **b.** 2.34 rabbits in group 1, 2.62 rabbits in group 2, 3.47 in group 3, and 4.56 in group 4   **c.** The long-range probability of rabbits in group 1 or 2 staying in group 1 or 2 is zero.

39. a. 43.6%, 22%, 10.2%, 23.2%, 1%   **b.** 54.57%, 21.39%, 6.65%, 17.11%, 0.27%

c. $\begin{bmatrix} 0.6387 & 0.2119 & 0.0418 & 0.1069 & 0.0007 \\ 0.5906 & 0.226 & 0.0485 & 0.1336 & 0.0013 \\ 0.53 & 0.2026 & 0.0866 & 0.1786 & 0.0022 \\ 0.5041 & 0.2107 & 0.0722 & 0.2089 & 0.0041 \\ 0.4653 & 0.2185 & 0.0834 & 0.2275 & 0.0053 \end{bmatrix}$ **d.** 46.53%   **e.** 22.75%   **41. a.** $\begin{bmatrix} 0.3484 & 0.3786 & 0.2730 \\ 0.3198 & 0.3782 & 0.3020 \\ 0.3164 & 0.3601 & 0.3235 \end{bmatrix}$ **b.** 0.3484   **c.** 0.3198

Exercises 10.2 (page 468)

For exercises . . .	1–6	7–10,20,25–27,32,40	11–19,28–31,33,35,37,38,41
Refer to example . . .	1	2	3

1. Regular   **3.** Not regular   **5.** Regular
7. $\begin{bmatrix} 2/5 & 3/5 \end{bmatrix}$ **9.** $\begin{bmatrix} 1/3 & 2/3 \end{bmatrix}$
11. $\begin{bmatrix} 5/31 & 19/93 & 59/93 \end{bmatrix}$ **13.** $\begin{bmatrix} 170/563 & 197/563 & 196/563 \end{bmatrix}$ **15.** $\begin{bmatrix} 2/7 & 19/42 & 11/42 \end{bmatrix}$ **17.** $\begin{bmatrix} 0 & 0 & 1 \end{bmatrix}$ **19.** $\begin{bmatrix} 1/2 & 7/20 & 3/20 \end{bmatrix}$
21. $\begin{bmatrix} (1-q)/(2-p-q) & (1-p)/(2-p-q) \end{bmatrix}$

25. $\begin{matrix} & \begin{matrix} \text{Works} & \text{Doesn't} \\ & \text{Work} \end{matrix} \\ \begin{matrix} \text{Works} \\ \text{Doesn't Work} \end{matrix} & \begin{bmatrix} 0.9 & 0.1 \\ 0.8 & 0.2 \end{bmatrix} \end{matrix}$; 8/9   **27. a.** $\begin{bmatrix} 0.4 & 0.6 \end{bmatrix}$; $\begin{bmatrix} 0.53 & 0.47 \end{bmatrix}$; $\begin{bmatrix} 0.5885 & 0.4115 \end{bmatrix}$; $\begin{bmatrix} 0.614825 & 0.385175 \end{bmatrix}$; $\begin{bmatrix} 0.626671 & 0.373329 \end{bmatrix}$; $\begin{bmatrix} 0.632002 & 0.367998 \end{bmatrix}$; $\begin{bmatrix} 0.634401 & 0.365599 \end{bmatrix}$; $\begin{bmatrix} 0.635480 & 0.364520 \end{bmatrix}$; $\begin{bmatrix} 0.635966 & 0.364034 \end{bmatrix}$; $\begin{bmatrix} 0.636185 & 0.363815 \end{bmatrix}$ **b.** 0.236364; 0.106364; 0.047864; 0.021539; 0.009693; 0.004362; 0.001963; 0.000884; 0.000398; 0.000179   **c.** Roughly 0.45   **d.** Each week, the difference between the probability vector and the equilibrium vector is slightly less than half of what it was the previous week.   **e.** $\begin{bmatrix} 0.75 & 0.25 \end{bmatrix}$; $\begin{bmatrix} 0.6875 & 0.3125 \end{bmatrix}$; $\begin{bmatrix} 0.659375 & 0.340625 \end{bmatrix}$; $\begin{bmatrix} 0.646719 & 0.353281 \end{bmatrix}$; $\begin{bmatrix} 0.641023 & 0.358977 \end{bmatrix}$; $\begin{bmatrix} 0.638461 & 0.361539 \end{bmatrix}$; $\begin{bmatrix} 0.637307 & 0.362693 \end{bmatrix}$; $\begin{bmatrix} 0.636788 & 0.363212 \end{bmatrix}$; $\begin{bmatrix} 0.636555 & 0.363445 \end{bmatrix}$; $\begin{bmatrix} 0.636450 & 0.363550 \end{bmatrix}$; 0.113636; 0.051136; 0.023011; 0.010355; 0.004659; 0.002097; 0.000943; 0.000424; 0.000191; 0.000086; roughly 0.45   **29.** 16.91% small, 28.47% medium, 54.62% large   **31.** Couples 0.4675; mother 0.3802; father 0.0748; relative 0.0515; other 0.0261   **33.** 0.6053 in type I, 0.2143 in type II, 0.0494 in type III, 0.1295 in type IV, 0.0013 in type V   **35. a.** $\begin{bmatrix} 1/3 & 1/3 & 1/3 \end{bmatrix}$ **37. a.** 1.80% in group 1, 23.68% in group 2, 38.47% in group 3, and 36.04% in group 4   **b.** 8

39. c. $\begin{bmatrix} 0 & 1 & 0 \\ 1/2 & 0 & 1/2 \\ 0 & 1 & 0 \end{bmatrix}$ **d.** Not a regular matrix   **e.** $\begin{bmatrix} 1/4 & 1/2 & 1/4 \end{bmatrix}$ **41. b.** The guard spends 3/7 of the time in front of the middle door and 2/7 of the time in front of each of the other doors.

Exercises 10.3 (page 477)

For exercises . . .	1–6	15,16,31	24a	7–14,22,23,24b,c,25–30,32
Refer to example . . .	1	2	3	4

1. State 3 is absorbing; matrix is that of an absorbing Markov chain.
3. State 1 is absorbing; matrix is not that of an absorbing Markov chain. **5.** States 2 and 4 are absorbing; matrix is that of an absorbing Markov chain. **7.** $F = [2]$; $FR = [0.3 \quad 0.7]$ **9.** $F = [3/2]$; $FR = [3/4 \quad 1/4]$

11. $F = \begin{bmatrix} 1 & 0 \\ 1/3 & 4/3 \end{bmatrix}$; $FR = \begin{bmatrix} 1/3 & 2/3 \\ 4/9 & 5/9 \end{bmatrix}$ **13.** $F = \begin{bmatrix} 25/17 & 5/17 \\ 5/34 & 35/34 \end{bmatrix}$; $FR = \begin{bmatrix} 4/17 & 15/34 & 11/34 \\ 11/34 & 37/68 & 9/68 \end{bmatrix}$

15. a. $\begin{bmatrix} 1 & 0 & 0 & 0 & 0 \\ 1/2 & 0 & 1/2 & 0 & 0 \\ 0 & 1/2 & 0 & 1/2 & 0 \\ 0 & 0 & 1/2 & 0 & 1/2 \\ 0 & 0 & 0 & 0 & 1 \end{bmatrix}$ **b.** $F = \begin{bmatrix} 3/2 & 1 & 1/2 \\ 1 & 2 & 1 \\ 1/2 & 1 & 3/2 \end{bmatrix}$; $FR = \begin{bmatrix} 3/4 & 1/4 \\ 1/2 & 1/2 \\ 1/4 & 3/4 \end{bmatrix}$ **c.** 3/4 **d.** 1/4 **17.** 0.8756

19.

p	0.1	0.2	0.3	0.4	0.5	0.6	0.7	0.8	0.9
x_a	0.9999999997	0.99999905	0.99979	0.98295	0.5	0.017046	0.000209	0.00000095	0.0000000003

21. A column matrix of all 1's **23. a.** $F = \begin{bmatrix} 14.2857 & 2.8571 \\ 8.5714 & 5.7143 \end{bmatrix}$; $FR = \begin{bmatrix} 1 \\ 1 \end{bmatrix}$ **b.** 1 **c.** 17.14 years

25. b. $F = \begin{bmatrix} 1.0625 & 0.1875 \\ 0.0625 & 1.1875 \end{bmatrix}$; $FR = \begin{bmatrix} 1 \\ 1 \end{bmatrix}$ **c.** 1 **d.** 1.25 **27. a.** $\begin{bmatrix} 0.80 & 0.15 & 0.05 \\ 0.05 & 0.80 & 0.15 \\ 0 & 0 & 1 \end{bmatrix}$; $F = \begin{bmatrix} 6.154 & 4.615 \\ 1.538 & 6.154 \end{bmatrix}$; $FR = \begin{bmatrix} 1.000 \\ 1.000 \end{bmatrix}$

b. 1 **c.** 10.77 yr **29. a.** $\begin{bmatrix} 0.4 & 0.3 & 0.2 & 0.1 & 0 \\ 0.2 & 0.1 & 0 & 0.6 & 0.1 \\ 0 & 0 & 1 & 0 & 0 \\ 0.1 & 0.1 & 0.4 & 0.1 & 0.3 \\ 0 & 0 & 0 & 0 & 1 \end{bmatrix}$; $F = \begin{bmatrix} 2.0436 & 0.7629 & 0.7357 \\ 0.6540 & 1.4441 & 1.0354 \\ 0.2997 & 0.2452 & 1.3079 \end{bmatrix}$; $FR = \begin{matrix} 3 \quad\quad 5 \\ \begin{matrix} 1 \\ 2 \\ 4 \end{matrix}\begin{bmatrix} 0.703 & 0.297 \\ 0.545 & 0.455 \\ 0.583 & 0.417 \end{bmatrix} \end{matrix}$

b. 0.297 **c.** 0.455 **d.** 0 **e.** 0.417 **f.** 2.04 **g.** 1.31 **31. a.**

	0	1	2	3	4	5	6	7
0	1	0	0	0	0	0	0	0
1	1/2	0	1/2	0	0	0	0	0
2	0	1/2	0	1/2	0	0	0	0
3	0	0	1/2	0	1/2	0	0	0
4	0	0	0	1/2	0	1/2	0	0
5	0	0	0	0	1/2	0	1/2	0
6	0	0	0	0	0	1/2	0	1/2
7	0	0	0	0	0	0	0	1

b. 3/7 **c.** 2/7

Chapter 10 Review Exercises (page 481)

For exercises . . .	1–4,13,15–26,43,45–57, 49a,50–58,60,67a,b	5–10,14,27–31,44,48, 59,67c,68	4–12,32–38,49b,c, 61–65,66
Refer to section . . .	1	2	3

1. False **2.** False **3.** True **4.** False
5. False **6.** True **7.** False **8.** False
9. False **10.** True **11.** True **12.** True

15. No **17.** Yes

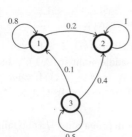

19. a. $C = \begin{bmatrix} 0.7 & 0.3 \\ 1 & 0 \end{bmatrix}$; $C^2 = \begin{bmatrix} 0.79 & 0.21 \\ 0.7 & 0.3 \end{bmatrix}$; $C^3 = \begin{bmatrix} 0.763 & 0.237 \\ 0.79 & 0.21 \end{bmatrix}$ **b.** 0.79

21. a. $E = \begin{bmatrix} 0.2 & 0.5 & 0.3 \\ 0.3 & 0.4 & 0.3 \\ 0 & 1 & 0 \end{bmatrix}$; $E^2 = \begin{bmatrix} 0.19 & 0.6 & 0.21 \\ 0.18 & 0.61 & 0.21 \\ 0.3 & 0.4 & 0.3 \end{bmatrix}$;

$E^3 = \begin{bmatrix} 0.218 & 0.545 & 0.237 \\ 0.219 & 0.544 & 0.237 \\ 0.18 & 0.61 & 0.21 \end{bmatrix}$ **b.** 0.219

23. $[0.377 \quad 0.623]$; $[5/13 \quad 8/13]$ **25.** $[0.492 \quad 0.264 \quad 0.244]$; $[43/80 \quad 19/80 \quad 9/40]$ **27.** Regular **29.** Not regular
35. State 3 is absorbing; matrix is that of an absorbing Markov chain. **37.** No absorbing states; hence, matrix is not that of an absorbing Markov chain.

39. $F = [5/4]$; $FR = [9/16 \quad 7/16]$ **41.** $F = \begin{bmatrix} 7/4 & 4/5 \\ 1 & 8/5 \end{bmatrix}$; $FR = \begin{bmatrix} 0.55 & 0.45 \\ 0.6 & 0.4 \end{bmatrix}$

43. a. $[0.5725 \quad 0.4275]$ **b.** $[0.6776 \quad 0.3224]$ **45.** $[0.428 \quad 0.322 \quad 0.25]$ **47.** $[0.4307 \quad 0.2839 \quad 0.2854]$

49. a. $\begin{bmatrix} 6.536 & 26.144 \\ 3.268 & 42.484 \end{bmatrix}$ **b.** 42.484 **c.** 26.144 **51.** 0.2 **53.** 0.4 **55.** 0.256 **57.** $[0.1945 \quad 0.5555 \quad 0.25]$

59. $[7/36 \quad 5/9 \quad 1/4]$ **61.** States 1 and 6 **63.** $F = \begin{bmatrix} 8/3 & 1/6 & 4/3 & 2/3 \\ 4/3 & 4/3 & 8/3 & 4/3 \\ 4/3 & 1/3 & 8/3 & 4/3 \\ 2/3 & 1/6 & 4/3 & 8/3 \end{bmatrix}$; $FR = \begin{bmatrix} 3/4 & 1/4 \\ 1/2 & 1/2 \\ 1/2 & 1/2 \\ 1/4 & 3/4 \end{bmatrix}$
65. 1/2 **67. a.** 0.1454 **b.** 0.2400
c. 0.2094, 0.2057, 0.2018, 0.1903, 0.1929

Chapter 11 Game Theory

Exercises 11.1 (page 493)

For exercises . . .	1–6	7,9–14,25,26,28,31c,32c,33,39	8,15–24,25d,26d,31c,32c 34,36–38,40
Refer to example . . .	1	2	3

1. $7 from B to A **3.** $3 from A to B
5. $1 from A to B **7.** Yes; column
2 dominates column 3. **9.** Column 3 dominates column 1; $\begin{bmatrix} 9 & -3 \\ -8 & -2 \end{bmatrix}$. **11.** Row 1 dominates rows 2 and 4; $\begin{bmatrix} 3 & 6 \\ 4 & -2 \end{bmatrix}$.

13. Column 1 dominates column 2; $\begin{bmatrix} 8 & -7 \\ -2 & 4 \end{bmatrix}$. **15.** $(2, 2)$; 4; strictly determined **17.** No saddle point; not strictly determined

19. $(3, 1)$; 3; strictly determined **21.** $(1, 3)$; 1; strictly determined **23.** No saddle point; not strictly determined

25. a. $\begin{bmatrix} 3 & 1 \\ 4 & -7 \\ 2 & 0 \\ -4 & -1 \end{bmatrix}$ **b.** $\begin{bmatrix} 3 & 1 \\ 4 & -7 \end{bmatrix}$ **c.** $\begin{bmatrix} 1 \\ -7 \end{bmatrix}$ **d.** $[1]$ **27. a.** No dominated columns **b.** No dominated rows

29. a. Set up in the stadium **b.** Set up in the gym **c.** Set up in both; $1560

31. a.

	Better	Not Better
Market	$50,000	−$25,000
Don't Market	−$40,000	−$10,000

b. $5000 if they market new product, −$22,000 if they don't; market the new product.
c. There are no dominated strategies. There is no saddle point.

33.

		B		
		1	2	3
A	1	15	−2	6
	2	7	15	9
	3	3	−3	15

; no **37.** The saddle point is −7 at $(3, 2)$, and the value of the game is −7.

39.

	Rock	Paper	Scissors
Rock	0	−1	1
Paper	1	0	−1
Scissors	−1	1	0

; not strictly determined

Exercises 11.2 (page 503)

For exercises . . .	1,2	3–20,29a,30–37	27,28
Refer to example . . .	1,2	3,4	3

1. a. −1 **b.** −0.84 **c.** −1.12 **d.** −0.88
3. Player A: 1: 1/7, 2: 6/7; player B: 1: 3/7, 2: 4/7; value 25/7 **5.** Player A: 1: 9/11, 2: 2/11; player B: 1: 4/11, 2: 7/11; value −8/11
7. Player A: 1: 10/17, 2: 7/17; player B: 1: 12/17, 2: 5/17; value 33/17 **9.** Strictly determined; saddle point at $(2, 1)$; value 3
11. Strictly determined; saddle point at $(2, 2)$; value −5/12 **13.** Player A: 1: 6/11, 2: 5/11; player B: 1: 7/11, 2: 4/11; value −12/11
15. Player A: 1: 1/14, 2: 0, 3: 13/14; player B: 1: 1/7, 2: 6/7; value 50/7 **17.** Player A: 1: 2/3, 2: 1/3; player B: 1: 0, 2: 1/9,
3: 8/9; value 10/3 **19.** Player A: 1: 0, 2: 3/4, 3: 1/4; player B: 1: 0, 2: 1/12, 3: 11/12; value 33/4

27. a. $E_1 = 4 - 3p_1$
$E_2 = 3 - p_1$
$E_3 = 1 + 2p_1$ **b.** **c.** 3/5 **29. a.** Each player should play heads with probability 3/8 and tails with probability 5/8, for a value of −1/8. No. **b.** Play heads and tails with probability 1/2 each. **c.** Always play tails, for a value of 0. **31.** Aim for the $4.9 million profit 1/4 of the time and the $4.75 million profit 3/4 of the time.

33. a.

Strain

		1	2
Medicine	1	0.75	0.4
	2	0	1

b. Prescribe medicine 1 about 20/27 of the time and medicine 2 about 7/27 of the time; effectiveness of 55.56%

35. a. $\begin{bmatrix} 1 & -1 \\ -1 & 1 \end{bmatrix}$ **b.** Player A: 1: 1/2, 2: 1/2; player B: 1: 1/2, 2: 1/2; value 0

Number of Fingers

		0	2
Number of Fingers	0	0	-2
	2	-2	4

37. a. (above) **b.** Player A: 1: 3/4, 2: 1/4; player B: 1: 3/4, 2: 1/4; value −1/2

Exercises 11.3 (page 510)

For exercises . . .	1–6,13,21,25	7–12,14–20,22–24,26
Refer to example . . .	1	2

1. Player A: 1: 3/4, 2: 1/4; player B: 1: 1/4, 2: 3/4; value 7/4
3. Player A: 1: 7/11, 2: 4/11; player B: 1: 8/11, 2: 3/11; value 10/11   **5.** Player A: 1: 2/7, 2: 5/7; player B: 1: 11/21, 2: 10/21; value −1/7   **7.** Player A: 1: 1/2, 2: 1/2; player B: 1: 0, 2: 3/10, 3: 7/10; value −1/2   **9.** Player A: 1: 5/7, 2: 2/7; player B: 1: 3/7, 2: 4/7, 3: 0; value 1/7   **11.** Player A: 1: 1/2, 2: 1/2, 3: 0; player B: 1: 1/6, 2: 2/3, 3: 1/6; value 0 **13.** The contractor should bid $30,000 with probability 9/29 and bid $40,000 with probability 20/29. The value of the game is $1396.55.

15. a. $\begin{bmatrix} 5000 & 10,000 & 10,000 \\ 8000 & 4000 & 8000 \\ 6000 & 6000 & 3000 \end{bmatrix}$ **b.** General Items should advertise in Atlanta with probability 4/9, in Boston with probability 5/9, and never in Cleveland. Original Imitators should advertise in Atlanta with probability 2/3, in Boston with probability 1/3, and never in Cleveland. The value of the game is $6666.67.   **17.** Merchant A should locate in cities 1, 2, and 3 with probability 27/101, 129/202, and 19/202, respectively. Merchant B should locate in cities 1, 2, and 3 with probability 39/101, 9/101, and 53/101, respectively. The value of the game is 885/101 ≈ 8.76 percentage points.   **19. a.** Fish inside with probability 0.6705 and fish a mixture with probability 0.3295. Have current with probability 0.3125 and no current with probability 0.6875. Payoff is 13.3125.   **b.** Fish outside, for a payoff of 14.35.   **21.** The student should choose the calculator with probability 3/8 and the book with probability 5/8.   **23.** The kicker should kick left with a probability of 0.4289, middle with a probability of 0.2375, and right with a probability of 0.3336. The goalie should move left with a probability of 0.5953, middle with a probability of 0.0922, and right with a probability of 0.3125. Payoff is 76.25.   **25.** The server should serve left with probability 0.533 and right with probability 0.467; the receiver should move left with probability 0.667 and right with probability 0.333. Payoff is 0.65.

Chapter 11 Review Exercises (page 513)

For exercises . . .	1–5,11–28,46,47–51, 55–59	6,7,29–36,52,60	8–10,37–45,53,54, 61,62
Refer to section . . .	1	2	3

1. False   **2.** True   **3.** True   **4.** True   **5.** False   **6.** False   **7.** False   **8.** True   **9.** True   **10.** True
13. $3 from A to B   **15.** $9 from B to A   **17.** Row 3 dominates row 1; column 1 dominates column 4.

19. $\begin{bmatrix} -11 & 6 \\ -10 & -12 \end{bmatrix}$ **21.** $\begin{bmatrix} -2 & 4 \\ 3 & 2 \\ 0 & 3 \end{bmatrix}$ **23.** (2, 2); value −3   **25.** (2, 2); value 0; fair game   **27.** (1, 4); value −4   **29.** Player A: 1: 5/6, 2: 1/6; player B: 1: 1/2, 2: 1/2; value 1/2   **31.** Player A: 1: 1/9, 2: 8/9; player B: 1: 5/9, 2: 4/9; value 5/9   **33.** Player A: 1: 1/5, 2: 4/5; player B: 1: 3/5, 2: 0, 3: 2/5; value −12/5   **35.** Player A: 1: 3/4, 2: 1/4, 3: 0; player B: 1: 1/2, 2: 1/2; value 1/2   **37.** Player A: 1: 4/7, 2: 3/7; player B: 1: 1/2, 2: 1/2; value −1   **39.** Player A: 1: 7/8, 2: 1/8; player B: 1: 1/2, 2: 1/2; value 1/2   **41.** Player A: 1: 1/2, 2: 1/2; player B: 1: 1/6, 2: 0, 3: 5/6; value −1/2   **43.** Player A: 1: 10/29, 2: 11/29, 3: 8/29; player B: 1: 7/29, 2: 16/29, 3: 6/29; value 2/29   **47.** Hostile   **49.** Friendly; $780   **51.** Labor and management should both always be friendly. The value of the game is $700.   **53.** Victor should invest in blue-chip stocks with probability 35/37 and growth stocks with probability 2/37. The value of the game is $2918.92.   **55.** Favor   **57.** Favor; gain of 1470 votes   **59.** Each candidate should favor the factory. The value of the game is 0.   **61.** Ravogna should defend installation No. 1 with probability 1/4 and installation No. 2 with probability 3/4. Rontovia should attack installation No. 1 with probability 3/4 and installation No. 2 with probability 1/4. The value of the game is 13/4.

Credits

Text Credits

page 33 Exercise 4: Copyright © Society of Actuaries. Used by permission.   **page 54** Exercise 46: Reprinted by permission of Suntex International, Inc.   **page 66** Exercise 48: Professor Nathan Borchelt, Clayton State University   **page 126** Exercise 17: Copyright © Society of Actuaries. Used by permission.   **page 133** Exercises 18-20: Copyright 1973-1991, American Institute of Certified Public Accountants, Inc. All Rights Reserved. Used with permission.   **page 150** Exercise 29: Professor Karl K. Norton, Husson College   **page 159** Exercises 31 and 32: Copyright 1973-1991, American Institute of Certified Public Accountants, Inc. All Rights Reserved. Used with permission.   **page 168** Exercise 17: Copyright © Society of Actuaries. Used by permission.   **page 199** Exercise 65: Copyright © Society of Actuaries. Used by permission.   **page 222** Exercise 80: Copyright © Society of Actuaries. Used by permission.   **page 232** Exercise 82: TNIV Bible   **page 240** Exercise 49: Milton Bradley   **page 256** Exercise 57: Milton Bradley   **page 262** Example 6: From the Complete Works of Lewis Carroll, Vintage Books, 1976   **page 264** Exercise 9: Quote from Aldous Huxley; Exercise 10: Quote from Sir Isaac; Exercise 12: Quote from Sir Winston Churchill   **page 266** Exercises 39-44: From the Complete Works of Lewis Carroll, Vintage Books, 1976   **page 274** Exercise 42: TNIV Bible   **page 278** Exercise 71: From the Complete Works of Lewis Carroll, Vintage Books, 1976; Exercises 69-70: Milton Bradley   **page 279** Logic puzzles: Copyright © 2011 Penny Publications. Reprinted by permission.   **pages 280-282** Exercises 1-5: Copyright © 2011 Penny Publications. Reprinted by permission.   **page 318** Exercise 46: Copyright © Society of Actuaries. Used by permission.   **page 319** Exercises 57-59: Copyright © Society of Actuaries. Used by permission.   **page 333** Exercise 69: Copyright © Society of Actuaries. Used by permission.   **page 340** Exercise 18: Copyright © Society of Actuaries. Used by permission.   **page 341** Exercises 19-20: Copyright © Society of Actuaries. Used by permission.   **page 342** Exercises 27-30: Copyright © Society of Actuaries. Used by permission.   **page 346** Exercise 78: Copyright © Society of Actuaries. Used by permission.   **page 347** Exercises 92-94: Copyright © Society of Actuaries. Used by permission.   **page 348** Exercise 97: Copyright © Society of Actuaries. Used by permission; Exercise 100: Data from Stanley Warner, "Randomized Response: A Survey Technique for Eliminating Evasive Answer Bias," *The Journal of the American Statistical Association*, Vol. 60, No. 309, March 1965, pp. 63-69   **page 350** Exercise 109: Copyright © Society of Actuaries. Used by permission; Extended Application: Robert M. Thrall   **page 361** Exercise 52: Milton Bradley   **page 387** Exercise 41: Professor Irvin R. Hentzel, Iowa State University; Exercise 42: Copyright © Society of Actuaries. Used by permission.   **page 388** Exercises 56-57: Copyright © Society of Actuaries. Used by permission.   **page 397** Exercises 32, 35: Copyright © Society of Actuaries. Used by permission; Exercise 54: James McDonald   **page 398** Exercise 38: Copyright © Society of Actuaries. Used by permission; Exercise 42: Data from Howard, R. A., J. E. Matheson, and D. W. North, "The Decision to Seed Hurricanes," *Science*, Vol. 176, No. 16, June 1972, pp. 1191-1202.   **page 402** Exercise 52: Copyright © The Mathematical Association of America 2010. All rights reserved. Reprinted by permission.   **page 403** Exercises 63-66: Copyright © Society of Actuaries. Used by permission   **page 404** Exercise 69: Adapted from "Media Clips," *Mathematics Teacher*, Vol. 92, No. 8, 1999. Copyright © 1999. Used with permission from the National Council of Teachers of Mathematics. All rights reserved.   **page 426** Exercise 26: Jeffery S. Berman   **page 491** Example 3: Professor Mary T. Treanor of Valparaiso University   **page 504** Exercise 31: Data from Brigham, Georges, "Pricing, Investment, and Games of Strategy," Management Sciences Models and Techniques, Vol. 1, 1960; Exercise 32: Williams, J. D., *The Compleat Strategyst*. New York: McGraw-Hill Book Company, 1966, pp. 56-57. Reprinted by permission from The Rand Corporation, Santa Monica, CA.

Photo Credits

page R-1 Ng Yin Chern/Shutterstock   **page 1** Stockbyte/Getty Images   **page 37** Comstock/Thinkstock   **page 42** Forster Forest/Shutterstock   **page 44** Kelly-Mooney Photography/Corbis   **page 76** Courtesy of Raymond N. Greenwell   **page 86** John and Karen Hollingsworth/U.S. Fish and Wildlife Service   **page 109** Photos/Thinkstock   **page 110** Tischenko Irina/Shutterstock   **page 112** F11 Photo/Shutterstock   **page 119** Neftall/Fotolia   **page 128** Ondrejschaumann/Dreamstime   **page 137** Jordache/Shutterstock   **page 142** Ewwwgenich1/Fotolia   **page 167** Courtesy of Raymond N. Greenwell   **page 169** Iko/Shutterstock   **page 174** Alma Sacra/Fotolia   **page 183** Semen Lixodeev/Shutterstock   **page 187** Mangostock/Shutterstock   **page 216** AP Images   **page 222** Hleib/Shutterstock   **page 224** Hightowernrw/Shutterstock   **page 283** Liquidlibrary/Thinkstock   **page 351** Avava/Shutterstock   **page 352** Stephen Coburn/Shutterstock   **page 406** Lisa F. Young/Shutterstock   **page 407** Kletr/Shutterstock   **page 452** Comstock/Thinkstock   **page 452** Comstock/Thinkstock   **page 484** Olly/Shutterstock   **page 486** Digital Vision/Thinkstock   **page 505** Courtesy of Raymond N. Greenwell

Chapter opener photos are repeated at a smaller size on table of contents and preface pp. v-vii, ix.

Index of Applications

Index

Note: A complete Index of Applications appears on pp. I-1 to I-4.

Sources

Chapter 1

Section 1.1

1. Example 10 from *Morbidity and Mortality Weekly Report*, Centers for Disease Control and Prevention, Vol. 58, No. 44, Nov. 13, 2009, p. 1227.
2. Example 14 from http://www.trends-collegeboard.com/college_ pricing/1_3_over_time_current_dollars.html.
3. Exercise 62 from *Time Almanac 2010*, p. 150.
4. Exercise 63 from *Time Almanac 2010*, pp. 637–638.
5. Exercise 64 from Alcabes, P., A. Munoz, D. Vlahov, and G. Friedland, "Incubation Period of Human Immunodeficiency Virus," *Epidemiologic Review*, Vol. 15, No. 2, The Johns Hopkins University School of Hygiene and Public Health, 1993, pp. 303–318.
6. Exercise 65 from Hockey, Robert V., *Physical Fitness: The Pathway to Healthful Living*, Times Mirror/ Mosby College Publishing, 1989, pp. 85–87.
7. Exercise 66 from *Science*, Vol. 253, No. 5017, July 19, 1991, pp. 306–308.
8. Exercise 67 from *Science*, Vol. 254, No. 5034, Nov. 15, 1991, pp. 936–938, and http://www.cdc.gov/nchs/data.
9. Exercise 68 from *World Health Statistics 2010*, World Health Organization, pp. 56–57.
10. Exercise 69 from *The New York Times*, Sept. 11, 2009, p. A12.
11. Exercise 70 from U. S. Census Bureau, http://www.census.gov/ population/socdemo/hh-fam/ms2.pdf.
12. Exercise 71 from *2008 Yearbook of Immigration Statistics*, Office of Immigration Statistics, Aug. 2009, p. 5.
13. Exercise 72 from *Science News*, June 23, 1990, p. 391.
14. Exercise 73 from Acker, A. and C. Jaschek, *Astronomical Methods and Calculations*, John Wiley & Sons, 1986; Karttunen, H. (editor), *Fundamental Astronomy*, Springer-Verlag, 1994.
15. Exercise 74 from http://www.stateofthemedia .org/2009/narrative_audio_audience.php?media=10&cat=2#1listeningtoradio.
16. Exercise 75 from http://www.trends-collegeboard.com/college_ pricing/1_3_over_time_current_dollars.html.

Section 1.2

1. Page 18 from http://www.agmrc.org/media/cms/oceanspray_4BB99D38246C8.pdf.
2. Exercise 46 from *Science News*, Sept. 26, 1992, p. 195, *Science News*, Nov. 7, 1992, p. 399.
3. Exercise 48 from http://www.calstate.edu/budget/fybudget/2009-2010/supportbook2/challenges-off-campus-costs.shtml.

Section 1.3

1. Page 25 from U.S. Dept. of Health and Human Services, National Center for Health Statistics, found in *New York Times 2010 Almanac*, p. 394.
2. Example 5 from *Public Education Finances 2007*, U.S. Census Bureau, July 2009, Table 8. http://www2.census.gov/govs/school/07f33pub .pdf; *The Nation's Report Card: Reading 2007*, National Center for Education Statistics, U.S. Department of Education, Sept. 2007, Table 11. http://nces.ed.gov/nationsreportcard/pdf/main2007/2007496.pdf.
3. Exercise 4 from "November 1989 Course 120 Examination Applied Statistical Methods" of the *Education and Examination Committee of The Society of Actuaries*. Reprinted by permission of The Society of Actuaries.
4. Exercise 10 from http://www.bea.gov/national/FA2004/SelectTable.asp.
5. Exercise 11 from http://www2.fdic.gov/hsob/hsobRpt.asp.
6. Exercise 12 from http://www.ncta.com/Stats/CableAvailableHomes.aspx.
7. Exercise 13 from http://www.federalreserve .gov/releases/g19/Current/.
8. Exercise 14 from http://www.nada.org/NR/rdonlyres/0FE75B2C-69F0-4039-89FE-1366B5B86C97/0/NADAData08_no.pdf.
9. Exercise 15 from American Airlines, http://www.aa.com.
10. Exercise 15 from *The New York Times*, Jan. 7, 2000.
11. Exercise 16 from www.nctm.org/wlme/wlme6/five.htm.
12. Exercise 17 from Stanford, Craig B., "Chimpanzee Hunting Behavior and Human Evolution," *American Scientist*, Vol. 83, May–June 1995, pp. 256–261, and Goetz, Albert, "Using Open-Ended Problems for Assessment," *Mathematics Teacher*, Vol. 99, No. 1, August 2005, pp. 12–17.
13. Exercise 18 from Pierce, George W., *The Songs of Insects*, Cambridge, Mass., Harvard University Press, Copyright © 1948 by the President and Fellows of Harvard College.
14. Exercise 19 from *Digest of Education Statistics 2006*, National Center for Education Statistics, Table 63.
15. Exercise 20 from *Historical Poverty Tables*, U.S. Census Bureau.
16. Exercise 21 from Lee, Grace, Paul Velleman, and Howard Wainer, "Giving the Finger to Dating Services," *Chance*, Vol. 21, No. 3, 2008, pp. 59–61.
17. Exercise 23 from data provided by Gary Rockswold, Mankato State University, Minnesota.
18. Exercise 25 from Carter, Virgil and Robert E. Machol, *Operations Research*, Vol. 19, 1971, pp. 541–545.

19. Exercise 26 from Whipp, Brian J. and Susan Ward, "Will Women Soon Outrun Men?" *Nature*, Vol. 355, Jan. 2, 1992, p. 25. The data are from Peter Matthews, *Track and Field Athletics: The Records*, Guinness, 1986, pp. 11, 44; from Robert W. Schultz and Yuanlong Liu, in *Statistics in Sports*, edited by Bennett, Jay and Jim Arnold, 1998, p. 189; and from *The World Almanac and Book of Facts 2006*, p. 880.
20. Exercise 27 from http://www.run100s.com/HR/.

Review Exercises

1. Exercises 56 and 57 from TradeStats Express™, http://tse.export.gov.
2. Exercise 58 from U.S. Census Bureau, Historical Income Tables–Households, Table H-6, 2008.
3. Exercise 59 from *Chicago Tribune*, Feb. 4, 1996, Sec. 5, p. 4, and NADA Industry Analysis Division, 2006.
4. Exercise 60 from Food and Agriculture Organization Statistical Yearbook, Table D1, Table G5, http://www.fao.org/economic/ess/publications-studies/statistical-yearbook/fao-statistical-yearbook-2009/en/.
5. Exercise 62 from http://www.ers.usda.gov/Data/FoodConsumption/spreadsheets/mtpcc.xls.
6. Exercise 63 from http://www.census.gov/population/socdemo/hh-fam/ms1.xls.
7. Exercise 64 from http://www.census.gov/hhes/www/poverty/histpov/famindex.html.
8. Exercise 65 from http://www.census.gov/popest/states/NST-ann-est.html; http://doa .alaska.gov/dop/fileadmin/socc/pdf/bkgrnd_socc23.pdf.
9. Exercise 66 from Moore, Thomas L., "Paradoxes in Film Rating," *Journal of Statistics Education*, Vol. 14, 2006, http://www.amstat.org/publications/jse/v14n1/datasets.moore.htm1

Extended Application

1. Page 43 from *Health, United States, 2009*, National Center for Health Statistics, U.S. Department of Health and Human Services, Table 24, http://www.cdc.gov/nchs/data/hus/hus09.pdf.

Chapter 2

Section 2.1

1. Exercise 44 from Goetz, Albert, "Basic Economics: Calculating Against Theatrical Disaster," *The Mathematics Teacher*, Vol. 89, No. 1, Jan. 1996, pp. 30–32.
2. Exercise 45 from Inouye, David, Billy Barr, Kenneth Armitage, and Brian Inouye, "Climate Change Is Affecting Altitudinal Migrants and Hibernating Species," *Proceedings of the National Academy of Science*, Vol. 97, No. 4, Feb. 15, 2000, pp. 1630–1633.

3. Exercise 46 from *National Traffic Safety Institute Student Workbook*, 1993, p. 7.
4. Exercise 47 from "Kobe's 81-Point Game Second Only to Wilt," http://sports.espn.go.com.
5. Exercise 48 from Suntex Int. Inc., Easton, PA, http://www.24game.com. Copied with permission. 24® is a registered trademark of Suntex International Inc., all rights reserved.

Section 2.2

1. Exercise 46 was provided by Prof. Nathan Borchelt, Clayton State University.
2. Exercise 55 from L. L. Bean, http://www.llbean.com.
3. Exercise 60 from Paredes, Miguel, Mohammad Fatehi, and Richard Hinthorn, "The Transformation of an Inconsistent Linear System into a Consistent System," *The AMATYC Review*, Vol. 13, No. 2, Spring 1992.
4. Exercise 61 from Dorrie, Heinrich, *100 Great Problems of Elementary Mathematics, Their History and Solution*, New York: Dover Publications, 1965, pp. 3–7.
5. Exercise 62 from *The New York Times 2010 Almanac*, p. 392.
6. Exercise 63 from Bellany, Ian, "Modeling War," *Journal of Peace Research*, Vol. 36, No. 6, 1999, pp. 729–739.
7. Exercise 65 from Szydlik, Stephen D., "The Problem with the Snack Food Problem," *The Mathematics Teacher*, Vol. 103, No. 1, Aug. 2009, pp. 18–28.
8. Exercise 66 from "Kobe's 81-Point Game Second Only to Wilt," http://sports.espn.go.com.
9. Exercise 68 from Guinard, J., C. Zoumas-Morse, L. Mori, B. Uatoni, D. Panyam, and A. Kilar, "Sugar and Fat Effects on Sensory Properties of Ice Cream," *Journal of Food Science*, Vol. 62, No. 4, Sept./Oct. 1997, pp. 1087–1094.
10. Exercise 69 from Anderson, Marlow and Todd Feil, "Turning Lights Out with Linear Algebra," *Mathematics Magazine*, Vol. 71, No. 4, 1998, pp. 300–303.
11. Exercise 70 from http://www.baseball-almanac.com.

Section 2.3

1. Exercise 44 from *Traffic Safety Facts Research Note*, NHTSA, December 2009.
2. Exercise 45 from *The New York Times 2010 Almanac*, p. 394.
3. Exercise 46 from U.S. Census Bureau Educational Attainment, Table A-2, http://www.census.gov/population/www/socdemo/educ-attn.html.
4. Exercise 47 from U.S. Census Bureau Educational Attainment, Table A-2, http://www.census.gov/population/www/socdemo/educ-attn.html.

Section 2.4

1. Exercise 52 from David I. Schneider, University of Maryland, based on the article "A Dynamic Analysis of Northern Spotted Owl

Viability in a Fragmented Forest Landscape," by Lamberson, R., R. McKelvey, B. Noon, and C. Voss, *Conservation Biology*, Vol. 6, No. 4, Dec. 1992, pp. 505–512.
2. Exercise 53 from http://www.census.gov/ipc/www/idb/region.php and https://www.census.gov/ipc/prod/wp02/tabA-01.xls

Section 2.5

1. Exercise 66 from Isaksen, Daniel, "Linear Algebra on the Gridiron," *The College Mathematics Journal*, Vol. 26, No. 5, Nov. 1995, pp. 358–360.

Section 2.6

1. Page 102 from http://www.bea.gov/industry/.
2. Exercise 19 from Leontief, Wassily, Input-Output Economics, 2nd ed., Oxford University Press, 1966, pp. 20–27.
3. Exercise 20 from Ibid, pp. 6–9.
4. Exercise 21 from Ibid, pp. 174–177
5. Exercise 22 from *Input-Output Tables of China, 1981*, China Statistical Information and Consultancy Service Centre, 1987, pp. 17–19.
6. Exercise 23 and 24 from Chase, Robert, Philip Bourque, and Richard Conway Jr., "The 1987 Washington State Input-Output Study," Report to the Graduate School of Business Administration, University of Washington, Sept. 1993.
7. Exercise 25 from an example created by Thayer Watkins, Department of Economics, San Jose State University, www.sjsu.edu/faculty/watkins/inputoutput.htm.

Review Exercises

1. Exercise 72 from Lamphear, F. Charles and Theodore Roesler, "1970 Nebraska Input-Output Tables," *Nebraska Economic and Business Report No. 10*, Bureau of Business Research, University of Nebraska-Lincoln, 1971.
2. Exercises 74 and 75 are based on the article "Medical Applications of Linear Equations" by David Jabon, Gail Nord, Bryce W. Wilson, and Penny Coffman, *The Mathematics Teacher*, Vol. 89, No. 5, May 1996, p. 398.
3. Exercise 76 from Benson, Brian, Nicholas Nohtadi, Sarah Rose, and Willem Meeuwisse, "Head and Neck Injuries Among Ice Hockey Players Wearing Full Face Shields vs. Half Face Shields," *JAMA*, Vol. 282, No. 24, Dec. 22/29, 1999, pp. 2328–2332.
4. Exercise 77 from Hibbeler, R., *Structural Analysis*, Prentice-Hall, 1995.
5. Exercise 79 from Atmospheric Carbon Dioxide Record from Mauna Loa, Scripps Institution of Oceanography http://cdiac.esd.ornl.gov/ftp/trends/co2/maunaloa.co2.
6. Exercise 80 from Alberty, Robert, "Chemical Equations Are Actually Matrix Equations," *Journal of Chemical Education*, Vol. 68, No. 12, Dec. 1991, p. 984.
7. Exercise 82 from http://www.baseball-reference.com.

Extended Application

1. Page 110 from Grossman, Stanley, "First and Second Order Contact to a Contagious Disease." *Finite Mathematics with Applications to Business, Life Sciences, and Social Sciences*, WCB/McGraw-Hill, 1993.

Chapter 3

Section 3.2

1. Exercise 17 from Problem 5 from "November 1989 Course 130 Examination Operations Research" of the *Education and Examination Committee of The Society of Actuaries*. Reprinted by permission of The Society of Actuaries.

Section 3.3

1. Page 133 from *Uniform CPA Examinations and Unofficial Answers*, copyright © 1973, 1974, 1975 by the American Institute of Certified Public Accountants, Inc.; reprinted with permission.
2. Exercise 25 from Reidhead, Van A., "Linear Programming Models in Archaeology," *Annual Review of Anthropology*, Vol. 8, 1979, pp. 543–578.

Review Exercise

1. Exercise 45 from Joy, Leonard, "Barth's Presentation of Economic Spheres in Darfur," in *Themes in Economic Anthropology*, edited by Raymond Firth, Tavistock Publications, 1967, pp. 175–189.

Chapter 4

Section 4.1

1. Exercise 29 was provided by Professor Karl K. Norton, Husson University.

Section 4.2

1. Exercise 31 from *Uniform CPA Examination Questions and Unofficial Answers*, copyright ©1973, 1974, 1975 by the American Institute of Certified Public Accountants, Inc., is reprinted with permission.
2. Exercise 35 from http://www.nutristrategy.com/activitylist4.htm.
3. Exercise 36 from http://www.nutristrategy.com/activitylist4.htm.

Section 4.3

1. Exercise 17 from Problem 2 from "November 1989 Course 130 Examination Operations Research" of the *Education and Examination Committee of The Society of Actuaries*. Reprinted by permission of The Society of Actuaries.
2. Exercise 27 from http://www.brianmac.demon.co.uk/energyexp.htm.

Section 4.4

1. Exercise 30 from http://www.nutristrategy .com/activitylist4.htm.

Review Exercise

1. Exercise 47 from http://www.nutristrategy .com/activitylist4.htm.

Extended Application

1. This application based on material from the following online sources: The website of the Optimization Technology Center at Northwestern University at http:// www.ece.nwu.edu/OTC/. There is a link to a thorough explanation of the stock-cutting problem. Home page of the Special Interest Group on Cutting and Packing at http://prodlog .wiwi.uni-halle.de/sicup/index.html. The linear programming FAQ at http://www.faqs.org/faqs/ linear-programming-faq/.

Chapter 5

Section 5.1

1. Exercise 54 from *The New York Times*, July 20, 1997, Sec. 4, p. 2.
2. Exercise 54 b from http://money.cnn.com.
3. Exercise 54 d from http://www.forbes.com/ lists/2006/10/BH69.html and http://www .forbes.com/lists/2010/10/billionaires-2010_ William-Gates-III_BH69.html.
4. Exercise 55 from SallieMae: http://www .salliemae.com/get_student_loan/find_student _loan/undergrad_student_loan/federal_student _loans/.
5. Exercise 58 from *The New York Times*, Dec. 31, 1995, Sec. 3. p. 5.
6. Exercise 65 adapted from Problem 5 from "Course 140 Examination, Mathematics of Compound Interest" of the *Education and Examination Committee of The Society of Actuaries.* Reprinted by permission of The Society of Actuaries.
7. Exercise 66 from *The New York Times*, March 30, 1995.
8. Exercise 67 from http://www.ibankmarine.com.
9. Exercise 68 from http://us.etrade.com.
10. Exercise 69 from https://www.bankrate.com.
11. Exercise 77 from *The New York Times*, March 11, 2007, p. 27.
12. Exercise 77 b from http://www.american .com/archive/2006/december/mitt-romney/

Section 5.2

1. Exercise 49 from http://articles.moneycentral .msn.com/Insurance/InsureYourHealth/High CostOfSmoking.aspx
2. Exercise 65 from *The Washington Post*, March 10, 1992, p. A1

Section 5.3

1. Exercise 39 from http://www.cars.com/go/ advice/incentives/incentivesAll.jsp, http://www.cars.com/go/buyIndex.jsp.

2. Exercise 40 from http://www.cars.com/go/ advice/incentives/incentivesAll.jsp, http://www.cars.com/go/buyIndex.jsp.
3. Exercise 41 from Gould, Lois, "Ticket to Trouble," *The New York Times Magazine*, April 23, 1995, p. 39.
4. Page 216 from http://www.direct.ed.gov/ RepayCalc/dlindex2.html.
5. Exercise 47 from *The New York Times,* Nov. 12, 1996, pp. A1, A22.

Review Exercises

1. Exercise 70 from "Pocket That Pension," *Smart Money*, Oct. 1994, p. 33.
2. Exercise 75 from http://www.bankrate.com
3. Exercise 77 from http://www.cars.com/go/ advice/incentives/incentivesAll.jsp, http://www.cars.com/go/buyIndex.jsp
4. Exercise 78 from http://www.cars.com/go/ advice/incentives/incentivesAll.jsp, http://www.cars.com/go/buyIndex.jsp
5. Exercise 80 from Problem 16 from "Course 140 Examination, Mathematics of Compound Interest" of the *Education and Examination Committee of The Society of Actuaries.* Reprinted by permission of The Society of Actuaries.
6. Exercise 81 from *The New York Times,* Sept. 27, 1998, p. BU 10.

Extended Application

1. Page 218 from COMAP, copyright COMAP "Consortium" 1991. COMAP, Inc. 57 Bedford Street #210, Lexington, MA 02420.

Chapter 6

Section 6.1

1. Page 225 from http://research.microsoft.com/
2. Page 231 from J.K. Lasser Institute, *Your Income Tax 2010,* New York: John Wiley & Sons, pp. 10, 562, 487, 435.
3. Page 232 from Goldman, D. R., ed., *American College of Physicians Complete Home Medical Guide*, 2nd ed., New York: DK Publishing, 2003, p. 57.
4. Page 232 from Ventura, John, *Law for Dummies*, Indianapolis, IN: Wiley Publishing, Inc. 2005, pp. 113, 20, 12, 55, 10.
5. Page 232 from Frost, S. E., ed., *Masterworks of Philosophy*, New York: McGraw-Hill, 1946.
6. Page 232 from *TNIV Bible*, Zondervan, 2005.

Section 6.2

1. Exercise 40 from J.K. Lasser Institute, *Your Income Tax 2010,* New York: John Wiley & Sons, p. 489.
2. Exercise 41 from "AppleCare Protection Plan for iPhone," Apple Inc., 2009, p. 11.
3. Exercise 42 from eBay.
4. Exercise 44 from Goldman, David R., ed., *American College of Physicians Complete Home Medical Guide,* 2nd ed., New York: DK Publishing, 2003, p. 223.

5. Exercise 46 from Ventura, John, *Law for Dummies*, Indianapolis, IN: Wiley Publishing, Inc. 2005, p. 21.
6. Exercise 47 from http://www.drmardy.com/ chiasmus/masters/kennedy1.shtml.
7. Exercise 49 from Milton Bradley Company, East Longmeadow, MA, 1996.
8. Exercise 50 from Smullyan, Raymond, *The Lady or the Tiger? And Other Logic Puzzles, Including a Mathematical Novel That Features Godel's Great Discovery*, New York: Knopf, 1982.

Section 6.3

1. Exercise 89 from J.K. Lasser Institute, *Your Income Tax 2010,* New York: John Wiley & Sons, pp. 13, 41, 576.
2. Exercise 90 from "AppleCare Protection Plan for iPhone," Apple Inc., 2009, p. 11.
3. Exercise 92 from Goldman, David R., ed., *American College of Physicians Complete Home Medical Guide,* 2nd ed., New York: DK Publishing, 2003, p. 59.
4. Exercise 93 from Ventura, John, *Law for Dummies*, Indianapolis, IN: Wiley Publishing, Inc. 2005, pp. 12, 110, 5.

Section 6.4

1. Exercise 41 from J.K. Lasser Institute, *Your Income Tax 2010,* New York: John Wiley & Sons, p. 55, 13, 435.
2. Exercise 42 from J.K. Lasser Institute, *Your Income Tax 2010,* New York: John Wiley & Sons, p. 271.
3. Exercise 43 from JP Morgan Chase & Co.
4. Exercise 44 from Goldman, David R., ed., *American College of Physicians Complete Home Medical Guide*, 2nd ed., New York: DK Publishing, 2003, p. 72.
5. Exercise 45 from Rosing, Norbert, "Bear Beginnings: New Life on the Ice," *National Geographic*, December 2000, p. 33.
6. Exercise 46 from Ventura, John, *Law for Dummies*, Indianapolis, IN: Wiley Publishing, Inc. 2005, p. 115.
7. Exercise 47 from Ventura, John, *Law for Dummies*, Indianapolis, IN: Wiley Publishing, Inc. 2005, p. 55, 190, 114.
8. Exercise 48 from Bartlett, John, *Bartlett's Familiar Quotations*, 15th ed., Boston: Little, Brown and Company, 1980.
9. Exercise 49 from Fisher, William E., "An Analysis of the Deutsch Sociocausal Paradigm of Political Integration," *International Organizational*, Vol. 23, No. 2, Spring 1969, pp. 254–290.
10. Exercise 50 from *The New York Times,* May 29, 2002, p. A1.
11. Exercise 51 from Lipset, Seymour Martin, *Political Man*, 1960, quoted by Hildebrand, David K., James D. Laing, and Howard Rosenthal, "Prediction Analysis in Political Research," *The American Political Science Review*, Vol. 70, No. 2, June 1976, pp. 509–535.

12. Exercise 52 from Rosenthal, Howard, "The Electoral Politics of Gaullists in the Fourth French Republic: Ideology or Constituency Interest?" *The American Political Science Review*, Vol. 63, June 1969, pp. 476–487.
13. Exercise 53 from Bower, Bruce, "Roots of Reason," *Science News*, Vol. 145, Jan. 29, 1994, pp. 72–73.
14. Exercise 57 from Directions for Scrabble® and Yahtzee®, Milton Bradley Company, East Longmeadow, MA.

Section 6.5

1. Page 62 and Exercises 39–44 from *The Complete Works of Lewis Carroll*, Vintage Books, 1976.

Section 6.6

1. Example 6 from www.fec.gov/pubrec/fe2000/prespop.htm.
2. Exercise 40 from *United States v. Brown*, 381 U.S. 437 (1965).
3. Exercise 42 from NIV Bible.

Review Exercise

1. Exercise 42 from *Formal Logic: Its Scope and Limits*, 2nd ed., by Richard Jeffrey, New York: McGraw-Hill, 1981.
2. From page 274 from J.K. Lasser Institute, *Your Income Tax 2010*, New York: John Wiley & Sons, pp. 561, 450, 399, 435.
3. Exercise 65 from Goldman, David R., ed., *American College of Physicians Complete Home Medical Guide*, 2nd ed., New York: DK Publishing, 2003, p. 55, 32, 45, 48.
4. Exercise 66 from Ventura, John, *Law for Dummies*, Indianapolis, IN: Wiley Publishing, Inc. 2005, p. 315, 348, 314.
5. Exercise 67 from Tilly, Charles, "Processes and Mechanisms of Democratization," *Sociological Theory*, Vol. 18, No. 1, March 2000, pp. 1–16.
6. Exercise 68 from Estling, Ralph, "It's a Good Thing Cows Can't Fly in Mobile," *Skeptical Inquirer*, Nov./Dec. 2002, pp. 57–58.
7. Exercise 68c from Chambers, Timothy, "On Venn Diagrams," *The Mathematics Teacher*, Vol. 97, No. 1, Jan. 2004, p. 3.
8. Page 278 from Milton Bradley Company.
9. Page 278 from *The Complete Works of Lewis Carroll*, Vintage Books, 1976.

Extended Application

1. From page 279 from *World-Class Logic Problems, Penny Press*, Autumn, 2003.
2. Exercise 1 from *World-Class Logic Problems Special*, Summer 2003, p. 6.
3. Exercise 2 from *World-Class Logic Problems Special*, October 2003, p. 23.
4. Exercise 3 from *World-Class Logic Problems Special,* Autumn 2003, p. 20.
5. Exercise 4 from *World-Class Logic Problems Special,* Autumn 2003, p. 6.
6. Exercise 5 from *World-Class Logic Problems Special*, Autumn 2003, p. 7.

Chapter 7

Section 7.1

1. Page 291 from *The Merck Manual of Diagnosis and Therapy*, 16th ed., Merck Research Laboratories, 1992, pp. 1075 and 1080.
2. Page 292 from *The New York Times 2010 Almanac*, page 408.
3. Exercise 77 from Stewart, Ian, "Mathematical Recreations: A Strategy for Subsets," *Scientific American*, Mar. 2000, pp. 96–98.
4. Page 292 from *The New York Times 2010 Almanac*, pp. 187–207.

Section 7.2

1. Example 7 from U.S. Fish and Wildlife Service, http://ecos.fws.gov.
2. Exercise 46 from *National Vital Statistics Reports*, Vol. 57, No. 14, April 17, 2009, p.18.
3. Exercise 47 from Benson, Brian, Nicholas Nohtaki, M. Sarah Rose, Willem Meeuwisse, "Head and Neck Injuries Among Ice Hockey Players Wearing Full Face Shields vs. Half Face Shields," *JAMA*, Vol. 282, No. 24, Dec. 22/29, 1999, pp. 2328–2332.
4. Exercise 48 from usa.gov.
5. Page 301 from *Population Projections of the United States by Age, Sex, Race, and Hispanic Origin: 1995 to 2050*, U.S. Bureau of the Census, Feb. 1996, pp. 16–17.
6. Page 301 from *The New York Times 2010 Almanac*, p. 296.

Section 7.3

1. Example 4 from www.epi.org.
2. Example 8 from National Safety Council, Itasca, IL, *Injury Facts*, 2007.
3. Exercise 50 from http://www.cartalk.com/content/puzzler/transcripts/200107/. Cartalk.com is a production of Dewey, Cheetham and Howe. Contents © 2007, Dewey, Cheetham and Howe.
4. Exercise 52 from www.nsf.gov/statistics.
5. Exercise 54 from www.bls.gov.
6. Exercise 57 from www.cdc.gov.
7. Exercise 58 from *Population Projections of the United States by Age, Sex, Race, and Hispanic Origin: 1995 to 2050*, Bureau of the Census, Feb. 1996, p. 12.
8. Exercise 59 from *Roll Call*, March 1, 2010.
9. Exercises 60 and 61 from Busey, John and David Martin, *Regimental Strengths and Losses at Gettysburg*, Hightstown, N.J., Longstreet House, 1986, p. 270.

Section 7.4

1. Example 8 from *The New York Times 2010 Almanac*, p. 365.
2. Exercise 24 from *Parade* magazine, Nov. 6, 1994, p. 11. © 1994 Marilyn vos Savant.

Initially published in *Parade* magazine. All rights reserved.
3. Exercise 26 from Staubach, Roger, *First Down, Lifetime to Go,* Word Incorporated, Dallas, 1976.
4. Exercise 45 from Menand, Louis, "Everybody's an Expert: Putting Predictions to the Test," *New Yorker*, Dec. 5, 2005, pp. 98–101.
5. Exercise 46 from Problem 3 from the 2005 Sample Exam P of the *Education and Examination Committee of the Society of Actuaries*. Reprinted by permission of the Society of Actuaries.
6. Exercise 50 from www.bls.gov.
7. Exercise 51 from www.bls.gov.
8. Exercise 53: The probabilities of a person being male or female are from *The World Almanac and Book of Facts*, 1995. The probabilities of a male and female being color-blind are from *Parsons' Diseases of the Eye* (18th ed.) by Stephen J. H. Miller, Churchill Livingstone, 1990, p. 269. This reference gives a range of 3 to 4% for the probability of gross color blindness in men; we used the midpoint of this range.
9. Exercise 56 from Wright, J. R., "An Analysis of Variability in Guinea Pigs," *Genetics,* Vol. 19, pp. 506–536.
10. Exercise 57 from Problem 2 from the 2005 Sample Exam P of the *Education and Examination Committee of the Society of Actuaries*. Reprinted by permission of the Society of Actuaries.
11. Exercise 58 from Problem 8 from the 2005 Sample Exam P of the *Education and Examination Committee of the Society of Actuaries*. Reprinted by permission of the Society of Actuaries.
12. Exercise 59 from Problem 15 from the 2005 Sample Exam P of the *Education and Examination Committee of the Society of Actuaries*. Reprinted by permission of the Society of Actuaries.
13. Exercise 60 from Safire, William, "The Henry Poll," *The New York Times*, June 25, 2001, p. A17 and http://abcnews.go.com.
14. Exercise 66 from usa.gov.
15. Exercises 67 and 68 from Gibson, J. L., "Putting Up with Fellow Russians: An Analysis of Political Tolerance in the Fledgling Russian Democracy," *Political Research Quarterly*, Vol. 51, No. 1, Mar. 1998, pp. 37–68.
16. Exercise 69 from *The New York Times*, Feb. 23, 2006, p. D3.
17. Exercise 70 from bookofodds.com.
18. Exercise 71 from bookofodds.com.

Section 7.5

1. Page 325 from Pringle, David, "Who's the DNA Fingerprinting Pointing At?" *New Scientist*, Jan. 29, 1994, pp. 51–52.
2. Exercise 31 from *Parade* magazine, June 12, 1994, p. 18. © 1994 Marilyn vos Savant. Initially published in *Parade* magazine. All rights reserved.

3. Exercise 32 from *Chance News* 10.01, Jan 16, 2001.

4. Exercise 35: This problem is based on the "Puzzler of the Week: Prison Marbles" from the week of Sept. 7, 1996, on National Public Radio's *Car Talk.*

5. Exercise 41 from airconsumer.dot.gov.

6. Exercise 43 from *Chicago Tribune,* Dec. 18, 1995, Sec. 4, p. 1.

7. Exercise 63 from American Medical Association, 2010.

8. Page 332 from Benson, Brian, Nicholas Nohtaki, M. Sarah Rose, and Willem Meeuwisse, "Head and Neck Injuries Among Ice Hockey Players Wearing Full Face Shields vs. Half Face Shields," *JAMA,* Vol. 282, No. 24, Dec. 22/29, 1999, pp. 2328–2332.

9. Exercise 69 from Problem 12 from the 2005 Sample Exam P of the *Education and Examination Committee of the Society of Actuaries.* Reprinted by permission of the Society of Actuaries.

10. Exercise 70 from Falk, Ruma, *Chance News,* July 23, 1995.

11. Exercise 73 from cdc.gov.

12. Exercise 74 from Matthews, Robert A. J., *Nature,* Vol. 382, Aug. 29, 1996, p. 3.

13. Exercise 75 from *Science News,* Vol. 169, April 15, 2006, pp. 234–236.

14. Exercise 77 from Takis, Sandra L., "Titanic: A Statistical Exploration," *Mathematics Teacher,* Vol. 92, No. 8, Nov. 1999, pp. 660–664.

15. Exercise 87 from Goodman, Terry, "Shooting Free Throws, Probability, and the Golden Ratio," Mathematics Teacher, Vol. 103, No. 7, March, 2010, pp. 482–487.

16. Exercise 88 from Goodman, Terry, "Shooting Free Throws, Probability, and the Golden Ratio," *Mathematics Teacher,* Vol. 103, No. 7, March, 2010, pp. 482–487.

17. Exercise 89 from Schielack, Vincent P., Jr., "The Football Coach's Dilemma: Should We Go for 1 or 2 Points First?" *The Mathematics Teacher,* Vol. 88, No. 9, Dec. 1995, pp. 731–733.

Section 7.6

1. Exercise 18 from Problem 8 from May 2003 Course 1 Examination of the *Education and Examination Committee of the Society of Actuaries.* Reprinted by permission of the Society of Actuaries.

2. Exercise 19 from Problem 20 from the 2005 Sample Exam P of the *Education and Examination Committee of the Society of Actuaries.* Reprinted by permission of the Society of Actuaries.

3. Exercise 20 from Problem 23 from the 2005 Sample Exam P of the *Education and Examination Committee of the Society of Actuaries.* Reprinted by permission of the Society of Actuaries.

4. Exercise 21 from Uniform CPA Examination, Nov. 1989.

5. Exercise 23 from Hoffrage, Ulrich, Samuel Lindsey, Ralph Hertwig, and Gerd Gigerenzer, *Science,* Vol. 290, Dec. 22, 2000, pp. 2261–2262.

6. Exercise 25 from "Mammography facility characteristics associated with interpretive accuracy of screening mammography," National Cancer Institute, www.cancer.gov.

7. Exercise 26 from www.cdc.gov.

8. Exercise 27 from Problem 31 from May 2003 Course 1 Examination of the *Education and Examination Committee of the Society of Actuaries.* Reprinted by permission of the Society of Actuaries.

9. Exercise 28 from Problem 21 from the 2005 Sample Exam P of the *Education and Examination Committee of the Society of Actuaries.* Reprinted by permission of the Society of Actuaries.

10. Exercise 29 from Problem 25 from the 2005 Sample Exam P of the *Education and Examination Committee of the Society of Actuaries.* Reprinted by permission of the Society of Actuaries.

11. Exercise 30 from Problem 26 from the 2005 Sample Exam P of the *Education and Examination Committee of the Society of Actuaries.* Reprinted by permission of the Society of Actuaries.

12. Exercise 31 from "Binge-Drinking Trends," Harvard School of Public Health, Vol 50, March 2002, p. 209.

13. Exercise 32 from Good, I. J., "When Batterer Turns Murderer," *Nature,* Vol. 375, No. 15, June 15, 1995, p. 541.

14. Page 342 from www.uscensus.gov.

15. Exercise 36 from "Factors Related to the Likelihood of a Passenger Vehicle Occupant Being Ejected in a Fatal Crash," National Highway Traffic Safety Administration, U.S. Department of Transportation, December, 2009.

16. Page 343 from "Cigarette Smoking Among Adults—United States, 2007," www.cdc.gov.

17. Exercise 39 from http://abcnews.go.com/ Technology/WhosCounting/story?id= 1560771.

18. Exercise 40 from Shimojo, Shinsuke, and Shin'ichi Ichikawa, "Intuitive Reasoning About Probability: Theoretical and Experimental Analyses of the 'Problem of Three Prisoners,'" *Cognition,* Vol. 32, 1989, pp. 1–24.

Review Exercises

1. Exercise 64 from *Parade* magazine, Sept. 9, 1990, p. 13. © 1990 Marilyn vos Savant. Initially published in *Parade* magazine. All rights reserved.

2. Exercise 78 from Problem 4 from the 2005 Sample Exam P of the *Education and Examination Committee of the Society of Actuaries.* Reprinted by permission of the Society of Actuaries.

3. Exercise 91 from Problem 5 from May 2003 Course 1 Examination of the *Education and Examination Committee of the Society of Actuaries.* Reprinted by permission of the Society of Actuaries.

4. Exercise 92 from Problem 5 from the 2005 Sample Exam P of the *Education and Examination Committee of the Society of Actuaries.* Reprinted by permission of the Society of Actuaries.

5. Exercise 93 from Problem 11 from the 2005 Sample Exam P of the *Education and Examination Committee of the Society of Actuaries.* Reprinted by permission of the Society of Actuaries.

6. Exercise 94 from Problem 18 from May 2003 Course 1 Examination of the *Education and Examination Committee of the Society of Actuaries.* Reprinted by permission of the Society of Actuaries.

7. Exercise 96 from Young, Victoria, "A Matter of Survival," *The Mathematics Teacher,* Vol. 95, No. 2, Feb. 2002, pp. 100–112.

8. Exercise 97 from Problem 13 from the 2005 Sample Exam P of the *Education and Examination Committee of the Society of Actuaries.* Reprinted by permission of the Society of Actuaries.

9. Exercise 98 from www.uscensus.gov and "Election Results 2008," *The New York Times,* November 5, 2008.

10. Exercise 100 from Warner, Stanley, "Randomized Response: A Survey Technique for Eliminating Evasive Answer Bias," *The Journal of the American Statistical Association,* Vol. 60, No. 309, Mar. 1965, pp. 63–69.

11. Exercise 101 from John Allen Paulos, "Coins and Confused Eyewitnesses: Calculating the Probability of Picking the Wrong Guy," *Who's Counting,* Feb. 1, 2001. http://more.abcnews .go.com/sections/science/ whoscounting_ index/whoscounting_index.html.

12. Exercise 102 from *Science,* Vol. 309, July 22, 2005, p. 543.

13. Exercise 103 from *The San Francisco Chronicle,* June 8, 1994, p. A1.

14. Exercise 104 from Carter, Virgil and Robert Machols, "Optimal Strategies on Fourth Down," *Management Science,* Vol. 24, No. 16, Dec. 1978. Copyright © 1978 by The Institute of Management Sciences.

15. Exercise 107 from http://www.cartalk.com/ content/puzzler/2002.html.

16. Exercise 108 from Clancy, Tom, *Debt of Honor,* New York: G. P. Putnam's Sons, 1994, pp. 686–687.

17. Exercise 109 from Problem 1 from May 2003 Course 1 Examination of the *Education and Examination Committee of the Society of Actuaries.* Reprinted by permission of the Society of Actuaries.

Extended Application

1. Page 350 from Wright. Roger, "Probabilistic Medical Diagnosis," *Some Mathematical Models in Biology,* rev. ed., Robert M. Thrall, ed., University of Michigan, 1967. Used by permission of Robert M. Thrall.

Chapter 8

Section 8.1

1. Exercise 51 from Sets, copyright © Marsha J. Falco.
2. Exercise 52 from Scattergories, copyright © Milton Bradley Company.

Section 8.2

1. Exercise 27 from *Weekend Edition*, National Public Radio, Oct. 23, 1994.
2. Exercise 39 from *Mathematics Teacher*, Vol. 96, No. 3, March 2003.
3. Exercise 52 from Heiser, Joseph F., "Pascal and Gauss Meet Little Caesars," *Mathematics Teacher*, Vol. 87, Sept. 1994, p. 389.
4. Exercise 53 from Weber, Griffin and Glenn Weber, "Pizza Combinatorics Revisited," *The College Mathematics Journal*, Vol. 37, No. 1, Jan. 2006, pp. 43–44.
5. Exercise 56 from www.slate.com
6. Exercise 58 from *The New York Times*, Feb. 23, 2006, p. D3.

Section 8.3

1. Exercise 34 from *Weekend Edition*, September 7, 1991.
2. Exercise 35 from *Parade* magazine, Apr. 30, 1995, p. 8. © 1995 Marilyn vos Savant. All rights reserved.
3. Exercise 46 from letter to the editor, *Mathematics Teacher*, Vol. 92, No. 8, Nov. 1999.
4. Exercise 68 from Gould, Lois, "Ticket to Trouble," *The New York Times Magazine,* Apr. 23, 1995, p. 39.
5. Exercise 69 from *Parade* magazine, Dec. 10, 2000, p. 11. © 2000 Marilyn vos Savant. All rights reserved.
6. Exercise 70 from Furst, Randy, "Advice from 2 number crunchers: Don't spend that $340 million just yet," *Minneapolis Star Tribune*, Oct. 18, 2005 and *Chance News 8*, Oct. 15–30, 2005.
7. Exercise 71 from Helman, Danny, "Reversal of Fortunes," *Chance*, Vol. 18, No. 3, Summer 2005, pp. 20–22.
8. Exercise 72 from *Science*, Vol. 258, Oct. 16, 1992, p. 398.
9. Exercise 73 from Madsen, Richard, "On the Probability of a Perfect Progression," *The American Statistician*, Aug. 1991, Vol. 45, No. 3, p. 214.
10. Exercise 74 from All Things Considered, NPR, October 16, 2009.
11. Exercise 75 from Bay, Jennifer M., Robert E. Reys, Ken Simms, and P. Mark Taylor, "Bingo Games: Turning Student Intuitions into Investigations in Probability and Number Sense," *Mathematics Teacher*, Vol. 93, No. 3, Mar. 2000, pp. 200–206.
12. Exercise 76 is based on Problem 1.6 in *40 Puzzles and Problems in Probability and Mathematical Statistics* by Wolfgang Schwarz, Springer, 2008, p. 5.

Section 8.4

1. Example 1 from *Harper's Magazine*, Mar. 1996, p. 13.
2. Example 2 from "Education Life," *The New York Times*, April 23, 2006, p. 24.
3. Example 5 from www.ftc.gov.
4. Example 7 from Grofman, Bernard, "A Preliminary Model of Jury Decision Making as a Function of Jury Size, Effective Jury Decision Rule, and Mean Juror Judgmental Competence," *Frontiers in Economics*, 1979, pp. 98–110.
5. Exercise 22 from *Science*, Vol. 309, July 22, 2005, p. 543.
6. Exercise 23 from Snell, J. Laurie, "40-Million-Dollar Thursday," *Chance News 9.04*, Mar. 7–April 5, 2000.
7. Page 386 from *Statistical Abstract of the United States*, 2010, Table 1152.
8. Exercise 41 submitted by Irvin R. Hentzel, Iowa State University.
9. Exercise 42 from Problem 39 from the 2005 Sample Exam P of the *Education and Examination Committee of the Society of Actuaries*. Reprinted by permission of the Society of Actuaries.
10. Page 387 from *Harper's Index*, April 2006.
11. Exercise 52 from advertisement in *Time*, July 17, 2000, for Pravachol®, developed and marketed by Bristol-Myers Squibb Company.
12. Exercise 53 from "Genetic Fingerprinting Worksheet," Centre for Innovation in Mathematics Teaching, http://www.ex.ac.uk/cimt/resource/fgrprnts.htm.
13. Exercise 54 from Rolly, Paul and JoAnn Jacobsen-Wells, "Bureaucrat's Math Makes Dizzy Dozen," *The Salt Lake Tribune*, Oct. 11, 2002.
14. Exercise 55 from De Smet, Peter A.G.M., "Drug Therapies: Herbal Remedies," *The New England Journal of Medicine*, Vol. 347, Dec. 19, 2002, pp. 2046–2056.
15. Exercise 56 from Problem 28 from the 2005 Sample Exam P of the *Education and Examination Committee of the Society of Actuaries*. Reprinted by permission of the Society of Actuaries.
16. Exercise 57 from Problem 41 from the 2005 Sample Exam P of the *Education and Examination Committee of the Society of Actuaries*. Reprinted by permission of the Society of Actuaries.
17. Exercise 58 from "Fewer Mothers Prefer Full-Time Work," Pew Research Center, July 12, 2007.
18. Page 388 from "Education Life," *The New York Times*, April 23, 2006, p. 7.
19. Exercise 63 from U.S. Department of Education, National Center for Education Statistics. (2009). *Digest of Education Statistics*, 2008. Table 226.
20. Exercise 64 from Kleiner, Carolyn and Mary Lord, "The Cheating Game," *U.S. News and World Report*, Nov. 22, 1999, pp. 55–66.
21. Exercise 65 from Hack, Maureen, M.B., Ch.B., Daniel J. Flannery, Ph.D., Mark Schluchter, Ph.D., Lydia Cartar, M.A., Elaine. Borawski, Ph.D., and Nancy Klein Ph.D. "Outcomes In Young Adulthood for Very-Low-Birth-Weight Infants." *New England Journal of Medicine*, 2002, Vol. 346, pp. 149–157.
22. Exercise 66 from Tillman, Jodie, "The Price They Paid: By Several Measures, Vermont Bears Heavy War Burden," *Valley News*, Jan. 30, 2005.
23. Exercise 67b from Morrison, Donald G. and David C. Schmittlein, "It Takes a Hot Goalie to Raise the Stanley Cup," *Chance*, Vol. 11, No. 1, 1998, pp. 3–7.
24. Exercise 67c from Groeneveld, Richard A. and Glen Meeden, "Seven Game Series in Sports," *Mathematics Magazine*, Vol. 48, No. 4, Sept. 1975, pp. 187–192.

Section 8.5

1. Example 7 from nces.ed.gov.
2. Exercise 32 was supplied by James McDonald, Levi Strauss and Company, San Francisco.
3. Exercise 34 from Problem 48 from the 2005 Sample Exam P of the *Education and Examination Committee of the Society of Actuaries*. Reprinted by permission of the Society of Actuaries.
4. Exercise 35 from Problem 96 from the 2005 Sample Exam P of the *Education and Examination Committee of the Society of Actuaries*. Reprinted by permission of the Society of Actuaries.
5. Exercise 37 from Weiss, Jeffrey and Shoshana Melman, based on "Cost Effectiveness in the Choice of Antibiotics for the Initial Treatment of Otitis Media in Children: A Decision Analysis Approach," *Journal of Pediatric Infectious Disease*, Vol. 7, No. 1, 1988, pp. 23–26.
6. Exercise 38 from Problem 36 from May 2003 Course 1 Examination of the *Education and Examination Committee of the Society of Actuaries*. Reprinted by permission of the Society of Actuaries.
7. Exercise 39 from Hack, Maureen M.B., Ch.B., Daniel J. Flannery, Ph.D., Mark Schluchter, Ph.D., Lydia Cartar, M.A., Elaine Borawski, Ph.D., and Nancy Klein, Ph.D. "Outcomes in Young Adulthood for Very-Low-Birth-Weight Infants." *New England Journal of Medicine*, 2002, Vol. 346, pp. 149–157.
8. Exercise 40 from Kleiner, Carolyn and Mary Lord, "The Cheating Game," *U.S. News and World Report*, Nov. 22, 1999, pp. 55–66.
9. Exercise 41 from http://caselaw.lp.findlaw.com/scripts/getcase.pl?court=3rd&navby=case&no=989009v3&exact=1.
10. Exercise 41d from *Chance News 9*, Nov. 1–27, 2005.
11. Exercise 42 from Howard, R. A., J. E. Matheson, and D. W. North, "The Decision to Seed Hurricanes," *Science*, Vol. 176, No. 16, June 1972, pp. 1191–1202. Copyright © 1972 by The American Association for the Advancement of Science. Reprinted with permission from AAAS.
12. Exercise 44 from Seligman, Daniel, "Ask Mr. Statistics," *Fortune*, July 24, 1995, pp. 170–171.
13. Exercise 55 from Bohan, James and John Shultz, "Revisiting and Extending the Hog

Game," *Mathematics Teacher*, Vol. 89, No. 9, Dec. 1996, pp. 728–733.
14. Exercise 56 from nfl.com.
15. Exercise 57 from baseball-reference.com.

Review Exercises

1. Exercise 52 from "Japanese University Entrance Examination Problems in Mathematics," by Ling-Erl Eileen T. Wu, ed., Mathematical Association of America, 1993, p. 5. Copyright © 1993 from Wu's *Japanese University Entrance Examination Problems in Mathematics*, published by The Mathematical Association of America.
2. Page 403 from *The New York Times*, May 30, 2006, p. A1.
3. Exercise 63 from Problem 16 from the 2005 Sample Exam P of the *Education and Examination Committee of the Society of Actuaries*. Reprinted by permission of the Society of Actuaries.
4. Exercise 64 from Problem 24 from the 2005 Sample Exam P of the *Education and Examination Committee of the Society of Actuaries*. Reprinted by permission of the Society of Actuaries.
5. Exercise 65 from *Course 130 Examination*, Operations Research, Nov. 1989. Reprinted by permission of the Society of Actuaries.
6. Exercise 66 from Ibid.
7. Exercise 68 from advertisement in *The New England Journal of Medicine*, Vol. 338, No. 9, Feb. 26, 1998, for Prozac®, developed and marketed by Eli Lilly and Company.
8. Exercise 69 from "Media Clips," *Mathematics Teacher*, Vol. 92, No. 8, 1999. Copyright 1999. Used with permission from the National Council of Teachers of Mathematics. All rights reserved.
9. Exercise 70 from www.mms.com.
10. Exercise 71 from www.bigten.org.
11. Exercise 76 from www.palottery.state.pa.us.
12. Exercise 77 from www.palottery.state.pa.us.
13. Exercise 79 from Matthews, Robert, "Why Does Toast Always Land Butter-Side Down?" *Sunday Telegraph*, March 17, 1996, p. 4.
14. Exercise 80 from J. Laurie Snell's report of Hal Stern's analysis in *Chance News 7.05*, Apr. 27–May 26, 1998.

Extended Application

1. Page 405 from Smith, Stephen, John Chambers, and Eli Shlifer, "Optimal Inventories Based on Job Completion Rate for Repairs Requiring Multiple Items," *Management Science*, Vol. 26, No. 8, Aug. 1980. © 1980 by The Institute of Management Sciences.

Chapter 9

Section 9.1

1. Example 3 from http://www.abiworld.org/AM/AMTemplate.cfm?Section=Home&TEMPLATE=/CM/ContentDisplay.cfm&CONTENTID=60229.

2. Example 9 from *The Handy Science Answer Book, 2nd ed.*, The Carnegie Library of Pittsburgh, 1997, p. 247.
3. Page 416 from http://www.ers.usda.gov/Data/Wheat/Yearbook/WheatYearbookTable01-Full.htm.
4. Exercise 38 from http://www.huffingtonpost.com/2010/05/10/highest-paid-ceos-2009-ex_n_569695.html#s89202.
5. Exercise 39 from http://www.census.gov/hhes/www/cpstables/032009/hhinc/new06_000.htm.
6. Exercise 40 from http://www.census.gov/hhes/www/cpstables/032009/hhinc/new06_000.htm.
7. Exercise 41 from U.S. Department of Transportation, *Air Travel Consumer Report*, Aug, 2010, p. 47.
8. Exercise 43 from *The Handy Science Answer Book, 2nd ed.*, Carnegie Library of Pittsburgh, 1997, p. 264.
9. Exercise 44 from *The New York Times*, July 31, 1996, p.B4, and www.accuweather.com.
10. Exercise 46 from *New York Times 2010 Almanac*, p. 209, and International Olympic Committee, private communication
11. Exercise 47 from http://discover.npr.org/features/feature html?wfld=1318509.
12. Exercise 48 from Schwartz, J., "Mean Statistics: When Is Average Best?" *The Washington Post*, Jan, 11, 1995, p. H7.
13. Exercise 49 from "http://baseball.about.com/od/newsrumors/a/2010baseballteampayrolls.htm.
14. Exercise 50 from *The New York Times*, Feb. 7, 1997, p. A1.

Section 9.2

1. Example 5 from http://en.wikipedia.org/wiki/Nathan's_Hot_Dog_Eating_Contest.
2. Example 6 from *Statistical Process Control* by Leonard A. Doty, Industrial Press, Inc., 1996.
3. Exercise 26 from Jeffery S. Berman, Senior Analyst, Marketing Information, Quaker Oats Company.
4. Exercise 30 from http://www.bls.gov/cps/cpsatabs.htm.
5. Exercise 32 from *The Handy Science Answer Book, 2nd ed.*, Carnegie Library of Pittsburgh, 1997, p. 264.
6. Exercise 33 from Collins, Vincent, R. Kenneth Lodffer, and Harold Tivey, "Observations on Growth Rates of Human Tumors," *American Journal of Roentgen*. Vol. 76, No. 5, Nov. 1956, pp. 988–1000.
7. Exercise 34 from http://www.the-movie-times.com/thrsdir/actors/actorProfiles.mv?wsmith.
8. Exercise 36 from Revak, Marie, and Jihan Williams, "The Double Stuf Dilemma, *Mathematics Teacher*, Vol. 92, No. 8, Nov. 1999, pp.674–675.

Section 9.3

1. Exercise 39 from Tang, Thomas Li-Ping, David Shin-Hsiung Tang, and Cindy Shin-Yi Tang, "Factors Related to University

Presidents' Pay: An Examination of Private College and Universities," *Higher Education*, Vol. 39, No. 4, June 2000, pp. 393–415.
2. Page 437 from http://www.ibiblio.org/rdu/sl-irrel.html.
3. Exercise 59 from Rothstein. R. "How Tests Can Drop the Ball." *The New York Times*, Sept. 13, 2000, p. B11.
4. Exercises 61 and 62 from Deppisch, Ludwig, Jose Centeno, David Gemmel, and Norca Torres, "Andrew Jackson's Exposure to Mercury and Lead," *JAMA*, Vol. 282, No. 6, Aug. 11, 1999, pp. 569–571.
5. Exercise 61a from Weiss, D., B. Whitten, and D. Leddy, "Lead Content of Human Hair (1871–1971)," *Science*, Vol. 178, 1972, pp. 69–70.
6. Exercise 61b from Iyengar, V. and J. Woittiez, "Trace Elements in Human Clinical Specimens," *Clinical Chemistry*, Vol. 34, 1988, pp. 474–481.
7. Exercise 62a from Suzuki, T., T. Hongo, M. Morita, and R. Yamamoto, "Elemental Contamination of Japanese Women's Hair from Historical Samples," *Sci. Total Environ.*, Vol. 39, 1984, pp. 81–91.
8. Exercise 63 from Norton, Kevin, Timothy Olds, Scott Olive, and Stephen Dank, *Sex Roles*. Vol. 34, Nos. 3/4, Feb.1996, pp. 287–294.

Section 9.4

1. Page 439 from http://www.laseratlanta.com/Archive%20Content/pr08223_speedmotorist.htm.
2. Exercise 15 from Seligman, Daniel and Patty De Llosa, "Ask Mr. Statistics," *Fortune*, May 1, 1995, p. 141.
3. Exercise 20 from http://www.bls.gov/cps/minwage2008tbls.htm#1.
4. Exercise 21 from Wilbur, H. M., "Propagule Size, Number, and Dispersion Pattern in *Ambystoma* and *Asclepias*," *The American Naturalist*, Vol. 111, No. 977, Jan.–Feb. 1977, pp. 43–68.
5. Exercise 22 from DeJong, G., "A Model of Competition for Food. I. Frequency-Dependent Viabilities," *The American Naturalist*, Vol. 110, No. 976, Nov.–Dec. 1976, pp. 1013–1027.
6. Exercise 23 from Barss, Peter, "Injuries Due to Falling Coconuts," *The Journal of Trauma*, Vol. 24, No. 11, 1984, pp. 990–991.
7. Exercise 26 from *The Handy Science Answer Book, 2nd ed.*, The Carnegie Library of Pittsburgh, 1997, p. 332.
8. Exercise 27 from Motorcycle Riders Killed or Injured by Time of Day and Day of Week, 2008, U.S. Department of Transportation, National Highway Traffic Safety Administration, http://www.iii.org/media/hottopics/insurance/motorcycle/.
9. Exercise 29 from http://www.cdc.gov/mmwr/pdf/ss/ss5905.pdf.
10. Exercise 30 from http://www.cdc.gov/mmwr/pdf/ss/ss5905.pdf.
11. Exercise 31 from http://www.historycentral.com/elections/2000state.html.
12. Exercise 32 from Harper's Index," *Harper's*, Sept. 1996, p. 15.

13. Exercise 34 from Reilly, R., "King of the Hill," *Sports Illustrated*, June 1989, pp. 20–25.
14. Exercise 34a from Litwiller, Bonnie and David Duncan, "The Probability of a Hole in One," *School Science and Mathematics*, Vol. 91, No. 1, Jan. 1991, p. 30.

Review Exercises

1. Exercise 57 from The *New York Times 2010 Almanac*, p. 903.
2. Exercise 58 from Evans, Randolph, Richard Evans, Scott Carvajal, and Susan Perry. "A Survey of Injuries Among Broadway Performers." *American Journal of Public Health.* Vol. 86, No. 1, Jan. 1996, pp. 77–80.
3. Exercise 59 from Evans, Randolph, Richard Evans, Scott Carvajal, and Susan Perry. "A Survey of Injuries Among Broadway Performers." *American Journal of Public Health.* Vol. 86, No. 1, Jan. 1996, pp.77–80.

Chapter 10

Section 10.1

1. Exercise 35 from Colins, Lyndhurst. "Estimating Markov Transition Probabilities from Micro-Unit Data," *Applied Statistics,* Vol. 23, No. 3, 1974, pp. 355–371.
2. Exercise 37 from McGilchrist, C.A., C.W. Aisbett, and S. Cooper, "A Markov Transition Model in the Analysis of the Immune Response." *Journal of Theoretical Biology,* Vol. 138, 1989, pp. 17–21.
3. Exercise 39 from Brown, Lawrence A and David B. Longbrake, " Migration Flows in Intraurban Space: Place Utility Considerations." *Annals of the Association of American Geographers.* Vol. 60, No. 2. June 1970, pp. 368–384.
4. Exercise 41 from Colwell, Derek, Brian Jones, and Jack Gillett, "A Markow Chain in Cricket." *The Mathematical Gazette,* Vol. 75, No. 472, June 1991, pp. 183–185.
5. Exercise 42 from Jones, Chris. "A Look at Overtime in the NFL." *Mathematics in Sports,* ed. Joseph Gallian, MAA, 2010.

Section 10.2

1. Exercise 28 from "Fixed-Rate Mortgages Dominant Among Refinancing Homeowners," Freddie Mac, Feb 15, 2010.
2. Exercise 29 from Collins, Lyndhurst. "Estimating Markov Transition Probabilities from Micro-Unit Data," *Applied Statistics,* Vol. 23, No. 3, 1974, pp. 355–371.
3. Exercise 31 from Hooper, Zol A, and E. A. Eff, "Social Mobility in the United States as a Markov Process," *Journal for Economic Educators,* Vol. 8, No. 1, Spring 2008.
4. Exercise 32 from Hertz, Tom, "Understanding Mobility in America," *Center for American Progress,* April 26, 2010.
5. Exercise 33 from Brown, Lawrence A, and David B., Longbrake, "Migration Flows in Intraurban Space: Place Utility Considerations," *Annals of the Association of American Geographers,* Vol. 60, No. 2, June 1970, pp. 368–384.
6. Exercise 34 from Stander, Julian et al., "Markov Chain Analysis and Specialization in Criminal Careers," *The British Journal of Criminology,* Vol. 29, No. 4, Autumn 1989, pp. 317–335.
7. Exercise 37 from Gunzenhauser, Georg W. and Raymond G. Taylor, "Concept Mastery and First Passage Time," *National Forum of Teacher Education Journal,* Vol. 1, No. 1, 1991–1992, pp. 29–34.

Section 10.3

1. Exercise 24 from Beck, J. Robert and Stephen G. Paulker, "The Markov Process in Medical Prognosis," *Medical Decision Making,* Vol. 4, No. 3, 1983, pp. 419–458.
2. Exercise 28 from Taylor, Raymond G., "Forecasting Teacher Shortages," *National Forum of Educational Administration and Supervision Journal,* Vol. 7, No. 2, 1990.
3. Exercise 30 from Hodgson, Ted R, and Maurice J. Burke, "Tennis, Anyone?" *The Mathematics Teacher,* Vol. 98, No. 9, May 2005, pp. 586–592.
4. Exercise 32 from Jones, Chris. "A Look at Overtime in the NFL." *Mathematics in Sports,* ed. Joseph Gallian, MAA, 2010.

Review Exercises

1. Exercise 49 from Chen, Pai-Lien, Estrada J. Bernard, and Pranab K. Sen, "A Markov Chain Model Used in Analyzing Disease History Applied to a Stroke Study," *Journal of Applied Statistics,* Vol. 26, No. 4, 1999, pp. 413–422.
2. Exercise 67 from Jäntti, M., Bratsberg, B., Roed, K., Raaum, O., Naylor, R. Österbacka, E., Björklund, A., and Eriksson, T., "American Exceptionalism in a New Light: A Comparison of Intergenerational Earnings Mobility in the Nordic Countries, the United Kingdom, and the United States," *The Institute for the Study of Labor,* June 2006.
3. Exercise 68 from Abbott, Stephen D, and Matt Richey, "Take a Walk on the Boardwalk," *The College Mathematics Journal,* Vol. 28, No. 3, May 1997, pp. 162–171.

Extended Application

1. Page 484 from Reid, W. M. and R. G. Taylor, "An Application of Absorbing Markov Analysis to Human Resource Issues in Public Administration," *Review of Public Personnel Administration,* Vol. 10, 1989, pp. 69–74.

Chapter 11

Section 11.1

1. Exercise 35 from Craven, Greg, "The Most Terrifying Video You'll Ever See," June 8, 2007, http://www.youtube.com/watch?v=zORv8wwiadQ.

Section 11.2

1. Exercise 29 from *Parade* magazine, March 31, 2002, p. 11 © 1995 by Marilyn vos Savant. Initially published in *Parade* magazine. All rights reserved.
2. Exercise 31 from Brigham, Georges, "Pricing, Investment, and Games of Strategy," *Management Sciences Models and Techniques,* Vol. 1, Copyright © 1960 by Pergamon Press, Ltd. Reprinted with permission.
3. Exercise 32 from Williams, J. D., *The Compleat Strategyst,* New York: McGraw-Hill Book Company, 1966, pp. 56–57. Reprinted by permission from The RAND Corporation, Santa Monica, CA. This is an excellent non-technical book on game theory.

Section 11.3

1. Exercise 18 from Duckett, Stephen, "Nurse Rostering with Game Theory," *Journal of Nursing Administration,* Jan. 1977, pp. 58–59.
2. Exercise 19 from Davenport, William, "Jamaican Fishing: A Game Theory Analysis," *Papers in Caribbean Anthropology,* Vol. 59, Compiled by Sidney W. Mintz, Yale University Publications in Anthropology, 1960.
3. Exercise 20 from Bernard, Jessie, "The Theory of Games of Strategy as a Modern Sociology of Conflict," *American Journal of Sociology,* Vol. 59, No. 5, 1954, pp. 411–424.
4. Exercise 22 from Karlin, Samuel, *Mathematical Methods and Theory in Games, Programming, and Economics,* Addison-Wesley, 1959.
5. Exercise 23 from Chiappori, P.-A., S. Levitt, and T. Groseclose, "Testing Mixed-Strategy Equilibria When Players Are Heterogeneous: The Case of Penalty Kicks in Soccer," *The American Economic Review,* Vol. 92, No. 4, Sept. 2002, pp. 1138–1151.
6. Exercise 24 from Thomas, L. C., *Games, Theory, and Applications,* Halsted Press, 1986.
7. Exercise 25 from Walker, M. and J. Wooders, "Minimax Play at Wimbledon," *The American Economic Review,* Vol. 91, No. 5, December 2001, pp. 1521–1538.

KEY DEFINITIONS, THEOREMS, AND FORMULAS

1.1 Point-Slope Form

If a line has slope m and passes through the point (x_1, y_1), then an equation of the line is given by

$$y - y_1 = m(x - x_1),$$

the point-slope form of the equation of a line.

2.2 Row Operations

For any augmented matrix of a system of equations, the following operations produce the augmented matrix of an equivalent system:

1. interchanging any two rows;

2. multiplying the elements of a row by any nonzero real number;

3. adding a nonzero multiple of the elements of one row to the corresponding elements of a nonzero multiple of some other row.

2.4 Product of Two Matrices

Let A be an $m \times n$ matrix and let B be an $n \times k$ matrix. To find the element in the ith row and jth column of the product matrix AB, multiply each element in the ith row of A by the corresponding element in the jth column of B, and then add these products. The product matrix AB is an $m \times k$ matrix.

2.5 Finding a Multiplicative Inverse Matrix

To obtain A^{-1} for any $n \times n$ matrix A for which A^{-1} exists, follow these steps.

1. Form the augmented matrix $[A|I]$, where I is the $n \times n$ identity matrix.

2. Perform row operations on $[A|I]$ to get a matrix of the form $[I|B]$ if this is possible.

3. Matrix B is A^{-1}.

3.2 Solving a Linear Programming Problem Graphically

1. Write the objective function and all necessary constraints.

2. Graph the feasible region.

3. Identify all corner points.

4. Find the value of the objective function at each corner point.

5. For a bounded region, the solution is given by the corner point producing the optimum value of the objective function.

6. For an unbounded region, check that a solution actually exists. If it does, it will occur at a corner point.

4.2 Simplex Method for Standard Maximization Problems

1. Determine the objective function.

2. Write all necessary constraints.

3. Convert each constraint into an equation by adding a slack variable in each.

4. Set up the initial simplex tableau.

5. Locate the most negative indicator. If there are two such indicators, choose the one farther to the left.

6. Form the necessary quotients to find the pivot. Disregard any quotients with 0 or a negative number in the denominator. The smallest nonnegative quotient gives the location of the pivot. If all quotients must be disregarded, no maximum solution exists. If two quotients are both equal and smallest, choose the pivot in the row nearest the top of the matrix.

7. Use row operations to change all other numbers in the pivot column to zero by adding a suitable multiple of the pivot row to a positive multiple of each row.

8. If the indicators are all positive or 0, this is the final tableau. If not, go back to Step 5 and repeat the process until a tableau with no negative indicators is obtained.

9. Read the solution from this final tableau.

5.1 Compound Amount

$$A = P(1 + i)^n$$

where $i = \dfrac{r}{m}$ and $n = mt$.

A is the future (maturity) value;
P is the principal;
r is the annual interest rate;
m is the number of compounding periods per year;
t is the number of years;
n is the number of compounding periods;
i is the interest rate per period.

5.2 Future Value of an Ordinary Annuity

$$S = R\left[\frac{(1 + i)^n - 1}{i}\right] \qquad \text{or} \qquad S = Rs_{\overline{n}|i}$$

where
S is the future value;
R is the payment;
i is the interest rate per period;
n is the number of periods.

5.3 Present Value of an Ordinary Annuity

The present value P of an annuity of n payments of R dollars each at the end of each consecutive interest period, with interest compounded at a rate of interest i per period, is

$$P = R\left[\frac{1 - (1 + i)^{-n}}{i}\right] \qquad \text{or} \qquad P = Ra_{\overline{n}|i}.$$

6.1, 6.3 Truth Tables

The following truth table defines the logical operators in this chapter.

p	q	$\sim p$	$p \wedge q$	$p \vee q$	$p \rightarrow q$	$p \leftrightarrow q$
T	T	F	T	T	T	T
T	F	F	F	T	F	F
F	T	T	F	T	T	F
F	F	T	F	F	T	T

7.3 Basic Probability Principle

Let S be a sample space of equally likely outcomes, and let event E be a subset of S. Then the probability that event E occurs is

$$P(E) = \frac{n(E)}{n(S)}.$$

7.4 Union Rule

For any two events E and F from a sample space S,

$$P(E \cup F) = P(E) + P(F) - P(E \cap F).$$

7.4 Odds

If $P(E') \neq 0$. the odds in favor of an event E are defined as the ratio of $P(E)$ to $P(E')$, or

$$\frac{P(E)}{P(E')}.$$

7.4 Properties of Probability

Let S be a sample space consisting of n distinct outcomes, $s_1, s_2, \ldots, s_n$. An acceptable probability assignment consists of assigning to each outcome s_i a number p_i (the probability of s_i) according to these rules.

1. The probability of each outcome is a number between 0 and 1.

$$0 \leq p_1 \leq 1, \quad 0 \leq p_2 \leq 1, \ldots, \quad 0 \leq p_n \leq 1$$

2. The sum of the probabilities of all possible outcomes is 1.

$$p_1 + p_2 + p_3 + \cdots + p_n = 1$$

7.5 Product Rule

If E and F are events, then $P(E \cap F)$ may be found by either of these formulas.

$$P(E \cap F) = P(F) \cdot P(E \,|F) \quad \text{or} \quad P(E \cap F) = P(E) \cdot P(F|E)$$

7.6 Bayes' Theorem

$$P(F_i|E) = \frac{P(F_i) \cdot P(E|F_i)}{P(F_1) \cdot P(E|F_1) + P(F_2) \cdot P(E|F_2) + \cdots + P(F_n) \cdot P(E|F_n)}$$

8.1 Multiplication Principle

Suppose n choices must be made, with

m_1 ways to make choice 1,

m_2 ways to make choice 2,

and so on, with

m_n ways to make choice n.